RADIO ACCESS NETWORKS FOR UMTS

RADIO ACCESS NETWORKS FOR UMTS

PRINCIPLES AND PRACTICE

Chris Johnson

Nokia Siemens Networks, UK

John Wiley & Sons, Ltd

Other Wiley Editorial Offices

John Wiley & Sons Inc., 111 River Street, Hoboken, NJ 07030, USA

Jossey-Bass, 989 Market Street, San Francisco, CA 94103-1741, USA

Wiley-VCH Verlag GmbH, Boschstr. 12, D-69469 Weinheim, Germany

John Wiley & Sons Australia Ltd, 42 McDougall Street, Milton, Queensland 4064, Australia

John Wiley & Sons (Asia) Pte Ltd, 2 Clementi Loop #02-01, Jin Xing Distripark, Singapore 129809

John Wiley & Sons Canada Ltd, 6045 Freemont Blvd, Mississauga, ONT, L5R 4J3, Canada

Wiley also publishes its books in a variety of electronic formats. Some content that appears in print may not be
available in electronic books.

Library of Congress Cataloging-in-Publication Data

Johnson, Chris (Chris W.)
   Radio access networks for UMTS : principles and practice / Chris Johnson.
      p. cm.
   Includes index.
   ISBN 978-0-470-72405-7 (cloth)
1. Mobile communication systems. I. Title.
   TK6570.M6J63 2008
   621.384–dc22   2007040535

British Library Cataloguing in Publication Data

A catalogue record for this book is available from the British Library

ISBN 978-0-470-72405-7 (HB)

Contents

Preface   **ix**

Acknowledgements   **xi**

Abbreviations   **xiii**

1   Introduction   **1**
   1.1   Network Architecture   1
   1.2   Radio Access Technology   4
   1.3   Standardisation   10

2   Flow of Data   **13**
   2.1   Radio Interface Protocol Stacks   13
      2.1.1   *Radio Interface Control Plane*   14
      2.1.2   *Radio Interface User Plane*   19
   2.2   RRC Layer   27
      2.2.1   *RRC States*   28
      2.2.2   *RRC Procedures*   54
      2.2.3   *RRC Messages*   56
      2.2.4   *UE RRC Timers, Counters and Constants*   61
      2.2.5   *Other Functions*   67
   2.3   RLC Layer   74
      2.3.1   *Transparent Mode*   75
      2.3.2   *Unacknowledged Mode*   77
      2.3.3   *Acknowledged Mode*   81
   2.4   MAC Layer   101
      2.4.1   *Architecture of the MAC Layer*   102
      2.4.2   *Format of MAC PDU*   109
      2.4.3   *Other Functions*   112
   2.5   Frame Protocol Layer   112
      2.5.1   *Dedicated Channels - Data Frames*   113
      2.5.2   *Dedicated Channels - Control Frames*   118
      2.5.3   *Common Channels - Data Frames*   121
      2.5.4   *Common Channels - Control Frames*   126
   2.6   Physical Layer   127
      2.6.1   *Physical Layer Processing*   128
      2.6.2   *Spreading, Scrambling and Modulation*   144
      2.6.3   *Other Functions*   153

3   Channel Types   **155**
   3.1   Logical Channels   155
   3.2   Transport Channels   158
   3.3   Physical Channels   166
          3.3.1   Common Pilot Channel (CPICH)   168
          3.3.2   Synchronisation Channel (SCH)   172
          3.3.3   Primary Common Control Physical Channel (P-CCPCH)   174
          3.3.4   Secondary Common Control Physical Channel (S-CCPCH)   176
          3.3.5   Paging Indicator Channel (PICH)   182
          3.3.6   MBMS Indicator Channel (MICH)   186
          3.3.7   Acquisition Indicator Channel (AICH)   188
          3.3.8   Physical Random Access Channel (PRACH)   191
          3.3.9   Dedicated Physical Channel (DPCH)   204
          3.3.10   Fractional Dedicated Physical Channel (F-DPCH)   228

4   Non-Access Stratum   **231**
   4.1   Concepts   231
   4.2   Mobility Management   233
   4.3   Connection Management   239
   4.4   PLMN Selection   244

5   Iub Transport Network   **249**
   5.1   Protocol Stacks   249
          5.1.1   Radio Network Control Plane   251
          5.1.2   Transport Network Control Plane   253
          5.1.3   Transport Network User Plane   257
   5.2   Architecture   260
   5.3   Overheads   264
   5.4   Service Categories   268

6   HSDPA   **273**
   6.1   Concept   273
   6.2   HSDPA Bit Rates   278
   6.3   PDCP Layer   283
   6.4   RLC Layer   284
   6.5   MAC-d Entity   287
   6.6   Frame Protocol Layer   288
          6.6.1   HS-DSCH Data Frame   289
          6.6.2   HS-DSCH Control Frames   292
   6.7   Iub Transport   294
          6.7.1   ATM Transport Connections   294
          6.7.2   Transport Overheads   296
   6.8   MAC-hs Entity   300
          6.8.1   Flow Control   301
          6.8.2   Scheduler   304
          6.8.3   Adaptive Modulation and Coding   307
          6.8.4   Hybrid Automatic Repeat Request (HARQ)   313
          6.8.5   Generation of MAC-hs PDU   320

6.9   Physical Channels   322
       6.9.1   *High Speed Shared Control Channel (HS-SCCH)*   323
       6.9.2   *High Speed Physical Downlink Shared Channel (HS-PDSCH)*   329
       6.9.3   *High Speed Dedicated Physical Control Channel (HS-DPCCH)*   332
6.10   Mobility   337

7   HSUPA   **343**
7.1   Concept   343
7.2   HSUPA Bit Rates   349
7.3   PDCP Layer   355
7.4   RLC Layer   355
7.5   MAC-d Entity   357
7.6   MAC-es/e Entity (UE)   358
       7.6.1   *E-TFC Selection*   359
       7.6.2   *Hybrid Automatic Repeat Request (HARQ)*   368
       7.6.3   *Generation of MAC-es PDU*   371
       7.6.4   *Generation of MAC-e PDU*   372
7.7   Physical Channels   374
       7.7.1   *E-DCH Dedicated Physical Control Channel (E-DPCCH)*   376
       7.7.2   *E-DCH Dedicated Physical Data Channel (E-DPDCH)*   378
       7.7.3   *E-DCH Hybrid ARQ Indicator Channel (E-HICH)*   387
       7.7.4   *E-DCH Relative Grant Channel (E-RGCH)*   390
       7.7.5   *E-DCH Absolute Grant Channel (E-AGCH)*   392
7.8   MAC-e Entity (Node B)   394
       7.8.1   *Packet Scheduler*   395
       7.8.2   *De-multiplexing*   399
7.9   Frame Protocol Layer   399
       7.9.1   *E-DCH Data Frame*   400
       7.9.2   *Tunnel Congestion Indication Control Frame*   401
7.10   MAC-es Entity (RNC)   402
7.11   Mobility   402

8   Signalling Procedures   **405**
8.1   RRC Connection Establishment   405
8.2   Speech Call Connection Establishment   429
       8.2.1   *Mobile Originated*   430
       8.2.2   *Mobile Terminated*   454
8.3   Video Call Connection Establishment   459
       8.3.1   *Mobile Originated and Mobile Terminated*   460
8.4   Short Message Service (SMS)   469
       8.4.1   *Mobile Originated*   470
       8.4.2   *Mobile Terminated*   474
8.5   PS Data Connection Establishment   477
       8.5.1   *Mobile Originated*   478
8.6   Soft Handover   501
       8.6.1   *Inter-Node B*   501
       8.6.2   *Intra-Node B*   512
8.7   Inter-System Handover   514
       8.7.1   *Speech*   515

9   Planning   **533**
      9.1   Link Budgets   533
            9.1.1   DPCH   535
            9.1.2   HSDPA   545
            9.1.3   HSUPA   547
      9.2   Radio Network Planning   548
            9.2.1   Path Loss based Approach   550
            9.2.2   3G Simulation based Approach   553
      9.3   Scrambling Code Planning   556
            9.3.1   Downlink   557
            9.3.2   Uplink   560
      9.4   Neighbour Planning   561
            9.4.1   Intra-Frequency   563
            9.4.2   Inter-Frequency   564
            9.4.3   Inter-System   566
            9.4.4   Maximum Neighbour List Lengths   567
      9.5   Antenna Subsystems   572
            9.5.1   Antenna Characteristics   573
            9.5.2   Dedicated Subsystems   576
            9.5.3   Shared Subsystems   578
      9.6   Co-siting   578
            9.6.1   Spurious Emissions   581
            9.6.2   Receiver Blocking   583
            9.6.3   Intermodulation   585
            9.6.4   Achieving Sufficient Isolation   586
      9.7   Microcells   587
            9.7.1   RF Carrier Allocation   588
            9.7.2   Sectorisation   588
            9.7.3   Minimum Coupling Loss   589
            9.7.4   Propagation Modelling   590
            9.7.5   Planning Assumptions   591
      9.8   Indoor Solutions   592
            9.8.1   RF Carrier Allocation   593
            9.8.2   Sectorisation   593
            9.8.3   Active and Passive Solutions   593
            9.8.4   Minimum Coupling Loss   594
            9.8.5   Leakage Requirements   595
            9.8.6   Antenna Placement   596

References   **597**

Index   **599**

Preface

This book provides a comprehensive description of the *Radio Access Networks for UMTS*. It is intended to address the requirements of both the beginner and the more experienced mobile telecommunications engineer. An important characteristic is the inclusion of sections from example log files. More than 180 examples have been included to support the majority of explanations and to reinforce the reader's understanding of the key principles. Another important characteristic is the inclusion of summary bullet points at the start of each section. The reader can use these bullet points either to gain a high-level understanding prior to reading the main content or for subsequent revision. The main content is based upon the release 6 version of the 3GPP specifications. Changes since the release 99 version are described while some of the new features appearing within the release 7 version are introduced.

Starting from the high-level network architecture, the first sections describe the flow of data between the network and end user. The functionality and purpose of each protocol stack layer is explained while the corresponding structure and content of packets are studied. A section is dedicated to describing and contrasting the sets of logical, transport and physical channels. The increasing importance of the bandwidth offered by the transport network connecting the population of Node B to the RNC justifies the inclusion of a dedicated section describing the Iub interface and the associated transport solutions. Dedicated sections are also included for both HSDPA and HSUPA. The bit rates and functionality associated with these technologies are described in detail. A relatively large section is used to describe some of the most important signalling procedures. These include RRC connection establishment, speech call connection establishment, video call connection establishment, PS data connection establishment, SMS data transfer, soft handover and inter-system handover. The accompanying description provides a step-by-step analysis of both the signalling flow and message content. Other sections focus upon the more practical subjects of link budgets and radio network planning. Topics include scrambling code planning, neighbour list planning, antenna subsystem design, co-siting, microcells and indoor solutions.

The content of this book represents the understanding of the author. It does not necessarily represent the view nor opinion of the author's employer. Descriptions are intended to be generic and do not represent the implementation of any individual vendor.

Acknowledgements

The author would like to acknowledge his employer, Nokia Siemens Networks UK Limited for providing the many opportunities to gain valuable project experience. The author would also like to thank his managers from within Nokia Siemens Networks UK Limited for supporting participation within projects which have promoted continuous learning and development. These include Andy King, Peter Love, Aleksi Toikkanen, Stuart Davis, Mike Lawrence and Chris Foster. The author would also like to thank Florian Reymond for providing the opportunities to work on global projects within Nokia Siemens Networks.

The author would like to acknowledge colleagues from within Nokia Siemens Networks who have supported and encouraged the development of material for this book. These include Poeti Boedhi-hartono, Simon Browne, Gareth Davies, Martin Elsey, Benoist Guillard, Terence Hoh, Harri Holma, Steve Hunt, Sean Irons, Phil Pickering, Kenni Rasmussen, Mike Roche, Lorena Serna Gonzalez, Ian Sharp, Achim Wacker, Volker Wille and Nampol Wimolpitayarat. In addition, the author would like to thank the managers and colleagues from outside Nokia Siemens Networks who have also supported the development of this book. These include Mohamed AbdelAziz, Paul Clarkson, Tony Conlan, Patryk Debicki, Nathan Dyson, Gianluca Formica, Dave Fraley, Ian Miller, Balan Muthiah, Pinaki Roychowdhury, Adrian Sharples and Ling Soon Leh.

The author would also like to offer special thanks to his parents who provided a perfect working environment during the weeks spent in Scotland. He would also like to thank them for their continuous support and encouragement.

The author would like to thank the team at John Wiley & Sons Limited who have made this publication possible. This team has included Mark Hammond, Sarah Hinton, Katharine Unwin and Brett Wells.

Comments regarding the content of this book can be sent to ran4umts@yahoo.co.uk. These will be considered when generating material for future editions.

Abbreviations

16QAM	16 Quadrature Amplitude Modulation
3GPP	3rd Generation Partnership Project
4PAM	4 Pulse Amplitude Modulation
64QAM	64 Quadrature Amplitude Modulation
AAL2	ATM Adaptation Layer 2
AAL5	ATM Adaptation Layer 2
ABR	Available Bit Rate
AC	Access Class
ACIR	Adjacent Channel Interference Ratio
ACLR	Adjacent Channel Leakage Ratio
ACS	Adjacent Channel Selectivity
AI	Access Indicator
AICH	Access Indicator Channel
ALCAP	Access Link Control Application Part
AM	Acknowledged Mode
AMC	Adaptive Modulation and Coding
AMR	Adaptive Multi Rate
APN	Access Point Name
ARFCN	Absolute Radio Frequency Channel Number
AS	Access Stratum
ASC	Access Service Class
ASN	Abstract Syntax Notation
ATM	Asynchronous Transfer Mode
BCC	Base station Colour Code
BCCH	Broadcast Control Channel
BCD	Binary Coded Decimal
BCH	Broadcast Channel
BER	Bit Error Rate
BFN	Node B Frame Number
BLER	Block Error Rate
BMC	Broadcast/Multicast Control
BSIC	Base Station Identity Code
CAC	Connection Admission Control
CBC	Cell Broadcast Centre
CBR	Constant Bit Rate

CBS	Cell Broadcast Services
CC	Call Control
CCCH	Common Control Channel
CCTrCh	Coded Composite Transport Channels
CDMA	Code Division Multiple Access
CDVT	Cell Delay Variation Tolerance
CFN	Connection Frame Number
CGI	Cell Global Identity
CI	Cell Identity
CID	Channel Identifier
CIO	Cell Individual Offset
CLP	Cell Loss Priority
CLR	Cell Loss Ratio
CM	Compressed Mode
COI	Code Offset Indicator
CPCH	Common Packet Channel
CPCS	Common Part Convergence Sublayer
CPI	Common Part Indicator
CPICH	Common Pilot Channel
CPS	Common Part Sublayer
CQI	Channel Quality Indicator
CRC	Cyclic Redundancy Check
C-RNTI	Cell Radio Network Temporary Identity
CS	Circuit Switched
CTCH	Common Traffic Channel
CTD	Cell Transfer Delay
CTFC	Calculated Transport Format Combination
DAS	Distributed Antenna System
DCCH	Dedicated Control Channel
DCH	Dedicated Channel
DDI	Data Description Indicator
DPCCH	Dedicated Physical Control Channel
DPCH	Dedicated Physical Channel
DPDCH	Dedicated Physical Data Channel
DRT	Delay Reference Time
DRX	Discontinous Receive
DSAID	Destination Signaling Association Identifier
DSCH	Downlink Shared Channel
DTCH	Dedicated Traffic Channel
DTX	Discontinuous Transmit
E-AGCH	E-DCH Absolute Grant Channel
Eb/No	Energy per bit/Noise spectral density
ECF	Establish Confirm
E-DCH	Enhanced Dedicated Channel
E-DPCCH	E-DCH Dedicated Physical Control Channel
E-DPDCH	E-DCH Dedicated Physical Data Channel
EGPRS	Enhanced General Packet Radio Service
E-HICH	E-DCH Hybrid ARQ Indicator Channel

EIRP	Effective Isotropic Radiated Power
E-RGCH	E-DCH Relative Grant Channel
ERQ	Establish Request
E-TFC	E-DCH Transport Format Combination
E-TFCI	E-DCH Transport Format Combination Indicator

FACH	Forward Access Channel
FBI	Feedback Information
FDD	Frequency Division Duplex
F-DPCH	Fractional Dedicated Physical Channel
FSN	Frame Sequence Number
FTP	File Transfer Protocol

GFR	Guaranteed Frame Rate
GGSN	Gateway GPRS Support Node
GMM	GPRS Mobility Management
GMSK	Gaussian Minimum Shift Keying
GPRS	General Packet Radio Service
GRAKE	Generalised RAKE
GSMS	GPRS Short Message Service
GTP-U	User plane GPRS Tunnelling Protocol

HARQ	Hybrid Automatic Repeat Request
HCS	Hierarchical Cell Structure
HEC	Header Error Correction
HFN	Hyper Frame Number
HLBS	Highest Priority Logical Channel Buffer Status
HLID	Highest Priority Logical Channel Identity
HLR	Home Location Register
HLS	Higher Layer Scheduling
HPLMN	Home Public Land Mobile Network
H-RNTI	HS-DSCH Radio Network Temporary Identity
HSCSD	High Speed Circuit Switched Data
HSDPA	High Speed Downlink Packet Access
HS-DPCCH	High Speed Dedicated Physical Control Channel
HS-DSCH	High Speed Downlink Shared Channel
HS-PDSCH	High Speed Downlink Shared Channel
HS-SCCH	High Speed Shared Control Channel
HSUPA	High Speed Uplink Packet Access

ICP	IMA Control Protocol
IE	Information Element
IETF	Internet Engineering Task Force
IMA	Inverse Multiplexing for ATM
IMEI	International Mobile Equipment Identity
IMSI	International Mobile Subscriber Identity
IPDL	Idle Period Downlink
IPv4	Internet Protocol version 4
IPv6	Internet Protocol version 6
ITP	Initial Transmit Power

ITU	International Telecommunications Union

LAC	Location Area Code
LAI	Location Area Identity
LLC	Logical Link Control
LSN	Last Sequence Number

MAC	Medium Access Control
MAP	Mobile Application Part
MBMS	Multimedia Broadcast Multicast Services
MBS	Maximum Burst Size
MCC	Mobile Country Code
MCCH	MBMS Control Channel
MCL	Minimum Coupling Loss
MCR	Minimum Cell Rate
MDC	Macro Diversity Combination
MDCR	Minimum Desired Cell Rate
MFS	Maximum Frame Size
MHA	Mast Head Amplifier
MIB	Master Information Block
MICH	MBMS Indicator Channel
MIMO	Multiple Input Multiple Output
MLP	MAC Logical channel Priority
MM	Mobility Management
MNC	Mobile Network Code
MSCH	MBMS Scheduling Channel
MSS	Maximum Segment Size
MTCH	MBMS Traffic Channel
MTU	Maximum Transmission Unit
MUD	Multi User Detection

NAS	Non-access Stratum
NBAP	Node B Application Part
NCC	Network Colour Code
NI	Notification Indicator
NMO	Network Mode of Operation
NNI	Network to Network Interface
NRT	Non Real Time
NSAP	Network Service Access Point
NSAPI	Network layer Service Access Point Identifier

| OSAID | Originating Signalling Association Identifier |
| OTDOA | Observed Time Difference of Arrival |

PAP	Password Authentication Protocol
PCA	Power Control Algorithm
PCCH	Paging Control Channel
P-CCPCH	Primary Common Control Physical Channel
PCH	Paging Channel
PCR	Peak Cell Rate

PDCP	Packet Data Convergence Protocol
PDH	Plesiochronous Digital Hierarchy
PDU	Packet Data Unit
PER	Packed Encoding Rules
PI	Paging Indication
PICH	Paging Indication Channel
PLMN	Public Land Mobile Network
PRACH	Physical Random Access Channel
PS	Packet Switched
P-SCH	Primary Synchronisation Channel
PSTN	Public Switched Telephone Network
P-TMSI	Packet Temporary Mobile Subscriber Identity
PWE3	Psuedo Wire Emulation Edge to Edge
QoS	Quality of Service
QPSK	Quadrature Phase Shift Keying
RAB	Radio Access Bearer
RAC	Routing Area Code
RACH	Random Access Channel
RAI	Routing Area Identity
RAN	Radio Access Network
RANAP	Radio Access Network Application Part
RAT	Radio Access Technology
RB	Radio Bearer
RDI	Restricted Digital Information
RFN	RNC Frame Number
RIP	Radio Interface Protocol
RL	Radio Link
RLC	Radio Link Control
RM	Rate Matching
RNC	Radio Network Controller
RNS	Radio Network Sub-system
ROHC	Robust Header Compression
RPP	Recovery Period Power control
RRC	Radio Resource Control
RRM	Radio Resource Management
RSCP	Received Signal Code Power
RSN	Re-transmission Sequence Number
RSSI	Received Signal Strength Indicator
RT	Real Time
RV	Redundancy Version
SA	Service Area
SAC	Service Area Code
SAI	Service Area Identity
SAR	Segmentation and Reassembly
SAW	Stop and Wait
S-CCPCH	Secondary Common Control Channel
SCH	Synchronisation Channel

SCR	Sustainable Cell Rate
SDH	Synchronous Digital Hierarchy
SDU	Service Data Unit
SEAL	Simple and Efficient ATM Adaptation Layer
SF	Spreading Factor
SFN	System Frame Number
SGSN	Serving GPRS Support Node
SI	Scheduling Information
SIB	System Information Block
SID	Size Index Identifier
SIR	Signal to Interference Ratio
SM	Session Management
SM-AL	Short Message Application Layer
SM-RL	Short Message Relay Layer
SMS	Short Message Service
SM-TL	Short Message Transfer Layer
SONET	Synchronous Optical Networking
SRB	Signalling Radio Bearer
SRNS	Serving Radio Network Sub-system
S-RNTI	SRNC Radio Network Temporary Identity
SS	Supplementary Services
SSADT	Service Specific Assured Data Transfer
SSCF	Service Specific Coordination Function
S-SCH	Secondary Synchronisation Channel
SSCOP	Service Specific Connection Orientated Protocol
SSCS	Service Specific Convergence Sublayer
SSDT	Site Selection Diversity Transmit
SSSAR	Service Specific Segmentation and Reassembly
SSTED	Service Specific Transmission Error Detection
STTD	Space Time Transmit Diversity
SUFI	Super Field
TB	Transport Block
TBS	Transport Block Set
TCP	Transmission Control Protocol
TCTF	Target Channel Type Field
TDD	Time Division Duplex
TDMA	Time Division Multiple Access
TEBS	Total E-DCH Buffer Status
TF	Transport Format
TFC	Transport Format Combination
TFCI	Transport Format Combination Indicator
TFCS	Transport Format Combination Set
TFI	Transport Format Indicator
TFO	Tandem Free Operation
TFS	Transport Format Set
TGD	Transmission Gap Distance
TGL	Transmission Gap Length
TGPL	Transmission Gap Pattern Length
TGPRC	Transmission Gap Pattern Repetition Count

TGPS   Transmission Gap Pattern Sequence
TGPSI   Transmission Gap Pattern Sequence Identifier
TGSN   Transmission Gap Starting Slot Number
THP   Traffic Handling Priority
TM   Transparent Mode
TMSI   Temporary Mobile Subscriber Identity
toAWE   Time of Arrival Window End point
toAWS   Time of Arrival Window Start point
TPC   Transmit Power Control
TPDU   Transfer Protocol Data Unit
TR   Technical Report
TrFO   Transcoder Free Operation
TS   Technical Specification
TSN   Transmission Sequence Number
TSTD   Time Switched Transmit Diversity
TTI   Transmission Time Interval
TTL   Time To Live

UARFCN   UTRA Absolute Radio Frequency Channel Number
UBR   Unspecified Bit Rate
UDI   Unrestricted Digital Information
UE   User Equipment
UEA   UMTS Encryption Algorithm
UIA   UMTS Integrity protection Algorithm
UM   Unacknowledged Mode
UMTS   Universal Mobile Telecommunications System
UNI   User to Network Interface
UPH   UE Power Headroom
URA   UTRAN Registration Area
U-RNTI   UTRAN Radio Network Temporary Identity
USIM   Universal Subscriber Identity Module
UTRAN   UMTS Terrestrial Radio Access Network
UUI   User to User Indication

VBR   Variable Bit Rate
VCC   Virtual Channel Connection
VPC   Virtual Path Connection
VCI   Virtual Channel Identifier
VoIP   Voice over IP
VPI   Virtual Path Identifier
VPLMN   Visited Public Land Mobile Network

WCDMA   Wideband Code Division Multiple Access

1

Introduction

1.1 Network Architecture

- The RAN includes RNC, Node B and UE. RNC are connected to Node B using the Iub interface. Neighbouring RNC are connected using the Iur interface. UE are connected to Node B using the Uu interface. The RAN is connected to the CN using the Iu interface.
- Each Node B has a controlling RNC and each UE connection has a serving RNC. The serving RNC provides the Iu connection to the CN. Drift RNC can be used by UE connections in addition to the serving RNC.

The network architecture defines the network elements and the way in which those network elements are interconnected. Figure 1.1 illustrates a section of the network architecture for UMTS. This book focuses upon the Radio Access Network (RAN) rather than the core network. The RAN represents the section of the network which is closest to the end-user and which includes the air-interface.

The RAN includes the Radio Network Controller (RNC), the Node B and the User Equipment (UE). The MSC and SGSN are part of the core network. An example UMTS network could include thirty RNC, ten thousand Node B and five million UE. The UE communicate with the Node B using the air-interface which is known as the Uu interface. The Node B communicates with the RNC using a transmission link known as the Iub interface. The RNC communicates with the core network using a transmission link known as the Iu interface. There is an Iu interface for the Circuit Switched (CS) core network and an Iu interface for the Packet Switched (PS) core network. The capacity of the Iu interface is significantly greater than the capacity of the Iub interface because the Iu has to be capable of supporting a large quantity of Node B whereas the Iub supports only a single Node B. Neighbouring RNC can be connected using the Iur interface. The Iur interface is particularly important for UE which are moving from the coverage area of one RNC to the coverage area of another RNC.

Each Node B has a controlling RNC and each UE connection has a serving RNC. The controlling RNC for a Node B is the RNC which terminates the Iub interface. The serving RNC for a UE connection is the RNC which provides the Iu interface to the core network. Figure 1.2 illustrates an example for a packet switched connection and four Node B.

RNC 1 is the controlling RNC for Node B 1 and 2 whereas RNC 2 is the controlling RNC for Node B 3 and 4. The controlling RNC is responsible for managing its Node B. RNC 1 is the serving RNC for the packet switched connection because it provides the connection to the PS core network. The serving

Radio Access Networks for UMTS   Chris Johnson
© 2008 John Wiley & Sons, Ltd

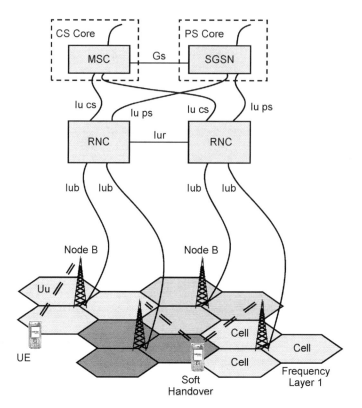

Figure 1.1   UMTS network architecture

RNC is responsible for managing its UE connections. As this example illustrates, an RNC can be categorised as both controlling and serving.

In the case of UE mobility, an RNC can also be categorised as a drift RNC. If a UE starts its connection within the coverage area of RNC 1 then that RNC becomes the serving RNC and will provide the connection to the core network. If the UE subsequently moves into the coverage area of the second RNC then the UE can be simultaneously connected to Node B controlled by both RNC 1 and RNC 2. This represents a special case of soft handover, i.e. inter-RNC soft handover. This scenario is illustrated in Figure 1.3. In general, soft handover allows UE to simultaneously connect to multiple Node B. This is in contrast to hard handover in which case the connection to the first Node B is broken before the connection to the second Node B is established. Soft handover helps to provide seamless mobility to active connections as UE move throughout the network and also helps to improve the RF conditions at cell edge where signal strengths are generally low and cell dominance is poor. In the case of inter-RNC soft handover, the UE is simultaneously connected to multiple RNC. The example illustrated in Figure 1.3 is based upon two RNC but it is possible for UE to be connected to more than two RNC if the RNC coverage boundaries are designed to allow it. In this example, RNC 1 is the serving RNC because it provides the Iu connection to the core network. RNC 2 is a drift RNC because it is participating in the connection, but it is not providing the connection to the core network. A single connection can have only one serving RNC, but can have more than one drift RNC.

Communication between the UE and the serving RNC makes use of the Iur interface when a drift RNC is involved. The Iur interface is an optional transmission link and is not always present. For

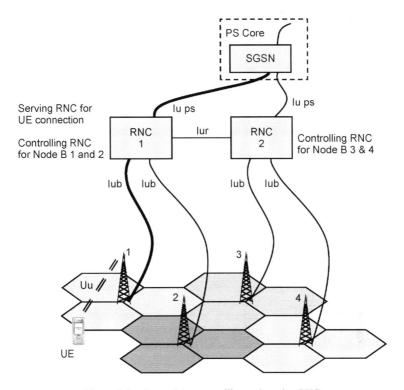

Figure 1.2 Categorising controlling and serving RNC

example, if a network is based upon RNC from two different network vendors then it is possible that those RNC are not completely compatible and the Iur interface is not deployed. If the Iur interface is not present then inter-RNC soft handover is not possible because there is no way to transfer information from the drift RNC to the serving RNC. In this case, the UE has to complete a hard handover when moving into the coverage area of the second RNC. The inter-RNC hard handover procedure allows the second RNC to become the serving RNC while the first RNC no longer participates in the connection.

Assuming that the Iur interface is present and that a UE continues to move into the coverage area of the drift RNC then it becomes inefficient to leave the original RNC as the serving RNC. There will be a time when the UE is not connected to any Node B which are controlled by the serving RNC and all information is transferred across the Iur interface. In this scenario it makes sense to change the drift RNC into the serving RNC and to remove the original RNC from the connection. This procedure of changing a drift RNC into the serving RNC is known as serving RNC relocation, or Serving Radio Network Subsystem (SRNS) relocation. A Radio Network Subsystem (RNS) is defined as an RNC and the collection of Node B connected to that RNC.

The radio network plan defines the location and configuration of the Node B. The density of Node B should be sufficiently great to achieve the target RF coverage performance. If the density of Node B is not sufficiently great then there may be locations where the UE does not have sufficient transmit power to be received by a Node B, i.e. coverage is uplink limited. Alternatively, there may be locations where a Node B does not have sufficient transmit power to be received by a UE, i.e. coverage is downlink limited. The connection from the UE to the Node B is known as the uplink or reverse link whereas the connection from the Node B to the UE is known as the downlink or forward link.

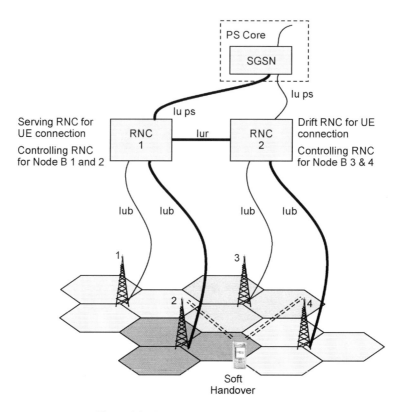

Figure 1.3 Categorising serving and drift RNC

1.2 Radio Access Technology

- The air-interface is based upon full duplex FDD with a nominal channel bandwidth of 5 MHz. Channel separations can be <5 MHz because the occupied bandwidth is <5 MHz.
- Operators are typically assigned between 2 and 4 UMTS channels.
- A frequency reuse of 1 is applied allowing both soft and hard handovers.
- Multiple access is based upon Wideband CDMA with a chip rate of 3.84 Mcps.
- The release 7 version of the 3GPP specifications defines 9 operating bands.
- The most common Node B configuration for initial network deployment is three sectors with 1 RF carrier, i.e. a 1+1+1 Node B configuration.
- HSDPA and HSUPA offer significantly increased throughput performance.

The UMTS air-interface makes use of separate RF carriers for the uplink and downlink. This approach is known as Frequency Division Duplexing (FDD) and is in contrast to technologies which use the same RF carrier for both the uplink and downlink. Using the same RF carrier for both the uplink and downlink requires time sharing, i.e. the RF carrier is assigned to the uplink for a period of time and then the RF carrier is assigned to the downlink for a period of time. This approach is known as Time Division Duplexing (TDD). A set of operating bands have been standardised for use by the UMTS air-interface. These operating bands are presented in Table 1.1.

Table 1.1 UMTS operating bands for the FDD air-interface

Operating Band	Uplink (MHz)	Downlink (MHz)	Duplex spacing (MHz)	Equivalent 2G band
I	1920–1980	2110–2170	190	
II	1850–1910	1930–1990	80	PCS 1900
III	1710–1785	1805–1880	95	DCS 1800
IV	1710–1755	2110–2155	400	
V	824–849	869–894	45	GSM 850
VI	830–840	875–885	45	
VII	2500–2570	2620–2690	120	
VIII	880–915	925–960	45	E-GSM
IX	1749.9–1784.9	1844.9–1879.9	95	

The availability of each operating band depends upon existing spectrum allocations and the strategy of the national regulator. The majority of countries deploying UMTS make use of operating band I as the core set of frequencies. The remaining operating bands can either be used as extension bands or can be used by countries where operating band I is not available. For example, operating band II is used in North America because operating band I is not available. Operating band II cannot be used as an extension for operating band I because the two sets of frequencies overlap with one another. Operating band VIII is commonly viewed as an extension band which benefits from improved coverage performance as a result of using lower frequencies. Operating band VIII is the same as the extended GSM 900 band and so its use for UMTS may require re-farming of any existing GSM 900 allocations.

Each operating band is divided into 5 MHz channels. Operating bands I and II have 12 uplink channels and 12 downlink channels. Operating band I has a frequency difference of 190 MHz between the uplink and downlink channels whereas operating band II has a frequency difference of 80 MHz. The difference between the uplink and downlink frequencies is known as the duplex spacing. Large duplex spacings cause more significant differences between the uplink and downlink path loss. The uplink is assigned the lower set of frequencies because the path loss is lower and link budgets are traditionally uplink limited. Small duplex spacings make it more difficult to implement transmit and receive filtering within the UE. Transmit and receive filtering is less of an issue within the Node B because larger and more expensive filters can be used. The uplink and downlink channels belonging to operating band I are illustrated in Figure 1.4.

National regulators award the 5 MHz channels to operators. Those operators then become responsible for deploying and operating UMTS networks. It is common to award between two and four channels to each operator. For example, a country which has four operators could have three channels assigned to each operator. It is possible that not all twelve channels are available and only a subset of the channels are allocated. Once an operator has been assigned a subset of the 5 MHz channels then the operator has some flexibility in terms of configuring the precise centre frequencies of its RF carriers. A UMTS RF carrier occupies less than 5 MHz and so the frequency separation between adjacent RF carriers can also be less than 5 MHz. An example deployment strategy is illustrated in Figure 1.4. In this example, three 5 MHz UMTS channels have been awarded to operator 2 while the adjacent channels have been awarded to operators 1 and 3. Adjacent channel interference mechanisms, e.g. non-ideal transmit filtering and non-ideal receive filtering are less significant when RF carriers are co-sited, or at least coordinated. Operator 2 is likely to co-site adjacent RF carriers which are assigned to the macrocell network (multiple RF carriers assigned to the same Node B) and is likely to coordinate adjacent RF carriers which are assigned to the microcell layer or to any indoor solutions. The Node B belonging to operators 1 and 3 may be neither co-sited nor coordinated with the Node B belonging to operator 2. Operator 2 can help to reduce the potential for any adjacent channel interference by reducing the frequency separation between its own RF carriers. This allows an increased frequency

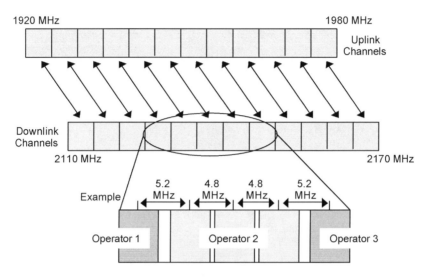

Figure 1.4   UMTS FDD operating band I

separation from the adjacent operators. The RF carriers within operating band I have been standardised using a 200 kHz channel raster. This means that the centre frequency of each RF carrier can be adjusted with a resolution of 200 kHz.

The Node B configuration defines characteristics such as the number of sectors and the number of RF carriers. The most common configuration for initial network deployment is three sectors with one RF carrier. This is known as a 1+1+1 Node B configuration. It requires at least three antennas to be connected to the Node B cabinet, i.e. at least one antenna serving each sector. If uplink receive diversity or downlink transmit diversity is used then either six single element antennas or three dual element antennas are required. If six single element antennas are used then there should be spatial isolation between the two antennas belonging to each sector. This tends to be less practical than using three dual element antennas. It is common to use cross polar antennas which accommodate two antenna elements within each antenna housing. In this case, isolation is achieved in the polarisation domain rather than the spatial domain. Figure 1.5 illustrates an example 1+1+1 Node B configuration using cross polar antennas.

When diversity is used then a separate RF feeder is required for each diversity branch. A 1+1+1 Node B with uplink receive diversity requires six RF feeders to connect the antennas to the Node B cabinet. Likewise, if Mast Head Amplifiers (MHA) are used then six of them would be required. The 1+1+1 Node B configuration has three logical cells, i.e. a logical cell is associated with each sector of the Node B. When the capacity of a single RF carrier becomes exhausted then it is common to upgrade to a second RF carrier. The Node B configuration is then known as a 2+2+2. This configuration has three sectors, but now has two RF carriers and six logical cells. Alternatively, a six sector single RF carrier configuration could be deployed which would be known as a 1+1+1+1+1+1. This configuration also has six logical cells but has six sectors and 1 RF carrier.

When a UMTS operator deploys a single RF carrier then that carrier must be shared between all users of the network and the frequency re-use is 1, i.e. all cells make use of the same RF carrier. GSM networks make use of frequency re-use patterns to assign different RF carriers to neighbouring cells. For example, a frequency re-use of 12 means that the radio network is planned in clusters of 12 cells and each cell within a cluster can use 1/12th of the available RF carriers. This type of approach helps to reduce co-channel interference, but leads to a requirement for hard handovers and a relatively large

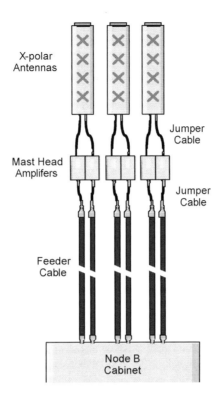

X-polar Antennas

Jumper Cable

Mast Head Amplifers

Jumper Cable

Feeder Cable

Node B Cabinet

Figure 1.5 Example 1+1+1 Node B configuration

number of RF carriers. GSM channels have a bandwidth of 200 kHz and so it is possible to place 25 GSM channels within the bandwidth of a single UMTS channel. The use of a wide bandwidth and a frequency re-use of 1 for UMTS provides benefits in terms of receiver sensitivity and spectrum efficiency. The air-interface of a single UMTS cell can support approximately 50 speech users when assuming the maximum Adaptive Multi-Rate (AMR) bit rate of 12.2 kbps. A single GSM RF carrier can support a maximum of 8 speech users when assuming Full Rate (FR) connections. This means that 5 MHz of GSM spectrum can support a maximum of 200 speech users (ignoring the impact of the broadcast channel which in practise would reduce the maximum number of GSM speech users). Assuming a frequency reuse of 10 reduces this figure to 20 speech users per 5 MHz in contrast to the 50 speech users supported by UMTS. The spectrum efficiency of GSM can approach that of UMTS when using small frequency re-use patterns which require more careful planning to avoid co-channel interference. Frequency hopping can also be used to improve the performance and spectrum efficiency of GSM. The number of speech users supported by both UMTS and GSM can be increased by decreasing the bit rates assigned to each connection. The UMTS AMR codec supports bit rates ranging from 4.75 to 12.2 kbps. The GSM Half Rate (HR) feature may be used to reduce the GSM speech bit rate.

GSM RF carriers are shared between multiple connections using Time Division Multiple Access (TDMA). A GSM radio frame is divided into eight time slots and these time slots can be assigned to different connections. An RF carrier belonging to a cell is never simultaneously assigned to more than one connection. A GSM speech connection is assigned different time slots within the radio frame for the uplink and downlink, i.e. the GSM MS does not have to simultaneously transmit and receive. This approach is known as half-duplex and tends to make the MS design easier and less expensive. GSM base stations have to simultaneously transmit and receive because they serve multiple connections and the uplink time slot of one connection can coincide with the downlink time slot of another connection. This is known as full-duplex operation.

UMTS RF carriers are shared between multiple connections using Code Division Multiple Access (CDMA). CDMA allows multiple connections to simultaneously use the same RF carrier. Instead of being assigned time slots, connections are assigned codes. These codes are used to mask the transmitted signal and allow the receiver to distinguish between signals belonging to different connections. The RNC assigns codes to both the uplink and downlink during the establishment of a connection. The version of CDMA used for UMTS is known as Wideband CDMA (WCDMA) because the bandwidth is relatively large compared with earlier CDMA systems. WCDMA connections are able to use all time slots in both the uplink and downlink directions. This means that WCDMA is full-duplex rather than half-duplex because UE must be capable of simultaneously transmitting and receiving.

The WCDMA air-interface makes use of two types of code in both the uplink and downlink. Channelisation codes are used to increase the bandwidth of the connection subsequent to physical layer processing at the transmitter. These codes are sometimes referred to as spreading codes. For example, a connection could have a bit rate of 240 kbps after physical layer processing. Each individual bit would then be multiplied by a 64 chip channelisation code. This would increase the bit rate of 240 kbps by a factor of 64 to a chip rate of 3.84 Mcps. The chip rate of 3.84 Mbps is standardised for WCDMA and all connections have the same chip rate after spreading. If the bit rate after layer 1 processing had been 480 kbps then each individual bit would have been multiplied by a 32 chip channelisation code. This would have increased the bit rate of 480 kbps by a factor of 32 to a chip rate of 3.84 Mcps. The chip rate of 3.84 Mcps defines the approximate bandwidth of the WCDMA signal in the frequency domain, i.e. the approximate bandwidth after baseband filtering is 3.84 MHz. Once the transmitted signal has been spread by a channelisation code then it is multiplied by a scrambling code. Scrambling codes have a chip rate of 3.84 Mcps and do not change the chip rate of the already spread signal. In the downlink direction, channelisation codes are used to distinguish between different connections and scrambling codes are used to distinguish between different cells, i.e. each connection within a cell is assigned a different channelisation code and each cell within the same geographic area is assigned a different scrambling code. In the uplink direction, channelisation codes are used to distinguish between the different physical channels transmitted by a single UE and scrambling codes are used to distinguish between different UE.

Table 1.2 summarises some of the most important characteristics of the GSM and UMTS air-interfaces.

Table 1.2   Comparison of GSM and UMTS air-interfaces

	GSM	WCDMA
Duplexing	FDD	FDD
Multiple access	TDMA	CDMA
MS transmit and receive	Half-duplex	Full-duplex
Handover	Hard	Hard and soft
Frequency re-use	4–18	1
Channel bandwidth	200 kHz	5 MHz
RF carrier bandwidth	200 kHz	3.84 MHz
Typical maximum bit rates	GSM 9.6 kbps	DPCH 403.2 kbps
	HSCSD 43.2 kbps	HSDPA 7.2 Mbps
	GPRS 62.4 kbps	HSUPA 1.44 Mbps
	EGPRS 179.2 kbps	
Power control rate	2 Hz or lower	1500 Hz
Typical maximum uplink transmit power	33 dBm	24 dBm
Typical minimum uplink transmit power	5 dBm	−50 dBm

The maximum bit rates represent typical figures rather than theoretical maxima and they are based upon the bit rates achieved at the top of the physical layer rather than at the application layer. The GSM bit rate of 9.6 kbps corresponds to 192 bits of data coded every 20 ms and transferred across the air-interface using a single time slot within four 4.615 ms radio frames. High Speed Circuit Switched Data (HSCSD) is the circuit switched evolution of GSM which allows a single connection to use multiple time slots from each radio frame. The HSCSD bit rate of 43.2 kbps corresponds to using three time slots and increasing the coding rate to allow the bit rate per time slot to increase from 9.6 to 14.4 kbps.

The General Packet Radio Service (GPRS) represents the packet switched evolution of GSM. GPRS supports four channel coding schemes and also allows the use of multiple time slots from each radio frame. The bit rate of 62.4 kbps corresponds to Coding Scheme 3 (CS-3) and the use of four time slots within each radio frame. CS-3 has a coding rate of approximately 0.75 and applies channel coding to 312 bits per 20 ms for every time slot that is used. Enhanced GPRS (EGPRS) supports nine coding schemes and allows the use of 8-PSK modulation in addition to GMSK modulation. When the signal to noise ratio conditions are relatively good, 8-PSK is able to triple the air-interface throughput relative to GMSK by transferring 3 bits per modulated symbol rather than 1 bit per modulated symbol. The EGPRS bit rate of 179.2 kbps corresponds to Modulation and Coding Scheme 7 (MCS-7) and the use of four time slots within each radio frame. MCS-7 has a coding rate of approximately 0.75 and applies channel coding to 896 bits per 20 ms for every time slot that is used.

The bit rates quoted for WCDMA are for the Dedicated Physical Channel (DPCH), High Speed Downlink Packet Access (HSDPA) and High Speed Uplink Packet Access (HSUPA). The DPCH bit rate of 403.2 kbps corresponds to channel coding 12 blocks of 336 bits every 10 ms. This bit rate can be supported in both the uplink and downlink directions. 403.2 kbps at the top of the physical layer corresponds to 384 kbps at the top of the RLC layer, i.e. each block of 336 bits includes 16 bits of RLC header information.

The HSDPA bit rate of 7.2 Mbps can be achieved when a 16QAM modulation scheme is used in combination with 10 channelisation codes with a spreading factor of 16 and a channel coding rate of 0.75. Based upon the chip rate of 3.84 Mcps, the 16QAM symbol rate is 240 ksps prior to spreading. There are 4 bits of information represented by each 16QAM symbol and so the bit rate per channelisation code before channel coding is 720 kbps. The use of 10 channelisation codes increases this figure to 7.2 Mbps. Higher HSDPA bit rates are possible if more than 10 channelisation codes are used or if the coding rate is increased. The maximum theoretical bit rate is 14.4 Mbps when the coding rate is increased to 1 and 15 channelisation codes with a spreading factor of 16 are used.

The HUSPA bit rate of 1.44 Mbps can be achieved when using two channelisation codes with a spreading factor of 4 and a channel coding rate of 0.75. Based upon the chip rate of 3.84 Mcps, the QPSK symbol rate is 960 ksps prior to spreading. Each QPSK branch uses one channelisation code and transfers 1 bit per symbol. This results in the bit rate of 1.44 Mbps before channel coding. Higher HSUPA bit rates are possible if channelisation codes with a spreading factor of 2 are used or if the coding rate is increased. The maximum theoretical bit rate is 5.76 Mbps when the coding rate is increased to one and two channelisation codes with a spreading factor of 2 are used in addition to two channelisation codes with a spreading factor of 4.

Table 1.2 indicates that the rate of power control for WCDMA is significantly greater than that for GSM. The uplink transmit power of a WCDMA UE and the downlink transmit power of a Node B can be changed 1500 times per second. This is in contrast to GSM which changes its transmit power at a maximum rate of twice per second. The combination of CDMA and a frequency re-use pattern of 1 increases the importance of power control for WCDMA. All users are simultaneously transmitting and receiving on the same RF carrier. If any user transmits with more power than necessary then levels of interference are increased unnecessarily for all other users. All other users would then have to increase their transmit powers to compensate. The power control for WCDMA helps to ensure that target quality criteria are achieved with minimum transmit power. Changing the transmit power 1500 times per second allows relatively rapid changes in the propagation channel to be tracked and compensated. For example, if the path loss suddenly decreases as a result of a user moving into an area which has line-of-sight propagation to a serving cell then power control is able to rapidly decrease both the uplink and downlink transmit powers. Power control is less important for GSM because neighbouring cells use different RF carriers and users served by the same cell and the same RF carrier are separated in time by the TDMA frame structure. Co-channel interference becomes more important for GSM when the frequency re-use is decreased and there is less isolation between cells using the same RF carrier.

The importance of power control for WCDMA is also illustrated by the large dynamic range between the maximum and minimum uplink transmit powers. A typical GSM mobile has a transmit power dynamic range of 28 dB whereas a typical WCDMA UE has a transmit power dynamic range of 74 dB. This large dynamic range allows a UE to decrease its transmit power when in close proximity to a Node B and to avoid generating unnecessary increases in the uplink interference floor of the Node B receiver. Assuming that the minimum coupling loss between a Node B receiver and a UE is 60 dB then a UE transmitting with a power of −50 dBm will be received with a power of −110 dBm. The thermal noise floor of a Node B receiver is typically −105 dBm and so the received signal will have a relatively small impact upon the thermal noise floor. The maximum transmit power of a WCDMA UE is typically less than that of a GSM MS, but this is compensated by WCDMA receivers being more sensitive than GSM receivers.

1.3 Standardisation

- 3GPP has standardised UMTS. 3GPP is a collaboration between a number of telecommunications standards organisations from around the world.
- 3GPP generates both Technical Reports and Technical Specifications.
- The Technical Specifications continue to evolve from the release 99 version to release 7 and newer versions.

The 3rd Generation Partnership Project (3GPP) was formally established in December 1998 and has been responsible for generating the technical specifications which define the UMTS protocols and performance requirements. 3GPP is a collaboration between a number of telecommunications standards organisations from around the world. These include ETSI, ARIB, CCSA, ATIS, TTA and

Table 1.3 Evolution of UMTS with 3GPP releases

	Release 99	Release 4	Release 5	Release 6	Release 7
DPCH	√	√	√	√	√
Soft handover	√	√	√	√	√
Hard handover	√	√	√	√	√
Inter-system handover	√	√	√	√	√
HSDPA			√	√	√
HSUPA				√	√
MBMS				√	√
CPCH	√	√			
DSCH	√	√			
Operating band I	√	√	√	√	√
Operating band II	√	√	√	√	√
Operating band VIII					√

TTC. 3GPP generates Technical Reports (TR) in addition to Technical Specifications (TS). An example TS is 25.331 which specifies the RRC signalling protocol between a UE and an RNC. An example TR is 25.816 which discusses the deployment of UMTS within operating band VIII.

The set of 3GPP TS and TR have been and continue to be published using a series of releases. The first version of the UMTS specifications was release 99. Early UMTS network deployments were based upon the release 99 version of the 3GPP specifications. The second, third and forth versions of the specifications are release 4, release 5 and release 6. Releases 7 and 8 are also under development. In general, the newer releases of the specifications include additional functionality and performance requirements. There are also examples of functionality being removed from the newer releases. Table 1.3 presents an example subset of the functionality which appears in each 3GPP release.

The release 99 and release 4 versions of the specifications define the use of Dedicated Physical Channels (DPCH) in combination with soft, hard and inter-system handover. Operating bands I and II are specified at this stage. Release 5 introduces HSDPA functionality whereas the Common Packet Channel (CPCH) and Downlink Shared Channel (DSCH) are removed. The CPCH and DSCH were removed because equipment manufacturers were not implementing them. Release 6 of the specifications introduces HSUPA and Multimedia Broadcast Multicast Services (MBMS) whereas release 7 introduces operating band VIII.

References [1–36] are the 3GPP TS most relevant to the main content of this book. These TS are referenced throughout the remainder of the text. References [37–44] are ITU-T Recommendations which are also relevant. The majority of these describe sections of the Iub protocol stack. Others specify the higher layer protocols. References [45–48] are documents generated by the ATM Forum and Internet Engineering Task Force (IETF). These are also relevant to the design and deployment of a Radio Access Network for UMTS.

2

Flow of Data

2.1 Radio Interface Protocol Stacks

- The control plane protocol stack is used to transfer signalling information whereas the user plane protocol stack is used to transfer application data.
- The control plane protocol stack includes RRC, RLC, MAC, Frame Protocol and Physical layers. SRB are associated with the control plane protocol stack.
- The control plane can be used to transfer signalling messages belonging to the higher layer Non-Access Stratum, e.g. call control and mobility management messages.
- The user plane protocol stack includes BMC, PDCP, RLC, MAC, Frame Protocol and Physical layers. The PDCP layer is particularly important for applications like VoIP where the overhead generated by the higher layers is relatively large.
- Logical channels are used to transfer information between the RLC and MAC layers. Transport channels are used to transfer information between the MAC and Physical layers.
- The HSDPA and HSUPA protocol stacks increase the functionality of the Node B. The MAC-hs layer is used by HSDPA, whereas the MAC-e layer is used by HSUPA.
- Key 3GPP specifications: TS 25.301.

Protocol stacks belonging to the radio interface can be categorised as belonging to the control plane or the user plane. Figure 2.1 presents the principle of control plane and user plane protocol stacks.

Control plane protocol stacks are used to transfer signalling. The RRC protocol is used to signal between an RNC and a UE, i.e. all of the messages which are used to establish, maintain and release connections make use of the RRC signalling protocol. When an RNC or a UE sends any RRC message then the control plane protocol stack is used. User plane protocol stacks are used to transfer actual application data. For example, when a UE downloads a file then a user plane protocol stack is used. Likewise, the audio belonging to a speech call makes use of a user plane protocol stack.

Radio interface protocol stacks define the way in which data is transferred between an RNC and a UE. The radio interface protocol stacks treat the Iub as a general transport layer which is able to transfer both control plane and user plane data between the RNC and the Node B. There is an additional protocol stack for the Iub interface which is described in Section 5.

Radio Access Networks for UMTS   Chris Johnson
© 2008 John Wiley & Sons, Ltd

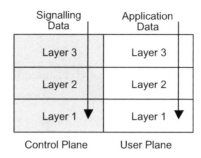

Figure 2.1   Principle of control plane and user plane protocol stacks

2.1.1 Radio Interface Control Plane

The control plane protocol stack used for RRC signalling is illustrated in Figure 2.2. This forms part of the radio interface protocol architecture specified in 3GPP TS 25.301.

Signalling messages are generated at the RRC layer. For example, when a UE wants to move from RRC Idle mode to RRC Connected mode then the RRC layer within the UE generates an RRC Connection Request message. This message is coded into a binary string and is passed down through the RLC, MAC and physical layers of the UE. Each layer processes the message as it passes through. The RLC and MAC layers add headers so the size of the message increases as it is passed down to the physical layer. This means that the bit rate at the bottom of the MAC layer is greater than the bit rate at the top of the RLC layer. The physical layer further increases the size of the message by adding redundancy to help protect the contents of the message as it is transferred across the air-interface. The message becomes an RF signal at the bottom of the physical layer and is transmitted towards the Node B. The Node B receives the RF signal and reverses the physical layer processing which was completed at the transmitter. The message is then passed across to the Frame Protocol layer which packages the message to allow transfer across the Iub transmission link. The transport layer is an ATM connection which can be provided over either a physical cable connection or a microwave link. Once the Frame Protocol layer belonging to the RNC receives the message, the data is extracted and passed up to the MAC layer. The MAC layer removes its header and extracts the payload. Likewise the RLC layer removes its header and passes the resulting payload to the receiving RRC layer. Finally, the RRC layer is able to decode the binary string to re-generate the original message. The message flow is reversed for downlink messages which are generated by the RRC layer of the RNC and received by the RRC layer of the UE.

Figure 2.3 illustrates an example of the processing completed by a serving RNC when a downlink signalling message is generated within the RRC layer. A drift RNC is not required to complete this processing and acts only as a means to route the message to the active set cells. In this example, the

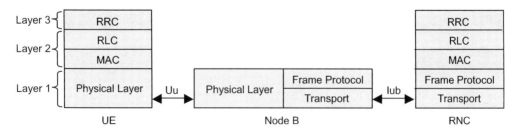

Figure 2.2   Radio interface control plane protocol stack

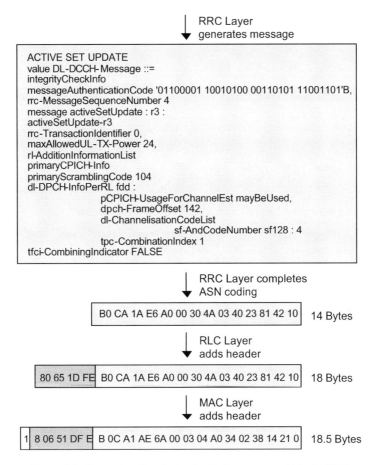

RRC Layer
generates message

ACTIVE SET UPDATE
value DL-DCCH-Message ::=
integrityCheckInfo
messageAuthenticationCode '01100001 10010100 00110101 11001101'B,
rrc-MessageSequenceNumber 4
message activeSetUpdate : r3 :
activeSetUpdate-r3
rrc-TransactionIdentifier 0,
maxAllowedUL-TX-Power 24,
rl-AdditionInformationList
primaryCPICH-Info
primaryScramblingCode 104
dl-DPCH-InfoPerRL fdd :
                    pCPICH-UsageForChannelEst mayBeUsed,
                    dpch-FrameOffset 142,
                    dl-ChannelisationCodeList
                                  sf-AndCodeNumber sf128 : 4
                    tpc-CombinationIndex 1
tfci-CombiningIndicator FALSE

RRC Layer completes
ASN coding

B0 CA 1A E6 A0 00 30 4A 03 40 23 81 42 10   14 Bytes

RLC Layer
adds header

80 65 1D FE B0 CA 1A E6 A0 00 30 4A 03 40 23 81 42 10   18 Bytes

MAC Layer
adds header

1 8 06 51 DF E B 0C A1 AE 6A 00 03 04 A0 34 02 38 14 21 0   18.5 Bytes

Figure 2.3 Processing of an Active Set Update message within the RNC

RRC message is an Active Set Update which is being used to instruct a UE to add a new cell into its active set, i.e. the UE is to enter soft handover and start communicating with an additional cell. The RRC message is relatively small and can be coded into 14 bytes. Abstract Syntax Notation (ASN) coding is used to translate the message into a binary string. Figure 2.3 illustrates this binary string in hexadecimal notation. ASN coding is specified by the ITU-T within recommendations X.680, X.681 and X.691. In this example, the RLC layer adds a 4 byte header before passing the message to the MAC layer which adds a 0.5 byte header. The message is then ready to be packaged by the Frame Protocol layer and to be sent across the Iub to the Node B for physical layer processing.

Different channel types are used to transfer messages between the RLC, MAC and physical layers. Logical channels are used to transfer messages between the RLC and MAC layers, whereas transport channels are used to transfer messages between the MAC and physical layers. Physical channels are used to transfer messages across the air-interface between the physical layer of the Node B and the physical layer of the UE. These three channel types are illustrated in Figure 2.4. This figure also illustrates the use of the terms Packet Data Unit (PDU) and Service Data Unit (SDU). The packet at the top of a layer is known as an SDU whereas the packet at the bottom of a layer is known as a PDU. In the control plane protocol stack, the size of a PDU is always greater than the size of the corresponding SDU because each layer adds an overhead.

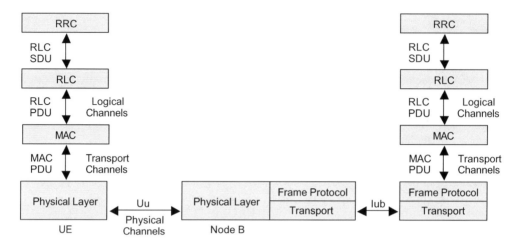

Figure 2.4   Interfaces between protocol stack layers

Some RRC messages are triggered by higher layer procedures within the UE. For example, when an end-user dials a phone number and presses the call button, the higher layers within the UE request the lower layers to provide a connection to the core network. If the UE is in RRC Idle mode, the higher layers trigger the UE to send an RRC Connection Request message to allow the UE to move into RRC Connected mode. The higher layers of the control plane are known as the Non-Access Stratum (NAS). This is in contrast to the RRC, RLC, MAC and physical layers which belong to the Access Stratum (AS). The AS provides a service to the NAS in terms of providing connectivity. When the NAS requires a connection then the AS is responsible for providing that connection. The concepts of AS and NAS are presented in Figure 2.5.

The NAS layer includes Call Control (CC) functionality for circuit switched connection establishment and release, and Session Management (SM) functionality for packet switched connection establishment and release. The NAS also includes mobility management for both the circuit switched and packet switched core network domains. Mobility management procedures are used by the core network to keep track of a UE's location as it moves throughout the network. NAS functionality appears within the UE and the core network, but does not appear within the RNC nor the Node B. The RNC and Node B are limited to being part of the AS.

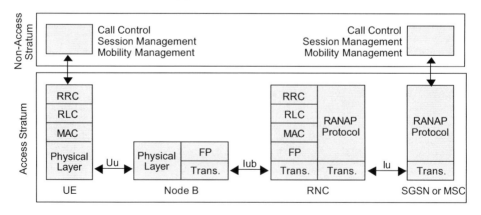

Figure 2.5   Interaction of Access Stratum and Non-Access Stratum

Once the AS has established a connection between the UE and the core network then the NAS is able to transfer messages. NAS messages originating from the UE are passed down to the RRC layer. A NAS message is packaged into an RRC message before being passed between a UE and an RNC. An example of a NAS message packaged into an RRC message is presented in Log File 2.1. This example uses an RRC Initial Direct Transfer message to package a NAS Service Request message. The NAS service request message is visible as a hexadecimal string within the RRC message.

The AS does not decode nor make use of the NAS message in any way. The AS is only responsible for transferring the NAS message. The RRC Downlink Direct Transfer and Uplink Direct Transfer messages are also able to package NAS messages for transfer between the UE and RNC. The RRC System Information Block 1 (SIB1) also includes NAS messages which are broadcast across the air-interface.

Transferring NAS messages between an RNC and the core network makes use of the RANAP signalling protocol. The RANAP signalling protocol provides a packaging service across the Iu similar to that provided by the Frame Protocol across the Iub. The RANAP Initial UE Message and Direct Transfer messages are used to transfer NAS messages across the Iu. Transfering NAS messages forms only a small part of the RANAP protocol. The RANAP protocol includes a large number of other messages which can be used by the control plane of the Iu interface.

The concept of Signalling Radio Bearer (SRB) is used to define the logical signalling connection between the RLC layer in the UE and the peer RLC layer in the RNC. An SRB is a control plane version of a Radio Bearer (RB). Every radio bearer has an identity and radio bearer identities 1 to 4 are reserved for SRB. It is possible to configure more than a single SRB to simultaneously link a UE to an RNC. Each SRB has its own logical channel between the RLC and MAC layers. The MAC layer is used to multiplex the multiple SRB onto a single transport channel which is then processed by the physical layer. This multiplexing operation requires the MAC layer to include a header to specify which SRB is using the transport channel at any point in time. Figure 2.6 illustrates the concept of multiplexing SRB onto a single transport channel.

3GPP TS 25.331 specifies that all messages sent on the Common Control Channel (CCCH) logical channel make use of SRB0. SRB0 is always encapsulated by the RACH transport channel in the uplink direction and the FACH transport channel in the downlink direction. The different types of logical and transport channels are presented in Sections 3.1 and 3.2 respectively. UE in RRC Idle mode are limited to using only SRB0. This SRB is used when establishing an RRC Connection and making the transition from RRC Idle mode to RRC Connected mode. SRB0 uses Transparent Mode (TM) RLC in the uplink direction and Unacknowledged Mode (UM) RLC in the downlink direction. Both of these RLC modes rely upon re-transmissions being provided by layer 3 rather than layer 2, i.e. by the RRC layer rather than the RLC layer. This is in contrast to Acknowledged Mode (AM) RLC which allows re-transmissions from the RLC layer. The use of TM and UM RLC means that entire RRC messages have to be re-transmitted rather than only the individual transport blocks which have been received in error. Re-transmissions may be required if the air-interface conditions are relatively poor.

```
INITIAL DIRECT TRANSFER
UL-DCCH-Message ::=
initialDirectTransfer :
cn-DomainIdentity cs-domain,
intraDomainNasNodeSelector
        version release99 :
        cn-Type gsm-Map-IDNNS :
        routingbasis localPTMSI :
        routingparameter '00000000 10'B
nas-Message '052461034F188005F40100A3BD'H,
v3a0NonCriticalExtensions
        initialDirectTransfer-v3a0ext
        start-Value '00000000 00001011 0110'B
```

Log File 2.1   Example RRC message used to transfer a NAS message

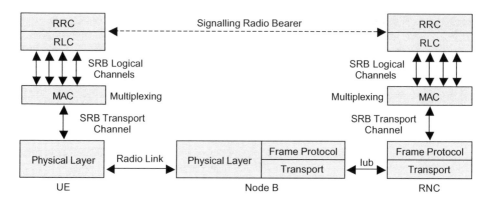

Figure 2.6   Concept of Signalling Radio Bearer

3GPP TS 25.331 specifies that SRB1, 2, 3 and 4 make use of Dedicated Control Channel (DCCH) logical channels. If all four of these SRB are used then four DCCH logical channels are multiplexed onto a single transport channel within the MAC layer. These DCCH SRB are configured during RRC Connection establishment within the RRC Connection Setup message. They can also be configured when making RRC Connected mode state changes, e.g. within a Radio Bearer Reconfiguration message when moving from CELL_DCH to CELL_FACH. Table 2.1 summarises the main characteristics of these four DCCH SRB and the CCCH SRB. When a UE is in RRC Connected mode SRB1, 2 and 3 are configured as a minimum while SRB4 is optional. SRB1 is used for all messages sent on the DCCH which use Unacknowledged Mode (UM) RLC. SRB2 is used for all messages sent on the DCCH which use Acknowledged Mode (AM) RLC and which do not contain NAS messages. SRB3 and SRB4 are used for all messages sent on the DCCH which contain NAS messages. SRB3 and 4 use AM RLC in the same way as SRB2. SRB3 is used rather than SRB 4 either when the NAS indicates that a message has high priority, or when SRB4 has not been configured. SRB4 is used when it has been configured and the NAS indicates that a message has low priority. The priority levels are used within the MAC layer when multiplexing the set of SRB onto a single transport channel, i.e. the priority determines which messages are sent first.

Table 2.1   Summary of signalling radio bearers

		SRB0	SRB1	SRB2	SRB3	SRB4
General	Transfers NAS messages	No	No		Yes	
	High priority	Yes	Yes			No
Uplink	RLC mode	TM	UM	AM		
	Logical channel	CCCH	DCCH			
	Transport channel	RACH	RACH, DCH, E-DCH			
	Physical channel	PRACH	PRACH, DPCH, E-DPCH			
Downlink	RLC mode	UM	UM	AM		
	Logical channel	CCCH	DCCH			
	Transport channel	FACH	FACH, DCH, HS-DSCH			
	Physical channel	S-CCPCH	S-CCPCH, DPCH, HS-PDSCH			

Table 2.2 RRC messages associated with each SRB

Always SRB0	Cell Update, RRC Connection Request, RRC Connection Setup, RRC Connection Reject, URA Update
Always SRB1	None
Always SRB2	Active Set Update Complete, Active Set Update Failure, Assistance Data Delivery, Cell Change Order From UTRAN, Cell Change Order from UTRAN Failure, Counter Check, Counter Check Response, Handover from UTRAN Command, Handover from UTRAN Failure, Handover to UTRAN Complete, Measurement Control, Measurement Control Failure, Paging Type 2, Physical Channel Reconfiguration Complete, Physical Channel Reconfiguration Failure, Radio Bearer Reconfiguration Complete, Radio Bearer Reconfiguration Failure, Radio Bearer Release Complete, Radio Bearer Release Failure, Radio Bearer Setup Complete, Radio Bearer Setup Failure, RRC Connection Setup Complete, RRC Status, Security Mode Command, Security Mode Complete, Security Mode Failure, Signalling Connection Release, Signalling Connection Release Indication, Transport Channel Reconfiguration Complete, Transport Channel Reconfiguration Failure, Transport Format Combination Control Failure, UE Capability Information, UTRAN Mobility Information Confirm, UTRAN Mobility Information Failure
Always SRB3	Initial Direct Transfer
Always SRB4	None
SRB0 or SRB1	Cell Update Confirm, RRC Connection Release, URA Update Confirm
SRB1 or SRB2	Active Set Update, Measurement Report, Physical Channel Reconfiguration, Radio Bearer Reconfiguration, Radio Bearer Release, Radio Bearer Setup, RRC Connection Release Complete, Transport Channel Reconfiguration, Transport Format Combination Control, UE Capability Enquiry, UE Capability Information Confirm, UTRAN Mobility Information
SRB3 or SRB4	Downlink Direct Transfer, Uplink Direct Transfer

Table 2.2 associates the set of RRC messages with each of the five SRB. These RRC messages have been extracted from 3GPP TS 25.331.

The SRB associated with each RRC message can be deduced from the logical channel type, RLC mode and whether or not the RRC message includes a NAS message. The majority of RRC messages make use of SRB2, i.e. the DCCH logical channel with AM RLC and without including a NAS message. There are no messages which always use SRB1 or SRB4. The RRC Initial Direct Transfer message is always sent with high priority whereas the priority of the Uplink and Downlink Direct Transfer messages depends upon the type of encapsulated NAS message.

2.1.2 Radio Interface User Plane

The radio interface user plane protocol stack is used to transfer application data between an RNC and a UE. The generic version of this protocol stack is illustrated in Figure 2.7.

The radio interface section of the user plane protocol stack includes only layers 1 and 2. The RRC layer is only associated with the control plane and does not appear as part of the user plane. All user plane data passes through the RLC, MAC and Physical layers. In addition, some types of data may pass through the Broadcast/Multicast Control (BMC) layer or the Packet Data Convergence Protocol (PDCP) layer.

The BMC layer is used for Cell Broadcast Services (CBS) and is transparent for all other services. The BMC layer operates only in the downlink direction making use of the CTCH logical channel and

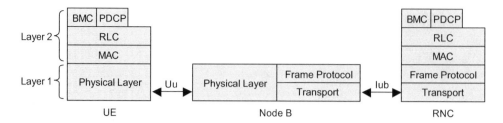

Figure 2.7   Radio interface user plane protocol stack

the FACH transport channel. The use of CBS requires an additional core network domain known as the Broadcast (BC) domain. The BC domain represents a third core network which is used in addition to the CS and PS core networks. The BC core network domain makes use of the Iu-bc interface to provide connectivity between a Cell Broadcast Center (CBC) and the RNC. The RNC receives user plane CBS messages from the CBC. The CBC specifies which Service Areas (SA) the messages are to be broadcast. SA defined for the BC domain include only a single cell. This is in contrast to SA defined for the CS and PS core network domains which may include more than a single cell. There is a single BMC entity within the RNC for each cell that supports CBS. The BMC layer is able to store, schedule and transfer CBS messages to the RLC layer for transmission to one or more UE. CBS always uses Unacknowledged Mode (UM) RLC. The corresponding BMC layer in the UE receives the messages and passes them to the higher layers. UE which support CBS are capable of receiving BMC messages in RRC Idle mode and the RRC Connected mode states CELL_PCH and URA_PCH. The BMC layer within the RNC communicates with the RRC layer within the control plane protocol stack. The BMC layer measures the quantity of cell broadcast traffic and informs the RRC layer of the result. The BMC layer also informs the RRC layer of CBS scheduling information. The RRC layer informs the BMC layer of the configuration of the CTCH logical channel being used for CBS.

The PDCP layer is optional and only applicable to PS services. The PDCP layer is able to provide header compression for IP data streams. This is particularly important for services where the payload is relatively small and the IP header represents a significant percentage of the total data. The user plane protocol stack for Voice over IP (VoIP) is illustrated in Figure 2.8. This figure illustrates both the radio interface section of the protocol stack and the higher layers. Only a single end-user is shown. The peer layers for the codec and RTP/UDP/IP layers are located at the second end-user. The second end-user

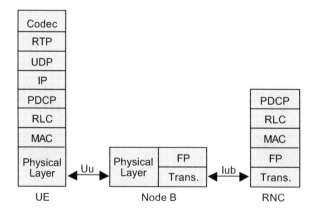

Figure 2.8   User plane protocol stack for Voice over IP (VoIP)

may not be connected using the UMTS network, e.g. the second end-user could be connected directly to the public internet.

In the case of the 12.2 kbps AMR speech codec, the payload is 244 bits per 20 ms speech frame. The IP header is 160 bits when using IP version 4 (IPv4) or 320 bits when using IP version 6 (IPv6). The UDP header is 64 bits and the RTP header is 96 bits. This results in a total RTP/UDP/IP header size of 320 bits (40 bytes) when IPv4 is used or 480 bits (60 bytes) when IPv6 is used. In addition, the RLC layer is operating in Unacknowledged Mode (UM) and adds a header of up to 16 bits. Without header compression the 12.2 kbps AMR speech bit rate would be increased to 29 kbps with IPv4 or 37 kbps with IPv6, i.e. the spectrum efficiency would be reduced by up to a factor of three relative to a CS speech connection which does not require any RTP/UDP/IP nor RLC headers. The overheads are even more significant during periods of speech DTX when the AMR speech codec generates a comfort noise payload of only 56 bits. Transfering such a large overhead across the air-interface would have a significant negative impact upon spectrum efficiency. The PDCP layer is able to increase the feasibility of VoIP across the air-interface by reducing the size of the RTP/UDP/IP header. The release 99 version of 3GPP TS 25.323 specifies the use of IETF RFC 2507 for header compression. The release 4 version introduces the use of IETF RFC 3095 which is also known as Robust Header Compression (ROHC). Both RFC are specifically designed to work well over links with significant packet loss ratios. The main principle of header compression is to avoid sending fields which do not change between consecutive packets. ROHC is able to reduce an RTP/UDP/IPv6 header from 60 bytes to 3 or 4 Bytes. This means that the 12.2 kbps AMR speech bit rate would be increased to 14.6 kbps with IPv6, i.e. the VoIP connection could use a 16 kbps PS data connection.

It is less important for some other PS data connections to make use of the PDCP layer. An example payload size for a typical TCP/IP session is 1420 bytes. This packet size increases to 1460 bytes once a TCP/IPv4 header has been included. In this case, the header represents less than 3% of the payload and header compression would have a less significant impact. The user plane protocol stack for a File Transfer Protocol (FTP) session is illustrated in Figure 2.9. This figure illustrates both the radio interface section of the protocol stack and the higher layers. The FTP application makes use of the TCP/IP protocol stack. A similar protocol stack can be used for internet browsing by swapping the FTP application layer with the HTTP application layer.

In contrast to the VoIP application, FTP sessions and internet browsing are categorised as non-real time applications. This means that the RLC layer can be used in Acknowledged Mode (AM) to allow re-transmissions and an increased reliability of data transfer.

Figure 2.10 illustrates the processing completed by the RNC during a downlink PS data TCP/IP file transfer. In this example, the RNC receives a 1460 Byte TCP/IP packet from the PS core network. This packet includes both the payload and the TCP/IP header. The size of the payload is known as the Maximum Segment Size (MSS). The MSS is negotiated between the TCP client and server during TCP

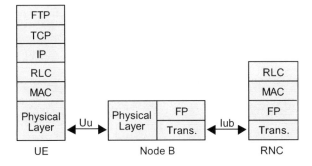

Figure 2.9   User plane protocol stack for a File Transfer Protocol (FTP) session

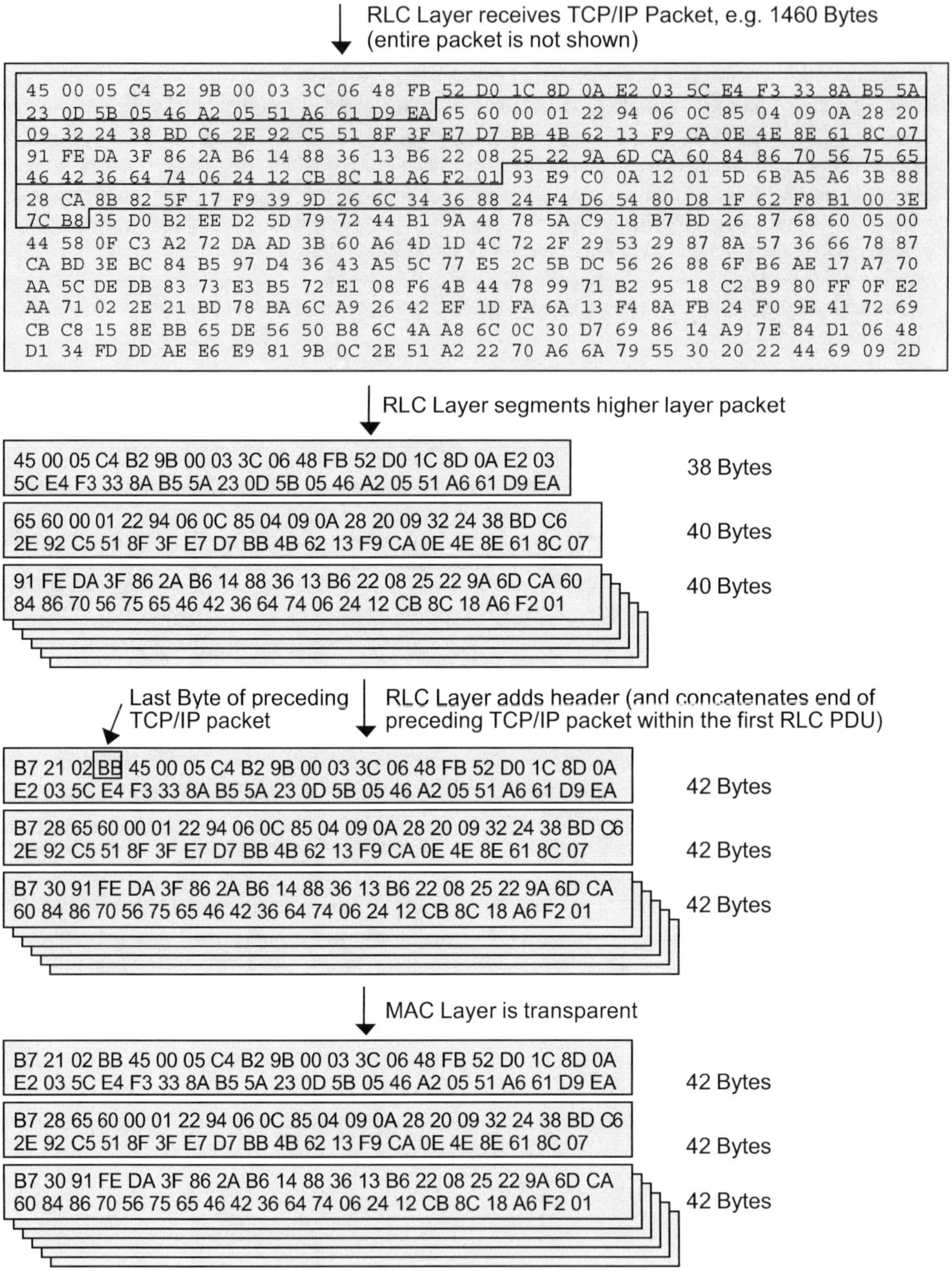

RLC Layer receives TCP/IP Packet, e.g. 1460 Bytes
(entire packet is not shown)

```
45 00 05 C4 B2 9B 00 03 3C 06 48 FB 52 D0 1C 8D 0A E2 03 5C E4 F3 33 8A B5 5A
23 0D 5B 05 46 A2 05 51 A6 61 D9 EA 65 60 00 01 22 94 06 0C 85 04 09 0A 28 20
09 32 24 38 BD C6 2E 92 C5 51 8F 3F E7 D7 BB 4B 62 13 F9 CA 0E 4E 8E 61 8C 07
91 FE DA 3F 86 2A B6 14 88 36 13 B6 22 08 25 22 9A 6D CA 60 84 86 70 56 75 65
46 42 36 64 74 06 24 12 CB 8C 18 A6 F2 01 93 E9 C0 0A 12 01 5D 6B A5 A6 3B 88
28 CA 8B 82 5F 17 F9 39 9D 26 6C 34 36 88 24 F4 D6 54 80 D8 1F 62 F8 B1 00 3E
7C B8 35 D0 B2 EE D2 5D 79 72 44 B1 9A 48 78 5A C9 18 B7 BD 26 87 68 60 05 00
44 58 0F C3 A2 72 DA AD 3B 60 A6 4D 1D 4C 72 2F 29 53 29 87 8A 57 36 66 78 87
CA BD 3E BC 84 B5 97 D4 36 43 A5 5C 77 E5 2C 5B DC 56 26 88 6F B6 AE 17 A7 70
AA 5C DE DB 83 73 E3 B5 72 E1 08 F6 4B 44 78 99 71 B2 95 18 C2 B9 80 FF 0F E2
AA 71 02 2E 21 BD 78 BA 6C A9 26 42 EF 1D FA 6A 13 F4 8A FB 24 F0 9E 41 72 69
CB C8 15 8E BB 65 DE 56 50 B8 6C 4A A8 6C 0C 30 D7 69 86 14 A9 7E 84 D1 06 48
D1 34 FD DD AE E6 E9 81 9B 0C 2E 51 A2 22 70 A6 6A 79 55 30 20 22 44 69 09 2D
```

RLC Layer segments higher layer packet

| 45 00 05 C4 B2 9B 00 03 3C 06 48 FB 52 D0 1C 8D 0A E2 03
5C E4 F3 33 8A B5 5A 23 0D 5B 05 46 A2 05 51 A6 61 D9 EA | 38 Bytes |

| 65 60 00 01 22 94 06 0C 85 04 09 0A 28 20 09 32 24 38 BD C6
2E 92 C5 51 8F 3F E7 D7 BB 4B 62 13 F9 CA 0E 4E 8E 61 8C 07 | 40 Bytes |

| 91 FE DA 3F 86 2A B6 14 88 36 13 B6 22 08 25 22 9A 6D CA 60
84 86 70 56 75 65 46 42 36 64 74 06 24 12 CB 8C 18 A6 F2 01 | 40 Bytes |

Last Byte of preceding / TCP/IP packet RLC Layer adds header (and concatenates end of preceding TCP/IP packet within the first RLC PDU)

| B7 21 02 BB 45 00 05 C4 B2 9B 00 03 3C 06 48 FB 52 D0 1C 8D 0A
E2 03 5C E4 F3 33 8A B5 5A 23 0D 5B 05 46 A2 05 51 A6 61 D9 EA | 42 Bytes |

| B7 28 65 60 00 01 22 94 06 0C 85 04 09 0A 28 20 09 32 24 38 BD C6
2E 92 C5 51 8F 3F E7 D7 BB 4B 62 13 F9 CA 0E 4E 8E 61 8C 07 | 42 Bytes |

| B7 30 91 FE DA 3F 86 2A B6 14 88 36 13 B6 22 08 25 22 9A 6D CA
60 84 86 70 56 75 65 46 42 36 64 74 06 24 12 CB 8C 18 A6 F2 01 | 42 Bytes |

MAC Layer is transparent

| B7 21 02 BB 45 00 05 C4 B2 9B 00 03 3C 06 48 FB 52 D0 1C 8D 0A
E2 03 5C E4 F3 33 8A B5 5A 23 0D 5B 05 46 A2 05 51 A6 61 D9 EA | 42 Bytes |

| B7 28 65 60 00 01 22 94 06 0C 85 04 09 0A 28 20 09 32 24 38 BD C6
2E 92 C5 51 8F 3F E7 D7 BB 4B 62 13 F9 CA 0E 4E 8E 61 8C 07 | 42 Bytes |

| B7 30 91 FE DA 3F 86 2A B6 14 88 36 13 B6 22 08 25 22 9A 6D CA
60 84 86 70 56 75 65 46 42 36 64 74 06 24 12 CB 8C 18 A6 F2 01 | 42 Bytes |

Figure 2.10 Processing of TCP/IP packet within the RNC (transparent PDCP)

connection establishment. The minimum of the values configured at the client and server is adopted. The total size of the TCP/IP packet is known as the Maximum Transmission Unit (MTU). The MTU is equal to the MSS plus the TCP/IP header size, i.e. the MSS is 1420 bytes and the MTU is 1460 bytes.

The RLC layer receives the TCP/IP packet and recognises that it is too large to send as a single unit of data. The RLC layer segments the TCP/IP packet into smaller data units which can be accommodated by single RLC PDU. A single RLC PDU for a downlink file transfer using a dedicated channel is typically 336 bits (42 bytes). The control plane RRC layer informs the user plane RLC layer of this size during the connection establishment procedure. The RLC layer accounts for the size of the RLC header when it segments the TCP/IP packet to ensure that the size of the RLC PDU is always equal to the value instructed. The example shown in Figure 2.10 illustrates that the steady state RLC header size is 2 bytes. The first RLC PDU has a larger header because this PDU includes a section of the preceding TCP/IP packet. The RLC header includes an additional 1 byte to indicate the quantity of data belonging to the preceding TCP/IP packet. In this example there is only a single byte belonging to the preceding TCP/IP packet. Section 2.3 describes the fields within the RLC header in greater detail.

Once the RLC headers have been added, the RLC PDU are passed down to the MAC layer. The number of RLC PDU passed to the MAC layer depends upon the bit rate of the service. If the service has been configured to support 384 kbps then 12 RLC PDU are passed to the MAC layer every 10 ms. This generates a bit rate at the bottom of the RLC layer of $12 \times 336/0.01 = 403.2$ kbps. The figure of 384 kbps represents the bit rate at the top of the RLC layer, i.e. an RLC throughput of $320 \times 12/0.01 = 384$ kbps. Figure 2.10 illustrates the case of 8 RLC PDU which corresponds to an RLC throughput of 128 kbps when the those PDU are delivered once every 20 ms.

In the case of PS data transfer across a dedicated channel, the MAC layer is transparent and does not add any header information. This means that the MAC PDU are equal to the RLC PDU. The set of MAC PDU are transferred across the Iub interface towards the physical layer within the Node B.

The processes are reversed in the receiving UE user plane protocol stack, i.e. the RLC layer removes its header and concatenates the payload to generate complete TCP/IP packets before passing to the higher layers. In the case of a downlink TCP/IP file transfer, the UE is required to send uplink TCP acknowledgements. These acknowledgements are 40 bytes in length when using IPv4 and can fit within a single RLC PDU. In the case of an uplink file transfer, the processing is reversed, i.e. the RLC layer within the UE segments the uplink TCP/IP packets, generates the RLC PDU, passes them through the MAC layer and transmits them towards the Node B. The RLC layer within the RNC is then responsible for removing the RLC header and reassembling the TCP packets before forwarding them towards the PS core network.

The user plane protocol stack for a circuit switched speech connection is illustrated in Figure 2.11. This figure illustrates both the radio interface section of the protocol stack and the higher layers. Only a single end-user is shown. The second end-user may not be connected using the UMTS network, e.g. the second end-user could be connected using the Public Switched Telephone Network (PSTN).

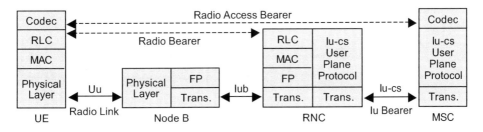

Figure 2.11   User plane protocol stack for the CS speech service

In the case of a CS speech connection, the higher layers are represented by the AMR speech codec. Figure 2.11 illustrates the concepts of Radio Access Bearer (RAB), Radio Bearer (RB) and Radio Link (RL). The RAB represents the logical connection between the UE and the core network. RAB are service specific and so a UE which is simultaneously using multiple services can have multiple RAB, e.g. a UE which has a CS speech connection while completing a PS file transfer has one RAB to the MSC and a second RAB to the SGSN. It is also possible to have multiple RAB to the same core network domain, e.g. a UE which is browsing the internet while downloading emails could have two RAB to the PS core network. Each RAB has a particular Quality of Service (QoS) profile associated with it and this can influence the way in which radio access resources are assigned. A RAB can have subflows which allows differential treatment of multiple bit streams belonging to the same RAB.

The RB represents the logical connection between the UE and the RNC. The concept of a user plane RB is the same as the concept of a control plane SRB. RB identities 1 to 4 are reserved for SRB and so user plane RB have identities which are greater than 4. The RL represents the physical channel connection between the UE and the Node B. If a UE is in soft handover then it will have multiple radio links, i.e. one radio link for each active set cell. If active set cells belong to the same Node B then radio links belong to the same radio link set.

An AMR speech connection makes use of one RAB with three RAB subflows, three logical channels, three transport channels and a single physical channel. The SRB are configured in parallel to the speech connection and make use of a further four logical channels and one transport channel. It is possible that only three SRB are configured requiring three rather than four logical channels. The complete set of logical, transport and physical channels are illustrated in Figure 2.12.

The four SRB logical channels are multiplexed onto a single transport channel within the MAC layer. The SRB transport channel and the three speech transport channels are multiplexed onto a single physical channel within the physical layer. The speech logical channels are not multiplexed within the MAC layer because each is processed differently within the physical layer. The three RAB subflows belonging to the speech service represent three bit streams of differing importance – class A, B and C bits. Class A bits have the greatest importance and are provided with the greatest error protection by the physical layer before transmitting across the air-interface. Class B bits have less importance than class A bits but greater importance than class C bits.

Figure 2.13 illustrates the processing completed by the RLC and MAC layers for the 12.2 kbps speech service. The RLC layer receives data units for each RAB subflow once every 20 ms. 20 ms represents the duration of a speech frame. The data unit sizes are 81 bits for the class A subflow, 103 bits for the class B subflow and 60 bits for the class C subflow. The AMR speech bit rate is then defined

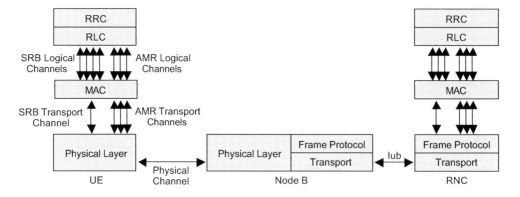

Figure 2.12 Logical, transport and physical channels used by the speech service

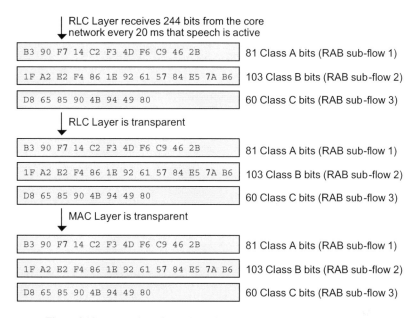

RLC Layer receives 244 bits from the core
network every 20 ms that speech is active

| B3 90 F7 14 C2 F3 4D F6 C9 46 2B | 81 Class A bits (RAB sub-flow 1) |

| 1F A2 E2 F4 86 1E 92 61 57 84 E5 7A B6 | 103 Class B bits (RAB sub-flow 2) |

| D8 65 85 90 4B 94 49 80 | 60 Class C bits (RAB sub-flow 3) |

RLC Layer is transparent

| B3 90 F7 14 C2 F3 4D F6 C9 46 2B | 81 Class A bits (RAB sub-flow 1) |

| 1F A2 E2 F4 86 1E 92 61 57 84 E5 7A B6 | 103 Class B bits (RAB sub-flow 2) |

| D8 65 85 90 4B 94 49 80 | 60 Class C bits (RAB sub-flow 3) |

MAC Layer is transparent

| B3 90 F7 14 C2 F3 4D F6 C9 46 2B | 81 Class A bits (RAB sub-flow 1) |

| 1F A2 E2 F4 86 1E 92 61 57 84 E5 7A B6 | 103 Class B bits (RAB sub-flow 2) |

| D8 65 85 90 4B 94 49 80 | 60 Class C bits (RAB sub-flow 3) |

Figure 2.13 Processing of speech service data by the RLC and MAC layers

by the total number of bits divided by the period of the speech frame, i.e. 244 / 0.02 = 12.2 kbps. This corresponds to the bit rate during periods when the end-user is active and there is speech to be transmitted. The bit rate is reduced during periods of speech DTX and the AMR codec generates only comfort noise. Comfort noise data is used to generate audio background noise at the receiver rather than allowing total silence during periods of speech DTX. This provides a more natural sounding conversation. Comfort noise data is transmitted once every eight speech frames during periods of speech DTX, i.e. once every 160 ms. Comfort noise data occupies 39 bits of the class A subflow resulting in an instantaneous bit rate of 1.95 kbps and an average bit rate of 244 bps.

Figure 2.13 illustrates that the RLC layer is used in Transparent Mode (TM) and does not add any header to the higher layer data. Likewise, the MAC layer is transparent and does not add any header. The MAC PDU are forwarded to the physical layer where layer 1 processing is completed to provide error protection and detection across the air-interface.

The generic user plane protocol stack presented in Figure 2.7 and the service-specific protocol stacks presented in Figure 2.8, Figure 2.9 and Figure 2.11 assume that dedicated transport channels are used for data transfer. The release 99 version of the 3GPP specifications defines the use of dedicated channels. The release 5 version of the specifications introduces the use of High Speed Downlink Packet Access (HSDPA). HSDPA is able to offer high bit rate PS data connections in the downlink direction. The HS-DSCH transport channel and HS-PDSCH physical channel are introduced for this purpose. The generic radio interface protocol stack for HSDPA is illustrated in Figure 2.14.

The use of HSDPA does not change the RLC layer, but introduces an additional MAC sublayer known as the MAC-hs. The MAC-hs layer is located within both the Node B and the UE. Including the MAC-hs layer within the Node B allows the system to become significantly more responsive. It also increases the intelligence and responsibilities of the Node B relative to the RNC. The MAC-hs layer within the Node B is responsible for flow control of downlink HSDPA data between the RNC and the Node B; scheduling and priority handling; Hybrid Automatic Repeat Request (HARQ); and transport format selection, i.e. bit rate selection. The MAC-hs layer within the UE is responsible for HARQ; reordering queue distribution; reordering; and disassembly.

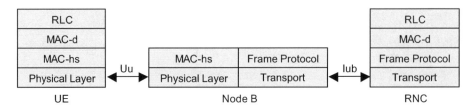

Figure 2.14   HSDPA protocol stack

These MAC-hs functions are described in detail within section 6. The Frame Protocol layer used for HSDPA across the Iub is different to that used for a dedicated channel. The HSDPA Frame Protocol forms part of the common channel Frame Protocol specification and includes control frames to support flow control functionality. The Frame Protocol layers for both dedicated channels and HSDPA are described in Section 2.5. The HSDPA physical layer is also different to that used for a dedicated channel. HSDPA allows the use of 16QAM modulation in addition to QPSK. It also allows multi-code transmission with a channelisation code spreading factor of 16. These differences are described in Section 2.6.

The release 6 version of the 3GPP specifications introduces the use of High Speed Uplink Packet Access (HSUPA). HSUPA is able to offer high bit rate PS data connections in the uplink direction. The E-DCH transport channel and E-DPCH physical channel are introduced for this purpose. HSUPA does not have to be paired with HSDPA, but in general the two are used in combination. The generic radio interface protocol stack for HSUPA is illustrated in Figure 2.15.

Similar to HSDPA, the use of HSUPA does not change the RLC layer, but introduces additional MAC sublayers known as the MAC-es and MAC-e. On the network side, the MAC-es is located within the RNC whereas the MAC-e is located within the Node B. The MAC-es and MAC-e layers are combined within the UE to form the MAC-es/e layer. Including the MAC-e layer within the Node B allows the system to become significantly more responsive. It also further increases the intelligence and responsibilities of the Node B relative to the RNC. The MAC-e layer within the Node B is responsible for scheduling; HARQ; E-DCH Control; de-multiplexing. The MAC-es/e layer within the UE is responsible for HARQ; transport format selection, i.e. bit rate selection; multiplexing and transmission sequence number setting. The MAC-es layer within the RNC is responsible for reordering queue distribution; reordering; macro diversity selection; and disassembly.

These MAC-es and MAC-e functions are described in detail within Section 7. The Frame Protocol layer used for HSUPA across the Iub is different to that used for a dedicated channel. The Frame Protocol layers for both dedicated channels and HSUPA are described in Section 2.5. The HSDPA Physical layer is also different from that used for a dedicated channel. HSUPA allows the use of spreading factor 2 channelisation codes and additional multi-code transmission combinations. These differences are described in Section 2.6.

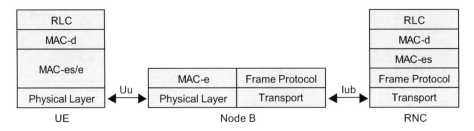

Figure 2.15   HSUPA protocol Stack

2.2 RRC Layer

- The RRC layer represents layer 3 of the control plane and is located within the RNC and UE. It is responsible for all signalling messages transferred between the RNC and UE. It is able to both generate its own messages and package messages originating from the non-access stratum layer, e.g. call control and mobility management messages.
- RRC messages are used to establish, release and maintain connections. RRC messages are encoded using unaligned PER ASN.1 specified by ITU-T recommendation X.691.
- The RRC layer uses a state machine which includes RRC Idle mode and RRC Connected mode. RRC Connected mode includes CELL_DCH, CELL_FACH, CELL_PCH and URA_PCH.
- UE in RRC Idle mode have minimal resource requirements. Their main tasks are cell selection and cell reselection, reading system information and using a DRX cycle to receive paging indicators and potentially paging type 1 messages. Paging messages are broadcast across entire location areas or routing areas. UE may trigger location area and routing area updates if registered with the CS and PS CN respectively. UE are able receive CBS and MBMS services. UE are identified by their IMSI, TMSI or P-TMSI.
- UE in CELL_DCH have the greatest resource requirement, but have the potential to transfer large quantities of data. Dedicated channels, HSDPA and HSUPA can be used in CELL_DCH. Dedicated channels are applicable to both CS and PS connections whereas HSDPA and HSUPA are applicable to PS connections. Soft handover, inter-frequency handover and inter-system handover are possible. UE can receive paging type 2 messages without a DRX cycle. UE are identified by their U-RNTI, H-RNTI or E-RNTI.
- UE in CELL_FACH have a lower resource requirement than UE in CELL_DCH. UE are required to receive continuously, but not transmit continuously. UE have the potential to transfer small quantities of PS data. The release 7 version of the 3GPP specifications introduces HSDPA in CELL_FACH. UE complete cell reselection and both inter-frequency and inter-system handover are possible. UE can receive paging type 2 messages without a DRX cycle. UE are identified by their U-RNTI and C-RNTI.
- UE in CELL_PCH have a lower resource requirement than UE in CELL_FACH, but cannot transfer uplink data. Only downlink CBS and MBMS services are possible. The RNC tracks the location of the UE on a per cell basis and paging type 1 messages can be forwarded to individual cells using a DRX cycle. UE complete cell reselection for intra-frequency, inter-frequency and inter-system mobility. UE are identified by their U-RNTI.
- URA_PCH is similar to CELL_PCH except the RNC tracks the location of the UE on a per URA basis rather than on a per cell basis.
- The RRC layer uses a set of timers, counters and constants to supervise specific procedures, e.g. T300, V300 and N300 are used to supervise the transmission of RRC Connection Request messages.
- The RRC layer uses a set of measurement reporting events to identify when UE experience certain conditions. These reporting events are categorised as intra-frequency, inter-frequency, inter-RAT, traffic volume, quality, UE internal and UE positioning. The most common reporting events are 1a, 1b and 1c for soft handover radio link addition, deletion and replacement respectively.
- Key 3GPP specifications: TS 25.331, TS 25.304.

The Radio Resource Control (RRC) layer represents layer 3 of the control plane belonging to the radio interface protocol stack. The RRC layer is located within the UE and the RNC. 3GPP TS 25.331

specifies the RRC protocol whereas TS 25.304 specifies some of the procedures which are directly applicable to the RRC layer. These include cell selection and cell reselection.

The transmit side of the RRC layer is responsible for generating all signalling messages transferred between an RNC and a UE, i.e. all messages transferred by the Signalling Radio Bearers (SRB). The RLC layer is located below the RRC layer and is responsible for providing the data transfer service. The receive side of the RRC layer is responsible for interpreting and acting upon signalling messages. Signalling messages may originate from either the RRC layer or the Non-Access Stratum (NAS) layer, e.g. an Active Set Update message originates from the RRC layer whereas a Service Request message originates from the NAS layer. When the message originates from the NAS layer, an RRC message is used to package the NAS message before transferring it between the RNC and UE. In the downlink direction, the RNC may receive NAS messages from either the CS or PS core network domains. In the uplink direction, the RNC is responsible for routing NAS messages to the appropriate core network domain. RRC messages which are used to package NAS messages include an information element to specify the relevant core network domain.

RRC messages are used to establish, maintain and release connections, e.g. when an end-user wishes to make a speech call then a series of RRC messages are used to establish the connection. Likewise, when an end-user wishes to end a speech call a series of RRC messages are used to release the connection. In general, specific RRC messages belong to specific RRC procedures, e.g. the Active Set Update and Active Set Update Complete messages belong to the Active Set Update procedure.

2.2.1 RRC States

The RRC protocol includes a state machine to define the procedures that a UE should complete and also the type of connection that a UE can be assigned. The RRC state machine is illustrated in Figure 2.16. The starting point within the state machine is RRC Idle mode. A UE enters RRC Idle mode once it has been switched on. Making the transition to RRC Connected mode involves establishing an RRC connection. A UE always has one RRC connection between itself and the serving RNC while in RRC Connected mode. It is not possible for a UE to have more than one RRC connection. Making the transition back to RRC Idle mode involves releasing the RRC connection.

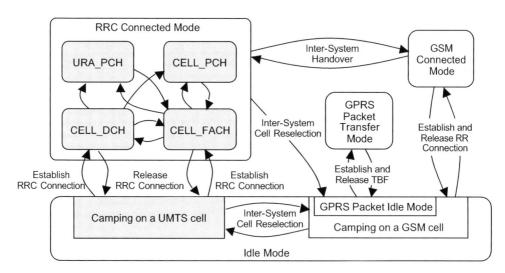

Figure 2.16 RRC state machine including transitions to and from GSM/GPRS

Establishing an RRC connection always requires the UE to transmit an RRC Connection Request message and for the RNC to respond with an RRC Connection Setup message. The RRC Connection Setup message can instruct the UE to move into either CELL_DCH or CELL_FACH. The UE may subsequently move into CELL_PCH or URA_PCH from either CELL_DCH or CELL_FACH. A UE can only return to CELL_DCH via CELL_FACH because it is necessary for a UE to be in CELL_FACH to complete the reconfiguration procedure required to move into CELL_DCH. A UE cannot signal with the network while in CELL_PCH nor URA_PCH. The transition from RRC Connected mode to RRC Idle mode involves the RNC transmitting an RRC Connection Release message while the UE is in either CELL_DCH or CELL_FACH. Alternatively if a UE moves out of coverage for a sufficiently long period of time then it will return to RRC Idle mode.

The GSM and GPRS state machines are presented adjacent to the UMTS RRC state machine. It is possible to complete an inter-system handover between RRC Connected mode and GSM connected mode, i.e. speech calls can be switched between the two systems. This may be done if a UE moves out of coverage of one system or if the load of one system becomes higher than the load of the other system. The release 99 to release 6 versions of the 3GPP specifications do not support inter-system handover for packet switched connections. Instead they support inter-system cell reselection which allows a UE to move between the two systems and subsequently request resources on the target system, i.e. resources are not reserved prior to making the transition. The release 7 version of the 3GPP specifications modifies the interaction between the RRC and GPRS state machines to include inter-system handover for packet switched connections. A UE will then be able to move directly between RRC Connected mode and GPRS Packet Transfer mode.

2.2.1.1 RRC Idle Mode

Battery life is maximised while the UE is in RRC Idle mode because there are relatively few procedures to complete and in general those procedures are based upon reception rather than transmission. The only procedure initiated from RRC Idle mode which involves transmission is RRC connection establishment. The set of RRC Idle mode procedures is summarised in Table 2.3.

When a UE is switched on, it enters RRC Idle mode and completes PLMN selection prior to selecting a cell and registering with the network. PLMN selection requires the reception of System Information messages to identify the PLMN to which a specific cell belongs, i.e. the PLMN identity is broadcast within the Master Information Block (MIB). The registration procedure requires signalling and is completed in RRC Connected mode. Figure 2.17 presents two possible sets of RRC state transitions for completing the registration procedure. In the first set, the UE moves into the RRC Connected mode state CELL_DCH, completes the registration procedure and then returns to RRC Idle mode. In the second set, the UE moves into the RRC Connected mode state CELL_FACH, completes the registration procedure and then returns to RRC Idle mode. Both sets of transitions are allowed by the 3GPP specifications.

Table 2.3   UE procedures in RRC Idle mode

Reception of System Information messages
PLMN selection
Cell selection
Registration (requires RRC connection establishment)
Reception of paging messages (may trigger RRC connection establishment)
Cell reselection
Location and routing area updates (requires RRC connection establishment)
Reception of Cell Broadcast and MBMS services if enabled

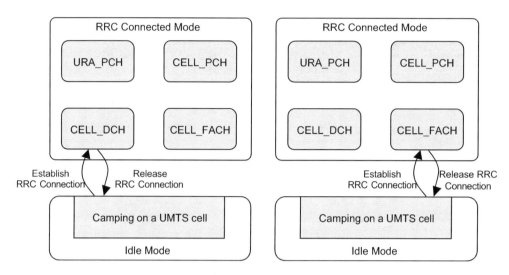

Figure 2.17 Two possible sets of RRC state transitions for the registration procedure

A UE may register with only the CS core network and become IMSI attached, or may register with both the CS and PS core networks and become both IMSI and GPRS attached. If a UE does not become GPRS attached when switched on, the PS connection establishment delay is increased because the UE has to complete the registration procedure as part of the connection establishment procedure. The drawback of becoming GPRS attached when the UE is switched on is an increased PS core network load. The registration procedures inform the core network of the UE location. The Mobility Management (MM) section of the CS core network NAS is informed of the current location area, whereas the GPRS Mobility Management (GMM) section of the PS core network NAS is informed of the current routing area. The core network is then aware of the UE location. The RNC does not maintain a record of the UE location while the UE is in RRC Idle mode.

If a UE is IMSI attached, the Call Control (CC) section of the CS core network NAS can trigger the UE to establish an RRC connection by broadcasting a paging message across the appropriate location area. This procedure would be used if a UE has an incoming speech or video call. The UE checks for a paging message once every Discontinuous Reception (DRX) cycle. The use of a DRX cycle avoids the requirement for the UE to continuously listen for paging messages and helps to increase battery life. Large DRX cycle lengths increase battery life, but also increase the connection establishment delay. Once a paging message has been received the UE will initiate the transition to RRC Connected mode by transmitting an RRC Connection Request message. Likewise, if a UE is IMSI attached, the CC section of the UE NAS can trigger RRC connection establishment for a mobile originated CS service. Figure 2.18 presents two sets of RRC state transitions for establishing and completing a CS service connection. The first set illustrates that both the call establishment and call completion phases are completed in CELL_DCH. The RRC connection is subsequently released and the UE returns to RRC Idle mode. The second set of state transitions illustrates the case where part of the call establishment signalling is completed in CELL_FACH. The UE is only moved to CELL_DCH once the user plane radio bearer has been established. The CS service is subsequently connected in CELL_DCH and the UE returned to RRC Idle mode once the connection has been completed.

If a UE is GPRS attached, the Session Management (SM) section of the PS core network NAS can trigger a UE to establish an RRC connection by broadcasting a paging message across the appropriate routing area. If a UE is IMSI attached, but not GPRS attached, all paging messages must be sent from the CS core network. Likewise, if a UE is GPRS attached, the SM section of the UE NAS can

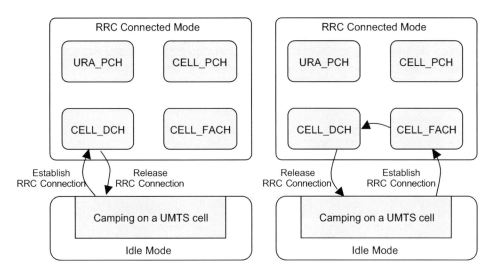

Figure 2.18   Two sets of RRC state transitions for a CS service connection

trigger RRC connection establishment for any mobile originated PS services. If a UE is not GPRS attached when the end-user initiates a PS service then the GMM section of the UE NAS must first trigger the UE to register with the PS core network before the SM section of the UE NAS can establish a connection.

Figure 2.19 presents two sets of RRC state transitions for establishing and completing a PS service connection. PS services have many more combinations of possible state transitions than CS services. CS services are limited to using the CELL_DCH connected mode state whereas PS services are able to use all four of the RRC Connected mode states. The first example illustrates a PS connection being established and completed in CELL_DCH. Once the data transfer has been completed the UE is moved to CELL_FACH and subsequently to CELL_PCH before being returned to RRC Idle mode. The

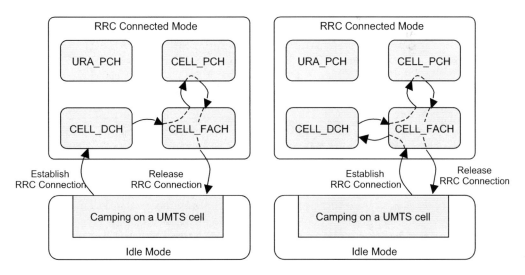

Figure 2.19   Two sets of RRC state transitions for a PS service connection

CELL_FACH and CELL_PCH states are used after the data transfer to avoid an immediate release of all resources. The traffic profile of PS data services is often characterised by multiple data transfers with intermediate periods of inactivity. If all resources were released after every data transfer then the end-user would experience an increased delay when starting to transfer the next block of data. Web browsing and email applications are often characterised by a reading time between downloads. These applications benefit from maintaining the RRC connection during the reading time. If multiple data transfers are completed then the UE can be moved backwards and forwards between CELL_DCH, CELL_FACH and CELL_PCH. The CELL_FACH and CELL_PCH states allow a reduced set of resources to be maintained for the UE until the start of the next data transfer. If the period of inactivity is relatively long then the RRC connection can be released and the UE can be returned to RRC Idle mode. This example does not make use of URA_PCH although URA_PCH can be used in a similar way to CELL_PCH, i.e. a UE can wait for any further PS data transfer in URA_PCH.

The second example presented in Figure 2.19 illustrates a similar set of state transitions, but the PS connection establishment is completed in CELL_FACH. If only small quantities of data are transferred then the UE can remain in CELL_FACH and avoid the requirement for CELL_DCH. If larger quantities of data are transferred then the UE can be moved into CELL_DCH. Once the data transfer has been completed the UE is moved to CELL_FACH and subsequently to CELL_PCH before being returned to RRC Idle mode.

If a UE is mobile while in RRC Idle mode then it completes intra-system cell reselections without informing the network unless the UE is IMSI attached and a location area boundary is crossed, or the UE is GPRS attached and a routing area boundary is crossed. If a UE is IMSI attached and a location area boundary is crossed the MM section of the UE triggers a location update procedure. This procedure informs the MM section of the CS core network of the new and old location areas. If a UE is GPRS attached and a routing area boundary is crossed the GMM section of the UE triggers a routing area update procedure. This procedure informs the GMM section of the PS core network of the new and old routing areas. A UE can also complete an inter-system cell reselection while in RRC Idle mode. An inter-system cell reselection is completed if a UE moves out of UMTS coverage, but remains within GSM coverage. The UE then registers with the GSM system and camps upon GSM in Idle mode. Likewise a UE can complete an inter-system cell reselection from GSM to UMTS requiring a registration procedure on the UMTS system.

Both the cell selection and cell reselection procedures make use of CPICH RSCP and CPICH Ec/Io measurements. In RRC Idle mode, a UE has to periodically monitor the CPICH RSCP and CPICH Ec/Io of the cell upon which it is camped. The cell selection criteria have to be checked at least once every DRX cycle. The RRC Idle mode DRX cycle can be configured to have a period of 640, 1280, 2560 or 5120 ms. If intra-system neighbouring cell measurements are triggered then the CPICH RSCP and CPICH Ec/Io of those cells must also be measured. Likewise, if inter-system cell reselection measurements are triggered the RSSI of the neighbouring GSM cells must be measured.

Whenever a UE completes cell reselection it must read system information messages broadcast on the Primary Common Control Physical Channel (P-CCPCH). The majority of system information has a scope of a single cell and becomes invalid once a UE changes cell. A UE has to re-read this system information as part of the cell reselection procedure. Other system information has a scope of a PLMN. This system information only has to be re-read if the UE changes PLMN or if the value tag associated with that system information changes. Value tags are used to indicate when the content of some system information changes. A value tag is broadcast for the Master Information Block (MIB), each Scheduling Block (SB) and each System Information Block (SIB). A change in the value tag indicates that the content of the relevant MIB, SB or SIB has changed and that the content must be re-read. This allows value tags to be used to force UE to re-read system information with a scope of PLMN even when changing cells within the same PLMN. Value tags may also change while stationary within a single cell and this forces the UE to re-read the relevant section of system information. If a network optimisation engineer decides to change a network parameter, e.g. the neighbour list broadcast in System Information Block (SIB) 11, then the value tag must also be changed to indicate that the system information must be re-read. The value tag for system information which has a scope of a single cell ranges from 1 to 4 whereas the value tag for system information which has a scope of a PLMN ranges from 1 to 256. The range is significantly less for system information which has a scope of a single cell because UE will re-read this system information when changing cell irrespective of the value tag. SIB 7 represents an exception and does not use a value tag. SIB 7 includes a measurement of the uplink interference floor which is relatively dynamic and changes over time. SIB 7 includes an expiration time factor which defines how frequently a UE should re-read its contents. This replaces the requirement for a value tag for SIB 7.

UE in RRC Idle mode are able to receive Cell Broadcast Services (CBS) and Multimedia Broadcast Multicast Services (MBMS) without the requirement to be in RRC Connected mode. The reception of MBMS services may require a short-term transition to RRC Connected mode to complete the MBMS counting procedure. This procedure allows the network to count the number of UE receiving an MBMS service and to subsequently allocate radio resources appropriately, i.e. if there is only a few MBMS UE then it may be more efficient to use dedicated physical channels rather than a common physical channel.

A UE in RRC Idle mode can be identified by its International Mobile Subscriber Identity (IMSI), Temporary Mobile Subscriber Identity (TMSI) or Packet Temporary Mobile Subscriber Identity (P-TMSI). An IMSI is fixed for a specific USIM and is not allocated by the network during registration. A TMSI can be allocated by the CS core network whereas a P-TMSI can be allocated by the PS core network. A TMSI is unique within a location area. The MCC, MNC and Location Area Code (LAC) are specified in combination with the TMSI to ensure that the UE identity is unique globally. A P-TMSI is unique within a routing area. The MCC, MNC, LAC and Routing Area Code (RAC) are specified in combination with the P-TMSI to ensure that the UE identity is unique globally. The TMSI and P-TMSI are used in preference to the IMSI to help protect the confidentiality of the UE identity. The TMSI and P-TMSI are less vulnerable to malicious attacks because they are not fixed. It is usual to reallocate the TMSI as part of the location area update procedure and to reallocate the P-TMSI as part of the routing area update procedure.

3GPP TS 25.331 specifies that when a UE transmits an RRC Connection Request message to enter RRC Connected mode then it should include an initial UE identity. The order of preference for the initial UE identity is specified as TMSI, P-TMSI, IMSI and International Mobile Equipment Identity (IMEI). The IMEI is only used if none of the other three identities are available. The IMEI identifies the UE rather than the USIM. It would be necessary to include the IMEI when a UE without a USIM is used to make an emergency call. The RNC echoes the same initial UE identity within the RRC Connection Setup or RRC Connection Reject messages. Other UE identities, e.g. U-RNTI, H-RNTI, E-RNTI, C-RNTI are assigned once the UE is in RRC Connection mode.

2.2.1.2 RRC Connected Mode – CELL_DCH

CELL_DCH is the most important RRC state in terms of data transfer. CELL_FACH allows relatively small quantities of PS data to be transferred whereas CELL_DCH allows large quantities of both CS and PS data to be transferred. CELL_DCH is also the RRC state which consumes the greatest quantity of UE and network resources. The main characteristics of CELL_DCH are summarised in Table 2.4.

A UE can be instructed to enter CELL_DCH from either RRC Idle mode or CELL_FACH. These state transitions are illustrated in Figure 2.20.

The state transition from RRC Idle mode to CELL_DCH is always completed using the RRC connection establishment procedure. The RRC Connection Setup message is used to instruct the UE to move into CELL_DCH. An example of the relevant section from an RRC Connection Setup message is presented in Log File 2.2.

Table 2.4 Main characteristics of CELL_DCH

Possible to transfer large quantities of uplink and downlink data
Dedicated channels can be used for both CS and PS connections
HSDPA and HSUPA can be used for PS connections
UE and network resource requirement is relatively high
Soft handover possible for dedicated channels and HSUPA
Inter-system and inter-frequency handover possible
Radio access network uses U-RNTI, H-RNTI, E-RNTI to identify UE
Paging Type 2 messages without a DRX cycle are used for paging purposes

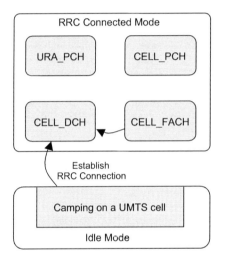

Figure 2.20   RRC state transitions used to enter CELL_DCH

```
RRC_CONNECTION_SETUP
DL-CCCH-Message
rrcConnectionSetup-r3
initialUE-Identity
    tmsi-and-LAI
        tmsi
            Bin: 01 13 F6 C8
        lai
            plmn-Identity
                mcc
                mcc value: 234
                mnc
                mnc value: 99
            lac
                Bin: 27 83
rrc-TransactionIdentifier: 0
new-U-RNTI
    srnc-Identity
        Bin: 04 3
    s-RNTI
        Bin: 10 69 8
rrc-StateIndicator: cell-DCH
```

Log File 2.2   Section from an RRC Connection Setup message

```
RADIO_BEARER_RECONFIGURATION
DL-DCCH-Message
integrityCheckInfo
messageAuthenticationCode
      Bin: C3 72 A0 59
rrc-MessageSequenceNumber: 12
radioBearerReconfiguration-r3
rrc-TransactionIdentifier: 1
rrc-StateIndicator: cell-DCH
```

Log File 2.3 Section from a Radio Bearer Reconfiguration message

The RRC state indicator instructs the UE to move into CELL_DCH rather than CELL_FACH. This section of the RRC Connection Setup message also assigns a new UTRAN Radio Network Temporary Identity (U-RNTI) to the UE. The U-RNTI is a concatenation of the SRNC identity and an SRNC RNTI (S-RNTI). The inclusion of the TMSI and LAI indicates that these were included within the RRC Connection Request message as the initial UE identity.

A UE can be instructed to make the transition from CELL_FACH to CELL_DCH using a Radio Bearer Reconfiguration or Transport Channel Reconfiguration message. It is necessary for the RNC to use a Radio Bearer Reconfiguration message if the RLC layer is reconfigured. If only the transport channels and physical channels are reconfigured then a Transport Channel Reconfiguration message can be used. Alternatively, a Radio Bearer Setup or Physical Channel Reconfiguration message could be used. An example of the relevant section from a Radio Bearer Reconfiguration message is presented in Log File 2.3.

Similar to the RRC Connection Setup message presented in Log File 2.2, the RRC state indicator instructs the UE to move into CELL_DCH. In this case, it is not necessary to include a U-RNTI because the UE is already in RRC Connected mode and has already been assigned a U-RNTI.

Table 2.5 presents the channel types which can be used in CELL_DCH. Dedicated channels, HSDPA and HSUPA make use of CELL_DCH. For the release 99 to release 6 versions of the 3GPP specifications, these channel types are not able to use any other RRC state. The release 7 version of the 3GPP specifications introduces the concept of HS-FACH which allows HSDPA to be used in CELL_FACH. Nevertheless, it is not intended for HSDPA to transfer large quantities of data in CELL_FACH. Dedicated channels can be used for both CS and PS connections whereas HSDPA and HSUPA can only be used for PS connections. CELL_DCH is the only RRC state which allows CS connections, i.e. UE with ongoing CS speech or video calls must be in CELL_DCH. CS connections are released when leaving CELL_DCH. PS connections are able to use all four RRC connected mode states although CELL_PCH and URA_PCH do not allow the transfer of any user plane data other than that generated by the CBS and MBMS services.

The relatively significant resource requirement means that UE are not held in CELL_DCH for longer than necessary. The release 7 version of the 3GPP specifications introduces the concept of Continuous Packet Connectivity which defines a set of features aimed at reducing the resource requirement and consequently allowing more UE to remain in CELL_DCH for longer. Figure 2.18 illustrates that UE

Table 2.5 Channel types used in CELL_DCH

	Logical channel	Transport channel	Physical channel	Direction
Dedicated	DTCH, DCCH	DCH	DPCH	Uplink and downlink
HSDPA	DTCH, DCCH	HS-DSCH	HS-PDSCH	Downlink
HSUPA	DTCH, DCCH	E-DCH	E-DPCH	Uplink

are returned to RRC Idle mode when CS connections are released, e.g. when a user disconnects a speech or video call. This transition to RRC Idle mode releases the complete set of connected mode resources. This scenario is not applicable to multi-RAB connections. If a UE has an ongoing PS connection when the CS connection is released, the resources associated with the CS connection are released, but the UE remains in CELL_DCH for the PS connection. Figure 2.19 illustrates that UE are moved to CELL_FACH when PS connections experience a period of inactivity, e.g. after a user has finished downloading an internet page or sending an email. A Radio Bearer Reconfiguration message can be used to instruct the UE to make this transition. The period of inactivity which triggers the transition to CELL_FACH is typically 3 to 5 seconds. A shorter period of inactivity may be applied to connections which consume larger quantities of resource, e.g. a 384 kbps dedicated channel connection could have a shorter inactivity time than a 64 kbps dedicated channel connection. When PS connections are moved to CELL_FACH the majority of resources are released, but the Iu-ps signalling connection is maintained.

Figure 2.16 illustrates that the network can instruct UE to move into CELL_PCH and URA_PCH from CELL_DCH. These transitions may be used for PS connections during periods of congestion. The CELL_PCH and URA_PCH states require less resource than CELL_DCH and CELL_FACH and so they can be useful in terms of allowing UE to maintain their RRC connections while minimising the resource requirement. The precise set of triggering mechanisms used to initiate the transition from CELL_DCH to other RRC connected mode states is implementation dependant. 3GPP has standardised the relevant signalling procedures which can be used when necessary, e.g. the Radio Bearer Reconfiguration procedure can be used to move a UE from CELL_DCH to CELL_PCH.

From the end-user perspective, the most significant resource used in CELL_DCH is the UE transmit power. Irrespective of the connection type, a UE has to continuously transmit the uplink DPCCH while in CELL_DCH. This transmit requirement has a significant impact upon UE power consumption and battery life. One of the features associated with Continuous Packet Connectivity in release 7 of the 3GPP specifications is uplink gating. This feature reduces UE power consumption by allowing transmission gaps during periods of inactivity, i.e. the uplink DPCCH is no longer transmitted during every time slot. If an uplink dedicated channel has been configured, the UE also has to transmit the DPDCH whenever there is data to send. Likewise, if an E-DPDCH has been configured, the UE has to transmit the E-DPDCH whenever there is data to send. Transmission of the E-DPDCH requires transmission of the E-DPCCH. If HSDPA has been configured then the UE also has to transmit the HS-DPCCH to provide channel quality reports and MAC-hs acknowledgements. The uplink gating feature associated with Continuous Packet Connectivity allows HS-DPCCH transmission gaps as well as DPCCH transmission gaps during periods of inactivity.

UE in CELL_DCH also have to receive continuously and this has an impact upon UE power consumption. The network transmits either the downlink DPCCH or the F-DPCH during every time slot. If a downlink dedicated channel has been configured, the UE has to continuously monitor for the reception of any data on the DPDCH. Likewise, if HSDPA has been configured, the UE has to continuously monitor the HS-SCCH to detect whether or not any data is about to be scheduled on the HS-PDSCH. If HSUPA has been configured, the UE has to monitor the AGCH, RGCH and HICH. UE in CELL_DCH also have to measure the CPICH for the purposes of triggering soft handover and potentially inter-system or inter-frequency handover. UE can also use CPICH measurements for channel estimation to help receive other physical channels. Another feature associated with release 7 Continuous Packet Connectivity is discontinuous reception during periods of CELL_DCH inactivity. Similar to discontinuous transmission, this helps to reduce UE power consumption.

Continuous UE transmit power also consumes resources from the network perspective. The reception of uplink transmit power generates an increase in the uplink interference floor. Air-interface capacity in the uplink direction is defined by a maximum allowed increase in uplink interference. There is a limit to the number of UE which can be in CELL_DCH without exceeding the maximum allowed increase in interference. UE achieving high uplink throughputs are likely to generate larger increases in the uplink interference, but they may spend less time in CELL_DCH as a result of their data transfers taking less time.

Air-interface capacity in the downlink direction is defined by the maximum downlink transmit power capability. A static transmit power allocation is made for the CPICH, P-SCH, S-SCH, AICH, PICH, P-CCPCH and S-CCPCH. The remainder of the downlink transmit power is available to support the connections belonging to UE in CELL_DCH. As presented in Table 2.5, these connections can be configured as either dedicated channels or HSDPA. If a UE has been configured with a dedicated channel, the active set cells must continuously transmit either the DPCCH or F-DPCH. The active set cells must also transmit the DPDCH whenever there is data to send. If a UE has been configured with HSDPA, the serving cell must transmit the HS-SCCH and HS-PDSCH whenever there is data to send. The release 7 Continuous Packet Connectivity features which help to increase the downlink air-interface capacity are F-DPCH gating (discontinuous transmission during periods of inactivity) and HSDPA without an HS-SCCH.

Dedicated channels require the allocation of uplink and downlink channelisation codes. The allocation of uplink codes is not significant in terms of network capacity because each UE has access to its own channelisation code tree. The allocation of downlink codes is more significant because each cell has a single channelisation code tree to share amongst all UE connected in CELL_DCH and CELL_FACH (assuming a single scrambling code has been configured). UE in CELL_FACH typically use a relatively small and fixed allocation from the code tree, i.e. one or more codes are allocated to the S-CCPCH. Fixed allocations are also made for the CPICH, P-CCPCH, AICH and PICH. The remainder of the code tree is available for dedicated channels, HSDPA and HSUPA. HSUPA requires a relatively small allocation to support the RGCH, AGCH and HICH. The channelisation code requirement for the dedicated channels and HSDPA depends upon the allocated bit rates. In both cases, larger sections of the code tree are used when high bit rates are allocated. If dedicated channels are maintained for longer than necessary then there will be an increased probability of blocking new connection requests. The use of HSDPA helps to increase the efficiency of the code tree by sharing a set of codes between all HSDPA connections. However, each HSDPA connection also requires a dedicated channelisation code to support the downlink DPCCH. This can become an issue if there is a large number of HSDPA connections, e.g. a population of voice over IP users. The F-DPCH helps to reduce the requirement by replacing the downlink DPCCH and allowing up to 10 UE to share the shame channelisation code.

UE in CELL_DCH also require network resources in terms of Node B and RNC processing capability. In general, larger throughputs require greater processing capabilities. The precise requirement is implementation dependant. The processing capability of a Node B is often shared between cells. This is in contrast to each cell having a dedicated processing capability.

Sharing the processing capability provides a multiplexing gain assuming that not all cells peak in activity simultaneously, i.e. a busy cell can use a larger share of the Node B processing capability while one or more other cells are less busy. Similarly, the processing capability of an RNC can be shared between all controlled Node B and there is a multiplexing gain if it is assumed that not all Node B peak in activity simultaneously. These multiplexing gains are usually evaluated during system dimensioning. The multiplexing gain at an RNC will be significantly greater than the multiplexing gain at a Node B because the number of Node B controlled by an RNC is much larger than the number of cells at a Node B.

UE in CELL_DCH also require resources from the Iub transport network. Section 5 describes the Iub transport network and the dedicated channel bandwidth requirement. The precise bandwidth requirement is implementation dependant and depends upon the quality of service being offered. In general, if a high quality of service is offered then it is necessary to assign a relatively large bandwidth to help protect the connection from congestion causing increased delay and potential packet loss. If a lower quality of service is offered then a lower bandwidth can be allocated. Real time connections typically require greater Iub bandwidth than non-real time connections. In addition to the physical bandwidth requirement, UE in CELL_DCH require logical Iub transport resources. If AAL2 and ATM are used as part of the Iub protocol stack then each connection requires one or more Channel Identifiers (CID). Section 5.1.3 describes the limitations associated with assigning a large number of CID.

CELL_DCH is the only RRC state which allows UE to establish soft and softer handover connections. Both dedicated channels and HSUPA support soft and softer handover. HSDPA does not support soft nor softer handover. The signalling procedures for soft and softer handover are presented in Section 8.6. The use of soft handover helps to improve air-interface performance towards the cell edge, but also creates an overhead in terms of Iub transport, Node B and air-interface resources.

Inter-system and inter-frequency handovers can be triggered from CELL_DCH. In general, these handovers require the use of compressed mode to generate transmission gaps during which the UE is able to measure the quality of neighbouring cells. Compressed mode may not be necessary if a UE has two receivers. The triggering mechanisms used for inter-system and inter-frequency handover are implementation dependant. Both types of handover are typically triggered using coverage based criteria, e.g. CPICH Ec/Io, CPICH RSCP, uplink transmit power and downlink transmit power. They can also be triggered using load or service based criteria. Load based handover can make use of downlink transmit power and uplink interference power triggering mechanisms. Service based handover triggers handovers based upon service type. Preferred network layers can be associated with specific service types, e.g. speech calls could be moved onto GSM while data connections are maintained on a WCDMA RF carrier.

UE in CELL_DCH can be paged using Paging Type 2 messages. These messages are transferred using the SRB on the DCCH. Paging could be necessary if there is an incoming CS speech call during an ongoing PS data session. Alternatively, push data services, e.g. push email could generate PS paging during an ongoing CS speech call. An example Paging Type 2 message is presented in Log File 2.4.

This paging message does not include any form of UE identity. If the UE is using a dedicated channel for the SRB the paging message is sent directly to the UE using the dedicated physical channel resource. If the UE is using HSDPA for the SRB the paging message is also sent directly to the UE although in this case an H-RNTI is used to indicate when the HSDPA resources are scheduled for the UE. The Paging Type 2 message includes a specification of the core network domain from where the paging originated and also the type of UE identity used by the core network when signalling the RANAP paging message to the RNC. Knowledge of this identity type allows the UE to use the same identity type when responding to the core network.

If the UE has an HSUPA connection an E-RNTI is assigned as a UE identity for scheduling purposes. In contrast to the U-RNTI and H-RNTI, the same E-RNTI can be assigned to more than a single UE. In addition, a UE can be assigned both a primary and secondary E-RNTI. The E-DCH Absolute Grant Channel (E-AGCH) uses the E-RNTI to schedule HSUPA air-interface resources to either a single UE or a group of UE.

```
PAGING_TYPE_2
DL-DCCH-Message
integrityCheckInfo
messageAuthenticationCode
    Bin: 31 30 31 31
rrc-MessageSequenceNumber: 6
pagingType2
rrc-TransactionIdentifier: 2
pagingCause: terminatingConversationalCall
cn-DomainIdentity: cs-domain
pagingRecordTypeID: tmsi-GSM-MAP-P-TMSI
```

Log File 2.4 Paging Type 2 message received while in CELL_DCH

Table 2.6   Main characteristics of CELL_FACH

Possible to transfer small quantities of uplink and downlink PS data
CS connections require transition to CELL_DCH
Release 7 of the 3GPP specifications introduces HSDPA for CELL_FACH
UE and network resource requirement relatively low
Cell re-selections completed when UE is mobile
Inter-system and inter-frequency handover possible
Radio access network uses U-RNTI and C-RNTI to identify UE
Paging Type 2 messages without a DRX cycle are used for paging purposes

2.2.1.3 RRC Connected Mode – CELL_FACH

CELL_FACH supports control plane signalling and the transfer of relatively small quantities of PS data. Transferring larger quantities of data or establishing a CS connection requires the transition to CELL_DCH. UE in CELL_FACH consume significantly less resources than UE in CELL_DCH, but greater resources than UE in CELL_PCH, URA_PCH and RRC Idle mode. The main characteristics of CELL_FACH are summarised in Table 2.6.

A UE can be instructed to enter CELL_FACH from any other RRC state. These transitions are illustrated in Figure 2.21.

A UE may enter CELL_FACH from RRC Idle mode to complete a signalling procedure or to establish a user plane connection. A UE is instructed to enter CELL_FACH rather than CELL_DCH within the RRC Connection Setup message. An example of the relevant section from an RRC Connection Setup message is presented in Log File 2.5.

UE in CELL_FACH are assigned a Cell Radio Network Temporary Identifier (C-RNTI) in addition to a UTRAN RNTI (U-RNTI). C-RNTI are assigned by the controlling RNC to address the UE within a specific cell.

The delay associated with establishing an RRC connection can be reduced by using CELL_FACH rather than CELL_DCH because it avoids the requirement for dedicated channel air-interface

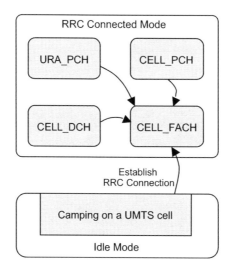

Figure 2.21   RRC state transitions used to enter CELL_FACH

```
new-U-RNTI
  srnc-Identity
    Bin: 00 B
  s-RNTI
    Bin: 02 1C C
new-C-RNTI
  Bin: 00 01
rrc-StateIndicator: cell-FACH
```

Log File 2.5   Section from an RRC Connection Setup message

synchronisation. However, establishing an RRC connection in CELL_FACH means that the RRC Connection Setup Complete message has to be transferred using the RACH rather than a DCH. A typical RACH SRB bit rate is 8.4 kbps compared with typical DCH SRB bit rates of 3.6 kbps and 14.4 kbps. The RRC Connection Setup Complete message is relatively large and is likely to require multiple random access procedures when transferred on the RACH.

Completing a user plane connection establishment procedure in CELL_FACH can help to reduce call setup delays. CELL_FACH is able to offer higher downlink SRB bit rates than CELL_DCH. A typical FACH SRB bit rate is 33.6 kbps compared with the typical DCH SRB bit rates of 3.6 and 14.4 kbps. However, as noted above, the RACH SRB bit rate may be less than the equivalent DCH SRB bit rate. CELL_FACH offers a potentially significant gain by removing the requirement for a synchronised reconfiguration procedure which is necessary in CELL_DCH when a user plane connection is added to an existing signalling connection. The signalling associated with RAB establishment in CELL_DCH is presented in Figure 2.22.

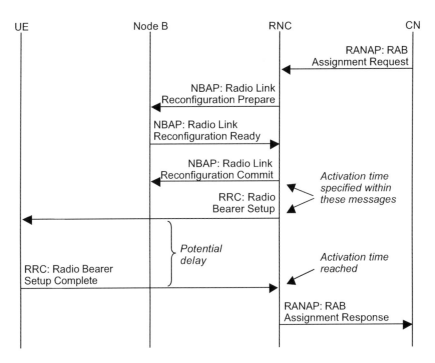

Figure 2.22   RAB establishment from CELL_DCH

At the start of RAB establishment the UE has been configured with a set of SRB multiplexed onto a single DCH transport channel. The core network initiates RAB establishment by forwarding a RANAP RAB Assignment Request message to the RNC. The RNC proceeds to reconfigure the Node B using a synchronised reconfiguration procedure. This starts by preparing the Node B in terms of specifying the new configuration. The RNC completes the procedure by using the NBAP Radio Link Reconfiguration Commit message to specify the activation Connection Frame Number (CFN) during which the new configuration is to be applied. The RNC has to calculate this CFN taking into account that the UE must be ready to apply the new configuration at the same time as the Node B. The UE has not yet been informed of the new configuration and so the CFN must allow sufficient time for the Radio Bearer Setup message to be transferred to the UE. In general, some margin is included when calculating the activation CFN to allow for one or more RLC re-transmissions which may or may not be necessary. Once the activation CFN has been reached both the Node B and UE apply the new configuration simultaneously. If the RNC specifies an activation CFN which passes after only a short delay there is a danger that the UE does not receive the Radio Bearer Setup message before the Node B applies the new configuration. This would result in a connection establishment failure. If the activation CFN is specified to pass after a long delay the UE is more likely to receive the Radio Bearer Setup message prior to the Node B applying the new configuration but the connection setup delay is increased.

Establishing a connection in CELL_FACH removes the requirement for the synchronised reconfiguration procedure because both the SRB and user plane transport channels are established simultaneously when the UE is instructed to move into CELL_DCH. The signalling associated with CS RAB establishment in CELL_FACH is illustrated in Figure 2.23 (PS RAB may be established completely in CELL_FACH without the transition to CELL_DCH).

At the start of RAB establishment the UE has been configured with a set of SRB multiplexed onto a single FACH transport channel. The core network initiates RAB establishment by forwarding a RANAP RAB Assignment Request message to the RNC. In this case, the UE does not have any dedicated resources to reconfigure. The RNC requests a new radio link at the Node B which is capable of supporting both the SRB and user plane connections. The Node B starts transmitting the new radio link as soon as it has received the request. The UE is then informed of the new radio link and is instructed to enter CELL_DCH. The UE immediately moves into CELL_DCH and achieves

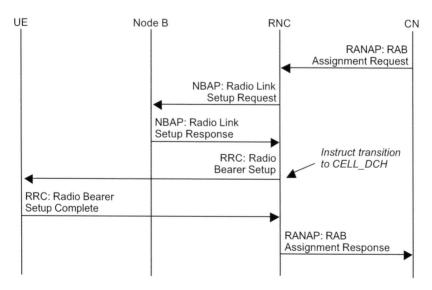

Figure 2.23   RAB establishment from CELL_FACH

air-interface synchronisation. The UE is then able to start transmitting and complete the radio bearer setup procedure. This approach does not require an activation CFN and so it is not necessary to balance the trade-off between connection setup delay and connection setup success rate.

A UE may be moved into CELL_FACH from CELL_DCH if it has a PS connection which is experiencing a period of inactivity. This could be inactivity at the end of a file transfer or after downloading an email or internet page. Inactive PS connections are not maintained in CELL_DCH for long periods of time due to the relatively high resource requirement. An example section from a Radio Bearer Reconfiguration message used to move a UE from CELL_DCH to CELL_FACH is presented in Log File 2.6.

The RRC state indicator instructs the UE to enter CELL_FACH and the controlling RNC assigns a C-RNTI. It is not necessary to specify a U-RNTI because the UE is already in RRC connected mode and has already been assigned a U-RNTI. It is likely that once the UE has been moved into CELL_FACH then it will be configured with measurement reporting event 4a. This reporting event allows the UE to inform the RNC if the occupancy of its RLC buffers exceeds a specific threshold. The RNC can then take the decision to move the UE back into CELL_DCH. Measurement reporting event 4a is described in greater detail within Section 2.2.5. In parallel, the RNC will be monitoring the occupancy of its own RLC buffers and will move the UE back into CELL_DCH if the quantity of downlink data becomes significant.

UE in CELL_PCH make use of CELL_FACH to complete cell updates. Cell updates can be triggered by cell reselection, periodic updates, an incoming paging message, the requirement to transfer uplink data, re-entering coverage, MBMS reception or an MBMS point-to-point radio bearer request. Cell Update messages are always transferred using the CCCH logical channel and transparent mode RLC (SRB 0 is used). Using the CCCH means that the UE identity must be included within the RRC message and the U-RNTI is used for this purpose. The RLC layer is unable to complete re-transmissions to provide reliable data transfer and so, if necessary, re-transmissions are generated from the RRC layer using timer T302, counter V302 and constant N302. A Cell Update can also be completed in CELL_FACH by a UE which was previously in CELL_DCH, but has experienced an unrecoverable RLC error. In this case, the UE moves to CELL_FACH to inform the RNC of the unrecoverable error. In addition, a UE which was previously in CELL_DCH can also complete a cell update procedure in CELL_FACH if it experiences dedicated channel radio link failure. UE in URA_PCH can make use of CELL_FACH to complete URA updates triggered by cell reselection across a URA boundary or by a periodic update. UE in URA_PCH can also make use of CELL_FACH for cell updates, e.g. triggered by an incoming paging message or the requirement to transfer uplink data.

UE in CELL_FACH do not have access to any dedicated physical channels nor any dedicated transport channels. They do however have access dedicated logical channels. Uplink data is always transferred using a PRACH physical channel and a RACH transport channel. The logical channel for user plane data is always a DTCH whereas the logical channel for control plane signalling can be either

```
RADIO_BEARER_RECONFIGURATION
DL-DCCH-Message
integrityCheckInfo
messageAuthenticationCode
     Bin: 44 96 68 44
rrc-MessageSequenceNumber: 9
radioBearerReconfiguration-r3
rrc-TransactionIdentifier: 1
new-C-RNTI
     Bin: 00 09
rrc-StateIndicator: cell-FACH
```

Log File 2.6   Section from a Radio Bearer Reconfiguration message

a DCCH or a CCCH. The choice between DCCH and CCCH depends upon the RRC message, e.g. Radio Bearer Reconfiguration Complete messages are sent on the DCCH whereas Cell Update messages are sent on the CCCH. An example bit rate for the DTCH multiplexed onto the RACH is 18 kbps. Downlink data is always transferred using a S-CCPCH physical channel and a FACH transport channel. The logical channel for user plane data can be either a DTCH, CTCH or MTCH. The majority of applications make use of the DTCH logical channel. An example bit rate for the DTCH multiplexed onto the FACH is 36 kbps. The CTCH logical channel can be used for Cell Broadcast Services (CBS) whereas the MTCH can be used for MBMS services. The logical channel for control plane signalling can be either a DCCH, CCCH, MCCH or MSCH. In general, the DCCH and CCCH are used, depending upon the RRC message whereas the MCCH and MSCH are used for MBMS services. Table 2.7 summarises the channel types which can be used in CELL_FACH.

Neither HSDPA nor HSUPA can be used in CELL_FACH when using release 6 or earlier versions of the 3GPP specifications. The release 7 version of the 3GPP specifications introduces HSDPA within CELL_FACH, but nevertheless it is not intended for HSDPA to transfer significant quantities of data in CELL_FACH. It is just intended to improve the efficiency of signalling and the transfer of small quantities of data.

The BCCH is used in CELL_FACH to read system information, e.g. cell reselection and neighbour information. A UE in CELL_FACH has to read system information messages broadcast on the P-CCPCH in the same way as in RRC Idle mode, i.e. based upon the scope of the system information and the value tag. The principles of cell reselection in CELL_FACH are the same as those for RRC Idle mode although it is possible to apply a different set of parameters. The cell reselection parameters for RRC Idle mode are broadcast in SIB 3. These parameters are also used for RRC Connected mode unless SIB 4 is broadcast. If SIB 4 is broadcast then it defines the cell reselection parameters for RRC Connected mode. Likewise SIB 11 broadcasts the neighbour list for use in RRC Idle mode. This neighbour list is also used for CELL_FACH unless SIB 12 is broadcast. If SIB 12 is broadcast then it defines the neighbour list for use in CELL_FACH. The controlling RNC keeps track of the cell serving each UE in CELL_FACH. The cell update procedure is used to update the controlling RNC whenever the UE completes a cell reselection. T305 can be used to trigger periodic cell updates while a UE is in CELL_FACH. Periodic cell updates can be used as a supervision mechanism to confirm to the RNC that a UE is still connected in CELL_FACH and has not moved out of coverage. However, in general periodic cell updates are not applicable to CELL_FACH because UE typically remain in CELL_FACH for relatively short periods of time, i.e. of the order of 10 s, whereas the minimum value that can be assigned to T305 is 5 minutes.

It is not necessary for the UE to monitor the PCCH for paging messages in CELL_FACH because the RNC can forward paging information on the DCCH using a Paging Type 2 message. A Paging Type 2 message in CELL_FACH is the same as a Paging Type 2 message in CELL_DCH with the exception that in CELL_FACH the UE identity is included as part of the MAC header. This allows all UE receiving the same FACH transport channel to isolate their appropriate messages.

Inter-frequency and inter-system cell reselections in CELL_FACH are more complex than those in RRC Idle mode, CELL_PCH and URA_PCH. A UE in CELL_FACH has to continuously receive one of more FACH transport channels. This means that the receiver does not normally have time to record

Table 2.7 Channel types used in CELL_FACH

Logical channel	Transport channel	Physical channel	Direction
DCCH, CCCH, DTCH	RACH	PRACH	Uplink
DCCH, CCCH, MCCH, MSCH, DTCH, CTCH, MTCH	FACH	S-CCPCH	Downlink
BCCH	BCH	P-CCPCH	Downlink

measurements from a different RF carrier. Compressed mode is not supported in CELL_FACH. UE implementations with multiple receivers may be able to simultaneously receive the FACH transport channels and record measurements from a different RF carrier. The 3GPP solution to recording inter-frequency and inter-system measurements while in CELL_FACH is to use FACH measurement occasions. A FACH measurement occasion is a period of time during which the RNC will not send any downlink data to a specific UE in CELL_FACH. This allows the UE to record measurements from other RF carriers. The duration of the measurement occasion is equal to the transmission time interval (TTI) of the FACH having the largest TTI on the S-CCPCH selected by the UE. The measurement occasion is repeated every $N \times 2^k$ radio frames, where N is the duration of the measurement occasion in radio frames and k is the FACH measurement occasion cycle length coefficient. The value assigned to the cycle length coefficient can be broadcast within SIB 11 or SIB 12. If a value is not broadcast then it indicates that the network does not support CELL_FACH measurement occasions and UE cannot complete inter-frequency nor inter-system cell reselections from CELL_FACH unless they have multiple receivers. The timing of the measurement occasions belonging to a specific UE is calculated from the C-RNTI. This helps to ensure that measurement occasions belonging to different UE are distributed in time rather than coinciding with one another.

The power consumption of a UE in CELL_FACH is less than in CELL_DCH, but greater than in CELL_PCH, URA_PCH and RRC Idle mode. The power consumption is less than CELL_DCH because the UE is not required to continuously transmit. It is greater than CELL_PCH, URA_PCH and RRC Idle mode because the UE is required to continuously receive. There is a requirement to receive continuously because control plane signalling can be sent by the RNC at any time without prior notification. Likewise, if a user plane connection has been established, user plane data can be sent by the RNC at any time. The UE has to continuously search for its identity appearing within messages transmitted on the FACH. In the case of the DTCH and DCCH, the UE identity is included as part of the MAC header. In the case of the CCCH, the UE identity is included as part of the RRC message.

2.2.1.4 RRC Connected Mode — CELL_PCH

CELL_PCH allows UE to remain in RRC Connected mode for relatively long periods of time while minimising the network and UE resource requirements. UE can be paged in CELL_PCH, but the response must be sent in CELL_FACH. UE in CELL_PCH make use of a DRX cycle to avoid the requirement for continuous reception. Similar to CELL_FACH, the RNC keeps track of the individual cell upon which the UE is camped. The main characteristics of CELL_PCH are summarised in Table 2.8.

A UE can be moved into CELL_PCH either from CELL_DCH or from CELL_FACH. In both cases a Physical Channel Reconfiguration message can be used to instruct the UE to make the transition (other messages such as Radio Bearer Reconfiguration and Transport Channel Reconfiguration can also

Table 2.8 Main characteristics of CELL_PCH

Uplink data transfer is not possible
Downlink MBMS and CBS data can be received
Paging Type 1 messages with a DRX cycle are used for paging purposes
Paging UE in CELL_PCH is more efficient than paging UE in RRC Idle mode
Transitions to CELL_FACH use the Cell Update procedure for more than one reason
The UE and network resource requirement is less than CELL_FACH
Cell reselections and cell updates completed when a UE is mobile
Inter-system and inter-frequency cell reselections possible
Radio access network uses U-RNTI to identify UE

```
PHYSICAL_CHANNEL_RECONFIGURATION
DL-DCCH-Message
integrityCheckInfo
messageAuthenticationCode
Bin: FC C5 CF 04
rrc-MessageSequenceNumber: 14
physicalChannelReconfiguration-r3
rrc-TransactionIdentifier: 0
  rrc-StateIndicator: cell-PCH
  utran-DRX-CycleLengthCoeff: 5
  frequencyInfo
    modeSpecificInfo fdd
    uarfcn-DL: 10637
  modeSpecificInfo fdd:
  dl-InformationPerRL-List
  dl-InformationPerRL-List value 1
    modeSpecificInfo fdd
      primaryCPICH-Info
      primaryScramblingCode: 88
```

Log File 2.7   Physical channel reconfiguration instructing a transition to CELL_PCH

be used). Log File 2.7 provides an example Physical Channel Reconfiguration message instructing a UE to make the transition to CELL_PCH.

The UE is instructed to move into CELL_PCH and use the RF carrier at 2127.4 MHz to camp upon scrambling code 88. The RF carrier frequency is derived by dividing the UTRA Absolute Radio Frequency Number (ARFCN) by 5 when using operating band I. The UE is also configured with a UTRAN Discontinuous Reception (DRX) cycle length of $2^5 = 32$ radio frames. When a UE is in RRC Connected mode the DRX cycle length is the minimum of the UTRAN DRX cycle length and the DRX cycle length belonging to any core network domains with which the UE is registered but does not have a signalling connection. In the case of CELL_PCH, the UE has a signalling connection across the Iu-ps and is normally registered with both the CS and PS core network domains. This means that the DRX cycle length is the minimum of the UTRAN DRX cycle length and the CS core network DRX cycle length. If it is assumed in this example that the CS core network DRX cycle length is 64 radio frames (read from SIB 1) then the UTRAN DRX cycle length dominates and the UE has to listen for paging messages once every 320 ms.

CELL_PCH does not allow any uplink data transfer. If a UE has either control plane or user plane data to send then it must make the transition to CELL_FACH. In the downlink direction, paging messages can be received on the PCCH logical channel and system information can be read from the BCCH logical channel. In addition MBMS information can be received on the MCCH and MTCH logical channels, while Cell Broadcast Service (CBS) information can be received on the CTCH logical channel. These channel types are summarised in Table 2.9. The DCCH and DTCH logical channels cannot be used in CELL_PCH.

Table 2.9   Channel types used in CELL_PCH

Channel type	Logical channel	Transport channel	Physical channel	Direction
Common	BCCH	BCH	P-CCPCH	Downlink
Common	PCCH	PCH	S-CCPCH	Downlink
Common	MCCH	FACH	S-CCPCH	Downlink
Common	MTCH	FACH	S-CCPCH	Downlink
Common	CTCH	FACH	S-CCPCH	Downlink

The BCCH makes use of the BCH transport channel and P-CCPCH physical channel whereas the PCH makes use of the PCH transport channel and the S-CCPCH physical channel. The MCCH, MTCH and CTCH make use of the FACH transport channel and the S-CCPCH physical channel.

UE in CELL_PCH complete the cell update procedure each time a cell boundary is crossed and cell reselection is completed. This allows the RNC to track the location of the UE on a per cell basis. Having knowledge of the cell upon which the UE is camped allows the RNC to direct paging messages to specific cells rather than across an entire location area or routing area. An example Cell Update message which has been triggered by cell reselection is presented in Log File 2.8. The UE is not able to transfer any uplink signalling in CELL_PCH and so has to move into CELL_FACH to send this message. The CCCH logical channel, RACH transport channel and PRACH physical channel are used to send this message.

The UE identifies itself using its UTRAN Radio Network Temporary Identifier (U-RNTI) which is the concatenation of the SRNC identity and the SRNC RNTI (S-RNTI). The cause of the Cell Update message is signalled as cell reselection and the UE informs the RNC of the CPICH Ec/Io experienced on the new cell. The CPICH Ec/Io has a signalled value of 30 which maps to an actual value between −9.5 and −9.0 dB. This mapping is presented in Table 8.5. The message does not include any indication of cell identity. This is not necessary because the RNC can deduce the cell identity by observing which cell received the RACH message (the Iub AAL2 CID allocated to the RACH identifies the cell belonging to a specific Node B).

The RNC records the new cell upon which the UE is camped and responds using a Cell Update Confirm message. The corresponding Cell Update Confirm message is presented in Log File 2.9.

```
CELL_UPDATE
UL-CCCH-Message
integrityCheckInfo
messageAuthenticationCode
Bin: AD 23 BA FD
rrc-MessageSequenceNumber: 1
cellUpdate
u-RNTI
  srnc-Identity
    Bin: 04 3
  s-RNTI
    Bin: 0C 22 9
startList
startList value 1
  cn-DomainIdentity: cs-domain
    start-Value
      Bin: 00 00 2
startList value 2
  cn-DomainIdentity: ps-domain
    start-Value
      Bin: 00 01 E
am-RLC-ErrorIndicationRb2-3or4: false
am-RLC-ErrorIndicationRb5orAbove: false
cellUpdateCause: cellReselection
rb-timer-indicator
t314-expired: false
t315-expired: false
measuredResultsOnRACH
currentCell
  modeSpecificInfo fdd
  measurementQuantity
    cpich-Ec-N0: 30 (-9.5 to -9.0 dB)
```

Log File 2.8 Cell Update message triggered by cell reselection

```
CELL_UPDATE_CONFIRM
DL-DCCH-Message
integrityCheckInfo
messageAuthenticationCode
Bin: F0 BD 15 11
rrc-MessageSequenceNumber: 1
cellUpdateConfirm-r3
rrc-TransactionIdentifier: 0
rrc-StateIndicator: cell-PCH
utran-DRX-CycleLengthCoeff: 5
rlc-Re-establishIndicatorRb2-3or4: false
rlc-Re-establishIndicatorRb5orAbove: false
```

Log File 2.9   Cell Update Confirm message triggered by cell reselection

The Cell Update Confirm message simply instructs the UE to return to CELL_PCH. This example cell update procedure has been based upon UE mobility, i.e. a UE moving from one cell to another and completing cell re-selection. UE in CELL_PCH also have to run a periodic cell update timer. This UE timer is called T305 and is reset whenever the UE enters CELL_PCH. This could be after a data transfer or after a cell update procedure resulting from mobility. When T305 expires the UE has to complete a periodic cell update procedure. In this case, the cause value within the Cell Update message is changed to 'periodic cell update'. Periodic cell updates serve as a keep-alive mechanism to keep the RNC informed that the UE remains connected in CELL_PCH. T305 is explained in greater detail within Section 2.2.4.

If a UE is in CELL_PCH and has uplink data to transfer then it is not permitted to simply move into CELL_FACH and start transmitting on the RACH. The UE must start by using the cell update procedure to inform the RNC that it has data to transfer. In this case, the Cell Update message appears the same as that presented within Log File 2.8, but the cause value is changed to uplink data transmission. The UE is not able to send this message from CELL_PCH and has to move to CELL_FACH to send it on the RACH. The RNC responds to the Cell Update message using a Cell Update Confirm. In this case, the content of the message is slightly different. An example Cell Update Confirm message associated with a cause value of uplink data transfer is presented in Log File 2.10.

In this case, the UE is instructed to remain in CELL_FACH and is assigned a Cell RNTI (C-RNTI). This version of the Cell Update message excludes the DRX cycle length because discontinuous reception is not used in CELL_FACH and so the UE has to receive the downlink continuously. The UE acknowledges the new C-RNTI using a UTRAN Mobility Information Confirm Message. An example is presented in Log File 2.11.

```
CELL_UPDATE_CONFIRM
DL-DCCH-Message
integrityCheckInfo
messageAuthenticationCode
Bin: 00 5C 2F 33
rrc-MessageSequenceNumber: 3
cellUpdateConfirm-r3
rrc-TransactionIdentifier: 2
new-C-RNTI
Bin: 00 09
rrc-StateIndicator: cell-FACH
rlc-Re-establishIndicatorRb2-3or4: false
rlc-Re-establishIndicatorRb5orAbove: false
```

Log File 2.10   Cell Update Confirm message triggered by uplink data transfer

```
UTRAN_MOBILITY_INFORMATION_CONFIRM
UL-DCCH-Message
integrityCheckInfo
messageAuthenticationCode
Bin: 76 0D EE F3
rrc-MessageSequenceNumber: 14
utranMobilityInformationConfirm
rrc-TransactionIdentifier: 2
```

Log File 2.11   UTRAN Mobility Information Confirm acknowledging a new C-RNTI

The UE may send other messages after receiving a Cell Update Confirm message, depending upon the content. For example, if the Cell Update Confirm message reconfigures radio bearers then the Radio Bearer Reconfiguration Complete message is sent. Likewise, Transport Channel Reconfiguration Complete, Physical Channel Reconfiguration Complete or Radio Bearer Release Complete messages may be sent after a Cell Update Confirm message. The Iu-ps signalling connection is maintained while the UE is in CELL_PCH and so the UE can start transferring data immediately it has completed the cell update procedure.

It is likely that the RNC will configure measurement reporting event 4a after a cell update procedure resulting from uplink data transfer. This allows the UE to inform the RNC if it has a significant quantity of uplink data to transfer in which case the RNC would subsequently instruct the UE to make the transition to CELL_DCH. Event 4a is described within Section 2.2.4.

A UE in CELL_PCH can be paged using a Paging Type 1 message. Paging Type 1 messages are transferred on the PCCH logical channel and are used to page UE in RRC Idle mode, CELL_PCH and URA_PCH. This is in contrast to Paging Type 2 messages which are used to page UE in CELL_DCH and CELL_FACH. Paging Type 2 messages do not include any indication of UE identity because they are transferred using the DCCH logical channel. An example Paging Type 1 message is presented in Log File 2.12.

This example includes a single paging record although 3GPP TS 25.331 allows up to eight paging records to be included within the paging message. A UE doesn't know whether or not it is being paged until it has read the contents of the paging records and determined whether or not its identity is listed.

In general, Paging Type 1 messages for UE in CELL_PCH are directed towards only a single cell because the RNC has knowledge of which cell each UE is camped upon. This means that the network paging load can be decreased by allowing UE to remain in CELL_PCH rather than RRC Idle mode. This reduction in network paging is maximised if location areas are not distributed across multiple RNC. If location areas are distributed across multiple RNC the CS core network forwards paging messages to all RNC which control cells belonging to the appropriate location area, i.e. the CS core

```
PAGING_TYPE_1
PCCH-Message
pagingType1
pagingRecordList
pagingRecordList value 1
  utran-Identity
  u-RNTI
    srnc-Identity
    Bin: 04 3
    s-RNTI
    Bin: 02 D4 E
```

Log File 2.12   Paging Type 1 message received while in CELL_PCH

network knows within which location area the UE is located, but does not know the individual cell. This scenario is illustrated in Figure 2.24.

The serving RNC receives the paging message from the CS core and directs it towards the appropriate cell. However the second RNC also receives the paging message and does not have any knowledge of where the UE is located. This RNC directs the paging message to all cells belonging to the location area. This scenario provides justification for avoiding location areas which span multiple RNC.

Once a UE in CELL_PCH has received a Paging Type 1 message it moves to CELL_FACH to send a Cell Update message. The Cell Update messages uses the UTRAN paging response cause value, but otherwise appears the same as the cell update message used for cell reselection, periodic cell update and uplink data transfer. The RNC responds with a Cell Update Confirm message which instructs the UE to remain in CELL_FACH. If the paging message was for a CS service then the UE will be moved into CELL_DCH. The UE may also be moved to CELL_DCH for PS services, depending upon the quantity of data to transfer.

UE can be maintained in CELL_PCH for relatively long periods of time because the resource requirement is relatively small. UE power consumption is low because UE are only required to listen for paging messages once every DRX cycle. The DRX cycle length in CELL_PCH can be smaller than the DRX cycle length in RRC Idle mode. This can have some impact upon the UE power consumption, but the impact is very small when compared with the power requirements of CELL_DCH, i.e. periodically receiving a signal requires significantly less power than transmitting and receiving continuously. The UE also has to monitor broadcast system information messages and measure the

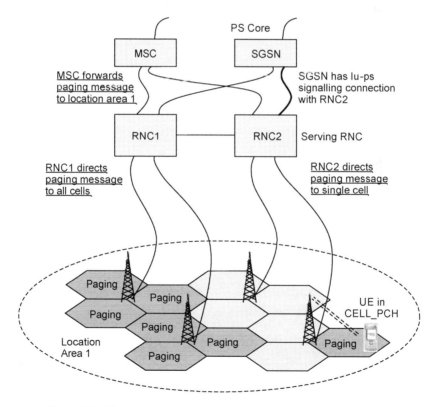

Figure 2.24   CELL_PCH paging when location areas are distributed across RNC

CPICH of the serving and neighbouring cells when in CELL_PCH. 3GPP TS 25.133 specifies that UE have to evaluate the cell selection minimum quality requirements and the cell reselection criteria at least once every DRX cycle. The UE power consumption in CELL_PCH can be increased if the UE starts to receive MBMS services. In this case, power is consumed by both receiving and displaying application data.

From the network perspective, UE in CELL_PCH require the Iu-ps signalling connection to be maintained and to be allocated logical RNC resources in terms of a U-RNTI. There are no dedicated resources on the air-interface, at the Node B nor on the Iub.

Inactive UE are typically maintained in CELL_PCH for 30 minutes before returning them to RRC Idle mode via CELL_FACH. If UE are maintained in CELL_PCH for only a short period of time there is an increased probability that the end-user initiates a subsequent connection from RRC Idle mode. This would then involve a more complex and time-consuming connection establishment procedure, i.e. connection establishment delay is increased.

2.2.1.5 RRC Connected Mode – URA_PCH

URA_PCH is similar to CELL_PCH with the exception that the RNC knows the location of the UE with a resolution of a UTRAN Registration Area (URA) rather than with a resolution of a cell. It is intended that a URA includes a cluster of cells. If a UE is mobile, it is only required to inform the RNC when it moves to a cell which triggers a change of the URA registered with the UE. The RNC may move a UE from CELL_PCH to URA_PCH if it is detected to have high mobility. This helps to reduce the signalling requirement in terms of cell updates with cause cell reselection. The drawback of moving a UE from CELL_PCH to URA_PCH is that paging messages have to be broadcast across a URA rather than a single cell, i.e. the paging load is greater. The main characteristics of URA_PCH are summarised in Table 2.10.

Similar to CELL_PCH, a UE can be moved into URA_PCH from either CELL_DCH or CELL_FACH. An RNC may decide that a UE in CELL_PCH should be moved into URA_PCH, but this transition has to be completed via CELL_FACH. It would be necessary for the RNC to page the UE in CELL_PCH and subsequently instruct the UE to move into URA_PCH after the UE has moved into CELL_FACH to respond to the paging message. There is no mechanism for the RNC to instruct the UE to go directly from CELL_PCH to URA_PCH. Figure 2.25 illustrates the transition from CELL_PCH to URA_PCH via CELL_FACH.

Figure 2.25 also illustrates the transition from CELL_DCH. The RNC may make use of this transition if the RNC becomes overloaded. This is similar to the RNC moving UE directly to CELL_PCH during periods of congestion. It is intended that URA_PCH is used for UE which have relatively high mobility. UE in CELL_PCH are required to complete the Cell Update procedure after every cell reselection. UE in URA_PCH are only required to complete the Cell Update procedure if a

Table 2.10 Main characteristics of URA_PCH

Uplink data transfer is not possible

Downlink MBMS and CBS data can be received

Paging Type 1 messages with a DRX cycle are used for paging purposes

Paging UE in URA_PCH is less efficient than paging UE in CELL_PCH

Transitions to CELL_FACH use URA and Cell Update procedures for more than one reason

The UE and network resource requirement is potentially less than CELL_PCH

Cell reselections and URA updates completed when a UE is mobile

Inter-system and inter-frequency cell reselections possible

Radio access network uses U-RNTI to identify UE

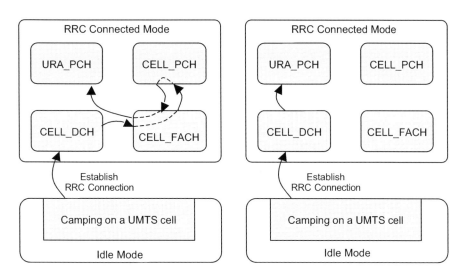

Figure 2.25 Example RRC state transitions to enter URA_PCH

cell reselection results in the UE camping upon a cell which does not belong to the URA currently registered with the UE.

UE are assigned a URA identity when instructed to enter URA_PCH. An example section from a Physical Channel Reconfiguration message instructing a UE to enter URA_PCH and assigning a URA identity is presented in Log File 2.13.

In this example, the UE is assigned URA identity 1. This URA identity should belong to the cell upon which the UE is camped. Radio network planners are able to assign up to eight URA identities to each cell. The set of URA identities belonging to a cell is broadcast on the BCCH within System Information Block (SIB) 2. Log File 2.14 presents an example of SIB 2.

In this example, the radio network planner has assigned 3 URA identities to the cell. Multiple URA identities are assigned to allow URA to overlap with one another and to help avoid ping-pong scenarios at URA boundaries. Figure 2.26 illustrates a UE moving from cell 1 to cell 2 and then returning to cell 1. Prior to moving into cell 2, the UE must have been assigned either URA identity 1 or 2. In this example it is assumed that the RNC assigned URA identity 1. After completing cell reselection onto cell 2, the UE reads SIB 2 to determine the set of URA. Cell 2 has only been assigned URA identity 2 whereas the UE has previously been assigned URA identity 1. This triggers the UE to complete a URA

```
PHYSICAL_CHANNEL_RECONFIGURATION
DL-DCCH-Message
integrityCheckInfo
messageAuthenticationCode
Bin: 31 30 31 30
rrc-MessageSequenceNumber: 3
physicalChannelReconfiguration-r3
rrc-TransactionIdentifier: 2
  rrc-StateIndicator: ura-PCH
  utran-DRX-CycleLengthCoeff: 5
  ura-Identity
     Bin: 00 01
```

Log File 2.13 Physical channel reconfiguration instructing a transition to URA_PCH

```
SYSTEM_INFORMATION_BLOCK_TYPE_2
SysInfoType2
ura-IdentityList
ura-IdentityList value 1
  Bin: 00 01
ura-IdentityList value 2
  Bin: 00 02
ura-IdentityList value 3
  Bin: 00 03
```

Log File 2.14   System Information Block (SIB) 2

Update procedure which allows the RNC to assign URA identity 2. If the UE then moves back across the cell boundary it will read SIB2 and determine that cell 1 has been assigned URA identities 1 and 2. In this case the UE does not need to trigger a URA Update procedure because its URA identity is listed within SIB 2. The UE can now move back and forth across the cell boundary without triggering a URA update because both cells and the UE have been assigned URA identity 2.

Assigning URA identities to each cell is part of the radio network planning process. The simplest approach is to assign each cell with a single URA identity which is equal to the cell identity. However, this approach defeats the purpose of URA_PCH and the characteristics of URA_PCH become the same as those for CELL_PCH. The benefits of URA_PCH are obtained if URA are defined to include clusters of cells rather than single cells and if URA overlap with one another. Associating clusters of cells with a single URA identity reduces the rate at which URA updates have to be completed by UE which are mobile. Allowing URA to overlap with one another avoids ping-pong scenarios. Figure 2.27 illustrates an example allocation of URA identities to a group of hexagonal cells.

This example is based upon clusters of seven cells belonging to the same URA. The URA overlap with one another with a maximum of two URA identities assigned to each cell. In practice, the radio network plan is not a perfect hexagonal grid and it is likely that there will be scenarios where more than two URA identities are necessary to avoid URA Update ping-pongs. In addition, in practice URA may be larger than seven cells to help reduce the complexity of the URA planning process. Planning large URA also has the benefit of reducing the quantity of URA Updates and the associated signalling load. The drawback of planning large URA is that the paging procedure becomes less efficient. The RNC has knowledge of which URA the UE is located within, but does not have knowledge of the individual cell. This means that paging messages must be broadcast by all cells belonging to the appropriate URA. Increasing the number of cells belonging to a URA increases the number of cells which have to broadcast the same paging message, i.e. the paging load increases.

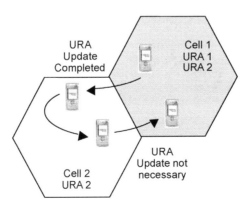

Figure 2.26   Mobility across a URA boundary

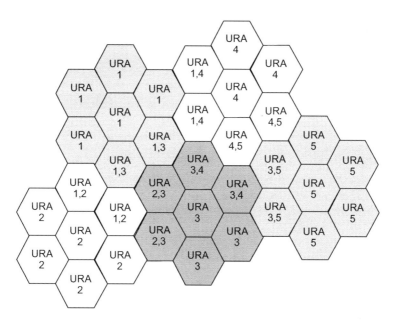

Figure 2.27   CELL_PCH paging when location areas are distributed across RNC

The decision to move a UE into URA_PCH is implementation dependent. However, in general it is intended that high mobility UE make use of URA_PCH while low mobility UE make use of CELL_PCH. This means that the signalling generated by Cell Updates is reduced at the cost of an increased paging load. If UE are initially moved into CELL_PCH by default the RNC can subsequently monitor the mobility of the UE by counting the number of cell reselection Cell Updates within a specific time interval. If the number of Cell Updates exceeds a threshold the RNC can decide to move the UE into URA_PCH via CELL_FACH.

The URA Update procedure has relatively few cause values when compared with the Cell Update procedure. The only cause values for the URA Update procedure are 'change of URA' and 'periodic URA update'. The T305 timer is used to trigger periodic URA updates in the same way as for CELL_PCH and CELL_FACH. The URA Update message is always transferred using the uplink CCCH logical channel in the same way as the Cell Update message. UE in URA_PCH can also trigger the Cell Update procedure. For example, a UE responding to a paging message moves to CELL_FACH and sends a Cell Update message with cause value 'paging response'. Alternatively, if a UE in URA_PCH has uplink data to transfer then it moves to CELL_FACH and sends a Cell Update message with cause value 'uplink data transmission'.

Table 2.11 summarises the channel types which can be used in URA_PCH. These are the same as those which can be used in CELL_PCH. System information can be read from the BCCH and the UE checks for paging messages on the PCCH once per DRX cycle. MBMS information can be received on the MCCH and MTCH logical channels, while Cell Broadcast Service (CBS) information can be received on the CTCH logical channel. The DCCH and DTCH logical channels cannot be used in URA_PCH.

Similar to CELL_PCH, UE can be maintained in URA_PCH for relatively long periods of time because the resource requirement is relatively small. UE power consumption is low because UE are only required to listen for paging messages once every DRX cycle. UE also have to evaluate the cell selection minimum quality requirements and the cell reselection criteria at least once every DRX cycle. This involves CPICH measurements from the serving and neighbouring cells. The UE power

Table 2.11 Channel types used in URA_PCH

Channel type	Logical channel	Transport channel	Physical channel	Direction
Common	BCCH	BCH	P-CCPCH	Downlink
Common	PCCH	PCH	S-CCPCH	Downlink
Common	MCCH	FACH	S-CCPCH	Downlink
Common	MTCH	FACH	S-CCPCH	Downlink
Common	CTCH	FACH	S-CCPCH	Downlink

consumption in URA_PCH can be increased if the UE starts to receive MBMS services. In this case, power is consumed by both receiving and displaying application data. From the network perspective, UE in URA_PCH require the Iu-ps signalling connection to be maintained and to be allocated logical RNC resources in terms of a U-RNTI. There are no dedicated resources on the air-interface, at the Node B nor on the Iub.

2.2.2 RRC Procedures

RRC procedures can be categorised into those which require signalling to the peer RRC layer and those which are completed within the RRC layer. The active set update procedure is an example of an RRC procedure which requires signalling to the peer RRC layer. The RRC layer within the RNC sends an Active Set Update message to the RRC layer within the UE. The UE then responds with either an Active Set Update Complete or Active Set Update Failure message. The DPCCH uplink open loop power control calculation is an example of an RRC procedure which is completed within the RRC layer without any signalling. The RRC layer within the UE calculates the initial transmit power of the DPCCH and provides the result to the physical layer.

Table 2.12 presents the set of RRC procedures which involve signalling to the peer RRC layer. These are further categorised as being relevant to RRC connection management, radio bearer control, RRC

Table 2.12 RRC procedures which involve signalling to the peer RRC layer

RRC Connection Management	Broadcast of System Information, Paging, RRC Connection Establishment, RRC Connection Release, RRC Connection Release Requested by Upper Layers, Transmission of UE Capability Information, UE Capability Enquiry, Initial Direct Transfer, Downlink Direct Transfer, Uplink Direct Transfer, UE Dedicated Paging, Security Mode Control, Signalling Connection Release, Signalling Connection Release Indication, Counter Check, Inter-RAT handover Information Transfer
Radio Bearer Control	Radio Bearer Establishment, Radio Bearer Reconfiguration, Radio Bearer Release, Transport Channel Reconfiguration, Transport Format Combination Control, Physical Channel Reconfiguration, Physical Channel Reconfiguration Failure
RRC Conection Mobility	Cell and URA Update, UTRAN Mobility Information, Active Set Update, Hard Handover, nter-RAT Handover to UTRAN, Inter-RAT Handover from UTRAN, Inter-RAT Cell Re-Selection to UTRAN, Inter-RAT Cell Re-Selection from UTRAN, Inter-RAT Cell Change Order to UTRAN, Inter-RAT Cell Change Order from UTRAN
Measurement	Measurement Control, Measurement Report, Assistance Data Delivery
MBMS	Reception of MBMS Control Information, MCCH Acquisition, MBMS Notification, MBMS Counting, MBMS p-t-m Radio Bearer Configuration, MBMS Modification Request, MBMS Service Scheduling

connection mobility, measurement and MBMS. Each of these procedures is specified individually within 3GPP TS 25.331. Some procedures make use of a combination of RRC messages whereas others make use of only a single RRC message. The downlink direction transfer procedure is an example which requires only a single RRC message, i.e. the Downlink Direct Transfer message.

RRC connection management procedures support the establishment and release of RRC connections. They also support the transfer of NAS messages between the RNC and UE. Radio bearer control procedures support the establishment, reconfiguration and release of radio bearers. Radio bearers represent logical connections between layer 3 of the RNC and layer 3 of the UE. Radio bearers make use of the lower layer transport and physical channels. This leads to transport channel and physical channel reconfigurations being categorised as radio bearer control procedures. RRC mobility procedures support the movement of a UE between cells and between Radio Access Technologies (RAT), i.e. between UMTS and GSM. Measurement procedures are used to control the measurements completed by a UE and also to allow those measurements to be reported to the RNC. MBMS procedures support the establishment and release of MBMS connections. The majority of MBMS procedures are based upon downlink signalling. The MBMS Modification Request procedure is the only MBMS procedure which includes uplink signalling. The MBMS Counting procedure is based upon downlink signalling but can trigger other procedures which include uplink signalling, e.g. it can trigger the cell Update procedure for UE in CELL_FACH, CELL_PCH and URA_PCH.

Table 2.13 presents the set of RRC procedures which do not involve signalling to the peer RRC layer. These are grouped into a single general category. Examples of these procedures are Radio Link Failure Criteria and Actions upon Radio Link Failure, PRACH Selection, Secondary CCPCH Selection and FACH Measurement Occasion Calculation. The Radio Link Failure Criteria and Actions upon Radio Link Failure procedure is applicable to UE in CELL_DCH. It is based upon the constants N313 and N315 as well as the timer T313. If a radio link failure is detected by the RRC layer all dedicated physical channel resources are released and the UE attempts a Cell Update procedure on the RACH using a 'radio link failure' cause.

PRACH selection is based upon generating a list of candidate PRACH and then selecting one at random. Secondary CCPCH selection is based upon generating a list of candidate S-CCPCH and then selecting one based upon the U-RNTI. The selection of a S-CCPCH has to be deterministic rather than random because both the RNC and UE must generate the same result, i.e. the RNC must know which S-CCPCH the UE is monitoring when scheduling downlink data transfer.

The FACH Measurement Occasion Calculation procedure is used to schedule radio frames during which the UE can stop receiving the FACH and record measurements either from a different RF carrier or a different RAT. This calculation is based upon the C-RNTI, the FACH measurement occasion cycle length and the longest FACH TTI belonging to the selected S-CCPCH.

Table 2.13   RRC procedures which do not involve signalling to the peer RRC layer

General	Selection of Initial UE Identity, Actions when Entering Idle mode from Connected mode, Open Loop Power Control upon Establishment of DPCCH, Physical Channel Establishment Criteria, Actions in 'out of service area' and 'in service area', Radio Link Failure Criteria and Actions upon Radio Link Failure, Open Loop Power Control, Maintenance of Hyper Frame Numbers, START value calculation, Integrity Protection, FACH Measurement Occasion Calculation, Establishment of Access Service Classes, Mapping of Access Classes to Access Service Classes, PLMN Type Selection, Neighbour Cells List Narrowing for Cell Reselection, CFN Calculation, Configuration of CTCH Occasions, PRACH Selection, Selection of RACH TTI, Secondary CCPCH Selection, Secondary CCPCH and FACH Selection for MCCH Reception, Unsupported Configuration, Actions Related to Radio Bearer Mapping, Actions when Entering another RAT from Connected Mode, Service Prioritisation, MBMS Frequency Selection, Actions Related to E-DCH Transmission Variable

2.2.3 RRC Messages

The RRC layer is responsible for generating the messages required by each of the RRC procedures involving signalling between the RNC and UE. The set of messages specified by 3GPP TS 25.331 is presented within Table 3.2. These messages are transferred using the BCCH, PCCH, CCCH, DCCH, MCCH and MSCH logical channels. The RLC, MAC and Physical layers provide a data transfer service for the RRC layer. The RRC layer encodes messages into binary format before passing them to the RLC layer for transmission. Packed Encoding Rules (PER) for Abstract Syntax Notation One (ASN.1) are used to translate RRC messages into binary strings. PER for ASN.1 is specified within ITU-T Recommendation X.691. This Recommendation specifies both aligned and unaligned versions of PER. 3GPP TS 25.331 specifies that the RRC layer applies the unaligned version. This approach increases decoding complexity, but reduces the size of the resultant binary strings. PER is based upon the definition of different encoding rules for different data types, e.g. there are different encoding rules for Integer, Bit String, Choice, Sequence and Enumerated data types. Once these encoding rules have been applied and a bit string has been generated, padding is added to ensure that the encoded message occupies an integer number of octets.

As an example of the PER ASN.1 encoding and decoding processes consider the RRC Connection Request message presented in Log File 2.15.

Encoding this message using the rules specified for unaligned PER ASN.1 generates the binary string presented in Log File 2.16. This log file also includes the hexadecimal version of the binary string. RRC messages are typically recorded as hexadecimal strings within UE and network log files.

The RRC Connection Request message is an uplink message and so the RRC layer within the UE is responsible for encoding while the RRC layer within the RNC is responsible for decoding. The logical channel type is signalled within the MAC header and the MAC layer within the RNC is able to pass this information to the RRC layer. This means that the RRC layer is provided with both the binary string representing the RRC message and the associated logical channel type. In this example the logical

```
RRC_CONNECTION_REQUEST
UL-CCCH-Message
   rrcConnectionRequest
      initialUE-Identity
        tmsi-and-LAI
          tmsi
            Bin: 00 00 84 4F
          lai
            plmn-Identity
              mcc
                mcc value: 206
              mnc
                mnc value: 99
            lac
            Bin: 0B B8
      establishmentCause: registration
      protocolErrorIndicator: noError
      measuredResultsOnRACH
        currentCell
          modeSpecificInfo fdd
            measurementQuantity
              cpich-Ec-N0: 41  (-4.0 to -3.5 dB)
```

Log File 2.15 RRC Connection Request message

```
Binary Format
0011 0001 0000 0000 0000 0000 1000 0100 0100 1111 0010 0000
0110 0000 1000 0000 0101 1101 1100 0011 0000 0001 0100 1000

Hexadecimal Format
31 00 00 84 4F 20 60 80 5D C3 01 48
```

Log File 2.16   Encoded RRC Connection Request message (binary and hexadecimal)

channel type is the CCCH. 3GPP TS 25.331 specifies the PER ASN.1 structure of an uplink CCCH message as:

```
UL-CCCH-Message ::= SEQUENCE {
integrityCheckInfo IntegrityCheckInfo OPTIONAL,
message UL-CCCH-MessageType}
```

Uplink CCCH messages are encoded as a Sequence which means that the message is preceded by a series of flags to indicate which optional information elements are present. In this example there is only a single optional information element and it has been excluded. This means that a 0 flag is inserted at the start of the message. This is illustrated in Figure 2.28.

The optional information element flag would be followed by the integrity check information if it had been included. In this example, it has been excluded and so the flag is followed by the uplink CCCH message type. 3GPP TS 25.331 specifies that the message type is encoded as a Choice according to:

```
UL-CCCH-MessageType ::= CHOICE {
cellUpdate CellUpdate,
rrcConnectionRequest RRCConnectionRequest,
uraUpdate URAUpdate,
spare NULL}
```

There are three possibilities for the uplink CCCH message and so it is necessary to use two bits to indicate which message type is being transferred. The RRC Connection Request message is the second in the list and indexing starts at 0 so the relevant index for this example is 1. The remainder of the binary string represents the contents of the RRC Connection Request message. Similar to the uplink DCCH message, the RRC Connection Request message is encoded as a SEQUENCE and so the main message content is preceded by a series of flags to indicate which optional information elements are

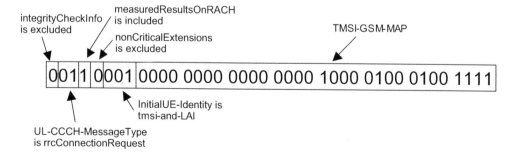

Figure 2.28   First section of an ASN.1 encoded RRC Connection Request message

present. 3GPP TS 25.331 specifies the PER ASN.1 structure of an RRC Connection Request message as:

```
RRCConnectionRequest ::= SEQUENCE {
initialUE-Identity InitialUE-Identity,
establishmentCause EstablishmentCause,
protocolErrorIndicator ProtocolErrorIndicator,
measuredResultsOnRACH MeasuredResultsOnRACH OPTIONAL,
nonCriticalExtensions SEQUENCE {} OPTIONAL}
```

This example is based upon the release 99 version of the RRC protocol. The same concepts apply to newer releases of the protocol, but additional information elements have been included. A release 99 example has been selected to help simplify the analysis. The RRC Connection Request structure indicates that there are two optional information elements and so it is necessary to precede the main information content by two flags. In this example, the flags indicate that the RACH measurement results are included but the non-critical extensions are excluded.

The first part of the RRC Connection Request message is the initial UE identity. 3GPP TS 25.331 specifies that this is encoded as a Choice based upon the following list of identities:

```
InitialUE-Identity ::= CHOICE {
imsi IMSI-GSM-MAP,
tmsi-and-LAI TMSI-and-LAI-GSM-MAP,
p-TMSI-and-RAI P-TMSI-and-RAI-GSM-MAP,
imei IMEI,
esn-DS-41 ESN-DS-41,
imsi-DS-41 IMSI-DS-41,
imsi-and-ESN-DS-41 IMSI-and-ESN-DS-41,
tmsi-DS-41 TMSI-DS-41}
```

There is a choice of eight different initial UE identities and so three bits are required to encode the selection. Indexing starts at 0 which means that a value of 1 corresponds to the Temporary Mobile Subscriber Identity (TMSI) and Location Area Identification (LAI) combination. This combination has the following structure:

```
TMSI-and-LAI-GSM-MAP ::= SEQUENCE {
tmsi TMSI-GSM-MAP,
lai LAI}
```

The TMSI and LAI are encoded as a SEQUENCE, but in this case there are no optional information elements and so it is not necessary to precede the main information content by any flags. The TMSI is inserted directly as a string of 32 bits. These 32 bits are illustrated in Figure 2.28. The structure of the LAI is specified by 3GPP TS 25.331 as:

```
LAI ::= SEQUENCE {
plmn-Identity PLMN-Identity,
lac BIT STRING (SIZE (16))}
```

The LAI is a concatenation of the PLMN identity and the Location Area Code (LAC). The LAC is inserted directly as a string of 32 bits, but the PLMN identity is defined as a Sequence based upon the Mobile Country Code (MCC) and Mobile Network Code (MNC):

```
PLMN-Identity ::= SEQUENCE {
mcc MCC,
mnc MNC}
```

Encoding for the MCC is relatively simple, i.e. the Binary Coded Decimal (BCD) version of the MCC is included as part of the encoded message. Encoding for the MNC is more complex because the length of the MNC can be either two or three decimal digits:

```
MCC ::= SEQUENCE (SIZE (3)) OF
Digit
MNC ::= SEQUENCE (SIZE (2..3)) OF
Digit
```

The MNC information element is preceded by an index which is used to indicate the length of the subsequent MNC. In this example, the MNC length is 2 and so an index of 0 is included.

An index of 1 would have been included if the MNC length had been 3. The section of the encoded RRC Connection Request message which includes the MCC, MNC and LAC is presented in Figure 2.29.

The PER ASN.1 structure of the RRC Connection Request message indicates that the initial UE identity is followed by the establishment cause. The structure of the establishment cause for the release 99 version of the RRC Connection Request message is:

```
EstablishmentCause ::= ENUMERATED {
originatingConversationalCall,
originatingStreamingCall,
originatingInteractiveCall,
originatingBackgroundCall,
originatingSubscribedTrafficCall,
terminatingConversationalCall,
terminatingStreamingCall,
terminatingInteractiveCall,
terminatingBackgroundCall,
emergencyCall,
interRAT-CellReselection,
interRAT-CellChangeOrder,
registration,
detach,
originatingHighPrioritySignalling,
originatingLowPrioritySignalling,
callRe-establishment,
terminatingHighPrioritySignalling,
terminatingLowPrioritySignalling,
terminatingCauseUnknown,
spare12, spare11, spare10, spare9, spare8, spare7,
spare6, spare5, spare4, spare3, spare2, spare1 }
```

The release 6 version of the establishment cause structure includes two additional cause values — MBMS reception and MBMS PTP RB Request. With the inclusion of the spare positions there are a

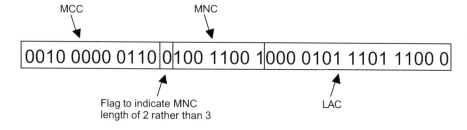

Figure 2.29   Second section of an ASN.1 encoded RRC Connection Request message

total of 32 cause values which can be encoded using 5 bits. This example is based upon a cause value of registration which appears 13th within the list. Indexing starts at 0 and so the registration cause value is referenced using an index of 12. This index is illustrated within Figure 2.30.

The establishment cause is followed by the protocol error indicator which has the following structure:

```
ProtocolErrorIndicator ::= ENUMERATED {
noError, errorOccurred}
```

There are only two possible values and so one bit is sufficient for encoding. Figure 2.30 illustrates that the first value is signalled which indicates no protocol error. The final section of the RRC Connection Request message provides the RACH measurement results. The structure of this section is shown below:

```
MeasuredResultsOnRACH ::= SEQUENCE {
    currentCell SEQUENCE {
        modeSpecificInfo CHOICE {
        fdd SEQUENCE {
            measurementQuantity CHOICE {
                cpich-Ec-N0 CPICH-Ec-N0,
                cpich-RSCP CPICH-RSCP,
                pathloss Pathloss}},
        tdd SEQUENCE {
            timeslotISCP TimeslotISCP-List OPTIONAL,
            primaryCCPCH-RSCP PrimaryCCPCH-RSCP OPTIONAL}}},
    monitoredCells MonitoredCellRACH-List OPTIONAL}
```

The structure is based upon a Sequence with one mandatory information element (current cell measurements) and one optional information element (monitored cell measurements). A flag is included to indicate whether or not the optional monitored cell measurements are included. In this example the flag is set to 0 and the measurements are excluded. The current cell measurements are also a Sequence, but with only a single mandatory information element. The mode specific information element is a Choice between FDD and TDD. This allows the RRC Connection Request message to be used by both FDD and TDD radio access technologies. This example is based upon FDD so a value of 0 is included to index the first entry in the list. The measurement quantity itself is signalled as a Choice between three entries and requiring two bits for encoding. A value of 0 indicates that the subsequent measurement is based upon CPICH Ec/Io. The structure of a CPICH Ec/Io measurement is

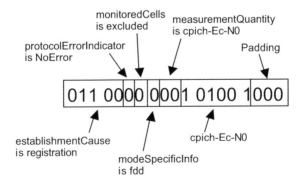

Figure 2.30   Third section of an ASN.1 encoded RRC Connection Request message

an integer value between 0 and 49 which requires 6 bits for encoding. These 6 bits allow a maximum range from 0 to 63, but values above 49 are spare and should not be used:

```
CPICH-Ec-N0 ::= INTEGER (0..63)
```

In this example the CPICH Ec/Io has a signalled value of 41 which maps onto an actual value of between -4.0 and -3.5 dB. This mapping between signalled and actual value is specified by 3GPP TS 25.133. The RRC Connection Request message is then completed with the inclusion of padding to generate an integer number of bytes.

2.2.4 UE RRC Timers, Counters and Constants

The RRC protocol specified within 3GPP TS 25.331 defines a set of UE timers, counters and constants. Each timer is configured with a maximum value which triggers an action upon expiry. 3GPP TS 25.331 defines a set of default values for the maximum of each timer. UE have knowledge of these default values without any signalling from the network. The RNC is able to override the defaults using either broadcast system information or dedicated UTRAN Mobility Information messages. The set of UE timers is presented in Table 2.14.

A subset of the UE timers presented in Table 2.14 have associated UE counters, i.e. a counter is incremented when the timer expires. These counters are presented in Table 2.15.

Each of the counters presented in Table 2.15 has a maximum value. An action is triggered if the counter exceeds the maximum value. The maximum values are defined using a set of constants. With the exception of N308, each constant has a default value which is known to the UE. Similar to the default maximum timer values, these defaults may be overridden by the RNC using either broadcast system information or dedicated UTRAN Mobility Information messages. The set of UE constants is presented in Table 2.16.

An example of timers and constants being used in CELL_DCH is illustrated in Figure 2.31. This example is based upon radio link failure, i.e. the UE loses synchronisation on the air-interface. If a UE is configured with a dedicated channel the pilot bits belonging to the DPCCH are used to evaluate whether or not the UE is synchronised. If a UE is configured with HSDPA and the F-DPCH has replaced the DPCCH then the TPC bits belonging to the F-DPCH are used to evaluate the synchronisation status.

The physical layer of the UE is able to generate an 'in-sync' or 'out-of-sync' primitive once every 10 ms radio frame. It is also possible that conditions are such that neither primitive is generated during a radio frame. The 'in-sync' or 'out-of-sync' primitives are reported to the RRC layer which is then responsible for acting appropriately. If N313 (default 20) consecutive 'out-of-sync' primitives are received then the UE starts timer T313 (default 3 s). If N315 (default 1) successive 'in-sync' indicators are received while T313 is running then the UE stops T313 and the UE remains synchronised.

If T313 expires before N315 'in-sync' indicators have been received the UE recognises that radio link failure has occurred. At this stage the UE releases its dedicated physical channel configuration and attempts to complete a cell update using the random access channel. Either or both of timers T314 (default 12 s) and T315 (default 180 s) are started. The early versions of 3GPP TS 25.331 stated that T314 should be associated with transparent and unacknowledged mode radio bearers whereas T315 should be associated with acknowledged mode radio bearers. This definition was subsequently removed, allowing the RNC to select which Radio Access Bearers (RAB) to associate with each timer. The relevant timer is now signalled as part of the RAB establishment procedure. An example is presented in Log File 2.17.

This example illustrates T315 being associated with a packet switched RAB. If a RAB has not been configured at the time that radio link failure is detected then T314 is applied for the signalling radio

Table 2.14 UE timers

Timer	Signalled within	Start	Stop	Upon expiry
T300 (default: 1 s) RRC Idle	SIB 1	When MAC layer indicates success or failure of sending an RRC Connection Request (non-MBMS).	When an RRC Connection Setup is received.	Re-transmit the RRC Connection Request if V300 $\leq$N300. Otherwise return to RRC Idle mode.
T302 (default: 4 s) RRC Connected	SIB 1 and UTRAN Mobility Information	When MAC layer indicates success or failure of sending either a Cell Update or a URA Update.	When a Cell Update Confirm or URA Update Confirm is received.	Re-transmit the Cell Update or URA update if V302 $\leq$N302. Otherwise enter RRC Idle mode.
T304 (default: 2 s) RRC Connected	SIB 1 and UTRAN Mobility Information	When the UE Capability Information has been delivered to the lower layers.	When a UE Capability Information Confirm is received.	Re-transmit UE Capability Information if V304 $\leq$N304. Otherwise initiate cell update procedure.
T305 (default: 30 min) RRC Connected	SIB 1 and UTRAN Mobility Information	When CELL_FACH, CELL_PCH, URA_PCH is entered. Or after Cell/URA Update Confirm.	When entering another RRC state.	Transmit Cell Update if T307 isn't running and UE is in coverage. Otherwise, if it isn't running start T307.
T307 (default: 30 s) RRC Connected	SIB 1 and UTRAN Mobility Information	When T305 expires and UE is out of coverage.	When UE has returned to coverage.	Make the transition to RRC Idle mode.
T308 (default: 160 ms) RRC Connected	SIB 1 and UTRAN Mobility Information	After transmission of RRC Connection Release Complete on air-interface.	T308 is not stopped.	Transmit RRC Connection Release Complete if V308 $\leq$N308, else go to RRC Idle mode.
T309 (default: 5 s) RRC Connected	SIB 1 and UTRAN Mobility Information	After reception of Cell Change Order From UTRAN.	Successful response to a connection establishment request in the new cell.	Resume the original connection with UTRAN.
T312 (default: 1 s) RRC Idle and Connected	SIB 1 and UTRAN Mobility Information	When the UE starts to establish a DPCH.	When UE detects N312 'in sync' indications from L1.	Criteria for physical channel establishment failure is fulfilled.

Timer			
T313 (default: 3 s) RRC Connected	SIB 1 and UTRAN Mobility Information	When UE detects N313 consecutive 'out of sync' indications from L1.	When UE detects N315 consecutive 'in sync' indications from L1.	The criteria for radio link failure is fulfilled.
T314 (default: 12 s) RRC Connected	SIB 1 and UTRAN Mobility Information	When the criteria for radio link failure is fulfilled, if RB associated with T314 exist or if only RRC conn. exists.	When the Cell Update procedure has been completed.	If T302 is running then await Cell/URA Update Confirm. Else, release RB associated with T314, and enter RRC Idle if T314 is not running.
T315 (default: 180 s) RRC Connected	SIB 1 and UTRAN Mobility Information	When the criteria for radio link failure is fulfilled, if RB associated with T315 exist.	When the Cell Update procedure has been completed.	If T302 is running then await Cell/URA Update Confirm. Else, release RB associated with T315, and enter RRC Idle if T314 is not running.
T316 (default: 30 s) RRC Connected	SIB 1 and UTRAN Mobility Information	When the UE is out of coverage in URA_PCH or CELL_PCH states.	When UE has returned to coverage.	Initiate cell update if in coverage, else start T317, move to CELL_FACH and initiate cell update when in coverage.
T317 (default: infinity) RRC Connected	SIB 1 and UTRAN Mobility Information	When the T316 expires or when out of coverage in CELL_FACH.	When UE has returned to coverage.	T317 does not expire.
T318 (default: 1s) RRC Idle	SIB 5 and MBMS Access Information	When MAC layer indicates success or failure of sending an RRC Connection Request (MBMS).	When an RRC Connection Setup is received.	Return to RRC Idle mode.

Table 2.15 UE counters

Counter	Reset	Incremented	Action
V300	At the start of RRC connection establishment	Expiry of T300	If V300 > N300, the UE returns to RRC Idle mode
V302	At the start of a cell or URA update	Expiry of T302	If V302 > N302, the UE returns to RRC Idle mode
V304	When sending the first UE capability information	Expiry of T304	If V304 > N304, the UE starts the Cell Update procedure
V308	When sending the first RRC Connection Release Complete	Expiry of T308	If V308 > N308, the UE stops re-transmitting the RRC Connection Release Complete

bearers. Unless a UE is configured with a multi-RAB, only one of T314 and T315 will be running. If the UE manages to complete a cell update while either T314 or T315 is running the UE remains in RRC Connected mode and maintains its connection. The RNC can reconfigure the connection as part of the Cell Update Confirm message. If T315 or T314 expires while the UE is attempting to camp on a cell to complete the cell update procedure then the radio bearers associated with the relevant timer are released. If only one timer is running, the UE returns to RRC Idle mode. If both timers are running, the UE waits for the second timer to expire before releasing the associated radio bearers and returning to RRC Idle mode.

Figure 2.32 illustrates the timers associated with periodic cell and URA updates for UE in CELL_FACH, CELL_PCH and URA_PCH. Periodic updates are used as a keep-alive mechanism to ensure that the network maintains knowledge of whether or not a UE remains connected. If a periodic update is missed then the RNC can deduce that the UE has moved out of coverage and is no longer in RRC connected mode. The most important timer for this procedure is T305 which has a default value of 30 min.

T305 is started when a UE moves into CELL_FACH, CELL_PCH or URA_PCH. If a UE moves to a new cell in either CELL_FACH or CELL_PCH then a cell update is completed and T305 is reset. Likewise, if a UE moves to a new URA in URA_PCH then a URA update is completed and T305 is reset. The UE attempts a periodic update when T305 expires. If the UE is in coverage then the update should be successful and T305 is reset for the next periodic update. If the UE is out of coverage when T305 expires, T307 is started (default 30 s). If the UE returns to coverage while T307 is running then an update can be completed successfully. Otherwise, if T307 expires the UE returns to RRC Idle mode. Although T305 is a UE timer, the RNC runs an equivalent supervision timer in parallel. This means that the RNC knows when to expect a periodic update from the UE. If the update does not arrive the RNC can deduce that the UE has returned to RRC Idle mode. The paging procedure provides an example of a procedure which would change after the RNC has deduced that a UE has returned to RRC

Table 2.16 UE constants

Constant	Default	Definition
N300	3	Maximum number of re-transmissions of RRC Connection Request
N302	3	Maximum number of re-transmissions of Cell/URA Update
N304	2	Maximum number of re-transmissions of UE Capability Information
N308	No default	Maximum number of re-transmissions of RRC Connection Release Complete
N312	1	Maximum number of 'in sync' primitives required from L1
N313	20	Maximum number of consecutive 'out of sync' received from L1
N315	1	Maximum number of consecutive 'in sync' received from L1 while T313 is active

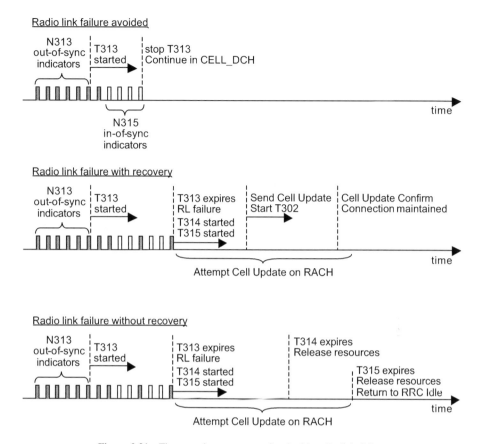

Figure 2.31 Timers and counters associated with radio link failure

Idle mode. If a UE in CELL_PCH moved out of coverage and returned to RRC Idle mode as a result of failing to complete a periodic cell update then the RNC would page the UE on a location area basis rather than on a cell basis. The RNC would also address the UE using its TMSI, P-TMSI or IMSI rather than its U-RNTI which the UE would have deleted after returning to Idle mode.

Figure 2.33 illustrates the timers which are associated with a loss of coverage while in CELL_FACH, CELL_PCH and URA_PCH. T316 has a default of 30 s whereas T317 has a fixed value of infinity.

If a UE moves out of coverage while in CELL_PCH then timer T316 is started. If the UE returns to coverage on the same cell while T316 is running then no further action is necessary. If the UE returns to coverage on a different cell then a cell update is completed with a cause value of 'cell reselection'. If T316 expires then the UE moves to CELL_FACH and T317 is started. T317 is also used by UE which

```
rab-InformationSetupList value 1
   rab-Info
   rab-Identity
     gsm-MAP-RAB-Identity
       Bin: 05
   cn-DomainIdentity:        ps-domain
   re-EstablishmentTimer: useT315
```

Log File 2.17   Example of associating T315 with a radio access bearer

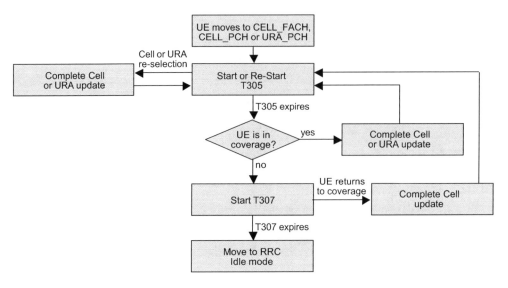

Figure 2.32   Timers associated with periodic cell and URA updates

have moved out of coverage while in CELL_FACH. T317 has a maximum value of infinity and so never expires. If the UE returns to coverage while T317 is running then a cell update is completed with cause 're-entering service area'. If the periodic cell or URA update timer T305 expires while T317 is running then the UE returns to RRC Idle mode.

The definition of T317 was changed by 3GPP during 2003. In earlier versions of TS 25.331 T317 was allowed to expire and had a default value of 180 s. It was discovered that this could cause a loss of synchronisation between the RRC state machines maintained by the UE and RNC and that this could

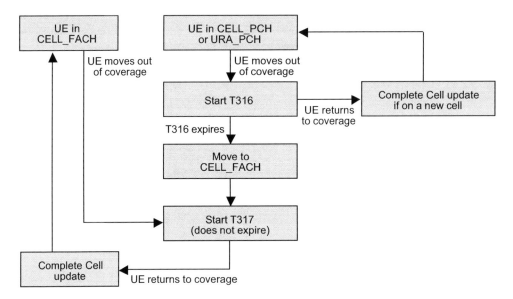

Figure 2.33   Timers associated with loss of coverage in CELL_FACH, CELL_PCH and URA_PCH

subsequently cause issues with the paging procedure. When T317 expired the UE returned to RRC Idle mode and could return to coverage and camp upon a cell without informing the RNC. One proposed solution was to force the UE to complete a routing area update after returning to coverage in RRC Idle mode. However the solution adopted was to prevent T317 expiring and so prevent the UE from returning to Idle mode.

2.2.5 Other Functions

3GPP TS 25.331 specifies a set of reporting events which can be used to trigger the UE to send measurement reports to the RNC. These measurement reports inform the RNC of conditions that the UE is experiencing and allows the RNC to react appropriately. The set of reporting events specified by

Table 2.17   RRC Reporting events

Measurement	Reporting events
Intra-frequency	1a-primary CPICH enters the reporting range 1b-primary CPICH leaves the reporting range 1c-non-active primary CPICH becomes better than active primary CPICH 1d-change of best cell 1e-primary CPICH becomes better than an absolute threshold 1f-primary CPICH becomes worse than an absolute threshold 1j-non-active E-DCH, but active DCH primary CPICH becomes better than an active E-DCH primary CPICH
Inter-frequency	2a-change of best frequency 2b-estimated quality of the current frequency is below a threshold and estimated quality of a non-used frequency is above a threshold 2c-estimated quality of a non-used frequency is above a threshold 2d-estimated quality of the current frequency is below a threshold 2e-estimated quality of a non-used frequency is below a threshold 2f-estimated quality of the current frequency is above a threshold
Inter-RAT	3a-estimated quality of the current frequency is below a threshold and estimated quality of other system is above a threshold 3b-estimated quality of other system is below a threshold 3c-estimated quality of other system is above a threshold 3d-change of best cell in other system
Traffic volume	4a-transport channel traffic volume becomes larger than a threshold 4b-transport channel traffic volume becomes smaller than a threshold
Quality	5a-a defined number of bad CRC are received
UE internal	6a-UE transmit power becomes larger than a threshold 6b-UE transmit power becomes less than a threshold 6c-UE transmit power reaches its minimum value 6d-UE transmit power reaches its maximum value 6e-UE RSSI reaches the UE's receiver dynamic range 6f-UE Rx-Tx time difference becomes larger than a threshold 6g-UE Rx-Tx time difference becomes less than a threshold
UE positioning	7a-UE position changes more than a threshold 7b-SFN to SFN measurement changes more than a threshold 7c-GPS time and SFN time have drifted apart more than a threshold

TS 25.331 is presented within Table 2.17. These reporting events are categorised according to their measurement type. The RNC uses Measurement Control messages to configure these reporting events.

The set of intra-frequency reporting events are typically used for handover procedures. The measurement quantity configured to trigger these events can be CPICH Ec/Io, CPICH RSCP or path loss. In the case of path loss, the UE estimates the result by subtracting the CPICH RSCP from the CPICH transmit power. Events 1a, 1b and 1c can be used to trigger soft handover radio link addition, deletion and replacement respectively. It is usual to trigger these events using CPICH Ec/Io rather than CPICH RSCP or path loss. CPICH Ec/Io results are generally more accurate because CPICH Ec/Io is calculated as a ratio of CPICH RSCP to RSSI. Measurement errors in CPICH Ec/Io and RSSI tend to be correlated and so cancel one another at least to some extent. Events 1a, 1b and 1c are described in greater detail within Section 8.6. Event 1d can be used to trigger an HSDPA serving cell change. This procedure is described within Section 6.10. Events 1e and 1f can be used to cancel and trigger inter-system and inter-frequency handover attempts. Both CPICH Ec/Io and CPICH RSCP can be used to trigger inter-system and inter-frequency handovers. If both measurement quantities are used then it is necessary to configure two event 1f measurements and two event 1e measurements. The UE and RNC distinguish between the two using the measurement identity, e.g. event 1e and 1f based upon CPICH Ec/Io could be assigned measurement identity 10, while event 1e and 1f based upon CPICH RSCP could be assigned measurement identity 11. An example of events 1e and 1f applied to the inter-system handover procedure is presented in Section 8.7. Events 1g, 1h and 1i are excluded from Table 2.17 because they are used for the TDD variant of UMTS rather than for the FDD variant. Reporting event 1j was introduced with the release 6 version of TS 25.331. This reporting event is applicable to HSUPA connections and allows cells in the DCH active set to replace cells in the E-DCH active set. This event is similar to event 1c in terms of comparing non-active set cells with the weakest active set cell.

The set of inter-frequency reporting events can also be used for handover procedures. The measurement quantity configured to trigger these events can be CPICH Ec/Io or CPICH RSCP, i.e. in this case path loss is not used. Events 2a, 2b, 2c and 2e involve measurements on other RF carriers and so typically require the use of compressed mode. Compressed mode can be avoided if the UE has a dual receiver. The requirement for compressed mode means that these reporting events are not used very often in practice. Events 2d and 2f involve measurements on the current RF carrier and do not require compressed mode. These reporting events can be used in a similar way to events 1e and 1f, i.e. both pairs of reporting events can be used to trigger inter-frequency and inter-system handovers. The main difference between the two pairs of reporting events is that events 1e and 1f are evaluated and reported on a per cell basis whereas events 2d and 2f are evaluated and reported based upon all cells within the active set. If the active set includes three cells and event 1f is being used to trigger inter-system handover then it would be normal to require all three cells to trigger event 1f before the inter-system handover procedure is initiated. This would then indicate that all cells within the active set have become relatively weak. If the active set includes three cells and event 2d is being used to trigger inter-system handover then it would be normal to allow a single event 2d measurement report based upon all three cells to initiate the inter-system handover procedure. Events 2d and 2f make use of a quality measure defined by the equation below.

$$\text{Quality Measure} = W \times 10 \times \log \left(\sum_{i=1}^{N} M_i \right) + (1 - W) \times 10 \times \log(M_{\text{best}})$$

Where W is a weighting coefficient provided by the RNC, N is the number of cells in the active set, M_i is the measurement result from the ith cell in the active set and M_{best} is the measurement result from the best cell within the active set. Measurement results are expressed in linear units. If the weighting coefficient is configured with a value of 0 then the quality measure will be based upon only the best cell within the active set, whereas if the weighting coefficient is configured with a value of 1 then the quality measure will be based upon a sum of the active set cells. A intermediate weighting coefficient allows a combination of the two scenarios. Events 1a and 1b make use of a similar weighting coefficient, but events 1e and 1f do not.

The set of inter-RAT (Radio Access Technology) reporting events can also be used for handover procedures. The measurement quantities configured to trigger these events can be either CPICH Ec/Io or CPICH RSCP on the UMTS system and RSSI on the GSM system. All four events involve measurements on other RF carriers and so typically require the use of compressed mode. The requirement for compressed mode means that these reporting events are not used very often in practice.

The traffic volume reporting events 4a and 4b can be used by the RNC to allocate appropriate uplink bit rates. Whereas the intra-frequency, inter-frequency and inter-RAT reporting events are based upon physical layer measurements, the traffic volume reporting events are based upon MAC layer measurements. The MAC layer is responsible for receiving buffer occupancy results from the RLC layer at least once every Transmission Time Interval (TTI). These buffer occupancy results are expressed in terms of bytes. Event 4a is triggered if the buffer occupancy becomes larger than a first threshold whereas event 4b is triggered if the buffer occupancy becomes smaller than a second threshold. An example Measurement Control message used to configure reporting event 4a is presented in Log File 2.18.

```
MEASUREMENT_CONTROL
DL-DCCH-Message
integrityCheckInfo
messageAuthenticationCode
Bin: EC 8F 80 96
rrc-MessageSequenceNumber: 3
measurementControl-r3
rrc-TransactionIdentifier: 0
measurementIdentity:           16
measurementCommand
setup
trafficVolumeMeasurement
trafficVolumeMeasurementObjectList
    trafficVolumeMeasurementObjectList value 1
    dch: 1
trafficVolumeMeasQuantity:                  rlc-BufferPayload
    trafficVolumeReportingQuantity
    rlc-RB-BufferPayload:                   true
    rlc-RB-BufferPayloadAverage:            false
    rlc-RB-BufferPayloadVariance:           false
    measurementValidity
    ue-State:                               all-States
    reportCriteria
trafficVolumeReportingCriteria
    transChCriteriaList
        transChCriteriaList value 1
        ul-transportChannelID
        dch: 1
    eventSpecificParameters
        eventSpecificParameters value 1
        eventID:                            e4a
        reportingThreshold:                 th8
        timeToTrigger:                      ttt240
        pendingTimeAfterTrigger:            ptat2
measurementReportingMode
    measurementReportTransferMode:          acknowledgedModeRLC
    periodicalOrEventTrigger:               eventTrigger
```

Log File 2.18 Measurement control message used to configure reporting event 4a

```
MEASUREMENT_REPORT
UL-DCCH-Message
integrityCheckInfo
messageAuthenticationCode
Bin: F5 65 B6 A3
rrc-MessageSequenceNumber: 4
measurementReport
measurementIdentity: 16
measuredResults
    trafficVolumeMeasuredResultsList
    trafficVolumeMeasuredResultsList value 1
        rb-Identity:            5
        rlc-BuffersPayload:     pl1024
        eventResults
        trafficVolumeEventResults
        ul-transportChannelCausingEvent: dch: 1
        trafficVolumeEventIdentity: e4a
```

Log File 2.19   Measurement report message triggered by reporting event 4a

In this example, the reporting event is configured with a measurement identity of 16. This identity is included within any Measurement Reports associated with the reporting event. The reporting event is configured to focus upon the DCH transport channel with identity 1. The traffic volume measurement quantity is signalled to be the RLC buffer payload. However, 3GPP TS 25.331 specifies that this information element should be ignored by the UE. Earlier versions of the specifications allowed the measurement quantity to be either RLC buffer payload, average RLC buffer payload or variance of the RLC buffer payload. TS 25.331 was revised such that the measurement quantity is always RLC buffer occupancy, irrespective of the quantity signalled within the Measurement Control message. The Measurement Control message presented in Log File 2.19 proceeds to specify that when a Measurement Report is triggered, the UE should include a measurement of RLC buffer payload for DCH transport channel 1. It is not necessary for the UE to include average and variance measurements. The Measurement Control also specifies that this reporting event is applicable to all RRC states.

The traffic volume reporting criteria section defines the conditions which trigger the UE to send a Measurement Report to the RNC. The measurement event is identified as type 4a while a reporting threshold of 8 bytes is configured with a time-to-trigger of 240 ms. This means that if the RLC buffer occupancy for DCH transport channel 1 exceeds 8 bytes for 240 ms then the UE will provide the RNC with a Measurement Report including the relevant RLC buffer occupancy measurement. A pending time after trigger of 2 s is specified. This defines the minimum interval between consecutive Measurement Reports for this reporting event. Consecutive Measurement Reports are only sent if the RLC buffer occupancy falls below the triggering threshold for event 4a and then returns to a level above the triggering threshold.

Log File 2.19 presents an example Measurement Report which has been triggered by the event 4a configured using the Measurement Control message presented in Log File 2.19.

The relevant reporting event is identified within the message using the measurement identity of 16. The main part of the Measurement Report then specifies that radio bearer 5 which belongs to DCH transport channel 1 has triggered event 4a with an RLC buffer occupancy of 1024 bytes (1 kbyte). The RNC can use this information to trigger an upgrade of the allocated uplink bit rate.

Reporting event 5a is based upon the downlink transport channel Block Error Rate (BLER). The event is triggered if the number of bad CRC experienced from a specific total number of CRC exceeds a threshold. For example, event 5a could be configured to trigger when 5 bad CRC are experienced after a total of 100 CRC, i.e. the downlink BLER becomes worse than 5%. The RNC could use this information to identify instances when the UE struggles to receive downlink data and could benefit from a decrease in the allocated bit rate to provide greater margin in the link budget. The UE will not necessarily evaluate event 5a periodically. If there is a break in the downlink data transfer then the UE will have to wait for longer before accumulating the configured total number of CRC. This is similar to the reason why UE logging tools report downlink BLER with a variable time interval, i.e. the BLER is reported after a specific number of CRC rather than after a specific time interval.

UE internal reporting events 6a, 6b, 6c and 6d are based upon UE transmit power measurements. Events 6a and 6b are based upon configurable thresholds whereas events 6c and 6d are based upon the minimum and maximum UE transmit power. The maximum UE transmit power may not necessarily be based upon the UE power class, but could also be based upon the maximum allowed transmit power signalled by the RNC. For example, a power class 3 UE which is capable of transmitting 24 dBm of uplink power could be informed within System Information Block (SIB) 3 that the maximum allowed uplink transmit power is 21 dBm. This means that triggering event 6d would be based upon 21 dBm rather than 24 dBm. Both events 6a and 6d are related to the UE transmit power becoming high. These events can be used to indicate that coverage has become uplink limited. If the uplink coverage continues to become worse then the connection is likely to fail as a result of the UE having insufficient transmit power Events 6a and 6d can be used to trigger inter-system and inter-frequency handovers. Event 6a provides greater flexibility because the triggering threshold can be configured. If inter-system or inter-frequency handover is triggered using event 6d then there may not be sufficient time to complete the necessary handover measurements before the connection fails. Event 6a can be configured with a threshold below the maximum and so the RNC has greater time to instruct the UE to complete handover measurements prior to the UE transmit power becoming completely exhausted. The drawback associated with configuring a lower threshold for handovers based upon event 6a is that the handover procedure is likely to be triggered more frequently and some handovers may be completed unnecessarily.

Event 6e provides a mechanism for detecting whether or not UE receive too much downlink power. If event 6e is configured then a measurement report is triggered when the total received downlink power reaches the maximum that the UE is capable of receiving. In general, a well-planned radio network should not allow UE to receive more power than they are capable of receiving, i.e. the minimum coupling loss between the Node B and the UE should provide sufficient isolation to protect the UE receiver from high power levels. Indoor solutions and picocells are more likely to generate minimum coupling loss problems despite having lower downlink transmit powers. Event 6e can be used to detect whether or not there are minimum coupling loss problems in the network. If UE start to report significant numbers of event 6e measurement reports then there is justification to re-optimise the radio network plan around the location where the measurement reports are triggered.

Events 6f and 6g are only applicable to UE with connections in soft handover, i.e. with more than one downlink radio link. Section 3.3.9.1 explains that when a Dedicated Physical Channel (DPCH) is first established the uplink slot timing follows the downlink slot timing by 1024 chips. When a UE enters soft handover the RNC attempts to synchronise the two downlinks from the perspective of the UE receiver. If the propagation delay from each cell is different then the slot timing should be synchronised from the UE perspective, but unsynchronised from the Node B perspective. The RNC makes use of timing information provided by the UE within the event 1a Measurement Report to determine the timing of the new radio link. An example of this timing information provided by the UE is presented in Log File 2.20.

The UE measures the time offset between the P-CCPCH of the candidate cell and its existing DPCH connection. This represents the System Frame Number (SFN) to Connection Frame Number (CFN) observed time difference. The time difference is defined by OFF $\times$ 38400 + Tm, where OFF can range from 0 to 255 radio frames and Tm can range from 0 to 38399 chips. This information allows the RNC to provide the new Node B with appropriate instructions regarding the timing of its new downlink DPCH transmission.

In practice, the synchronisation of downlink signals originating from different active set cells will not be perfectly synchronised and they will appear as multi-path signals for the UE RAKE receiver to combine. If the UE is mobile after the new radio link has been added to the active set, the

```
intraFreqMeasuredResultsList value 2
  cellSynchronisationInfo
  modeSpecificInfo fdd
      countC-SFN-Frame-difference
      countC-SFN-High:        2
      off:                    5
      tm:                     35327
  modeSpecificInfo fdd
      primaryCPICH-Info
      primaryScramblingCode:  34
      cpich-Ec-N0:            32    (-8.5 to -8.0 dB)
```

Log File 2.20 Example section from an event 1a Measurement Report

synchronisation between the downlink signals will change as the UE moves away from one active set cell and towards the other, i.e. one propagation delay increases while the other propagation delay decreases. The UE will then start to receive one downlink signal before the other and will have to buffer that signal before being able to combine the two signals within the RAKE receiver. This means that an increased time difference between the downlink radio links corresponds to an increased buffering requirement for the UE. It can also have an impact upon inner loop power control. If one radio link is received very late then the UE will not have time to account for its Transmit Power Control (TPC) command when transmitting the subsequent uplink time slot.

Event 6f allows the UE to inform the RNC that one of its radio links is arriving too early. This corresponds to experiencing a time difference between the received downlink and transmitted uplink of greater than 1024 chips. Figure 2.34 illustrates this scenario.

The RNC configures event 6f using a Measurement Control message. Configuring a triggering threshold of 1280 chips means that the UE uses a Measurement Report to inform the RNC if the time difference between a received downlink and the transmitted uplink exceeds 1280 chips. This represents a downlink which is arriving 256 chips early. The value of 1280 represents the maximum value which can be configured according to 3GPP TS 25.331. Smaller values between 1280 and 1024 chips can also be configured for event 6f. If the RNC receives an event 6f Measurement Report then it can remove the radio link from the active set allowing the UE to re-trigger event 1a for radio link addition. The radio link would then be re-added to the active set with corrected timing. The 256 chip threshold corresponds to a distance of 20 km ($256 \times 3 \times 10^8 / 3.84 \times 10^6$). This represents a large distance compared with the size of most macrocells and so event 6f should not be triggered very frequently when configured with a threshold of 1280 chips.

Event 6g allows the UE to inform the RNC that one of its radio links is arriving too late. This corresponds to experiencing a time difference between the received downlink and transmitted uplink of less than 1024 chips. Figure 2.35 illustrates this scenario.

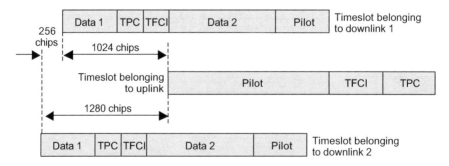

Figure 2.34 Event 6f triggered - time difference between uplink and downlink becomes too great

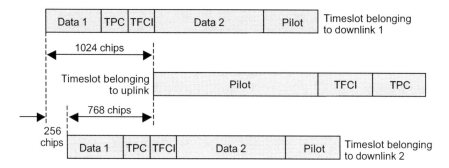

Figure 2.35 Event 6g triggered — time difference between uplink and downlink becomes too small

Configuring a triggering threshold of 768 chips means that the UE uses a Measurement Report to inform the RNC if the time difference between a received downlink and the transmitted uplink becomes less than 768 chips. This represents a downlink which is arriving 256 chips late. The value of 768 represents the minimum value which can be configured according to 3GPP TS 25.331. Larger values between 768 and 1024 chips can also be configured for event 6g. If the RNC receives an event 6g Measurement Report then it can remove the radio link from the active set allowing the UE to re-trigger event 1a for radio link addition. The radio link would then be re-added to the active set with corrected timing. Similar to event 6f, the 256 chip threshold corresponds to a distance of 20 km and so event 6g should not be triggered very frequently when configured with a threshold of 768 chips.

Log File 2.21 presents an example Measurement Report generated after event 6g has been triggered using a threshold of 874 chips (a drift of 150 chips relative to target of 1024 chips). In this example,

```
Measurement Report (UL-DCCH)
integrityCheckInfo
messageAuthenticationCode
Binary string (Bin) : 11110110000011101010001001001001
rrc-MessageSequenceNumber:        12
measurementIdentity:              6
measuredResults
MeasuredResults: ue-InternalMeasuredResults
    ue-InternalMeasuredResults
    modeSpecificInfo: fdd
        ue-RX-TX-ReportEntryList
        UE-RX-TX-ReportEntryList:
        [0]
            primaryCPICH-Info
            primaryScramblingCode:        236
            ue-RX-TX-TimeDifferenceType1: 768
        [1]
            primaryCPICH-Info
            primaryScramblingCode:        368
            ue-RX-TX-TimeDifferenceType1: 982
eventResults
EventResults : ue-InternalEventResults
    ue-InternalEventResults
    UE-InternalEventResults: event6g
    event6g
    primaryScramblingCode:    236
```

Log File 2.21 Example section from an event 6g Measurement Report

there are two cells in the active set. The Measurement Report specifies that the cell with scrambling code 236 has triggered event 6g. The Measurement Report also provides time offset measurements for both cells in the active set. The cell with scrambling code 368 has an acceptable time offset of 982 chips, i.e. this is less than 150 chips away from the target of 1024 chips. The cell with scrambling code 236 has an unacceptable time offset of 768 chips. The value of 768 chips represents the minimum value which can be reported, i.e. 3GPP TS 25.331 specifies a reporting range from 768 to 128 chips for both events 6f and 6g. In practice, the actual time offset could be less than 768 chips.

Reporting events 7a, 7b and 7c are relevant to location services. Event 7a is based upon the UE triggering a Measurement Report if its position changes by more than a specific threshold. Triggering thresholds ranging from 10 m to 100 km can be configured. Event 7b is based upon the UE triggering a Measurement Report if the relative SFN timing of any two measured cells changes by more than a specific threshold. Triggering thresholds ranging from 0.25 to 5000 chips can be configured. Event 7c is based upon the UE triggering a Measurement Report if the GPS Time Of the Week (TOW) and the SFN have drifted apart by more than a specific threshold. Triggering thresholds ranging from 1 to 100 ms can be configured.

2.3 RLC Layer

- The RLC layer can be used in three different modes: Transparent Mode (TM), Unacknowledged Mode (UM) and Acknowledged Mode (AM).
- TM is suitable for real time services, e.g. the speech and video call services. It is also used by SRB 0 in the uplink direction, e.g. RRC Connection Request message. The release 99 version of the 3GPP specifications does not allow TM to be used for DCCH SRB.
- TM does not add any header information and so does not introduce any overhead. Padding is not supported and so higher layer packet sizes must be a multiple of the RLC PDU size. If higher layer packets are segmented they must be transferred during the same TTI and no other higher layer packet can be transferred during that TTI. Multiple higher layer packets can be transferred during each TTI if segmentation is disabled.
- TM RLC does not support ciphering and so ciphering is completed by the MAC layer when necessary. Timer based SDU discard without explicit signalling is supported.
- UM RLC is suitable for real time services which have unpredictable higher layer packet sizes, e.g. the Voice over IP service. It is also used by SRB 0 in the downlink direction, e.g. RRC Connection Setup message, and by SRB 1 in the uplink and downlink directions.
- UM adds a header of at least 1 byte to attach a sequence number to each PDU. Length indicators may also be included to indicate the number of bytes between the RLC header and the end of an SDU. Segmentation, concatenation and padding are supported to allow one or more SDU to be packed within a single PDU.
- UM RLC supports ciphering and timer based SDU discard without explicit signalling. Additional UM functionality has been added to the release 6 version of the 3GPP specifications for the purposes of MBMS.
- AM RLC is suitable for non-real time services which require reliable data transfer, e.g. file transfers, email and internet browsing. It is also used by SRB 2, 3 and 4.
- AM adds a header of at least 2 bytes to attach a sequence number to each PDU. Length indicators may also be included to indicate the number of bytes between the RLC header and the end of an SDU. Segmentation, concatenation and padding are supported to allow one or more SDU to be packed within a single PDU. Re-transmissions are supported to provide reliable data transfer.

- AM uses transmit and receive windows to avoid sequence number ambiguity and to allow flow control. Window sizes should be configured according to the bit rate, round trip time and maximum number of re-transmissions.
- AM RLC Status PDU are used to inform the transmitter of which PDU have been received successfully. They can also be used for flow control. Status PDU can be triggered periodically, by the receiver detecting a missing PDU or by the transmitter polling the receiver. Polling can be triggered by: last PDU in buffer, last PDU in re-transmission buffer, poll timer, every nth PDU, every mth SDU, window-based and timer-based.
- AM RLC status PDU, RLC headers and RLC padding generate an overhead which is visible as an increased MAC throughput. TCP acknowledgements may result in PDU which include only header information and padding.
- AM RLC supports ciphering and SDU discard based upon: timer-based with explicit signalling, discard after maximum number of transmissions and no discard after maximum number of transmissions.
- Key 3GPP specifications: TS 25.322 and TS 25.331.

The Radio Link Control (RLC) layer of the radio interface protocol stack is specified within 3GPP TS 25.322. The RLC layer is used to transfer both control plane signalling and user plane data, i.e. RRC messages as well as application data pass through the RLC layer. The RLC layer can be used in three different modes: Transparent Mode (TM), Unacknowledged Mode (UM) and Acknowledged Mode (AM). 3GPP TS 25.331 specifies which RLC modes can be used to transfer each RRC message. The RNC is responsible for selecting the RLC mode when more than single option exists, e.g. Measurement Reports can be transferred using either AM or UM RLC. The RNC is also responsible for selecting an appropriate RLC mode for user plane connections. Real time applications tend to use either TM or UM RLC whereas non-real time applications tend to use AM RLC.

2.3.1 Transparent Mode

Transparent Mode (TM) RLC is suitable for real time services which have a requirement to minimise delay, e.g. it is used by the speech and video call services. TM RLC is able to minimise delay by offering only simple and basic functionality. In many cases TM RLC does not modify the higher layer packets in any way and those packets are able to pass directly through the RLC layer. In other cases TM RLC may modify the higher layer packets in terms of segmentation. TM RLC does not add any header information and so the bit rate at the top of the RLC layer is equal to the bit rate at the top of the MAC layer. TM RLC does not offer any sequence numbering nor does it offer any support for re-transmissions. The functionality offered by TM RLC is presented in Table 2.18.

TM RLC requires the higher layer packet (RLC SDU) size to be a multiple of the RLC PDU size. This is because the RLC PDU must be a fixed size and TM RLC does not support padding. If an SDU is larger than a PDU, the RLC layer at the transmitter is able to segment the SDU into multiple PDU. In

Table 2.18 Functionality offered by Transparent Mode RLC

	Downlink transmit	Downlink receive	Uplink transmit	Uplink receive
Segmentation	√		√	
SDU discard	√		√	
Transfer of data	√	√	√	√
Reassembly		√		√

```
rb-InformationSetupList
rb-InformationSetupList value 1
rb-Identity: 5
rlc-InfoChoice
rlc-Info
   ul-RLC-Mode
      ul-TM-RLC-Mode
      segmentationIndication: false
   dl-RLC-Mode
      dl-TM-RLC-Mode
      segmentationIndication: false
```

Log File 2.22 Example configuration for Transparent Mode RLC

this case, all PDU generated from the same SDU must be transferred during the same Transmission Time Interval (TTI) and no PDU generated from any other SDU can be transferred during the same TTI. This requirement is necessary because TM RLC does not add any header information to help reassembly at the receiver.

Log File 2.22 illustrates an example section from a Radio Bearer Setup message used to configure a radio bearer with TM RLC. In this example, segmentation of RLC SDU is not permitted. This means that the RLC SDU size must always equal the RLC PDU size. Multiple RLC SDU can be transferred during each TTI because the receiver knows that each PDU represents a separate and complete SDU.

Log File 2.22 does not include any information regarding the SDU discard function. The SDU discard options for TM RLC are summarised in Table 2.19.

RRC messages can include information regarding the uplink SDU discard function to configure the UE. RRC messages do not include information regarding the downlink SDU discard function because it is not necessary for the UE to know about the discard function within the RNC. Excluding uplink SDU discard information from within RRC messages (SDU discard not configured) means that by default the UE RLC layer discards buffered SDU belonging to any previous TTI when receiving SDU belonging to a new TTI. Alternatively, the RNC can configure the UE with a discard timer which defines the maximum duration that an SDU can remain within the RLC layer (timer based SDU discard without explicit signalling). The discard timer can be assigned a value between 10 and 100 ms. It is started whenever an RLC SDU is received from the higher layers. An SDU is discarded if the timer expires before it has been passed to the MAC layer. Similarly, the RNC can configure a downlink discard timer for itself.

Neither the speech nor video call services make use of RLC segmentation, i.e. the SDU size is equal to the PDU size. In addition, it is not necessary for these services to use an SDU discard timer. This means the RLC layer can be used in its simplest and most efficient form.

Data transfer between the RLC and MAC layers makes use of logical channels. TM RLC PDU can be transferred using the BCCH, PCCH, CCCH, DCCH and DTCH logical channels. With the exception of the DTCH, these logical channels transfer RRC messages. The RRC messages transferred using TM RLC are presented in Table 2.20.

Signalling Radio Bearers (SRB) are used to transfer RRC messages on the CCCH and DCCH logical channels. SRB 0 is used to transfer CCCH messages. This SRB uses TM RLC in the uplink direction and unacknowledged mode RLC in the downlink direction. Segmentation is not permitted for this SRB in the uplink direction and so the encoded RRC messages must occupy exactly one RLC PDU. The

Table 2.19 SDU discard methods (TM RLC)

Timer-based without explicit signalling
Not configured

Table 2.20 RRC messages able to use Transparent Mode RLC

Logical channel	Messages
BCCH	System Information Change Indication
PCCH	Paging Type 1
CCCH	Cell Update, URA Update, RRC Connection Request
DCCH	Transport Format Combination Control (3GPP release 4 onwards)

RRC layer is able to include padding to ensure that RLC SDU have an appropriate size. The use of TM RLC for DCCH logical channels is not supported by the release 99 version of the 3GPP specifications. Support was removed to help simplify that version of the specifications. Subsequent versions of the specifications allow DCCH radio bearers with identities between 5 and 32 to be configured with TM RLC (identity 0 is used for SRB 0 transferring CCCH information whereas identities 1 to 4 are used for SRB 1 to 4 transferring DCCH information with unacknowledged and acknowledged mode RLC). A DCCH using TM RLC is able to transfer the Transport Format Combination Control message. This message can be used to control the bit rate of a multi-rate speech connection. This message can also be transferred using unacknowledged mode and acknowledged mode RLC. It is intended that TM RLC is used when a UMTS to GSM speech connection makes use of either Tandem Free Operation (TFO) or Transcoder Free Operation (TrFO).

The use of TFO and TrFO means there are no speech transcoding operations between two UE. If one UE changes its speech service bit rate then the other UE must follow that change. In the case of GSM, speech service bit rates are changed according to the C/I conditions. These conditions are likely to change both rapidly and frequently. If a UMTS UE is connected to a GSM UE there is a requirement to inform the UMTS UE of the bit rate changes triggered by the GSM UE. The Transport Format Combination Control message can be used for this purpose. TM RLC offers a way to transfer this message both rapidly and efficiently. The message is relatively small, in the order of 5 bytes and so including a 1 or 2 byte RLC header represents a relatively significant overhead. The use of TM RLC avoids the requirement to include an RLC overhead. It is less important to use TM RLC when TFO and TrFO speech calls are between two UMTS UE because bit rate changes are likely to be triggered less frequently, i.e. UMTS UE benefit from inner loop power control to help maintain constant C/I conditions.

TM RLC does not offer any ciphering capability and so ciphering, if necessary has to be completed by the MAC layer.

2.3.2 Unacknowledged Mode

Unacknowledged Mode (UM) RLC is suitable for real time services which have unpredictable higher layer packet sizes, e.g. it can be used for the voice over IP (VoIP) service. UM RLC is able to offer greater functionality than TM RLC by including header information. The inclusion of header information means that the bit rate at the top of the MAC layer is greater than the bit rate at the top of the RLC layer. UM RLC functionality has been enhanced to help reduce the RLC overhead for services such as VoIP.

The functionality offered by UM RLC is presented in Table 2.21. In the case of UM RLC, the SDU size does not have to be a multiple of the PDU size. Segmentation, concatenation and padding is supported to allow one or more SDU to be packed within a single PDU. Length indicators within the RLC header can be used to indicate the number of bytes between the RLC header and the end of an SDU. The RLC header also includes a sequence number to allow the receiver to detect when PDU are missing. In contrast to TM RLC which relies upon the MAC layer to complete ciphering, UM RLC is capable of ciphering. If enabled, ciphering is applied to the RLC payload and length indicators, but not to the sequence number, i.e. the first byte of the RLC PDU remains unciphered.

Table 2.21   Functionality offered by Unacknowledged Mode RLC

	Downlink transmit	Downlink receive	Uplink transmit	Uplink receive
Segmentation	√		√	
Concatenation	√		√	
Padding	√		√	
Ciphering	√		√	
SDU discard	√		√	
Transfer of data	√	√	√	√
Deciphering		√		√
Reassembly		√		√
Sequence number check		√		√
Out of sequence SDU delivery		√		
Duplicate avoidance and reordering		√		

The SDU discard function for UM RLC is similar, but not identical. to that for TM RLC. The options are summarised in Table 2.22, i.e. the same options as for TM RLC.

A discard timer can be configured in the same way as for TM RLC and in that case, the behaviour of the discard function is the same as for TM RLC, i.e. an SDU is discarded at the transmitter if the discard timer expires before it has been forwarded to the MAC layer. The functionality differs from TM RLC if a discard timer has not been configured. In that case, UM RLC only discards SDU if the RLC buffer becomes full. This means that the size of the RLC buffer has an impact upon the probability of discarding RLC SDU.

UM RLC was updated for the release 6 version of the 3GPP specifications to include the 'out of sequence SDU delivery' and 'duplicate avoidance and reordering' functionality. These functions are used by MBMS in the downlink direction. The 'out of sequence SDU delivery' function is applicable to the MCCH logical channel whereas the 'duplicate avoidance and reordering' function is applicable to the MTCH logical channel.

UM RLC PDU include a header of at least 1 byte. The structure of an UM RLC PDU is illustrated in Figure 2.36. There are two possibilities depending upon whether 7 or 15 bits are assigned to the length indicators.

The 7 bit sequence number allows a range from 0 to 127 and is always included within an UM RLC PDU. This sequence number is initialised with a value of 0 and is incremented by 1 every time a new PDU is generated. The sequence number is always followed by a 1 bit Extension (E) field. The interpretation of this extension field can be configured by the RNC during radio bearer establishment. The interpretation can be configured to be either normal or alternative. These two configurations are presented in Table 2.23.

The release 6 version of the 3GPP specifications introduces the alternative interpretation to improve the efficiency of UM RLC for VoIP and other similar services for which the majority of higher layer packets have a fixed length. The alternative interpretation avoids the requirement to include length indicators when the UM RLC PDU size is configured to equal the RLC SDU size plus 1 byte, i.e. one byte of header information for the sequence number and extension field. The normal interpretation requires length indicators for each RLC SDU. The alternative interpretation does not prevent length indicators being used when an SDU does not fit exactly into an RLC PDU.

Table 2.22   SDU discard methods (UM RLC)

Timer based without explicit signalling
Not configured

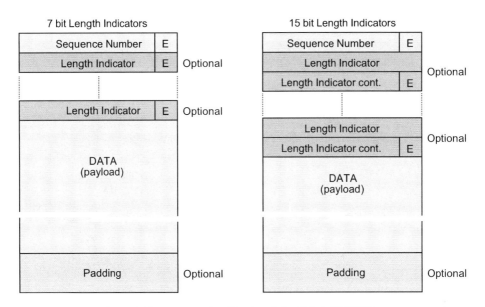

Figure 2.36 Structure of an Unacknowledged Mode RLC PDU

Length indicators are included to inform the receiver of the location of the last byte belonging to an SDU. Multiple length indicators can be included if multiple SDU are included within a single PDU. Length indicators configured with reserved values can also be included to signal other information. The reserved values for a 7 bit length indicator are presented in Table 2.24.

These reserved values allow the receiver to be informed of where RLC SDU start and finish. Log File 2.23 provides an example of a single RLC SDU occupying a series of UM RLC PDU. This example is based upon an RRC Connection Setup message with 7 bit length indicators and normal interpretation of the extension field.

The first UM RLC PDU includes a reserved length indicator value to inform the receiver that the first byte of data belongs to the start of a new SDU. The next five PDU do not include any length indicators whereas the seventh PDU includes two length indicators. The first indicator uses a non-reserved value to inform the receiver that the RLC SDU ends after a further 6 bytes of data. The second indicator uses a reserved value to inform the receiver that the remainder of the PDU is padding. This example demonstrates that the overhead generated by the RLC layer is not constant. The first PDU has an RLC overhead of 11%, i.e. 2 bytes of header compared with 18 bytes of payload. The next five PDU have an RLC overhead of 5%, i.e. 1 byte of header compared with 19 bytes of payload. The final PDU has a high RLC overhead because there is insufficient data to fully occupy the fixed size PDU. In this case the RLC overhead is 230%, i.e. 14 bytes of header and padding compared with 6 bytes of payload.

Table 2.23 Normal and alternative interpretations of the extension field

	Normal interpretation	Alternative interpretation
0	the next byte includes data or padding	the next byte includes a complete SDU
1	the next byte includes a length indicator	the next byte includes a length indicator

Table 2.24   Reserved length indicator values for UM RLC PDU (7 bit length indicator)

Reserved value	Interpretation
0000 000	The previous RLC PDU was exactly occupied by the last segment an of RLC SDU and an associated length indicator was not included.
1111 100	The first byte of data in the RLC PDU is the first byte of an RLC SDU.
1111 101	The first byte of data in the RLC PDU is the first byte of an RLC SDU and the last byte of the PDU is the last byte of the same SDU.
1111 110	The RLC PDU contains a segment of an SDU, but neither the first nor last byte of the SDU.
1111 111	The remainder of the RLC PDU is padding. The padding length can be zero.

UM RLC does not support re-transmissions and receiving a single PDU in error results in the complete set of PDU associated with the same SDU being discarded. The receiver then has to rely upon the higher layers to generate a re-transmission, e.g. the RRC layer within the RNC can re-transmit the RRC Connection Setup message if it does not receive an RRC Connection Setup Complete message within a certain time window.

```
RLC PDU 1 - 71 F8 30 EF 20 20 16 88 64 68 10 75 3B 01 FC 03 B4 45 83 03
- Sequence Number: 56
- E: Next field is LI and E bit
- LI: The first data octet is the first octet of a RLC SDU
- E: Next field is data
Data:  30 EF 20 20 16 88 64 68 10 75 3B 01 FC 03 B4 45 83 03

RLC PDU 2 -72 49 D2 E0 84 B8 6A 10 00 14 21 67 B4 51 B1 2E 8A 28 27 4B
- Sequence Number: 57
- E: Next field is data
        - Data :  49 D2 E0 84 B8 6A 10 00 14 21 67 B4 51 B1 2E 8A 28 27 4B

RLC PDU 3 -74 8A 12 E3 A8 C0 00 51 89 9E D1 46 C4 BA 28 A0 9D 2E 48 4B
- Sequence Number: 58
- E: Next field is data
- Data :  8A 12 E3 A8 C0 00 51 89 9E D1 46 C4 BA 28 A0 9D 2E 48 4B

RLC PDU 4 - 76 96 A5 00 01 4A 36 7B 35 1B 12 E8 A2 62 74 B9 B1 2E 7A 9C
- Sequence Number: 59
- E: Next field is data
- Data :  96 A5 00 01 4A 36 7B 35 1B 12 E8 A2 62 74 B9 B1 2E 7A 9C

RLC PDU 5 - 78 00 45 32 00 01 99 36 BF CA 02 E4 09 00 88 C0 19 82 5C 41
- Sequence Number: 60
        - E: Next field is data
- Data :  00 45 32 00 01 99 36 BF CA 02 E4 09 00 88 C0 19 82 5C 41

RLC PDU 6 - 7A 00 00 92 00 46 00 EB 50 91 8E 08 01 54 35 BC 01 9D 10 80
- Sequence Number: 61
- E: Next field is data
- Data :  00 00 92 00 46 00 EB 50 91 8E 08 01 54 35 BC 01 9D 10 80

RLC PDU 7 - 7D 0D FE 81 94 00 C0 C1 00 00 00 00 00 00 00 00 00 00 00 00
- Sequence Number: 62
- E: Next field is LI and E bit
- LI: 6
- E: Next field is LI and E bit
- LI: Rest of RLC PDU is padding
    - E: Next field is data
    - Data :  81 94 00 C0 C1 00
    - Padding:  00 00 00 00 00 00 00 00 00 00 00
```

Log File 2.23   Example RRC message packaged within a series of UM RLC PDU

Table 2.25   RRC messages able to use Unacknowledged Mode RLC

Logical channel	Messages
CCCH	RRC Connection Reject, RRC Connection Setup
DCCH	Active Set Update, MBMS Modification Request, Measurement Report, Physical Channel Reconfiguration, Radio Bearer Reconfiguration, Radio Bearer Release, Radio Bearer Setup, RRC Connection Release Complete, Transport Channel Reconfiguration, Transport Format Combination Control, UE Capability Enquiry, UE Capability Information Confirm, UTRAN Mobility Information
DCCH or CCCH	Cell Update Confirm, RRC Connection Release, URA Update Confirm
DCCH or MCCH	MBMS Modified Services Information
MCCH	MBMS Access Information, MBMS Common p-t-m rb Information, MBMS Current Cell p-t-m rb Information, MBMS General Info, MBMS Neighbouring Cell p-t-m rb Information, MBMS Unmodified Services Information
MSCH	MBMS Scheduling Information

In the uplink direction, 7 bit length indicators are used if the largest RLC PDU size is $\leq$ 125 bytes. A 7 bit length indicator has a range from 0 to 127. There are 5 reserved values which result in the range becoming 1–123. The maximum value of 123 corresponds to the maximum PDU size of 125 bytes minus 1 byte for the sequence number and 1 byte for the length indicator. If the largest uplink RLC PDU $>$ 125 bytes then a 15 bit length indicator is used. The same rule is applied for the downlink when using 3GPP specifications prior to release 5. The release 5 and newer versions of the specifications allow the RNC to explicitly signal the number of bits assigned to the downlink length indicators. This change was introduced to avoid any ambiguity during radio bearer reconfigurations.

UM RLC PDU can be transferred using the CCCH, CTCH, DCCH, DTCH, MCCH, MTCH and MSCH logical channels. The CCCH, DCCH, MCCH and MSCH are used to transfer RRC messages whereas the CTCH, DTCH and MTCH are used to transfer application data. The set of RRC messages which can be transferred using UM RLC is presented in Table 2.25.

The CCCH messages are associated with SRB 0. These are downlink messages and are in contrast to uplink SRB 0 messages which are transferred using TM RLC. The DCCH messages are associated with SRB 1. This SRB uses UM RLC in both the uplink and downlink directions.

2.3.3 Acknowledged Mode

Acknowledged Mode (AM) RLC is suitable for non-real time services which require reliable data transfer, e.g. it can be used for file transfers, email and internet browsing. AM RLC is able to offer reliable data transfer by allowing re-transmissions of RLC PDU. This avoids the requirement for the higher layers to re-transmit entire RLC SDU. This can be particularly important for typical data services which can have RLC SDU in the order of 1500 bytes. The payload belonging to a typical RLC PDU is 320 bits. This means that a bit error requires only 320 bits to be re-transmitted rather than 1500 bytes. The functionality offered by AM RLC is presented in Table 2.26.

Similar to UM RLC, the SDU size for AM RLC does not have to be a multiple of the PDU size. Segmentation, concatenation and padding are supported to allow one or more SDU to be packed within a single PDU. Length indicators within the RLC header can be used to indicate the number of bytes between the RLC header and the end of an SDU. The RLC header also includes a sequence number to allow the receiver to detect when PDU are missing. This information can be used to trigger re-transmissions.

Table 2.26 Functionality offered by Acknowledged Mode RLC

	Downlink transmit	Downlink receive	Uplink transmit	Uplink receive
Segmentation	√		√	
Concatenation	√		√	
Padding	√		√	
Ciphering	√		√	
SDU discard	√	√	√	√
Flow control	√	√	√	√
Error correction	√	√	√	√
Transfer of data	√	√	√	√
Protocol error detection and recovery	√	√	√	√
Deciphering		√		√
Reassembly		√		√
In-sequence delivery		√		√
Duplicate detection		√		√

AM RLC makes use of transmit and receive windows to avoid sequence number ambiguity and to provide a flow control capability. The general principles of the transmit and receive windows are illustrated in Figure 2.37.

The lower and upper edges of the transmit window are defined by the variables VT(A) and VT(MS) respectively. VT(A) represents the sequence number following the last in-sequence acknowledged

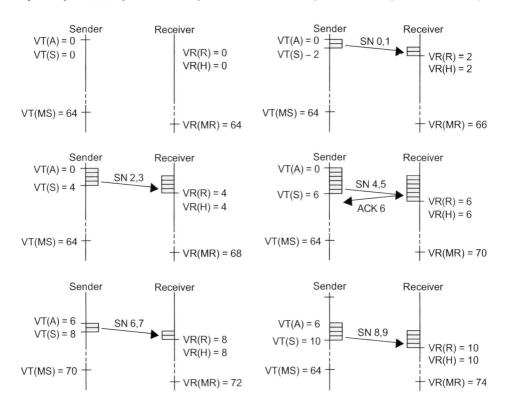

Figure 2.37 AM RLC transmit and receive windows (assuming no re-transmissions)

PDU. VT(MS) represents the sequence number of the first PDU which can be rejected by the receiver. The sender should not transmit any PDU with sequence numbers $\geq$ VT(MS). The sender also makes use of a VT(S) variable to record the sequence number of the next PDU to be transmitted for the first time.

The lower and upper edges of the receive window are defined by the variables VR(R) and VR(MR) respectively. VR(R) represents the sequence number following the last in-sequence received PDU. VR (MR) represents the sequence number of the first PDU which would be rejected by the receiver. The receiver also makes use of a VR(H) variable to record the highest expected sequence number of the next PDU. Although the transmit and receive windows have similar definitions they do not remain perfectly synchronised because the sender does not have instantaneous information regarding successfully received PDU.

The example presented in Figure 2.37 is based upon two RLC PDU transmitted during each Transmission Time Interval (TTI). This could represent a 32 kbps PS data connection if the RLC PDU size is 336 bits and the TTI is 20 ms. Data transfer starts when the sender transmits the first two PDU. These PDU are not discarded by the sender after transmission, but are stored within a re-transmission buffer. VT(S) is incremented for each PDU that has been sent. This provides the sender with a record of the sequence number to include within the next PDU. In this example, the first two PDU are successfully received and the receive window slides onwards by two. The receive window is now unsynchronised relative to the transmit window because the sender does not know whether or not the receiver has successfully received the two PDU.

The sender proceeds to send the next two PDU during the subsequent TTI. VT(S) is again incremented for each PDU that has been sent. The transmit window remains stationary because the sender has not yet received any acknowledgement from the receiver. The receiver successfully receives both PDU and the receive window slides onwards by a further two. The forth and fifth PDU are sent during the third TTI. In this example, it is assumed that the first RLC SDU, i.e. higher layer packet, occupies the first 6 RLC PDU. The sender polls the receiver when transmitting the sixth PDU (using a poll flag within the RLC PDU header). This triggers the receiver to return a status report indicating that the first 6 PDU have been successfully received. The transmitter is then able to update its transmit window and delete the acknowledged PDU from the re-transmission buffer. The transmit window moves closer to the receive window, but remains behind due to the PDU which have been sent while waiting for the response to the poll. Data transfer then continues and the next two PDU are transmitted.

The transmit and receive window sizes are configured by the RRC layer. The example presented in Figure 2.37 assumes transmit and receive window sizes of 64 RLC PDU. In practice, the window size should be configured according to the bit rate, round trip time and maximum number of re-transmissions. High bit rate connections tend to require large windows because PDU are transferred at a high rate relative to the round trip time. Increasing the number of re-transmissions also increases the requirement for a large window. The RLC layer can stall if the window size is not sufficiently large, i.e. the sender will not be able to transmit any new PDU while waiting for an acknowledgment from the receiver. The concept of the RLC layer stalling is illustrated in Figure 2.38

The large transmit window allows PDU with relatively high sequence numbers to be sent while the sender waits for an acknowledgement from a re-transmission. The small transmit window limits the number of PDU which can be sent while waiting for the acknowledgement. In this case there is a period of DTX while the sender waits. The transmission of PDU recovers once the acknowledgement has been received and the transmit window is able to slide forwards.

Based upon an RLC PDU size of 336 bits, a 384 kbps connection can transfer 12 PDU during a 10 ms TTI. If the RLC round trip time is assumed to equal 160 ms then the sender is able to transmit $12 \times 16 = 192$ PDU during the time required for the sender to receive a negative acknowledgement. If an original transmission and up to three re-transmissions are allowed then the total number of PDU becomes $4 \times 192 = 768$. This means that a window size of 768 PDU is appropriate. Similarly, a 128 kbps connection can transfer 4 PDU during a 10 ms TTI. In this case, the sender is able to transmit $4 \times 16 = 64$ PDU during the time required for the sender to receive a negative acknowledgement. If

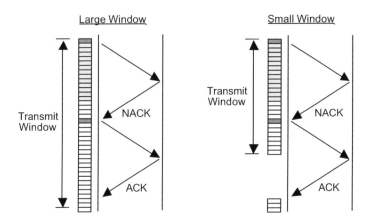

Figure 2.38 AM RLC layer stalling when small transmit window is used

an original transmission and up to three re-transmissions are allowed then the total number of PDU becomes $4 \times 64 = 256$ and in this case, a window size of 256 PDU is appropriate. The round trip time will increase for larger TTI and so the window size requirement will also increase. However, in general larger TTI are associated with lower bit rates and so fewer PDU are transferred during the round trip time and so the requirement for a large window size decreases. The appropriate window size can also be influenced by memory size and the associated buffering capability. If a UE has only limited memory then it may not be able to support the large window sizes associated with high data rates.

The example presented in Figure 2.37 is based upon completely successful transmission, i.e. 0% Block Error Rate (BLER). In that case, VT(S) does not approach VT(MS) and there is no danger of the RLC layer stalling. Figure 2.39 presents an example which includes a re-transmission.

Both Figure 2.37 and Figure 2.39 are based upon very short round trip times, i.e. the number of PDU transferred while waiting for an acknowledgement is small. This approach has been adopted only for illustrative purposes to help keep the figures and explanation relatively simple. In practice, the round trip time will be greater, e.g. 160 ms and the number of PDU transferred while waiting for an acknowledgement will be greater.

Data transfer starts in Figure 2.39 when the sender transmits the first two PDU. It is assumed that the first two PDU are received successfully and the receive window slides onwards by two. The sender proceeds to send the next two PDU during the subsequent TTI. In this example, it is assumed that the third PDU is received in error. The receiver recognises that it has received sequence number 3 without receiving sequence number 2. This triggers the receiver to return a status report to the sender (assuming that this status report triggering mechanism has been configured). The PDU received in error prevents the receiver window from advancing in the usual way, i.e. PDU have been received out-of-sequence whereas VR(R) is defined as the sequence number following the last in-sequence PDU. Receiving the PDU in error causes the values belonging to VR(R) and VR(H) to differ, i.e. the highest expected sequence number is greater than the lower end of the receive window. The sender continues to transmit the forth and fifth PDU while the status report is returned. Once the status report has been received, the PDU which was received in error is re-transmitted together with a new PDU. The status report indicates that all PDU with sequence numbers preceding 2 have been received successfully. This allows the sender to advance its transmit window by 2. In this example, it is assumed that the receiver receives the re-transmission successfully. This allows the receive window to advance to the highest successfully received in-sequence PDU. The receiver returns a status report to inform the sender that the re-transmission was successful and that further re-transmissions are not necessary. This status report allows the sender to advance its transmit window.

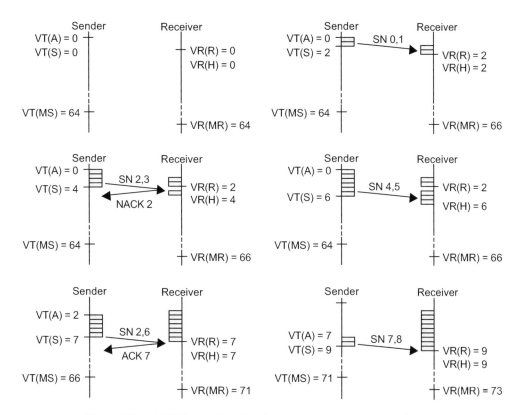

Figure 2.39 AM RLC transmit and receive windows (assuming re-transmissions)

Returning to the set of AM RLC functions presented in Table 2.26, AM RLC supports ciphering. If enabled, ciphering is applied to the RLC payload and length indicators, but not to the sequence number, i.e. the first two bytes of the RLC PDU remain unciphered.

The SDU discard function for AM RLC has greater flexibility than the equivalent function for TM and UM RLC. In the case of TM and UM RLC, the SDU discard function only involves the sender. In the case of AM RLC, it can involve both the sender and receiver. The RRC layer can configure AM RLC to use any one of the methods presented in Table 2.27. In contrast the TM and UM RLC, it is mandatory to configure an SDU discard method for AM RLC, i.e. the 'not configured' option does not appear within Table 2.27.

The 'timer-based with explicit signalling' method relies upon a discard timer in the same way as the 'timer-based without explicit signalling' method used by TM and UM RLC. The RRC layer configures the discard timer which defines the maximum duration that an SDU can remain within the RLC layer. Each RLC SDU has its own timer which is started when the SDU is received from the higher layers. An SDU is discarded if the timer expires before it has been completely passed to the

Table 2.27 SDU discard methods (AM RLC)

Timer based with explicit signalling
Discard after maximum number of transmissions
No discard after maximum number of transmissions

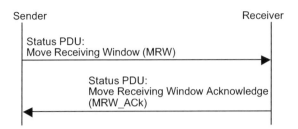

Figure 2.40   Move receiving window signalling after RLC SDU discard timer has expired

MAC layer. In this case, the sender discards the SDU and initiates the signalling procedure illustrated in Figure 2.40.

The sender issues a status PDU which instructs the receiver to discard any PDU which have already been received and are associated with the discarded SDU. The status PDU also instructs the receiver to move its receive window to avoid waiting for any further PDU associated with the discarded SDU. The structure of a status PDU is described later in this section. The receiver uses another status PDU to acknowledge receipt of the PDU from the sender.

The 'discard after maximum number of transmissions' method triggers an SDU discard after a PDU containing either part of the SDU, or a length indicator signalling the end of the SDU has been transmitted the maximum number of times. The maximum number of transmissions is defined by MaxDat -1, where MaxDat is a parameter configured by the RRC layer. If an SDU is discarded as a result of this method, the sender also initialises the signalling procedure illustrated in Figure 2.40, i.e. informing the receiver to discard any PDU associated with the SDU and to advance its receive window.

The 'no discard after maximum number of transmissions' method is the same as the 'discard after maximum number of transmissions' method except that the RLC reset procedure is triggered. The signalling associated with the RLC reset procedure is illustrated in Figure 2.41.

The RLC reset procedure causes both RLC peers to discard all RLC PDU in their receivers and all RLC SDU that were previously sent by their transmitters. Discarding these PDU and SDU means that it is necessary for the higher layers to complete re-transmissions if a reliable data transfer service is to be maintained.

Figure 2.42 illustrates the structure of an AM RLC PDU. The PDU starts with a Data/Control (D/C) flag represented using a single bit. This flag informs the receiver of the PDU type. A value of 1 indicates that the subsequent PDU has the structure illustrated in Figure 2.42, i.e. it is a data PDU which may or may not include a piggybacked status PDU. A value of 0 indicates that the subsequent PDU could be either a status PDU, a reset PDU or a reset Acknowledge PDU.

In the case of AM RLC, the sequence number has a length of 12 bits. This allows a range from 0 to 4095. This is in contrast to the range from 0 to 127 used for UM RLC. AM RLC requires a significantly greater range to avoid cycling the range while re-transmissions are completed.

The polling (P) bit is used to request a status report from the receiver. The RRC layer can configure one or more triggering mechanisms to set the polling bit at the transmitter. These triggering mechanisms are presented in Table 2.28.

Figure 2.41   RLC reset procedure

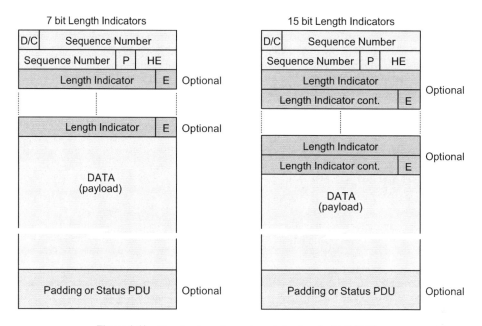

Figure 2.42   The structure of an acknowledged mode RLC PDU

The first triggering mechanism polls the receiver when an RLC PDU is passed to the lower layers for the first time and there is no further data from the higher layers. This triggering mechanism also polls the receiver if an RLC PDU is passed to the lower layers for the first time and the sender is not allowed to transfer any further PDU because the upper limit of the transmit window has been reached. The second triggering mechanism is similar to the first except that it is based upon RLC PDU being re-transmitted rather than PDU transmitted for the first time.

The third polling mechanism is used to repeat a poll rather than to trigger an initial poll. If configured, a poll timer is started when a PDU containing a poll is submitted to the lower layers. The sender repeats the poll if the poll timer expires before receiving a status report from the receiver. The poll timer is re-started if a poll is triggered by any other mechanism while it is running.

The forth polling mechanism is used to trigger a poll after every nth PDU has been transferred to the lower layers. Both original transmissions and re-transmissions are counted. The value of n is defined by the Poll_PDU parameter which can be configured by the RRC layer. Similarly, the fifth polling mechanism is used to trigger a poll after every mth SDU. In this case, polling is triggered when the PDU containing the length indicator defining the end of the mth SDU is transferred to the lower

Table 2.28   Triggering mechanisms for polling the receiver

1. Last PDU in buffer
2. Last PDU in re-transmission buffer
3. Poll timer
4. Every nth PDU
5. Every mth SDU
6. Window based
7. Timer based

layers. The value of m is defined by the Poll_SDU parameter which can be configured by the RRC layer.

The sixth polling mechanism is based upon reaching a certain percentage of the transmit window. For example, if the polling window is configured with a value of 70% then polling is triggered once the sequence number reaches 70% of the range allowed by the transmit window.

The final polling mechanism allows the RRC layer to configure periodic polling. Periodic polling is triggered if there are PDU to send when the periodic timer expires.

The RRC layer can also configure a guard timer to avoid sending polls too frequently. If configured, this timer is started when a poll is sent and no subsequent polls are allowed until the timer has expired. Polls are delayed until the timer has expired if they are triggered while the timer is running. The guard timer is configured using the Timer_Poll_Prohibit parameter.

Returning to the AM RLC PDU structure illustrated in Figure 2.42, the pair of Header Extension (HE) bits are used to indicate whether the following byte includes data or a length indicator. A value of 00 indicates data whereas a value of 01 indicates a length indicator. The 10 and 11 values are not used by the release 6 and previous versions of the specifications.

The AM RLC PDU may contain one or more length indicators. Similar to UM RLC, both 7 bit and 15 bit length indicators can be used. In the case of AM RLC PDU the number of bits allocated to the length indicator is determined by the size of the RLC PDU. If the RLC PDU size is greater than 126 bytes then 15 bit length indicators are used. Otherwise 7 bit length indicators are used. This is in contrast to UM RLC where the RRC layer can configure the number of bits allocated to the length indicators used in the downlink direction. Similar to UM RLC, length indicators are included to inform the receiver of the location of the last byte belonging to an SDU. If multiple SDU are included within a single PDU then multiple length indicators can be included. Length indicators configured with reserved values can also be included to signal other information. The set of reserved values for a 7 bit length indicator is presented in Table 2.29. The set of reserved values is the same as for UM RLC, but their interpretation is different. In the case of AM RLC, only three of the five reserved values are used and their meaning is relatively straightforward.

Length indicators are followed by an extension (E) bit to signal whether the following byte includes data or a further length indicator. In the case of AM RLC, normal interpretation is always applied to the extension bit and the alternative interpretation which can be configured for UM RLC is not applicable.

The payload of the RLC PDU may be followed by a piggybacked status PDU. Status PDU can be triggered periodically, by the receiver detecting a missing PDU or by the transmitter polling the receiver. The structure of a piggybacked status PDU is illustrated in Figure 2.43.

The Reserved 2 (R2) bit is not used and always has a value of 0. Likewise, the PDU type bits are not used and always have a value of 000. The piggybacked status PDU can then include one or more Super Fields (SUFI). The set of SUFI is presented in Table 2.30.

The 'no more data' SUFI is used to indicate the last SUFI within a status PDU. Any content after this SUFI is considered to be padding. An example of the 'no more data' SUFI is presented in Log File 2.24.

Table 2.29 Reserved length indicator values for AM RLC PDU (7 bit length indicator)

Reserved value	Interpretation
0000 000	The previous RLC PDU was exactly occupied by the last segment an of RLC SDU and an associated length indicator was not included.
1111 100	Reserved, but not used
1111 101	Reserved, but not used
1111 110	The remainder of the RLC PDU contains a piggybacked status PDU.
1111 111	The remainder of the RLC PDU is padding. The padding length can be zero.

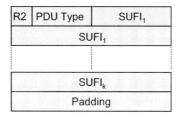

Figure 2.43   The structure of a piggybacked Status PDU

Table 2.30   Status PDU Super Fields

1. No more data
2. Window size
3. Acknowledgement
4. List
5. Bitmap
6. Relative list
7. Move receiving window
8. Move receiving window acknowledgement

```
SUFI:   No More Data
```

Log File 2.24   No more data SUFI

```
SUFI:   Window Size
WSN:    128
```

Log File 2.25   Window Size SUFI

The 'window size' SUFI can be used by the receiver to control the size of the transmit window used by the sender. This represents a form of flow control. If the receiver wishes to reduce the rate at which data is sent then the transmit window size can be reduced. An example of the window size SUFI is presented in Log File 2.25. In this example, the receiver instructs the sender to apply a transmit window size of 128.

The 'acknowledgement' SUFI is used to return an acknowledgement to the sender. This SUFI includes a Last Sequence Number (LSN) to indicate that with the exception of any erroneous PDU signalled within the same status PDU, all PDU with sequence numbers less than the LSN have been received successfully. Erroneous PDU can be signalled using the 'list', 'bitmap' or 'relative list' SUFI. The 'acknowledgment' SUFI is always included last within the status PDU and when included it is not necessary to include the 'no more data' SUFI. An example of the acknowledgement SUFI is presented in Log File 2.26.

```
SUFI:   Acknowledgement
LSN:    37
```

Log File 2.26   Acknowledgement SUFI

```
SUFI:   List
Length: the number of SN, L pairs:    2
sequence number of AMD PDU:           164
number of consecutive AMD PDUs not correctly received after SN: 2
sequence number of AMD PDU:           171
number of consecutive AMD PDUs not correctly received after SN: 1
```

Log File 2.27   List SUFI

The 'list' SUFI is used to specify a list of RLC PDU which have been received in error. Both individual and bursts of errors can be specified. An example of the list SUFI is presented in Log File 2.27.

The list SUFI starts with a length indicator which specifies the number of elements belonging to the list. Each element represents a burst of one or more PDU which have been received in error. Each burst is specified using the sequence number of the first PDU belonging to the burst followed by the number of successive PDU belonging to the same burst. In this example, PDU with sequence numbers 164, 165, 166, 171 and 172 have been received in error. This SUFI informs the sender that these PDU require re-transmission.

The 'bitmap' SUFI is also used to specify a list of RLC PDU which have been received in error. An example of the bitmap SUFI is presented in Log File 2.28.

The 'bitmap' SUFI starts with a length indicator to specify the number of bytes used to represent the subsequent bitmap. The number of bytes is equal to the value of the length indicator plus 1. A maximum of 16 bytes can be used to represent the bitmap. This would correspond to a span of $16 \times 8 = 128$ PDU. A single byte is used to represent the bitmap in this example, i.e. providing a span of 8 PDU. The sequence number corresponds to the first bit belonging to the bitmap. Bitmap values of 0 indicate PDU which have been received in error whereas bitmap values of 1 indicate PDU which have been received successfully. In this example, PDU with sequence numbers 473, 474, 477 and 479 have been received in error. This SUFI informs the sender that these PDU require re-transmission.

The 'relative list' SUFI provides another way to specify a list of RLC PDU which have been received in error. An example of the relative list SUFI is presented in Log File 2.29.

The length indicator specifies the number of codewords (CW) included within the SUFI. A maximum of 15 codewords can be included. The length indicator can be set to 0 if only a single PDU has been received in error. In that case, the sequence number specifies the single PDU received in error. Otherwise, the sequence number specifies the first PDU which has been received in error and the

```
SUFI:   Bitmap
Length: the size of the bimap:        0
sequence number of AMD PDU:           473
Bitmap:                               00110101
```

Log File 2.28   Bitmap SUFI

```
SUFI:   Relative List
Length: the number of codewords:      4
sequence number of AMD PDU:           37
CW1:                          0111
CW2:                          1101
CW3:                          0010
CW4:                          0101
```

Log File 2.29   Relative list SUFI

```
SUFI:   Move Receiving Window
Length: the number of SN_MRW fields:  2
SN_MRW1:                              36
SN_MRW2:                              48
N_Length                              1
```

Log File 2.30   Move receiving window SUFI

subsequent sequence of codewords define the remaining PDU which have been received in error. The codewords are used to specify the sequence number of the next erroneous PDU relative to the previous erroneous PDU. Each codeword has a length of 4 bits and the least significant bit is used for control purposes. If the least significant bit has a value of 0 then this indicates that the remaining 3 bits form part of a number defining the relative sequence number. If the least significant bit has a value of 1 then this indicates that the remaining 3 bits terminate the number defining the relative sequence number. The first two codewords in Log File 2.29 have least significant bits equal to 1 and so their remaining 3 bits represent entire numbers, i.e. 3 and 6. The third codeword has a least significant bit equal to 0 and so represents only part of a number. The forth codeword has a least significant bit equal to 1 and so pairs with the third codeword to generate the number 17. This means that the four codewords indicate values of 3, 6 and 17 and the PDU received in error are 37, 40, 46 and 63. Bursts of PDU received in error can be signalled using a codeword value of 0001 followed by one or more codewords defining the number of PDU within the burst.

The 'move receiving window' SUFI is used to instruct the receiver to advance its receive window. It can also be used to inform the receiver about one or more RLC SDU which have been discarded at the sender. An example of the move receiving window SUFI is presented in Log File 2.30.

The length indicator specifies the number of sequence numbers included within the SUFI. A maximum of 15 sequence numbers can be included. A length indicator equal to 0 means that a single sequence number is included and that the SDU to be discarded at the receiver extends beyond the upper edge of the transmit window at the sender. Each sequence number defines the end of an SDU which has been discarded at the sender. If N_length is equal to 0 then the last sequence number, SN_last is used to instruct the receiver to discard all SDU which have not yet been fully received and which have PDU with sequence numbers less than SN_last. Otherwise the value of N_length specifies the length indicator within the PDU with sequence number SN_last which defines the last byte of the last SDU to be discarded. In either case, the receiver advances the receive window to avoid waiting for PDU associated with discarded SDU.

The 'move receiving window acknowledge' SUFI is used to acknowledge a 'move receiving window' SUFI. The associated handshake is illustrated in Figure 2.40. An example of the 'move receiving window acknowledge' SUFI is presented in Log File 2.31.

The acknowledged sequence number SN_ACK informs the sender of the updated value of VR(R) after acting upon the move receiving window SUFI. If VR(R) has been updated to a value equal to SN_last then N is set equal to N_length. Otherwise N is set to 0.

A standalone status PDU can be used instead of a piggybacked status PDU. The structure of a standalone status PDU is illustrated in Figure 2.44. The structure is the same as that for a piggybacked status PDU except that the R2 bit is replaced by a Data/Control (D/C) flag which is set to 0 to indicate that the subsequent PDU is either a status PDU, reset PDU or reset acknowledge PDU. The PDU type field is used to indicate a status PDU.

```
SUFI:   Move Receiving Window Acknowledge
N:          1
SN_ACK:     48
```

Log File 2.31   Move receiving window acknowledge SUFI

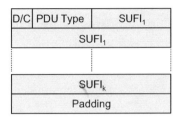

Figure 2.44   The structure of a Status PDU

A drawback associated with using the standalone version of the status PDU is that the RLC PDU size is typically large relative to the quantity of information content belonging to the status PDU. This means that a significant quantity of padding has to be transferred through the MAC layer, across the Iub and across the air-interface. An example of a standalone status PDU is presented in Log File 2.32. This example illustrates the acknowledge SUFI being used to inform the sender that all PDU up to sequence number 2440 have been received successfully. The sender could have requested this status PDU using any of the polling mechanisms presented in Table 2.28, e.g. after sending the last PDU belonging to an RLC SDU.

The information content of the status PDU occupies only 20 bits whereas the RLC PDU is 336 bits. The significance of this overhead depends upon the set of polling mechanisms and consequently the rate at which status PDU are transferred. If timer-based periodic polling is used then the overhead becomes more significant for low data rate connections. If polling is triggered after every mth RLC SDU then the overhead becomes dependent upon the size of the higher layer packets.

Data applications based upon TCP/IP may use higher layer packets with a maximum size of 1460 bytes. Each of these packets requires 37 RLC PDU when assuming an RLC payload of 320 bits. If a single standalone status PDU is transferred after every higher layer packet then the overhead is equal to $1/37 = 3\%$. This overhead increases for smaller higher layer packets. The overhead is transferred in the opposite direction to the application data. For example, the RLC status PDU acknowledging a downlink data transfer are sent in the uplink direction. This overhead is in addition to the acknowledgements generated by the higher layers. The TCP layer also requires acknowledgments which could be transferred after every TCP packet or after every second TCP packet (depending upon whether or not the delayed acknowledgement TCP feature is used). The combination of RLC and TCP acknowledgements for a downlink data transfer is illustrated in Figure 2.45.

RLC acknowledgements are returned to the RNC whereas TCP acknowledgments are returned to the application server. This means that TCP acknowledgements are visible as RLC, MAC and physical layer throughput whereas RLC acknowledgements are visible as MAC and physical layer throughput. Both types of acknowledgement load the Iub interface. A TCP acknowledgement is 320 bits when using IP version 4 and requires 2 RLC PDU when using an RLC PDU size of 336 bits. The 320 bits of

```
02 98 80 00 00 00 00 00 00 00 00 00 00 00 00 00 00 00 00 00 00
00 00 00 00 00 00 00 00 00 00 00 00 00 00 00 00 00 00 00 00 00
      ACKNOWLEDGED MODE DATA PDU
      - D/C: Control PDU
      - PDUType: STATUS
      - SUFI type: ACK
      - LSN: 2440
      - Padding:
        0 00 00 00 00 00 00 00 00 00 00 00 00 00 00 00 00 00 00 00 00 00 00 00 00
          00 00 00 00 00 00 00 00 00 00 00 00 00 00 00 00
```

Log File 2.32   Status PDU (acknowledgment SUFI)

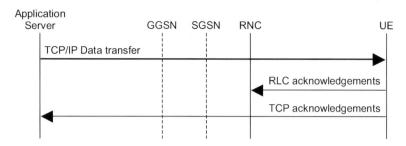

Figure 2.45 Combination of RLC and TCP acknowledgments (downlink data transfer)

TCP/IP acknowledgement fit exactly within the payload of the first RLC PDU. A second RLC PDU is required to accommodate the length indicator used to signal the end of the higher layer packet. An example of 2 RLC PDU transferring a TCP acknowledgement is presented in Log File 2.33.

The first RLC PDU includes a 2 byte header and a 40 byte payload. The 40 byte payload accommodates the TCP acknowledgement and generates the RLC throughput. The second RLC PDU includes a 4 byte header and 38 bytes of padding. This PDU generates MAC and physical layer throughput but does not generate any RLC throughput, i.e. none of the content is passed to the higher layers. If a TCP acknowledgement is returned for every TCP packet the corresponding overhead is $2/37 = 5\%$ from the perspective of the MAC layer and $320/(1460 \times 8) = 3\%$ from the perspective of the RLC layer.

Log File 2.34 presents a set of RLC PDU belonging to an uplink TCP/IP file transfer. 4 uplink RLC PDU are transferred during each TTI. This corresponds to a 64 kbps connection based upon an RLC PDU size of 336 bits and a TTI of 20 ms. A single downlink status PDU and two downlink data PDU are included. The status PDU is acknowledging the end of an RLC SDU whereas the two data PDU are

```
RLC PDU 1
CC 98 45 00 00 28 BE 36 40 00 80 06 BE FE 0A E2 03 5C 52 D0 1C 8D 0B 0E 00 50
6C A0 B7 A8 76 7B C5 A6 50 10 F7 90 CE DF 00 00
ACKNOWLEDGED MODE DATA PDU
- D/C: Data PDU
- Sequence Number: 2451
- P: Status report not requested
- HE: Succeeding octet contains data
- Data 45 00 00 28 BE 36 40 00 80 06 BE FE 0A E2 03 5C 52 D0 1C 8D 0B 0E 00 50
6C A0 B7 A8 76 7B C5 A6 50 10 F7 90 CE DF 00 00

RLC PDU 2
CC A5 01 FE 55 55 55 55 55 55 55 55 55 55 55 55 55 55 55 55 55 55 55 55 55 55
55 55 55 55 55 55 55 55 55 55 55 55 55 55 55 55
ACKNOWLEDGED MODE DATA PDU
- D/C: Data PDU
- Sequence Number: 2452
- P: Status report requested
- HE: Succeeding octet contains 7bit LI and E bit
- LI: Prev RLC PDU exactly filled with last segment of RLC SDU
- E: Next field is LI and E bit
- LI: Rest of RLC PDU is padding
- E: Next field is data
- Padding 55 55 55 55 55 55 55 55 55 55 55 55 55 55 55 55 55 55 55 55 55 55 55
55 55 55 55 55 55 55 55 55 55 55 55 55 55 55 55
```

Log File 2.33 RLC PDU transferring a TCP/IP acknowledgement

```
RLC UL AM PDU Data PDU seq number:3567,3568,3569,3570
RLC UL AM PDU Data PDU seq number:3571,3572,3573,3574
RLC DL AM PDU Control PDU status (RLC Ack)
RLC UL AM PDU Data PDU seq number:3575,3576,3577,3578
RLC UL AM PDU Data PDU seq number:3579,3580,3581,3582
RLC UL AM PDU Data PDU seq number:3583,3584,3585,3586
RLC UL AM PDU Data PDU seq number:3587,3588,3589,3590
RLC UL AM PDU Data PDU seq number:3591,3592,3593,3594
RLC DL AM PDU Data PDU seq number:623,624 (TCP Ack)
RLC UL AM PDU Control PDU/Data PDU status,seq number:3595,3596,3597
RLC UL AM PDU Data PDU seq number:3598,3599,3600,3601
RLC UL AM PDU Data PDU seq number:3602,3603,3604,3605
```

Log File 2.34   Sample of RLC PDU for an uplink TCP/IP file transfer

transferring a TCP acknowledgement. The two downlink data PDU are followed by an uplink status PDU because the TCP acknowledgement represents an RLC SDU and polling has been configured to trigger after every higher layer packet.

Combining the overheads generated by the TCP and RLC acknowledgements results in a 3% overhead from the perspective of the RLC layer and an 8% overhead from the perspective of the MAC layer. These overheads would be reduced if the delayed acknowledgement TCP feature is enabled and only every second TCP packet is acknowledged by the TCP layer.

In practice, the overhead from the perspective of the MAC layer is increased by status PDU which are triggered when RLC PDU are received in error. The rate at which these status PDU are generated depends upon the BLER. Similar to standalone status PDU containing only the acknowledgement SUFI, standalone status PDU informing the sender of PDU received in error typically include a large quantity of padding. An example standalone status PDU which includes the relative list and acknowledgement SUFI is presented in Log File 2.35.

This status PDU has been triggered by the receiver because a burst of RLC PDU have been received in error. The first PDU received in error has a sequence number of 735. The subsequent three codewords specify that 24 consecutive PDU have been received in error, i.e. PDU with sequence numbers 735 to 758. The subsequent ACK SUFI indicates that no PDU with sequence numbers greater that 734 have been received successfully. The information content of the status PDU occupies 52 bits which results in a requirement for 284 bits of padding.

If the BLER is 1% and if RLC PDU are received in error individually rather than in bursts the additional overhead generated by these status PDU would be 1%, i.e. on average, one status PDU

```
05 32 DF 10 72 2D F0 00 00 00 00 00 00 00 00 00 00 00 00 00 00
00 00 00 00 00 00 00 00 00 00 00 00 00 00 00 00 00 00 00 00 00
   ACKNOWLEDGED MODE DATA PDU
   - D/C: Control PDU
   - PDUType: STATUS
   - SUFI type: RLIST
   - LENGTH: 3
   - FSN: 735
   - CW1: 0001 (Error Burst Indicator)
   - CW2: 0000 (first part of number)
   - CW3: 0111 (final part of number 011000)
   - SUFI type: ACK
   - LSN: 735
   - Padding:
     0 00 00 00 00 00 00 00 00 00 00 00 00 00 00 00 00 00 00 00 00 00
     00 00 00 00 00 00 00 00 00 00 00 00
```

Log File 2.35   Status PDU (relative list and acknowledgment SUFI)

reporting an erroneous data PDU would be transferred after every 100 data PDU. RLC PDU are likely to be received in error individually when the propagation environment is static and bit errors are distributed with a uniform random distribution. A fading propagation environment is more likely to generate bursts of erroneous RLC PDU, i.e. bursts of errors are experienced during fades. This means that for a fixed BLER, the requirement for status PDU is reduced because each status PDU reports multiple erroneous data PDU. Consequently, the overhead generated by the status PDU would decrease. The overhead would also decrease if piggybacked status PDU are used instead of standalone status PDU, i.e. for the example presented in Log File 2.35 the overhead would be reduced from 336 to 52 bits.

AM RLC PDU can be transferred using the DCCH and DTCH logical channels. The DCCH is used to transfer RRC signalling whereas the DTCH is used to transfer application data. Other logical channels are not applicable to AM RLC. The set of RRC messages which can be transferred using AM RLC is presented in Table 2.31.

The vast majority of these messages are transferred using SRB 2. SRB 2 uses AM to transfer RRC messages associated with the access stratum. The content of these messages is transferred between the RNC and UE. The Initial Direct Transfer, Uplink Direct Transfer and Downlink Direct Transfer messages make use of either SRB 3 or SRB 4. These SRB use AM to transfer RRC messages which include information content for the non-access stratum.

Log File 2.36 presents an example RLC configuration for SRB 2. The SDU discard function has been configured to use the 'no discard after maximum number of transmissions' method. This means that the RLC reset procedure is triggered after the maximum number of re-transmissions have been completed. MaxDat is configured with a value of 20 and so the maximum number of transmissions is 19 while the maximum number of re-transmissions is 18. A relatively high figure is used in this example to reflect the importance of the control plane signalling transferred by SRB 2.

The uplink transmit window size is configured with a value of 64. This represents a reasonable figure when considering that, in this example, the SRB has been configured with a 40 ms TTI and that a single

Table 2.31 RRC messages able to use Acknowledged Mode RLC

Logical channel	Messages
DCCH	Active Set Update, Active Set Update Complete, Active Set Update Failure, Assistance Data Delivery, Cell Change Order From UTRAN, Cell Change Order from UTRAN Failure, Counter Check, Counter Check Response, Downlink Direct Transfer, Handover from UTRAN Command, Handover from UTRAN Failure, Handover to UTRAN Complete, Initial Direct Transfer, MBMS Modification Request, MBMS Modified Services Information, Measurement Control, Measurement Control Failure, Measurement Report, Paging Type 2, Physical Channel Reconfiguration, Physical Channel Reconfiguration Complete, Physical Channel Reconfiguration Failure, Radio Bearer Reconfiguration, Radio Bearer Reconfiguration Complete, Radio Bearer Reconfiguration Failure, Radio Bearer Release, Radio Bearer Release Complete, Radio Bearer Release Failure, Radio Bearer Setup, Radio Bearer Setup Complete, Radio Bearer Setup Failure, RRC Connection Release Complete, RRC Connection Setup Complete, RRC Status, Security Mode Command, Security Mode Complete, Security Mode Failure, Signalling Connection Release, Signalling Connection Release Indication, Transport Channel Reconfiguration, Transport Channel Reconfiguration Complete, Transport Channel Reconfiguration Failure, Transport Format Combination Control, Transport Format Combination Control Failure, UE Capability Enquiry, UE Capability Information, UTRAN Mobility Information Confirm, Uplink Direct Transfer, UTRAN Mobility Information, UE Capability Information Confirm, UTRAN Mobility Information Failure

```
srb-InformationSetupList value 2
rb-Identity: 2
rlc-InfoChoice
  rlc-Info
    ul-RLC-Mode
      ul-AM-RLC-Mode
        transmissionRLC-Discard
          noDiscard:                    dat20
        transmissionWindowSize:         tw64
        timerRST:                       tr300
        max-RST:                        rst1
        pollingInfo
          timerPollProhibit:                       tpp120
          timerPoll:                               tp300
          poll-SDU:                                sdu1
          lastTransmissionPDU-Poll:                false
          lastRetransmissionPDU-Poll:              true
          pollWindow:                              pw70
          dl-RLC-Mode
            dl-AM-RLC-Mode
              inSequenceDelivery:                  true
              receivingWindowSize:                 rw64
              dl-RLC-StatusInfo
              missingPDU-Indicator:                true
```

Log File 2.36 Example AM RLC configuration for a signalling radio bearer

RLC PDU can be transferred during each TTI, i.e. it is likely that the transmitter can send a maximum of three or four PDU during each round trip time. Allowing a maximum of 19 transmissions corresponds to 19 round trip times and 76 PDU (assuming that 4 PDU are sent during each round trip time). 3GPP allows a finite set of window sizes to be configured rather than allowing any size. The set of allowed sizes is: 1, 8, 16, 32, 64, 128, 256, 512, 768, 1024, 1536, 2047, 2560, 3072, 3584 and 4095. A window size of 64 represents a rounding down of the figure of 76. In practice, it is unlikely that an SRB would require such a large window because RRC messages are not sufficiently large nor are they sent sufficiently frequently to generate transmissions during every TTI, i.e. SRB typically experience large periods of DTX and have relatively low activity factors.

In general, an uplink reset timer of 300 ms means that the UE spends 300 ms waiting for a reset acknowledge PDU after sending a reset PDU. The reset PDU is re-transmitted if the timer expires before receiving an acknowledgement assuming that the maximum number of transmissions has not been reached. However, the maximum number of transmissions is defined by Max-RST −1, and in this case the maximum number of transmissions is 0. This means that the reset PDU is never sent and the RRC layer is immediately informed of an unrecoverable error. The RRC layer within the UE proceeds to inform the RNC using TM RLC and a Cell Update message. The RNC subsequently releases the RRC connection. 3GPP specifies that AM SRB should always use a Max-RST value of 1 (3GPP RP-010549 change requests 1029 and 1030).

Log File 2.36 illustrates that three triggering mechanisms are configured for polling: 'last PDU in re-transmission buffer', 'every mth SDU' and 'window-based'. The value of m is configured with a value of 1 and so polling is triggered after every RLC SDU. The window-based mechanism is configured with a threshold of 70% and so polling is triggered if the transmit sequence number reaches 70% of the range defined by the transmit window. The poll timer is configured with a value of 300 ms and so polls are re-transmitted if a status report has not been received after 300 ms. The poll prohibit timer is configured with a value of 120 ms and so the minimum interval between consecutive polls is 120 ms.

The UE is instructed that downlink RLC SDU should be delivered to the higher layers in-sequence. The UE is also configured with a downlink receive window of 64, i.e. equal to the uplink transmit window. The missing PDU indicator instructs the UE to generate a status report every time a missing PDU is identified. This allows the RNC to schedule a re-transmission.

Figure 2.46 illustrates the AM RLC PDU transferred during the first phase of a speech call establishment procedure. These PDU are transferred using a combination of SRB 2 and 3.

The RRC Connection Request and RRC Connection Setup messages are not included within Figure 2.46 because they are transferred on SRB 0 using TM and UM RLC respectively. The RRC Connection Setup Complete message is transferred using SRB 2. This message typically occupies three RLC PDU. These are the first PDU transferred on SRB 2 and so the sequence numbers start from 0. The third PDU includes a poll requesting the RNC to respond with a status report. This poll is triggered by the end of the RLC SDU, i.e. using the 'every mth SDU' polling mechanism with $m = 1$. The RNC responds to the poll by sending a status report including the acknowledgement SUFI and specifying an LSN of 3 to indicate that PDU with sequence numbers 0, 1 and 2 have been received successfully.

The UE proceeds to send an Initial Direct Transfer message including a CM Service Request. The CM Service Request is directed towards the non-access stratum within the core network. Initial Direct Transfer, Uplink Direct Transfer and Downlink Direct Transfer messages are sent using either SRB 3 or 4. This example assumes that SRB 3 is used. Sending the CM Service Request message is not dependent upon receiving the acknowledgement for the RRC Connection Setup Complete message. Log files may show the CM Service Request message being sent prior to receiving the

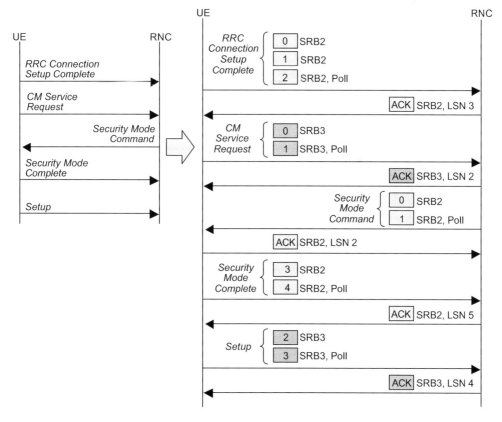

Figure 2.46 AM RLC PDU transferred during the first phase of speech call establishment

acknowledgement. Each SRB uses its own sequence numbering and so the RLC PDU associated with the CM Service Request message start from 0. The message occupies two RLC PDU and the second PDU includes a poll according to the 'every mth SDU' polling mechanism. The RNC successfully receives both PDU and the poll. A status report is returned including the acknowledgement SUFI and specifying an LSN of 2.

The CM Service Request triggers the security mode procedure and the RNC sends a Security Mode Command message using SRB 2. Uplink and downlink sequence numbering is independent and so the sequence numbers for this message start from 0. The message is transferred using 2 RLC PDU and the RLC layer within the UE responds with a status report. The RRC layer within the UE responds with a Security Mode Complete message which occupies a further two RLC PDU on SRB 2. The sequence numbers continue from those used for the RRC Connection Setup Complete message. The RLC layer within the RNC responds with a status report and the UE proceeds by sending the non-access stratum Setup message within an Uplink Direct Transfer message.

Figure 2.47 illustrates an example based upon the transfer of a Radio Bearer Setup message. This message typically occupies 8 RLC PDU when establishing a speech connection and is transferred using SRB 2. For the purposes of this example it is assumed that one of the PDU is received in error.

Sequence numbers start from 8 because other downlink messages have already been transferred using SRB 2. RLC PDU with sequence numbers 8 and 9 are received correctly, but the PDU with sequence number 10 is received in error. The UE successfully receives PDU 11 which allows it to recognise that it has missed PDU 10. The UE returns a status report including both the 'list' and 'acknowledgement' SUFI (alternatively the 'relative list' or 'bitmap' SUFI could be used in place of the 'list' SUFI). The list SUFI informs the RNC that PDU 10 has been received in error. The RNC reacts by scheduling PDU 10 for re-transmission. In this example, the UE successfully receives the

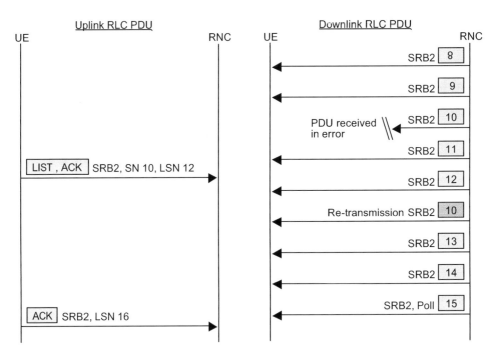

Figure 2.47   RRC message with intermediate AM RLC PDU received in error

re-transmission and no further status PDU are generated until the UE is polled by the final PDU belonging to the RRC message.

Figure 2.48 illustrates an example based upon receiving the final PDU in error rather than an intermediate PDU. In this case, the receiver does not receive the poll attached to the final PDU nor is it able to detect a missing PDU as a result of out-of-sequence numbering. This means the UE does not return a status report requesting the erroneous PDU to be re-transmitted. Instead, the RNC has to rely upon using its poll timer to re-transmit the poll attached to the final PDU. The RNC expects a status report from the UE after sending the poll for the first time. When the poll timer expires and a status report has not been received the RNC deduces that the PDU has been lost and a re-transmission is scheduled. The poll timer should be at least as large as the round trip time otherwise re-transmissions may be sent unnecessarily. In this example, the UE successfully receives the re-transmission and returns a status PDU with a positive acknowledgement.

Log File 2.37 presents an example RLC configuration for a user plane radio bearer. In this case, the SDU discard function has been configured to use the 'discard after maximum number of transmissions' method. This means that an SDU is discarded at the transmitter if the receiver fails to receive any of its PDU after the maximum number of re-transmissions. MaxDat is configured with a value of 8 and so the maximum number of transmissions is seven and the maximum number of re-transmissions is six. This SDU discard method also triggers the move receiving window signalling procedure illustrated in Figure 2.40. The timerMRW parameter defines the period of time that the UE should allow for the RNC to respond before re-transmitting the status PDU including the move receiving window SUFI. The MaxMRW parameter defines the corresponding maximum number of re-transmissions.

In this example, the uplink transmit window size is configured with a value of 128. If it is assumed that this user plane radio bearer is established to support a 64 kbps uplink data rate then 4 RLC PDU could be transferred during a 20 ms TTI (based upon an RLC PDU size of 336 bits). If the round trip time for a 20 ms TTI is assumed to be 180 ms then 36 PDU could be transferred during each round trip time. Allowing a maximum of seven transmissions means that a total of $7 \times 36 = 252$ PDU could be transferred while completing re-transmissions. These figures indicate that the RLC layer could stall if more than three transmissions are required. The probability of requiring more than three transmissions is likely to be relatively small. If the BLER is 5% and it is assumed that the probability of receiving each transmission in error is statistically independent then the probability of not receiving a PDU

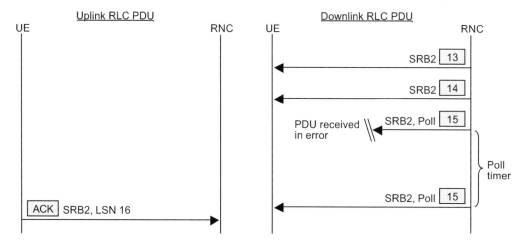

Figure 2.48 RRC message with final AM RLC PDU received in error

```
            rb-InformationSetupList value 1
              rb-Identity: 5
              pdcp-Info
                losslessSRNS-RelocSupport:       notSupported
                pdcp-PDU-Header :                absent
              rlc-InfoChoice
                rlc-Info
                  ul-RLC-Mode
                    ul-AM-RLC-Mode
                      transmissionRLC-Discard
                        maxDAT-Retransmissions
                          maxDAT:               dat8
                          timerMRW:             te300
                          maxMRW:               mm12
                      transmissionWindowSize:   tw128
                      timerRST:                 tr250
                      max-RST:                  rst12
                      pollingInfo
                        timerPollProhibit:              tpp120
                        timerPoll:                      tp240
                        poll-SDU:                       sdu1
                        lastTransmissionPDU-Poll:       false
                        lastRetransmissionPDU-Poll:     true
                        pollWindow:                     pw70
                  dl-RLC-Mode
                    dl-AM-RLC-Mode
                      inSequenceDelivery:       true
                      receivingWindowSize:      rw512
                      dl-RLC-StatusInfo
                        missingPDU-Indicator:   false
```

Log File 2.37   Example AM RLC configuration for a user plane radio bearer

successfully after three transmissions is 0.0125%. In practice, it is unlikely that each transmission is statistically independent because the propagation channel has time coherence, but nevertheless the probability of requiring more than three transmissions is relatively small.

The uplink reset timer has been configured with a value of 250 ms. This means that the UE spends 250 ms waiting for a reset acknowledge PDU after sending a reset PDU. The reset PDU is re-transmitted if the timer expires before receiving an acknowledgement assuming that the maximum number of transmissions has not been reached. The maximum number of transmissions is defined by Max-RST -1, i.e. in this example the maximum number of transmissions is 11.

Three triggering mechanisms have been configured for polling: 'last PDU in re-transmission buffer', 'every mth SDU' and 'window-based'. The value of m is configured with a value of 1 and so polling is triggered after every RLC SDU. The window based mechanism is configured with a threshold of 70% and so polling is triggered if the transmit sequence number reaches 70% of the range defined by the transmit window. The poll timer is configured with a value of 240 ms and so polls are re-transmitted if a status report has not been received after 240 ms. The poll prohibit timer is configured with a value of 120 ms and so the minimum interval between consecutive polls is 120 ms.

The UE is instructed that downlink RLC SDU should be delivered to the higher layers in-sequence. The UE is also configured with a downlink receive window of 512. This value is greater than the uplink transmit window and so indicates that the downlink bit rate is likely to be greater than the uplink bit rate, i.e. a greater number of RLC PDU can be transferred during each round trip time. The missing PDU indicator is disabled and so the UE is not instructed to generate status reports when missing PDU are detected, i.e. the RNC relies upon status reports generated in response to polls to trigger re-transmissions.

2.4 MAC Layer

- The RNC includes MAC-d, MAC-c/sh/m and MAC-es entities. The Node B includes MAC-hs, MAC-e and MAC-b entities. The UE includes MAC-d, MAC-c/sh/m, MAC-hs, MAC-e/es, MAC-m and MAC-b entities.
- MAC-d entities are responsible for the DCCH and DTCH logical channels. These may be mapped directly onto DCH transport channels or they may be forwarded to either the MAC-c/sh/m, MAC-hs or MAC-e/es.
- MAC-d entities can multiplex a group of DCCH or a group of DTCH. This requires the inclusion of a C/T header to identify which logical channel is sent at any point in time.
- MAC-d entities allow transport channel type switching, e.g. from FACH to DCH.
- MAC-d entities provide TFC selection for DCH transport channels based upon the transmit power capability and the quantity of data to be sent.
- MAC-d entities are able to cipher logical channels which use transparent mode RLC.
- MAC-c/sh/m entities are responsible for the PCCH, CCCH, CTCH, MCCH, MTCH and MSCH logical channels. They are also responsible for the BCCH logical channel when transferred using the FACH. In addition, MAC-c/sh/m entities can receive the DCCH and DTCH logical channels from MAC-d entities.
- MAC-c/sh/m entities are able to multiplex different logical channel types onto the same transport channel. This requires the inclusion of a TCTF header to identify which logical channel type is sent at any point in time
- MAC-c/sh/m entities are responsible for including a UE identity header when a common transport channel is used and the payload does not already include the UE identity. MAC-c/sh/m entities are also responsible for adding an MBMS service type header when transferring MTCH information.
- MAC-c/sh/m entities provide TFC selection for the RACH and FACH transport channels. Channel priorities are used to ensure that the information with greatest importance is transferred first.
- MAC-hs entities are used for HSDPA. They receive MAC-d PDU from MAC-d entities. MAC-hs entities provide flow control, scheduling, TFRC selection, generation of MAC-hs PDU and HARQ.
- MAC-e/es entities are used for HSUPA. They receive MAC-d PDU from MAC-d entities. MAC-e/es entities provide scheduling, E-TFC selection, generation of MAC-e/es PDU, macro diversity selection and HARQ.
- MAC-m entities are used for MBMS. They only exist in the UE and are used for selective combining when receiving MTCH transmissions from more than one cell.
- MAC-b entities within the Node B are used to schedule BCCH transmissions on the BCH.
- The MAC layer is responsible for providing traffic volume measurements from the RLC layer. These measurements can be used to trigger capacity requests and capacity releases.
- The MAC layer is responsible for the determining the ASC from the MAC logical channel priorities when in RRC connected mode. It also controls RACH transmissions in terms of monitoring the number of PRACH preamble cycles, running a back-off timer after receiving an AICH negative acknowledgement and completing persistency checks.
- Key 3GPP specifications: TS 25.321.

The Medium Access Control (MAC) layer of the radio interface protocol stack is specified within 3GPP TS 25.321. The MAC layer is used to transfer both control plane signalling and user plane data, i.e. RRC messages as well as application data pass through the MAC layer. RRC messages are transferred down from the RLC layer at the transmitter using the CCCH, DCCH, BCCH, PCCH, MCCH and MSCH logical channels. Similarly, application data is transferred using the CTCH, DTCH and MTCH logical channels.

2.4.1 Architecture of the MAC Layer

Figure 2.49 illustrates the high-level architecture of the MAC layer from the network perspective. The MAC layer is distributed between the RNC and Node B. It includes a set of MAC entities which are categorised according to the logical channels they process. Figure 2.49 illustrates a single instance of each MAC entity whereas in practice the number of entities depends upon the number of connections and the number of cells connected to the RNC, e.g. there is one MAC-d entity for every UE that has one or more dedicated logical channel connections to the RNC, and there is one MAC-c/sh/m entity for every cell connected to the RNC

Figure 2.50 illustrates the high-level architecture of the MAC layer from the UE perspective. The MAC layer within the UE includes the peer entities for the MAC layer within the RNC and Node B. The MAC-e and MAC-es entities are merged within the UE and the UE can include a MAC-m entity which does not exist on the network side.

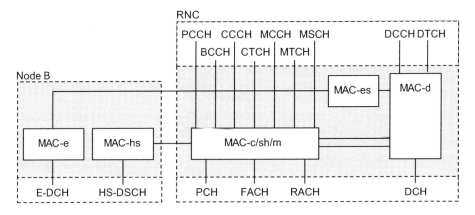

Figure 2.49   Architecture of the MAC layer for the network (MAC-b not shown)

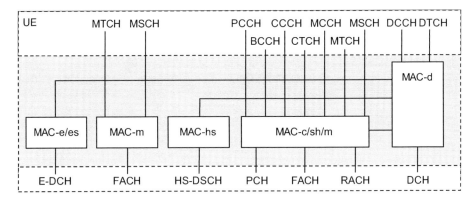

Figure 2.50   Architecture of the MAC layer for the UE (MAC-b not shown)

Table 2.32 DCCH and DTCH logical channel to transport channel mappings

	CELL_DCH	CELL_FACH
Uplink	DCH via MAC-d or E-DCH via MAC-e/es	RACH via MAC-c/sh/m
Downlink	DCH via MAC-d or HS-DSCH via MAC-hs	FACH via MAC-c/sh/m

2.4.1.1 MAC-d

MAC-d entities are located within the RNC and UE. The RNC has one MAC-d entity for every UE that has one or more dedicated logical channel connections. The UE has one MAC-d entity for all of its dedicated logical channel connections. The MAC-d entity is responsible for receiving the DCCH and DTCH logical channels from the RLC layer. It may either map these logical channels directly onto DCH transport channels, or it may pass them to other MAC entities to be mapped onto other transport channel types. Table 2.32 presents the logical channel to transport channel mappings for the DCCH and DTCH. This table is applicable to the release 6 version of the 3GPP specifications. Release 7 introduces the potential to use the HS-DSCH transport channel in CELL_FACH.

A MAC-d entity can be instructed to multiplex a group of logical channels onto a single transport channel. The set of Signalling Radio Bearers (SRB) represents a common example of MAC-d multiplexing. Figure 2.51 illustrates the set of logical channels and transport channels belonging to a circuit switched speech connection.

The speech information makes use of three logical channels and three transport channels. This information is not multiplexed at the MAC layer. The SRB data makes use of four logical channels, but only one transport channel. The MAC layer is responsible for multiplexing these four logical channels onto a single transport channel. This represents a scheduling task, e.g. the MAC layer must decide which data is transferred first if data arrives simultaneously for both SRB 1 and SRB 2. Prioritisation is guided by the MAC logical channel priority which is configured during connection establishment. Multiplexing a group of logical channels onto a single transport channel requires the inclusion of MAC header information to signal which logical channel is being sent at any point in time. The C/T header field is used for this purpose. This header field is illustrated in section 2.4.2.3.

The MAC-d entity controls access to the MAC-c/sh/m, MAC-hs and MAC-e/es entities for the DCCH and DTCH. A C/T header is always included when DCCH or DTCH information is forwarded to the MAC-c/sh/m for transmission on the RACH or FACH. Figure 2.52 illustrates the example of a MAC-d entity directing DTCH information towards the MAC-hs entity for mapping onto the

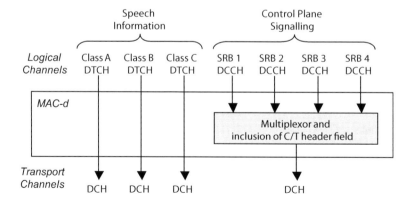

Figure 2.51 Mapping of logical channels onto transport channels for a speech connection

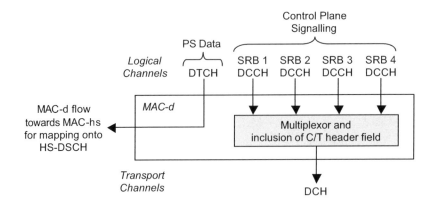

Figure 2.52 Mapping of logical channels onto transport channels for an HSDPA data connection

HS-DSCH transport channel. In this case, a C/T header is only included if the MAC-d entity completes logical channel multiplexing. This example assumes that the SRB are transferred using a DCH. Alternatively, the multiplexed SRB with the C/T header could also be directed towards to the MAC-hs entity.

The MAC-d entity is able to provide transport channel type switching when instructed by the RRC layer. A common example of transport channel type switching is encountered when a UE in CELL_FACH is moved into CELL_DCH. The RNC MAC-d entity stops directing the logical channels towards the MAC-c/sh/m and starts either mapping them directly onto a DCH or directing them towards the MAC-hs. Similarly, the UE MAC-d entity stops directing the logical channels towards the MAC-c/sh/m and starts either mapping them directly onto a DCH or directing them towards the MAC-e/es.

The MAC-d entity is responsible for selecting the appropriate DCH Transport Format Combination (TFC) from the Transport Format Combination Set (TFCS). The concepts of TFC and TFCS are described in greater detail within Section 3.2. Figure 2.53 illustrates the Transport Format Set (TFS) for an uplink 384 kbps packet switched DCH data connection. This example is kept simple by considering the selection of a Transport Format (TF) from a TFS. The same principles apply when selecting a TFC from a TFCS.

The MAC-d entity within the UE evaluates each of the TF at the start of each Transmission Time Interval (TTI). TF can be placed in one of three states: supported, excess power or blocked. These three states are illustrated in Figure 2.54.

A TF is moved to the excess power state if the UE calculates that it has required more than the maximum UE transmit power capability for at least 15 of the previous 30 measurement periods. A TF is

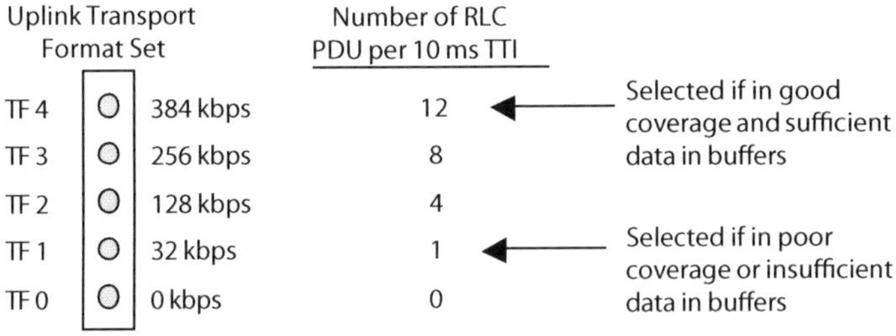

Uplink Transport Format Set			Number of RLC PDU per 10 ms TTI	
TF 4	O	384 kbps	12	← Selected if in good coverage and sufficient
TF 3	O	256 kbps	8	data in buffers
TF 2	O	128 kbps	4	
TF 1	O	32 kbps	1	← Selected if in poor coverage or insufficient
TF 0	O	0 kbps	0	data in buffers

Figure 2.53 Uplink transport format selection from the transport format set

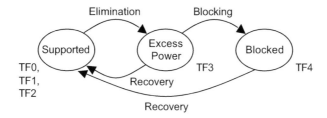

Figure 2.54 State machine for DCH transport formats belonging to the transport format set

moved to the blocked state if it remains in the excess power state for a prolonged period of time. The definition of a prolonged period of time depends upon the service type. In the case of data services, it is defined as 30 ms + longest TTI belonging to the transport format combination. In the case of speech services using multi-rate codecs, it accounts for the delay associated with changing codec rates. A TF is moved back into the supported state if the UE calculates that it has not required more than the maximum UE transmit power capability for the previous 30 measurement periods. The example in Figure 2.54 illustrates TF 4 in the blocked state and TF3 in the excess power state. This indicates that the UE is towards the edge of coverage and has insufficient transmit power to support the higher bit rates.

TF in the blocked state are excluded from the TF selection process. TF in the excess power state may also be excluded from the TF selection process. The UE also defines a minimum set which includes TF which cannot be excluded from the selection process. This ensures that the UE is always able to select at least one TF with a non-zero bit rate. In the case of logical channels using acknowledged mode RLC, the minimum set includes the TF based upon a single RLC PDU. The TF selection process identifies the TF capable of transferring the largest number of RLC PDU that are ready for transmission. The MAC layer will select a relatively low TF if the number of RLC PDU ready for transmission is relatively small, e.g. when an application does not have large quantities of data to transfer. The MAC layer will select the highest available transport format when the number of RLC PDU ready for transmission is high.

The MAC-d entity is also responsible for ciphering RLC PDU belonging to logical channels using transparent mode RLC. Instructions for ciphering are provided by the RRC layer. When ciphering is requested, the MAC-d entity ciphers the RLC PDU, but leaves the MAC header unciphered.

2.4.1.2 MAC-c/sh/m

MAC-c/sh/m entities are located within the RNC and UE. The RNC has one MAC-c/sh/m entity for every cell that it controls. The UE has a single MAC-c/sh/m entity. The MAC-c/sh/m entity always handles the PCCH, CCCH, CTCH, MCCH, MTCH and MSCH logical channels. The MAC-c/sh/m also handles the BCCH when it is transferred using the FACH transport channel. In addition, the MAC-c/sh/m entity can receive the DCCH and DTCH logical channels from the MAC-d entity.

The MAC-c/sh/m entity is able to multiplex different logical channel types onto a single transport channel. This is in contrast to the MAC-d entity which is able to multiplex different logical channels of the same type. Figure 2.55 illustrates an example of the MAC-c/sh/m entity multiplexing different logical channel types onto a single transport channel. This example is based upon a UE multiplexing the CCCH, DCCH and DTCH logical channel types onto a RACH transport channel.

Multiplexing a group of logical channel types onto a single transport channel requires the inclusion of MAC header information to signal which logical channel type is being sent at any point in time. The Target Channel Type Field (TCTF) header is used for this purpose. This header field is illustrated in section 2.4.2. DCCH and DTCH data coming from the MAC-d entity will already include a C/T header

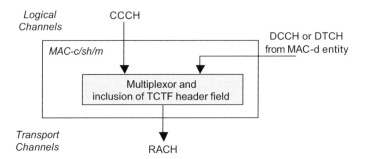

Figure 2.55 Multiplexing CCCH, DCCH and DTCH information onto a single transport channel

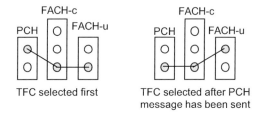

Figure 2.56 TFC selection when PCH and FACH-u information arrive simultaneously

field if logical channel multiplexing has been completed by the MAC-d entity. In this case, the TCTF header field is included in addition to the C/T header field.

The MAC-c/sh/m entity is responsible for adding UE identity type and UE identity header fields when necessary. These header fields are necessary when a common transport channel is used and the payload does not already include the UE identity. Paging Type 1 messages are transferred using the PCH transport channel (a common transport channel), but do not require a UE identity within the MAC header because the UE identity is already included within the RRC message. DTCH information transferred on the FACH (a common transport channel) does not include the UE identity within the payload and so the UE identity must be included as part of the MAC header.

The MAC-c/sh/m entities within the RNC are also responsible for adding an MBMS service identity header when transferring MTCH information on the FACH transport channel. This identity allows the UE to separate the information belonging to different MBMS services.

Similar to the MAC-d entity selecting an appropriate TFC from the TFCS for the DCH transport channel, the MAC-c/sh/m entity is responsible for selecting an appropriate TFC for the RACH and FACH transport channels. This function accounts for the logical channel priority to ensure that the most important information is transferred first. Figure 2.56 illustrates an example based upon the PCH, FACH-c and FACH-u belonging to the TFCS. In this example it is assumed that PCH and FACH-u data arrive simultaneously. The MAC-c/sh/m entity recognises that the PCH message has the higher priority and so starts by selecting the TFC which allows the PCH message to be transferred. The TFC which allows the FACH-u information to be transferred is selected after the PCH message has been sent.

2.4.1.3 MAC-hs

The MAC-hs entity is used for HSDPA. MAC-hs entities are located within the Node B and UE. The MAC-hs entity is placed within the Node B rather than within the RNC to increase its responsiveness to changes in the RF channel conditions and to reduce the round trip time for triggering re-transmissions

Table 2.33 Functionality offered by the MAC-hs entity

Flow control: managing the flow of MAC-d PDU from the RNC to the Node B. Flow control ensures that the Node B has sufficient data while avoiding Node B buffer overflow.

Scheduling: identifying when air-interface resources are assigned to a specific priority queue belonging to a specific MAC-d flow. If code multiplexing is used, multiple priority queues can access the air-interface resources at any point in time.

Transport Format and Resource Combination (TFRC) selection: identifying the HS-DSCH transport block size and determining the number of channelisation codes and modulation scheme.

Generation of MAC-hs PDU: concatenation of MAC-d PDU and the inclusion of a MAC-hs header and potentially, the inclusion of padding. The MAC-hs PDU represents the HS-DSCH transport block.

Hybrid Automatic Repeat Request (HARQ): supporting the re-transmission protocol for MAC-hs re-transmissions between the Node B and UE. Based upon parallel stop and wait processes with soft bit combining of re-transmissions at the UE.

of erroneous data. The Node B has one MAC-hs entity for every HSDPA cell. An HSDPA UE has a single MAC-hs entity.

The MAC-hs entity within the Node B receives DCCH and DTCH information from the MAC-d entity within the RNC, i.e. MAC-d PDU are transferred across the Iub interface. These MAC-d PDU will include a C/T header field if the MAC-d entity has completed logical channel multiplexing. Otherwise, the MAC-d PDU will be the same as the RLC PDU. The MAC-hs entity maps the DCCH and DTCH MAC-d PDU onto the HS-DSCH transport channel.

The functionality offered by the MAC-hs entity is described in detail within Chapter 6. A summary of this functionality is presented in Table 2.33.

2.4.1.4 MAC-e/es

The MAC-e/es entity is used for HSUPA. MAC-e entities are located within the Node B whereas MAC-es entities are located within the RNC. The MAC-e and MAC-es entities are combined within the UE to form a MAC-e/es entity. The MAC-e entity is placed within the Node B rather than within the RNC to increase its responsiveness to changes in the RF channel conditions and to reduce the round trip time for triggering re-transmissions of erroneous data. The RNC has one MAC-es entity for every UE with an HSUPA connection. Likewise, the Node B has one MAC-e entity for every UE with an HSUPA connection. A UE with an HSUPA connection has a single MAC-e/es entity.

The MAC-e/es entity within the UE receives DCCH and DTCH information from the MAC-d entity. These MAC-d PDU will include a C/T header field if the MAC-d entity has completed logical channel multiplexing. Otherwise, the MAC-d PDU will be the same as the RLC PDU. The MAC-e/es entity maps the DCCH and DTCH MAC-d PDU onto the E-DCH transport channel.

The functionality offered by the MAC-e/es entity is described in detail within Chapter 7. A summary of this functionality is presented in Table 2.34.

2.4.1.5 MAC-m

MAC-m entities are located within the UE and do not exist on the network side. An MBMS UE makes use of MAC-m entities when completing selective combining for the downlink MTCH. In this case, there is one MAC-c/sh/m for the current cell and one MAC-m entity for each neighbouring cell. Selective combining is completed at the RLC layer based upon CRC results and unacknowledged mode RLC sequence numbers. The MAC-m entities combined with the MAC-c/sh/m entity allow the UE to transfer parallel streams of MTCH information to the RLC layer.

Table 2.34 Functionality offered by the MAC-e/es entity

MAC-e/es (UE)	E-DCH Transport Format Combination (E-TFC) selection: identifying the transport block size based upon scheduling grants provided by the Node B, the UE transmit power capability and the quantity of data to be sent.
	Generation of MAC-e/es PDU: concatenation of MAC-d PDU, the inclusion of both MAC-es and MAC-e headers and, potentially, the inclusion of padding and scheduling information. The MAC-e/es PDU represents the E-DCH transport block.
	Hybrid Automatic Repeat Request (HARQ): supporting the re-transmission protocol for MAC-e/es re-transmissions between the Node B and UE. Based upon parallel stop and wait processes with soft bit combining of re-transmissions at the Node B.
MAC-e (Node B)	Scheduler: assigns the E-DCH resources across the set of HSUPA UE. The scheduler grants uplink capacity to each UE using absolute and relative grants.
	Extracting MAC-es PDU: uses the MAC-e header to extract the MAC-es PDU and forwarding both the MAC-es PDU and the MAC-e header to the RNC.
	Hybrid Automatic Repeat Request (HARQ): the peer HARQ entity for the UE.
MAC-es (RNC)	Macro diversity selection: selecting the best MAC-es PDU when UE are in soft handover and PDU are received from more than one Node B.
	Reordering Queue distribution: routing the MAC-es PDUs to the correct reordering buffer.
	Reordering: reordering MAC-es PDUs according to the sequence number and Node B tagging. MAC-es PDUs with consecutive sequence numbers are delivered to the disassembly function.
	Disassembly. extracting MAC d PDU from the MAC-es PDU and delivering them to the MAC-d entity for transfer to the RLC layer.

The MAC-m entity reads the Target Channel Type Field (TCTF) from the MAC header to identify which transport blocks include MTCH information. The MAC-m entity also reads the MBMS service identity from MTCH transport blocks to differentiate between services.

2.4.1.6 MAC-b

MAC-b entities are located within the Node B and UE. The MAC-b entity within the Node B maps BCCH logical channel information onto the BCH transport channel before passing it to the Physical layer. Placing the MAC-b layer within the Node B rather than the RNC avoids the requirement to continuously signal BCCH messages across the Iub. The MAC-b layer is responsible for scheduling BCCH transmissions according to the instructions provided by the RRC layer. The BCCH data and scheduling information is signalled to the Node B using NBAP System Information Update Request messages. These messages include the data provided by the RRC layer. An NBAP System Information Update Request message is sent to a Node B when the Node B is switched on. Further update requests are sent if there is a requirement to change any of the system information. Some system information content can be generated by the Node B rather than the RNC, e.g. the uplink RSSI measurement broadcast in SIB 7 can be generated by the Node B. In this case, the RRC layer provides instructions for scheduling the system information message, but the Node B generates the content.

The MAC-b layer is transparent in terms of overheads and does not add any header information. On the network side there is one MAC-b entity for every cell. On the UE side there is one MAC-b entity for every cell from which BCCH information is read.

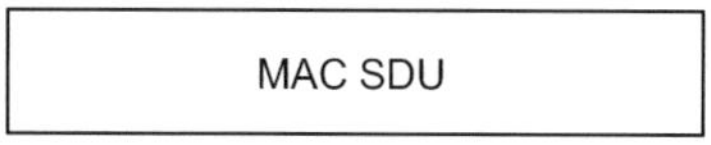

Figure 2.57   MAC PDU for BCCH mapped onto BCH and PCCH mapped onto PCH

TCTF	MAC SDU

Figure 2.58   MAC PDU for BCCH mapped onto FACH

2.4.2 Format of MAC PDU

MAC PDU represent the transport blocks which are passed between the MAC and physical layers. The format of a MAC PDU depends upon the source logical channel, the target transport channel and whether or not logical channel multiplexing is used.

2.4.2.1 BCCH and PCCH

The MAC layer does not add any header information when the BCCH is mapped onto the BCH. In this case, the MAC layer is transparent for the BCCH and the MAC PDU is equal to the MAC SDU. Likewise, the MAC layer does not add any header information when the PCCH is mapped onto the PCH. These mappings are illustrated in Figure 2.57.

If the BCCH is mapped onto the FACH, a 2 bit TCTF is included by the MAC-c/sh/m entity. This header is used by the receiving UE to identify that BCCH information has been included within the FACH. A TCTF value of 00 indicates that the BCH is present. The corresponding MAC PDU is illustrated in Figure 2.58.

2.4.2.2 CCCH and CTCH

When the CCCH is mapped onto the RACH in the uplink direction or onto the FACH in the downlink direction, the MAC-c/sh/m entity adds a TCTF. The length of this header information is 2 bits for the RACH and 8 bits for the FACH, i.e. TCTF values of 00 and 0100 0000 respectively. Similarly, when the CTCH is mapped onto the FACH in the downlink direction, an 8 bit TCTF header is added by the MAC-c/sh/m entity. In this case the TCTF has a value of 1000 0000. The corresponding MAC PDU is illustrated in Figure 2.59.

Log File 2.38 presents an example MAC PDU and SDU for the CCCH logical channel mapped onto the RACH transport channel.

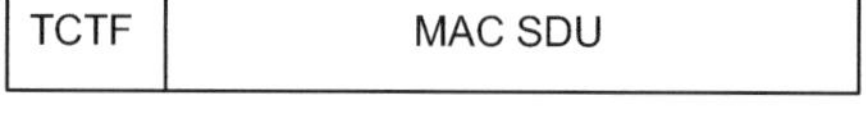

TCTF	MAC SDU

Figure 2.59   MAC PDU for CCCH mapped onto RACH/FACH and CTCH mapped onto FACH

```
MAC PDU: 0E B5 1D AD 9B 14 80 30 00 28 28 C0 5B 44 00 00 00 00 00 00 00
CCCH mapped to RACH/FACH
- TCTF: CCCH
- MAC SDU: 3A D4 76 B6 6C 52 00 C0 00 A0 A3 01 6D 10 00 00 00 00 00 00 00
```

Log File 2.38   Example MAC PDU for CCCH mapped onto RACH

The first 2 bits of the MAC PDU are 00 which indicates that the PDU is used to transfer the CCCH logical channel. The remainder of the MAC PDU represents the MAC SDU. The MAC SDU can be derived by translating the PDU into binary, truncating the TCTF bits, realigning the bytes and translating back to hexadecimal, i.e. the first five bytes of the MAC PDU are:

```
0000 1110 1011 0101 0001 1101 1010 1101 1001 1011
```

Truncating the first two bits and re-aligning the bytes leads to:

```
0011 1010 1101 0100 0111 0110 1011 0110 0110 11
```

Which, when converted back to hexadecimal becomes:

```
3A D4 76 B6 6
```

This hexadecimal string matches the start of the MAC SDU. If the header length had been a multiple of 4 bits then the SDU could have been obtained from the PDU by inspection, i.e. by truncating the header directly in hexadecimal format.

2.4.2.3 DCCH and DTCH

When the DCCH or DTCH is mapped onto an uplink or downlink DCH the requirement for a MAC header depends upon whether or not the MAC-d entity is completing any logical channel multiplexing. If the MAC-d entity is not completing any multiplexing then a MAC header is not included. If the MAC-d entity is completing multiplexing then the MAC-d entity includes a 4 bit C/T header. These scenarios are illustrated in Figure 2.60.

The C/T header informs the receiver of the identity of the logical channel within the MAC SDU. For example, if four SRB have been configured for the DCH then the C/T header would have a decimal value between 1 and 4 and the receiver would know which SRB is being sent at any point in time.

When the DCCH or DTCH is mapped onto the RACH in the uplink direction or onto the FACH in the downlink direction the MAC-d entity adds a 4 bit C/T header before passing the MAC-d PDU onto the MAC-c/sh/m entity. The C/T header informs the receiver of the identity of the logical channel within the MAC SDU. The MAC-c/sh/m entity subsequently adds a header which includes the TCTF, UE identity type flag and UE identity. The TCTF is used to inform the receiver of the logical channel type within the MAC SDU. A value of 01 is used to identify the DCCH and DTCH on the RACH whereas a value of 11 is used on the FACH. The TCTF does not differentiate between the DCCH and DTCH. The receiver uses the C/T header to identify the specific logical channel. The UE identity type flag is added to inform the receiver of the UE identity type which follows within the subsequent header field. A value of 00 indicates that a U-RNTI has been used whereas a value of 01 indicates that a C-RNTI has been used. The FACH can use either UE identity but the RACH must always use the C-RNTI. The UE identity has a length of 32 bits when the U-RNTI is used and a length of 16 bits when

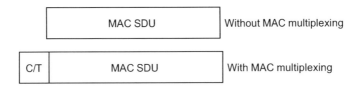

Figure 2.60   MAC PDU for the DCCH or DTCH mapped onto the DCH

TCTF	ID Type	UE ID	C/T	MAC SDU

Figure 2.61 MAC PDU for DCCH or DTCH mapped onto RACH/FACH

```
MAC PDU: D0 00 11 80 F8 F3 9D 62 8C CE D6 78 3A 67 D3 99 51 2E 8B 71 C6
DTCH/DCCH mapped to RACH/FACH
- TCTF: DCCH or DTCH over FACH
- UE-Id Type: C-RNTI
- UE-Id: 1
- C/T Field: Logical channel 2
- MAC SDU:  80 F8 F3 9D 62 8C CE D6 78 3A 67 D3 99 51 2E 8B 71 C6
```

Log File 2.39 Example MAC PDU for DCCH mapped onto FACH

the C-RNTI is used. The UE identity is included in the downlink direction to indicate which UE should decode the MAC SDU. The UE identity is included in the uplink direction to inform the receiving Node B of which UE sent the MAC PDU. A MAC PDU corresponding to the DCCH mapped onto the RACH or FACH is presented in Figure 2.61.

Log File 2.39 presents an example MAC PDU and SDU for the DCCH logical channel mapped onto the FACH transport channel. The MAC SDU can be extracted from the MAC PDU by truncating the header which has a length of 3 bytes. The TCTF header information indicates that the MAC SDU originates from either the DCCH or DTCH logical channels. The UE identity is provided in the form of a C-RNTI with a value of 1. Finally, the logical channel identity indicates that the MAC SDU originates from SRB 2, i.e. an access stratum RRC message from the RNC using acknowledged mode RLC.

Chapter 6 describes the MAC PDU format when the DTCH and DCCH logical channels are mapped onto the HS-DSCH transport channel. Chapter 7 describes the MAC PDU format when the DTCH and DCCH logical channels are mapped onto the E-DCH transport channel.

2.4.2.4 MCCH, MSCH and MTCH

When the MCCH or MSCH is mapped onto a FACH, the requirement for a MAC header depends upon whether or not multiple logical channels are multiplexed onto that FACH. A MAC header is not required if the MAC-c/sh/m entity is not completing any multiplexing. The MAC-c/sh/m entity includes an 8 bit TCTF header if the MAC-c/sh/m entity is completing multiplexing. These scenarios are illustrated in Figure 2.62.

Both TCTF and MBMS Identifier header fields are included when the MTCH is mapped onto a downlink FACH. The structure of the resulting PDU is presented in Figure 2.63.

In this case the TCTF has a length of 4 bits. The MBMS identity also has a length of 4 bits and is used to signal an identifier for the MBMS service. This allows the receiving UE to associate transport blocks with specific services.

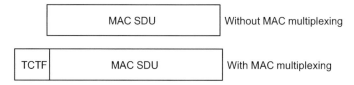

Figure 2.62 MAC PDU for MCCH and MSCH mapped onto FACH

Figure 2.63 MTCH mapped to FACH

2.4.3 Other Functions

The MAC layer is responsible for receiving buffer occupancy results from the RLC layer at least once every Transmission Time Interval (TTI). These buffer occupancy results can be used to trigger capacity requests when the buffer occupancy becomes high, or capacity releases when the buffer occupancy becomes low. The use of these measurements in the context of UE measurement reporting events 4a and 4b is described in section 2.2.5. These events can be used to help manage the allocation of uplink resources. The RNC can complete similar measurements internally to help manage the allocation of downlink resources without the requirement for any signalling.

The MAC layer also has some control over RACH transmissions. The MAC layer is responsible for determining the Access Service Class (ASC) from the MAC logical channel priorities when a UE is in RRC Connected mode (the RRC layer determines the ASC from the AC when a UE is in RRC Idle mode). The MAC layer is also responsible for managing the number of PRACH preamble cycles, running a back-off timer after receiving an AICH negative acknowledgement and completing persistency checks. These procedures are described in greater detail within Section 3.3.8.

2.5 Frame Protocol Layer

- The Frame Protocol layer is responsible for packaging both control plane signalling and user plane data for transfer across the Iub and Iur interfaces.
- The Frame Protocol layer is specified separately for dedicated channels (DCH, E-DCH) and common channels (RACH, FACH, PCH, HS-DSCH). It is not necessary for Frame Protocol to support the transfer of PCH data frames across the Iur.
- Uplink DCH and RACH data frames include CRC check results to inform the RNC of which transport blocks have been received in error. A quality estimate may also be included within uplink DCH data frames. RACH data frames include a measurement of the propagation delay between the UE and Node B.
- FACH data frames include a transmit power which provides the potential for power control. This would require the RNC to have some knowledge of the downlink coverage conditions experienced by individual UE.
- PCH data frames are used in pairs. The first data frame includes a bitmap to define the content of the PICH. The second data frame includes the paging message itself.
- E-DCH (HSUPA) data frames are used to package MAC-es PDU rather than transport blocks because the Node B includes the MAC-e layer which extracts the MAC-es PDU.
- HS-DSCH (HSDPA) data frames are used to package MAC-d PDU rather than transport blocks because the Node B includes the MAC-hs layer which generates transport blocks.
- The Frame Protocol layer is also responsible for providing a set of control functions, e.g. Uplink Outer Loop Power Control control frames can be used to signal uplink SIR target changes to the Node B, Capacity Allocation and Capacity Request control frames can be used for HSDPA flow control, and Radio Interface Parameter Update control frames can be used to reconfigure the Node B.
- Key 3GPP specifications: TS 25.427, TS 25.435 and TS 25.425

The Frame Protocol layer is responsible for packaging both control plane signalling (RRC messages) and user plane data (application data) for transfer across the Iub and Iur interfaces. The Frame Protocol layer also provides support for various control functions. For example, an uplink outer loop power control frame is defined which allows the RNC to signal changes in the uplink SIR target. The frame protocol layer also includes control frames to support the flow control of HSDPA data frames from the RNC to the Node B.

The Frame Protocol layer is specified separately for the dedicated channels and the common channels. 3GPP TS 25.427 specifies the Frame Protocol layer for transferring the dedicated channels across both the Iub and Iur, whereas 3GPP TS 25.435 specifies the Frame Protocol layer for transferring the common channels across the Iub. 3GPP TS 25.425 specifies the Frame Protocol layer for transferring the common channels across the Iur. The HS-DSCH (HSDPA) is categorised as a common channel whereas the E-DCH (HSUPA) is categorised as a dedicated channel.

2.5.1 Dedicated Channels - Data Frames

Dedicated channels include the DCH and E-DCH. The DCH is applicable to both the uplink and downlink whereas the E-DCH is only applicable to the uplink. In the case of the downlink DCH, the MAC-d entity within the RNC is able to transfer a transport block set to the Frame Protocol layer once every Transmission Time Interval (TTI). For example, the MAC-d entity can transfer up to 12 MAC PDU of 336 bits during each 10 ms TTI for a 384 kbps PS data connection. The Frame Protocol layer packages these transport blocks within a data frame and forwards them to the Iub transport network, e.g. AAL2/ATM layers. The Iub transport network is then responsible for delivering the data frame to the peer Frame Protocol layer within the Node B. If the connection is using a drift RNC then the Frame Protocol layer is also responsible for forwarding the data frame to the Iur transport network.

Figure 2.64 illustrates the structure of the Frame Protocol data frame for a downlink DCH. The data frame starts by including a 7 bit Cyclic Redundancy Check (CRC) for the header. These CRC bits are used by the Node B to detect whether or not the header information has been received correctly. If the header has been received in error then the entire data frame can be discarded. The CRC bits are

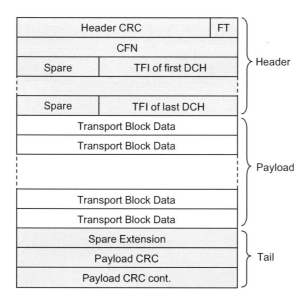

Figure 2.64   Structure of an Iub/Iur Frame Protocol data frame for a downlink DCH

followed by a 1 bit Frame Type (FT) flag used to indicate whether the subsequent frame is a data frame
or a control frame. A value of 0 indicates a data frame whereas a value of 1 indicates a control frame.
Figure 2.64 illustrates a data frame and so the FT flag would have a value of 0.

The second byte of the header includes an 8 bit Connection Frame Number (CFN) used to indicate
the first radio frame belonging to the TTI during which the set of transport blocks are to be transmitted
across the air-interface. The CFN is followed by a series of 5 bit Transport Format Indicators (TFI). A
TFI value is included for each transport channel and is used to indicate which transport format within
the transport format set is included within the data frame. This informs the Node B of how many
transport blocks to expect for each transport channel belonging to the transport format combination.

The payload section of the data frame includes the set of transport blocks. The transport blocks are
arranged such that those belonging to the first transport channel are sent first. Padding can be used at
the end of a transport block to ensure that the start of each transport block is positioned at the start of a
new byte. If there are multiple transport channels then the transport blocks belonging to the remaining
transport channels follow those belonging to the first transport channel.

The final part of the downlink data frame includes an optional CRC for the contents of the payload.
These CRC bits can be used by the Node B to detect whether or not the payload information has been
received without error.

An example Frame Protocol downlink data frame for a PS data connection is presented in
Log File 2.40. The header specifies that the Node B should start transmitting this set of transport
blocks during the radio frame with CFN 71.

```
DCH DL DATA FRAME
Header CRC: 62
Conn. Frame number: 71
TFI Block of DCH
1. TFI: 4 {12 blocks, 336 bits/block}
Transport Block Sets
1. Transport Block Set
1.    Transport Block F6 00 86 1A 2A 23 B7 2F A9 1B 8A C3 8A 08 C2 FA 9D 2B 12
      47 62 D1 98 12 42 3A 3D 58 85 42 AA C6 D0 F6 8C CD CD E1 2F C4 2A 43
2.    Transport Block F6 08 54 BB 54 44 64 8A 13 C5 41 03 4B CE DC 2C A2 75 4C
      67 BC 76 F1 29 E4 37 A3 21 A0 EA F1 3C 9C 38 AA 8E 48 24 F9 6C AC D9
3.    Transport Block F6 10 94 45 13 22 C8 57 2A 38 9E 8D 62 F5 65 85 0E 2C B9
      92 0A 95 37 5F 02 A5 88 97 A3 65 62 2C 8C AC 25 94 D0 8E 8A 81 E9 F3
4.    Transport Block F6 18 44 A7 9E 8B 16 28 5C 75 D6 56 43 B1 65 7D 0B 4F B7
      17 A1 A7 BA F3 83 08 96 1E 1E AD 4B 01 82 64 23 C7 61 BB BA CB 8B 19
5.    Transport Block F6 20 75 42 DC B3 7B 49 DD F8 2E F3 9A 96 F1 3E 00 24 A0
      0A 8C 00 17 69 89 18 58 26 E2 CA 59 A0 04 22 EF 08 34 A5 20 09 84 4D
6.    Transport Block F6 28 8C 93 CF 07 45 65 32 41 60 00 1A 88 84 66 4B C1 D9
      A8 55 44 D4 FF FB B2 6C 78 00 05 FB 63 5A EB 2C 63 72 4B 85 1B 5D 64
7.    Transport Block F6 30 C9 3C 15 F5 77 6B ED 31 2D C1 25 10 2D 31 A6 24 E0
      13 A6 E4 D7 5C AB 98 C6 5E C9 E3 55 E3 0D D3 82 B4 D3 21 C6 53 F9 36
8.    Transport Block F6 38 BA C7 29 3A C8 B2 42 02 44 41 25 23 6E 18 53 C7 6E
      50 76 74 BB 1A 1A EF 99 05 A4 5C 90 61 DB A7 97 DE 8C 68 78 3A 31 48
9.    Transport Block F6 40 64 E8 3C 4E 58 12 3F 88 00 60 BE 7C 89 7A 36 8E 97
      FA 0C 30 17 59 11 14 8E 1E 53 65 E5 34 1D 92 2D 3F 3B 16 89 07 0B 97
10.   Transport Block F6 48 A1 2C E5 F7 44 2B DA 3C 75 66 55 41 35 1C 30 1A 57
      08 52 27 5D 88 A2 7A E7 12 42 42 50 34 58 D0 45 0C D1 2C 81 52 16 52
11.   Transport Block F6 50 34 88 F0 7B C0 92 74 70 25 40 1C 50 F2 57 08 A4 70
      11 98 8A D8 3C 9E 52 A4 46 E9 37 F3 2B 29 A9 F0 64 2E AD 23 B4 EC 92
12.   Transport Block F6 59 1C 6C 21 81 D8 92 2B D9 E4 FA A2 19 AF E7 25 45 00
      00 2C B3 49 20 00 3C 06 2D E8 52 D0 1C 8D 0A E2 03 5C 00 50 0B 0E 76
Spare Extension
Payload CRC: 10272
```

Log File 2.40   Example Frame Protocol data frame for a downlink DCH PS data connection

The header also specifies a TFI of 4, which in this example corresponds to 12 transport blocks each with a size of 336 bits. The Node B has already been informed of the transport format set and TTI during connection establishment, i.e. using an NBAP Radio Link Setup Request or NBAP Radio Link Reconfiguration Prepare message. This example is based upon the transport format set illustrated in Figure 2.65.

The TFI within Log File 2.40 is followed by the data frame payload which includes the set of 12 transport blocks. The PS data service also makes use of a control plane transport channel used by the Signalling Radio Bearers (SRB). The SRB transport channel makes use of separate Frame Protocol data frames. The user plane data and the SRB are then transferred across the Iub and Iur using separate logical ATM connections.

In the uplink direction, a Node B can receive transport blocks once every TTI. For example, a Node B which has a 64 kbps DCH PS data connection can receive four transport blocks of 336 bits during each 20 ms TTI. Node B which have multiple active set cells combine the uplink signals within the Physical layer, e.g. using a RAKE receiver, to generate a single set of uplink transport blocks. These transport blocks are then transferred from the Node B to the RNC after packaging them within a Frame Protocol uplink data frame. A data frame is transferred for every TTI that data is received from the UE. Figure 2.66 illustrates the structure of the Frame Protocol data frame for an uplink DCH.

The header of the uplink data frame includes the same fields as the downlink data frame, i.e. a 7 bit CRC for the header, a 1 bit FT flag, an 8 bit CFN and a series of 5 bit TFI. In this case, the CFN indicates the radio frame during which the set of transport blocks were received. If the TTI is greater than 10 ms then the set of transport blocks will be received across multiple CFN. The CFN included within the data frame header indicates the first CFN during which the transport blocks were received.

Similar to the downlink data frame, the payload section includes the set of transport blocks. The payload also includes a single 8 bit quality estimate and a series of 1 bit CRC indicators. The quality estimate can be used by the serving RNC to select the highest quality set of transport blocks when a connection is in soft handover and the CRC indicators do not provide any differentiation. The quality estimate is a Bit Error Rate (BER) measurement which can be measured from either the physical channel or from one of the transport channels. In the case of a physical channel measurement, the DPCCH pilot bits can be used by comparing the received bit sequence with the expected bit sequence. In the case of a transport channel measurement, the BER can be estimated from the DPDCH bits during the channel decoding procedure. The choice between physical and transport channel BER is configured during connection establishment. The QE-selector within the NBAP Radio Link Setup Request and NBAP Radio Link Reconfiguration Prepare messages can be used to instruct a Node B which quality estimate to include. If one of the transport channels included within the Frame Protocol data frame has a QE-selector value of 'selected' then the BER should be measured from that transport channel. If the transport channel BER is not available, or if none of the transport channels have a QE-selector value of 'selected' then the BER should be measured from the physical channel.

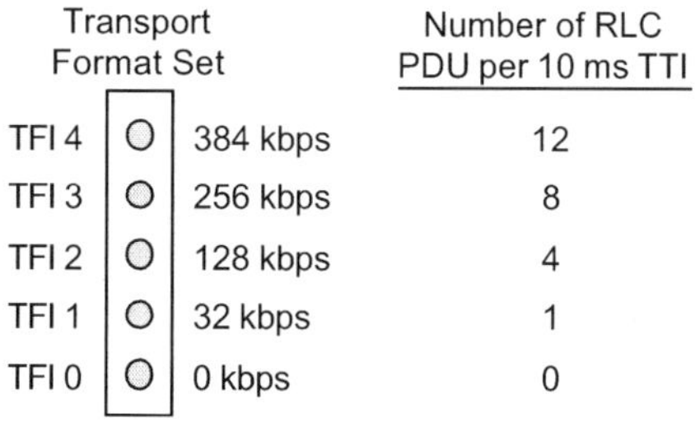

Transport Format Set			Number of RLC PDU per 10 ms TTI
TFI 4	O	384 kbps	12
TFI 3	O	256 kbps	8
TFI 2	O	128 kbps	4
TFI 1	O	32 kbps	1
TFI 0	O	0 kbps	0

Figure 2.65 Example transport format set belonging to a 384 kbps PS data connection

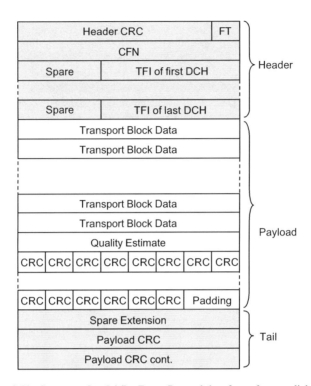

Figure 2.66   Structure of an Iub/Iur Frame Protocol data frame for an uplink DCH

The CRC indicators represent the results of any CRC checks completed by the physical layer of the Node B. The physical layer of the UE can attach CRC bits to transport blocks prior to channel coding and transmitting across the air-interface. The Node B receiver can then use those CRC bits to determine whether or not the received transport blocks include any errors after channel decoding. CRC bits are attached to transport blocks belonging to transport channels which use Acknowledged Mode (AM) RLC. The CRC indicators can then be used to determine whether or not an RLC re-transmission is necessary for a specific transport block. CRC bits can also be attached to transport blocks belonging to transport channels using either Unacknowledged Mode (UM) or Transparent Mode (TM) RLC. RRC signalling messages sent using UM RLC can be discarded if a CRC check indicates an error in one or more of the associated transport blocks. The CRC check results can also be used to calculate the uplink Block Error Rate (BLER) for a specific transport channel. This information can be used within the uplink outer loop power control procedure. The speech service transport channel belonging to the class A bits uses TM RLC, but includes a CRC check which can be used to help ensure that the target BLER performance is achieved. If the result of a CRC check is positive then a value of 0 is included within the Frame Protocol data frame. Otherwise a value of 1 is inserted. There is a CRC indicator included for every transport block within the data frame. If a specific transport block did not include any CRC bits when transmitted across the air-interface then the CRC indicator is set to 0.

The final part of the Frame Protocol uplink data frame includes an optional CRC field for the contents of the payload. These CRC bits can be used by the RNC to detect whether or not the payload information has been received correctly.

An example uplink data frame for the speech service is presented in Log File 2.41. The speech service makes use of a 20 ms TTI and so the CFN of 72 corresponds to the first 10 ms radio frame belonging to the TTI.

```
DCH UL DATA FRAME
Header CRC: 55
Conn. Frame number: 72
TFI Block of DCH
1. TFI: 2 {1 blocks, 81 bits/block}
2. TFI: 1 {1 blocks, 103 bits/block}
3. TFI: 1 {1 blocks, 60 bits/block}
Transport Block Sets
1. Transport Block Set
1.    Transport Block:  BF AF 02 96 37 A8 CC D6 51 D5 61
2. Transport Block Set
1.    Transport Block:  3E 88 9A 7C 7C EB 62 1C 7A 36 E2 9A 28
3. Transport Block Set
1.    Transport Block:  DE 13 11 23 EB EA E1 70
Quality Estimate: 1
CRC Indicators: 00
Spare Extension
Payload CRC: 42186
```

Log File 2.41 Example Frame Protocol data frame for the uplink DCH speech service

The user plane of the speech service makes use of three transport channels and so has three transport format sets, i.e. the class A, B and C transport format sets. These transport format sets are illustrated in Figure 2.67. The class A transport format set includes three transport formats whereas the class B and C transport format sets include two transport formats. Log File 2.41 indicates TFI 2 is being used for the class A transport channel whereas TFI 1 is being used for the class B and C transport channels. This combination of transport formats results in a bit rate of 12.2 kbps which implies that speech is active during this TTI.

The speech service also makes use of a control plane transport channel used by the SRB. The SRB transport channel makes use of separate Frame Protocol data frames and is transferred across the Iub and Iur using separate logical ATM connections.

The quality estimate included in Log File 2.41 has a decimal value of 1. Table 2.35 presents the mapping between the signalled value and the actual value. This mapping has been extracted from 3GPP TS 25.133. A signalled value of 1 means that the BER is less than 0.86%. The Frame Protocol data frame gives no indication of whether this BER has been measured from a transport channel or from the physical channel.

All of the CRC indicators in Log File 2.41 are 0. Transport blocks belonging to the class A transport channel include CRC bits for the air-interface and so it can be deduced that the class A transport block has been received without error. The class B and C transport blocks do not include CRC bits for the air-interface and so their CRC indicators are automatically set to 0 and it is not known whether or not they include any errors. If this had been a file transfer data service then all transport blocks would include CRC bits for the air-interface because it becomes important to ensure that all higher layer packets are received without error.

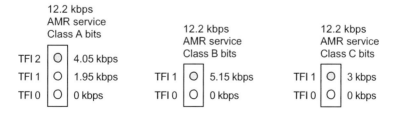

Figure 2.67 User plane transport format sets belonging to the 12.2 kbps AMR speech service

Table 2.35 Mapping between signalled BER value and actual BER value

Signalled value	Actual value
0	$BER = 0$
1	$-\infty \leq LOG10(BER) < -2.06375$
2	$-2.06375 \leq LOG10(BER) < -2.055625$
3	$-2.055625 \leq LOG10(BER) < -2.04750$
. . .	. . .
254	$-0.016250 \leq LOG10(BER) < -0.008125$
255	$-0.008125 \leq LOG10(BER) < 0$

Transport blocks belonging to the E-DCH transport channel are also packaged by Frame Protocol uplink data frames although using a different data frame structure. In the case of the E-DCH there is only one transport block per TTI and the Node B includes the MAC-e layer which completes some processing of the transport blocks prior to forwarding their content to the RNC, i.e. each transport block is forwarded as a series of one or more MAC-es PDU. Also, in the case of the E-DCH, the Node B may forward the MAC-es PDU less frequently than once per TTI, e.g. if HSUPA has been configured to use a 2 ms TTI the Node B could buffer the MAC-es PDU over a period of 10 ms before transferring them towards the RNC. Frame Protocol data frames for the E-DCH are described in greater detail within Section 7.9.

2.5.2 Dedicated Channels - Control Frames

The set of control frames belonging to the dedicated channel Frame Protocol is presented in Table 2.36.

Each control frame has an identifier which is included as part of the control frame header to indicate the type of control frame being sent. The generic structure of a control frame is illustrated in Figure 2.68. The structure of the header is the same for all control frames, but the structure of the payload is dependent upon the control frame type.

Outer Loop Power Control frames are used by the serving RNC to instruct SIR target changes at the active set Node B, i.e. this control frame is only used in the downlink direction. The outer loop power control functionality within the serving RNC adjusts the SIR Target to help ensure that the target uplink quality criteria is achieved. The uplink quality criteria can be based upon a BLER target and compared against the BLER measured from the CRC indicators within the uplink data frame. An example Outer Loop Power Control frame is presented in Log File 2.42. The actual value of the SIR target can be derived from the signalled value by dividing by 10 and subtracting 8.2. This mapping is specified in

Table 2.36 Control frame types belonging to the dedicated channel Frame Protocol

Control frame type	Identifier
Outer Loop Power Control	0000 0001
Timing Adjustment	0000 0010
DL Synchronisation	0000 0011
UL Synchronisation	0000 0100
DL Node Synchronisation	0000 0110
UL Node Synchronisation	0000 0111
Radio Interface Parameter Update	0000 1001
TNL Congestion Indication	0000 1011

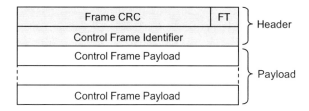

Figure 2.68   Structure of an Iub/Iur dedicated channel control frame

```
55 01 6E
OUTER LOOP PWR CTRL
Control Frame CRC: 42
Control Frame Type: 1
UL SIR TARGET: 2.8 dB
```

Log File 2.42   Example Outer Loop Power Control control frame

3GPP TS 25.427. In this example, the signalled value in hexadecimal format is 6E which equals 110 in decimal. Applying the mapping results in a SIR target of 2.8 dB.

Timing Adjustment control frames are used to help ensure that active set Node B always receive downlink data from the serving RNC within the correct time window to allow transmission across the air-interface. This control frame is only used in the uplink direction. If a downlink data frame is received outside the allowed time window then the Node B can return a Timing Adjustment control frame which specifies the CFN of the relevant downlink data frame and also the time of arrival for that data frame. The time of arrival is measured as the time difference between the end point of the allowed time window and the actual arrival time. A negative value indicates that the data frame was received outside the allowed time window. The actual value of the time of arrival in ms can be derived from the signalled value by applying 2's complement binary arithmetic and dividing by 8.

The Downlink and Uplink Synchronisation frames are used to either achieve or restore DCH synchronisation in the downlink direction. These control frames are also used as a keep-alive to maintain activity on the transport connection. The serving RNC initiates the procedure by sending a Downlink Synchronisation control frame. This control frame includes a target CFN value. After receiving the downlink control frame, the Node B responds with an Uplink Synchronisation control frame which echoes the target CFN and adds a time of arrival value. The RNC can then compensate its downlink timing based upon the time of arrival. An example pair of Downlink and Uplink Synchronisation frames is presented in Log File 2.43.

```
37 03 A4
DOWNLINK SYNC
Control Frame CRC: 27
Control Frame Type: 3
Conn. Frame number: 164

85 04 A4 00 4F
UPLINK SYNC
Control Frame CRC: 66
Control Frame Type: 4
Conn. Frame number: 164
Time of Arrival: 9.875 ms
```

Log File 2.43   Example Downlink and Uplink Synchronisation control frames

In this example the downlink CFN is 164 and the time of arrival is 9.875 ms. The value of 9.875 ms has been derived from the hexadecimal value of 00 4F which equals a decimal value of 79. Dividing by 8 leads to the value of 9.875 ms. 2's complement negative values are flagged by the left most binary digit having a value of 1, i.e. the left most hexadecimal value would be greater than 7.

The Downlink and Uplink Node Synchronisation frames are used by the serving RNC to obtain Node B timing information. The serving RNC sends a Downlink Node Synchronisation frame containing the RNC Frame Number (RFN) corresponding to the time that the frame was sent. After receiving the downlink control frame, the Node B responds with an Uplink Node Synchronisation control frame which echoes the RFN and adds the Node B Frame Numbers (BFN) corresponding to the time that the downlink frame was received and the time that the uplink frame was returned.

The Radio Interface Parameter Update control frame is used to update radio interface parameters linked to a specific UE. The set of parameters which can be updated has increased with each release of the 3GPP specifications. Table 2.37 presents the set of parameters for release 99 through to release 6.

The first parameter to be included was the power offset for the inner loop power control commands which are transmitted as part of the downlink DPCCH. This parameter was included to allow the serving RNC to instruct power offset changes as the downlink interference conditions change. The next parameter to be included was the downlink transmit power control mode. The argument for including this parameter is based upon being able to change the downlink power control mode from 0 to 1 when a UE enters soft handover, and from 1 to 0 when a UE leaves soft handover. DPC mode 0 increases the reliability of the downlink power control commands by repeating them over three time slots. This increase in reliability can help to avoid transmit power differences between active set Node B. The multiple radio sets indicator was added to allow Node B to be informed of when a UE is in soft handover with another Node B. This information can be used by Node B to influence the procedure when synchronisation is lost. The maximum UE transmit power was included to help increase the quantity of information available to an HSUPA serving Node B when making scheduling decisions. If the Radio Interface Parameter Update control frame does not include a valid CFN then parameter changes are applied as soon as possible. Otherwise, changes are applied at the specified CFN.

The TNL Congestion Indication control frame is used by the serving RNC to inform a Node B that transport network congestion has been detected on either the Iub or Iur. This control frame is only used in the downlink direction and is targeted at controlling the uplink throughput generated by HSUPA. The header of an HSUPA data frame includes a Frame Sequence Number (FSN) and CFN to allow the RNC to detect congestion resulting from either lost frames or increased delay. If a Node B receives a TNL Congestion Indication control frame with a cause value of either increased delay or lost frames then it should reduce the uplink bit rate on its Iub. This reduction should be applied to at least the HSUPA connection for which congestion was detected. The TNL Congestion Indication frame can also be used to indicate no congestion to allow Node B to return to normal operation subsequent to experiencing congestion.

Table 2.37 Parameters which can be updated using the Radio Interface Parameter Update

3GPP release	Parameters
release 99	TPC Power Offset
release 4	TPC Power Offset, DPC Mode
release 5	TPC Power Offset, DPC Mode, Multiple Radio Sets Indicator
release 6	TPC Power Offset, DPC Mode, Multiple Radio Sets Indicator, Maximum UE Transmit Power

2.5.3 Common Channels - Data Frames

Common channels include the RACH, FACH, PCH and HS-DSCH. The RACH is applicable to the uplink whereas the FACH, PCH and HS-DSCH are applicable to the downlink. Frame Protocol for the common channels is specified separately for the Iub and Iur. With the exception of the PCH transport channel, all common channels are specified to be transferred across both the Iub and Iur. The PCH is only specified to be transferred across the Iub because the core network forwards Paging Type 1 messages to the appropriate RNC and it is not necessary to transfer them across the Iur.

Figure 2.69 illustrates the structure of the Iub data frame used to package uplink RACH transport blocks before transferring them to the controlling RNC. The header of the RACH data frame is similar to the header of the DCH data frame. The CFN corresponds to the first radio frame occupied by the set of transport blocks, i.e. the first radio frame belonging to the Transmission Time Interval (TTI). The TTI for the RACH transport channel can be either 10 or 20 ms. The header of the RACH data frame can only include a single TFI. This indicates that the data frame can package only a single transport channel. That transport channel could be used to transfer either control plane signalling or user plane data.

The header of the RACH data frame also includes a propagation delay measurement. This measurement is completed by the Node B and is specified within 3GPP TS 25.215. The measurement is based upon the time difference between the downlink frame timing of the PRACH message at the Node B and the uplink frame timing of the PRACH message at the Node B, i.e. the time difference between when the PRACH message is expected and when the PRACH message is received. The measurement result represents the one-way propagation delay rather than the round trip time. The RACH data frame does not include a quality estimate, but CRC indicators are included for the set of transport blocks to reflect the result of any CRC checks across the air-interface.

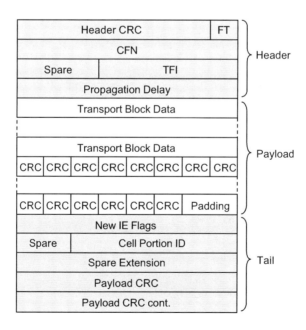

Figure 2.69 Structure of an Iub RACH uplink data frame

```
RACH DATA FRAME
Header CRC: 106
Conn. Frame number: 238
Transport Format Ind: 0 {1 blocks, 168 bits/block}
FDD - Propagation Delay: 9 chips
Transport Blocks
1. Transport Block 0C 51 57 9C 41 88 D5 28 07 10 00 5A 00 00 00 00 00 00 00
00 00
CRC Indicators 00
Spare Extension
Payload CRC: 40355
```

Log File 2.44 Example RACH uplink data frame

The new Information Element (IE) flags and cell portion ID sections have been introduced for the release 6 version of 3GPP TS 25.435. The new IE flags section has been introduced to allow any subsequent fields to be optional rather than mandatory. Each bit within the new IE flags section indicates whether or not a specific new IE is present. The new IE flags section is only necessary when there is at least one new IE present. The release 6 version of the data frame includes only a single new IE (Cell Portion ID) and so the flags section is only necessary when that new IE is present. The right-most bit within the new IE flags section is used to indicate that the cell portion ID is present. The Cell Portion ID has been introduced for the purposes of beamforming. A cell portion represents an area of a cell covered by a specific antenna beam. A Node B with beamforming capability is able to detect from which cell portion the RACH message has been received and can subsequently inform the RNC using the Frame Protocol data frame. An example RACH data frame is presented in Log File 2.44.

The TTI for the RACH transport channel is broadcast within System Information Block (SIB) 5 and optionally within SIB 6. In this example, the TTI is 20 ms and the CFN of 238 corresponds to the first radio frame belonging to the TTI. The transport format set of the RACH transport channel is also broadcast within SIB 5 and optionally within SIB 6. The transport format set for this example is illustrated in Figure 2.70.

The transport format set for the RACH differs from the transport format set for a DCH by excluding the 0 kbps transport format. The RACH transport channel does not require a 0 kbps transport format because it is a common transport channel and UE only make use of it when there is data to transfer. Log File 2.44 indicates that TFI 0 is being used which corresponds to a single transport block of size 168 bits.

The propagation delay measurement is signalled with a value of 3. 3GPP TS 25.435 specifies that the actual value is equal to the signalled value multiplied by 3 and that the actual value is quantified in terms of chips. This means that the actual value is 9 chips which can be translated into a delay by dividing by the chip rate of 3.84 Mcps. The delay can then be translated into a propagation distance by multiplying by the speed of light ($3 \times 10^8 \text{ ms}^{-1}$). Applying these calculations results in a propagation delay of 2.3 µs and a propagation distance of 703 m.

As indicated by the TFI, there is only one transport block within the data frame which in this example can be decoded as an RRC Connection Request message sent on the CCCH logical channel. The CRC indicator is 0, meaning that the transport block has been received without error. The new IE

Figure 2.70 Example transport format set belonging to the RACH transport channel

flags and cell portion ID sections are not present, which indicates that the data frame is based upon a version of the specifications prior to release 6, or that beamforming is not supported by the serving Node B.

Figure 2.71 illustrates the structure of the Iub Frame Protocol data frame used to package downlink FACH transport blocks before transferring them to the serving Node B. The header of the FACH data frame is similar to the header of the RACH data frame. The CFN corresponds to the first radio frame to be occupied by the set of transport blocks, i.e. the first radio frame belonging to the TTI. Similar to the RACH data frame, there is only a single TFI which indicates that the FACH data frame accommodates only a single transport channel. Separate FACH transport channels are configured for the control plane and user plane, i.e. FACH-c and FACH-u. These transport channels can be multiplexed onto the same S-CCPCH physical channel, but they are transferred across the Iub using separate Frame Protocol data frames and separate logical transport connections. The propagation delay measurement which appears in the RACH data frame is replaced by a downlink transmit power. This value offers the potential to power control the relevant S-CCPCH physical channel.

The FACH data frame then includes the set of transport blocks belonging to the specified transport format. There are no CRC indicators because this is a downlink data frame and the message has not yet been transferred across the air-interface. The final part of the data frame is a CRC check for the payload. An example FACH data frame is presented in Log File 2.45.

The TTI for the FACH transport channel is broadcast within SIB 5 and optionally within SIB 6. In this example, the TTI is 10 ms and the CFN of 12 corresponds to the only radio frame belonging to the TTI. The transport format set of the FACH-c and FACH-u transport channels are also broadcast within SIB 5 and optionally within SIB 6. This example is based upon the FACH-c and the corresponding transport format set is presented in Figure 2.72.

The transport format set for the FACH-c transport channel includes a 0 kbps transport format despite being a common channel because multiple transport channels are multiplexed onto the same physical channel and it necessary to have a 0 kbps transport format when defining the transport format combination set. This allows the transport format combination set to include transport format combinations where one transport channel is inactive while other transport channels are active.

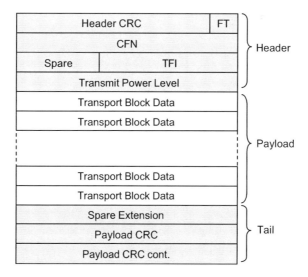

Figure 2.71 Structure of an Iub FACH downlink data frame

```
FACH DATA FRAME
Header CRC: 6
Conn. Frame number: 12
Transport Format Ind: 2 {2 blocks, 168 bits/block}
Transmit Power Level: 5.5 dB
Transport Blocks
1. Transport Block 40 1B F8 30 EF 28 AB CE 20 C4 6A 94 03 88 00 04 08 B3 85
   83 03
2. Transport Block 40 1C 49 D2 E0 84 B8 6A 10 00 14 21 67 B4 51 B1 6E 8A 28
   27 4B
Spare Extension
Payload CRC: 4418
```

Log File 2.45   Example FACH downlink data frame

Log File 2.45 indicates a TFI of 2 which corresponds to two transport blocks of 168 bits. In this example, these transport blocks form part of an RRC Connection Setup message sent on the CCCH logical channel. The RRC Connection Setup message is relatively large and cannot be transferred within a single TTI. The RNC makes use of multiple Frame Protocol data frames and multiple TTI to transfer the message to the Node B.

The FACH transmit power level is signalled using a value of 37 in hexadecimal which is equal to 55 in decimal. 3GPP TS 25.435 specifies that the actual value is equal to the signalled value divided by 10, and that the actual value represents the transmit power relative to the maximum FACH transmit power. The maximum FACH transmit power is configured during the cell start-up procedure within the NBAP Common Transport Channel Setup Request message. In this example, the Node B is instructed to transmit the FACH with a power which is 5.5 dB less than the maximum. Use of this information element is implementation dependant and some Node B may ignore the value instructed by the RNC, or the RNC may signal a fixed value. In either case, the Node B would then use a fixed transmit power for the FACH.

Figure 2.73 illustrates the structure of the Iub data frame used to package both downlink PCH transport blocks and paging channel indicators for the PICH physical channel. PCH data frames are used in consecutive pairs. The first data frame includes the PICH bitmap used to trigger the relevant UE to listen to the PCH during the subsequent radio frame. The second data frame includes the corresponding PCH message. The PCH data frame header includes a CFN which occupies 12 bits rather than 8 bits because the CFN cycle used for the PCH ranges from 0 to 4095 rather than from 0 to 255. There is a 1 bit PI flag to indicate whether or not there is a paging indicator bitmap within the payload. Similar to the RACH and FACH data frames, there is a single TFI to indicate the PCH transport format within the payload.

The first part of the payload includes the PICH bitmap when it is included. The length of the PICH bitmap depends upon the number of paging groups configured for the paging channel.

The number of paging groups can be either 18, 36, 72 or 144 and the number of bits in the PICH bit map equals the number of paging groups. If a paging message is to be scheduled during the next radio frame for a UE belonging to a specific paging group then the bit within the PICH bitmap associated

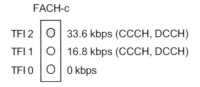

Figure 2.72   Example transport format set belonging to the FACH-c transport channel

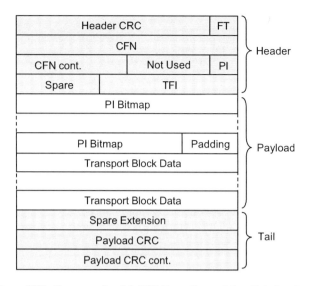

Figure 2.73   Structure of an Iub PCH Frame Protocol downlink data frame

with that paging group is set to 1. Otherwise the PICH bitmap is populated with 0. If the PCH has been configured to accommodate only a single paging record then there can be a maximum of 1 positive indication within the PICH bitmap. The second part of the payload includes the PCH transport blocks when they are included.

An example pair of PCH data frames is presented in Log File 2.46. The first data frame is used to instruct the Node B to broadcast a positive paging indication for a specific paging group while the second data frame is used to transfer the corresponding paging message. The first data frame has a

```
5E 25 D1 00 00 00 00 00 02 00 00 00 00 80 F3
PCH DATA FRAME
Header CRC: 47
Conn. Frame number: 605
Paging Indication: PI-bitmap in payload
Transport Format Ind: 0 {0 blocks, 80 bits/block}
PI-bitmap:  00 00 00 00 02 00 00 00 00
Transport Blocks: No Transport Blocks
Spare Extension
Payload CRC: 33011

46 25 E0 01 40 C1 24 68 40 00 02 00 8A E6 13 71
PCH DATA FRAME
Header CRC: 35
Conn. Frame number: 606
Paging Indication: no PI-bitmap in payload
Transport Format Ind: 1 {1 blocks, 80 bits/block}
PI-bitmap: not present
Transport Blocks
1. Transport Block:  40 C1 24 68 40 00 02 00 8A E6
Spare Extension
Payload CRC: 4977
```

Log File 2.46   Example pair of downlink PCH data frames

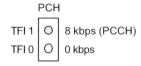

Figure 2.74 Example transport format set belonging to the PCH transport channel

CFN of 605 and the paging indication flag within the header is positive to indicate that the payload includes a PICH bitmap. The TFI of the first data frame is 0 which indicates that the payload does not include a paging message. The PICH bitmap in this example has a length of 9 bytes (72 bits) which indicates that the network has been configured to use 72 paging groups. Expanding the PICH bitmap from hexadecimal to binary illustrates that only a single paging group has a positive indication and so the subsequent radio frame can only include paging records belonging to that single paging group.

The second data frame has a CFN of 606, i.e. corresponding to the consecutive radio frame. The payload does not include a PICH bitmap, but includes the paging message which was indicated by the preceding data frame. The TFI has a value of 1 which corresponds to a single transport block with a size of 80 bits. The transport format set of the PCH transport channel is broadcast within SIB 5 and optionally within SIB 6. The transport format set used for this example is presented in Figure 2.74.

In this case, the maximum bit rate of the PCH is relatively low and only a single paging record can be accommodated within each paging message. Nevertheless, this allows up to 100 paging messages to be sent every second when considering that the PCH uses a TTI of 10 ms. The transport block included as the payload of the second data frame can be decoded as an RRC Paging Type 1 message.

2.5.4 Common Channels - Control Frames

The set of control frames belonging to the Iub common channel Frame Protocol is presented in Table 2.38. Each control frame has an identifier which is included as part of the control frame header to indicate the type of control frame being sent.

The first five of these control frames are the same as those used for the dedicated channel Frame Protocol described in section 2.5.2. The HS-DSCH Capacity Request and HS-DSCH Capacity Allocation control frames are used for HSDPA. These control frames are described within Section 6.6.

Table 2.38 Control frames types belonging to the Iub common channel Frame Protocol

Control frame type	Identifier
Timing Adjustment	0000 0010
DL Synchronisation	0000 0011
UL Synchronisation	0000 0100
DL Node Synchronisation	0000 0110
UL Node Synchronisation	0000 0111
HS-DSCH Capacity Request	0000 1010
HS-DSCH Capacity Allocation	0000 1011

2.6 Physical Layer

- The physical layer is responsible for processing the transport blocks generated by the MAC layer and subsequently using them to modulate the RF carrier.
- CRC bits can be attached to allow error detection at the receiver after channel decoding. Longer CRC attachments reduce the probability of undetected errors.
- Channel coding can be 1/2 rate convolutional coding, 1/3 rate convolutional coding or 1/3 rate turbo coding. 1/3 rate coding provides greater redundancy than 1/2 rate coding. Turbo coding provides improved performance for high data rate connections whereas convolutional coding allows relatively simple Viterbi decoding. Turbo coding generates systematic, parity 1 and parity 2 bits. The systematic bits are equal to the input bits.
- The first stage of interleaving is completed prior to transport channel multiplexing and is used to randomise the order of the bits across the radio frames belonging to the TTI. The second stage of interleaving is completed after transport channel multiplexing and is used to randomise the order of the bits across the time slots belonging to each radio frame. Interleaving improves the performance of channel coding by helping to ensure that the locations of bit errors are randomised rather than grouped in bursts.
- Rate matching involves either repetition or puncturing. The uplink uses dynamic rate matching to ensure that the channel-coded data always exactly occupies the capacity offered by the DPDCH. The repetition or puncturing ratio can change depending upon the transport format combination and spreading factor. The downlink uses static rate matching and the repetition or puncturing ratio remains constant irrespective of the transport format combination. The downlink uses a fixed spreading factor with periods of DTX when lower transport format combinations are active. Rate matching attributes are used to balance the SIR requirements of each transport channel. This improves efficiency when transferring multiple transport channels using a single physical channel.
- Physical layer processing in the downlink direction includes the insertion of DTX indication bits. The first stage of inserting DTX indication bits is applicable when the position of the transport channel data within a radio frame is fixed. Fixed transport channel positions allows blind detection of the transport format combination and avoids the requirement for TFCI bits within the DPCCH. The second stage of inserting DTX indication bits is applicable when the position of the transport channel data within a radio frame is flexible. DTX indication bits act as spacers which are not transferred across the air-interface.
- A RAKE receiver may be used to compensate for phase variations introduced by the propagation channel. A RAKE receiver also allows delay-spread components to be detected and combined based upon the principles of maximum ratio combining. RAKE receivers may be replaced by more advanced receiver structures which are capable of interference cancellation.
- Spreading is used to increase the bit rate generated by the Physical layer to the chip rate of 3.84 Mcps. Chip sequences known as channelisation codes are used for spreading. The length of a channelisation code is equal to its spreading factor. Spreading factors of 2, 4, 8, 16, 32, 64, 128 and 256 are specified for the uplink (spreading factor of 2 is only applicable to HSUPA). Spreading factors of 4, 8, 16, 32, 64, 128, 256 and 512 are specified for the downlink.
- Channelisation codes are organised in a tree hierarchy. They are orthogonal as long as they are synchronised and one code is not the parent of the other code. Scrambling codes are not orthogonal to one another.
- The uplink DPCCH is mapped onto the quadrature branch of the constellation whereas the DPDCH is mapped onto the in-phase branch. The DPCCH always uses a spreading factor of 256. Different amplitude offsets can be applied to each branch and so the constellation can become rectangular rather than square. Complex scrambling results in rotations of the constellation.

- The downlink DPCCH and DPDCH are time multiplexed onto both the in-phase and quadrature branches of the constellation. This involves a serial to parallel conversion which halves the combined DPCCH and DPDCH bit rate. This means that downlink spreading factors are typically twice their equivalent uplink spreading factors. The downlink DPDCH and DPCCH share the same spreading factor.
- In the uplink direction both channelisation codes and scrambling codes are specified to help reduce the peak to average ratio and improve amplifier efficiency.
- Modulation is used to map the complex chip sequences after spreading and scrambling onto a carrier which is subsequently mixed to the appropriate RF frequency.
- The physical layer is also responsible for inner loop power control, air-interface synchronisation, some sections of the random access procedure and for completing measurements on the air-interface.
- Key 3GPP specifications: TS 25.212, TS 25.213, TS 25.214, TS 25.215 and TS 25.944.

The physical layer is used to process the data belonging to each transport channel prior to mapping onto a physical channel. The physical layer is also responsible for subsequent spreading and scrambling prior to modulating a carrier for transfer across the air-interface.

2.6.1 Physical Layer Processing

2.6.1.1 Uplink

The MAC layer within the UE can forward an uplink transport block set to the physical layer once every Transmission Time Interval (TTI). The physical layer processes the transport block set before using it to modulate the uplink RF carrier. Table 2.39 presents the physical layer processing applied to the transport block set belonging to a 12.2 kbps speech service. This example is based upon an active speech connection with an inactive SRB. The speech service uses a 20 ms TTI and so the MAC layer can forward a transport block set to the physical layer once every 20 ms. The transport block set includes three transport blocks which belong to three transport channels. Summing the quantity of data within each transport block generates a result of 244 bits which corresponds 12.2 kbps.

Table 2.39 Uplink 12.2 kbps speech service transport block processing (inactive SRB)

	Class A speech bits	Class B speech bits	Class C speech bits
Transport block size	81	103	60
CRC attachment	$81 + 12 = 93$	$103 + 0 = 103$	$60 + 0 = 60$
TrBk concatenation/code block segmentation	Not necessary	Not necessary	Not necessary
Channel coding	$(93 + 8) \times 3 = 303$	$(103 + 8) \times 3 = 333$	$(60 + 8) \times 2 = 136$
Radio frame equalisation	$303 + 1 = 304$	$333 + 1 = 334$	$136 + 0 = 136$
1st interleaving	304	334	136
Radio frame segmentation	2×152	2×167	2×68
Rate matching	2×221	2×250	2×129
TrCh multiplexing		2×600	
Physical channel segmentation		Not necessary	
2nd interleaving		2×600	
Physical channel mapping		2×600	

The physical layer starts by attaching Cyclic Redundancy Check (CRC) bits. These bits are used by the receiver to detect whether or not a transport block includes any errors after the channel decoding process. 3GPP TS 25.212 specifies CRC lengths of 0, 8, 12, 16 and 24 bits. Longer CRC lengths result in larger overheads, but are able to detect lower Bit Error Rates (BER). Short CRC lengths have an increased probability of leaving undetected errors. The minimum detected BER associated with each CRC length is presented in Table 2.40.

The minimum detected BER figures for CRC lengths of 8, 16 and 24 correspond to the residual BER figures which can be configured as part of the Quality of Service (QoS) profile belonging to the interactive and background traffic classes. The 16 bit CRC is based upon the $x^{16} + x^{12} + x^5 + 1$ polynomial which is known as the CRC-CCITT polynomial (CCITT was formerly a group within the ITU). This polynomial is also used by the X.25, Bluetooth, IrDA, and PPP standards. The first versions of 3GPP TS 25.212 only included the 16 bit CRC. It was recognised that a 16 bit CRC could represent a significant overhead for connections using relatively small transport blocks. Consequently, smaller CRC lengths of 8 and 12 bits were introduced. The CRC length of 24 was also introduced for services which require particularly low undetected error rates. In the case of the speech service, only the transport block belonging to the class A transport channel has CRC bits attached, i.e. error detection is not applied to the class B and C transport channels. This means that the speech service Block Error Rate (BLER) is calculated from the class A transport channel. Outer loop power control makes use of this BLER information to adjust the SIR target.

The physical layer evaluates the requirement for transport block concatenation and code block segmentation after the CRC bits have been attached. Transport block concatenation involves concatenating transport blocks belonging to the same transport channel. Transport blocks belonging to different transport channels are not concatenated at this stage. The speech service has only one transport block belonging to each transport channel and so concatenation is not necessary. The transport blocks are known as code blocks once transport block concatenation has been completed. Code block segmentation is necessary if the resulting code blocks are too large for the subsequent channel coding procedure. The maximum code block size for convolutional coding is 504 bits whereas the maximum code block size for turbo coding is 5114 bits. Turbo coding also requires a minimum code block size of 40 bits and so padding may be added if very small code blocks are generated. Code block segmentation is not necessary for the speech service because convolutional coding is used and the code blocks are smaller than the maximum of 504 bits.

Channel coding is completed after code block segmentation. 3GPP TS 25.212 specifies 1/3 rate convolutional coding, 1/2 rate convolutional coding and 1/3 rate turbo coding. The coding rate reflects the quantity of redundancy introduced to help protect the code blocks as they are transferred across the air-interface. A coding rate of 1/3 means that the number of bits after channel coding is 3 times the number of bits before channel coding. A coding rate of 1/3 provides greater protection than a coding rate of 1/2 but generates a larger overhead. Turbo coding generates three streams of output data - systematic, parity 1 and parity 2. The systematic bits are equal to the input bits whereas the parity 1 and parity 2 bits are calculated from the input bits to generate redundancy. Turbo coding provides improved performance for high data rate connections whereas convolutional coding allows relatively simple Viterbi decoding. In the case of convolutional coding, 8 tail bits are added to the code block prior to channel coding. These tail bits reset the shift register used for coding and prevent error propagation at the receiver. In the case of turbo coding, 12 tail bits are added after the channel coding process. The speech service uses 1/3 rate convolutional coding for the class A and B transport channels, and 1/2 rate

Table 2.40 Minimum detectable BER for each CRC length

CRC length	8 bits	12 bits	16 bits	24 bits
Minimum detected BER	3.9×10^{-3}	2.4×10^{-4}	1.5×10^{-5}	6.0×10^{-8}

convolutional coding for the class C transport channel. This reflects the importance of the data transferred by the class A and B transport channels. The speech service bit rate increases from 12.2 kbps to 38.6 kbps after channel coding.

Radio frame equalisation is used to ensure that the channel coded data can be equally distributed between the 10 ms radio frames belonging to the TTI, e.g. if the TTI is 40 ms then radio frame equalisation ensures that the number of coded bits is a multiple of 4. Padding is added if necessary. The speech service uses a 20 ms TTI and so the number of bits belonging to each transport channel must be a multiple of 2. The class A and B transport channels each require 1 bit of padding to generate blocks of data which contain an even number of bits.

The blocks of data are then interleaved using a combination of block interleaving with column reordering. Interleaving is completed separately for each transport channel. This represents the first of two interleaving stages – the first stage is applied prior to transport channel multiplexing and is used to randomise the order of the bits across the set of radio frames belonging to the TTI. The second stage is applied after transport channel multiplexing and is used to randomise the order of the bits belonging to different transport channels within a radio frame. The general principle of the first interleaving stage is presented in Figure 2.75.

The data is read into an array which has a number of columns equal to the number of 10 ms radio frames within the TTI. The speech service uses two columns because the TTI is 20 ms. Figure 2.75 illustrates four columns which correspond to a 40 ms TTI and could represent the interleaving applied to a 3.4 kbps SRB. The data is read into the array row by row. The data always exactly occupies the array because the preceding radio frame equalisation has made sure that the quantity of data is a multiple of the number of radio frames within the TTI. The columns of the array are reordered if the TTI is either 40 or 80 ms. If the TTI is 40 ms the second and third columns are swapped. If the TTI is 80 ms the column order changes from (0,1,2,3,4,5,6,7) to (0,4,2,6,1,5,3,7). Data is then read out from the array column by column. If the TTI is 10 ms there is only a single radio frame per TTI and a single column in the interleaving array. In this case, the first interleaving stage does not have any impact.

Interleaving is used to help improve the performance of channel decoding after the data has experienced a fading radio environment. The performance of channel decoding is improved if bit errors are randomly distributed throughout the data. Fading radio environments tend to generate bursts of

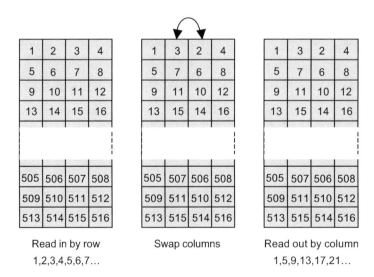

Read in by row   Swap columns   Read out by column
1,2,3,4,5,6,7...   1,5,9,13,17,21...

Figure 2.75   First stage interleaving for a transport channel using a 40 ms TTI

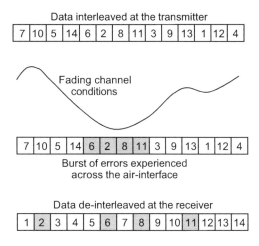

Figure 2.76   Principle of interleaving

errors rather than randomly distributed errors, i.e. a burst of errors may occur during a deep fade. The combination of interleaving at the transmitter and de-interleaving at the receiver is able to redistribute bit errors to help remove any bursts. This general principle is illustrated in Figure 2.76. In this example, the de-interleaving process redistributes a burst of four bit errors. The effectiveness of interleaving improves as the interleaving time (interleaving depth) increases, e.g. interleaving across 40 ms is more effective than interleaving across 10 ms. Using a short interleaving depth increases the probability of bursts of errors remaining after de-interleaving. The drawback associated with an increased interleaving depth is an increased system delay, i.e. the receiver has to wait for longer before being able to start decoding. This drawback is visible when completing PINGs to an application server. Smaller round trip times can be measured for shorter TTI.

Radio frame segmentation is completed after the first interleaving stage. This divides the blocks of interleaved data into the equal sized sections to be sent during each radio frame. Rate matching is then applied to the data associated with each radio frame. Rate matching involves either repetition or puncturing to ensure that the blocks of data fit exactly into the capacity offered by the DPDCH. The capacity offered by the DPDCH depends upon the spreading factor and so the UE must identify the spreading factor prior to completing rate matching. The UE identifies the spreading factor using a combination of the minimum allowed spreading factor signalled by the RNC and the algorithm specified within 3GPP TS 25.212. The minimum allowed spreading factor is typically signalled within a Radio Bearer Setup message or a Radio Bearer Reconfiguration message. An example of the relevant section from a Radio Bearer Setup message is presented in Log File 2.47.

This example is applicable to the speech service and the minimum spreading factor is specified as 64. This corresponds to a maximum DPDCH capacity of 600 bits per radio frame. The algorithm defined in 3GPP TS 25.212 specifies that if puncturing can be avoided for all transport channels then

```
ul-DPCH-Info
   modeSpecificInfo fdd
         scramblingCodeType:       longSC
         scramblingCode:           417359
         spreadingFactor:          sf64
         tfci-Existence:           true
         puncturingLimit:          pl0-68
```

Log File 2.47   Example minimum spreading factor signalled by the RNC

the maximum spreading factor which avoids puncturing should be selected. Otherwise, if it is necessary to puncture at least one transport channel then the minimum allowed spreading factor should be selected to help minimise the level of puncturing. Identifying the requirement for puncturing requires knowledge of the transport channel Rate Matching attributes (RM). The puncturing ratio is defined by the following equation:

$$Puncturing\ ratio_n = N_{\text{data}} \times \frac{RM_n}{\sum\limits_{i=1}^{K} (RM_i \times N_i)}$$

where, N_{data} is the number of bits offered by the DPDCH per 10 ms radio frame, RM_n is the rate matching attribute for the nth transport channel, N_i is the number of bits per radio frame after radio frame segmentation for the ith transport channel, and K is the number of transport channels. This equation generates the results presented in Table 2.41.

Puncturing ratios greater than 1 indicate that repetition is possible whereas puncturing ratios less than 1 indicate that puncturing is required. Repetition involves repeating specific bits to increase the total number of bits occupying the DPDCH. Repetition increases redundancy although it is less powerful than the redundancy introduced by channel coding. Puncturing involves removing specific bits to reduce the number of bits occupying the DPDCH. Both repetition and puncturing use specific bit patterns to allow the receiver to determine which bits have been repeated or which bits have been punctured.

The quantity of repetition or puncturing applied to a specific transport channel depends upon the relative values assigned to the rate matching attributes. The absolute values have little importance other than larger values allowing a greater range of relative values. If a transport channel is assigned a relatively small rate matching attribute then it is more likely to be punctured. Rate matching attributes are only of interest when there are multiple transport channels to be multiplexed into a single radio frame. The rate matching attribute has no impact if there is only a single transport channel, i.e. $K = 1$ in the previous equation and the numerator and denominator rate matching attributes cancel one another.

The RNC is responsible for configuring the rate matching attributes during connection establishment. The objective of configuring specific values is to ensure that each transport channel multiplexed onto the same physical channel has an equal DPCCH SIR target. The SIR target is used by inner loop power control and is adjusted by outer loop power control. Outer loop power control can define only one SIR target. If each transport channel has a different SIR requirement then outer loop power control must adopt the highest. This means that the other transport channels would be received with a higher quality than necessary. It would be more efficient to apply greater puncturing to the transport channels with a lower SIR requirement, and apply less puncturing to the transport channel with the highest SIR requirement. This would help to equalise the SIR requirements for each transport channel. The general principle of allocating rate matching attributes to a set of transport channels is illustrated in Figure 2.77.

Table 2.41 Uplink puncturing ratios for the speech service (inactive SRB)

	Class A TrCh	Class B TrCh	Class C TrCh
Number of bits after radio frame segmentation	152	167	68
Rate matching attribute	196	202	256
Puncturing ratio for SF256 ($N_{\text{data}} = 150$ bits)	0.36	0.37	0.47
Puncturing ratio for SF128 ($N_{\text{data}} = 300$ bits)	0.73	0.75	0.95
Puncturing ratio for SF64 ($N_{\text{data}} = 600$ bits)	1.45	1.50	1.90

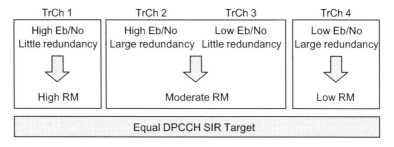

Figure 2.77 Principle of rate matching attribute allocation

If a transport channel has a high Eb/No requirement and little redundancy has been added by the physical layer, e.g. rate 1/2 convolutional coding has been applied, then that transport channel requires a high rate matching attribute to avoid puncturing the data too heavily. If a transport channel has a low Eb/No requirement and a lot of redundancy has been added by the physical layer then that transport channel can be allocated a relatively low rate matching attribute to allow an increased quantity of puncturing. The maximum rate matching attribute specified by 3GPP is 256. Rate matching attributes tend to be defined relative to this value.

The speech service example presented in Table 2.41 illustrates that SF64 allows repetition whereas SF128 and SF256 require puncturing. This means that SF64 is selected for this transport format combination. SF64 offers 600 DPDCH bits per 10 ms radio frame. Rate matching must be used to ensure that the number of bits generated the set of transport channels also equals 600 bits. This results in the speech service bit rate increasing to 60 kbps. The number of transport channel bits after rate matching is defined by the following equation:

$$N_{i,\text{afterRM}} = N_i \times Puncturing \ ratio$$

where, $N_{i,\text{afterRM}}$ is the number of bits per radio frame for the ith transport channel after rate matching, N_i is the number of bits per radio frame prior to rate matching, and the puncturing ratio is the value defined by the previous equation.

Transport channel multiplexing is completed after rate matching. This involves concatenating the blocks of data from each transport channel which are to be transferred during the same TTI. From this point onwards, the physical layer processes the combination of transport channels rather than processing each transport channel individually. Physical channel segmentation is only necessary for DPCH connections when more than a single DPDCH is used, i.e. it becomes necessary to distribute the data between multiple physical channels. In practise only single DPDCH are used for DPCH connections. Physical channel segmentation is more likely to be necessary for HSDPA and HSUPA connections. This is described in Sections 6.9.2 and 7.7.2, respectively.

Physical channel segmentation is followed by a second stage of interleaving. Similar to the first stage, block interleaving with column reordering is used. Interleaving is completed separately for each 10 ms radio frame and each physical channel. This second stage of interleaving is used to randomise the order of the bits belonging to each transport channel. The general principle of the second interleaving stage is presented in Figure 2.78.

The array used for the second stage of interleaving always has 30 columns. The number of columns is a multiple of the number of time slots within a radio frame (15 time slots per radio frame) to ensure a uniform distribution of the bits across the time slots. Data is read into the array row by row. The columns are reordered and the data is read out column by column. The resultant data stream is then mapped onto the physical channel ready for spreading and scrambling prior to modulating the RF carrier.

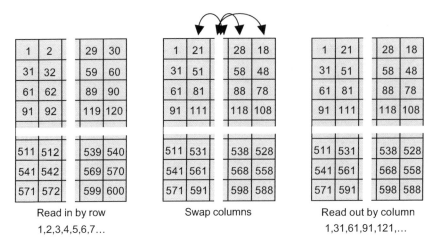

Figure 2.78 Second stage interleaving for a 10 ms radio frame

Table 2.42 presents the physical layer processing completed for the speech service when the SRB is active. The SRB uses a single transport block of 148 bits and a TTI of 40 ms. These figures correspond to a throughput of 3.7 kbps. A 16 bit CRC is attached to the SRB transport block before completing 1/3 rate convolutional coding. The coded data is divided into four blocks rather than two because the TTI is 40 ms rather than 20 ms.

Rate matching is then completed across all four transport channels. This reduces the potential for repetition relative to when the SRB was inactive, i.e. the SRB requires some of the DPDCH capacity. The puncturing ratios for an active SRB are presented in Table 2.43. An uplink spreading factor of 64 is selected because it avoids the requirement for puncturing.

Table 2.42 Uplink speech service transport block processing (active SRB)

	Class A speech bits	Class B speech bits	Class C speech bits	SRB bits
Transport block size	81	103	60	148
CRC attachment	$81 + 12 = 93$	$103 + 0 = 103$	$60 + 0 = 60$	$148 + 16 = 164$
TrBk concatenation/ Code block segmentation	Not necessary	Not necessary	Not necessary	Not necessary
Channel coding	$(93 + 8) \times 3 = 303$	$(103 + 8) \times 3 = 333$	$(60 + 8) \times 2 = 136$	$(164 + 8) \times 3 = 516$
Radio frame equalisation	$303 + 1 = 304$	$333 + 1 = 334$	$136 + 0 = 136$	$516 + 0 = 516$
1st interleaving	304	334	136	516
Radio frame segmentation	2×152	2×167	2×68	4×129
Rate matching	2×166	2×187	2×97	4×150
TrCh multiplexing	4×600 (includes a 2nd speech transport block set)			
Physical channel segmentation	Not necessary			
2nd interleaving	4×600			
Physical channel mapping	4×600			

Table 2.43   Uplink puncturing ratios for the speech service (active SRB)

	Class A TrCh	Class B TrCh	Class C TrCh	SRB
Number of bits after radio frame segmentation	152	167	68	129
Rate matching attribute	196	202	256	210
Puncturing ratio for SF256 (N_{data} = 150 bits)	0.27	0.28	0.36	0.29
Puncturing ratio for SF128 (N_{data} = 300 bits)	0.54	0.56	0.71	0.58
Puncturing ratio for SF64 (N_{data} = 600 bits)	1.09	1.12	1.42	1.17

Transport channel multiplexing is used to combine the speech and SRB data. Two speech transport block sets are paired with a single SRB transport block set. Figure 2.79 illustrates this combination of transport channels with different TTI. The second stage of interleaving is applied to the multiplexed transport channels. The interleaved transport channel data is then mapped onto the DPDCH physical channel.

Table 2.44 presents the physical layer processing applied to the transport block set belonging to a 384 kbps PS data connection. This example is based upon an active data connection with an inactive SRB. The data service uses a 10 ms TTI and so the MAC layer can forward a transport block set to the physical layer once every 10 ms. The transport block set includes 12 transport blocks which include a total of 4032 bits and correspond to a bit rate of 403.2 kbps.

CRC bits are attached to each of the 12 transport blocks. This allows errors to be detected independently for each transport block and so the RLC layer can complete re-transmissions on a per transport block basis. Rate 1/3 turbo coding is applied after the set of transport blocks have been concatenated. Code block segmentation is not necessary because the code block size is less than the maximum of 5114 bits. The bit rate increases from 403.2 kbps at the top of the Physical layer to 1.27 Mbps after channel coding. Radio frame segmentation is not necessary because the TTI is 10 ms and the entire block of data is transferred during a single radio frame. Rate matching is applied to match the size of the data block to the capacity of the DPDCH. In this case, there is only a single transport channel and so the rate matching attribute has no impact. A single DPDCH has a capacity of 9600 bits when the spreading factor is 4. This means that puncturing must be applied rather than repetition unless multiple DPDCH are used. Table 2.45 presents the puncturing ratios assuming a single

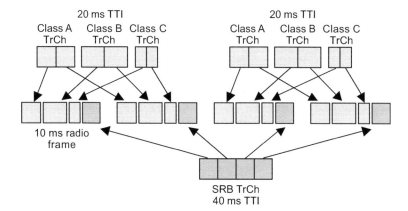

Figure 2.79   Speech service transport channel multiplexing with an active SRB

Table 2.44 Uplink 384 kbps data service transport block processing (inactive SRB)

	Data bits
Transport block size	336×12
CRC attachment	$(336 + 16) \times 12 = 4224$
TrBk concatenation/code block segmentation	4224
Channel coding	$(4224 \times 3) + 12 = 12684$
Radio frame equalisation	$12684 + 0 = 12684$
1st interleaving	12684
Radio frame segmentation	Not necessary
Rate matching	9600
TrCh multiplexing	Not necessary
Physical channel segmentation	Not necessary
2nd interleaving	9600
Physical channel mapping	9600

Table 2.45 Uplink puncturing ratios for the uplink 384 kbps data service (SRB inactive)

	Data TrCh
Number of bits after radio frame segmentation	12684
Rate matching attribute	176
Puncturing ratio for SF8 ($N_{\text{data}} = 4800$ bits)	0.38
Puncturing ratio for SF4 ($N_{\text{data}} = 9600$ bits)	0.76

DPDCH. The UE selects a spreading factor of 4 because it requires the least puncturing. The puncturing ratio of 0.76 indicates that rate matching removes 24% of the channel coded data. Only the parity 1 and parity 2 bits are punctured when puncturing is applied to turbo coded data. The systematic bits are left unpunctured.

Table 2.46 presents the physical layer processing completed for the 384 kbps packet switched data service when the SRB is active. The rate matching attributes become important because there is more

Table 2.46 Uplink 384 kbps data service transport block processing (active SRB)

	Data bits	SRB bits
Transport block size	336×12	148
CRC attachment	$(336 + 16) \times 12 = 4224$	$148 + 16 = 164$
TrBk concatenation/Code block segmentation	4224	Not necessary
Channel coding	$(4224 \times 3) + 12 = 12684$	$(164 + 8) \times 3 = 516$
Radio frame equalisation	$12684 + 0 = 12684$	$516 + 0 = 516$
1st interleaving	12684	516
Radio frame segmentation	1×12684	4×129
Rate matching	9460	4×140
TrCh multiplexing	4×9600 (includes 2nd, 3rd and 4th data TBS)	
Physical channel segmentation	Not necessary	
2nd interleaving	4×9600	
Physical channel mapping	4×9600	

than a single transport channel to map onto the DPDCH. This example assumes rate matching attributes of 176 and 256 for the data and SRB transport channels, respectively. The quantity of data bits is significantly greater than the quantity of SRB bits and so the inclusion of the SRB has little impact upon the puncturing ratio applied to the data transport channel, i.e. it reduces from 0.76 to 0.75. Transport channel multiplexing is used to combine the data and SRB information after rate matching. Four data transport block sets are combined with each SRB transport block set as a result of the 10 and 40 ms TTI.

The spreading, scrambling and modulation processes applied after physical channel mapping are described in Section 2.6.2. The receiver is responsible for reversing the physical layer processing completed by the transmitter, e.g. de-interleaving, transport channel de-multiplexing, channel decoding and error detection. It is necessary for the receiver to generate a bit stream from the received signal prior to completing these processes. It is common for a RAKE receiver to be used for this purpose. The structure of a typical RAKE receiver is illustrated in Figure 2.80.

The input signal originates from the Node B antenna. This signal is filtered, mixed to baseband and fed through an analogue to digital (A/D) converter before arriving at the RAKE receiver. A/D conversion generates a stream of digital samples. It is usual for the sampling rate to be greater than the chip rate, e.g. a sampling rate of 7.68 Mega samples per second would generate two samples per chip. This is known as oversampling. Oversampling improves the time resolution of the signal processed by the RAKE receiver. If delay spread components generated by the radio propagation channel have a spacing of 1.5 chips and oversampling is not used then the receiver will be unable to capture the full power of both components. The drawback of oversampling is an increased baseband processing requirement. It is typical for receivers to use oversampling rates of between 2 and 4. Oversampling by a factor of 4 provides a 0.25 chip time resolution.

The correlator which is positioned at the front end of the RAKE receiver is used to identify the timing of the strongest delay spread components. The correlator compares a filtered version of the DPCCH pilot bit sequence with the received signal. Peaks in the correlator output indicate the presence of a delay spread component. Figure 2.81 illustrates the concept of the correlator. The reference samples represent the DPCCH pilot bit sequence and remain fixed. The samples from the received signal are input to the correlator using a sliding window.

If uplink receive diversity is configured at the Node B, correlations have to be completed separately for each of the received signals. The output from the correlator is used to allocate the strongest delay

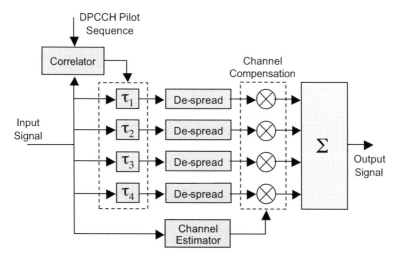

Figure 2.80 Typical structure of a RAKE receiver

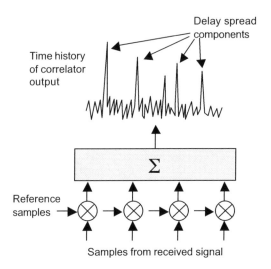

Figure 2.81 Correlator used to search for delay spread components

spread components to the RAKE fingers. If dual branch receive diversity is used then, on average, half of the RAKE fingers would be allocated to each branch although instantaneously more or less fingers may be allocated to any one branch. For simplicity Figure 2.80 illustrates a RAKE receiver with four fingers. In practice, RAKE receivers are likely to have between six and eight fingers. The output from the correlator also defines the delays τ_1, τ_2, τ_3 and τ_4. These delays are used to time align the delay spread components processed by each RAKE finger.

The set of RAKE fingers de-scramble and de-spread the time delayed versions of the input sequence. De-scrambling and de-spreading changes the received signal from having a chip rate of 3.84 Mcps to a symbol rate defined by the spreading factor, e.g. a spreading factor of 64 generates a symbol rate of 60 ksps. Channel compensation is applied after de-scrambling and de-spreading. This process compensates for any phase rotations experienced by the received signal. These phase rotations could be generated by Doppler frequencies resulting from the UE mobility, or they could be generated by frequency offsets between the transmitter and receiver, e.g. the UE could transmit at 1922.6 MHz + 25 Hz while the Node B receives at 1922.6 MHz − 25 Hz. The impact of phase rotations upon a QPSK constellation is illustrated in Figure 2.82.

The channel estimator is responsible for tracking these rotations and providing the channel compensation block with information which allows the received signal to be rotated back to its original position. The channel compensation block also scales the amplitude of the symbols belonging

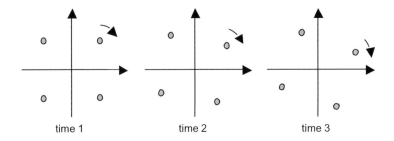

Figure 2.82 Phase rotations generated by a frequency offset

to each RAKE finger. Scaling the amplitude follows the rules of maximum ratio combining, i.e. the RAKE fingers with the highest signal to noise ratios are provided with greater weightings than the fingers with poor signal to noise ratios. The outputs from the RAKE fingers are then summed to generate the data stream which is ready for de-interleaving and de-coding.

More advanced receivers may be used instead of the RAKE receiver. Advanced receivers may encorporate interference cancellation techniques. Multi-user Detection (MUD) provides a solution for reducing the interference generated by other active connections. Generalised RAKE (GRAKE) receivers provide a solution for whitening the coloured (frequency selective) interference usually experienced by a RAKE receiver.

2.6.1.2 Downlink

A Node B can receive a downlink transport block set from the RNC once every TTI. Transport block sets are transferred across the Iub using dedicated channel Frame Protocol data frames. Table 2.47 presents the physical layer processing applied to the transport block set belonging to a 12.2 kbps speech service. This example assumes that the SRB is inactive.

The downlink transport block set is the same as the uplink transport block set, i.e. three transport blocks associated with three transport channels. CRC attachment in the downlink direction is also the same as that in the uplink direction. Likewise, transport block concatenation, code block segmentation and channel coding are the same as in the uplink direction. Uplink and downlink Physical layer processing starts to differ after channel coding. Rate matching follows channel coding in the downlink direction whereas radio frame equalisation and the first stage of interleaving follow channel coding in the uplink direction. Rate matching in the downlink direction is static whereas rate matching in the uplink direction is dynamic. Dynamic rate matching means that the puncturing ratio can change for each transport format combination. This is illustrated by comparing the speech service puncturing ratios in Table 2.41 with those in Table 2.43. The uplink puncturing ratio decreases when the transport format combination is changed by introducing the SRB. Dynamic rate matching allows the puncturing ratio to adapt to ensure that the capacity offered by the DPDCH is always exactly occupied by the data

Table 2.47 Downlink 12.2 kbps speech service transport block processing (inactive SRB)

	Class A speech bits	Class B speech bits	Class C speech bits	SRB bits
Transport block size	81	103	60	0
CRC attachment	$81 + 12 = 93$	$103 + 0 = 103$	$60 + 0 = 60$	0
TrBk concatenation/Code block segmentation	Not necessary	Not necessary	Not necessary	Not necessary
Channel coding	$(93 + 8) \times 3 = 303$	$(103 + 8) \times 3 = 333$	$(60 + 8) \times 2 = 136$	0
Rate matching	280	318	164	0
1st insertion of DTX indication bits	Not necessary	Not necessary	Not necessary	516 DTX
1st interleaving	280	318	164	516 DTX
Radio frame segmentation	2×140	2×159	2×82	4×129 DTX
TrCh multiplexing	$4 \times 381 + 4 \times 129$ DTX (includes a 2nd speech transport block set)			
2nd insertion of DTX indication bits	Not applicable			
Physical channel segmentation	Not necessary			
2nd interleaving	$4 \times 381 + 4 \times 129$ DTX			
Physical channel mapping	$4 \times 381 + 4 \times 129$ DTX			

belonging to the transport format combination. In the uplink direction, the spreading factor can also change to help match the capacity offered by the DPDCH to the quantity of data. Static rate matching means that the puncturing ratio remains constant for each transport format combination. The puncturing ratio is defined by the largest transport format combination and periods of Discontinuous Transmission (DTX) are introduced when smaller transport format combinations are used. The spreading factor remains constant in the downlink direction and so the capacity offered by the DPDCH is fixed.

The transport format combination presented in Table 2.47 does not represent the largest combination because the SRB is inactive. This means there is a period of DTX during the section of the DPDCH that the SRB could occupy. The maximum transport format combination is presented in Table 2.48. It is assumed that downlink DPCH slot format 8 which offers 510 data bits per radio frame is used (shown in Table 3.33).

This maximum transport format combination defines the fixed puncturing ratios for each transport channel. The same puncturing ratios are applied in Table 2.47 when the SRB is inactive. The same puncturing ratios would also be applied during periods of speech inactivity when only Silence Descriptor (SID) frames are transferred using the class A transport channel. These frames generate 39 bit transport blocks which become 177 bits after channel coding. Static rate matching maintains the 0.92 puncturing ratio and reduces the number of channel coded data bits to 163.

Rate matching is followed by the first stage of inserting DTX indication bits. This stage is only applicable if the connection has been configured to use fixed transport channel locations. The use of fixed transport channel locations means that the data belonging to each transport format always occupies the same section of the DPDCH. This allows the UE to complete blind detection of the downlink transport format combination and avoids the requirement to include Transport Format Combination Indicator (TFCI) bits as part of the DPCCH, i.e. the capacity of the DPDCH is increased. The UE completes blind detection by recognising which sections of the DPDCH include data and which sections are using DTX. DTX indication bits act as spacers to avoid the data bits moving together during Physical layer processing when smaller transport formats are used. They are removed after the processed data is mapped onto the DPDCH, i.e. they are not transmitted across the air-interface.

Log File 2.48 presents an example section from a Radio Bearer Setup message which configures the downlink of a speech connection for fixed transport channel locations and blind detection. The log file also specifies a spreading factor of 128 and the inclusion of four DPCCH pilot bits. This allows the UE to deduce that DPCH slot format 8 should be used.

The first stage of inserting DTX indication bits is completed separately for each transport channel. For the example in Table 2.47, the speech transport channels are using their maximum transport formats but the SRB transport channel is inactive. This means that DTX indication bits are not necessary for the speech transport channels, but are necessary for the SRB transport channel. The SRB transport channel would usually occupy 516 bits when active and so 516 DTX indication bits are inserted. If the speech connection becomes inactive and only SID frames are generated on the class A

Table 2.48 Downlink puncturing ratios for the speech service (SRB active)

	Class A TrCh	Class B TrCh	Class C TrCh	SRB
Number of bits after channel coding	303	333	136	516
Rate matching attribute	196	202	256	210
Bits allocated	280	318	164	516
Bits per 10 ms radio frame	140	159	82	129
Puncturing ratio for SF128 ($N_{data} = 510$ bits)	0.92	0.95	1.21	1.00

```
modeSpecificInfo fdd
        powerOffsetPilot-pdpdch: 0
        spreadingFactorAndPilot
        sfd128:                   pb4
        positionFixedOrFlexible: fixed
        tfci-Existence:           false
```

Log File 2.48   Example configuration of fixed transport channel locations

transport channel, the 163 bits associated with the SID frame would be increased to 280 bits using a series of DTX indicators. All other transport channels would be fully occupied by DTX indication bits.

The first stage of interleaving is the same as that applied to the uplink. It is completed separately for each transport channel and is used to randomise the order of the bits across the radio frames belonging to the TTI. If the data includes DTX indication bits then these are also distributed across the TTI. Table 2.47 does not include any transport channels which have a combination of both data and DTX indication bits. Figure 2.83 illustrates the example of speech service class A transport channel when SID frames are generated during periods of speech inactivity. The interleaving array has two columns because the TTI is 20 ms. The DTX indication bits are positioned towards the end of the TTI at the input of the interleaver. The interleaving process redistributes the bits such that the DTX indicators become evenly distributed towards the end of each radio frame belonging to the TTI.

Radio Frame segmentation and transport channel multiplexing are used to combine the transport channel data to be transferred during each radio frame. Subsequent physical layer processing operates on a radio frame basis rather than on a transport channel basis.

The second stage of inserting DTX indication bits is only applicable if the connection has been configured to use flexible transport channel locations. This example is based upon fixed transport channel locations and so it is not applicable. Physical channel segmentation is not necessary because a single DPDCH is used to transfer the data. The second stage of interleaving is the same as that used for the uplink, i.e. it is used to randomise the order of the bits across each radio frame. If the radio frame includes DTX indication bits they become distributed across the time slots belonging to the radio frame. The resultant data stream is then mapped onto the physical channel ready for spreading and scrambling prior to modulating the RF carrier.

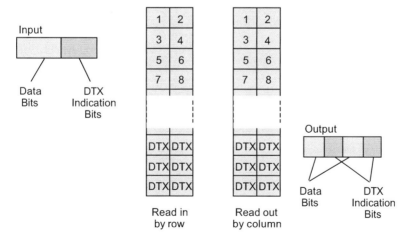

Figure 2.83   First stage interleaving with DTX indication bits

Table 2.49 Downlink 12.2 kbps speech service transport block processing (active SRB)

	Class A speech bits	Class B speech bits	Class C speech bits	SRB bits
Transport block size	81	103	60	148
CRC attachment	$81 + 12 = 93$	$103 + 0 = 103$	$60 + 0 = 60$	$148 + 16 = 164$
TrBk concatenation/code block segmentation	Not necessary	Not necessary	Not necessary	Not necessary
Channel coding	$(93 + 8) \times 3 = 303$	$(103 + 8) \times 3 = 333$	$(60 + 8) \times 2 = 136$	$(164 + 8) \times 3 = 516$
Rate matching	280	318	164	516
1st insertion of DTX indication bits	Not necessary	Not necessary	Not necessary	Not necessary
1st interleaving	280	318	164	516
Radio frame segmentation	2×140	2×159	2×82	4×129
TrCh multiplexing	4×510 (includes a 2nd speech transport block set)			
2nd insertion of DTX indication bits	Not applicable			
Physical channel segmentation	Not necessary			
2nd interleaving	4×510			
Physical channel mapping	4×510			

Table 2.49 presents the downlink physical layer processing completed for the speech service when the SRB is active. The processing applied to the speech transport channels does not change because the rate matching is static rather than dynamic and each transport channel has a fixed share of the DPDCH capacity. The SRB data replaces the DTX indication bits which were previously included. Table 2.49 represents the maximum transport format combination and so it is not necessary to include any DTX indication bits.

Table 2.50 presents the downlink physical layer processing applied to a 384 kbps packet switched data connection with an inactive SRB. The data service uses a 10 ms TTI and so the RNC can forward a transport block set to the Node B once every 10 ms.

Table 2.50 Downlink 384 kbps data service transport block processing (inactive SRB)

	Data bits	SRB bits
Transport block size	336×12	0
CRC attachment	$(336 + 16) \times 12 = 4224$	0
TrBk concatenation/Code block segmentation	4224	0
Channel coding	$(4224 \times 3) + 12 = 12684$	0
Rate matching	8987	0
1st insertion of DTX indication bits	Not applicable	Not applicable
1st interleaving	8987	0
Radio frame segmentation	8987	0
TrCh multiplexing	8987	
2nd insertion of DTX indication bits	8987 + 133 DTX	
Physical channel segmentation	Not necessary	
2nd interleaving	8987 + 133 DTX	
Physical channel mapping	8987 + 133 DTX	

Table 2.51   Downlink puncturing ratios for the 384 kbps data service (SRB active)

	Data TrCh	SRB
Number of bits after channel coding	12684	516
Rate matching attribute	176	256
Bits allocated	8987	532
Bits per 10 ms radio frame	8987	133
Puncturing Ratio for SF128 ($N_{data} = 510$ bits)	0.71	1.03

Similar to the speech service, Physical layer processing in the downlink direction is the same as that in the uplink direction until channel coding has been completed. Static rate matching is used in the downlink direction and so the puncturing ratio is based upon the largest transport format combination. The largest transport format combination and the associated puncturing ratios are presented in Table 2.51. It is assumed that downlink DPCH slot format 15 which offers 9120 data bits per radio frame is used (shown in Table 3.33)

The first stage of inserting DTX indication bits is only applicable if fixed transport channel locations have been configured. This example is based upon flexible transport channel locations and so it is not applicable. Log File 2.49 presents an example section from a Radio Bearer Setup message which configures the downlink of a 384 kbps data connection for flexible transport channel locations. The log file also specifies a spreading factor of 8 which allows the UE to deduce that DPCH slot format 15 should be used. The inclusion of TFCI bits is specified whereas it is not necessary to specify the number of DPCCH pilot bits because there are always 16 pilot bits when the downlink spreading factor is 8 and compressed mode is not active.

Neither the first stage of interleaving nor radio frame segmentation have any impact because there is only a single radio frame within the TTI. Transport channel multiplexing does not have any impact because there is only a single transport channel which is active. The second stage of inserting DTX indication bits is applicable because flexible transport channel locations have been configured. DTX indication bits are added to the end of the data belonging to each radio frame to ensure that the total number of bits is equal to the capacity offered by the DPDCH. In this case, the DPDCH offers 9120 bits and so 133 DTX indicators are added. These DTX indicators occupy the capacity reserved for the SRB. The number of DTX indicators would increase if the data transport channel used a smaller transport format. For example, the number of DTX indicators would increase to 6112 if a transport format which includes 4 rather than 12 transport blocks is selected. Prior to the second stage of interleaving, DTX indicators always appear towards the end of the radio frame when flexible transport channel locations are used. If fixed transport channel locations are used then the DTX indicator bits are added prior to transport channel multiplexing and they can be distributed throughout the radio frame depending upon which transport channels require them.

The second stage of interleaving is used to distribute the DTX indication bits across the radio frame as well as randomise the order of the data bits. The resultant data stream is then mapped onto the physical channel ready for spreading and scrambling prior to modulating the RF carrier.

```
modeSpecificInfo fdd
        powerOffsetPilot-pdpdch: 0
        spreadingFactorAndPilot
          sfd8:
        positionFixedOrFlexible: flexible
        tfci-Existence:          true
```

Log File 2.49   Example configuration of flexible transport channel locations

2.6.2 Spreading, Scrambling and Modulation

2.6.2.1 Uplink

Spreading is used to increase the bit rate generated by the physical layer to the chip rate of 3.84 Mcps. Table 2.39 illustrates that the speech service can generate a bit rate of 60 kbps after physical layer processing. Spreading increases this bit rate by a factor of 64. Figure 2.84 illustrates an example based upon a bit rate of 240 kbps which could represent an uplink 64 kbps data connection. Spreading increases this bit rate by a factor of 16. Each bit generated by the physical layer is multiplied by the same sequence of 16 chips.

If the physical layer had generated a bit rate of 480 kbps then it would be necessary to increase the bit rate by a factor of 8. Figure 2.85 illustrates an example based upon a spreading factor of 8. This example could represent an uplink 128 kbps data connection.

The chip rate of the sequences used for spreading is always 3.84 Mcps. The preceding examples illustrate that the length of the chip sequence is equal to the spreading factor which is dependent upon the bit rate generated by the Physical layer.

Chip sequences provide a way for CDMA systems to multiplex connections onto the same RF carrier without the requirement for time division multiplexing. 3GPP TS 25.213 specifies the set of chip sequences which can be used for spreading. These chip sequences are known as channelisation codes or spreading codes. Uplink channelisation code lengths of 4, 8, 16, 32, 64, 128 and 256 are specified for

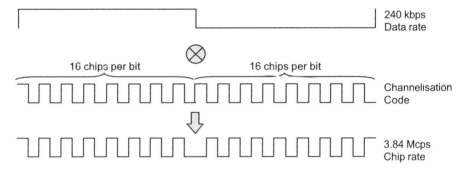

Figure 2.84   Spreading by a factor of 16 (requires a chip sequence with a length of 16 chips)

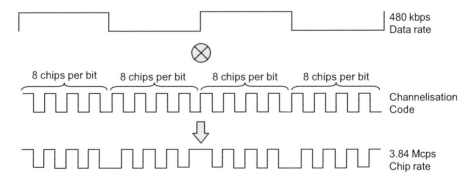

Figure 2.85   Spreading by a factor of 8 (requires a chip sequence with a length of 8 chips)

DPCH connections (HSUPA can also use a channelisation code length of 2). The number of codes with a specific length (spreading factor) is equal to the length of the code, e.g. there are 4 channelisation codes with a spreading factor of 4, 32 channelisation codes with a spreading factor of 32 and 256 channelisation codes with a spreading factor of 256. The set of channelisation codes is organised in a tree hierarchy. This hierarchy is illustrated in Figure 2.86 for spreading factors 4, 8 and 16. Channelisation codes with higher spreading factors extend the tree to the right.

The chip sequences which define the set of channelisation codes have been selected to have specific mathematical properties. Channelisation codes which belong to different branches of the code tree have the potential to be orthogonal to one another. Channelisation codes are only orthogonal when they are synchronised. Orthogonality means that multiple connections can be established without those connections interfering with one another. In practice, the requirement for synchronisation means that orthogonality is not achieved in the uplink direction because the transmitters belonging to different UE are not synchronised. Orthogonality is more achievable in the downlink direction in which case the transmission is point to multi-point. HSUPA connections take advantage of orthogonality when a single UE makes use of multiple channelisation codes to transfer its data. Not every code within the code tree is orthogonal to every other code. Figure 2.87 illustrates the relative position of codes which are orthogonal and non-orthogonal to one another. Channelisation codes are orthogonal as long as one code is not the parent of the other code.

Figure 2.88 illustrates the general pattern used to construct the set of channelisation codes. Each code is generated by concatenating two versions of its parent code. The first version is a duplicate of the parent code whereas the second version is either a duplicate or an inversion. Even numbered codes are based upon two duplicates whereas odd numbered codes are based upon a duplicate and an inversion.

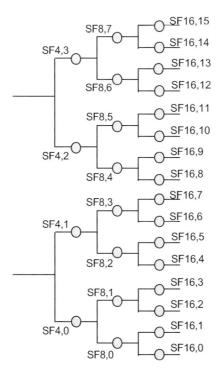

Figure 2.86   Channelisation code tree for spreading factors 4, 8 and 16

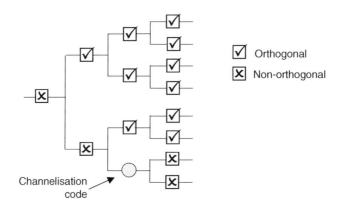

Figure 2.87   Orthogonal and non-orthogonal channelisation codes

Table 2.52 illustrates the process used to multiplex signals onto the same RF carrier and subsequently extract the original data at the receiver. This example is based upon two blocks of input data, $(1,1,-1,1,-1)$ and $(-1,1,1,-1,-1)$. Both blocks of data are assumed to have a bit rate of 960 kbps and so channelisation codes with a spreading factor of 4 are applied. The first block of data is spread using channelisation code (SF4,1) whereas the second block of data is spread using channelisation code (SF4,2). The spreading operation is a simple multiplication of the input data by the channelisation code. Both the input data and channelisation codes are represented using -1 and 1 rather than 1 and 0. This mapping is specified within 3GPP TS 25.213. The two spread signals can then be summed and used to modulate the RF carrier before transmission across the air-interface.

The receiver is required to extract the original blocks of data from the sum of the spread signals. If the transmission is in the downlink direction then the two blocks of data could belong to two separate UE. In this case, each UE would apply its allocated channelisation code to extract its own data. Extracting the original data requires a combination of multiplication and integration. The received signal is first multiplied by the allocated channelisation code and the result is integrated across the bit period. This example is based upon a spreading factor of 4 and so the bit period corresponds to four chips. The sign of the result corresponds to the sign of the original bit. The spreading factor of 4 generates results which are ±4. If the spreading factor had been 16 then the receiver would generate

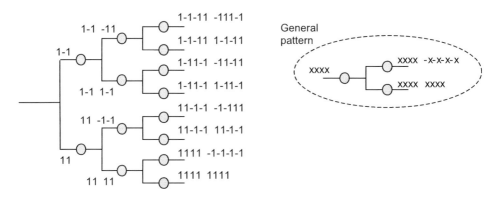

Figure 2.88   Pattern of channelisation code structure

Table 2.52 Extracting original data from the sum of spread signals

Input data 1	1				1				−1				1				−1			
Spreading code 1	1	1	−1	−1	1	1	−1	−1	1	1	−1	−1	1	1	−1	−1	1	1	−1	−1
Spread data 1	1	1	−1	−1	1	1	−1	−1	−1	−1	1	1	1	1	−1	−1	−1	−1	1	1
Input data 2	−1				1				1				−1				−1			
Spreading code 2	1	−1	1	−1	1	−1	1	−1	1	−1	1	−1	1	−1	1	−1	1	−1	1	−1
Spread data 2	−1	1	−1	1	1	−1	1	−1	1	−1	1	−1	−1	1	−1	1	−1	1	−1	1
Sum of spread data	0	2	−2	0	2	0	0	−2	0	−2	2	0	0	2	−2	0	−2	0	0	2
De-spread data 1	0	2	2	0	2	0	0	2	0	−2	−2	0	0	2	2	0	−2	0	0	−2
Integration 1				4				4				−4				4				−4
De-spread data 2	0	−2	−2	0	2	0	0	2	0	2	2	0	0	−2	−2	0	−2	0	0	−2
Integration 2				−4				4				4				−4				−4

results which are ±16. This demonstrates that higher spreading factors allow easier decisions when the signal to noise ratio is relatively poor, i.e. there is a larger difference between the two values and so the result is less susceptible to noise.

The spreading operation increases the bandwidth of the signal by a factor equal to the spreading factor. Figure 2.89 illustrates examples for spreading factors of 64 and 16. It is assumed that the bandwidth occupied by the data is approximately equal to its bit rate. In this context, the spreading factor is referred to as the spreading gain. The spreading gain is often quoted in dB and so spreading gains of 64 and 16 would be quoted as 18 dB and 12 dB respectively. The spreading gain represents the ratio between the chip rate and the bit rate after physical layer processing. This is in contrast to the processing gain which represents the ratio between the chip rate and the bit rate at the top of the RLC layer. The Eb/No figures used throughout system dimensioning and network planning are based upon RLC layer bit rates and so processing gain is used rather than spreading gain.

Larger spreading gains have the benefit of smaller noise bandwidths at the receiver after de-spreading. Figure 2.89 illustrates that a spreading gain of 64 has a noise bandwidth of 60 kHz after de-spreading whereas a spreading gain of 16 has a noise bandwidth of 240 kHz after de-spreading. This means that a receiver with a spreading gain of 16 captures four times more noise than a receiver with a spreading gain of 64. The drawback associated with selecting a large spreading gain is a smaller

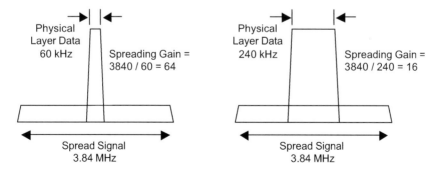

Figure 2.89 Increase in signal bandwidth after spreading

physical layer bit rate, i.e. a smaller DPDCH capacity, and a corresponding increased requirement for puncturing.

In the uplink direction, spreading is completed separately for the DPDCH and DPCCH. The uplink DPCCH is spread using channelisation code (SF256,0) whereas the DPDCH is spread using channelisation code (SFx,x/4), e.g. channelisation code index 16 is used if the DPDCH is spread using a spreading factor of 64. Both the DPDCH and DPCCH have chip rates of 3.84 Mcps subsequent to spreading. The content of the DPCCH is described in Section 3.3.9.

The DPDCH and DPCCH may be transmitted using either equal or different transmit powers. The relative transmit power of each physical channel is controlled using an amplitude gain factor. The amplitude gain factor applied to the DPDCH is known as β_d whereas the amplitude gain factor applied to the DPCCH is known as β_c. These amplitude gain factors can be assigned values between 1/15 and 1, in steps of 1/15. At any point in time at least one of β_c or β_d must have a value equal to 1. Figure 2.90 illustrates the impact of these gain factors upon the shape of the complex constellation. The DPCCH is mapped onto the quadrature branch of the constellation whereas the DPDCH is mapped onto the in-phase branch of the constellation.

β_c and β_d define the relative amplitude of each physical channel rather than the relative power. The relative power is defined by the square of the relative amplitudes. For example, if $\beta_c = 11/15$ while $\beta_d = 15/15$ then the DPDCH is transmitted with 2.7 dB greater power than the DPCCH, i.e. $10 \times \log(225/121) = 2.7$ dB.

Scrambling is completed once the amplitude gain offsets have been applied. 3GPP TS 25.213 specifies 2^{24} long scrambling codes and 2^{24} short scrambling codes. This means there are more than 16 million of each type of scrambling code. Each long scrambling code has a length of 38400 chips (10 ms) whereas each short scrambling code has a length of 256 chips (67 μs). Short scrambling codes are intended to be used by advanced receivers which are capable of interference cancellation and Multi-user Detection (MUD). Long scrambling codes represent the default for less advanced receivers. Similar to channelisation codes, scrambling codes are based upon a chip rate of 3.84 Mcps and so scrambling does not increase the bandwidth of the spread signal. A single uplink scrambling code is assigned to the DPCCH and DPDCH. This scrambling code is used by the Node B to identify the connection, i.e. every connection is assigned a different uplink scrambling code. The RNC is responsible for assigning the uplink scrambling code and for ensuring that the same scrambling code is not assigned to multiple UE.

Figure 2.91 illustrates the combination of uplink spreading and scrambling. Scrambling codes are complex rather than real and so require a complex multiplication. Complex multiplications cause phase rotations and so the scrambling process causes the DPCCH and DPDCH constellation to rotate. Figure 2.92 illustrates the impact of these phase rotations upon the constellation. A scrambling code

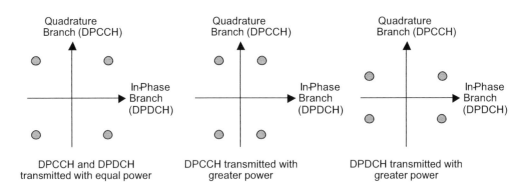

Figure 2.90   Impact of DPCCH and DPDCH amplitude gain factors upon the constellation

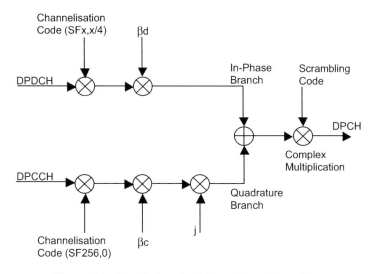

Figure 2.91   Combination of uplink spreading and scrambling

can have values of either $1 + j$, $1 - j$, $-1 + j$ or $-1 - j$ and so the rotation can be either $45°$, $135°$, $225°$ or $315°$.

Both the channelisation codes and the scrambling code have an impact upon the phase changes experienced by the DPCH. Large phase changes tend to increase the peak to average power ratio and result in less efficient amplifier utilisation. Uplink channelisation and scrambling codes have been specified to help limit the number of large phase changes [49]. The uplink DPCCH is spread using channelisation code (SF256,0) which is a continuous stream of 1s. The uplink DPDCH is spread using channelisation code (SFx,x/4) which is a series of alternating 1,1s with $-1,-1$s. These channelisation codes mean that within the period of a single DPDCH bit the phase of the spread signal only changes after every second chip. Scrambling codes are generated from two chip sequences, $c_{1,n}$ and $c_{2,n}$. These chip sequences are combined to generate a scrambling code using the following equation.

$$C_n(i) = c_{1,n}(i)(1 + j(-1)^i c_{2,n}(2\lfloor i/2 \rfloor))$$

This equation allows the value associated with $c_{1,n}$ to change after every chip period, but only allows the value associated with $c_{2,n}$ to change after every second chip period. This definition limits the

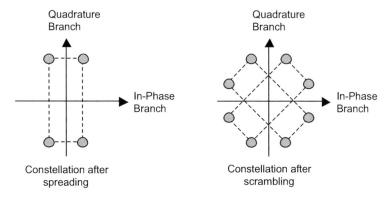

Figure 2.92   Example constellations after spreading and scrambling

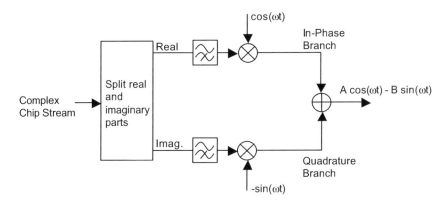

Figure 2.93   Uplink modulation process

scrambling code phase change to $\pm90°$ when making the transition from an even-numbered chip to an odd-numbered chip. The phase change can be $\pm90°$, $0°$ or $180°$ when making the transition from an odd-numbered chip to an even-numbered chip. The combined effect of the channelisation and scrambling codes is that within the period of a DPDCH bit, phase changes are limited to $\pm90°$ when making the transition from an even-numbered chip to an odd-numbered chip whereas larger phase changes are possible when making the transition from an odd-numbered chip to an even-numbered chip.

Modulation provides a way of mapping the resultant chip stream onto a carrier. The carrier can then be mixed to the appropriate RF frequency and transmitted across the air-interface. Figure 2.93 illustrates the uplink modulation process. The complex chip stream generated by spreading and scrambling is split into its real and imaginary components. The real component is processed by a low-pass filter and subsequently multiplied by a $\cos(\omega t)$ function. The imaginary component is processed by a low-pass filter and subsequently multiplied by a $-\sin(\omega t)$ function. The resultant signals from the two branches are then summed. Summing the $\cos(\omega t)$ and $-\sin(\omega t)$ functions modulates the phase onto the carrier, e.g. $\cos(\omega t) - \sin(\omega t) = A\cos(\omega t + \pi/4)$.

2.6.2.2 Downlink

The general concept of spreading in the downlink direction is the same as in the uplink direction, i.e. spreading is used to increase the bit rate generated by the physical layer to the chip rate of 3.84 Mcps. The downlink DPDCH and DPCCH are time multiplexed onto the in-phase and quadrature branches of the constellation. This is in contrast to mapping the uplink DPDCH onto the in-phase branch and the uplink DPCCH onto the quadrature branch. Figure 2.94 illustrates the format of the resultant downlink DPCH. Power offsets can be configured to specify the transmit power of the TPC, TFCI and pilot bits

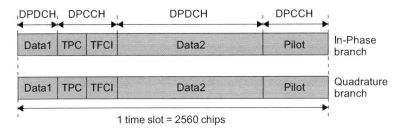

Figure 2.94   Mapping of DPDCH and DPCCH onto the in-phase and quadrature branches

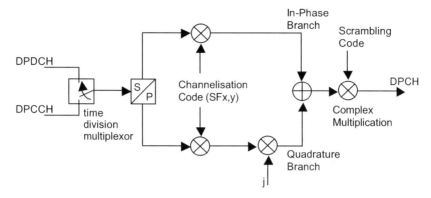

Figure 2.95 Combination of downlink spreading and scrambling

relative to the transmit power of the DPDCH. These power offsets are described in greater detail within Section 3.3.9.

Mapping the DPDCH and DPCCH onto both branches requires a serial to parallel conversion. This conversion is illustrated in Figure 2.95 as part of the spreading and scrambling process.

Each pair of in-phase and quadrature bits represents one constellation point (modulation symbol). There are two downlink DPDCH bits per symbol while the DPDCH is active, and two downlink DPCCH bits per symbol while the DPCCH is active. Table 2.53 presents two examples of the bit rates before serial to parallel conversion with the corresponding symbol rates after conversion. These examples continue from those presented as part of the explanation to the downlink physical layer processing.

Downlink speech connections can generate 510 bits per radio frame when the SRB is active. This represents the DPDCH data and corresponds to a bit rate of 51 kbps. The DPCCH belonging to slot format 8 generates 90 bits of data per radio frame which corresponds to 9 kbps. The sum of the two physical channels results in a combined bit rate of 60 kbps which becomes a symbol rate of 30 ksps after serial to parallel conversion. The symbol rate is half of the bit rate and so the downlink spreading factor is twice as large as the equivalent uplink spreading factor, i.e. the uplink speech service generates a DPDCH bit rate of 60 kbps and requires a spreading factor of 64 whereas the downlink speech service generates a DPDCH bit rate of 51 kbps and requires a spreading factor of 128. The uplink DPCCH is always sent at a relatively low bit rate (15 kbps) and spread using a spreading factor of 256. The downlink DPCCH is spread using the same spreading factor as the DPDCH and so can be sent at either a high or low bit rate, depending upon the requirements of the DPDCH.

Figure 2.95 illustrates that the same channelisation code is used for both the in-phase and quadrature branches. This code re-use is important in the downlink direction because the code tree is shared between all connections using the same scrambling code. This corresponds to all connections within a cell when assuming a single scrambling code per cell. In the uplink direction, each UE has its own scrambling code and its own channelisation code tree. Uplink channelisation codes do not become a

Table 2.53 Bit rates and symbol rates for the downlink DPDCH and DPCCH

	12.2 kbps speech (slot format 8)	384 kbps data (slot format 15)
DPDCH bit rate prior to serial/parallel	51 kbps	192 kbps
DPCCH bit rate prior to serial/parallel	9 kbps	48 kbps
Total bit prior to serial/parallel	60 kbps	240 kbps
Symbol rate subsequent to serial/parallel	30 ksps	120 ksps

limiting factor unless multi-code transmission is supported. Channelisation codes can become a limiting factor for HSUPA when two spreading factor 2 codes plus two spreading factor 4 codes are used simultaneously. Re-using the same channelisation code for both the in-phase and quadrature branches does not result in a loss of orthogonality because the in-phase and quadrature branches are orthogonal themselves, i.e. the two branches are transparent to one another.

Channelisation code orthogonality can be achieved in the downlink direction because the set of radio links transmitted by a specific cell are synchronised. Radio links transmitted by neighbouring cells are not synchronised and so orthogonality is lost. In addition, neighbouring cells use different scrambling codes and scrambling codes are not orthogonal to one another. UE can benefit from high orthogonality when located in areas of good dominance and neighbouring cells have little impact upon the downlink interference floor. UE at cell edge are more likely to experience increased levels of non-orthogonal intercell interference. The orthogonality between connections belonging to a specific cell would be perfect if the radio propagation conditions were ideal, i.e. 100% orthogonality. In practice, radio propagation introduces delay spread components which have the effect of making the received signal appear partially unsynchronised. Delay spread components are generated when the signal transmitted by a Node B follows more than a single path to reach the UE. The chip rate of 3.84 Mcps means that the duration of a single chip is 0.26 µs. Multiplying by the speed of light results in an equivalent distance of 78 m. Receivers are typically capable of resolving delay spread components with a 0.5 chip time difference. This means that path differences of more than 39 m generate delay spread components which cause a reduction in orthogonality. Orthogonality is higher at locations which have line-of-sight propagation conditions. In general, these tend to be locations close to the cell while locations towards the cell edge are more likely to have increased numbers of delay spread components.

The scrambling process illustrated in Figure 2.95 is complex and results in rotations of the constellation. Similar to the uplink, the rotation can be either 45°, 135°, 225° or 315°. The transmit power allocated to the downlink in-phase and quadrature branches is equal and so the constellation is square rather than rectangular. The constellations before and after scrambling are illustrated in Figure 2.96.

A gain factor is applied to the complex signal after scrambling. This gain factor is influenced by inner loop power control and is used to determine the transmit power of the DPCH. Each DPCH has its own gain factor and so can be power controlled individually. This gain factor can also be used to control the transmit power offset between the DPDCH and DPCCH. Figure 2.97 illustrates these gain factors and the subsequent summing of individual downlink physical channels.

The downlink physical channels combined as part of the first summing stage include the majority of the common channels, i.e. CPICH, P-CCPCH, S-CCPCH, PICH and AICH. The primary and secondary synchronisation channels are also summed subsequent to multiplication by their gain factors. These channels are shown as using a separate summing stage because they do not make use of

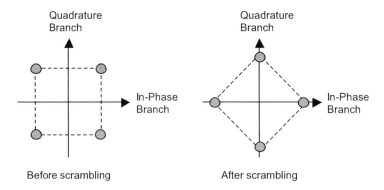

Figure 2.96   Constellation before and after scrambling (different DPDCH/DPCCH transmit powers)

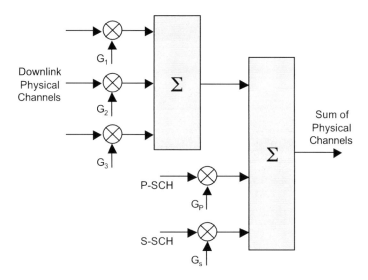

Figure 2.97   Combination of downlink physical channels

orthogonal channelisation codes from the code tree. The sum of the physical channels is passed to the modulator to map the resultant chip stream onto a carrier. The carrier can then be mixed to the appropriate RF frequency and transmitted across the air-interface. The uplink modulation process illustrated in Figure 2.93 is also applicable to the downlink.

2.6.3 Other Functions

The physical layer is responsible for providing inner loop power control functionality. This involves completing SIR measurements and subsequently generating Transmit Power Control (TPC) commands for transmission as part of the DPCCH. It also involves obeying received TPC commands.

The physical layer is also responsible for completing the measurements specified within 3GPP TS 25.215. For example, the UE may be instructed to complete CPICH measurements to support the cell selection, cell re-selection and soft handover processes. The Node B may be instructed to complete downlink transmit power and uplink received power measurements to support admission control and load control processes.

Air-interface synchronisation is achieved and maintained by the physical layer. This can be initial cell synchronisation when a UE is switched on. Alternatively, it could be synchronisation with neighbouring cells to complete cell reselections or handovers. It could also be dedicated or common channel synchronisation. The Physical layer generates the dedicated channel in-sync and out-of-sync primitives which are used by the higher layers to determine when synchronisation has been achieved or lost.

Some parts of the random access procedure are completed by the physical layer. The physical layer selects the signature used to mask the PRACH preambles. It also transmits the preambles and detects acknowledgements on the AICH. In addition, the physical layer ensures that the maximum number of preambles per preamble cycle is not exceeded.

In the case of an HSDPA connection, the Physical layer within the UE is responsible for generating the CQI values used to report the downlink channel conditions to the Node B. The use of CQI values is described in greater detail within Chapter 6. In the case of an HSUPA connection, the physical layer within the UE is responsible for detecting acknowledgements on the E-HICH physical channel, relative grants on the E-RGCH and absolute grants on the E-AGCH. The use of HSUPA acknowledgements and grants is described in greater detail within Chapter 7.

3

Channel Types

3.1 Logical Channels

- Logical channels are used to transfer data between the RLC and MAC layers. The MAC layer provides the mapping between logical channels and transport channels.
- Logical channels define the type of information being transferred. They are categorised as either control or traffic.
- Control channels include the BCCH, PCCH, CCCH, DCCH, MCCH and MSCH. These logical channels are responsible for transferring RRC signalling messages.
- Traffic channels include the CTCH, DTCH and MTCH. These logical channels are responsible for transferring application data. The DTCH is the only logical channel able to transfer application data in both the uplink and downlink directions.
- Key 3GPP specifications: TS 25.301 and TS 25.331

Logical channels are used to transfer data between the RLC and MAC layers. The logical channel defines the type of information being transferred. There are two categories of logical channel - control channels and traffic channels. Control channels are used to transfer control plane signalling whereas traffic channels are used to transfer user plane data. Table 3.1 presents the complete set of logical channels. All logical channels can be used in the downlink direction whereas only the CCCH, DCCH and DTCH logical channels can be used in the uplink direction.

The RRC protocol defines the control plane signalling messages transferred by the control channels. The majority of RRC messages have a one-to-one mapping between message type and logical channel type, e.g. the Active Set Update message is always sent using the DCCH logical channel and the RRC Connection Request message is always sent using the CCCH logical channel. There is a relatively small set of messages which can be sent on one of two logical channels. In these cases, the choice of logical channel depends upon the scenario. The relationship between RRC message type and logical channel is presented in Table 3.2.

The BCCH logical channel is responsible for transferring all downlink system information whereas the PCCH logical channel is responsible for transferring Paging Type 1 messages. In general, the CCCH logical channel is used to transfer control plane messages when the DCCH is not available. The information necessary for a UE to use the CCCH logical channel is broadcast as part of the system information on the BCCH. UE are able to read this information and use the CCCH before establishing an RRC connection. This is the reason why the RRC Connection Request, RRC Connection Setup and

Table 3.1 Logical channels

Logical channel	Type	Uplink	Downlink
Broadcast Control Channel (BCCH)	Control		√
Paging Control Channel (PCCH)	Control		√
Common Control Channel (CCCH)	Control	√	√
Dedicated Control Channel (DCCH)	Control	√	√
MBMS Control Channel (MCCH)	Control		√
MBMS Scheduling Channel (MSCH)	Control		√
Common Traffic Channel (CTCH)	Traffic		√
Dedicated Traffic Channel (DTCH)	Traffic	√	√
MBMS Traffic Channel (MTCH)	Traffic		√

RRC Connection Reject messages are always sent using the CCCH logical channel. DCCH logical channels are configured during the RRC connection establishment procedure and become available after entering RRC Connected mode. The CCCH logical channel must also be used in the RRC Connected mode states CELL_PCH and URA_PCH because the DCCH is not available in those states,

Table 3.2 RRC messages transferred by each logical channel

Logical channel	RRC message
BCCH	System Information, System Information Change Indication
PCCH	Paging Type 1
CCCH	RRC Connection Request, RRC Connection Setup, RRC Connection Reject, Cell Update, URA Update
DCCH	Active Set Update, Active Set Update Complete, Active Set Update Failure, Assistance Data Delivery, Cell Change Order From UTRAN, Cell Change Order from UTRAN Failure, Counter Check, Counter Check Response, Downlink Direct Transfer, Handover from UTRAN Command, Handover from UTRAN Failure, Handover to UTRAN Complete, Initial Direct Transfer, MBMS Modification Request, Measurement Control, Measurement Control Failure, Measurement Report, Paging Type 2, Physical Channel Reconfiguration, Physical Channel Reconfiguration Complete, Physical Channel Reconfiguration Failure, Radio Bearer Reconfiguration, Radio Bearer Reconfiguration Complete, Radio Bearer Reconfiguration Failure, Radio Bearer Release, Radio Bearer Release Complete, Radio Bearer Release Failure, Radio Bearer Setup, Radio Bearer Setup Complete, Radio Bearer Setup Failure, RRC Connection Release Complete, RRC Connection Setup Complete, RRC Status, Security Mode Command, Security Mode Complete, Security Mode Failure, Signalling Connection Release, Signalling Connection Release Indication, Transport Channel Reconfiguration, Transport Channel Reconfiguration Complete, Transport Channel Reconfiguration Failure, Transport Format Combination Control, Transport Format Combination Control Failure, UE Capability Enquiry, UE Capability Information, UE Capability Information Confirm, Uplink Direct Transfer, UTRAN Mobility Information, UTRAN Mobility Information Confirm, UTRAN Mobility Information Failure
CCCH or DCCH	RRC Connection Release, Cell Update Confirm, URA Update Confirm
MCCH	MBMS Access Information, MBMS Common P-T-M RB Information, MBMS Current Cell P-T-M RB Information, MBMS General Information, MBMS Neighbouring Cell P-T-M RB Information, MBMS Unmodified Services Information
MCCH or DCCH	MBMS Modified Services Information
MSCH	MBMS Scheduling Information

i.e. Cell Update and URA Update messages are sent using the CCCH logical channel. An exception to the rule of only using the CCCH logical channel when the DCCH is not available is when the Cell Update message is sent while a UE is in CELL_FACH. In this case, the UE can be configured with DCCH logical channels, but the Cell Update message is still sent on the CCCH logical channel.

The RRC Connection Release, Cell Update Confirm and URA Update Confirm messages can be sent using either the CCCH or DCCH logical channels. The RRC Connection Release message only uses the CCCH if a DCCH is not available. Likewise the choice of logical channel for the Cell Update Confirm and URA Update Confirm messages depends upon the scenario. A benefit of selecting the DCCH is that the RLC layer can provide ciphering to help maintain privacy whereas messages sent on the CCCH are not ciphered. Sending these messages on the CCCH also means that the UE identity has to be included as part of the RRC message. If the messages are sent using the DCCH and a common transport channel the UE identity is included as part of the MAC header. If the DCCH logical channel is used in combination with a dedicated transport channel then the UE identity is not necessary because the transport channel itself identifies the UE.

The DCCH logical channel is responsible for transferring the majority of RRC messages. These messages include those which are used to package higher layer NAS messages, i.e. the Initial Direct Transfer, Uplink Direct Transfer and Downlink Direct Transfer messages. A benefit of using the DCCH logical channel is that acknowledged mode RLC is supported. Acknowledged mode RLC increases the reliability and efficiency of transmission by allowing re-transmissions of individual PDU from the RLC layer. The CCCH logical channel is limited to using transparent mode RLC in the uplink and unacknowledged mode RLC in the downlink. Both of these RLC modes require re-transmissions of entire messages from the RRC layer when an error is detected.

The MCCH and MSCH are point-to-multi-point logical channels used by UE that support the Multimedia Broadcast/Multicast Service (MBMS). The majority of MBMS control plane information is sent on the MCCH. The MSCH is used to send scheduling information to indicate when specific MBMS services are transmitted. System information on the BCCH informs UE of the information necessary to read the MCCH and MSCH. Both logical channels can be read from either RRC Idle mode or RRC Connected mode. MBMS procedures can also make use of other logical channels and more general RRC messages, e.g. the MBMS counting procedure which allows the network to estimate the number of MBMS UE within a cell can use the Cell Update message and the CCCH or DCCH logical channels. The MBMS Modified Services Information message can be sent on either the MCCH or DCCH. This message is sent on the DCCH when the MBMS counting procedure determines that a point-to-point MBMS service connection should be established.

The DTCH represents the only traffic channel which can be used in both the uplink and downlink directions. The CTCH and MTCH are limited to the downlink direction, meaning that uplink user plane data is always sent using the DTCH. The vast majority of downlink user plane data also makes use of the DTCH. The DTCH provides a point-to-point connection dedicated to a single UE. The CTCH and MTCH provide downlink point-to-multi-point connections which can be used by a group of UE. The Cell Broadcast Service (CBS) makes use of the CTCH whereas MBMS services make use of the MTCH.

The MAC layer is responsible for mapping logical channels onto transport channels at the transmitter. Transport channels are then used to transfer data between the MAC and Physical layers. The set of transport channels available to each logical channel is presented in Table 3.3.

Some logical channels have a one-to-one relationship with a specific transport channel. Others have a choice of transport channels. The RRC layer and the radio resource management algorithms at layer 3 dictate which transport channel should be used at any point in time when there is a choice of transport channels, i.e. the MAC layer is responsible for completing the mapping under the instructions of the higher layers. There is some dependence upon the RRC state of the UE. For example, if a UE is in CELL_FACH then uplink DTCH and DCCH data has to be transferred using the RACH while downlink DTCH and DCCH data has to be transferred using the FACH. If a UE is in CELL_DCH then uplink DTCH and DCCH data can be transferred using either the DCH or E-DCH, while downlink DTCH and DCCH data can be transferred using either the DCH or HS-DSCH.

Table 3.3 Logical channels mapped onto transport channels

Logical channel	Transport channel
BCCH	BCH, FACH
PCCH	PCH
CCCH	FACH, RACH
DCCH	FACH, RACH, DCH, HS-DSCH, E-DCH
MCCH	FACH
MSCH	FACH
CTCH	FACH
DTCH	FACH, RACH, DCH, HS-DSCH, E-DCH
MTCH	FACH

Table 3.4 Logical channels extracted from transport channels

Transport channel	Logical channel
BCH	BCCH
PCH	PCCH
RACH	CCCH, DCCH, DTCH
FACH	BCCH, CCCH, DCCH, MCCH, MSCH, CTCH, DTCH, MTCH
DCH	DTCH, DCCH
HS-DSCH	DTCH, DCCH
E-DCH	DTCH, DCCH

The MAC layer is responsible for extracting logical channel data from the transport channels at the receiver. This represents the inverse of the mapping completed at the transmitter. The set of logical channels which can be extracted from each transport channel is presented in Table 3.4.

The receiver must be able to deduce which logical channel has been sent on a specific transport channel. In the case of the BCH and PCH there is only a single possibility, i.e. the BCH only transfers the BCCH and the PCH only transfers the PCCH. The remaining transport channels are able to transfer more than a single type of logical channel. In the case of the RACH and the FACH, the Target Channel Type Field (TCTF) can be included as part of the MAC header to indicate which logical channel type has been received at any point in time. In addition, if multiple instances of the same logical channel type are multiplexed onto the same transport channel the C/T MAC header field can be used to signal the logical channel identifier. A common example of this occurs when multiple signalling radio bearer DCCH are multiplexed onto a single transport channel.

3.2 Transport Channels

- Transport channels are used to transfer data between the MAC and Physical layers. The physical layer provides the mapping between transport channels and physical channels.
- Transport channels define how information is transferred. They are categorised as either common or dedicated. Common transport channels can be used by more than a single UE.
- Common channels include the BCH, PCH, RACH, FACH and HS-DSCH. These transport channels are mapped onto common physical channels.
- Dedicated channels include the DCH and E-DCH. These transport channels are mapped onto dedicated physical channels.
- All transport channels have associated Transport Formats (TF) and Transport Format Sets (TFS). A transport format defines the rate at which a specific transport channel can transfer data between

the MAC and physical layers. It also defines some of the processing completed by the physical layer. A transport format set is the collection of all transport formats for a specific transport channel.

- A transport format for the RACH, FACH, BCH, PCH and DCH transport channels is defined by a combination of semi-static and dynamic parameters. The semi-static parameters are TTI, channel coding, CRC size and rate matching attribute. The dynamic parameters are transport block size and transport block set size. Semi-static parameters are equal for all transport formats and cannot change unless reconfigured by the RRC layer. Dynamic parameters define the TF within the TFS and can be selected by the MAC layer.
- A transport format for the HS-DSCH transport channel is defined by a combination of static and dynamic parameters. The static parameters are fixed by 3GPP and include the TTI, channel coding and CRC size. The dynamic parameters can be selected by the MAC layer and include transport block size, transport block set size, redundancy version and modulation scheme.
- A transport format for the E-DCH transport channel is defined by a combination of static, semi-static and dynamic parameters. Static parameters include the channel coding and CRC size. The only semi-static parameter is the TTI. Dynamic parameters include transport block size, transport block set size and redundancy version.
- The bit rate at the top of the physical layer is defined by the TBS size and TTI. The TBS size is always a multiple of the TB size. The TBS size always equals the TB size for the HS-DSCH and E-DCH transport channels.
- A specific combination of transport formats across different transport format sets is known as a transport format combination. The collection of all allowed transport format combinations is known as the transport format combination set.
- Key 3GPP specifications: TS 25.211, TS 25.301 and TS 25.302

Transport channels are used to transfer data between the MAC and Physical layers. The transport channel defines how information is transferred. There are two categories of transport channel - common and dedicated. Common channels can be used by more than a single UE whereas dedicated channels can be used by only a single UE. Table 3.5 presents the complete set of transport channels. The BCH, PCH, FACH, DCH and HS-DSCH can be used in the downlink direction whereas the RACH, DCH and E-DCH can be used in the uplink direction. The DCH is the only transport channel which can be used in both the uplink and downlink directions.

In general, the common channels require in-band identification of the UE. The UE identification is included within the RRC message when the RACH or FACH transport channels are used by the CCCH logical channel. The UE identification is also included within the RRC message when the PCH transport channel is used by the PCCH. The UE identification is included within the MAC header when the RACH or FACH transport channels are used by either the DCCH or DTCH logical channels. In the case of the HS-DSCH transport channel which is transferred using the HS-PDSCH physical channel,

Table 3.5 Transport channels.

Transport channel	Type	Uplink	Downlink
Broadcast Channel (BCH)	Common		√
Paging Channel (PCH)	Common		√
Random Access Channel (RACH)	Common	√	
Forward Access Channel (FACH)	Common		√
High Speed Downlink Shared Channel (HS-DSCH)	Common		√
Dedicated Channel (DCH)	Dedicated	√	√
Enhanced Dedicated Channel (E-DCH)	Dedicated	√	

the UE identity is provided on a separate physical channel, i.e. the HS-SCCH physical channel indicates when HSDPA time slots and codes have been scheduled for a specific UE. Dedicated transport channels do not require in-band identification of the UE because each connection is identified by its physical resource allocation, i.e. by RF carrier, scrambling code and channelisation code.

The BCH transport channel is used to broadcast system information messages. The BCH is always mapped onto the P-CCPCH and is broadcast across the entire area of a cell. A UE accesses the BCH whenever it selects or re-selects a new cell. This could be in RRC Idle mode or CELL_FACH, CELL_PCH or URA_PCH.

The PCH transport channel is used to broadcast paging messages to UE which are in either RRC Idle mode or CELL_FACH, CELL_PCH or URA_PCH. The PCH transport channel is always mapped onto a S-CCPCH physical channel. The MAC layer may multiplex the PCH and FACH onto a shared S-CCPCH, or alternatively onto dedicated S-CCPCH. Similar to the BCH, the PCH must be broadcast across the entire area of a cell. The PCH is used in combination with the PICH physical channel. The PICH physical channel is used to inform the set of UE belonging to a specific paging group that a PCH message belonging to that paging group will be transmitted during the next radio frame.

The RACH transport channel is used when establishing a connection from RRC Idle mode and when signalling or transferring relatively small quantities of user plane data while in CELL_FACH. A single RACH transport channel is used for both control plane signalling and user plane data. The RACH is always mapped onto the PRACH physical channel which makes use of a contention based access procedure, i.e. there is a probability that collisions occur when multiple UE attempt to make use of the PRACH.

The FACH transport channel is the downlink equivalent of the RACH. It is also used when establishing a connection from RRC Idle mode and when signalling or transferring relatively small quantities of user plane data while in CELL_FACH. The FACH transport channel is always mapped onto the S-CCPCH physical channel. Separate FACH transport channels are configured for different purposes. The FACH-c is configured to transfer control plane signalling, whereas the FACH-u is configured to transfer user plane data. The FACH-s is used for cell broadcast services. Transmissions in the downlink direction are point-to-multi-point rather than multi-point-to-point. This allows transmissions on the S-CCPCH to be scheduled and it is not necessary to use a contention based access procedure. The MAC layer is responsible for scheduling each FACH on the S-CCPCH.

The HS-DSCH transport channel is used for HSDPA. Release 6 and earlier versions of the 3GPP specifications limit the use of the HS-DSCH to UE in CELL_DCH. The release 7 version of the 3GPP specifications introduces the possibility of using the HS-DSCH in CELL_FACH. There is a single HS-DSCH within each cell that supports HSDPA, i.e. the HS-DSCH is time shared between all active HSDPA connections. The HS-DSCH is always mapped onto one or more HS-PDSCH. The HS-DSCH can be used to transfer both control plane signalling and user plane data. If the HS-DSCH is only used for user plane data then a DCH may be used in parallel to transfer control plane signalling.

The DCH transport channel is used in both the uplink and downlink directions. The DCH can only be used when a UE is in CELL_DCH. DCH transport channels can be used to transfer both control plane signalling and user plane data. The MAC layer can multiplex the set of Signalling Radio Bearers (SRB) onto a single DCH. If separate DCH are used for control plane signalling and user plane data, the physical layer is responsible for multiplexing the transport channels onto the same physical channel. Alternatively, the MAC layer could multiplex both control plane and user plane logical channels onto a single DCH. The DCH transport channel is always mapped onto a DPDCH physical channel. The DCH is able to transfer greater quantities of traffic and at higher bit rates than the RACH and FACH.

The E-DCH transport channel is used for HSUPA. The E-DCH can only be used when a UE is in CELL_DCH. Each UE that has an active HSUPA connection has a single E-DCH transport channel. The E-DCH is always mapped to one or more E-DPDCH. The E-DCH can be used to transfer both control plane signalling and user plane data. The MAC-e/es layer within the UE is able to multiplex both DTCH and DCCH logical channels onto a single E-DCH. Logical channel identifiers are included

Table 3.6 Transport channels mapped onto physical channels

Transport channel	Physical channel
RACH	Physical Random Access Channel (PRACH)
FACH	Secondary Common Control Physical Channel (S-CCPCH)
BCH	Primary Common Control Physical Channel (P-CCPCH)
PCH	Secondary Common Control Physical Channel (S-CCPCH)
HS-DSCH	High Speed Physical Downlink Shared Channel (HS-PDSCH)
DCH	Dedicated Physical Data Channel (DPDCH)
E-DCH	Enhanced Dedicated Physical Data Channel (E-DPDCH)

as part of the MAC-e header. If the E-DCH is only used for user plane data then a DCH may be used in parallel for control plane signalling.

The physical layer is responsible for mapping transport channels onto physical channels at the transmitter. Physical channels are then used to transfer the data across the air-interface. The set of physical channels available to each transport channel is presented in Table 3.6. With the exception of the S-CCPCH, there is a one-to-one relationship between transport channel and physical channel. The S-CCPCH is able to transfer both the FACH and PCH transport channels.

All transport channels have associated Transport Formats (TF) and Transport Format Sets (TFS). A transport format defines the rate at which a specific transport channel can transfer data between the MAC and Physical layers. It also defines some of the processing completed by the Physical layer. A transport format set is the collection of all transport formats for a specific transport channel. The concepts of transport format and transport format set are illustrated in Figure 3.1.

This example is based upon the 128 kbps PS data service using a DCH transport channel. The DCH has four transport formats within the transport format set. Each transport format is associated with a different bit rate. The MAC layer is responsible for selecting the appropriate transport format at any point in time.

The parameters used to define a transport format for the RACH, FACH, BCH, PCH and DCH are presented in Table 3.7. These parameters are categorised as either semi-static or dynamic. Semi-static parameters are equal for all transport formats within the transport format set. These parameters remain fixed unless they are reconfigured by the RRC layer. Dynamic parameters can be different for each transport format within the transport format set. These parameters are selected by the MAC layer and are used to achieve the different bit rates associated with each transport format.

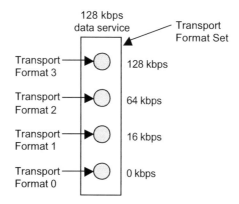

Figure 3.1 Concepts of transport format and transport format set

Table 3.7 Transport format parameters for the RACH, FACH, BCH, PCH and DCH

Semi-static	Transmission Time Interval (TTI)
	Channel coding
	CRC size
	Rate matching attribute
Dynamic	Transport Block (TB) size
	Transport Block Set (TBS) size

A transport block (TB) is equivalent to a MAC PDU and represents the basic unit of data transferred between the MAC and Physical layers. A TB includes the RLC and MAC headers. The number of bits within a TB is defined by the TB size. A transport block set (TBS) is a collection of one or more TB. The number of bits within a TBS is defined by the TBS size. The TBS size is always a multiple of the TB size. The Transmission Time Interval (TTI) defines the inter-arrival time of TBS passed between the MAC and Physical layers. It also defines the rate at which TBS can be transferred across the air-interface.

The combination of the TBS size and TTI defines the bit rate at the top of the physical layer. For example, if the TBS size is 4032 bits and the TTI is 10 ms then the bit rate at the top of the physical layer is 403.2 kbps. A TBS size of 4032 bits could be obtained using 12 transport blocks with a TB size of 336 bits. These figures correspond to the 384 kbps PS data service.

Each TF within the TFS has a different configuration for the combination of TB size and TBS size. Either, or both of these parameters can be changed to define a different transport format. The MAC layer is able to change the transport format and consequently change the bit rate once every TTI. It is common for the TB size to remain constant while the TBS size changes, e.g. the number of TB included for a 64 kbps TF would be half the number of TB included for a 128 kbps TF.

In the downlink direction, the TF selected by the RNC MAC layer is signalled to the Node B using the header of the Frame Protocol data frame (TFI field described in Section 2.5). The TF can be signalled across the air-interface using the TFCI field within the DPCCH, or the receiving UE can be left to complete blind detection of which TF has been used within each TTI. The latter avoids the requirement to transmit TFCI bits on the DPCCH. In the uplink direction, the TF selected by the UE MAC layer is signalled to the UE Physical layer using internal communication between the layers. Blind detection of the TF is not supported in the uplink direction and so TFCI bits are always transmitted on the DPCCH to inform the Node B receiver of which TF is used during each TTI.

3GPP TS 25.212 specifies that the channel coding completed by the Physical layer can be either convolution coding or turbo coding. Convolution coding rates of 1/3 and 1/2 are specified, whereas a turbo coding rate of 1/3 is specified. The CRC size is specified to be either 0, 8, 12, 16 or 20 bits. Rate matching attributes are required when multiple transport channels are multiplexed onto the same physical channel. Rate matching attributes define the share of the DPDCH bits which are allocated to each transport channel. Rate matching is described in greater detail within Section 2.6. Table 3.8 presents the TTI, channel coding, CRC size, TB size and TBS size configurations allowed for each transport channel type (excluding the HS-DSCH and E-DCH).

The configuration of the BCH transport channel is fully specified and there is only a single TF within the TFS. All UE have knowledge of this TF and so are able to extract BCH transport blocks without any further information. The configurations used by the RACH, FACH and PCH are broadcast on SIB 5 and optionally on SIB 6. These transport channels are specified to use 1/2 rate convolutional coding. The range of values specified for the TB size and the TBS size indicates that the TBS can include up to 40 TB. The DCH transport channel is fully configurable and the RRC layer is able to select from the full set of parameters.

When multiple transport channels are multiplexed onto the same physical channel there is a requirement to identify the combination of transport formats selected from each transport format set. It may also be necessary to avoid certain combinations, e.g. to avoid exceeding a maximum bit rate capability. A specific combination of transport formats across different transport format sets is known as a

Table 3.8 Transport format parameters for the RACH, FACH, BCH, PCH and DCH

Transport channel	TTI	Channel coding	CRC size	TB size	TBS size
RACH	10, 20 ms	1/2 rate Conv.	0, 8, 12, 16, 24	0 to 5000 bits	0 to 200 000 bits
FACH	10, 20, 40, 80 ms	1/2 rate Conv. 1/3 rate Conv. 1/3 rate Turbo	0, 8, 12, 16, 24	0 to 5000 bits	0 to 200 000 bits
BCH	20 ms	1/2 rate Conv.	16	246 bits	246 bits
PCH	10 ms	1/2 rate Conv.	0, 8, 12, 16, 24	0 to 5000 bits	0 to 200 000 bits
DCH	10, 20, 40, 80 ms	1/2 rate Conv. 1/3 rate Conv. 1/3 rate Turbo	0, 8, 12, 16, 24	0 to 5000 bits	0 to 200 000 bits

Transport Format Combination (TFC). The collection of all allowed TFC is known as the Transport Format Combination Set (TFCS). The concepts of TFC and TFCS are illustrated in Figure 3.2.

In this example, there are two transport channels multiplexed onto the same physical channel. The first transport channel has four transport formats, whereas the second transport channel has two transport formats. The TFCS includes two TFC. In practice, the TFCS would include more than two combinations otherwise it would not be possible to use the lower bit rates within the first transport format set. The RRC layer within the RNC is responsible for configuring the TFCS. The MAC layer is able to select an appropriate TFC at any point in time. Each transport channel can be configured to use a different set of transport format characteristics. For example, transport channel 1 could use a 20 ms TTI while transport channel 2 could use a 40 ms TTI. This means that the physical layer would receive the transport block set for transport channel 1 once every 20 ms and the transport block set for transport channel 2 once every 40 ms. There is a one-to-one correspondence between the DPCCH TFCI value and a certain TFC. The TFCI is used to inform the receiver of the current TFC.

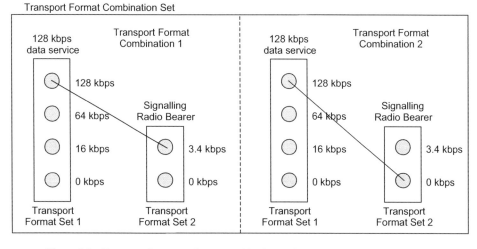

Figure 3.2 Concepts of transport format combination and transport format combination set

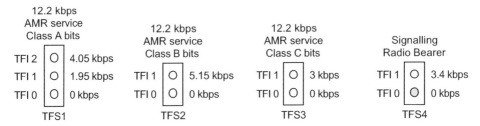

Figure 3.3 Transport format sets used by the 12.2 kbps AMR speech service

Figure 3.3 presents the transport format sets used by the 12.2 kbps AMR speech service. The user plane protocol stack requires three transport channels to transfer the class A, B and C bit streams generated by the AMR codec. The transport format set for the class A bits includes three transport formats whereas the transport format sets for the class B and C bits include two transport formats. The intermediate transport format belong to the class A transport channel is used for transferring comfort noise parameters during periods of speech DTX. The control plane protocol stack requires one transport channel. This transport channel accommodates multiple signalling radio bearers which have already been multiplexed into a single data stream by the MAC layer.

Figure 3.4 presents the transport format combination set used by the 12.2 kbps AMR speech service. There is a total of six transport format combinations within the transport format combination set. The first combination is applied when there is neither user plane data nor signalling to transfer and all transport format sets are assigned 0 kbps. The second combination is applied when there is comfort noise data to transfer from the AMR codec, but there is no speech nor any control plane signalling. The third combination is applied when there is speech data to transfer, but no control plane signalling. The remaining combinations follow the same pattern, but with control plane signalling included.

Figure 3.5 presents an example pair of transport format sets used by the 128 kbps PS data service. The user plane protocol stack requires one transport channel to transfer the application data. The associated transport format set includes four transport formats. Similar to the AMR speech connection, the control plane protocol stack requires a single transport channel.

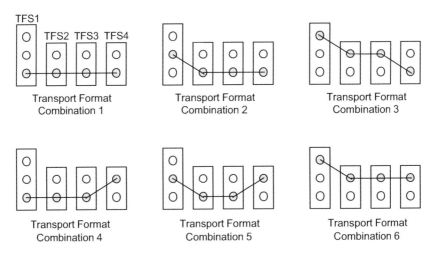

Figure 3.4 Transport format combination set used by the 12.2 kbps AMR speech service

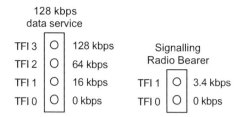

Figure 3.5 Transport format sets used by the 128 kbps data service

Figure 3.6 presents an example transport format combination set used by the 128 kbps PS data service.

The parameters used to define an HS-DSCH transport format are presented in Table 3.9. These parameters are categorised as either static or dynamic. The HS-DSCH does not use semi-static parameters. Static parameters are fully specified by 3GPP and are not configurable. Dynamic parameters are selected by the MAC-hs layer of the Node B, but must remain within the limits defined by the RRC layer.

The HS-DSCH TBS size is always equal to the TB size and only a single transport block is transferred to between the MAC and physical layers during each TTI. An appropriate TB size is selected by the Node B every time a UE is scheduled. The Node B also selects the appropriate redundancy version and modulation scheme. The dynamic parameters selected by the Node B are signalled to the receiving UE using the HS-SCCH. These parameters are always signalled and blind detection is not used for the HS-DSCH.

The set of parameters used to define an E-DCH transport format is presented in Table 3.10. An E-DCH transport format includes static, semi-static and dynamic parameters. The static parameters are fully specified by 3GPP and are not configurable. The TTI represents the only semi-static parameter for the E-DCH. This parameter is configured by the RNC RRC layer and subsequently signalled to the UE RRC layer, e.g. using a Radio Bearer Setup or Radio Bearer Reconfiguration message. The RRC layers of the RNC and UE are then responsible for informing the lower layers.

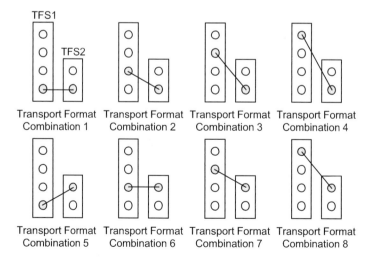

Figure 3.6 Transport format combination set used by the 128 kbps data service

Table 3.9   Transport format parameters for the HS-DSCH

Static	Transmission Time Interval (TTI)	2 ms
	Channel coding	1/3 rate Turbo
	CRC size	24
Dynamic	Transport Block (TB) size	1 to 200 000 bits
	Transport Block Set (TBS) size	1 to 200 000 bits
	Redundancy Version (RV)	1 to 8
	Modulation Scheme	QPSK, 16QAM

Table 3.10   Transport format parameters for the E-DCH

Static	Channel coding	1/3 rate Turbo
	CRC size	24
Semi-Static	Transmission Time Interval (TTI)	2, 10 ms
Dynamic	Transport Block (TB) size	1 to 200 000 bits
	Transport Block Set (TBS) size	1 to 200 000 bits
	Redundancy Version (RV)	0 to 3

The dynamic parameters are selected by the MAC-e/es layer of the UE, but must remain within the limits defined by the RRC layer. Similar to the transport format for the HS-DSCH, the TBS size is always equal to the TB size and only a single transport block is transferred between the MAC and Physical layers during each TTI. The dynamic parameters selected by the UE are signalled to the receiving Node B using the E-DPCCH. These parameters are always signalled and blind detection is not used for the E-DCH.

3.3 Physical Channels

- Physical channels are used to transfer data across the air-interface. They are categorised as either common or dedicated. Common channels can be used by more than a single UE.
- Common channels include the SCH, CPICH, P-CCPCH, S-CCPCH, PICH, AICH, PRACH, HS-PDSCH, HS-SCCH, E-AGCH and MICH.
- Dedicated channels include the DPDCH, DPCCH, F-DPCH, HS-DPCCH, E-DPDCH, E-DPCCH, E-RGCH and E-HICH.
- Every cell in the network must be capable of transmitting the CPICH, SCH, P-CCPCH, S-CCPCH, PICH, AICH and PRACH. The DPDCH and DPCCH are used for DPCH connections. The HS-PDSCH, HS-SCCH and HS-DPCCH are used for HSDPA. The E-DPDCH, E-DPCCH, E-AGCH, E-RGCH and E-HICH are used for HSUPA. The MICH is used for MBMS and the F-DPCH can replace the downlink DPCCH.
- The Primary CPICH allows the UE to identify the primary scrambling code belonging to a cell. CPICH measurements are used for cell selection and cell reselection. Soft handover is also based upon CPICH measurements. Likewise, inter-system and inter-frequency handover can be based upon CPICH measurements. CPICH RSCP can be measured directly whereas CPICH Ec/Io and path loss are calculated. CPICH Ec/Io measurements are based upon RSSI and exclude the benefit of downlink orthogonality.
- The Primary and Secondary SCH are used during initial cell synchronisation. The Primary SCH is a sequence of 256 chips. It is repeated at the start of every time slot and is used to achieve slot synchronisation. The Secondary SCH is a series of 15 sequences of 256 chips which repeat every radio frame. Identifying three consecutive chip sequences allows the UE to deduce radio frame timing and the primary scrambling code group.

- The P-CCPCH is responsible for broadcasting system information messages using the BCH transport channel and BCCH logical channel. The spreading factor is fixed at 256.
- The S-CCPCH is used to transfer the PCH and FACH transport channels. It can be used for SAB and MBMS services. Multiple S-CCPCH can be configured to increase the PCH and FACH capacity. The spreading factor can range from 4 to 256.
- The PICH is used to broadcast paging indicators which trigger groups of UE to decode subsequent paging messages. A DRX cycle length is used to avoid the requirement for continuous reception. Configuring a large number of paging indicators per radio frame decreases the length of each indicator and potentially reduces its reliability. It also decreases the number of UE within each paging group so UE receive positive indicators less frequently. The IMSI determines which paging indicator is received by each UE.
- The MICH is used to broadcast MBMS notification indicators which trigger groups of UE to decode the MCCH during the subsequent modification period. A DRX cycle is used to avoid the requirement for continuous reception. UE can receive the MICH at any time during the MBMS modification period. One or more MBMS services can be associated with each notification indicator.
- The AICH is used to acknowledge the reception of a PRACH preamble. A negative acknowledgement indicates that the Node B has received the preamble, but a random access collision has occurred. A positive acknowledgement indicates that the UE can proceed to transmit the PRACH message. The AICH transmission timing can be configured with a value of 1 to relax the delay requirements for larger cell ranges.
- The PRACH is separated into a preamble part and a message part. Preambles are used to gain access to a PRACH message time slot and to ensure that the message is transmitted with sufficient uplink power. A PRACH physical channel is defined by its scrambling code. 16 orthogonal preamble signatures can be used to allow multiple UE to transmit during the same access slot. Subchannels define which access slots can be used by each UE. UE identify their subchannels using their ASC. This is derived from the AC when in RRC Idle mode and from the MAC Logical Channel Priorities when in RRC Connected mode. The PRACH message can use a 10 or 20 ms TTI and a spreading factor between 32 and 256. The preamble signature determines the channelisation code index.
- The DPDCH provides the main data transfer capability when HSDPA and HSUPA are not used. The DPDCH is always accompanied by a DPCCH. These physical channels can only be used in CELL_DCH. They support both soft handover and inner loop power control. The uplink DPDCH supports spreading factors between 4 and 256 whereas the downlink supports spreading factors between 4 and 512. The uplink DPCCH always uses spreading factor 256 whereas the downlink uses the same spreading factor as the DPDCH. The scrambling code is used to identify a connection in the uplink whereas the channelisation code is used in the downlink.
- The F-DPCH represents a special case of the downlink DPCCH which helps to conserve channelisation codes by time multiplexing the TPC commands belonging to a maximum of 10 UE. The F-DPCH can be used when HSDPA replaces the DPDCH.
- Key 3GPP specifications: TS 25.211 and TS 25.213

Physical channels are used to transfer data across the air-interface. There are two categories of physical channel - common and dedicated, i.e. the same categories as used for transport channels. Common channels can be used by more than a single UE whereas dedicated channels can be used by only a single UE. Table 3.11 presents the set of physical channels.

The physical channels shown in shaded rows are used to transfer transport channels. The remaining physical channels only exist at the physical layer. There is a total of 14 downlink physical channels and 7 uplink physical channels. The DPDCH and DPCCH are the only physical channels which can be used

Table 3.11 Physical channels

Physical channel	Type	Uplink	Downlink
Synchronisation Channel (SCH)	Common		√
Common Pilot Channel (CPICH)	Common		√
Primary Common Control Physical Channel (P-CCPCH)	Common		√
Secondary Common Control Physical Channel (S-CCPCH)	Common		√
Paging Indication Channel (PICH)	Common		√
Access Indication Channel (AICH)	Common		√
High Speed Physical Downlink Shared Channel (HS-PDSCH)	Common		√
E-DCH Absolute Grant Channel (E-AGCH)	Common		√
MBMS Indicator Channel (MICH)	Common		√
Physical Random Access Channel (PRACH)	Common	√	
High Speed Shared Control Channel (HS-SCCH)	Common	√	
Dedicated Physical Data Channel (DPDCH)	Dedicated	√	√
Dedicated Physical Control Channel (DPCCH)	Dedicated	√	√
Fractional Dedicated Physical Channel (F-DPCH)	Dedicated		√
E-DCH Relative Grant Channel (E-RGCH)	Dedicated		√
E-DCH Hybrid ARQ Indicator Channel (E-HICH)	Dedicated		√
High Speed Dedicated Physical Control Channel (HS-DPCCH)	Dedicated	√	
E-DCH Dedicated Physical Data Channel (E-DPDCH)	Dedicated	√	
E-DCH Dedicated Physical Control Channel (E-DPCCH)	Dedicated	√	

in both the uplink and downlink directions. The set of physical channels directly associated with HSDPA, i.e. the HS-PDSCH, HS-SCCH and HS-DPCCH are described within Chapter 6. The set of physical channels directly associated with HSUPA, i.e. the E-DPDCH, E-DPCCH, E-HICH, E-RGCH and E-AGCH are described within Chapter 7.

3.3.1 Common Pilot Channel (CPICH)

There are two types of Common Pilot Channel (CPICH): Primary CPICH and Secondary CPICH. Both CPICH only exist at the physical layer and neither is used to transfer higher layer information. The Primary CPICH is a mandatory downlink common channel broadcast from every cell in the network. 3GPP TS 25.213 specifies that the Primary CPICH is always spread using channelisation code (SF256,0) and always scrambled using the primary scrambling code assigned to the cell. Channelisation code (SF256,0) is a series of 256 consecutive 1s and so spreading does not have any impact upon the original bit sequence. In addition, when transmit diversity is not used the original bit sequence is a continuous stream of 0s. These 0s map to +1 during the modulation process and so after scrambling the transmitted signal is effectively just the primary scrambling code. If two-branch open or closed loop transmit diversity is used, the bit sequence applied to the first antenna element is the same as when transmit diversity is not used but the bit sequence applied to the second antenna element is a continuous repetition of four 1s followed by four 0s. The bit rate is 30 kbps which generates a modulation symbol rate of 15 ksps, i.e. 2 bits per modulation symbol. The spreading factor of 256 increases the symbol rate to 3.84 Mcps. Only one Primary CPICH is broadcast by a cell and its transmit power should be sufficient to provide a high probability of reception across the entire target coverage area. The primary CPICH can be used as a phase reference for the SCH, P-CCPCH, S-CCPCH, DPCH, PICH, AICH and MICH. This means that it can be used as a reference for timing and propagation channel estimation.

The Secondary CPICH is an optional downlink common channel which can be introduced for the purposes of beamforming. In contrast to the Primary CPICH, the Secondary CPICH does not have to be broadcast across the entire area of a cell. Adopting the fixed beam approach to beamforming divides the cell into a number of segments. Relatively narrow fixed beams are then used to provide coverage across those segments. For example, a cell belonging to a three-sector site typically provides 120° of coverage which could be divided between four fixed beams of 30°. The signals belonging to individual connections are then transmitted and received over a relatively narrow beamwidth. This helps to reduce the level of interference in the system, increasing both service coverage and system capacity. The Primary CPICH cannot be used to estimate the propagation channel for signals transmitted by an individual fixed beam because a different antenna gain pattern has been used. The Primary CPICH always has to use an antenna gain pattern which will provide coverage across the full range of azimuths. A set of Secondary CPICH can be assigned to the cell allowing each fixed beam to transmit its own CPICH. These Secondary CPICH can then be used as a phase reference for individual fixed beams. 3GPP TS 25.213 specifies that the spreading factor used for a Secondary CPICH is always 256 but the channelisation code can be chosen by the RNC. It also specifies that a Secondary CPICH can be scrambled using either the primary scrambling code assigned to the cell or one of the 15 secondary scrambling codes associated with the primary scrambling code.

The timing of the CPICH is illustrated in Figure 3.7. The CPICH is transmitted with constant power, i.e. without power control, and has a 100% activity factor. The CPICH transmission is divided into 10 ms radio frames. Each radio frame is equivalent to the duration of a single downlink scrambling code. A scrambling code has a length of 38400 chips which occupies 10 ms after accounting for the WCDMA chip rate of 3.84 Mcps. This means that the Primary CPICH repeats the primary scrambling code once per 10 ms radio frame.

The Primary CPICH and any Secondary CPICH belonging to the same cell have the same timing. CPICH belonging to different cells of the same Node B have a constant, but configurable, time offset. Each Node B has its own time reference used to define the Node B Frame Number (BFN). The BFN has a range from 0 to 4095 and so cycles once every 40.96 s. The start of a BFN frame is used as a reference for the start of each CPICH frame. CPICH frames are numbered using the Cell System Frame Number (SFN) which also has a range from 0 to 4095. The time offset between the start of a BFN frame and the start of a CPICH frame is defined by the T_Cell parameter. The RNC sends this parameter to the Node B when a cell is switched on using the NBAP Cell Setup Request message. The signalled value of T_Cell can range from 0 to 9 whereas the actual value is equal to the signalled value multiplied by 256 chips. This means that the time offset between the BFN and the SFN can range from 0 to 2304 chips (0 to 0.6 ms). The relevance of 256 chips and the range from 0 to 9 is explained in Section 3.3.2 which addresses the Synchronisation Channels (SCH). It is common practice to assign T_Cell with values of 0, 256 and 512 chips for the first, second and third cells belonging to a Node B. If a Node B has more than three cells on the same RF carrier then each additional cell can be assigned a T_Cell value incremented by 256 chips. Cells belonging to a different RF carrier can re-use the same T_Cell values. These time offsets for a Node B with 3 cells are presented in Figure 3.8.

Different Node B are not synchronised with one another and so there is no fixed timing relationship between their CPICH. The time difference observed by a UE is reported as part of the soft handover procedure. This helps new radio links to be synchronised with existing radio links when they are added to the active set.

Many of the measurement reporting events specified within 3GPP TS 25.331 are based upon CPICH measurements. Intra-frequency reporting events 1a, 1b and 1c make use of CPICH measurements for radio link addition, deletion and replacement. Event 1d indicates a change of the best cell based upon

Figure 3.7   Radio frame timing for the CPICH

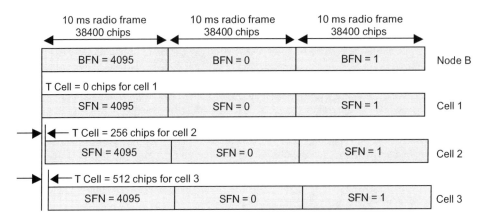

Figure 3.8   Example time offsets between the BFN and SFN

CPICH measurements and events 1e and 1f indicate when CPICH measurements become better than or worse than an absolute threshold. Inter-frequency reporting events 2a to 2f also make use of CPICH measurements. These reporting events may be used to trigger inter-system and inter-frequency handover as well as soft handover. In addition, CPICH measurements are used for initial cell selection and cell reselection.

The three most important measurements obtained from the CPICH are CPICH Ec/Io, CPICH Received Signal Code Power (RSCP) and path loss. CPICH RSCP is a direct measurement whereas the CPICH Ec/Io and path loss are calculated from other measurements. The CPICH Ec/Io is defined as the ratio between the received energy per chip and the received noise spectral density. This can be expressed using the equation below. Assuming the receiver bandwidth numerically equals the chip rate allows the equation to be simplified into its most common form, i.e. CPICH Ec/Io is calculated by dividing a CPICH RSCP measurement by a Received Signal Strength Indicator (RSSI) measurement. The equation below assumes linear rather than logarithmic units.

$$\text{CPICH Ec/Io} = \frac{\text{CPICH RSCP}}{\text{Chip rate}} \times \frac{\text{Bandwidth}}{\text{RSSI}} = \frac{\text{CPICH RSCP}}{\text{RSSI}}$$

CPICH Ec/Io tends to be more accurate than either CPICH RSCP or path loss because it is calculated as a ratio of two measurements. In general there is some correlation between the measurement error in the CPICH RSCP and the measurement error in the RSSI. The correlated parts of these measurement errors cancel when the CPICH Ec/Io is calculated.

The path loss is defined as the ratio between the CPICH transmit power and the CPICH RSCP. This can be expressed in linear units using the equation below. The same equation can be expressed in logarithmic units by subtracting the CPICH RSCP from the CPICH transmit power.

$$\text{Path loss} = \frac{\text{CPICH transmit power}}{\text{CPICH RSCP}}$$

The path loss equation defines the total loss between the UE antenna connector and the CPICH transmit power measurement reference point. In this context the path loss result includes more than just the air-interface propagation loss. The result is likely to include both UE and Node B antenna gains and may also include the downlink feeder loss and any Mast Header Amplifier (MHA) insertion loss. A CPICH transmit power value is broadcast on the P-CCPCH as part of System Information Block (SIB) 5. The measurement reference point of the CPICH transmit power should be the same as that for the uplink RSSI value broadcast in SIB 7. These two values are used in combination during the uplink open loop power control procedures. Inaccuracies will be introduced if different measurement reference points are used. If the Node B has not been configured with MHA then the measurement reference point should be at the Node B antenna connector. In this case, the value of CPICH transmit power broadcast in SIB 5 typically equals the RNC databuild parameter which has been used to configure it. If the Node B has been configured with MHA then the measurement reference point should be at the air-interface side of the MHA. This means that the CPICH transmit power value is reduced by the feeder loss and the MHA insertion loss, e.g. if the CPICH transmit power has been configured as 33 dBm at the Node B antenna connector, and the feeder loss is 2 dB while the MHA insertion loss is 0.5 dB then the CPICH transmit power broadcast in SIB 5 would be 30.5 dBm.

CPICH RSCP and path loss both provide a measure of absolute coverage, i.e. they quantify whether or not the absolute received power at a specific location is sufficient. CPICH Ec/Io provides a measure of relative coverage in terms of a signal to noise ratio. CPICH Ec/Io is often referred to as a measure of signal quality. In practice, both the CPICH RSCP and CPICH Ec/Io have to be sufficient for a UE to reliably establish and complete its connections. The cell selection procedure which determines whether or not a UE is able to camp upon a cell typically uses a CPICH RSCP threshold of -115 dBm and a CPICH Ec/Io threshold of -18 dB. These cell selection criteria are broadcast on the P-CCPCH as part of SIB 3. The CPICH Ec/Io threshold for cell selection is known as Qqualmin whereas the CPICH RSCP threshold used for cell selection is known as Qrxlevmin. The release 99 and release 4 versions of the 3GPP specifications limit the minimum value of Qrxlevmin to -115 dBm. The release 5 and newer versions of the 3GPP specifications allow Qrxlevmin to be decreased as far as -119 dBm. CPICH RSCP and path loss measurements are often compared with thresholds calculated from link budgets. Link budget analysis can be used to estimate the maximum allowed path loss or the minimum required CPICH RSCP. These values can then be used to define thresholds which trigger inter-system or inter-frequency handover. They can also be used to define thresholds for the radio network planning process and for initial RF optimisation.

CPICH Ec/Io measurements do not provide a direct indication of downlink coverage for the DPCH nor the HS-PDSCH. It is possible to experience the same CPICH Ec/Io at two locations and experience different DPCH and HS-PDSCH performance. CPICH Ec/Io measurements do not account for downlink orthogonality whereas the reception of DPCH and HS-PDSCH benefit from orthogonality. A UE could be close a heavily loaded cell and experience a relatively poor CPICH Ec/Io because the downlink RSSI is elevated by the downlink load. In this case the downlink DPCH and HS-PDSCH performance would be relatively good because the interference is dominated by orthogonal own cell power. Alternatively, the UE could experience the same CPICH Ec/Io towards the edge of an unloaded cell. In this case the interference floor is elevated by other cell power and the downlink DPCH and HS-PDSCH performance would be relatively poor because inter-cell interference is non-orthogonal.

UE also make use of CPICH Ec/Io measurements when generating Channel Quality Indicators (CQI) for HSDPA. In this case the UE is responsible for accounting for downlink orthogonality prior to generating the CQI value. If the benefit of orthogonality is ignored then the CQI values generated will be relatively low and the HSDPA bit rates assigned by the Node B are also likely to be relatively low. Differences between the CPICH transmit power and the HS-PDSCH transmit power also have to be taken into account when generating CQI values and assigning HSDPA bit rates.

It is common to assign 10% of the total downlink transmit power to the CPICH and a further 10% to the other common channels. These figures determine the maximum CPICH Ec/Io which can be achieved when the serving cell is unloaded. An accurate calculation must account for the activity factors associated with each of the common channels, but if it is assumed that the average total common channel transmit power sums to 20% whereas the CPICH has been assigned 10% then the maximum achievable CPICH Ec/Io is -3 dB. This figure is applicable to a UE which is close to the serving cell. The figure decreases as the UE moves towards cell edge and levels of intercell interference increase. Assigning 10% of the total downlink transmit power to the CPICH means that when the cell is fully loaded and transmitting at maximum power the CPICH Ec/Io can decrease to −10 dB when the UE is close to its serving cell. This figure would also decrease as the UE moves towards cell edge. During network deployment and initial RF optimisation it is common to target a minimum CPICH Ec/ Io of −9 dB. This figure allows the total downlink transmit power to increase by a factor of 4, i.e. 6 dB and still maintain an acceptable CPICH Ec/Io. Assigning less than 10% of the total downlink transmit power to the CPICH would mean that the CPICH Ec/Io can become relatively poor once the cell and its neighbours become loaded.

In general, neighbouring cells with the same physical configuration should be assigned equal CPICH transmit powers, e.g. two neighbouring macrocells with MHA and similar feeder lengths should be assigned equal CPICH transmit powers. Soft handover imbalances can be introduced if neighbouring cells are assigned different CPICH transmit powers. If the physical configuration of neighbouring cells is different then there can be justification for assigning different CPICH transmit powers, e.g. if neighbouring cells have MHA, but one cell has a significantly greater feeder loss then the CPICH transmit power can be increased to compensate. If neighbouring cells do not have MHA then differences in feeder loss are less important because they appear in both the uplink and downlink directions. Imbalances are likely to be more significant at the interface between different site types, i.e. between macrocells, microcells and indoor solutions.

The CPICH is also used in combination with the Synchronisation Channels (SCH) during the initial cell synchronisation procedure. Once a UE has achieved frame synchronisation and identified the scrambling code group from the SCH, the CPICH is used to identify the primary scrambling code.

3.3.2 Synchronisation Channel (SCH)

There are two types of Synchronisation Channel (SCH): Primary SCH (P-SCH) and Secondary SCH (S-SCH). Both exist only at the physical layer and neither is used to transfer higher layer information. Both synchronisation channels are mandatory downlink common channels broadcast from every cell in the network. The general timing of the P-SCH and S-SCH is illustrated in Figure 3.9.

The P-SCH is defined by a burst of 256 chips which is transmitted during the first 10% of every time slot. There are 15 time slots within each 10 ms radio frame and so the P-SCH chip sequence is repeated once every 0.67 ms. Every cell of every network uses the same chip sequence for the P-SCH. This allows a UE to achieve slot synchronisation relatively rapidly without the requirement to search for multiple sequences. A UE searches for the P-SCH as soon as it has been switched on or after returning from an area outside coverage. This task forms the first step of the initial cell synchronisation procedure. UE also search for the P-SCH whenever trying to identify neighbouring cells.

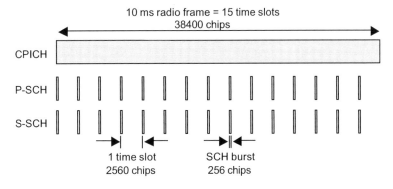

Figure 3.9   Timing of the P-SCH and S-SCH relative to the CPICH

The S-SCH is defined by a series of 15 consecutive bursts of 256 chips. Each burst occupies the first 10% of a time slot and the collection of 15 bursts is repeated once every 10 ms. Each of the 15 bursts is defined by a unique chip sequence. 3GPP TS 25.213 specifies 16 unique chip sequences which can be used by the S-SCH. The selection of which combination of chip sequences to transmit is dependent upon the scrambling code group associated with the primary scrambling code. There are 64 scrambling code groups and 64 allowed combinations of chip sequences, i.e. there is a one-to-one mapping between scrambling code group and the combination of chip sequences. Table 3.12 presents an example subset of the allowed combinations of chip sequences.

For example, if the primary scrambling code belongs to scrambling code group 2 the S-SCH transmits chip sequence 1, followed by chip sequence 2, followed by chip sequence 1, and so on. Once a UE has identified the combination of chip sequences on the S-SCH then it is able to deduce the scrambling code group associated with the primary scrambling code and also the radio frame timing. A UE does not have to receive the complete set of 15 consecutive chip sequences to deduce the scrambling code group and radio frame timing. The combinations of chip sequences have been defined so that it is possible for a UE to deduce both the scrambling code group and the radio frame timing after receiving only three consecutive chip sequences, i.e. each combination of three consecutive chip sequences is unique throughout the complete set of chip sequences for all scrambling code groups. This includes combinations of chip sequences which cross the boundary between the end of one radio frame and the start of another radio frame. In practice a UE may use more than three chip sequences to increase the reliability of the procedure.

The P-SCH and S-SCH burst length of 256 chips provides the reason for the T_Cell parameter introduced in Section 3.3.1 having a range from 0 to 2304 with a step size of 256 chips. The T_Cell parameter defines the time offset between the start of a Node B frame and the start of a cell frame. If a different T_Cell value is assigned to each cell belonging to a Node B then the P-SCH and S-SCH belonging to those cells do not overlap. This helps to distribute the quantity of non-orthogonal interference. The chip sequences defined for the P-SCH and S-SCH are not orthogonal to the channelisation codes used by the other downlink physical channels. If the P-SCH and S-SCH were transmitted from each cell at the same time then the bursts of non-orthogonal interference would become stronger. There is also an argument for assigning the same T_Cell value to all cells belonging to the same Node B. If the set of primary scrambling codes assigned to a Node B belong to the same scrambling code group then identical P-SCH and S-SCH are transmitted by each cell. If the difference in the propagation delays from each cell is relatively small then a UE will receive single combined versions of the P-SCH and S-SCH. This increases the received signal strength and potentially allows a more rapid synchronisation. The effectiveness of this strategy depends upon UE implementation. Some UE implementations may make the assumption that there is only a single primary scrambling code associated with each detected P-SCH. If the P-SCH from more than a single cell are received at the same time then the UE must check for more than a single scrambling code.

Table 3.12 S-SCH chip sequences used to identify the scrambling code group

Scrambling code group	Time slot														
	0	1	2	3	4	5	6	7	8	9	10	11	12	13	14
0	1	1	2	8	9	10	15	8	10	16	2	7	15	7	16
1	1	1	5	16	7	3	14	16	3	10	5	12	14	12	10
2	1	2	1	15	5	5	12	16	6	11	2	16	11	15	12
...	...	...	...	...	...	...	...	...	...	...	...	...	...	...	...
61	9	10	13	10	11	15	15	9	16	12	14	13	16	14	11
62	9	11	12	15	12	9	13	13	11	14	10	16	15	14	16
63	9	12	10	15	13	14	9	14	15	11	11	13	12	16	10

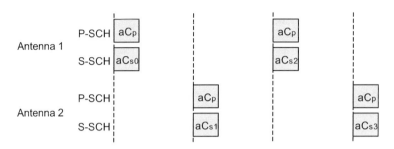

Figure 3.10 Time switched transmit diversity for the P-SCH and S-SCH

The P-SCH and S-SCH are typically transmitted with 3 dB less power than the CPICH. In the case of the P-SCH, a lower transmit power can be justified by the high repetition rate of the chip sequence. If a UE fails to receive one P-SCH chip sequence then the opportunity to receive another occurs after only 0.67 ms. Similarly, in the case of the S-SCH, a lower transmit power can be justified by the UE requiring only three consecutive chip sequences to deduce the scrambling code group and radio frame timing. If a UE fails to receive a S-SCH chip sequence then a subsequent combination of three consecutive chip sequences can be received with relatively little delay.

The P-SCH and S-SCH can make use of Time Switched Transmit Diversity (TSTD) to help improve performance. TSTD represents an open loop form of transmit diversity. It is mandatory for UE to support TSTD for the SCH but optional for the network. In this case, the bursts of chip sequences alternate between the two antenna elements. TSTD for the P-SCH and S-SCH is illustrated in Figure 3.10.

This figure also illustrates that the P-SCH and S-SCH chip sequences are multiplied by a factor a. This factor is applied irrespective of whether or not TSTD is used for the P-SCH and S-SCH. It allows the UE to deduce whether or not Space Time Transmit Diversity (STTD) is being applied to the P-CCPCH. If a has a value of -1 then STTD is not being applied to the P-CCPCH, whereas if a has a value of $+1$ then STTD is being applied to the P-CCPCH.

3.3.3 Primary Common Control Physical Channel (P-CCPCH)

The Primary Common Control Physical Channel (P-CCPCH) is responsible for broadcasting system information messages using the BCCH logical channel and BCH transport channel. The P-CCPCH is a mandatory downlink common channel broadcast from every cell in the network. 3GPP TS 25.213 specifies that the P-CCPCH is always spread using channelisation code (SF256,1) and always scrambled using the primary scrambling code assigned to the cell. The general timing of the P-CCPCH is illustrated in Figure 3.11.

The P-CCPCH is time multiplexed with the P-SCH and S-SCH, i.e. the P-CCPCH is active during the 90% of each time slot that the P-SCH and S-SCH are inactive. 90% of a single slot corresponds to 2304 chips which corresponds to 9 modulation symbols after accounting for the spreading factor of 256. There are 2 bits per modulation symbol and so there are 18 bits within each time slot or 270 bits within each 10 ms radio frame. This represents the number of bits after physical layer processing. The Transmission Time Interval (TTI) for the BCH transport channel is 20 ms and so the capacity offered by the P-CCPCH is 540 bits per 20 ms. The MAC-b entity within the Node B transfers 246 bits to the physical layer once every 20 ms. A 16 bit CRC field is added with 8 tail bits. This increases the total number of bits from 246 to 270 bits. Rate 1/2 convolution coding is then completed to increase the number of bits to 540 bits. In the case of the P-CCPCH, the number of channel-coded bits exactly

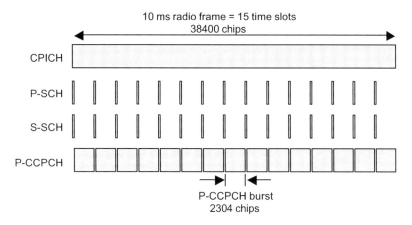

Figure 3.11 Timing for the P-CCPCH

equals the number of bits offered by the physical channel and so the rate matching procedure is not required to complete any puncturing nor any repetition. The transport format used by the BCH and the subsequent processing applied by the physical layer is standardised by 3GPP. It is not necessary for a UE to receive any information prior to decoding the contents of the P-CCPCH.

The P-CCPCH is the first physical channel decoded by the UE which includes higher layer information. The majority of information broadcast on the P-CCPCH originates from the RRC layer within the RNC. There is also a relatively small quantity of information which originates from the NAS layer within the core network. In addition, it is possible for a Node B to insert information, e.g. the uplink received signal strength measurement which appears within SIB 7.

The P-CCPCH is typically transmitted with 5 dB less power than the CPICH. This is a fixed transmit power and there is no power control. Neither the MAC nor RLC layers add any headers to the BCCH information. This means that the bit rate at the top of the RLC layer is equal to 12.3 kbps and the processing gain is equal to 25 dB. These figures are similar to those for the 12.2 kbps AMR speech service although the AMR speech service is typically assigned a maximum downlink transmit power which is comparable to the CPICH transmit power. The P-CCPCH can afford to have less transmit power assigned because the information is repeated and reception is relatively less service affecting. Nevertheless it is important that the P-CCPCH can be received across the entire cell.

The P-CCPCH can make use of Space Time Transmit Diversity (STTD) which represents an open loop transmit diversity scheme. It is mandatory for UE to support STTD, but it is optional for the network. The principle of STTD is illustrated in Figure 3.12. The original bit sequence is divided into blocks of four consecutive bits. The bit sequence for antenna 1 is not modified but the bit sequence for antenna 2 is modified by re-ordering all 4 bits and inverting 2 of the bits. Each bit from the original sequence is transmitted at a different time from each antenna. Inverting two of the bits means that a different modulation symbols are transmitted from each antenna.

The P-CCPCH has a capacity of 18 bits per time slot. 18×15 time slots = 270 bits which is not a multiple of 4. For the first 14 time slots belonging to a 10 ms radio frame, the last two bits of each even-numbered time slot are concatenated with the first two bits of each odd-numbered time slot. The last 2 bits of the fifteenth time slot are not STTD coded but are transmitted from both antennas. As described in the previous section, a UE can deduce whether or not STTD is being applied to the P-CCPCH by detecting the value of a from the P-SCH and S-SCH.

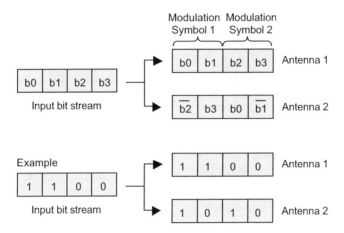

Figure 3.12   Principle of Space Time Transmit Diversity (STTD)

3.3.4 Secondary Common Control Physical Channel (S-CCPCH)

The Secondary Common Control Physical Channel (S-CCPCH) is responsible for broadcasting information sent on the PCH and FACH transport channels. The PCH is only used for paging messages, i.e. control plane signalling, whereas the FACH can be used for both control plane signalling and user plane data. MBMS and SAB services make use of the FACH transport channel and so they also make use of the S-CCPCH physical channel. In addition, the S-CCPCH can be used to send BCCH system information on the FACH. Each cell belonging to the network requires at least one S-CCPCH. Multiple S-CCPCH can be configured to increase PCH and FACH capacity. The drawback of configuring multiple S-CCPCH is an increased downlink transmit power requirement which reduces the transmit power available for other physical channels. MBMS services may be configured with a dedicated RF carrier in which case the majority of the downlink transmit power belonging to that carrier can be assigned to the S-CCPCH.

S-CCPCH which are configured to transfer PCH data are always scrambled using the primary scrambling code. Other S-CCPCH can be scrambled using either the primary scrambling code or a secondary scrambling code. The channelisation code used by the S-CCPCH can have a spreading factor within the range 4 to 256. The spreading factor has a direct impact upon the throughput offered by the S-CCPCH and the corresponding downlink transmit power requirement. Smaller spreading factors correspond to higher throughputs and higher downlink transmit power requirements. For example, a spreading factor of 128 corresponds to a modulation symbol rate of 30 ksps and a bit rate at the bottom of the physical layer of 60 kbps. This is equivalent to 600 bits per radio frame or 40 bits per time slot (15 time slots within each radio frame). In general, not all of the data bits per time slot are available for higher layer information. 3GPP TS 25.211 specifies a total of 18 slot formats for the S-CCPCH. Half of these slot formats include pilot bits whereas the release 6 and all preceding releases of the 3GPP specifications do not support the use of S-CCPCH pilot bits. This effectively reduces the number of slot formats to 9. S-CCPCH pilot bits may be supported in future releases of the 3GPP specifications to allow the use of a slow power control algorithm, i.e. the UE measures the signal to noise ratio from the downlink pilot bits and provides feedback to the network. This power control is likely to be slow relative to the inner loop power control used for the dedicated physical channels because the feedback would be provided to the RNC at layer 3 rather than to the Node B at layer 1. The Frame Protocol for FACH data frames already includes a field which allows the RNC to instruct the Node B to change the transmit power of the S-CCPCH. The set of S-CCPCH slot formats which exclude pilot bits is presented in Table 3.13.

Table 3.13 S-CCPCH slot formats which exclude pilot bits

Slot format	Spreading factor	Channel symbol rate (ksps)	Channel bit rate (kbps)	Bits per time slot	Data bits	TFCI bits
0	256	15	30	20	20	0
2	256	15	30	20	18	2
4	128	30	60	40	40	0
6	128	30	60	40	38	2
8	64	60	120	80	72	8
10	32	120	240	160	152	8
12	16	240	480	320	312	8
14	8	480	960	640	632	8
16	4	960	1920	1280	1272	8

The channel bit rates represent the bit rate after physical layer processing and are relatively high compared with the bit rates at the MAC and RLC layers. Slot format 8 has a channel bit rate of 108 kbps after accounting for the overhead generated by the Transport Format Combination Indication (TFCI) bits. This slot format typically has a maximum MAC layer throughput of 36 kbps, i.e. 33% of the channel bit rate. The majority of the slot formats include TFCI bits to instruct the UE which transport format combination is being used at any point in time. A UE is able to relate a specific TFCI value to a specific transport format combination using information broadcast on the P-CCPCH. System Information Block (SIB) 5 specifies both the slot format and the transport format combination set for each S-CCPCH.

An example section from SIB 5 is presented in Log File 3.1. This example is applicable to a cell which has been configured with a single S-CCPCH.

```
sCCPCH-SystemInformationList
        sCCPCH-SystemInformationList value 1
        secondaryCCPCH-Info
        modeSpecificInfo fdd
                sttd-Indicator:                 false
                sf-AndCodeNumber                sf64: 1
                pilotSymbolExistence:           false
                tfci-Existence:                 true
                positionFixedOrFlexible:        flexible
                timingOffset:                   30
        tfcs
        normalTFCI-Signalling
        complete
        ctfcSize
        ctfc4Bit
                ctfc4Bit value 1        ctfc4: 0
                ctfc4Bit value 2        ctfc4: 1
                ctfc4Bit value 3        ctfc4: 2
                ctfc4Bit value 4        ctfc4: 3
                ctfc4Bit value 5        ctfc4: 4
                ctfc4Bit value 6        ctfc4: 6
        fach-PCH-InformationList
        fach-PCH-InformationList value 1
        transportFormatSet
        commonTransChTFS
                tti     tti10
                tti10   value 1
```

```
                rlc-Size fdd
                octetModeRLC-SizeInfoType2
                        sizeType1:                      4
                        numberOfTbSizeList
                        numberOfTbSizeList value 1:     zero
                        numberOfTbSizeList value 2:     one
                        logicalChannelList:             allSizes
                        semistaticTF-Information
                                channelCodingType
                                convolutional:          half
                                rateMatchingAttribute: 210
                                crc-Size:               crc16
                transportChannelIdentity:       5
                ctch-Indicator:                 false
        fach-PCH-InformationList value 2
        transportFormatSet
        commonTransChTFS
                tti     tti10
                tti10   value 1
                rlc-Size fdd
                octetModeRLC-SizeInfoType2
                        sizeType1:                      15
                        numberOfTbSizeList
                        numberOfTbSizeList value 1:     zero
                        numberOfTbSizeList value 2:     one
                        numberOfTbSizeList value 3:     small1: 2
                        logicalChannelList:             allSizes
                        semistaticTF-Information
                                channelCodingType
                                convolutional:          half
                                rateMatchingAttribute: 200
                                crc-Size:               crc16
                transportChannelIdentity:       7
                ctch-Indicator:                 false
        fach-PCH-InformationList value 3
        transportFormatSet
        commonTransChTFS
                tti     tti10
                tti10   value 1
                rlc-Size fdd
                octetModeRLC-SizeInfoType2
                        sizeType1:                      15
                        numberOfTbSizeList
                        numberOfTbSizeList value 1:     zero
                        logicalChannelList:             allSizes
                        tti10 value 2
                rlc-Size fdd
                octetModeRLC-SizeInfoType2
                        sizeType2:                      3
                        numberOfTbSizeList
                        numberOfTbSizeList value 1:     one
                        logicalChannelList:             allSizes
                        semistaticTF-Information
                                channelCodingType:      turbo
                                rateMatchingAttribute: 110
                                crc-Size:               crc16
                transportChannelIdentity:       10
                ctch-Indicator:                 false
```

Log File 3.1 Example section from SIB 5 illustrating the S-CCPCH configuration

The first part of the message specifies whether or not Space Time Transmit Diversity (STTD) is being used. STTD represents an open loop form of transmit diversity and can be applied to the S-CCPCH in the same way as for the P-CCPCH. The first part of Log File 3.1 also specifies the relevant channelisation code and whether or not TFCI and pilot bits are included as part of the slot format. In this example, channelisation code number 1 with a spreading factor of 64 is used. There are TFCI bits, but no pilot bits. This information allows the UE to deduce that slot format 8 is being used. The position of the data bits within the physical channel is flexible rather than fixed and a time offset of 30 is specified. The time offset defines the time difference between the start of the CPICH radio frame and the start of the S-CCPCH radio frame. The time offset is specified in steps of 256 chips and so a value of 30 corresponds to an offset of 7680 chips. Figure 3.13 illustrates the general timing relationship between the CPICH and the S-CCPCH.

This figure also illustrates that the TFCI bits can be transmitted with a power which is greater than the data bits. A Node B is usually informed of the power offset during the start-up procedure after being switched on. The NBAP Common Transport Channel Setup Request message specifies the power offset to be applied. The power offset can range from 0 to 6 dB with a step size of 0.25 dB. A typical value for a S-CCPCH with a spreading factor of 64 is 4 dB. This means that if the S-CCPCH data bits are configured to have a transmit power of 33 dBm, i.e. equal to a typical CPICH transmit power, the TFCI bits generate a short-term increase in the S-CCPCH transmit power to 37 dBm. The example section from SIB 5 presented in Log File 3.1 does not specify a secondary scrambling code number and so the UE can deduce that the S-CCPCH is scrambled using the primary scrambling code.

The second part of the message specifies the set of allowed Calculated Transport Format Combinations (CTFC). In this example there are six CTFC values corresponding to six Transport Format Combinations (TFC). These CTFC values can be used in combination with the subsequent Transport Format Set (TFS) information to determine the six TFC. In this example, there are three TFS. These TFS are summarised in Table 3.14.

There is no explicit signalling to inform the UE of which transport channel is associated with the PCH and which transport channels are associated with the FACH. 3GPP TS 25.331 specifies that when a S-CCPCH includes a PCH, the PCH is always associated with the first transport channel in the list, i.e. the UE is able to deduce that transport channel 5 is used for the PCH while transport channels 7 and 10 are used for the FACH. A S-CCPCH cannot include more than one PCH transport channel. The example presented in Log File 3.1 includes only a single S-CCPCH. If a cell is configured with more

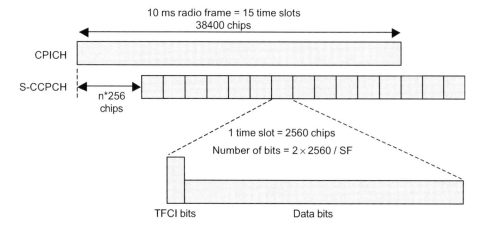

Figure 3.13   Frame timing of the S-CCPCH relative to the CPICH

Table 3.14 Example transport format sets for the S-CCPCH

	TrCh 5	TrCh 7	TrCh 10
Channel coding	1/2 rate Conv.	1/2 rate Conv.	1/3 rate Turbo
CRC size	16	16	16
Transmission Time Interval (TTI)	10 ms	10 ms	10 ms
Transport Block (TB) size	80 bits	168 bits	360 bits
Transport Block Set (TBS) size	0, 80 bits	0, 168, 336 bits	0, 360 bits
Rate matching attribute	210	200	110

than a single S-CCPCH then 3GPP TS 25.331 also specifies that the S-CCPCH responsible for the PCH should be listed first. It is possible to configure more than a single S-CCPCH to transfer PCH data. In this example, the transport block set size for the PCH is relatively small, i.e. 80 bits. This limits the number of paging records which can be included within the message. A transport block set size of 80 bits allows a maximum of one paging record to be included. In general, one paging record allows a single UE to be paged although it is also possible to use one paging record to page a group of RRC Connected mode UE using their UTRAN Radio Network Temporary Identities (U-RNTI). A group of RRC Connected mode UE may be paged to instruct them to release their RRC connections. 3GPP allows up to eight paging records within each paging message and so larger transport block set sizes can be used to increase the paging capacity of the network. Each S-CCPCH which includes a PCH also has the relevant PICH information specified within SIB 5. An example of this information is presented in Section 3.3.5.

The PCH transport channel is only able to encapsulate the PCCH logical channel and so the UE does not require any further information regarding the content. However, the FACH is able to encapsulate the BCCH, CCCH, DCCH, MCCH, MSCH, CTCH, DTCH and MTCH logical channels. In this case, the MAC header includes a flag to indicate which logical channel is included at any point in time. The MAC headers are not visible from the information presented in this section but transport channel 7 is being used for the DCCH, i.e. control plane signalling, whereas transport channel 10 is being used for the DTCH, i.e. user plane data. Examples of MAC headers are presented within Section 2.4.

The three TFS presented in Table 3.14 are illustrated at the top of Figure 3.14. The PCH and FACH-u TFS include two transport formats whereas the FACH-c TFS includes three transport formats. The set of six allowed TFC can be derived from the six CTFC values specified within Log File 3.1. The derivation requires the definition of two intermediate variables for each transport channel. The first variable L is equal to the number of transport formats within the transport format set. The second variable P is equal to the product of the L values associated with the preceding transport channels, with the exception that the P value associated with the first transport channel is always equal to 1. These rules lead to the L and P values presented in Table 3.15. The CTCF values are then calculated using a combination of the P variables and the TFI associated with the transport formats belonging to each transport format set. The CTFC value for a specific TFC is defined using the following equation.

$$\text{CTFC} = \sum^{\text{TrCh}} \text{TFI}_n \times P_n$$

The MAC layer is responsible for scheduling PCH and FACH data onto the S-CCPCH. This allows prioritisation between the different transport channels and the associated logical channels. The MAC layer is able to provide greatest priority to the PCH transport channel to help minimise the blocking probability for paging messages. The MAC layer must account for the DRX cycle length and the associated paging occasion when scheduling PCH messages. A UE listens to the PICH physical channel once per DRX cycle and subsequently listens for a corresponding PCH message if a positive

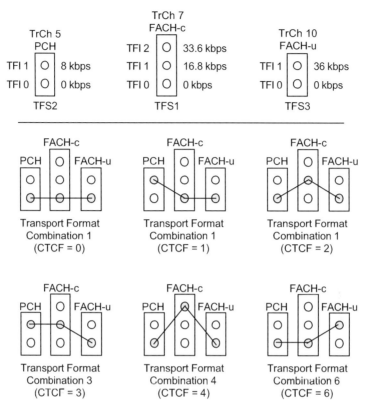

Figure 3.14   Example transport format sets and transport format combinations for the S-CCPCH

Paging Indication (PI) has been received. Paging occasions and the timing between a positive PI and the associated PCH message are described in section 3.3.5. The MAC layer can provide the FACH-c transport channel with the next highest priority. The set of TFC presented in Figure 3.14 illustrate that a FACH-c message can be sent simultaneously with a PCH message as long as an intermediate bit rate is used. Alternatively, if there are no PCH messages to be sent FACH-c messages can be sent using the

Table 3.15   Derivation of the CTFC values from the Transport Format Combinations (TFC)

TrCh 5	TrCh 7	TrCh 10	
$L_1 = 2$	$L_2 = 3$	$L_3 = 2$	
$P_1 = 1$	$P_2 = L_1 = 2$	$P_3 = L_1 \times L_2 = 6$	
TFI $\times$ P_1	TFI $\times$ P_2	TFI $\times$ P_3	CTFC
$0 \times P_1 = 0$	$0 \times P_2 = 0$	$0 \times P_3 = 0$	0
$1 \times P_1 = 1$	$0 \times P_2 = 0$	$0 \times P_3 = 0$	1
$0 \times P_1 = 0$	$1 \times P_2 = 2$	$0 \times P_3 = 0$	2
$1 \times P_1 = 1$	$1 \times P_2 = 2$	$0 \times P_3 = 0$	3
$0 \times P_1 = 0$	$2 \times P_2 = 4$	$0 \times P_3 = 0$	4
$0 \times P_1 = 0$	$0 \times P_2 = 0$	$1 \times P_3 = 6$	6

maximum bit rate within the TFS. These first two priorities result in the FACH-u transport channel having the lowest priority. The set of TFC presented in Figure 3.14 illustrate that FACH-u data can only be sent when there is neither PCH nor FACH-c messages to be sent.

3.3.5 Paging Indicator Channel (PICH)

The Paging Indicator Channel (PICH) is used to broadcast Paging Indicators (PI) across the entire coverage area of a cell. A positive PI is used to trigger a group of UE to decode a paging message from within the S-CCPCH. At least one of those UE should then find a relevant paging record. UE in RRC Idle mode or CELL_PCH and URA_PCH listen to the PICH once per DRX cycle. It is not necessary for UE in CELL_DCH to listen to the PICH because paging messages can be sent directly on the DPCH. The DRX cycle length used in RRC Idle mode can be different to that used in RRC Connected mode. In RRC Idle mode, the DRX cycle length is defined by the minimum of the DRX cycle lengths of the core networks with which the UE has registered. DRX cycle lengths for each core network are broadcast within SIB 1 on the P-CCPCH. An example of the relevant section of SIB 1 is presented within Log File 4.1 in section 4.2. If a UE has not registered with a specific core network domain then the DRX cycle length for that domain is disregarded. In RRC Connected mode, the DRX cycle length is defined by the minimum of the UTRAN DRX cycle length and the DRX cycle lengths belonging to any core network domains with which the UE has registered, but does not have a signalling connection, i.e. the UE is in the mobility management Idle state rather than Connected state. If a UE is in CELL_PCH with a signalling connection to the PS core network then the DRX cycle length is defined by the minimum of the UTRAN DRX cycle length and the CS core network DRX cycle length. The UTRAN DRX cycle length is signalled to the UE during the RRC connection establishment procedure. The RRC Connection Setup message includes a mandatory information element for the UTRAN DRX cycle length. Other RRC messages may be used to subsequently update the UTRAN DRX cycle length if necessary, e.g. the Transport Channel Reconfiguration or Radio Bearer Release messages can be used to update the UTRAN DRX cycle length. The UTRAN DRX cycle length can be configured with a value which is less than the core network DRX cycle lengths. This allows RRC Connected mode UTRAN originated paging to be more responsive. The UTRAN DRX cycle length can be configured with values of 8, 16, 32, 64, 128, 256, 512 radio frames, whereas the core network DRX cycle length can be configured with values of 64, 128, 256, 512 radio frames.

The PICH is always scrambled using the primary scrambling code and is always spread using a fixed spreading factor of 256. The 3GPP specifications do not specify which channelisation code is used to spread the PICH and so the RNC can assign any available code. UE are informed of the channelisation code within SIB 5 broadcast on the P-CCPCH. An example of the relevant section of SIB 5 is presented in Log File 3.2. In this case, channelisation code number 3 has been assigned to the PICH. Each S-CCPCH which includes a PCH transport channel has its own PICH and so SIB 5 may include multiple PICH definitions. Channelisation code numbering starts from 0 and so channelisation code 3 represents the forth code in the tree.

The spreading factor of 256 results in 150 modulation symbols per 10 ms radio frame which represent 300 bits (2 bits per modulation symbol). The PICH is active for 288 of the 300 bits and inactive for the remaining 12 bits. This generates an average activity factor of 96%. The 288 bits are grouped into a specific number of PI. The number of PI per radio frame can be 18, 36, 72 or 144. If

```
pich-Info
    channelisationCode256:    3
    pi-CountPerFrame:         e18
    sttd-Indicator:           false
```

Log File 3.2 Example section from SIB 5 illustrating the PICH configuration

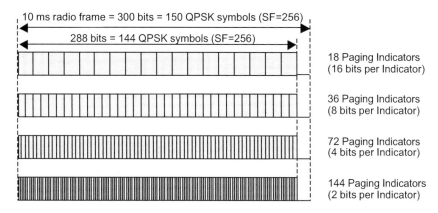

Figure 3.15 Number of paging indicators per 10 ms radio frame

there are 18 PI per radio frame then each PI has 16 bits (eight modulation symbols) whereas, if there are 144 PI per radio frame then each PI has 2 bits (one modulation symbol). This relationship between the number of PI per radio frame and the number of bits per PI is illustrated in Figure 3.15.

The example section from SIB 5 presented in Log File 3.2 indicates that the PICH has been configured with 18 PI per radio frame. Minimising the number of PI per radio frame increases the redundancy within the PI. This increased redundancy can be used to either increase reliability for a specific PICH transmit power or decrease the PICH transmit power for a specific reliability. A typical PICH transmit power for 18 PI per radio frame and a CPICH transmit power of 33 dBm is 28 dBm, i.e. the PICH is transmitted with 5 dB less power than the CPICH. The PICH is transmitted with a fixed power and does not have any form of dynamic power control. The UE is informed of the PICH transmit power relative to the CPICH transmit power within the first part of SIB 5. An example of this information is presented in Log File 3.3.

The UE is informed of the PICH transmit power to help determine a signal strength decision threshold which can be applied when identifying whether or not a positive PI has been received, i.e. to decide whether a positive PI has been received or if the received signal is just noise which appears similar to a positive PI. CPICH measurements can be used to determine the path loss, which can then be used to calculate the expected PICH received signal strength.

The drawback of minimising the number of PI per radio frame is that the number of UE belonging to each paging group will be greater. In this case, UE will receive positive PI more frequently and will then have to decode the contents of the S-CCPCH more frequently. The UE identity for a paging record has to be extracted from the paging message at layer 3. This means that a UE has to complete physical, MAC, RLC and RRC layer processing before being able to determine whether a paging record is targeted for itself or another UE. This processing requirement has an impact upon the UE battery life.

Increasing the number of PI per radio frame does not increase the capacity of the paging procedure. The capacity of the paging procedure is influenced by the throughput offered by the PCH transport channel, the size of location and routing areas and the number of UE in RRC Connected mode. Large location and routing areas decrease paging capacity because they contain increased numbers of UE.

```
SysInfoType5
        sib6indicator:          false
        pich-PowerOffset:       -5
```

Log File 3.3 Example section from SIB 5 illustrating the PICH transmit power

When a UE in RRC Idle mode is paged, the paging record is broadcast across the entire location area or routing area because the network does not know the precise location of the UE. Having a large number of UE within an area increases the number of paging messages which have to be broadcast by each cell. If a UE is in RRC Connected mode then the network knows the location of the UE in terms of either its cell or its UTRAN Registration Area (URA). A URA is typically a small cluster of cells and is only applicable when a UE is in the RRC Connected mode state URA_PCH. Paging a UE within a cell or within a URA avoids the requirement for every cell within the location or routing area to broadcast the same paging record.

If the DRX cycle length is 640 ms, each radio frame has 18 PI and there is a single S-CCPCH transferring paging messages then there are 1152 paging groups, i.e. the population of UE is divided into 1152 groups for paging purposes. If there are multiple S-CCPCH transferring paging messages then the UE selects one of the S-CCPCH using the calculation below.

$$\text{S-CCPCH index} = \text{IMSI mod} K$$

Where K equals the number of S-CCPCH used to transfer paging messages. If two S-CCPCH are used to transfer paging messages then all UE with an even-numbered IMSI will receive their paging messages on the first S-CCPCH while all UE with an odd-numbered IMSI will receive their paging messages on the second S-CCPCH. The RNC completes the same calculation when a paging message is to be transferred to ensure that the paging messages are sent on the expected S-CCPCH. The core network informs the RNC of the relevant IMSI within the RANAP paging message when requesting the paging procedure across the Iu interface. The IMSI represents a permanent Non-Access Stratum (NAS) UE identity.

Once a UE has identified the appropriate S-CCPCH then it must also identify the set of System Frame Numbers (SFN) during which its PI will be transmitted. These SFN are known as paging occasions. Paging occasions are calculated from the equation below.

$$\text{Paging occasion} = \{(\text{IMSI div} K) \bmod (\text{DRX cycle length})\} + n \times \text{DRX cycle length}$$

Where n is an integer which ranges from 0 to a maximum value which depends upon the DRX cycle length. If the DRX cycle length is 64 radio frames then the maximum value of n is 63. This maximum value can be deduced by considering that SFN values range from 0 to 4095. This means that the paging occasion result should never exceed 4095, i.e. $63 + (64 \times 63) = 4095$. It is apparent from the equation above that a paging occasion occurs once per DRX cycle. The RNC uses this equation to identify the next paging occasion for a UE when a paging message is to be transferred. This ensures that the PI and the subsequent paging message are sent during the expected SFN. The MAC layer within the RNC is responsible for scheduling the PI and the subsequent paging message during an appropriate paging occasion. This involves buffering while waiting for the next available paging occasion. The quantity of buffering will increase if the PCH is relatively congested.

After a UE has determined its set of paging occasions, it must identify which Paging Indicator (PI) to read from within each paging occasion. This procedure is completed in two steps. The first step determines an intermediate variable PI′ which is defined by the equation below.

$$\text{PI}' = (\text{IMSI div } 8192) \bmod N_p$$

Where N_p is the number of PI per radio frame. On the network side, this first step is completed by the RNC when a paging message is to be transferred and the result is signalled to the Node B within the appropriate Frame Protocol PCH data frame. Once PI′ has been determined, the actual PI for a specific paging occasion can be calculated using the equation below.

$$\mathrm{PI}_q = \left(\mathrm{PI}' + \left\lfloor ((18 \times (\mathrm{SFN} + \lfloor \mathrm{SFN}/8 \rfloor + \lfloor \mathrm{SFN}/64 \rfloor + \lfloor \mathrm{SFN}/512 \rfloor)) \bmod 144) \times \frac{N_p}{144} \right\rfloor \right) \bmod N_p$$

This second equation means that the PI changes as a function of the SFN, i.e. as a function of the paging occasion. For example, a UE may read the second PI during one paging occasion, but then read the fourth PI during the next paging occasion. On the network side, this calculation is completed by the Node B when a Frame Protocol PCH Data Frame indicates that a positive PI is to be broadcast.

A positive PI is represented by a series of consecutive 1s whereas a negative PI is represented by a series of consecutive 0s, e.g. if there are 72 PI per radio frame then each PI is represented by four consecutive 1s or four consecutive 0s. If a UE receives a positive PI then the UE has to decode the corresponding paging message from the S-CCPCH. This paging message can include one or more paging records. If none of the paging records address the UE then the UE continues to read its PI once per DRX cycle. If the paging message includes a paging record addressed to the UE then the UE has to react accordingly, e.g. if the UE is in RRC Idle mode then the reception of a paging record triggers the UE to initiate the RRC Connection establishment procedure.

The timing of the PICH relative to its corresponding S-CCPCH is illustrated in Figure 3.16. 3GPP TS 25.211 specifies that the timing of a PICH should always precede the timing of its S-CCPCH by 7680 chips (three time slots).

The time difference between the start of a PICH and its S-CCPCH helps to provide UE with some additional time to start receiving the S-CCPCH. If a UE receives a positive PI then the relevant paging message is within the S-CCPCH radio frame which has the next consecutive frame number. If the PI read by the UE is at the end of the PICH radio frame then it has relatively little time to start receiving the S-CCPCH.

Paging a single UE requires the use of two consecutive Iub Frame Protocol PCH data frames. These PCH data frames are generated by the RNC and sent to the Node B across the Iub interface. Frame Protocol is described in greater detail within Section 2.5. An example of two consecutive PCH data frames recorded from the Iub interface is presented in Log File 2.46 within Section 2.5. The first PCH Data Frame has a frame number of 605 and includes a positive PI. The Node B is informed of the positive PI using a PI bitmap. In this example there are 72 PI per radio frame and so the PI bitmap has a length of 9 bytes. The PI bitmap can be expanded into binary notation as:

00000000 00000000 00000000 00000000 00000010 00000000 00000000 00000000 00000000

This illustrates that the 38th PI (counting from 0) is positive whereas all other PI are negative. This PI represents PI' in the equations above rather than PI_q. The Node B is responsible for

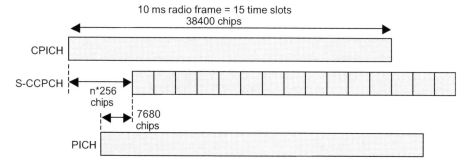

Figure 3.16 Frame timing of the PICH relative to the CPICH and S-CCPCH

calculating PI_q as:

$$PI_q = \left(38 + \left\lfloor ((18 \times (605 + \lfloor 605/8 \rfloor + \lfloor 605/64 \rfloor + \lfloor 605/512 \rfloor)) \bmod 144) \times \frac{72}{144} \right\rfloor \right) \bmod 72 = 56$$

The Node B then proceeds to broadcast PI number 56 as positive. All UE reading the 56th PI will proceed to decode the PCH message from within the subsequent S-CCPCH radio frame. The second PCH Data Frame within Log File 2.46 has a frame number of 606 and includes the paging message associated with the positive PI. The two PCH data frames are separated by 10 ms which corresponds to the duration of one radio frame on the air-interface and the PCH transport channel TTI.

3.3.6 MBMS Indicator Channel (MICH)

The MBMS Indicator Channel (MICH) is used to broadcast MBMS Notification Indicators (NI) across the entire coverage area of a cell. A positive NI is used to trigger UE which have activated a specific MBMS service to decode control information on the MCCH. The MCCH logical channel is always transferred using the FACH transport channel and the S-CCPCH physical channel. This means that the MICH is always associated with a S-CCPCH which includes at least one FACH transport channel. This is in contrast to the PICH which is always associated with a S-CCPCH which includes one PCH transport channel. A single cell can have a maximum of one FACH transport channel carrying MCCH information and so there is a maximum of one MICH per cell. 3GPP TS 25.331 specifies that the S-CCPCH carrying the MCCH is always the last S-CCPCH listed in SIB 5 and likewise the FACH transport channel carrying the MCCH is always the last FACH listed in SIB 5. If a UE has activated more than one MBMS service then it will monitor more than one NI.

UE can take advantage of Discontinuous Reception (DRX) to reduce power consumption when receiving the MICH. The MICH does not make use of specific occasions such as the paging occasions defined for the PICH. NI are repeated every 10 ms radio frame during each MBMS modification period. An MBMS modification period represents a period of time during which the majority of MBMS control information broadcast on the MCCH does not change. The MBMS modification period can be configured as 128, 256, 512 or 1024 radio frames, i.e. the shortest supported modification period is 1.28 s whereas the longest is 10.24 s. The only MBMS control information which can change during a modification period is the contents of the MBMS Access Information message. MBMS control information can be repeated during a modification period. The number of repetitions per modification period can be configured as 1, 2, 4 or 8. This value can be configured independently for the MBMS control information which does not change and the MBMS Access Information message. UE are informed of both the modification period and the number of repetitions per modification period within SIB 5.

If the network plans to change the contents of the MBMS control information for a specific service then the population of UE are informed by broadcasting a positive NI during the preceding modification period. The positive NI is repeated every 10 ms radio frame during that modification period so UE can decode the positive NI as long as they access the MICH at least once. The exact frequency with which a UE receives the MICH is dependent upon the UE implementation, but it must be at least once per modification period. If the modification period is greater than the paging DRX cycle length then the UE can receive the MICH at the same time as the PICH.

The MICH is always scrambled using the primary scrambling code and is always spread using a fixed spreading factor of 256. The 3GPP specifications do not specify which channelisation code is used to spread the MICH and so the RNC can assign any available code. UE are informed of the channelisation code being used within SIB 5. The physical layer structure of the MICH is the same as that of the PICH except that Paging Indicators (PI) are replaced by NI. The spreading factor of 256 results in 150 modulation symbols per 10 ms radio frame which represent 300 bits (2 bits per modulation symbol). The MICH is active for 288 of the 300 bits and inactive for the remaining 12 bits. This generates an average activity factor of 96%. The 288 bits are grouped into a specific number of NI. The number of NI per radio frame can be 18, 36, 72 or 144. If there are 18 NI per radio frame then each NI has 16 bits (8 modulation symbols) whereas, if there are 144 NI per radio frame then each NI has 2 bits (1 modulation symbol).

Minimising the number of NI per radio frame increases the redundancy within the NI. This increased redundancy can be used to either increase reliability for a specific MICH transmit power or decrease the MICH transmit power for a specific reliability. The MICH is transmitted with a fixed power and does not have any form of dynamic power control. The UE is informed of the MICH transmit power relative to the CPICH transmit power within SIB 5. The UE is informed of the MICH transmit power to help determine a signal strength decision threshold which can be applied when identifying whether or not a positive NI has been received, i.e. to decide whether a positive NI has been received or if the received signal is just noise which appears similar to a positive NI. CPICH measurements can be used to determine the path loss, which can then be used to calculate the expected MICH received signal strength.

The drawback associated with minimising the number of NI per radio frame is that the number of MBMS services associated with each NI will increase. The impact of this trend depends upon the number of MBMS services offered by the operator. If there is an increased number of MBMS services associated with each NI then UE will receive positive NI more frequently and will have to decode the contents of the S-CCPCH more frequently. The relevant MBMS service identity has to be extracted from the MBMS control information at layer 3. This means that a UE has to complete Physical, MAC, RLC and RRC layer processing before being able to determine whether the MBMS control information is relevant to itself or to UE using a different MBMS service. This processing requirement has an impact upon the UE battery life.

Each UE that has activated one or more MBMS services must determine which NI to read from the MICH. This procedure must be completed for each MBMS service that has been activated. The procedure is completed in two steps. The first step determines an intermediate variable NI$'$ which is defined by the equation below.

$$\mathrm{NI}' = (\mathrm{TMGI} + \lfloor \mathrm{TMGI}/65536 \rfloor) \bmod 65536$$

Where the Temporary Mobile Group Identity (TMGI) is calculated from an exclusive OR operation between the MBMS Service identity and the PLMN identity. On the network side, this first step is completed by the RNC when a positive NI is to be broadcast. Transferring MBMS NI$'$ information across the Iub to the Node B differs to transferring paging PI$'$ information. PI$'$ information is transferred using Frame Protocol PCH data frames whereas NI$'$ information is transferred using NBAP signalling. NBAP signalling can be used because it is only necessary to update the Node B once per modification period rather than once per 10 ms radio frame. The NBAP MBMS Notification Update Command message is used to inform the Node B of the NI$'$ result and the corresponding frame number for the start of the relevant modification period.

Once NI$'$ has been determined, the actual NI for a specific MBMS service can be calculated using the equation below.

$$\mathrm{NI}_q = \left\lfloor ((25033 \times (\mathrm{NI}' \oplus ((25033 \times \mathrm{SFN}) \bmod 65536))) \bmod 65536) \times \frac{N_n}{65536} \right\rfloor$$

Where the SFN is the frame number of the P-CCPCH radio frame during which the start of the MICH radio frame occurs, N_n is the number of notification indicators per radio frame and the $\oplus$ symbol indicates an exclusive OR operation. This second equation means that the NI_q changes as a function of the SFN. On the network side, this calculation is completed by the Node B when an NBAP MBMS Notification Update Command indicates that a positive NI is to be broadcast. A positive NI is represented by a series of consecutive 1s whereas a negative NI is represented by a series of consecutive 0s, e.g. if there are 72 NI per radio frame then each NI is represented by four consecutive 1s or four consecutive 0s. If a UE receives a positive NI then the UE has to decode the corresponding MBMS control information from the S-CCPCH.

The timing of the MICH relative to its corresponding S-CCPCH is the same as the time of the PICH relative to its corresponding S-CCPCH. This timing relationship is illustrated in Figure 3.16. 3GPP TS 25.211 specifies that the timing of a MICH should always precede the timing of its S-CCPCH by 7680 chips (three time slots). The time difference between the start of a MICH and its S-CCPCH helps to provide UE with some additional time to start receiving the S-CCPCH. If a UE receives a positive NI then the relevant MBMS control information is broadcast during the next modification period. If the NI read by the UE is at the end of the modification period then it has relatively little time to start receiving the S-CCPCH.

3.3.7 Acquisition Indicator Channel (AICH)

The Access Indicator Channel (AICH) is a downlink physical channel used to acknowledge the receipt of uplink PRACH preambles. The AICH exists only at the physical layer and is not used to transfer any higher layer information. The Node B is responsible for generating the content of the AICH. The AICH can include up to 16 Acquisition Indicators (AI) to simultaneously acknowledge up to 16 PRACH preambles. 3GPP TS 25.211 specifies that the AICH uses a fixed spreading factor of 256. The 3GPP specifications do not specify which channelisation code is used to spread the AICH and so the RNC can assign any available code. 3GPP TS 25.213 specifies that the AICH is always scrambled using the primary scrambling code. An AICH physical channel is always associated with a specific PRACH physical channel. If a cell is configured with multiple PRACH then it will also be configured with multiple AICH. In this case, each AICH will have a different channelisation code allocated from the code tree. 3GPP TS 25.211 specifies the AICH timing and structure illustrated in Figure 3.17.

The AICH has 15 access slots distributed across two 10 ms radio frames. The first access slot is time aligned with the start of every even-numbered radio frame. Each AICH access slot includes 16

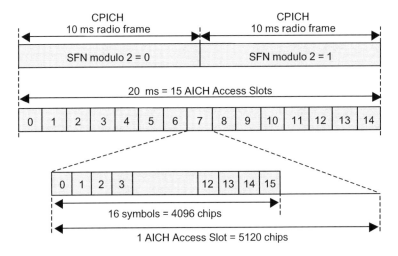

Figure 3.17   Frame timing and structure of the AICH relative to the CPICH

modulation symbols which correspond to 4096 chips after accounting for the spreading factor of 256. The duration of a complete AICH access slot includes 5120 chips and so the AICH is inactive for 1024 chips, i.e. the maximum activity factor of the AICH is 80%. If there are no AI to send then the AICH becomes inactive and the activity factor is 0%. This is in contrast to the PICH and MICH which have fixed activity factors of 96%. Each AI occupies the complete 4096 chips and so the activity factor becomes 80% as soon as there is a single AI to send. Multiple AI are sent simultaneously within a single access slot by superimposing orthogonal signatures. Each modulation symbol represents 2 bits and so there are 32 bits within each AICH access slot. These 32 bits are referred to as 32 AICH signals because they have values of $+1$, -1 and 0, rather than 1 and 0.

The set of 32 AICH signals are defined by the equation below. The index j ranges from 0 to 31, i.e. the 32 AICH signals are represented by $a_0, a_1, a_2, \ldots a_{31}$.

$$a_j = \sum_{s=0}^{15} \text{AI}_s b_{s,j}$$

Each AICH signal is generated by summing contributions from up to 16 AI. The variable s identifies the contribution from a specific AI. If there is only a single PRACH preamble to acknowledge then the variable s will have only a single value. In this case, the value of s will correspond to the PRACH preamble signature being acknowledged. There are 16 PRACH preamble signatures and so the value of s can range from 0 to 15. The value of AI_s can be $+1$, -1 and 0. The value of $+1$ indicates a positive acknowledgement whereas the value of -1 indicates a negative acknowledgement. The value of 0 indicates that the cell has not received a corresponding PRACH preamble and so it is not necessary to send an AI. For example, if a positive acknowledgement is to be sent for the 5th PRACH preamble signature then AI_4 will have a value of $+1$ whereas all other AI_s will have a value of 0. The variable $b_{s,j}$ corresponds to the jth signal within AICH signature pattern s. There are 16 AICH signatures which correspond to the 16 PRACH preamble signatures. An AICH signature has a length of 32 signals whereas a PRACH preamble signature has a length of 16 signals. The AICH signature patterns are generated by repeating each signal belonging to a specific PRACH preamble signature, e.g. a PRACH preamble signature starting as $+1$, -1, $+1$ becomes an AICH signature starting as $+1$, $+1$, -1, -1, $+1$, $+1$.

UE are informed of the AICH configuration within SIB 5 which is broadcast as part of the BCCH on the P-CCPCH. An example of the relevant section of SIB 5 is presented in Log File 3.4. This section of SIB 5 is associated with a specific PRACH physical channel. If multiple PRACH physical channels are configured then there will be multiple instances of this information.

In this example, the RNC has assigned channelisation code number 2. Channelisation code numbering starts from 0 and so channelisation code 2 represents the third code in the tree. This section of SIB 5 also indicates that Space Time Transmit Diversity (STTD) is not being used. STTD represents an open loop form of transmit diversity which can be applied to the AICH. If STTD is applied then the principle presented in section 3.3.3 is applied to each AICH signature pattern prior to combining with their associated AI values.

Log File 3.4 also indicates that the value of the AICH transmission timing is 0. This variable can have a value of either 0 or 1. A value of 0 means that the time offset between the AICH transmission timing and the PRACH transmission timing is equal to 1.5 AICH access slots (7680 chips). A value of 1

```
aich-Info
       channelisationCode256:        2
       sttd-Indicator:               false
       aich-TransmissionTiming:      e0
```

Log File 3.4   Example section from SIB 5 illustrating the AICH configuration

means that the time offset between the AICH transmission timing and the PRACH transmission timing is equal to 2.5 AICH access slots (12800 chips). These timing relationships are defined at the UE. The air-interface propagation delay means that the timing relationship at the Node B is different. The timing relationship from the UE perspective is illustrated in Figure 3.18.

The AICH transmission timing defines how long a UE has to wait before potentially receiving an acknowledgement to a PRACH preamble. PRACH preambles are transmitted as part of the random access procedure. A UE will not receive an acknowledgement to every PRACH preamble because the random access procedure is based upon transmitting the first PRACH preamble with relatively low power and subsequently increasing the transmit power between consecutive preambles until reception at the Node B is successful and a corresponding acknowledgement is returned. The AICH transmission timing has implications upon the delay between sending consecutive PRACH preambles as well as the delay between sending an acknowledged PRACH preamble and the PRACH message. Using an AICH transmission timing of 0 allows the system to be more responsive and reduces delay. The time difference between 1.5 AICH access slots and 2.5 AICH access slots is 1.33 ms. This time difference accumulates for each PRACH preamble transmitted by the UE. If a UE transmits 20 PRACH preambles prior to receiving an acknowledgement the time difference becomes 26.6 ms.

Figure 3.19 illustrates the impact of the air-interface propagation delay upon the AICH transmission timing. This example is based upon an AICH transmission timing of 0.

The AICH timing at the Node B is synchronised with the CPICH, as illustrated in Figure 3.17. Both the AICH and CPICH experience the same air-interface propagation delay and so the two physical channels remain synchronised when received by the UE. The UE determines the timing of its PRACH preambles based upon the timing of the CPICH. The PRACH preambles experience a propagation delay as they are transmitted across the air-interface. From the UE perspective, the time difference between the end of a PRACH access slot and the start of the corresponding AICH access slot is 0.5 access slots (2560 chips). From the Node B perspective, this time difference is reduced to (2560 chips − twice the air-interface propagation delay). If a cell has a range of 20 km then the air-interface propagation delay is 66 μs. This delay is equivalent to 256 chips after accounting for the chip rate of 3.84 Mcps. These figures indicate that the time difference between the end of a PRACH preamble and the start of the corresponding AICH access slot at the Node B is equivalent to 0.5 ms (2048 chips). This means that the Node B has 0.5 ms to generate and schedule the corresponding AI. If the propagation delay increases to the extent that the Node B no longer has sufficient time to generate and schedule the AI then the AICH transmission timing must be increased.

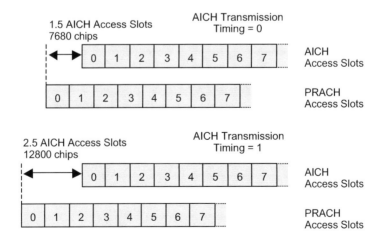

Figure 3.18   AICH and PRACH timing observed from the UE

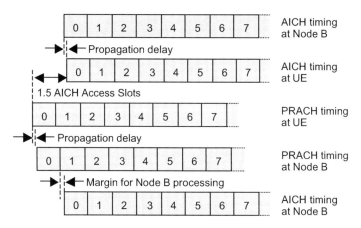

Figure 3.19 Impact of propagation delay upon requirement for AICH transmission timing

```
SysInfoType5
        sib6indicator:          false
        pich-PowerOffset:       -5
        modeSpecificInfo fdd
            aich-PowerOffset:   -8
```

Log File 3.5 Example section from SIB 5 illustrating the AICH transmit power

The AICH must have reliable coverage across the entire area of a cell otherwise the random access procedure is unlikely to have a high success rate. The AICH is typically assigned 8 dB less transmit power then the CPICH, i.e. if the CPICH has a transmit power of 33 dBm then the AICH would have a transmit power of 25 dBm. This represents the instantaneous transmit power when the AICH is active. The average transmit power depends upon the activity of the AICH, but will not exceed 80% of the instantaneous value, i.e. the maximum average transmit power is 24 dBm when the AICH is assigned an instantaneous transmit power of 25 dBm. A relatively low transmit power can be assigned as a result of the high spreading factor and the factor of 2 repetition within the AICH signature patterns. The UE is informed of the AICH transmit power relative to the CPICH transmit power within the first part of SIB 5. An example is presented in Log File 3.5.

The UE is informed of the AICH transmit power to help determine a signal strength decision threshold which can be applied when identifying whether or not a positive AI has been received, i.e. to decide whether a positive AI has been received or if the received signal is just noise which appears similar to a positive AI.

3.3.8 Physical Random Access Channel (PRACH)

The Physical Random Access Channel (PRACH) is an uplink physical channel separated into a preamble part and a message part. The preamble part exists only at the physical layer and is not used to transfer any higher layer information. This part is controlled by a combination of the RRC, MAC and physical layers. The message part is used to transfer higher layer information generated by either control plane signalling or user plane application data. The PRACH operates in combination with the AICH using the random access procedure. There are three general scenarios for the interaction of the PRACH with the AICH. These are illustrated in Figure 3.20. These three scenarios may belong to either the same or separate random access attempts.

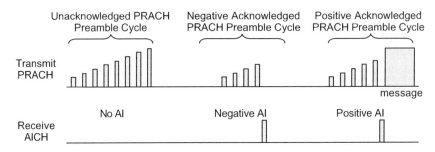

Figure 3.20   General scenarios for the interaction of the PRACH with the AICH

The first scenario illustrates an unacknowledged PRACH preamble cycle. PRACH preambles are transmitted with increasing power, but neither a positive nor negative Acquisition Indicator (AI) is received. This indicates that either the Node B has not received any of the preambles or the UE has not received the AI. In this case, the failed preamble cycle is reported to the MAC layer within the UE. The second scenario illustrates the reception of a negative AI. This could result from a random access collision with another UE. The failed preamble cycle is again reported to the MAC layer. The third scenario illustrates the successful transmission of a PRACH message after receiving a positive acknowledgement from the Node B.

The PRACH physical channel can be used while a UE is in either RRC Idle mode or CELL_FACH. The PRACH physical channel cannot be used in the RRC Connected mode states CELL_DCH, CELL_PCH nor URA_PCH. Each cell belonging to the network requires at least one PRACH physical channel.

3.3.8.1 PRACH Preamble

PRACH preambles are used to gain access to a random access opportunity for the transmission of a PRACH message. They are also used to help ensure that sufficient uplink transmit power is used for the PRACH message.

There are 15 PRACH access slots every 20 ms. As illustrated in Figure 3.21, these access slots precede the AICH access slots by either 1.5 or 2.5 slots, depending upon the value of the AICH transmission timing parameter. The AICH transmission timing also determines the minimum waiting period between transmitting consecutive PRACH preambles and the minimum waiting period between an acknowledged PRACH preamble and the transmission of the PRACH message. These dependencies are illustrated in Figure 3.21.

If the AICH transmission timing is configured with a value of 0 then the minimum waiting time between the start of consecutive preambles is three access slots. In addition, the minimum waiting time between an acknowledged preamble and the PRACH message is also three access slots. These figures represent only a minimum requirement. The waiting time can be greater and can be affected by the allocation of subchannels. Subchannels are described later in this section. If the AICH transmission timing is configured with a value of 1 then the minimum waiting time between consecutive preambles is four access slots. In addition, the minimum waiting time between an acknowledged preamble and the PRACH message is also four access slots. As described in Section 3.3.7 the increased time intervals associated with an AICH transmission timing of 1 allow for an increased delay resulting from either air-interface propagation or Node B processing.

Each PRACH access slot has a length of 5120 chips (1.3 ms). A PRACH preamble has a length of 4096 chips and occupies only 80% of the access slot. This is the same occupancy as an AI within an AICH access slot. A PRACH preamble is defined by the equation below, where k represents the chip

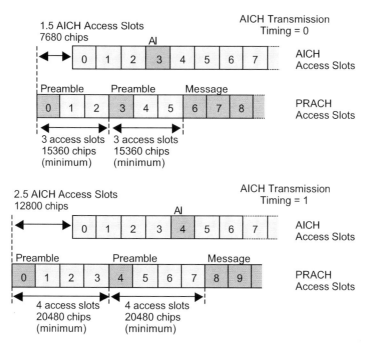

Figure 3.21 Waiting time between preambles and between preambles and PRACH message

number with a range from 0 to 4095.

$$C_{\text{pre},n,s}(k) = S_{\text{pre},n}(k) \times C_{\text{sig},s}(k) \times e^{j\left(\frac{\pi}{4} + \frac{\pi}{2}k\right)}$$

The variable n defines the selection of a specific PRACH scrambling code. There are 8192 uplink scrambling codes defined specifically for the preamble of the PRACH physical channel. These scrambling codes are not used by any other physical channel. The message part of the PRACH physical channel also uses a different set of scrambling codes. The complete set of 8192 PRACH preamble scrambling codes is divided into 512 groups of 16 codes. There is a one-to-one relationship between the group of 16 PRACH preamble scrambling codes and the downlink primary scrambling code. PRACH preamble scrambling codes have a length of 4096 chips and are used to define the PRACH physical channel. The variable $S_{\text{pre},n}(k)$ in the equation above represents the kth chip belonging to the nth PRACH preamble scrambling code. If a cell is configured with one PRACH physical channel then only 1 PRACH preamble scrambling code is configured. Likewise, if a cell is configured with two PRACH physical channels then two PRACH preamble scrambling codes are configured. Increasing the number of PRACH scrambling codes belonging to a cell increases the capacity of the PRACH. The number of PRACH scrambling codes configured for a specific cell can be read from SIB 5. When a cell is configured with more than a single PRACH, the UE selects a PRACH at random. The selected PRACH can change after every PRACH transmission and after every cell reselection.

The variable s defines the selection of a specific PRACH preamble signature. There are 16 preamble signatures each with a length of 4096 chips. Each preamble signature is defined by 256 repetitions of a shorter signature of length 16 chips. It is not necessary to allow the use of all preamble signatures within a cell. Increasing the availability of signatures increases the capacity of the PRACH physical channel, but also increases the Node B processing requirement, i.e. the Node B has to search for a larger set of signatures during each PRACH access slot. The variable $C_{\text{sig},s}(k)$ in the equation above represents the kth chip belonging to the sth PRACH signature.

Both the PRACH preamble scrambling codes and the PRACH signatures are specified by 3GPP TS 25.213. The product of the PRACH scrambling code and the PRACH preamble signature is multiplied by a complex phasor which rotates by 90° between consecutive chips.

The information necessary for a UE to start transmitting PRACH preambles is included within SIB 5. An example of a relevant section of SIB 5 is presented in Log File 3.6.

This example is applicable to a cell which has been configured with a single PRACH scrambling code. If a cell is configured with multiple PRACH scrambling codes then there will be multiple instances of this information. Log File 3.6 illustrates that the cell has been configured to use four of the 16 PRACH signatures, i.e. signature numbers 12, 13, 14 and 15. The specified spreading factor and puncturing limit are applicable to the PRACH message rather than the PRACH preambles. The preamble scrambling code associated with this PRACH physical channel is 0. This represents the first scrambling code from the group of 16 associated with the primary scrambling code. This section of SIB 5 also specifies that the cell has been configured to use subchannels 0 to 11, i.e. the complete set of subchannels.

Subchannels provide a way to prioritise access to the PRACH access slots. If a UE is provided access to a specific set of subchannels then those subchannels define which PRACH access slots may be used, i.e. high priority users could be granted access to a larger set of subchannels which would then allow access to a larger set of access slots. 3GPP TS 25.214 specifies the set of PRACH access slots associated with each of 12 subchannels. These 12 subchannels and their associated PRACH access slots are presented in Table 3.16.

The set of 12 subchannels generates an access slot pattern which is repeated every 80 ms. As an example, a UE which is granted access to subchannel 1 is allowed to use access slots 1 and 13 within the first 20 ms PRACH frame, but then only access slot 10 within the second frame, access slot 7 within

```
prach-SystemInformationList
prach-SystemInformationList value 1
    prac-RACH-Info
    modeSpecificInfo fdd
        availableSignatures:                    Bit12,Bit13,Bit14,Bit15
        availableSF:                            sfpr32
        preambleScramblingCodeWordNumber:       0
        puncturingLimit:                        pl1
                availableSubChannelNumbers:
                                                Bit0,Bit1,Bit2,Bit3,Bit4,
                                                Bit5,Bit6,Bit7,Bit8,Bit9,
                                                Bit10,Bit11
```

Log File 3.6 Example section from SIB 5 illustrating the PRACH preamble configuration

Table 3.16 PRACH access slots associated with each PRACH subchannel

	Subchannel number											
SFN modulo 8	0	1	2	3	4	5	6	7	8	9	10	11
0	0	1	2	3	4	5	6	7				
1	12	13	14						8	9	10	11
2				0	1	2	3	4	5	6	7	
3	9	10	11	12	13	14						8
4	6	7					0	1	2	3	4	5
5			8	9	10	11	12	13	14			
6	3	4	5	6	7					0	1	2
7						8	9	10	11	12	13	14

the third frame and access slot 4 within the fourth frame. In practice, UE are granted access to more than a single subchannel and so the number of candidate PRACH access slots is greater.

Although the section of SIB 5 presented in Log File 3.6 indicates that 4 signatures and 12 subchannels can be used within the cell, this does not indicate that every UE can make use of all 4 signatures and all 12 subchannels. A subsequent section of SIB 5 is used to partition these resources between groups of UE. An example of this partitioning is presented in Log File 3.7.

This example is relatively simple because it includes only a single PRACH partition. This partition grants all UE with access to all 4 signatures and all 12 subchannels. The set of signatures associated with the partition is referenced explicitly using start and end indices. This illustrates that the signatures associated with a PRACH partition are always consecutive from the perspective of the list provided earlier in the SIB. The start and end indices do not reference actual signature numbers, but reference positions from within the list, i.e. in this example the list includes signature numbers 12, 13, 14 and 15, whereas the partition references signature indices 0, 1, 2 and 3. The set of subchannels associated with the partition are signalled using a bit string of length 4 (b0,b1,b2,b3). The bit string illustrated in Log File 3.7 is equal to 1111, i.e. all four bits are signalled as being positive. The subsequent interpretation of this bit string depends upon the value assigned to the AICH transmission timing. If the AICH transmission timing has a value of 0 then the left most bit is truncated to leave b1,b2,b3. The remaining bit string is then repeated four times to generate b1,b2,b3,b1,b2,b3,b1,b2, b3,b1,b2,b3. If the AICH transmission timing has a value of 1 then the complete bit string is repeated three times to generate b0,b1,b2,b3,b0,b1,b2,b3,b0,b1,b2,b3. This approach generates a pattern of repetition which matches the minimum interval between consecutive PRACH preambles, i.e. a repetition pattern of 3 for an AICH transmission timing of 0 and a repetition pattern of 4 for an AICH transmission timing of 1. The resulting bit string of length 12 is then compared with the set of available subchannels provided earlier in the SIB. If both the earlier list and the PRACH partition have a subchannel indicated as positive then UE assigned to that PRACH partition are allowed to use that subchannel.

UE determine which PRACH partition they have been assigned using their Access Service Class (ASC). The first PRACH partition defined in SIB 5 corresponds to ASC 0. If further PRACH partitions are defined then these correspond to ASC 1, 2, 3 etc. The ASC associated with a specific UE and random access attempt is determined in one of two ways. If a UE is in RRC Idle mode and is using the PRACH to send an RRC Connection Request message then the ASC is derived from its Access Class (AC). If a UE is

```
prach-Partitioning
fdd value 1
     accessServiceClass-FDD
        availableSignatureStartIndex:   0
        availableSignatureEndIndex:     3
        assignedSubChannelNumber:       Bit0,Bit1,Bit2,Bit3
ac-To-ASC-MappingTable
        ac-To-ASC-MappingTable value:   0,0,0,0,0,0,0
modeSpecificInfo fdd
        primaryCPICH-TX-Power:          30
        constantValue:                  -25
        prach-PowerOffset
             powerRampStep:             1
             preambleRetransMax:        7
           rach-Transmiss ionParameters
                  mmax:                     16
                  nb01Min:                  0
                  nb01Max:                  50
```

Log File 3.7 Further PRACH preamble configuration information from SIB 5

in CELL_FACH then the ASC is derived from the MAC Logical channel Priority (MLP) associated with the logical channel used to generate the PRACH message.

UE in RRC Idle mode use the Access Class (AC) to Access Service Class (ASC) mapping table broadcast in SIB 5. This mapping table is visible in Log File 3.7. Use of this table requires that a UE has knowledge of its AC. A single subscriber may belong to more than one AC. The Universal Subscriber Identity Module (USIM) contains data specifying the AC or the set of AC. This data is stored within a USIM Elementary File (EF). 3GPP TS 31.102 specifies the contents of this file. The EF relevant to AC allocation is presented in Table 3.17.

The Access Control Class EF contains a payload with a length of 16 bits. These 16 bits (with the exception of the 11th bit) are used to flag which AC have been assigned to the USIM. There is a total of 15 AC and all subscribers belong to one of the first 10 which are numbered 0 to 9. This AC is assigned at random when programming the USIM. High-priority users may also be assigned an AC numbered from 11 to 15. These high-priority users and their corresponding AC are presented in Table 3.18.

The applicability of AC allocations depends upon the network with which the UE has registered. AC 0–9 are applicable to both the home and all visited PLMN. AC 11 and 15 are only applicable to the home PLMN, whereas AC 12, 13 and 14 are applicable to the home PLMN and any visited PLMN within the home country. AC 10 represents a special case which is used for the control of emergency calls. This AC is not programmed into the USIM, but can be sent across the air-interface. AC 10 indicates whether or not network access is permitted for emergency calls.

Once a UE in RRC Idle mode has identified its AC then it can use the AC to ASC mapping table to identify its corresponding ASC. If a UE has been assigned multiple AC then the mapping from AC to ASC makes use of the AC with the highest value. ASC are defined in the range 0–7 where ASC 0 represents the highest priority ASC. Table 3.19 presents the mapping of the Information Elements (IE) signaled in SIB 5 to the AC numbers.

For example, if the first IE belonging to the AC to ASC mapping table in SIB 5 has a value of 3 then this indicates that all UE with AC 0–9 have an ASC of 3. If the third IE belonging to the mapping table has a value of 2 then this indicates that all UE with AC 11 have an ASC of 2. The example presented in Log File 3.7 is relatively simplistic and all AC are associated with ASC 0, i.e. all IE have a value of 0. This means that all UE are associated with the first PRACH partition.

Table 3.17   USIM Elementary Files relevant to Access Class allocation

Elementary file (EF)
Access Control Class

Table 3.18   High-priority Access Classes

Users	Access Class
PLMN use	11
Security services	12
Public utilities	13
Emergency services	14
PLMN staff	15

Table 3.19   Access Classes associated with each information element within SIB 5 mapping table

AC	0–9	10	11	12	13	14	15
ASC	1st IE	2nd IE	3rd IE	4th IE	5th IE	6th IE	7th IE

If the PRACH is used by a UE in CELL_FACH then the MAC Logical Channel Priority (MLP) is used to determine the ASC. An MLP value is signalled to the UE for each logical channel when configuring the UE for CELL_FACH. A Radio Bearer Reconfiguration message can be used for this purpose. An example of the relevant section from a Radio Bearer Reconfiguration message is presented in Log File 3.8.

In this example, logical channel 2 has been assigned an MLP of 1. MLP values can be assigned within the range 1–8, where 1 represents the highest priority. The ASC is then set using the equation below.

$$ASC = Min(MLP, NumASC)$$

The value of NumASC corresponds to the maximum ASC number that has been configured where ASC numbering ranges from 0 to 7. The example presented in Log File 3.7 has only a single ASC and so the value of NumASC is 0, i.e. this would dominate over all MLP values and all PRACH transmissions would be associated with an ASC of 0. If the PRACH message includes multiple logical channels with different MLP values then the lowest MLP (highest priority) value is used in the equation above.

The section of SIB 5 presented in Log File 3.7 also informs the UE of the CPICH transmit power and a constant. These values are used by the UE during the open loop power control calculation which determines the transmit power of the first PRACH preamble. This open loop power control calculation is completed by the RRC layer. The UE is also informed of the transmit power increase to be applied between consecutive PRACH preambles. This transmit power increase can range from 1 to 8 dB. The example illustrated in Log File 3.7 configures the UE with a transmit power increase of 1 dB. The maximum number of PRACH preambles which can be transmitted as part of a single PRACH preamble cycle is also specified. This parameter has been configured with a value of 7 in Log File 3.7, but can be configured with any value between 1 and 64. Configuring a relatively large number of PRACH preambles per cycle and a relatively large transmit power step size allows UE to reach increased uplink transmit powers during each preamble cycle. This increases the probability of the Node B successfully receiving a preamble, but also increases the probability of the Node B experiencing short-term increases in its receiver interference floor. If the initial open loop power control calculation generates a result which is relatively low then there will be a requirement for either a greater number of PRACH preambles or a larger transmit power step size. Using a higher number of preambles increases the delay associated with the random access procedure.

The final section of SIB 5 illustrated in Log File 3.7 configures a set of parameters used by the MAC layer. The variable mmax defines the maximum number of preamble cycles which can belong to a single random access attempt. The MAC layer of the UE is informed if the physical layer reaches the maximum number of preambles per cycle without receiving an acknowledgement. The MAC layer is then responsible for instructing the physical layer whether or not a further preamble cycle can be

```
rb-MappingInfo
rb-MappingInfo value 1
        ul-LogicalChannelMappings
        oneLogicalChannel
                ul-TransportChannelType:        rach
                logicalChannelIdentity:         2
                rlc-SizeList
                        explicitList
                        explicitList value 1
                        rlc-SizeIndex:          1
                mac-LogicalChannelPriority:     1
```

Log File 3.8   Logical channel priority within a Radio Bearer Reconfiguration message.

Table 3.20 Persistency values associated with each Access Service Class

ASC	0	1	2	3	4	5	6	7
Persistency	1	$P(N)$	$s_2 P(N)$	$s_3 P(N)$	$s_4 P(N)$	$s_5 P(N)$	$s_6 P(N)$	$s_7 P(N)$

attempted. The minimum interval between the start of consecutive preamble cycles is 10 ms. The maximum number of PRACH preamble cycles can range from 1 to 32. The MAC layer is also informed if the UE receives a negative acknowledgement from the Node B during a preamble cycle. In this case the MAC layer can instruct a further preamble cycle if the maximum number of cycles has not been reached. After receiving a negative acknowledgement, and if a further preamble cycle is allowed, the MAC layer runs a back-off timer before instructing the next preamble cycle. This back-off timer is configured using the nb01Min and nb01Max parameters illustrated in Log File 3.7. The MAC layer selects an integer number at random between nb01Min and nb01Max. This value defines the number of 10 ms radio frames that the MAC layer must wait before instructing the physical layer to complete a further preamble cycle. Each of these parameters can be configured with a value between 0 and 50. If both parameters are configured with an equal value then the back-off time is fixed.

The MAC layer may also be required to complete a persistency check when instructing the physical layer to complete a PRACH preamble cycle. If only a single ASC has been configured, i.e. ASC 0, then a persistency check is not required. If more than a single ASC has been defined then persistency checks are completed for ASC 1 and upwards. Persistency checks are completed by generating a random number between 0 and 1. If the random number is less than the persistency value then the MAC layer can instruct the physical layer to start a PRACH preamble cycle. Otherwise the MAC layer must wait 10 ms before attempting the persistency check again with another random number. The persistency values for each ASC are presented in Table 3.20.

These persistency values are defined using a combination of information from SIB 5 and SIB 7. If SIB 5 specifies more than 2 PRACH partitions then the partitions associated with ASC 2 upwards can have an associated scaling factor defined. These scaling factors are represented by s_2, s_3, s_4, s_5, s_6, s_7 and each can have a value between 0.2 and 0.9 with a step size of 0.1. In general, the scaling factors should decrease for the higher ASC (lower priority ASC) to reduce their probability of starting a PRACH preamble cycle. However, scaling factors are optional and if they are not specified then a default value of 1 is assumed. SIB 7 specifies a dynamic persistence level for each PRACH physical channel. The dynamic persistence level, N can have a value from 1 to 8 and is used to calculate the variable $P(N)$ using the equation below.

$$P(N) = 2^{-(N-1)}$$

The MAC layer is responsible for informing the higher layers if the maximum number of preamble cycles is reached without receiving a positive acknowledgement to transmit the PRACH message. Likewise, if a positive acknowledgement is received then the MAC layer is responsible for informing the higher layers that a PRACH message can be sent.

3.3.8.2 PRACH Message

The PRACH message is used to transfer uplink information sent on the RACH transport channel. Only a single RACH transport channel can be mapped onto a PRACH physical channel. The RACH transport channel can be used for both control plane signalling and user plane data. The message part of the PRACH physical channel is divided into a data branch and a control branch. The data branch is used to transfer information from the RACH transport channel whereas the control branch is used to transfer Physical layer control bits. Figure 3.22 illustrates the data and control branches belonging to the PRACH message.

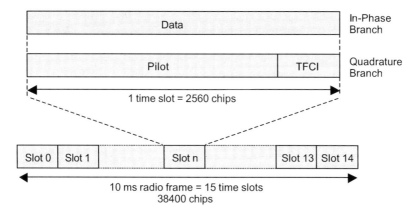

Figure 3.22 Data and control branches belonging to the PRACH message

The data branch forms the in-phase component of the uplink modulation whereas the control branch forms the quadrature component. PRACH messages are transmitted using the 10 ms radio frame structure with 15 time slots. PRACH messages can be configured to use either a 10 or 20 ms Transmission Time Interval (TTI) and so messages can span either 1 or 2 radio frames. The spreading factor of the channelisation code applied to the data branch can be 32, 64, 128 or 256. The spreading factor has a direct impact upon the throughput offered by the PRACH and the corresponding uplink transmit power requirement. Smaller spreading factors correspond to higher throughputs and higher uplink transmit powers. The four data branch spreading factors define the slot formats presented in Table 3.21.

The modulation symbol rate is defined by the chip rate of 3.84 Mcps divided by the spreading factor. There is a single information bit per symbol and so the bit rate is equal to the symbol rate. This represents the bit rate subsequent to physical layer processing, i.e. after channel coding redundancy has been added to help protect the higher layer information bits. An example configuration for the spreading factor to be applied to the data branch of the PRACH message is presented in Log File 3.6.

The spreading factor applied to the control branch is always 256 and there is only a single slot format. This slot format is presented in Table 3.22.

Table 3.21 Slot formats for the data branch of the PRACH message

Slot format	Spreading factor	Channel symbol rate (ksps)	Channel bit rate (kbps)	Bits per time slot
0	256	15	15	10
1	128	30	30	20
2	64	60	60	40
3	32	120	120	80

Table 3.22 Slot format for the control branch of the PRACH message

Slot format	Spreading factor	Channel symbol rate	Channel bit rate	Bits per time slot	Pilot bits	TFCI bits
0	256	15 ksps	15 kbps	10	8	2

The spreading factor of 256 provides a total of 10 bits per time slot. 8 bits are assigned to the pilot and 2 bits are assigned to the Transport Format Combination Indicator (TFCI). The pilot bits are used by the Node B receiver for synchronisation and channel estimation purposes. Channel estimation allows the receiver to provide coherent detection, i.e. account for rotations of the modulation constellation caused by frequency offsets and the radio propagation channel. 3GPP TS 25.211 specifies the set of pilot bits to be used for the PRACH message control branch. The pilot bit sequence is unique for each time slot belonging to a single radio frame. The TFCI bits are used by the Node B receiver to deduce which transport format has been sent during a specific TTI. 30 TFCI bits are distributed across each 10 ms radio frame. If a 20 ms TTI is used then the TFCI bits are repeated across each of the two 10 ms radio frames.

The channelisation codes used for the PRACH message data and control branches are defined by a combination of the signature which has been used for the preceding PRACH preambles and the spreading factor. The set of 16 PRACH preamble signatures is used to point to the set of 16 channelisation code tree branches with a spreading factor of 16. There is a one-to-one relationship between the signature used for the PRACH preambles and the channelisation code tree branch used for the PRACH message, e.g. if the fifth PRACH preamble signature has been used then the fifth branch of the channelisation code tree should be selected. The data branch is spread using the channelisation code with the highest index of the appropriate spreading factor. The control branch is spread using the channelisation code with the lowest index of spreading factor 256. Figure 3.23 illustrates the example of the channelisation code allocation when the second signature has been used for the PRACH preambles and the PRACH message data branch is transmitted using a spreading factor of 64. In this example, the channelisation code with the highest index for the data branch is (SF64,7) whereas the channelisation code with the lowest index for the control branch is (SF256,16).

There are 8192 uplink scrambling codes defined specifically for the message part of the PRACH physical channel. These scrambling codes are not used by any other physical channel. There is a one-to-one correspondence between the scrambling code used for the PRACH preamble and the scrambling code used for the PRACH message. PRACH message scrambling codes have a length of 38400 chips (10 ms).

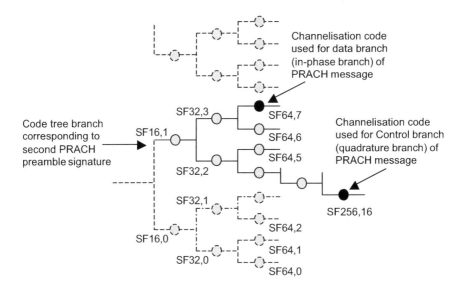

Figure 3.23   Allocation of channelisation codes for the PRACH message

```
transportChannelIdentity: 1              rach-TFCS
rach-TransportFormatSet                  normalTFCI-Signalling
commonTransChTFS                         complete
tti20 value 1                            ctfcSize
  rlc-Size fdd                           ctfc2Bit value 1
    octetModeRLC-SizeInfoType2             ctfc2: 0
      sizeType1: 15                        powerOffsetInformation
  numberOfTbSizeList                       gainFactorInformation
  numberOfTbSizeList value 1: one            signalledGainFactors
  logicalChannelList: configured             modeSpecificInfo fdd
tti20 value 2                                  gainFactorBetaC: 12
  rlc-Size fdd                                 gainFactorBetaD: 15
    octetModeRLC-SizeInfoType2                 powerOffsetPp-m: 2
      sizeType2: 3                       ctfc2Bit value 2
  numberOfTbSizeList                       ctfc2: 1
  numberOfTbSizeList value 1: one          powerOffsetInformation
  logicalChannelList: configured           gainFactorInformation
semistaticTF-Information                    signalledGainFactors
  channelCodingType                        modeSpecificInfo fdd
    convolutional: half                      gainFactorBetaC: 11
  rateMatchingAttribute: 1                   gainFactorBetaD: 15
  crc-Size: crc16                            powerOffsetPp-m: 2
```

Log File 3.9   Example section from SIB 5 illustrating the RACH transport format set

If a UE is in RRC Idle mode then the only message which can be sent to the network is the RRC Connection Request message. This message is always transmitted using the CCCH logical channel, the RACH transport channel and the PRACH physical channel. The UE is able to read the information necessary to use the CCCH from SIB 5. An example section from SIB 5 is presented in Log File 3.9.

This section of SIB 5 defines two transport block sizes which can be derived using the equations specified within 3GPP TS 25.331. Table 8.3 in Section 8.5.1 summarises these equations. There are three main groups of equations categorised by the RLC size information mode. Each equation within a group is referenced by its size type. Log File 3.9 indicates that both transport block sizes are defined using octet mode RLC size information type 2. This indicates that the third set of equations are applicable. Log File 3.9 also indicates that size type 1 is used for the first transport block size whereas size type 2 is used for the second transport block size. This indicates that the first transport block size can be calculated as $(8 \times 15) + 48 = 168$ bits, and the second transport block size can be calculated as $(16 \times 3) + 312 = 360$ bits.

The PRACH transport format combination set is relatively simple because there is only a single transport channel and so there is no transport channel multiplexing. Log File 3.9 illustrates that there are only two Calculated Transport Format Combination (CTFC) values. These two values correspond to the two transport block set sizes. This section of SIB 5 also defines the channel coding type, CRC size, TTI and rate matching attributes. The combination of these parameters defines both the static and dynamic sections of the transport format set. These are summarised in Table 3.23.

Table 3.23   Example transport format set for the PRACH

	TrCh 1
Channel coding	1/2 rate Convolutional
CRC size	16
Transmission Time Interval (TTI)	20 ms
Transport Block (TB) size	168, 360 bits
Transport Block Set (TBS) size	168, 360 bits
Rate matching attribute	1

This set of values is consistent with the set of allowed values presented for the RACH in Table 3.8. The value assigned to the rate matching attribute is arbitrary and has no impact because there is no transport channel multiplexing on the PRACH. The bit rate associated with each of the two transport formats can be calculated by dividing the transport block set size by the TTI, i.e. the bit rates are 8.4 and 18 kbps.

There is no explicit signalling to inform the UE of which transport format to use for the CCCH logical channel. 3GPP TS 25.331 specifies that the CCCH makes use of Signalling Radio Bearer (SRB) 0 and that SRB 0 uses the first transport format within the transport format set. This indicates that the RRC Connection Request message is transmitted using an 8.4 kbps connection. This represents the bit rate at the top of the MAC layer which is also equal to the bit rate at the top of the RLC layer because the uplink CCCH uses transparent mode RLC (the RLC layer does not add any header information). The MAC layer adds a 2 bit header which increases the bit rate to 8.6 kbps at the top of the physical layer.

Log File 3.9 also specifies the gain offsets which are to be applied to the data and control branches of the PRACH. The signalled values of these gain offsets must be divided by 15 to obtain the actual values. The signalled values can range from 0 to 15. Either the data branch or the control branch must have a signalled value of 15. In this example, the first transport format has a control branch gain offset of 12/15 and a data branch gain offset of 15/15. This indicates that the data branch is transmitted with 1.9 dB more power than the control branch, i.e. power difference $= 10 \times \log(15^2/12^2)$. The second transport format has a control branch gain offset of 11/15 and a data branch gain offset of 15/15. This indicates that the overhead generated by the control branch decreases as the bit rate of the data branch increases. The log file also specifies the power offset to be applied between the acknowledged PRACH preamble and the control part of the PRACH message. This power offset can be configured to have any integer value between -5 and 10 dB. In this example, both transport formats have a power offset of 2 dB. Table 3.24 presents a set of example transmit powers for the gain offsets specified in Log File 3.9 assuming that the acknowledged PRACH preamble had a transmit power of 5 dBm. These figures illustrate that the transport format with the higher bit rate makes use of a higher total transmit power although the transmit power allocated to the control branch remains constant.

If a UE is in CELL_FACH the RACH transport channel can be used to transfer information belonging to either the DTCH, CCCH or DCCH. In this case, RRC signalling is used to link each logical channel with one of the transport formats defined in SIB 5. The Radio Bearer Reconfiguration message can be used for this purpose when a UE is moved from CELL_DCH to CELL_FACH. An example of a relevant section from a Radio Bearer Reconfiguration message is presented in Log File 3.10.

This log file illustrates the configuration of a radio bearer with identity 5. The complete RRC message includes similar configuration information for radio bearer identities 1 to 4. TS 25.331 specifies that radio bearer identities 1 to 4 are reserved for DCCH Signalling Radio Bearers (SRB) using unacknowledged and acknowledged mode RLC. Identities 5 to 32 can only be used for SRB using transparent mode RLC. The first part of Log File 3.10 specifies the RLC configuration for acknowledged mode data transfer. This allows the UE to deduce that radio bearer 5 is intended for use by the DTCH logical channel. The second part of the log file specifies the radio bearer mapping

Table 3.24   Example PRACH transmit powers for a preamble to message power offset of 2 dB

Example preamble transmit power	Transport format	Control branch power (dBm)	Data branch power (dBm)	Total PRACH message power (dBm)
5 dBm	1	7	8.9	11.1
	2	7	9.7	11.6

```
          rb-Identity:        5
          rlc-Info
                  ul-RLC-Mode
                  ul-AM-RLC-Mode
                  transmissionRLC-Discard
                  noDiscard:                        dat8
                  transmissionWindowSize:           tw256
                  timerRST:                         tr1000
                  max-RST:                          rst12
                  pollingInfo
                          timerPollProhibit:                tpp140
                          timerPoll:                        tp1000
                          poll-SDU:                         sdu1
                          lastTransmissionPDU-Poll:         true
                          lastRetransmissionPDU-Poll:       true
                          pollWindow:                       pw50
                  dl-RLC-Mode
                  dl-AM-RLC-Mode
                  inSequenceDelivery:               true
                  receivingWindowSize:              rw768
                  dl-RLC-StatusInfo
                          missingPDU-Indicator:             false
      rb-MappingInfo
      rb-MappingInfo value 1
              ul-LogicalChannelMappings
              oneLogicalChannel
                      ul-TransportChannelType:          rach
                      logicalChannelIdentity:           5
                      rlc-SizeList
                      explicitList
                      explicitList value 1
                      rlc-SizeIndex:                    2
                      mac-LogicalChannelPriority:       8
              dl-LogicalChannelMappingList
          dl-LogicalChannelMappingList value 1
                  dl-TransportChannelType:          fach
                  logicalChannelIdentity:           5
```

Log File 3.10   Example section from a Radio Bearer Reconfiguration message

information. The radio bearer is mapped onto a RACH transport channel in the uplink direction and a FACH transport channel in the downlink direction. The RACH transport channel is linked with RLC size index 2 which indicates that the DTCH should use the second of the two transport formats presented in Log File 3.9, i.e. an RLC size of 360 bits. The DTCH is assigned the maximum allowed value of the MAC Logical channel Priority (MLP), i.e. the lowest priority. Other parts of the same RRC message assign an MLP value of 1 to radio bearers 1 to 3 and an MLP value of 7 to radio bearer 4. This indicates that the DCCH for SRB 1 to 3 are treated with the greatest priority whereas the DCCH for SRB 4 has a lower priority and the DTCH has the lowest priority. The MAC layer within the UE accounts for these priorities when scheduling information to be transferred on the RACH transport channel. Figure 3.24 summarises the example RACH transport format set for UE which are in CELL_FACH.

There is a single transport format set which includes two transport formats. The transport format with the higher bit rate is used by the DTCH while the transport format with the lower bit rate is shared between the CCCH and DCCH. The DCCH is configured with a higher priority than the DTCH. 3GPP TS 25.331 specifies that the CCCH always has a logical channel priority of 1, meaning that it also has a higher priority than the DTCH, i.e. control plane signalling is always transferred prior to user plane data.

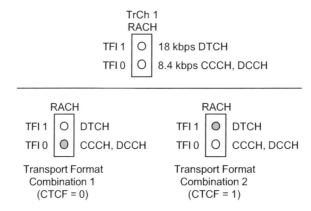

Figure 3.24   Example transport format set and transport format combinations for the RACH

3.3.9 Dedicated Physical Channel (DPCH)

The Dedicated Physical Channel (DPCH) can be used to transfer user plane data and control plane signalling in both the uplink and downlink directions. The DTCH and DCCH logical channels are mapped onto the DCH transport channel when using the DPCH physical channel. The DPCH and DCH can only be used when a UE is in CELL_DCH. The DTCH and DCCH logical channels can also be mapped onto the RACH and FACH transport channels and so these logical channels can also be used from CELL_FACH. The DPCH is able to make use of both soft and softer handover when a connection is located towards the boundary between two or more cells. The use of soft handover helps to improve the signal to noise ratio conditions at locations where the level of intercell interference would otherwise be relatively high. The DPCH also makes use of inner loop power control to help minimise the quantity of uplink and downlink interference.

3.3.9.1 Uplink

The uplink DPCH includes the Dedicated Physical Control Channel (DPCCH) and the Dedicated Physical Data Channel (DPDCH). It may also include the High Speed Dedicated Physical Control Channel (HS-DPCCH), the Enhanced Dedicated Physical Control Channel (E-DPCCH) and the Enhanced Dedicated Physical Data Channel (E-DPDCH). The HS-DPCCH is used for HSDPA connections whereas the E-DPCCH and E-DPDCH are used for HSUPA connections. Table 3.25 presents the various combinations of the physical channels which can belong to the uplink DPCH.

Table 3.25   Physical channels belonging to the uplink DPCH

Physical channel	R99 DPCH	HSDPA (SRB on DPDCH)	HSDPA and HSUPA (SRB on DPDCH)	HSDPA and HSUPA (SRB on E-DPDCH)
DPCCH	√	√	√	√
DPDCH	√	√	√	
HS-DPCCH		√	√	√
E-DPCCH			√	√
E-DPDCH			√	√

Release 99 DPCH connections require only the DPCCH and DPDCH. These represent the only two uplink DPCH physical channels specified by 3GPP prior to release 5. The release 5 version of the specifications introduces HSDPA and the uplink HS-DPCCH physical channel. The release 6 version of the specifications introduces HSUPA and the uplink E-DPCCH and E-DPDCH physical channels. The complete set of five physical channels is required when HSDPA is combined with HSUPA and the Signalling Radio Bearers (SRB) are transferred using the DPDCH. In this case, the E-DPDCH is only used to transfer user plane data. If the E-DPDCH is used to transfer both the SRB and the user plane data then the DPDCH is no longer required. Section 6.9 describes the HS-DPCCH whereas section 7.7 describes the E-DPCCH and E-DPDCH.

Figure 3.25 illustrates the combination of the DPDCH and DPCCH for a release 99 uplink DPCH. The are 15 DPDCH and 15 DPCCH time slots within each 10 ms radio frame. The DPCCH is used to transfer physical layer control information. An uplink DPCH always has one DPCCH and cannot have more than one DPCCH. Pilot bits are included to allow the Node B receiver to maintain synchronisation and to complete channel estimation. Channel estimation is necessary to complete coherent detection of the received modulation. Coherent detection compensates for phase changes introduced by both the radio propagation channel and any offset between the centre frequency of the transmitter and the centre frequency of the receiver. The pilot bits can also be used by the Node B receiver to complete Signal to Interference Ratio (SIR) measurements which are then compared with the uplink SIR target when generating downlink Transmit Power Control (TPC) commands. Downlink TPC bits are included within the uplink DPCCH for downlink inner loop power control, i.e. to instruct cells belonging to the active set to either increase or decrease their transmit powers. Both pilot bits and TPC bits are mandatory within every DPCCH time slot. Transport Format Combination Indicator (TFCI) bits may be included to inform the Node B of the transport format combination which is transmitted at any point in time. If TFCI bits are not included then the Node B receiver has to complete blind detection of the transport format combination. Feedback Information (FBI) bits are included when closed loop downlink transmit diversity is used.

Table 3.26 presents the set of uplink DPCCH slot formats. The first four slot formats exclude FBI bits whereas the second four slot formats include FBI bits. Slot formats 0A, 0B, 2A and 2B are only applicable to compressed mode in which case the UE transmits less than 15 time slots per radio frame. Slot formats 0A and 2A are applicable to the double frame compressed mode method which distributes the compressed mode transmission gap across two radio frames. Slot formats 0B and 2B are applicable to the single frame compressed mode method which limits the compressed mode transmission gap to a single radio frame. Slot formats 1 and 3 can be used for both compressed and normal frames.

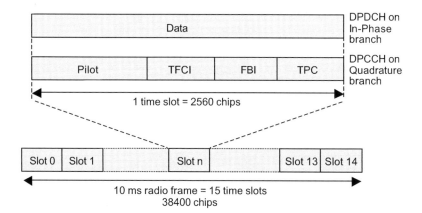

Figure 3.25   DPDCH and DPCCH belonging to a release 99 uplink DPCH

Table 3.26 Uplink DPCCH slot formats

Slot format	Channel symbol rate (ksps)	Channel bit rate (kbps)	Bits per radio frame	Bits per time slot	N_{pilot}	N_{TPC}	N_{TFCI}	N_{FBI}	Slots per radio frame
0	15	15	150	10	6	2	2	0	15
0A	15	15	150	10	5	2	3	0	10–14
0B	15	15	150	10	4	2	4	0	8–9
1	15	15	150	10	8	2	0	0	8–15
2	15	15	150	10	5	2	2	1	15
2A	15	15	150	10	4	2	3	1	10–14
2B	15	15	150	10	3	2	4	1	8–9
3	15	15	150	10	7	2	0	1	8–15

The uplink DPCCH is always spread using channelisation code (SF256,0), i.e. the first channelisation code with a spreading factor of 256. The spreading factor of 256 means that the channel symbol rate is equal to 15 ksps (after dividing the chip rate of 3.84 Mcps by 256). Each DPCCH symbol represents a single bit of physical layer control information and so the channel bit rate is equal to the channel symbol rate. The channel bit rate leads to the availability of 10 bits per time slot. These 10 bits are shared between the pilot, TPC, TFCI and FBI bits. The pilot bits represent the most important information and are allocated the largest share of the 10 DPCCH bits.

If a UE is in CELL_DCH then it has to transmit the DPCCH during every time slot. The only exception is when the UE is in compressed mode and transmission gaps are generated during which the UE stops transmitting. These transmission gaps are used to complete measurements on other UMTS RF carriers or other Radio Access Technologies (RAT), e.g. GSM. Transmitting the DPCCH during every time slot allows the Node B to maintain air-interface synchronisation. It also allows inner loop power control to continuously track changes in the propagation and interference conditions. Transmitting the DPCCH during every time slot has an impact upon both the UE power consumption and the uplink load at the Node B receiver. The release 7 version of the 3GPP specifications introduces the concept of uplink gating to reduce the number of time slots during which the DPCCH is transmitted. This has benefits in terms of increasing UE battery life and reducing the uplink interference floor at the Node B receiver.

The DPDCH is used to transfer higher layer information. This information could be either user plane data or control plane signalling. User plane data originates from the end-user application, e.g. a speech call or a file transfer. Control plane signalling originates from the RRC layer of the UE protocol stack. The 3GPP specifications allow the UE to transmit more than one DPDCH although in practice UE and network implementations tend to be limited to a single uplink DPDCH. Table 3.27 presents the set of uplink DPDCH slot formats.

Table 3.27 Uplink DPDCH slot formats

Slot format	Spreading factor	Channel symbol rate (ksps)	Channel bit rate (kbps)	Bits per radio frame	Bits per time slot
0	256	15	15	150	10
1	128	30	30	300	20
2	64	60	60	600	40
3	32	120	120	1200	80
4	16	240	240	2400	160
5	8	480	480	4800	320
6	4	960	960	9600	640

The DPDCH spreading factor can range from 256 to 4. The spreading factor has a direct impact upon the channel symbol rate and the corresponding channel bit rate. Decreasing the spreading factor increases the bit rate but also increases the UE transmit power requirement, i.e. the Node B receiver requires greater uplink power to decode the higher bit rates. The bit rates presented in Table 3.27 represent the bit rate at the bottom of the physical layer after redundancy has been added to help protect the higher layer information. A 12.2 kbps speech connection can use an uplink DPDCH spreading factor of 64, i.e. a channel bit rate of 60 kbps. In this case, the application bit rate is increased by a factor of 4.9 prior to insertion within the DPDCH. This allows a high quantity of redundancy to help protect the user plane speech information. The speech service uses transparent mode RLC and so re-transmissions are not allowed and it is important that information is received reliably with the first transmission. The drawback of decreasing the spreading factor to allow greater redundancy is a reduced spreading gain. Reducing the spreading gain increases the quantity of interference visible to the received signal after de-spreading. Decreasing the spreading gain by a factor of 2 increases the interference power by 3 dB. A 64 kbps uplink PS data connection can use a spreading factor of 16, i.e. a channel bit rate of 240 kbps. In this case, the application bit rate is increased by a factor of 3.75. A 384 kbps uplink PS data connection can use a spreading factor of 4, i.e. a channel bit rate of 960 kbps, and the application bit rate is increased by a factor of 2.5. These PS data connections have less redundancy, but make use of acknowledged mode RLC and so re-transmissions are allowed. Circuit switched video calls also use an application bit rate of 64 kbps and a channel bit rate of 240 kbps. In this case, transparent mode RLC is used and a higher transmit power operating point is necessary to ensure that data is received reliably. The transmit power operating point is defined by the inner loop power control SIR target.

The channelisation code used to spread the uplink DPDCH is specified by 3GPP TS 25.213. If only a single DPDCH is configured then the index of the channelisation code is equal to the spreading factor divided by 4. This concept is illustrated in Figure 3.26.

Figure 3.26 also illustrates the channelisation code used for the uplink DPCCH. 3GPP TS 25.213 specifies that if more than a single DPDCH is configured then all DPDCH must use a spreading factor

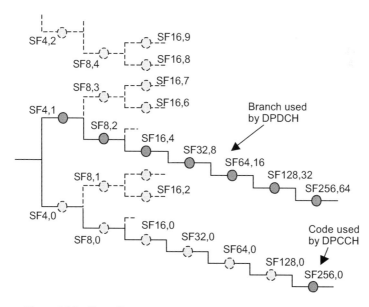

Figure 3.26   Channelisation codes used for the uplink DPCCH and DPDCH

of 4 and that a maximum of 6 DPDCH can be configured. The channelisation code tree includes only four channelisation codes with a spreading factor of 4 and so codes must be re-used by up to 2 DPDCH. When the same code is used by two DPDCH then those DPDCH must be on different branches of the modulation constellation, i.e. one DPDCH must be on the in-phase branch while the other DPDCH must be on the quadrature branch. This allows orthogonality between DPDCH to be maintained because the in-phase and quadrature branches are orthogonal to one another in the domain of the modulation constellation. The allocation of channelisation codes when multiple DPDCH are configured is presented in Table 3.28.

The uplink scrambling code assigned to the DPCCH and DPDCH is used to identify the connection at the Node B receiver, i.e. every connection is assigned a different uplink scrambling code. The RNC is responsible for assigning the uplink scrambling code and for ensuring that the same scrambling code is not assigned to multiple UE. 3GPP TS 25.213 specifies 2^{24} long scrambling codes and 2^{24} short scrambling codes. This means there are more than 16 million of each type of scrambling code. Each long scrambling code has a length of 38400 chips (10 ms) whereas each short scrambling code has a length of 256 chips (67 µs). Short scrambling codes are intended to be used by advanced receivers which are capable of interference cancellation and Multi-User Detection (MUD). Long scrambling codes represent the default for less advanced receivers. It is relatively simple for a single RNC to maintain a record of which uplink scrambling codes have been assigned and which are available for new connections. However, an RNC is not aware of the uplink scrambling codes assigned by its neighbouring RNC. In general, uplink scrambling codes are not reassigned during an active connection and there is a risk that neighbouring RNC assign the same scrambling code to different connections and then those connections move into the same geographic area and cause interference to one another. This risk can be avoided if each RNC is allocated a different subset of uplink scrambling codes. For example, if a network includes 20 RNC then each RNC can be assigned approximately 830 000 uplink scrambling codes. This represents a relatively large number and should be sufficient to allow every connection in CELL_DCH to be assigned a unique scrambling code. Uplink scrambling codes for the DPCCH and DPDCH are complex valued rather than real valued. This means that the scrambling process causes a rotation of the modulation constellation. This effect is discussed in greater detail within Section 2.6.2.

The DPDCH and DPCCH may be transmitted using either equal or different transmit powers. The relative transmit power of each physical channel is controlled by applying a pair of amplitude gain factors after channelisation code spreading. This process is illustrated in Section 2.6.2. The amplitude gain factor applied to the DPDCH is known as β_d whereas the amplitude gain factor applied to the DPCCH is known as β_c. The amplitude gain factors can have values of 1/15 to 1 in steps of 1/15. At any point in time at least one of the β_c and β_d values must be equal to 1. These gain factors define the relative amplitude of each physical channel rather than the relative power. The relative power is defined

Table 3.28 Channelisation code allocation when more than a single DPDCH is configured

Physical channel	Constellation branch	Channelisation code
DPCCH	Quadrature	(SF256,0)
DPDCH$_1$	In-Phase	(SF4,1)
DPDCH$_2$	Quadrature	(SF4,1)
DPDCH$_3$	In-Phase	(SF4,2)
DPDCH$_4$	Quadrature	(SF4,2)
DPDCH$_5$	In-Phase	(SF4,3)
DPDCH$_6$	Quadrature	(SF4,3)

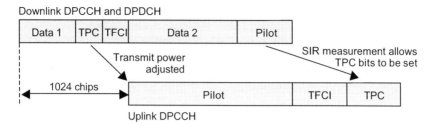

Figure 3.27   Timing of the uplink DPCCH relative to the downlink DPCH

by the square of the relative amplitudes. For example, if $\beta_c = 11/15$ while $\beta_d = 15/15$ then the DPDCH is transmitted with 2.7 dB greater power than the DPCCH ($=10 \times \log(225/121)$).

A UE derives the timing of the uplink DPCCH from the timing of the downlink DPCCH. The start of an uplink DPCCH time slot is transmitted 1024 chips after the start of the corresponding downlink DPCCH time slot. 1024 chips represents 40% of a single time slot. This timing relationship is defined at the UE and appears slightly differently at the Node B as a result of the air-interface propagation delay. The uplink and downlink DPCCH timing relationship at the UE is illustrated in Figure 3.27.

This timing relationship, as well as the structure of the uplink and downlink DPCCH have been defined to help ensure that inner loop power control can operate efficiently. The start of an uplink DPCCH time slot occurs after the UE has received the TPC bits from the corresponding downlink time slot. This allows the UE to adjust its transmit power according to the Node B instructions with very little delay. This justifies the TPC bits being positioned towards the front of the downlink DPCCH time slot. In addition, the UE receives the downlink pilot bits before having to define the TPC bits to be inserted within the uplink DPCCH. This allows the UE to define the TPC bits based upon very recent channel conditions.

Establishing a new DPCH involves NBAP signalling with the Node B and RRC signalling with the UE. The NBAP signalling is completed first to allow the Node B to start transmitting prior to the UE starting to receive. During RRC Connection establishment, the Node B is informed of the uplink DPCH information within an NBAP Radio Link Setup Request message. An example of the relevant section from an NBAP Radio Link Setup Request message is presented in Log File 3.11.

The Node B is informed that the UE will be instructed to start transmitting using the long scrambling code with number 1400276. The Node B is also informed that the minimum spreading factor for the uplink DPDCH will be 128. This indicates that the UE can use DPDCH slot formats 0 and 1. The uplink puncturing limit is signalled using a value of 7. This signalled value can be mapped to an actual

```
UL-DPCH-Information-RL-SetupReq
        ul-scramblingCode
                - ul-ScramblingCodeNumber: 1400276
                - ul-ScramblingCodeLength: long
        - minUL-ChannelisationCodeLength: len128
        - punctureLimit: 7
        tFCS
                cTFC
                - ctfc2bit: 0
                cTFC
                - ctfc2bit: 1
        ul-DPCCH-SlotFormat
                - non-extended: 0
                - ul-SIR-Target: 19
```

Log File 3.11   Example section from an NBAP Radio Link Setup Request message

value by multiplying by 0.04 and adding 0.4, i.e. the actual value is 0.68. The Node B can use the puncturing limit and the minimum DPDCH spreading factor to determine which spreading factor the UE will select for each transport format combination. The process for uplink spreading factor selection is described later in this section. The Node B is informed that there are two transport format combinations with Calculated Transport Format Combination (CTFC) values of 0 and 1. The number of transport format combinations is relatively low in this example because the DPCH is configured for only a signalling connection, i.e. a single transport channel. The number of transport format combinations will increase if the DPCH is reconfigured to transfer both signalling and user plane data, e.g. a DPCH configured for the 12.2 kbps speech service typically has six transport format combinations. The transport format set associated with the CTFC values of 0 and 1 is not shown in Log File 3.11, but is included elsewhere in the NBAP Radio Link Setup Request message. The uplink DPCCH slot format is specified to be 0 and the initial uplink SIR target has a signalled value of 19. The actual value of the SIR target can be obtained by dividing by 10, i.e. the actual value is 1.9 dB. This value for the uplink SIR target is used by inner loop power control within the Node B until outer loop power control within the RNC instructs a change.

The NBAP Radio Link Setup Request message is used to provide the Node B with DPCH information during initial connection establishment and during handover when the Node B does not already have any cells within the active set. If the Node B already has one or more cells within the active set then the NBAP Radio Link Addition Request message is used. This message includes less information because the Node B already has knowledge of the DPCH. If a DPCH is reconfigured then the Node B is informed using either an NBAP Radio Link Reconfiguration Prepare message or an NBAP Radio Link Reconfiguration Request message. The former is used for synchronised reconfiguration, i.e. the Connection Frame Number (CFN) during which the new configuration is to be applied is specified, whereas the latter is used for unsynchronised reconfiguration.

A UE is informed of the uplink DPCH configuration using RRC signalling. If the UE is making the transition from RRC Idle mode then all of the configuration information is included within the RRC Connection Setup message. If the UE is making the transition from the RRC Connected mode state CELL_FACH then the configuration information can be included within a Radio Bearer Reconfiguration message. An example of the relevant section from an RRC Connection Setup message is presented in Log File 3.12.

The log file starts by specifying that UE should not transmit more than 24 dBm. This corresponds to power class 3. Power class 4 corresponds to 21 dBm whereas power classes 1 and 2 correspond to 33 dBm and 27 dBm respectively. The log file then specifies a DPCCH power offset which is used by the UE for its open loop power control calculation of the initial DPCCH transmit power. The power

```
maxAllowedUL-TX-Power:           24
ul-ChannelRequirement
ul-DPCH-Info
ul-DPCH-PowerControlInfo fdd
        dpcch-PowerOffset:       -49
        pc-Preamble:             7
        sRB-delay:               7
        powerControlAlgorithm
                algorithm1:      0
modeSpecificInfo fdd
        scramblingCodeType:      longSC
        scramblingCode:          1400276
        spreadingFactor:         sf128
        tfci-Existence:          true
        puncturingLimit:         pl0-68
```

Log File 3.12 Example section from an RRC Connection Setup message

control preamble defines the duration of an initial transient period during which the UE transmits only the DPCCH. The preamble period is defined in terms of 10 ms radio frames and so the value of 7 presented in this example corresponds to 70 ms. These initial DPCCH radio frames allow time for the Node B receiver to achieve synchronisation prior to receiving any DPDCH information. 3GPP TS 25.331 allows the value of the power control preamble to range from 0 to 7. Assigning a large value allows more time for the Node B to achieve synchronisation, but increases the connection establishment delay. Additional margin for Node B synchronisation can be provided using the SRB delay parameter. This value defines an additional number of radio frames during which the DPDCH cannot be used for signalling purposes, i.e. SRB 0 to 4 cannot be transmitted until the number of 10 ms radio frames defined by the sum of the power control preamble and SRB delay have passed. The DPDCH can be used for user plane data during the radio frames defined by the SRB delay. This is applicable to the CELL_FACH to CELL_DCH transition rather than the RRC Idle mode to CELL_DCH transition. The former scenario allows user plane radio bearers to be configured at the same time as making the transition whereas the latter typically involves subsequent establishment of user plane radio bearers. The SRB delay can be configured with a value between 0 and 7. Similar to the power control preamble, a larger value allows more time for the Node B to achieve synchronisation, but also increases the connection establishment delay. This section of the log file also specifies the uplink power control algorithm to be applied for inner loop power control.

Log File 3.12 then specifies that long scrambling code number 1400276 should be applied after spreading. The spreading factor for the uplink DPDCH is specified as 128. This represents the minimum allowed spreading factor. The UE may use larger spreading factors if there is relatively little data to transmit. The minimum spreading factor defines the set of slot formats which the UE is allowed to use. A minimum spreading factor of 128 means that the UE can use slot formats 0 and 1. This is a relatively small set of allowed slot formats because this example is configuring the DPDCH for a signalling connection. If the DPDCH was being configured for a 64 kbps data connection then the minimum allowed spreading factor could be 16 which would mean that the UE could use slot formats 0 to 4. The uplink DPCCH slot format can be deduced from the indication of TFCI bits and the absence of any indication of FBI bits. This information indicates that slot format 0 should be used during normal frames and either slot format 0A or 0B during compressed frames, depending upon whether single or double frame compressed mode transmission gaps are configured.

Finally, this section of the log file specifies a DPDCH puncturing limit. This figure represents the minimum percentage of channel coded bits which must remain subsequent to rate matching. The puncturing limit can be configured with any value between 40% and 100% with a step size of 4%. This example illustrates a figure of 68%. The puncturing limit is used by the UE when selecting an appropriate spreading factor. The UE selects the uplink spreading factor by identifying whether or not any of the allowed spreading factors can be used without having to puncture any of the transport channels. If any spreading factors are identified which satisfy this criteria then the largest spreading factor is selected, i.e. the spreading factor with the largest spreading gain. Otherwise, the UE evaluates which spreading factors can be used without the requirement for any of the transport channels to be punctured beyond the puncturing limit. In this case, the UE selects the smallest spreading factor which satisfies the puncturing limit criteria for all transport channels, i.e. the spreading factor which requires the least puncturing.

Table 3.29 presents a section of the physical layer processing applied to the 12.2 kbps AMR speech service when all transport channels are at their maximum bit rate. This corresponds to a period of speech activity coinciding with the transmission of an RRC signalling message. In the case of the 12.2 kbps speech service, the UE is typically configured with a minimum uplink spreading factor of 64, i.e. the UE can select from spreading factors 64, 128 and 256. The number of bits after channel coding is calculated by adding the CRC bits and tail bits to the transport blocks and subsequently dividing by the coding rate. The puncturing required for each spreading factor is calculated for each transport

Table 3.29 Puncturing for 12.2 kbps AMR speech service (all transport channels at maximum bit rate)

	AMR Class A	AMR Class B	AMR Class C	SRB
Transport block size (bits)	81	103	60	148
Transport block set size (bits)	81	103	60	148
TTI (ms)	20	20	20	40
CRC bits	12	0	0	16
Tail bits	8	8	8	8
Coding rate	1/3	1/3	1/2	1/3
Data after channel coding per TTI (bits)	303	333	136	516
Data after channel coding per 10 ms (bits)	151.5	166.5	68	129
Rate matching attribute	195	201	256	210
Puncturing with SF64	1.09	1.12	1.43	1.17
Puncturing with SF128	0.54	0.56	0.71	0.58
Puncturing with SF256	0.27	0.28	0.36	0.29

channel using the equation below.

$$Puncturing\ ratio_n = N_{\text{data}} \times \frac{\text{RM}_n}{\sum_{i=1}^{N} (\text{RM}_i \times N_i)}$$

Where, n represents the transport channel being considered, N_{data} is the total number of DPDCH bits per 10 ms radio frame offered by a specific spreading factor, RM_i is the rate matching attribute for the ith transport channel, N_i is the number of bits per 10 ms after channel coding for the ith transport channel, and N is the total number of transport channels. The puncturing results in Table 3.29 indicate that only spreading factor 64 is able to accommodate all of the transport channels without any puncturing, i.e. all of the puncturing results are greater than 1. This means the UE will select spreading factor 64 when all four transport channels are active at their maximum bit rates. The result is the same if only the speech transport channels are active at their maximum bit rates while the SRB is inactive.

Table 3.30 presents the same section of physical layer processing during a period of speech inactivity when only comfort noise information is transmitted.

These puncturing results indicate that the UE can increase the uplink DPDCH spreading factor to 256 during periods of speech inactivity when only comfort noise information is transmitted.

Table 3.30 Puncturing for 12.2 kbps AMR speech service (only AMR comfort noise active)

	AMR Class A	AMR Class B	AMR Class C	SRB
Transport block size (bits)	39	0	0	0
Transport block set size (bits)	39	0	0	0
TTI (ms)	20	20	20	40
CRC bits	12	0	0	16
Tail bits	8	8	8	8
Coding rate	1/3	1/3	1/2	1/3
Data after channel coding per TTI (bits)	177	0	0	0
Data after channel coding per 10 ms (bits)	88.5	0	0	0
Rate matching attribute	195	201	256	210
Puncturing with SF64	6.78			
Puncturing with SF128	3.39			
Puncturing with SF256	1.69			

```
ul-CommonTransChInfo                        computedGainFactors: 0
modeSpecificInfo fdd                      ctfc6Bit value 4
ul-TFCS                                     ctfc6: 12
normalTFCI-Signalling                       powerOffsetInformation
complete                                    gainFactorInformation
ctfcSize                                      computedGainFactors: 0
  ctfc6Bit                                ctfc6Bit value 5
  ctfc6Bit value 1                          ctfc6: 13
    ctfc6: 0                                powerOffsetInformation
    powerOffsetInformation                  gainFactorInformation
    gainFactorInformation                     computedGainFactors: 0
      computedGainFactors: 0            ctfc6Bit value 6
  ctfc6Bit value 2                          ctfc6: 23
    ctfc6: 1                                powerOffsetInformation
    powerOffsetInformation                  gainFactorInformation
    gainFactorInformation                     signalledGainFactors
      computedGainFactors: 0                  modeSpecificInfo fdd
  ctfc6Bit value 3                            gainFactorBetaC: 10
    ctfc6: 11                                 gainFactorBetaD: 15
    powerOffsetInformation                    referenceTFC-ID: 0
    gainFactorInformation
```

Log File 3.13   Example section from a Radio Bearer Setup message

Log File 3.12 excludes the section of the RRC Connection Setup message which specifies the amplitude gain factors between the DPDCH and DPCCH. These are relatively simple in the case of an RRC Connection Setup message because there is only a single non-zero transport format combination and only a single pair of amplitude gain factors. Log File 3.13 illustrates the more complex example of the amplitude gain factors for a DPCH which is being configured for the 12.2 kbps AMR speech service. This example has been extracted from a Radio Bearer Setup message.

There are two possible approaches which can be used to inform the UE of the DPDCH and DPCCH gain factors. The first approach involves explicitly signalling the gain factors for each transport format combination. The second approach involves explicitly signalling the gain factors for one or more reference transport format combinations and then allowing the UE to calculate the gain factors for the remaining transport format combinations. Log File 3.13 illustrates the case where the gain factors are signalled for one reference transport format combination. The DPCCH amplitude gain factor has a signalled value of 10 whereas the DPDCH amplitude gain factor has a signalled value of 15. The actual value is derived from the signalled value by dividing by 15, i.e. the actual gain factors for the reference transport format combination are 10/15 and 15/15. The reference transport format combination has a CTFC value of 23 which corresponds to all four transport channels at their maximum bit rates. The derivation of the CTFC values is explained in Section 3.3.4.

The UE calculates the amplitude gain factors for the remaining transport format combinations using the equation below.

$$A_j = \frac{\beta_{d,ref}}{\beta_{c,ref}} \times \sqrt{\frac{L_{ref}}{L_j}} \times \sqrt{\frac{K_j}{K_{ref}}}$$

Where L_{ref} is the number of DPDCH used by the reference transport format combination and L_j is the number of DPDCH used for the jth transport format combination. In general, both L_{ref} and L_j have values of 1 and this part of the equation has no impact. K_{ref} is given by the equation below.

$$K_{ref} = \sum_{i=1}^{N} RM_i \times N_i$$

Table 3.31 DPDCH and DPCCH amplitude gain factors for the 12.2 kbps AMR speech service

	AMR Class A (bits)	AMR Class B (bits)	AMR Class C (bits)	SRB (bits)			Power Ratio
	RM 195	RM 201	RM 256	RM 210	β_c	β_d	
CTFC 0	0	0	0	0			
CTFC 1	88.5	0	0	0	15/15	10/15	−3.5 dB
CTFC 11	151.5	166.5	68	0	11/15	15/15	2.7 dB
CTFC 12	0	0	0	129	15/15	12/15	−1.9 dB
CTFC 13	88.5	0	0	129	15/15	15/15	0.0 dB
CTFC 23	151.5	166.5	68	129	10/15	15/15	3.5 dB

Where N is the total number of transport channels belonging to the transport format combination, RM_i is the rate matching attribute for the ith transport channel and N_i is the number of bits per 10 ms radio frame for the ith transport channel after channel coding and for the reference transport format combination. K_j is calculated in the same way as K_{ref} except that N_i is the number of bits per 10 ms radio frame for the ith transport channel after channel coding and for the jth transport format combination.

Once a value has been determined for A_j then the values of β_c and β_d are derived using the following rules. If $A_j > 1$ then $\beta_d = 1$ and β_c is the largest value of β_c for which $\beta_c \leq 1/A_j$. If $A_j \leq 1$ then $\beta_c = 1$ and β_d is the smallest value of β_d for which $\beta_d \geq A_j$. Table 3.31 illustrates the results for the 12.2 kbps AMR speech service example presented in Log File 3.13. The row associated with the reference transport format combination has been shaded.

The power ratio column quantifies the power difference between the DPDCH and DPCCH. A positive figure in this column indicates that the DPDCH is transmitted with greater power than the DPCCH. The results illustrate that the relative transmit power assigned to the DPDCH decreases as the bit rate of the DPDCH decreases.

3.3.9.2 Downlink

The downlink DPCH for release 99, release 4 and release 5 connections includes the Dedicated Physical Control Channel (DPCCH) and the Dedicated Physical Data Channel (DPDCH). The release 6 version of the 3GPP specifications introduces the Fractional Dedicated Physical Channel (F-DPCH). The F-DPCH can be used to replace the DPCCH and DPDCH when both the user plane data and control plane signalling are transferred using one or more HS-PDSCH. Table 3.32 summarises the use of these physical channels.

The F-DPCH reduces the requirement for downlink resources in terms of both transmit power and channelisation codes. Section 3.3.10 describes the F-DPCH. Figure 3.28 illustrates the combination of the downlink DPDCH and DPCCH.

Similar to the uplink there are 15 time slots within each 10 ms radio frame. The downlink DPDCH and DPCCH are time multiplexed onto both the in-phase and quadrature branches. This is in contrast to

Table 3.32 Physical channels belonging to the downlink DPCH.

Physical channel	R99 DPCH	HSDPA (SRB on DPDCH)	HSDPA & HSUPA (SRB on DPDCH)	HSDPA & HSUPA (SRB on HS-PDSCH)
DPCCH	√	√	√	
DPDCH	√	√	√	
F-DPCH				√

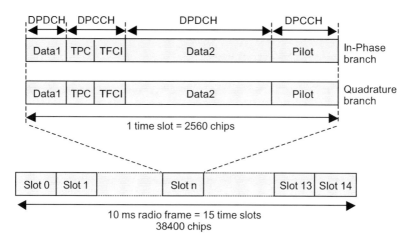

Figure 3.28 DPDCH and DPCCH belonging to a downlink DPCH

the uplink in which case the in-phase branch is dedicated to the DPDCH and the quadrature branch is dedicated to the DPCCH. The downlink DPCCH is used to transfer physical layer control information similar to the uplink DPCCH. A downlink DPCH always has one DPCCH and cannot have more than one DPCCH. Pilot bits are included to allow the UE receiver to maintain synchronisation and to complete channel estimation. Channel estimation is necessary to complete coherent detection of the received modulation. Coherent detection compensates for phase changes introduced by both the radio propagation channel and any offset between the centre frequency of the transmitter and the centre frequency of the receiver. Transmit Power Control (TPC) bits are included for inner loop power control, i.e. to instruct the UE to either increase or decrease its transmit power. Both pilot bits and TPC bits are mandatory within every time slot. Transport Format Combination Indication (TFCI) bits may be included to inform the UE of the transport format combination which is transmitted at any point in time. If TFCI bits are not included then the UE receiver has to complete blind detection of the transport format combination. The downlink DPCCH does not include FBI bits because uplink closed loop transmit diversity has not been standardised.

Table 3.33 presents a subset of the downlink DPCH slot formats. The slot formats for compressed mode have been excluded to keep the list relatively short.

The downlink DPCH slot formats combine the DPCCH and DPDCH. Downlink spreading factors range from 512 to 4 whereas the equivalent uplink spreading factors ranged from 256 to 4. In the case of the downlink, both the DPCCH and DPDCH are spread using the same spreading factor. The channel symbol rate is defined by the chip rate divided by the spreading factor. The channel bit rate equals twice the channel symbol rate because the DPCCH and DPDCH are mapped onto both the in-phase and quadrature branches and there are 2 bits per symbol. In this case the channel bit rate represents the instantaneous bit rate of either the DPCCH or DPDCH. Time multiplexing of these two physical channels means that the average bit rate is less than the figure presented in Table 3.33. For example, slot format 10 has an instantaneous channel bit rate of 60 kbps, but an average DPCCH channel bit rate of 15 kbps and an average DPDCH channel bit rate of 45 kbps. Slot format 15 has an instantaneous channel bit rate of 960 kbps, but an average DPCCH channel bit rate of 48 kbps and an average DPDCH channel bit rate of 912 kbps.

These channel bit rates represent the bit rate at the bottom of the physical layer after redundancy has been added to help protect the higher layer information. A 12.2 kbps speech call can use slot format 8 which has a spreading factor of 128. The equivalent uplink spreading factor for a 12.2 kbps speech call

Table 3.33 Downlink DPCH slot formats

Slot format	Spreading factor	Channel symbol rate (ksps)	Channel bit rate (kbps)	Bits per time slot	N_{Data1}	N_{Data2}	N_{TPC}	N_{TFCI}	N_{Pilot}
0	512	7.5	15	10	0	4	2	0	4
1	512	7.5	15	10	0	2	2	2	4
2	256	15	30	20	2	14	2	0	2
3	256	15	30	20	2	12	2	2	2
4	256	15	30	20	2	12	2	0	4
5	256	15	30	20	2	10	2	2	4
6	256	15	30	20	2	8	2	0	8
7	256	15	30	20	2	6	2	2	8
8	128	30	60	40	6	28	2	0	4
9	128	30	60	40	6	26	2	2	4
10	128	30	60	40	6	24	2	0	8
11	128	30	60	40	6	22	2	2	8
12	64	60	120	80	12	48	4	8[*]	8
13	32	120	240	160	28	112	4	8[*]	8
14	16	240	480	320	56	232	8	8[*]	16
15	8	480	960	640	120	488	8	8[*]	16
16	4	960	1920	1280	248	1000	8	8[*]	16

[*]TFCI bits are optional and may be replaced by a period of discontinuous transmission.

is 64. The downlink spreading factor is twice as large because the downlink DPDCH is distributed across both the in-phase and quadrature branches, i.e. the bit rate on either branch is half of the total bit rate and so each branch requires twice as much spreading to reach the chip rate of 3.84 Mcps. The average DPDCH channel bit rate for slot format 8 is 51 kbps. This means that the application bit rate is increased by a factor of 4.2 prior to insertion within the DPDCH. This allows a high quantity of redundancy to help protect the user plane speech information. Similar to the uplink, the speech service uses transparent mode RLC and so re-transmissions are not allowed and it is important that information is received reliably with the first transmission. The drawback of decreasing the spreading factor to allow greater redundancy is a reduced spreading gain. Reducing the spreading gain increases the quantity of interference visible to the received signal after de-spreading. Decreasing the spreading gain by a factor of 2 increases the interference power by 3 dB.

A downlink 64 kbps PS data connection can use slot format 13 which has a spreading factor of 32. Similar to the speech service, this spreading factor is twice as large as the equivalent uplink spreading factor, i.e. the spreading factor for an uplink 64 kbps PS data connection is 16. The average DPDCH channel bit rate for slot format 13 is 210 kbps and in this case, the application bit rate is increased by a factor of 3.3. A downlink 384 kbps PS data connection can use slot format 15 which has a spreading factor of 8 and an average DPDCH channel bit rate of 912 kbps. This means that the application bit rate is increased by a factor of 2.4 prior to insertion within the DPDCH.

If a UE is in CELL_DCH, and the F-DPCH from the release 6 version of the specifications is not used then the active set cells have to transmit the DPCCH during every time slot. The only exception is when the UE is in compressed mode and transmission gaps are generated during which the active set cells stop transmitting. These transmission gaps allow the UE to complete measurements on other UMTS RF carriers or other Radio Access Technologies (RAT), e.g. GSM. Transmitting the DPCCH during every time slot allows the UE to maintain air-interface synchronisation. It also allows inner loop power control to continuously track changes in the propagation and interference conditions. Transmitting the DPCCH during every time slot has an impact upon both the Node B power consumption and the utilisation of downlink channelisation codes. The F-DPCH described in Section 3.3.10 provides a solution which allows up to 10 UE to time share the same downlink channelisation code.

The channelisation code used to spread the downlink DPDCH and DPCCH is assigned by the RNC. A single channelisation code is used to spread the DPDCH and DPCCH on both the in-phase and quadrature branches. Each cell has a limited number of channelisation codes which are managed by the controlling RNC. The number of channelisation codes available depends upon the set of supported scrambling codes. Every cell in the network has one primary scrambling code which provides one channelisation code tree, i.e. a total of 1020 codes ($4 \times SF4$ codes, $8 \times SF8$ codes, $16 \times SF16$ codes,..., $512 \times SF512$ codes). If a cell supports the use of the left and right alternative scrambling codes then two additional channelisation code trees are introduced. However these additional code trees can only be used by connections which are in compressed mode using the SF/2 method. If a cell supports the use of secondary scrambling codes then a maximum of 15 additional code trees can be introduced, i.e. an additional code tree for each secondary scrambling code. If a cell supports both secondary and alternative scrambling codes then a maximum of 48 codes tress can be used, but 32 of these code trees are only available to connections in compressed mode using the SF/2 method, i.e. every primary and secondary scrambling code can have a left alternative scrambling code and a right alternative scrambling code. The number of channelisation code trees is summarised in Table 3.34.

DPCH which are spread using channelisation codes belonging to the same code tree are orthogonal as long as one code is not the parent of the other. This concept is described in Section 2.6.2. DPCH which are spread using channelisation codes belonging to different code trees are not orthogonal. This results from the DPCH having different scrambling codes, i.e. DPCH must have the same scrambling code to be orthogonal to one another. This means that increasing the number of scrambling codes increases the availability of channelisation codes, but also increases the level of downlink interference.

Assuming that a cell is configured to use only its primary scrambling code then the controlling RNC must assign DPCH channelisation codes from a single code tree and must account for the codes which are already used by the common channels and by HSDPA. Figure 3.29 illustrates an example of the channelisation codes used by the common channels and a subset of the channelisation codes which are available for the DPCH. This figure does not illustrate any channelisation codes assigned to HSDPA.

The common channels illustrated in Figure 3.29 make use of five channelisation codes and block a further 29, i.e. a total of 34 codes are occupied. This represents 3.3% of the total code tree. In this case, the channelisation code tree is unlikely to limit the cell capacity unless the quantity of traffic is high and the air-interface conditions are good. If HSDPA is active the probability of the channelisation code tree limiting cell capacity increases. HSDPA makes use of up to 15 HS-PDSCH, each of which requires a spreading factor 16 channelisation code. HSDPA also requires at least one HS-SCCH which uses spreading factor 128. If code multiplexing is used to simultaneously schedule multiple HSDPA connections then multiple HS-SCCH are required. Assuming five HS-PDSCH and one HS-SCCH means that the channelisation code tree occupancy increases from 3.3% to 35.6%. The occupancy

Table 3.34 The impact of scrambling code support upon the number of channelisation code trees

	Primary scrambling code (1 per cell)	Secondary scrambling codes (15 per cell)	Alternative scrambling codes (2 per primary and 2 per secondary SC)	
Case 1	√			1 code tree
Case 2	√		√	3 code trees
Case 3	√	√		16 code trees
Case 4	√	√	√	48 code trees

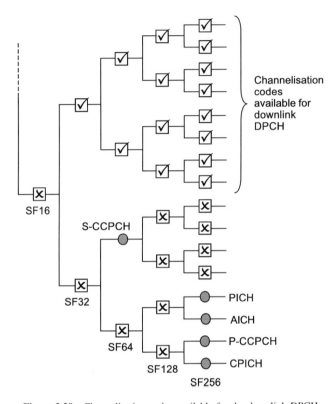

Figure 3.29 Channelisation codes available for the downlink DPCH

increases further as more codes are assigned to HSDPA. Table 3.35 summarises a set of example channelisation code tree occupancies when HSDPA is active.

These figures illustrate that the number of channelisation codes available for DPCH connections is severely limited when 15 HS-PDSCH are active. The example of 15 HS-PDSCH and 3 HS-SCCH leaves only one SF128 code, 2 SF256 and four SF512 codes for DPCH connections, i.e. only one 12.2 kbps speech connection could be supported. The channelisation code occupancy is also relatively high when HSDPA uses either five or ten HS-PDSCH. This indicates that cell capacity is more likely to become limited by the channelisation code tree when HSDPA is active.

Table 3.35 Channelisation code tree occupancy generated by common channels and HSDPA

Number of HS-PDSCH	Number of HS-SCCH	Channelisation code tree occupancy (%)
0	0	3.3
5	1	35.6
10	1	66.8
10	2	67.5
15	1	97.8
15	2	98.5
15	3	99.3

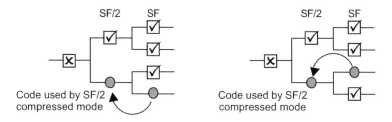

Figure 3.30 Channelisation codes used by the SF/2 compressed mode method

3GPP TS 25.213 specifies that when a DPCH makes use of the SF/2 compressed mode method then the parent channelisation code is always used to spread the signal. This relationship is illustrated in Figure 3.30.

It is only possible to assign the parent code for SF/2 compressed mode if the neighbouring channelisation code with the same parent is unassigned. Otherwise the parent channelisation code is blocked. The solution to this scenario is to make use of an alternative scrambling code. An alternative scrambling code guarantees that an SF/2 channelisation code will be available for compressed mode. 3GPP TS 25.213 specifies that channelisation codes belonging to the lower half of the tree can make use of the left alternative scrambling code whereas channelisation codes belonging to the upper half of the tree can make use of the right alternative scrambling code. This concept is illustrated in Figure 3.31.

The drawback of using the left and right alternative scrambling codes is that the DPCH becomes non-orthogonal and increases the quantity of downlink interference. This leads to a trade-off when the controlling RNC assigns new channelisation codes to DPCH connections. The trade-off balances the requirement to minimise downlink interference with the requirement to avoid fragmentation of the code tree. The probability of connections becoming blocked increases if the code tree becomes fragmented. The trade-off between having to use the alternative scrambling codes and code tree fragmentation is illustrated in Figure 3.32.

The code assignment strategy illustrated by scenario 1 avoids the requirement to use the alternative scrambling codes because both of the assigned codes have a parent code which is available for SF/2 compressed mode. However, the code tree is relatively fragmented and there are only two codes available for new connections. The strategy illustrated by scenario 2 avoids code tree fragmentation and an additional code is available for new connections. However, the parent code is not available for

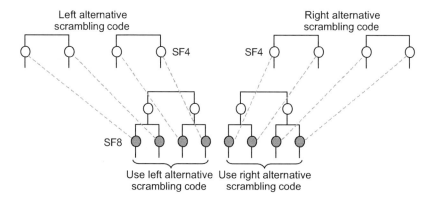

Figure 3.31 Left and right alternative scrambling codes for SF/2 compressed mode

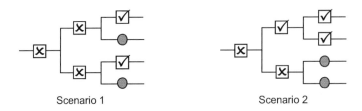

Scenario 1   Scenario 2

Figure 3.32   Example channelisation code assignments

SF/2 compressed mode and so an alternative scrambling code must be used. In practice, the requirement to avoid code tree fragmentation is given greater importance and the alternative scrambling codes are used when necessary. In general, networks are designed to minimise the requirement for compressed mode and some connections may be configured to use the Higher Layer Scheduling (HLS) method rather than the SF/2 method. The HLS compressed mode method does not require a change of channelisation code.

The RNC configures the frame timing of the downlink DPCH by defining a frame offset and a chip offset. The combination of the frame offset and chip offset defines the timing of the DPCH Connection Frame Number (CFN) relative to the timing of the cell System Frame Number (SFN). This timing difference is illustrated in Figure 3.33.

The SFN has a range from 0 to 4095 whereas the CFN has a range from 0 to 255. This means that there are 16 cycles of the CFN for a single cycle of the SFN. Modulo 256 arithmetic is used when deriving the CFN from the SFN. The frame offset and chip offset are signalled to the Node B using an NBAP Radio Link Setup Request message. An example of the relevant section of an NBAP Radio Link Setup Request message is presented in Log File 3.14.

3GPP TS 25.433 specifies that the frame offset can have any integer value between 0 and 255. When a UE makes the transition to CELL_DCH and the first DPCH radio link is established the range is limited to between 0 and 7. In this scenario, a range of eight radio frames is sufficient and corresponds to the maximum Transmission Time Interval (TTI) of 80 ms. This range of 8 radio frames is mirrored by the maximum range defined for the corresponding RRC signalling used to inform the UE of the equivalent information. The corresponding RRC signalling uses a range from 0 to 306688 chips, i.e. seven frames and 37888 chips. It is shown later in this section that the full range of frame offset values become applicable when a UE enters soft handover and the NBAP Radio Link Setup Request message is used to specify the timing of a new radio link. 3GPP TS 25.433 specifies that the chip offset can have any integer value between 0 and 38399. 3GPP TS 25.402 specifies that when the Node B receives the chip offset then it is rounded to the nearest multiple of 256 chips. When a UE makes the transition to CELL_DCH and the first DPCH radio link is established then the chip offset is rounded to the nearest multiple of 512 chips. This corresponds to the length of the longest downlink channelisation code. This step size is mirrored by the value defined for the corresponding RRC signalling, i.e. a step size of 512 is

Figure 3.33   Frame timing of the downlink DPCH relative to the P-CCPCH

```
RadioLinkInformationItem-RL-SetupReq
        - rL-ID: 1
        - c-ID: 11897
        - first-RLS-Ind: first-RLS
        - frameOffset: 7
        - chipOffset: 1536
        - propagationDelay: 3
        dL-ChannelisationCodeInformation-RL-SetupReq
                DL-ChannelisationCodeInformationItem-RL-SetupReq
                - dL-ScramblingCode: 0
                - dL-ChannelisationCodeNumber: 8
        - dL-TransmissionPower: -142
        - maxDL-Power: -51
        - minDL-Power: -180
```

Log File 3.14   Example section from an NBAP Radio Link Setup Request message

used when selecting a value from between 0 and 306688 chips. It is shown later in this section that the step size of 256 chips becomes applicable when a UE enters soft handover and the NBAP Radio Link Setup message is used to specify the timing of a new radio link.

Selecting the frame offset from the range 0 to 7 and the chip offset from the range 0 to 37888 (after accounting for the step size of 512 chips) allows the start of the DPCH CFN timing to be distributed across a period of eight radio frames. The RNC may select the frame offset and chip offset belonging to different DPCH at random to help distribute the CFN starting times. This helps to distribute the Iub and air-interface loads and reduces the probability of experiencing large peaks in activity. The TTI defines how frequently the RNC transfers downlink transport block sets belonging to a specific DPCH to the Node B, and how frequently the Node B transfers uplink transport block sets to the RNC. DPCH TTI can be configured with values of 10, 20, 40 and 80 ms. An 80 ms TTI is only allowed when the downlink spreading factor is 512. In this case, the bit rate is relatively low and the increased delay associated with an 80 ms TTI is less significant. Larger TTI help to improve physical layer performance by allowing a longer interleaving period which results in fewer consecutive bit errors after de-interleaving at the receiver. Distributing the values of the frame offset and chip offset means that the time instants when transport blocks are transferred between the RNC and Node B are also distributed. The quantisation step of 512 chips is applied to ensure that all downlink DPCH using the same scrambling code remain orthogonal. Channelisation codes are only orthogonal to one another when they are synchronised. The longest channelisation code length is 512 chips and so applying a chip offset step size of 512 chips ensures that channelisation code boundaries always coincide.

The CFN cycle of 256 radio frames corresponds to 2.56 seconds. It is not necessary for the Node B and UE to wait until the CFN equals 0 before starting to transmit. The Node B will start to transmit as soon as it has received the NBAP Radio Link Setup Request message irrespective of the CFN. Likewise, the UE will start to transmit once it has received the corresponding RRC message and has achieved air-interface synchronisation. The RRC message used to inform the UE of the CFN timing depends upon the RRC state transition. If the UE is making the transition from RRC Idle mode to CELL_DCH then the RRC Connection Setup message is used to inform the UE. If the UE is making the transition from CELL_FACH to CELL_DCH then the Radio Bearer Reconfiguration, Radio Bearer Setup or Transport Channel Reconfiguration messages can be used. An example of the relevant section of an RRC Connection Setup message is presented in Log File 3.15.

This log file signals the default DPCH offset using a value 528. The signalled value is mapped to the actual value by multiplying by 512 chips, i.e. the actual value is 270336 chips. The signalled value can range from 0 to 599 which corresponds to 0–306688 chips. The actual value of the default DPCH offset

```
dl-CommonInformation
dl-DPCH-InfoCommon
        cfnHandling: maintain
        modeSpecificInfo fdd
                powerOffsetPilot-pdpdch:          0
                spreadingFactorAndPilot
                        sfd256:                   pb4
                positionFixedOrFlexible:          flexible
                tfci-Existence:                   true
        modeSpecificInfo fdd
                defaultDPCH-OffsetValue:          528
dl-InformationPerRL-List
dl-InformationPerRL-List value 1
        modeSpecificInfo fdd
                primaryCPICH-Info
                primaryScramblingCode:            32
        dl-DPCH-InfoPerRL
                pCPICH-UsageForChannelEst:        mayBeUsed
                dpch-FrameOffset:                 6
                dl-ChannelisationCodeList
                dl-ChannelisationCodeList value 1
                sf-AndCodeNumber
                        sf256:                    8
                tpc-CombinationIndex:             0
```

Log File 3.15   Example section from an RRC Connection Setup message

provided in the RRC Connection Setup message is equal to the combination of the frame offset and chip offset provided to the Node B in the NBAP Radio Link Setup message. The example in Log File 3.14 has a frame offset of 7 and a chip offset of 1536. These figures combine to generate an offset of $(7 \times 38400) + 1536 = 270336$ chips. The UE is then able to derive the CFN timing from the SFN timing in the same way as the Node B. The UE reads the SFN from the BCCH information broadcast on the BCH transport channel and P-CCPCH physical channel. The BCH uses a TTI of 10 ms and the SFN is available to read during every radio frame. An example of the SFN within the decoded BCCH information is illustrated in Log File 3.16.

```
BCCH-BCH-Message
message
sfn-Prime: 561
payload
      completeAndFirst
      completeSIB-List
      completeSIB-List value 1
      sib-Type:        systemInformationBlockType1
              sib-Data-variable
              Bin: E4 9E 0D 08 48 09 16 40 10 00 11 92 03 D3 60 9A
      firstSegment
      sib-Type:        systemInformationBlockType3
              seg-Count       : 2
              sib-Data-variable
              Bin: 01 0C B2 DC 42 DA C0 09 44
```

Log File 3.16   Example BCCH message from the P-CCPCH

The main content of the BCCH is the system information blocks but the SFN is included within the RRC header. The UE has to apply modulo 256 arithmetic when deriving the CFN from the SFN in the same way as the Node B. 3GPP TS 25.331 specifies that when a UE makes a transition to CELL_DCH then the CFN is calculated using the equation below.

$$CFN = (SFN - (DOFF \, div \, 38400)) \, mod \, 256$$

Where DOFF is equal to the default DPCH offset specified in the RRC message used to instruct the UE to make the transition to CELL_DCH. Log File 3.15 also includes a radio link specific frame offset. In this example, the signaled value is equal to 6. The signalled value is mapped to the actual value by multiplying by 256 chips, i.e. the actual value is 1536 chips which equals the chip offset provided to the Node B in the NBAP Radio Link Setup Request message. The signalled value can have a range from 0 to 149 which corresponds to 0 to 38144 chips. In the case of the RRC Connection Setup message this radio link specific DPCH frame offset does not add any further information because the value of 1536 chips can be derived from the default DPCH offset by applying modulo 38400 arithmetic.

The radio link specific DPCH frame offset becomes more important when new radio links are added to the active set. In this case, the default DPCH offset is not required because it has already been used to initialise the CFN timing, but a new radio link specific DPCH frame offset is used to specify the DPCH radio frame timing of the new radio link relative to the P-CCPCH radio frame timing of the new cell. This scenario is illustrated in Figure 3.34.

The radio link specific DPCH frame offset informs the UE of where to search for the start of a radio frame belonging to the new radio link. This information is provided to the UE within the Active Set Update message. An example Active Set Update message illustrating this information is presented in Log File 3.17. In this example, the new radio link specific DPCH frame offset has a signalled value of 12 which corresponds to an actual value of 3072 chips.

The Node B providing the new radio link is informed of the corresponding frame offset and chip offset illustrated in Figure 3.34. If the new radio link is added using a Node B which does not already have any radio links within the active set then the NBAP Radio Link Setup Request message is used. In this case, it is necessary to use the full range of the frame offset parameter, i.e. 0 to 255 rather than the limited range of 0 to 7 which was used during the establishment of the first radio link. It is necessary to use the full range because individual Node B are not synchronised with one another and their SFN timing can have any offset. If the new radio link is added using a Node B which already has one or more radio links within the active set then the NBAP Radio Link Addition Request message is used. Cells belonging to the same

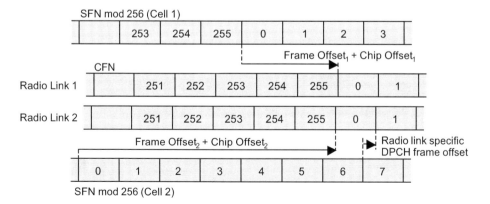

Figure 3.34 Frame timing of two DPCH radio links relative to their P-CCPCH

```
activeSetUpdate-r3
rrc-TransactionIdentifier:        0
maxAllowedUL-TX-Power:            24
rl-AdditionInformationList
rl-AdditionInformationList value 1
        primaryCPICH-Info
        primaryScramblingCode: 34
        dl-DPCH-InfoPerRL
                pCPICH-UsageForChannelEst:        mayBeUsed
                dpch-FrameOffset:                 12
                dl-ChannelisationCodeList
                dl-ChannelisationCodeList value 1
                        sf-AndCodeNumber
                        sf128:  5
                tpc-CombinationIndex:   0
        tfci-CombiningIndicator: false
```

Log File 3.17   Example Active Set Update message

Node B have SFN timings which are offset using the T_Cell parameter introduced in Section 3.3.1. The T_Cell parameter has a range from 0 to 2304 chips with a step size of 256 chips. This step size provides justification for the chip offset parameter having a step size of 256 chips.

This step size means that when the active set includes multiple radio links it is unlikely that all radio links will arrive at the UE completely synchronised. Instead the UE will be required to buffer within a certain time window until it has received the dominant multipath signals from all radio links. If the relative time difference between radio links becomes too great as a result of delays changing subsequent to radio link establishment then the UE can trigger events 6f and 6g to inform the RNC that the timing of a specific radio link requires adjustment. The step size of 512 chips which is used for the first DPCH radio link within the active set was justified in terms of it preserving orthogonality between all channelisation codes. The use of a 256 chip step size for the timing of subsequent radio links means that those radio links may not always be orthogonal if spreading factor 512 is used. However spreading factors 256 to 4 will remain orthogonal.

The section of the NBAP Radio Link Setup message presented in Log File 3.14 also specifies the radio link identifier. This identifier has a range from 0 to 31 and is UE specific rather than Node B specific. This indicates that the 3GPP specifications allow the controlling RNC to establish a maximum of 32 radio links for a specific UE. In practice, network implementations are likely to limit the number of radio links within the active set, e.g. a maximum active set size of 3 or 6. The cell identifier is also included to instruct the Node B to setup the new radio link at a specific cell. The cell identifier has a range from 0 to 65535 and must be unique for all cells connected to the controlling RNC. The first radio link set indication informs the Node B whether the new radio link is the first radio link for the UE or is a soft handover radio link. This information can influence the pattern of uplink Transmit Power Control (TPC) commands sent by the Node B prior to achieving uplink synchronisation. If the NBAP Radio Link Setup Request message indicates that the new radio link is the first radio link for the UE then the value of the parameter 'downlink TPC pattern 01 count' is recalled. This parameter is signalled to the Node B within the NBAP Cell Setup Request message during the cell start-up procedure. If the value assigned to the 'downlink TPC pattern 01 count' parameter is greater than 0 then the Node B transmits n instances of the TPC command pattern (0,1) followed by one instance of the TPC command (1), where n equals the value assigned to the 'downlink TPC pattern 01 count' parameter. This pattern is transmitted repeatedly with a reset after every fourth radio frame until uplink synchronisation has been achieved. Small values of n instruct UE to increase their average transmit powers more rapidly. If the value of n is equal to 0, or if the new radio link being established is not the first radio link for the UE then the Node B transmits a continuous stream of power-up (1) commands until uplink synchronisation has been achieved.

The propagation delay included within the NBAP Radio Link Setup Request message is a reflection of the value measured by the Node B during the random access procedure. This information is only available if the radio link establishment is for the first radio link. Propagation delay information is not included when the radio link establishment is for soft handover. The propagation delay information helps the Node B to achieve uplink synchronisation by providing guidance on where to search for the uplink DPCCH. The actual value of the propagation delay is equal to the signaled value multiplied by 3 chips. The example presented in Log File 3.14 has an actual value of 9 chips. Based upon the chip rate of 3.84 Mcps, 9 chips corresponds to a delay of $9/3.84 = 2.3$ µs. This can be translated to a distance by multiplying by the speed of light, i.e. 3×10^8 ms^{-1}. In this example the corresponding distance is 703 m. The delay associated with a single chip is equivalent to 78 m. The 3GPP specifications prior to release 7 support a maximum propagation delay of 765 chips which corresponds to 59.8 km. The release 7 version of the specifications increases this maximum propagation delay to 3069 chips which corresponds to 239.8 km.

The NBAP Radio Link Setup Request message presented in Log File 3.14 references the downlink scrambling code using a value of 0. This indicates that the primary downlink scrambling code should be used to scramble the DPCH. A value greater than 0 would indicate that a secondary scrambling code should be used. The downlink DPCH slot format is specified elsewhere within the NBAP Radio Link Setup Request message. The slot format defines the spreading factor of the channelisation code to be applied. The section of the message presented in Log File 3.14 specifies that the channelisation code number should be 8. Channelisation code numbering starts from 0 and so this represents the ninth code in the tree. The initial downlink transmit power is signaled using a value of -142. The actual value equals the signaled value divided by 10, and the actual value represents the initial transmit power of the DPCH relative to the transmit power of the CPICH. The interpretation is the same for the maximum and minimum downlink DPCH transmit powers. In this example, if the CPICH transmit power is 33 dBm the initial transmit power for the radio link is 18.8 dBm, while the maximum downlink transmit power is 27.9 dBm and the minimum downlink transmit power is 15 dBm. These powers are relatively low because this example radio link establishment is for a 3.4 kbps signaling connection. If the radio link was for a 12.2 kbps speech connection then the maximum downlink transmit power would be of the order of 33 dBm while the equivalent figure for a 384 kbps data connection would be of the order of 40 dBm.

Log File 3.18 presents another section from the same NBAP Radio Link Setup Request message. This section starts by defining two Calculated Transport Format Combinations (CTFC). The value of 0 corresponds to DTX for the DPDCH while the value of 1 corresponds to the single transport channel

```
DL-DPCH-Information-RL-SetupReq
tFCS
cTFC
- ctfc2bit: 0
cTFC
- ctfc2bit: 1
dL-DPCH-SlotFormat
- non-extended: 5
- multiplexingPosition: flexible-positions
powerOffsetInformation
        - pO1: 0
        - pO2: 12
        - pO3: 0
- tPC-DL-StepSize: one
- limitedPowerInc: not-used
```

Log File 3.18 Example section from an NBAP Radio Link Setup Request message

active at its maximum bit rate. The maximum bit rate is defined elsewhere within the NBAP Radio Link Setup Request message using the TTI and transport block set size.

The Node B is also informed that downlink DPCH slot format 5 should be used. This corresponds to a spreading factor of 256 with four DPCCH pilot bits, two TPC bits and two TFCI bits per time slot. This information complements the channelisation code number provided within the section of the NBAP Radio Link Setup Request message presented in Log File 3.14. In the case of the downlink DPCH, the spreading factor is fixed at the value specified within the NBAP Radio Link Setup Request message. This is in contrast to the uplink DPDCH which is configured using a minimum allowed spreading factor. The uplink DPDCH is able to change its spreading factor to suit the transport format combination being transmitted at any point in time. The uplink rate matching is then dynamic in terms of ensuring that the number of data bits at the bottom of the physical layer is always exactly equal to the number of DPDCH bits offered by the selected spreading factor. In the downlink direction, the spreading factor is selected based upon the transport format combination which generates the greatest bit rate. Other transport format combinations which have a lower bit rate use the same spreading factor but allow periods of DTX during which the DPDCH is not transmitted.

Log File 3.18 also illustrates that the Node B is informed that it should use flexible positions when multiplexing the transport channel data onto the DPDCH. This is in contrast to using fixed positions which means that the data belonging to each transport channel has a fixed number of DPDCH bits reserved and a fixed position within the DPDCH. The use of fixed positions allows the receiver to complete blind detection of the transport format combination and avoids the requirement to transmit TFCI bits. The Node B is then informed of the DPCCH power offsets relative to the DPDCH. These power offsets are fixed for all transport format combinations. This is in contrast to the uplink where the power difference between the DPDCH and DPCCH can change according to the transport format combination. The variables PO1, PO2 and PO3 correspond to the power offsets for the TFCI, TPC and pilot bits respectively. The actual value of the power offset is equal to the signaled value divided by 4. In this example, the TFCI and pilot bits are transmitted with an equal power to the DPDCH whereas the TPC bits are transmitted with 3 dB greater power. The NBAP Radio Link Setup Request message then instructs the Node B to use a 1 dB step size for inner loop power control, and not to use the limited power increase algorithm.

The section of the RRC Connection Setup message presented in Log File 3.15 informs the UE of the power offset between the downlink DPCCH pilot bits and the downlink DPDCH, i.e. PO3. The actual value is equal to the signaled value divided by 4. In this example, the actual value is equal to 0 dB. Knowledge of this power offset helps the UE to define received power thresholds when decoding the DPDCH. The UE is not informed of PO1 nor PO2. The UE is informed that the downlink spreading factor is 256, there are 4 pilot bits per time slot and TFCI bits are transmitted. This information allows the UE to deduce that slot format 5 is to be used for the downlink DPCH. The UE is also informed of the primary scrambling code belonging to the cell where the new connection is being established. In many cases, this information is not essential because the UE already has knowledge of the primary scrambling code. However, if the UE is being redirected to another cell potentially on a different RF carrier then the UE will not know the primary scrambling code of the target cell. The UE is informed that the primary CPICH can be used for channel estimation. It is usual to allow the primary CPICH to be used for channel estimation when both the primary CPICH and downlink DPCH are transmitted from the same antenna beam. If beamforming has been implemented then the primary CPICH and the downlink DPCH are likely to be transmitted using different antenna beams, i.e. the primary CPICH is broadcast using a wide antenna beam to provide coverage across the entire cell whereas the downlink DPCH is transmitted using a relatively narrow antenna beam directed towards the UE. In this case, the primary CPICH and the downlink DPCH experience different link loss. A secondary CPICH transmitted using the same antenna beam as the downlink DPCH can be used for channel estimation instead. Log File 3.15 also illustrates that the UE is informed of the channelisation code to be used for the downlink DPCH. The final element within this section of the RRC Connection Setup message is an indication that the new radio link belongs to a radio link set which has been assigned a TPC combination index of 0. Radio links are assigned the same TPC combination index when their TPC commands are always equal. This information informs the UE that the TPC commands can be combined prior to making a decision regarding which TPC command was transmitted. It is usual for radio links belonging to the same Node B to have the same TPC combination index and for those cells to always transmit equal TPC commands. When a Node B has more than a single cell within the active set then the uplink signals received by each cell are combined within the same receiver, i.e. within the RAKE receiver or advanced receiver. The Node B then measures a single SIR from the combined DPCCH pilot bits which is then used to derive an appropriate TPC command for the UE. This TPC command is subsequently transmitted by all active set cells belonging to that Node B. When active set cells belong to different Node B their TPC commands may be different and a different TPC combination index is assigned. This means that equal TPC combination indices are associated with softer handover whereas different TPC combination indices are associated with soft handover. When the active set cells have different TPC combination indices the UE has to decode multiple TPC commands for use by the inner loop power control algorithm.

Figure 3.35 illustrates the timing of the downlink DPCH time slot relative to the timing of the uplink DPCCH time slot. The timing relationship at the UE is fixed by the 1024 chip offset between the start for the received downlink DPCH time slot and the start of the transmitted uplink DPCCH time slot. The timing relationship at the Node B depends upon the propagation delay.

From the Node B perspective, the start of a transmitted downlink DPCH time slot follows the start of a received uplink DPCCH time slot by $2560 - (1024 + 2 \times$ number of chips corresponding to the propagation delay) chips. The figure of 2560 chips is equal to the number of chips within a single time slot. This expression indicates that the time difference becomes less as the propagation delay increases. If the propagation delay becomes too great then it will not be possible to use the TPC command at the end of the uplink time slot to determine the transmit power change at the start of the next downlink time slot. In this case, there will be a one time slot delay in the reaction of the inner loop power control. Assuming no processing delay within the Node B and that the inner loop power control delayed reaction is to be avoided then the maximum allowed propagation delay corresponds to 768 chips, i.e. 60 km. In practice, the Node B will have a processing delay and the maximum cell range will be less. If a one time slot inner loop power control delay can be accepted then the maximum propagation delay

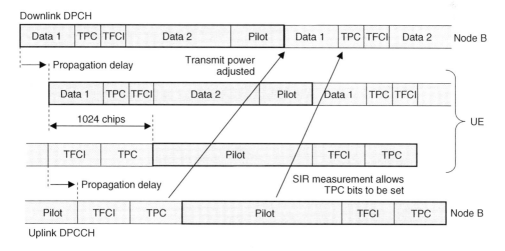

Figure 3.35 Timing of the downlink DPCH relative to the uplink DPCCH

can be increased to $768 + 2560/2 = 2048$ chips which is equivalent to 160 km. Larger cell ranges will have a greater impact upon inner loop power control. This impact is generally accepted as a consequence of supporting such large cell ranges.

3.3.10 Fractional Dedicated Physical Channel (F-DPCH)

The Fractional Dedicated Physical Channel (F-DPCH) is introduced by release 6 of the 3GPP specifications. It is a downlink physical channel intended for use when both the user plane and signalling radio bearers are transferred using the HS-PDSCH, i.e. there is no DPDCH. The F-DPCH represents a special case of the downlink DPCCH which helps to conserve downlink channelisation code resources. The conservation of downlink channelisation codes becomes particularly important for HSDPA applications which have relatively low bit rates and relatively low activity factors. In these cases, there can be a large number of simultaneously active connections while the set of HS-PDSCH occupies a significant percentage of the channelisation code tree, i.e. the set of HS-PDSCH leaves little code tree resource for the downlink DPCCH. Voice over IP (VoIP) represents an example application which justifies the introduction of the F-DPCH. A comparison of the F-DPCH and downlink DPCCH is presented in Table 3.36.

3GPP TS 25.211 specifies that the F-DPCH uses a spreading factor of 256 and has only a single slot format which includes two uplink Transmit Power Control (TPC) bits. These 2 TPC bits accommodate a single QPSK symbol. Uplink TPC bits are required for inner loop power control. Inner loop power control is necessary to help ensure that active set cells receive sufficient uplink power without receiving excessive uplink power. Connections which use the F-DPCH in the downlink direction are likely to use the DPCCH, E-DPCCH, E-DPDCH and HS-DPCCH in the uplink direction. These four physical channels are power controlled by the TPC bits within the F-DPCH. The uplink DPDCH is not necessary if both the user plane and signalling radio bearers are transferred using the E-DPDCH. The downlink TPC bits within the uplink DPCCH power control only the F-DPCH. The downlink DPDCH is no longer present and other downlink physical channels such as the HS-PDSCH and HS-SCCH have different power control mechanisms. These are described in Chapter 6.

Initial proposals for the F-DPCH included both pilot bits and TPC bits. The use of dedicated downlink pilot bits was subsequently removed during the specification process. Downlink pilot bits are used by the UE for inner loop power control purposes, i.e. the UE measures the downlink SIR from the pilot bits and compares the result with the SIR target defined by outer loop power control. This comparison can be used to generate the downlink TPC commands which instruct the Node B to either increase or decrease its transmit power. In the case of the F-DPCH, the UE requires an alternative method of generating the downlink TPC commands. This requirement is handled by the RNC providing the UE with a TPC command error rate target. This value is signalled during connection establishment, e.g. within a Radio Bearer Reconfiguration message if the UE is making the transition from CELL_FACH to CELL_DCH. Providing a TPC command error rate target is analogous to providing a BLER target when the DPDCH is used. The UE translates the TPC command error rate target into an initial SIR target, similar to how it would translate a BLER target into an initial SIR

Table 3.36 Comparison of the fields included within the downlink DPCCH and F-DPCH

	Downlink DPCCH	F-DPCH
TPC bits	Yes	Yes
Pilot bits	Yes	No
TFCI bits	Optional	No

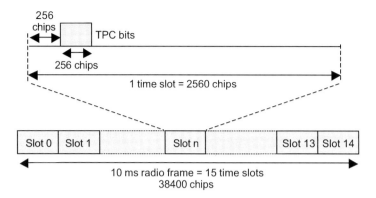

Figure 3.36 Structure of the F-DPCH

target. The UE then measures the SIR from the F-DPCH TPC bits and compares the result with the target SIR to generate the downlink TPC commands. If the UE is in soft handover then the TPC error rate target requirement is applied to the radio link associated with the HS-PDSCH (the HS-PDSCH is only ever transmitted from one active set cell). The UE may use an outer loop power control algorithm to adjust the target SIR if the estimated TPC command error rate differs from the target.

Downlink pilot bits belonging to the DPCCH are also used by the UE to maintain an air-interface synchronisation status, i.e. if the received signal quality becomes too poor then the UE deduces that it can no longer receive the uplink TPC commands reliably and switches its transmit power off. This out-of-synchronisation handling is specified by 3GPP TS 25.101. The procedure is similar for the F-DPCH except that the UE must use measurements from the TPC commands to determine whether or not the received signal quality has become too poor.

The F-DPCH does not require Transport Format Combination Indicator (TFCI) bits because the DPDCH is no longer present and the equivalent information for the HS-PDSCH is included within the HS-SCCH.

The structure of the F-DPCH is illustrated in Figure 3.36. The F-DPCH for a single connection is active for 10% of the time and inactive for the remaining 90% of the time. This means that if other connections are synchronised appropriately then up to ten connections can make use of the same downlink channelisation code.

4

Non-Access Stratum

4.1 Concepts

- The Non-Access Stratum (NAS) makes use of the Access Stratum (AS) to transfer data between the core network and the UE. The core network and UE include both AS and NAS layers whereas the RNC and Node B include only AS layers.
- The control plane of the NAS is divided into two sublayers: the Connection Management (CM) sublayer and the Mobility Management (MM) sublayer. The MM sublayer provides connectivity for the CM sublayer.
- Mobility Management and Call Control entities belong to the CS domain. GPRS Mobility Management and Session Management entities belong to the PS domain.
- Key 3GPP specifications: TS 23.122 and TS 23.221.

The Non-Access Stratum (NAS) represents a set of higher layer protocols which use the Access Stratum to transfer data between the core network and UE. The core network and UE include both AS and NAS layers whereas the RNC and Node B include only AS layers. Figure 4.1 illustrates the concept of the AS providing a data transfer service for the NAS.

The radio interface protocol stack provides AS connectivity between the UE and RNC, whereas the Iu interface protocol stack provides AS connectivity between the RNC and core network. When a UE establishes a connection from RRC Idle mode, the radio interface protocol stack is the first section of the AS to provide connectivity, i.e. Signalling Radio Bearers (SRB) are set up to allow control plane signalling between the UE and RNC. Once the SRB have been set up, the UE is able to use an RRC Initial Direct Transfer message to trigger the setup of a signalling connection between the RNC and core network. The RRC Initial Direct Transfer message also includes an initial NAS message for the core network, e.g. CM Service Request, or Location Updating Request. Once both the SRB and Iu signalling connection have been established, the control plane of the AS is able to transfer messages for the control plane of the NAS. The connection can be towards either the CS or PS core networks, depending upon the reason for establishing the connection. Further signalling from the NAS and from within the AS is required to establish a user plane connection between the UE and the core network. If the control plane connection has been established for a signalling procedure such as a location area update then a user plane connection is not required and the procedure is completed using only the control plane. If the control plane connection has been established for an end-user application such as a

---

Radio Access Networks for UMTS   Chris Johnson
© 2008 John Wiley & Sons, Ltd

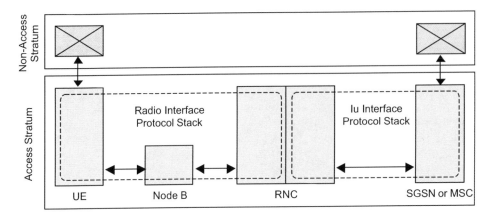

Figure 4.1   Access Stratum providing a data transfer service for the Non-Access Stratum

CS speech call or a PS data transfer then a user plane connection must be established within the AS to allow the NAS to transfer application data.

The control plane of the NAS belonging to a UE is presented in Figure 4.2. In the case of the core network, the CS and PS entities are separated between the two core network domains. The control plane of the NAS is divided into two sublayers: the Connection Management (CM) sublayer and the Mobility Management (MM) sublayer. Both of these sublayers belong to layer 3 of the overall protocol stack.

The MM sublayer provides connectivity for the CM sublayer. The NAS message which triggers the RRC Initial Direct Transfer message originates from the MM sublayer, i.e. the MM sublayer ensures that there is a signalling connection for either the MM or CM sublayer. If, for example, the connection has been established for a location area update then the CM sublayer is not involved and does not transfer any messages. If the connection has been established for a CS speech call or a PS data transfer then the MM sublayer provides the signalling connection for the CM messages which are subsequently

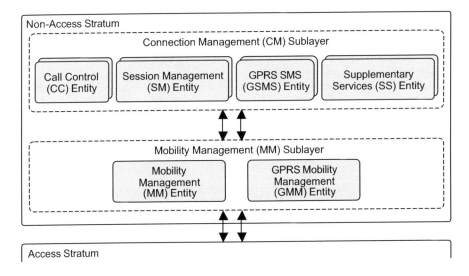

Figure 4.2   Control plane of the Non-Access Stratum (NAS) within a UE

used to establish the user plane connection. The signalling connection established by the MM sublayer is also known as a MM connection.

The MM entity within the MM sublayer and the Call Control (CC) entity within the CM sublayer are associated with CS domain services, whereas the GMM entity within the MM sublayer and the Session Management (SM) entity within the CM sublayer are associated with PS domain services. The CM sublayer also includes entities for the GPRS Short Message Service (GSMS) and for Supplementary Services (SS). The set of CM sublayer entities shown in Figure 4.2 is not necessarily exhaustive and other entities can be included. For example, the GSMS entity is used if SMS are transferred using the PS core network domain, but an SMS entity is used if SMS are transferred using the CS core network domain. In this case, an SMS entity would be included as part of the CM sublayer.

4.2 Mobility Management

- The Mobility Management sublayer is responsible for supporting UE mobility from the perspective of the CS and PS core networks.
- Separate CS and PS state machines are maintained in both the UE and core network. Detached, Idle and Connected states are used by each state machine.
- A Detached UE cannot be contacted by the network and the UE does not provide updates of its location. Registration allows the transition from Detached to Idle and the UE becomes attached to the core network. The UE is then required to provide location updates. Establishing an Iu signalling connection allows the transition from Idle to Connected. The UE and core network are then able to transfer NAS information.
- TMSI and P-TMSI are assigned as temporary identities when a UE registers with the CS and PS core networks respectively. Temporary identities help to improve confidentiality.
- The minimum size of a location area is one cell. The maximum size is the collection of cells connected to a single MSC. A location area can span multiple RNC. Routing areas are defined within location areas. There can be a single routing area within each location area.
- Key 3GPP specifications: TS 23.221, TS 24.007 and TS 24.008

The Mobility Management (MM) sublayer is responsible for supporting UE mobility from the perspective of the CS and PS core networks. It is also responsible for helping to maintain the confidentiality of the UE identity. A UE has to register with the core network before the MM sublayer starts to support mobility or identity confidentiality. Separate CS and PS state machines are maintained in both the UE and the core network. The CS and PS state machines run independently, but the CS state machine within the UE should remain synchronised with the CS state machine within the MSC. Likewise, the PS state machine within the UE should remain synchronised with the PS state machine within the SGSN. Figure 4.3 illustrates the concept of the CS and PS state machines. These state machines run in parallel to the RRC state machine described in Section 2.2.

The main CS service states are CS Detached, CS Idle and CS Connected. Likewise, the main PS service states are PS Detached, PS Idle and PS Connected. A UE starts in the CS and PS Detached states after being switched on. At that time the core network has no knowledge of the UE and has not provided a temporary UE identity. The UE cannot be accessed by the network and the UE does not provide any updates of its location. Under normal circumstances a UE will register with at least the CS core network. This registration procedure allows the UE to move into the CS Idle state and the UE becomes IMSI attached. The CS core network is informed of the UE location in terms of its Location Area (LA) and the CS core network assigns the UE with a temporary identity known as a Temporary Mobile Subscriber Identity (TMSI). A TMSI has a fixed length of 4 bytes and is unique within a LA.

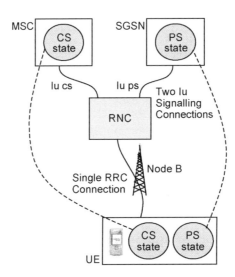

Figure 4.3   Mobility management state machines within the UE and core network

The same TMSI can be assigned within a neighbouring LA and so the TMSI is always referenced in combination with its parent LA Identity (LAI). The LAI is a concatenation of the Mobile Country Code (MCC), Mobile Network Code (MNC) and Location Area Code (LAC). This combination of values means that any TMSI concatenated with its LAI should be a unique value globally. The MCC is a 3 digit code assigned by the International Telecommunications Union (ITU), e.g. the MCC for the United Kingdom is 234. The MNC is a 2 or 3 digit code. The concatenation of the MCC and MNC defines the Public Land Mobile Network (PLMN) identifier. A LAC has a length of 2 bytes which allows 65536 combinations. However, LAC values of 0 and 65534 should not be assigned because they are reserved for UE to indicate that the LAC has been deleted from memory, i.e. if a UE reports that its LAC is 65534 then this indicates that it has deleted its LAC from memory. This can be triggered by a registration procedure being rejected by the network.

A UE may also register with the PS core network and move into the PS Idle state. A UE is known as GPRS attached once it has registered with the PS core network. The registration procedure informs the PS core network of the UE location in terms of its Routing Area (RA) and the PS core network assigns the UE with a temporary identity known as a Packet Temporary Mobile Subscriber Identity (P-TMSI). A P-TMSI has a fixed length of 4 bytes and is unique within a RA. The same P-TMSI can be assigned within a neighbouring RA and so the P-TMSI should always be referenced in combination with its RA Identity (RAI). The RAI is a concatenation of the LAI and Routing Area Code (RAC). This definition of RAI means that a RA is always a subset of LA. It also indicates that any P-TMSI concatenated with its RAI should be a unique value globally. A RAC has a length of 1 byte which allows 256 combinations, i.e. there can be a maximum of 256 RA within a single LA.

If a UE registers with the CS core network, but does not register with the PS core network then it is assigned a TMSI but not a P-TMSI. The TMSI and P-TMSI are used in preference to the IMSI to help protect the confidentiality of the UE identity. The TMSI and P-TMSI are less vulnerable to malicious attacks because they are not fixed. It is usual to reallocate the TMSI as part of the location area update procedure and to reallocate the P-TMSI as part of the routing area update procedure. These procedures can be triggered either by UE mobility or by a periodic update timer. If a UE crosses a LA boundary while in the CS Idle state then the MM entity within the UE triggers a location area update procedure. Likewise, if a UE crosses a RA boundary while in the PS Idle state then the GMM entity within the UE triggers a routing area update procedure. A UE is able to deduce that is has crossed a location area or

```
SysInfoType1
cn-CommonGSM-MAP-NAS-SysInfo
        lac       : 2783 Hex
cn-DomainSysInfoList
        cn-DomainSysInfoList value 1
        cn-DomainIdentity: cs-domain
        cn-Type
                gsm-MAP
                Hex: 0A01
                cn-DRX-CycleLengthCoeff: 6
        cn-DomainSysInfoList value 2
        cn-DomainIdentity: ps-domain
        cn-Type
                gsm-MAP
                Hex: 6401
                cn-DRX-CycleLengthCoeff: 6
```

Log File 4.1 Example section of System Information Block 1

routing area boundary by reading the contents of the broadcast channel Master Information Block (MIB) and System Information Block (SIB) 1. The MIB includes the MCC and MNC (PLMN Identity), whereas SIB 1 includes the LAC and RAC. SIB 1 also includes the timer T3212 which is used to trigger periodic location area updates. The equivalent timer for periodic routing area updates is T3312. This timer is not broadcast as part of the system information, but is provided to each UE individually during the PS registration procedure, i.e. within the GMM Attach Accept or the GMM Routing Area Update Accept messages.

An example of the relevant section of SIB 1 is presented in Log File 4.1. There are three NAS messages included within this example section of SIB1. The first NAS message is common to all core network domains, whereas the second NAS message is applicable to the CS core network and the third NAS message is applicable to the PS core network.

The coding for each of these NAS messages is specified by 3GPP TS 24.008 and is presented in Figure 4.4. The common NAS message specifies the LAC. The LAC is common to both the CS and PS core network domains because it forms part of both the LAI and RAI. The CS domain NAS message specifies the periodic location area update timer T3212 and the attach/detach flag. The value of T3212 is specified in units of decihours. This means that the timer can have a range of 0 to 1530 minutes, with

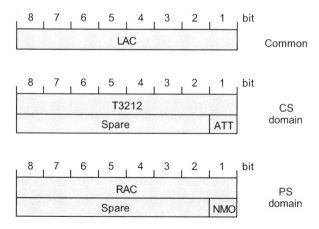

Figure 4.4 Coding for the NAS messages broadcast within SIB1

a step size of 6 minutes. The example presented in Log File 4.1 has a value of 0A in hexadecimal which corresponds to a decimal value of 10 and a T3212 value of 60 minutes. A value of 0 means that periodic location area updating is disabled. The attach/detach flag indicates whether or not a UE should complete the attach and detach procedures. A value of 0 indicates that the UE should not complete the attach and detach procedures, whereas a value of 1 indicates that the UE should complete the procedures.

The PS domain NAS message specifies the RAC and the Network Mode of Operation (NMO). The example presented in Log File 4.1 has a value of 64 in hexadecimal for the RAC and a value of 1 for the NMO. A NMO value of 0 maps to NMO I whereas a value of 1 maps to NMO II. NMO I indicates that the Gs interface is present between the SGSN and MSC. In this case, the UE is able to complete combined registration procedures via the SGSN. The UE includes a flag within the GMM Attach Request message to indicate whether or not a GPRS attach or a combined GPRS/IMSI attach is being requested. If NMO II is broadcast within SIB1 then the Gs interface is not present and separate CS and PS registration procedures must be completed.

A LA represents an area within which a UE can move while in the CS Idle state without having to update the CS core network. The minimum size of a LA is a single cell, whereas the maximum size is the collection of cells connected to a single MSC. The cells belonging to a LA can be connected to one or more RNC. If a UE crosses a LA boundary while in the CS Idle state then it must complete a location area update procedure to inform the CS core network. A RA represents an area within which a UE can move while in the PS Idle state without having to update the PS core network. The minimum size of a RA is a single cell and a RA is always contained within a single LA. The collection of cells belonging to a RA must be connected to the same SGSN although they can be connected to more than one RNC. If a UE crosses a RA boundary while in the PS Idle state then it must complete a routing area update procedure to inform the PS core network. Figure 4.5 illustrates the hierarchy of PLMN, location areas and routing areas.

This figure illustrates two example strategies for the definition of LA and RA. The PLMN belonging to operator 1 has multiple RA within each LA whereas the PLMN belonging to operator 2 has only a single RA within each LA. The second strategy is easier to manage and requires fewer routing area update procedures from UE in PS Idle mode. The same RAC can be assigned to all cells belonging to the PLMN. The number of UE in PS Idle mode depends upon whether UE register with only the CS core network or with both the CS and PS core networks. If UE register with only the CS core network they avoid the requirement for routing area updates, but must complete the PS core network registration procedure as part of the PS connection establishment procedure. The first strategy illustrated in Figure 4.5 reduces the PS paging load if paging messages are sent from the PS core network. When a UE is in RRC idle mode and CS-IDLE state then the CS core network domain knows the location of the UE with a resolution of its location area. When a CS-originated paging message needs to be sent, then

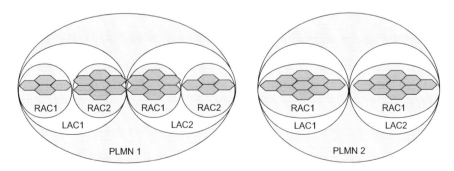

Figure 4.5 Hierarchy of PLMN, location areas and routing areas

it must be broadcast across the entire location area. If location areas are planned to be relatively large then they accommodate relatively large numbers of cells and UE. This tends to increase the quantity of paging traffic. If location areas are planned to be relatively small there will be an increase in the quantity of signalling generated by normal location area updates. This creates a trade-off between the quantity of paging traffic and the quantity of normal location area update signalling. The same arguments exist for routing areas, PS originated paging and routing area updates.

Both the RNC and core network require some knowledge of the LA and RA plan. The MSC requires a specification of which RNC handles cells belonging to a specific LA. This allows the MSC to route messages to the appropriate RNC. Likewise, the SGSN requires a specification of which RNC handles cells belonging to a specific RA. The RNC requires a specification of which cells belong to a specific LA and which cells belong to a specific RA. This allows the RNC to route messages to the appropriate cells.

If a UE is in CS Idle state then it can be paged by the CS core network and likewise, if a UE is in PS Idle state then it can be paged by the PS core network. If a UE is in CS Idle state, but PS Detached state, then all paging must be sent from the CS core network. The Discontinuous Reception (DRX) cycle length coefficients used by the paging procedure are broadcast as part of SIB1. These are visible within the section of SIB1 presented in Log File 4.1. The DRX cycle length in terms of the number of 10 ms radio frames is defined by 2 to the power of the cycle length coefficient. In this case, both the CS and PS core networks have a cycle length coefficient of 6 which corresponds to a DRX cycle length of 640 ms, i.e. the UE listens to a paging indication once every 640 ms. A short DRX cycle length tends to decrease UE battery life because the UE has to receive information more frequently. The benefit of a short DRX cycle length is a reduced average mobile terminated connection establishment delay.

A UE enters the CS Connected state by establishing a signalling connection to the CS core network and likewise a UE enters the PS Connected state by establishing a signalling connection to the PS core network. The uplink RRC Initial Direct Transfer message is used to trigger the establishment of an Iu signalling connection. This provides a connection between the UE and core network when combined with the SRB connection between the UE and RNC. The MM sublayer is responsible for ensuring that this signalling connection is established. Once a UE enters the Connected state, the UE and core network can exchange NAS messages to either complete a NAS signalling procedure or establish a user plane application connection. A UE does not update the relevant core network of its location when in the Connected state—location area and routing area updates are not triggered by the UE crossing a boundary nor by the periodic update timer expiring. Instead, UE mobility is handled by the Serving RNC which provides a fixed signalling connection to the core network unless SRNS relocation is completed to move the signalling connection to another RNC. If one core network domain is in the connected state while the other core network domain is in the Idle state, the UE only updates the core network which is in the Idle state.

The CS and PS state machines run in parallel to the RRC state machine, but there is some dependence between the two. If either of the CS or PS state machines are in the Connected state then the UE must be in RRC Connected mode. If neither of the CS nor PS state machines are in the Connected state then the UE must be in RRC Idle mode.

Table 4.1 presents the set of procedures associated with the MM entity within the MM sublayer, i.e. CS domain mobility management procedures.

Table 4.1 Procedures associated with the MM entity of the MM sublayer

Type	CS procedures
Common	TMSI reallocation, authentication, identification, IMSI detach, abort, MM information
Specific	Normal location updating, periodic location updating, IMSI attach
Connection management	MM connection establishment

Table 4.2   Messages associated with the MM entity of the MM sublayer

Type	CS messages
Registration	Location Updating Request, Location Updating Accept, Location Updating Reject, IMSI Detach Indication
Security	Authentication Request, Authentication Response, Authentication Reject, Authentication Failure, Identity Request, Identity Response, TMSI Reallocation Command, TMSI Reallocation Complete
Connection management	CM Service Request, CM Service Accept, CM Service Prompt, CM Service Reject, CM Service Abort, CM Re-establishment Request, Abort
Miscellaneous	MM Information, MM Status, MM Null

These procedures are categorised as common, specific and connection management. These categories determine when a procedure can be initiated. A common procedure can be initiated at any time that a CS signalling connection exists. A specific procedure can only be initiated if there are no other specific procedures ongoing, or if no MM connection exists. A MM connection establishment can only be completed if there are no specific procedures ongoing. For example, a normal location update procedure cannot be completed in parallel to an IMSI attach procedure, and a MM connection cannot be established while an IMSI attach or location update procedure is ongoing. The set of messages used by these MM procedures is presented in Table 4.2.

The content of these messages is specified within 3GPP TS 24.008. Similar to the NAS messages included within SIB1, these messages can be decoded by inspection using the rules specified by TS 24.008 and TS 24.007. These NAS messages are transferred between the UE and RNC using either an RRC Initial Direct Transfer, Uplink Direct Transfer or Downlink Direct Transfer message. RANAP signalling is used to transfer the messages between the RNC and Core Network. The contents of the RRC Initial Direct Transfer message is passed into a RANAP Initial UE message. The contents of RRC Uplink Direct Transfer and Downlink Direct Transfer messages is passed into RANAP Direct Transfer messages, i.e. RANAP Direct Transfer messages are used after an Iu signalling connection has been established.

Table 4.3 presents the set of procedures associated with the GMM entity within the MM sublayer, i.e. PS domain mobility management procedures. Similar to the CS domain mobility management procedures, these procedures are categorised as common, specific and connection management. These categories determine when a procedure can be initiated. A common procedure can be initiated at any time that a PS signalling connection exists. A specific procedure can only be initiated if there are no other specific procedures ongoing, or if no MM connection exists. A MM connection establishment can only be completed if there are no UE originated specific procedures ongoing. The set of messages used by these GMM procedures is presented in Table 4.4.

Similar to the CS domain messages, these NAS messages are transferred between the UE and RNC using either an RRC Initial Direct Transfer, Uplink Direct Transfer or Downlink Direct Transfer

Table 4.3   Procedures associated with the GMM entity of the MM sublayer

Type	PS procedures
Common	P-TMSI allocation and reallocation, GPRS authentication and ciphering, GPRS identification, GPRS information
Specific	GPRS attach and combined attach, GPRS detach and combined detach, normal routing area updating and combined routing area updating, periodic routing area updating
Connection management	MM connection establishment

Table 4.4   Messages associated with the GMM entity of the MM sublayer

Type of GMM message	Message
Registration	Attach Request, Attach Accept, Attach Complete, Attach Reject, Detach Request, Detach Accept, Routing Area Update Request, Routing Area Update Accept, Routing Area Update Complete, Routing Area Update Reject,
Security	P-TMSI Reallocation Command, P-TMSI Reallocation Complete, Authentication and Ciphering Request, Authentication and Ciphering Response, Authentication and Ciphering Failure, Authentication and Ciphering Reject, Identity Request, Identity Response,
Connection management	Service Request, Service Accept, Service Reject
Miscellaneous	GMM Information, GMM Status

message. Likewise, RANAP signalling is used to transfer the messages between the RNC and Core Network.

4.3 Connection Management

- The Connection Management sublayer includes entities for circuit switched Call Control (CC) and packet switched Session Management (SM).
- The CC entity is responsible for establishing, maintaining and terminating calls for the CS domain. The prerequisite for the CC entity to initiate any procedure is for the MM entity within the MM sublayer to provide a signalling connection between the RNC and MSC.
- Likewise, the SM entity is responsible for establishing, maintaining and terminating connections for the PS domain.
- Key 3GPP specifications: TS 23.221, TS 24.007 and TS 24.008

The Connection Management (CM) sublayer illustrated in Figure 4.2 includes entities for circuit switched Call Control (CC), packet switched Session Management (SM), the GPRS Short Message Service (GSMS) and Supplementary Services (SS). As mentioned previously, this set of entities is not exhaustive and other entities such as the SMS entity may also be included. Call-related supplementary services such as call forwarding and call holding are handled by the CC entity. The SS entity handles call-independent supplementary services.

The CC entity is responsible for establishing both mobile originated and mobile terminated calls for the CS core network, e.g. speech and video calls. Mobile originated calls can be either normal or emergency, whereas mobile terminated calls can only be normal, i.e. emergency calls terminate at a fixed line rather than at a mobile. The CC entity is also responsible for maintaining and terminating calls after they have been established. The prerequisite for the CC entity to initiate any procedure is for the MM entity within the MM sublayer to provide a signalling connection between the RNC and MSC. Once this signalling connection has been established then all of the necessary CC messages can be transferred. The set of procedures supported by the CC entity is presented in Table 4.5.

The set of messages used by these CC procedures is presented in Table 4.6. These messages have been extracted from 3GPP TS 24.008 which also provides a detailed specification of the content of each message.

The NAS MM and CC signalling associated with establishing a mobile-originated CS connection is presented in Table 4.7.

Table 4.5 Procedures associated with the CC entity of the CM sublayer

Type	Procedures
Call establishment	Mobile originating call establishment, mobile terminating call establishment, network-initiated mobile originating Call
Signalling during the active state	User notification, Call rearrangements, Codec change, support of dual services, user-initiated service level up- and downgrading, support of multimedia calls
Call clearing	Clearing initiated by the UE, clearing initiated by the network
Miscellaneous	In-band tones and announcements, call collisions, status procedures, call re-establishment, progress, DTMF protocol control

Table 4.6 Messages associated with the CC entity of the CM sublayer

Type of MM message	Procedures
Call establishment	Alerting, call confirmed, call proceeding, connect, connect acknowledge, emergency setup, progress, CC-establishment, CC-establishment confirmed, start CC, setup
Call information phase	Modify, modify complete, modify reject, user information
Call clearing	Disconnect, release, recall, release complete
Supplementary service control	Facility, hold, hold acknowledge, hold reject, retrieve, retrieve acknowledge, retrieve reject
Miscellaneous	Congestion control, notify, start DTMF, start DTMF acknowledge, start DTMF reject, status, status enquiry, stop DTMF, stop DTMF acknowledge

The uplink CM Service Request from the MM entity is used to establish a signalling connection to the CS core network. This message can be acknowledged by the core network in one of two ways. The core network can return a CM Service Accept message or it can trigger the security mode procedure. Successful completion of the security mode procedure is also interpreted as a positive acknowledgement for the CM Service Request. The example presented in Table 4.7 assumes that the security mode procedure is completed after the CM Service Request. The CM Service Request moves the UE from CS Idle mode to CS connected mode. The CC entity then starts the signalling required to establish the CS connection at the higher layers. The UE starts by sending a Setup message which includes the phone number being called. The CS core network returns the Call Proceeding message to acknowledge that the requested call establishment information has been received. Once the target phone has been reached and has started to ring then the CS core network returns the Alerting message to the UE. If the target person answers the phone then the Connect message is sent and finally the UE acknowledges connection using the Connect Acknowledge message.

Table 4.7 NAS message flow for a mobile-originated CS connection (UE in CS Idle)

Protocol	Direction	Message	End-user event
Mobility management	Uplink	CM service request	User dials number
Call control	Uplink	Setup	
Call control	Downlink	Call proceeding	
Call control	Downlink	Alerting	User hears ringing tone
Call control	Downlink	Connect	Target user answers phone
Call control	Uplink	Connect acknowledge	Connection established

Table 4.8 NAS message flow for a mobile terminated CS connection (UE in CS Idle)

Protocol	Direction	Message	End-user event
Radio resource management	Uplink	Paging response	
Call control	Downlink	Setup	
Call control	Uplink	Call confirmed	
Call control	Uplink	Alerting	Phone starts to ring
Call control	Uplink	Connect	User answers phone
Call control	Downlink	Connect acknowledge	Connection established

A similar set of messages for a mobile terminated CS connection is presented in Table 4.8. The mobile originated signalling uses the CM Service Request message to trigger the establishment of the signalling link whereas the mobile terminated case uses the Paging Response message. The CM Service Request message belongs to the MM protocol but the Paging Response message belongs to the Radio Resource Management (RRM) protocol. Messages belonging to the MM protocol are specified by 3GPP TS 24.008 whereas messages belonging to the RRM protocol are specified by 3GPP TS 44.018. The contents of the CM Service Request and Paging Response messages are very similar. Both messages include a ciphering key sequence, mobile station classmark II information and a mobile station identity. However the CM Service Request message also includes CM Service Type information to indicate the reason for requesting the signalling connection, and optionally can include a priority level.

Once the UE has triggered the establishment of a signalling connection using the RRM Paging Response message and has moved into CS Connected mode, CC signalling is used to establish the user plane speech connection at the higher layers. In the mobile terminating case, the Setup message is transferred in the downlink direction rather than the uplink direction. This message can include the phone number of the originating party. The UE acknowledges the Setup message with the Call Confirmed message. The Alerting message is sent once the phone starts to ring and the Connection message is sent if the phone is answered. The network completes the procedure by returning the Connect Acknowledge message.

Once a CS connection has been established the CC entity can modify the connection using the Modify and Modify Complete messages. At the end of a connection, the CC entity is responsible for completing the call clearing procedure. The CC signalling for mobile originated call clearing is presented in Table 4.9. In this case the UE sends the Disconnect message to the core network. If the other party had initiated the call clearing procedure then the Disconnect would be a downlink message from the core network.

The Disconnect message is used to clear the actual end-to-end connection whereas the Release message is used to clear the transaction identifier. Once the CC signalling presented in Table 4.9 has been completed, RANAP signalling is used to release the Iu signalling connection and RRC signalling is used to release the RRC connection. The UE then moves back to the CS Idle state.

The SM entity within the Connection Management sublayer is responsible for establishing, maintaining and terminating connections for the PS core network. The prerequisite for the SM entity to initiate any procedure is for the GMM entity within the MM sublayer to provide a signalling

Table 4.9 NAS message flow for mobile originated CS connection call clearing

Protocol	Direction	Message	End-user event
Call control	Uplink	Disconnect	User hangs up
Call control	Downlink	Release	
Call control	Uplink	Release complete	

Table 4.10 Procedures associated with the SM entity of the CM sublayer

Type	Procedures
PDP Context handling	PDP context activation, secondary PDP context activation, PDP context modification, PDP context deactivation
MBMS handling	MBMS context activation, MBMS context deactivation
Status	SM status enquiry

connection between the UE and the SGSN. Once this signalling connection has been established all of the necessary SM messages can be transferred. The set of procedures supported by the SM entity is presented in Table 4.10.

The set of messages used by these SM procedures is presented in Table 4.11. These messages have been extracted from 3GPP TS 24.008 which also provides a detailed specification of the content of each message.

The NAS GMM and SM signalling associated with establishing a mobile originated PS data connection is presented in Table 4.12.

The uplink Attach Request message from the GMM entity is used to establish a signalling connection to the PS core network and to initiate the registration procedure. Once the UE has registered, it moves from the PS Detached state to the PS Idle state. The UE then sends a Service Request message to the GMM entity within the SGSN. The Service Request message moves the UE from the PS Idle state to the PS Connected state. The UE is then able to start transferring the SM messages to establish the higher layer connection. The Activate PDP Context Request procedure is used to establish a connection between the UE and an Access Point Name (APN) within the Gateway GPRS Support Node (GGSN). A GGSN may be configured with one of more APN, e.g. there could be one APN for internet browsing and a second APN for video streaming or file download. The general architecture of the GGSN and its APN is illustrated in Figure 4.6.

An active PDP context represents a connection between the UE and an APN with a specific Quality of Service (QoS) profile. The QoS profile defines characteristics such as guaranteed bit rate and maximum transfer delay. It is possible to establish multiple PDP contexts to the same APN, but with different QoS profiles. The first PDP context is known as the primary PDP context whereas all other PDP contexts to the same APN are known as secondary PDP contexts. This may be necessary if the same APN supports a range of applications with different QoS requirements. It is also possible for a UE to have multiple PDP contexts to different APN, i.e. it is possible to have multiple primary PDP contexts.

Table 4.11 Messages associated with the SM entity of the CM sublayer

Type of MM message	Procedures
Session establishment	Activate PDP context request, activate PDP context accept, activate PDP context reject, activate secondary PDP context request, activate secondary PDP context accept, activate secondary PDP context reject, request PDP context activation, request PDP context activation reject
Session information Phase	Modify PDP context request, modify PDP context accept, modify PDP context reject
Session clearing	Deactivate PDP context request, deactivate PDP context accept
Miscellaneous	SM status
MBMS	Activate MBMS context request, activate MBMS context accept, activate MBMS context reject, request MBMS context activation, request MBMS context activation reject

Table 4.12   NAS messages for a mobile originated PS connection (UE in PS detached)

Protocol	Direction	Message
GPRS mobility management	Uplink	Attach request
GPRS mobility management	Downlink	Attach accept
GPRS mobility management	Uplink	Service request
Session management	Uplink	Activate PDP context request
Session management	Downlink	Activate PDP context accept

Once the Activate PDP Context Accept message has been received from the network, the application layer can start to transfer data between the UE and APN. In this example, it is assumed that the UE is in the PS Detached state at the start of the connection establishment procedure. This means that the UE has to complete registration and PDP context activation as part of the connection establishment procedure. It is possible for the UE to have already registered with the PS core network and to already be in the PS Idle state. Most UE allow the end-user to select whether or not the UE registers with the PS core network at all times or only when a PS service establishment is initiated, i.e. the 'when available/ when needed' option within the packet data connection settings of the UE. If the UE is already registered with the PS core network then it only has to complete PDP context activation as part of the connection establishment procedure. The NAS signalling for this scenario is presented in Table 4.13.

In this case, the connection establishment delay is reduced, but the network load is increased. The network load increases as a result of the UE being in the PS Idle state which requires the PS core network to keep track of the UE location, i.e. the UE has to complete routing area updates. It is also possible for the UE to have both registered with the PS core network and to have an active PDP context at the start of the connection establishment procedure. In this case, the UE only has to send the GMM Service Request message.

At the end of a connection, the SM entity is responsible for deactivating the PDP context whereas the GMM entity is responsible for clearing the UE's registration. The SM and GMM signalling for connection clearing is presented in Table 4.14.

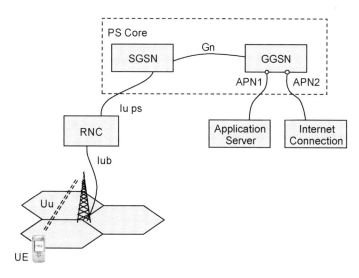

Figure 4.6   PDP context connection between a UE and APN

Table 4.13 NAS messages for a mobile-originated PS connection (UE in PS Idle).

Protocol	Direction	Message
GPRS mobility management	Uplink	Service request
Session management	Uplink	Activate PDP context request
Session management	Downlink	Activate PDP context accept

Table 4.14 NAS message flow for PS connection clearing.

Protocol	Direction	Message
Session management	Uplink	Deactivate PDP context request
Session management	Downlink	Deactivate PDP context accept
GPRS mobility management	Uplink	Detach request
GPRS mobility management	Downlink	Detach accept

If the UE has been configured to remain GPRS attached the GMM signalling is not necessary. Likewise, it is possible to maintain the active PDP connection and in that case the SM signalling is not necessary.

4.4 PLMN Selection

- PLMN selection can be completed using either an automatic or manual mode. In both cases, the NAS within the UE instructs the AS to complete a PLMN search. The access stratum reads PLMN identities from the system information and reports them to the NAS.
- In the case of automatic selection, the PLMN search can be stopped as soon as the home PLMN is found. If the home PLMN cannot be found then the PLMN which have been found are prioritised prior to selection. In the case of manual selection, the set of PLMN are presented to the end-user in order of their priority.
- Elementary Files within the USIM can support PLMN selection. These files can specify the interval between searches for higher priority PLMN when the home PLMN has not been selected in automatic mode. They can also record forbidden PLMN and network parameters, e.g. the previously used scrambling code, RF carrier and neighbour lists.
- Key 3GPP specifications: TS 23.122 and TS 31.102

Public Land Mobile Network (PLMN) selection can be completed using either an automatic or manual mode. The end-user is able to select between these two modes by configuring the phone settings within the UE user interface. In both cases, the UE NAS instructs the UE AS to complete a PLMN search. The AS scans specific RF carriers by identifying the strongest cell on each carrier and reading the PLMN identity broadcast on the P-CCPCH. The P-CCPCH is used to transfer the BCH transport channel which encapsulates the BCCH logical channel. The PLMN identity is included within the Master Information Block (MIB) which is broadcast as part of the system information within the BCCH. An example section of the MIB which includes the PLMN identity is presented in Log File 4.2. This log file illustrates that the PLMN identity is a concatenation of the Mobile Country Code (MCC) and Mobile Network Code (MNC).

Once the AS has read the PLMN identity from the MIB, its value is reported to the NAS. The AS also measures the CPICH RSCP from the same cell that the PLMN identity has been read. If the

```
MasterInformationBlock
    mib-ValueTag: 2
    plmn-Type
    gsm-MAP
    plmn-Identity
            mcc: 2,3,4
            mnc: 9,9
```

Log File 4.2 Example section of the MIB which includes the PLMN Identity

CPICH RSCP is less than −95 dBm then the measurement result is provided to the NAS with the PLMN identity. If the CPICH RSCP is greater than or equal to −95 dBm then the measurement result is not reported to the NAS, but the PLMN is flagged as being a high-quality PLMN. If the PLMN belongs to the GSM system then the GSM RSSI is measured rather than the CPICH RSCP and the threshold for a high-quality cell is −85 dBm.

The Universal Subscriber Identity Module (USIM) contains data to assist the PLMN search procedure. This data is stored within USIM Elementary Files (EF). 3GPP TS 31.102 specifies the contents of these files. The EF which are relevant to the PLMN search procedure are listed in Table 4.15.

The 'HPLMN Selector with Access Technology' EF specifies the PLMN identity of the Home PLMN (HPLMN) and also the set of associated radio access technologies. The Release 99 to release 6 versions of the specifications support only a single HPLMN. It is possible for the HPLMN to be associated with more than a single radio access technology, i.e. an operator can assign the same PLMN identity to both its GSM and UMTS networks. When more than a single radio access technology is associated with the HPLMN the choice between the access technologies is dependant upon the UE implementation. If this EF is not stored by the USIM then the UE is able to determine the HPLMN identity from its IMSI, i.e. the HPLMN identity forms the first five or six digits of the IMSI. In this case, the UE searches all radio access technologies, but provides priority to the GSM system. Once the initial PLMN selection procedure has been completed inter-system cell reselection can be used to provide mobility between the radio access technologies.

The 'Higher Priority PLMN Search Period' EF becomes applicable if the UE registers with a PLMN other than the HPLMN, i.e. the UE registers with a Visited PLMN (VPLMN) and if the UE is in automatic mode. This EF is not applicable if the UE has been configured for manual PLMN selection. The search period defines the frequency with which the UE searches for a higher priority PLMN. The search period has a range from 6 minutes to 8 hours with a step size of 6 minutes. Configuring a value of 0 indicates that periodic searches should not be completed. If this EF is not stored by the USIM then a default value of 60 minutes is assumed. This EF is applicable to the release 5 and 6 versions of the 3GPP specifications. The release 99 and 4 versions of the specifications include an EF with the same identity, but named as HPLMN search period, i.e. the HPLMN search period has been generalised to the higher priority search period in the newer versions of the 3GPP specifications.

The 'User-controlled PLMN Selector with Access Technology' EF allows the end-user to specify a set of preferred PLMN and radio access technologies. A USIM with this EF can store a minimum of

Table 4.15 USIM Elementary Files relevant to PLMN selection

Elementary file (EF)
HPLMN selector with access technology
Higher priority PLMN search period
User-controlled PLMN selector with access technology
Operator-controlled PLMN selector with access technology
Forbidden PLMN
Network parameters

eight PLMN with their associated radio access technologies. The PLMN are ordered according to their priority with the first PLMN treated as having the greatest priority. Each entry can be edited by the end-user, but cannot be updated by the operator across the air-interface.

The 'Operator-controlled PLMN Selector with Access Technology' EF allows the operator to specify a set of preferred PLMN and radio access technologies. A USIM with this EF can store a minimum of 8 PLMN with their associated radio access technologies. Similar to the user controlled list, the PLMN are ordered according to their priority.

The 'Forbidden PLMN' EF allows a minimum of 12 PLMN identities to be stored as forbidden PLMN. When the UE is configured for automatic PLMN selection then the UE does not attempt to register with any of the forbidden PLMN. A PLMN is stored within this EF if the network rejects a Location Updating Request with the cause 'PLMN not allowed'. If the list of forbidden PLMN becomes full then any further forbidden PLMN cause the oldest stored PLMN to be removed. The list of forbidden PLMN is retained when the UE is powered off and when the USIM is removed. PLMN are removed from the list of forbidden PLMN if manual PLMN selection results in a successful registration procedure. It is not possible for the HPLMN to be stored within the list of forbidden PLMN.

The 'Network Parameters' EF is able to store information which can be used to reduce the delay associated with selecting a cell and reading the associated PLMN identity. The previously used RF carrier and scrambling code can be stored in addition to the set of intra-frequency and inter-frequency neighbours. Intra-frequency neighbours are recorded in terms of their scrambling code whereas inter-frequency neighbours are recorded in terms of their scrambling code and RF carrier. It is possible to store information on 32 intra-frequency cells and 32 inter-frequency cells. The 32 inter-frequency cells can be distributed across a maximum of 3 RF carriers.

If a UE is operating in automatic PLMN selection mode then the NAS can select a PLMN and instruct the AS to stop searching at any time. This can happen relatively quickly if the RF carrier information stored within the 'Network Parameters' EF corresponds to the HPLMN, i.e. the HPLMN can be the first PLMN reported to the NAS and the AS can be instructed to stop searching. If the HPLMN is not found then the AS continues to search each RF carrier within its supported frequency bands. Once the NAS has received the set of PLMN identities associated with its set of supported RF carriers then it selects a PLMN for registration based upon the following order of priority:

1. HPLMN
2. Each PLMN/access technology combination within the 'User-controlled PLMN Selector with Access Technology' EF, accounting for the priority order within that EF and excluding any PLMN which appears within the 'Forbidden PLMN' EF.
3. Each PLMN/access technology combination within the 'Operator-controlled PLMN Selector with Access Technology' EF, accounting for the priority order within that EF and excluding any PLMN which appears within the 'Forbidden PLMN' EF.
4. Other PLMN reported to have high quality and not appearing within the 'Forbidden PLMN' EF. PLMN within this category are selected at random.
5. Other PLMN prioritised for a specific radio access technology in order of decreasing signal strength. CPICH RSCP is used for UMTS whereas RSSI is used for GSM. Prioritisation between radio access technologies is dependent upon the UE implementation.

If a UE is operating in manual PLMN selection mode then the AS stratum searches each RF carrier within its supported frequency bands and reports the associated set of PLMN identities to the NAS. The NAS displays those identities to the end-user in the following order:

1. HPLMN
2. Each PLMN/access technology combination within the 'User-controlled PLMN Selector with Access Technology' EF, accounting for the priority order within that EF.

3. Each PLMN/access technology combination within the 'Operator-controlled PLMN Selector with Access Technology' EF, accounting for the priority order within that EF.
4. Other PLMN reported to have high quality. PLMN within this category are ordered randomly.
5. Other PLMN prioritised for a specific radio access technology in order of decreasing signal strength. CPICH RSCP is used for UMTS whereas RSSI is used for GSM. Ordering between radio access technologies is dependant upon the UE implementation.

The end-user is then able to select the PLMN upon which registration is attempted. If registration fails then the end-user is presented with the list again to allow a further selection.

5

Iub Transport Network

5.1 Protocol Stacks

- The Iub transport network provides connectivity between each Node B and its controlling RNC. Iub traffic is categorised as either radio network control plane, transport network control plane or user plane. The release 4 and earlier versions of the specifications define protocol stacks based upon ATM. The release 5 version of the specifications introduces the option to use IP.
- The radio network control plane is used to transfer NBAP signalling between the RNC and Node B. NBAP signalling between the RNC and Node B is the equivalent of RRC signalling between the RNC and UE. AAL5 is used when the ATM protocol stack is adopted. AAL5 is suited to transferring large and variable length packets with relatively little overhead. It is the most commonly used AAL for data traffic.
- NBAP signalling procedures are categorised as common or dedicated. Common procedures are either not related to a specific UE, or involve the request of a new Node B communication context. Dedicated procedures are related to an existing Node B communication context. Dedicated NBAP messages include the Node B communication context identity as part of the message.
- The transport network control plane is used to establish and release AAL2 connections belonging to the user plane. The ALCAP signalling protocol specified within ITU-T recommendation Q.2630.1 defines the signalling procedures used to establish and release AAL2 connections. The establishment procedure assigns a CID to the new AAL2 connection.
- The user plane is used to transfer both control plane and user plane data belonging to the Radio Interface Protocol stack. This information is packaged using either common channel or dedicated channel Frame Protocol data frames. The user plane VCC is also responsible for transferring Frame Protocol control frames. AAL2 is used when the ATM protocol stack is adopted. AAL2 is intended for low rate, delay sensitive applications which generate short and variable length packets. It provides the ability to multiplex connections within the ATM cells belonging to a single VCC.
- The maximum number of AAL2 CID that can be assigned within a single VCC is 248. A subset of these CID are reserved for the RACH, PCH, FACH-c and FACH-u when a Node B is switched on. Each speech connection requires two CID. Likewise, video call connections and DCH data

connections require two CID. It is necessary to configure a second VCC if the set of CID becomes exhausted.
- Key 3GPP specifications: TS 25.426, TS 25.430, TS 25.433 and TS 25.434
- Key ITU-T recommendations: Q.2630.1, Q.2130, Q.2110, I.363.5 and I.366.1

The Iub transport network provides connectivity between each Node B and its controlling RNC. 3GPP TS 25.430 categorises Iub traffic as either radio network control plane, transport network control plane or user plane. The corresponding protocol stacks specified by the release 99 and release 4 versions of 3GPP TS 25.430 are presented in Figure 5.1.

The release 5 and release 6 versions of 3GPP TS 25.430 enhance these protocol stacks by allowing the optional use of IP for the radio network control plane and user plane. The enhanced version of these protocol stacks is presented in Figure 5.2.

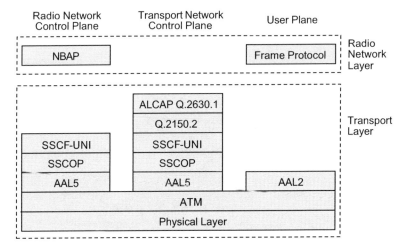

Figure 5.1 Iub transport network protocol stacks (3GPP releases 99 and 4)

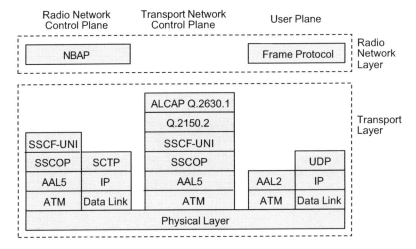

Figure 5.2 Iub transport network protocol stacks (3GPP Releases 5 and 6)

5.1.1 Radio Network Control Plane

The radio network control plane is used to transfer Node B Application Part (NBAP) signalling between the RNC and Node B. The NBAP signalling protocol is specified within 3GPP TS 25.433. NBAP signalling between the RNC and Node B is the equivalent of RRC signalling between the RNC and UE. For example, if the RNC decides that a radio link should be reconfigured then the UE is informed using RRC signalling whereas the Node B is informed using NBAP signalling. NBAP signalling procedures are categorised as common or dedicated. Common procedures are either not related to a specific UE, or involve the request of a new Node B communication context for a specific UE. Dedicated procedures are related to an existing Node B communication context. Dedicated NBAP messages include the Node B communication context identity as part of the message. The set of common and dedicated NBAP procedures is presented within Table 5.1.

A subset of the common NBAP procedures is typically used when a Node B is switched on, e.g. the cell setup, common transport channel setup, system information update, and common measurement initiation procedures. The cell setup procedure initialises a cell in terms of specifying its RF carrier, maximum downlink transmit power, primary scrambling code, timing information, and CPICH, P-SCH, S-SCH and P-CCPCH transmit powers. The common transport channel setup procedure configures the transport format sets for the RACH, FACH and PCH. It also specifies the transmit powers and channelisation codes for the S-CCPCH, PICH and AICH. The system information update procedure is used to provide the MAC-b entity within the Node B with instructions regarding the scheduling of system information. The Node B is also provided with copies of the Master Information Block (MIB) and each System Information Block (SIB). The MAC-b entity within the Node B then becomes responsible for broadcasting this information on the P-CCPCH. The common measurement initiation procedure is used to instruct the Node B to start reporting cell-specific measurements such as the total downlink transmit power, the uplink receiver interference floor and the number of acknowledged PRACH preambles. The most important common NBAP procedure related to dedicated channel establishment is radio link setup. This procedure is used when a UE enters CELL_DCH from either RRC Idle mode or from the RRC connected mode state CELL_FACH. It is also used when a soft handover radio link is added to the active set.

The most frequently used dedicated NBAP procedures tend to be radio link addition, radio link reconfiguration, radio link deletion, radio link restoration, radio link failure and compressed mode

Table 5.1 Common and dedicated NBAP procedures

Common NBAP procedures	Common transport channel setup, common transport channel reconfiguration, common transport channel deletion, block resource, unblock resource, audit required, audit, common measurement initiation, common measurement reporting, common measurement termination, common measurement failure, cell setup, cell reconfiguration, cell deletion, resource status indication, system information update, radio link setup, physical shared channel reconfiguration, reset, information exchange initiation, information reporting, information exchange termination, information exchange failure
Dedicated NBAP procedures	Radio link addition, synchronised radio link reconfiguration prepare, synchronised radio link reconfiguration commit, synchronised radio link reconfiguration cancellation, unsynchronised radio link reconfiguration, radio link deletion, downlink power control, dedicated measurement initiation, dedicated measurement reporting, dedicated measurement termination, dedicated measurement failure, radio link failure, radio link restoration, compressed mode command, radio link pre-emption, bearer rearrangement, radio link activation, radio link parameter update

command. The radio link addition procedure is used when a softer handover radio link is added to the active set. The quantity of signalling information transferred by the radio link addition procedure is less than that transferred by the radio link setup procedure because the Node B already has knowledge of the existing one or more radio links (more than one radio link if the UE is already in softer handover prior to triggering the radio link addition procedure). The radio link reconfiguration procedure is used when changing the configuration of an existing radio link. This can apply during the establishment of a speech call when the user plane transport channels are added to the existing control plane transport channel. Alternatively it could apply during a packet switched data transfer if the maximum uplink or downlink bit rate is changed, e.g. the maximum bit rate could be reduced if the UE is moving into an area of poor coverage. The radio link deletion procedure is used when a radio link is removed from the active set. This could occur during an active dedicated channel connection if either soft or softer handover radio links are removed. Alternatively, it could occur at the end of a dedicated channel connection when all radio links are removed from the active set. The radio link restoration and radio link failure procedures are used by the Node B to inform the RNC of when air-interface synchronisation has been achieved and lost respectively. The RNC can use this information to trigger other actions, e.g. the RNC will release all resources if air-interface synchronisation has been lost by all active set cells for a relatively long period of time. The compressed mode procedure is used to instruct a Node B to start a specific transmission gap sequence which allows the UE to complete either inter-frequency or inter-system measurements.

The release 99 and release 4 versions of the radio network control plane make use of ATM Adaptation Layer 5 (AAL5). AAL5 is a Simple and Efficient ATM Adaptation Layer (SEAL) and is the most commonly used AAL for data traffic. AAL5 is suited to transferring large and variable length packets with relatively little overhead. Similar to other AAL, AAL5 combines the use of service specific sublayers with common part sublayers. These sublayers are presented within Figure 5.3.

The Service Specific Coordination Function (SSCF) coordinates the transfer of data packets between the NBAP layer and the Service Specific Connection Orientated Protocol (SSCOP) sublayer. The SSCF sublayer is specified by ITU-T recommendation Q.2130 whereas the SSCOP sublayer is specified by ITU-T recommendation Q.2110. The SSCOP sublayer provides support for error correction by retransmission, in-sequence delivery of higher layer packets and flow control. Higher layer packets which form the SSCOP payload are specified to have a maximum size of 65528 bytes, i.e. significantly greater than the size of any NBAP message. The SSCOP sublayer adds 4 bytes of tail information and up to 3 bytes of padding. The 4 bytes of tail information are used to signal a sequence number, PDU

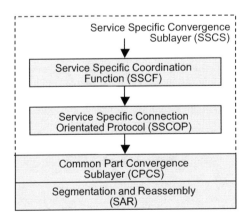

Figure 5.3 Sublayers belonging to the AAL5 protocol stack layer

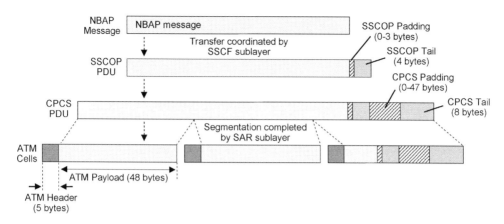

Figure 5.4 AAL5 processing of NBAP messages

type and an indication of the number of padding bytes. Between 0 and 3 padding bytes are included to ensure that the SSCOP PDU length is always a multiple of 4 bytes.

The SSCOP PDU is passed to the Common Part Convergence Sublayer (CPCS) which adds 8 bytes of tail information and up to 47 bytes of padding. Between 0 and 47 bytes of padding is added to ensure that the resultant CPCS PDU length is always a multiple of 48 bytes. The 8 bytes of tail information includes a 4 byte Cyclic Redundancy Check (CRC), a 2 byte length indicator, a 1 byte Common Part Indicator (CPI) and a 1 byte CPCS User to User (CPCS-UU) indication field. The CRC bytes provide an error detection capability at the receiver which can be used to trigger a retransmission from the SSCOP sublayer. The CPCS PDU is passed to the Segmentation and Reassembly (SAR) sublayer which is responsible for ensuring that the CPCS PDU is segmented into sections of 48 bytes. These sections of 48 bytes then form the payload belonging to a series of ATM cells. The CPCS and SAR sublayers are specified within ITU-T recommendation I.363.5.

The processing of an NBAP message by the AAL5 sublayers is summarised in Figure 5.4.

5.1.2 Transport Network Control Plane

The transport network control plane illustrated in Figure 5.1 is used to establish and release ATM Adaptation Layer 2 (AAL2) connections belonging to the user plane protocol stack. ITU-T recommendation Q.2630.1 is used for this purpose and represents an Access Link Control Application Part (ALCAP) signalling protocol. The two most important procedures defined within Q.2630.1 are those used to establish and release AAL2 connections. The signalling associated with these procedures is presented within Figure 5.5. Q.2630.1 does not include any procedures to reconfigure existing AAL2 connections. This means that reconfiguration requires the release of the existing connection and the establishment of a new connection.

The RNC Connection Admission Control (CAC) starts the AAL2 connection establishment procedure by identifying whether or not there are sufficient resources for the new connection. Assuming there are sufficient resources, the RNC proceeds to send an ALCAP Establish Request (ERQ) message to the Node B. An example ERQ message is presented within Log File 5.1.

The path identifier within the ERQ message is used to address the Virtual Channel Connection (VCC) within which the new AAL2 connection is to be established. In this example, the new AAL2

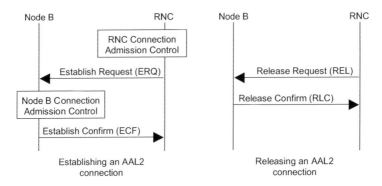

Figure 5.5 ALCAP signalling used to establish and release AAL2 user plane connections

connection is assigned a Channel Identifier (CID) of 20 which allows it to be distinguished from other AAL2 connections belonging to the same VCC. The general hierarchy of AAL2 connections, VCC and Virtual Path Connections (VPC) is illustrated in Figure 5.6.

A single physical connection can be configured to transfer data belonging to one or more VPC. Each VPC using the same physical connection must be identified using a unique Virtual Path Identifier (VPI). A VPC can be configured to transfer data belonging to one or more VCC and each VCC using the same VPC must be identified using a unique Virtual Channel Identifier (VCI). A VCC can transfer data belonging to one or more AAL2 connections. Each AAL2 connection within a specific VCC must be identified using a unique CID.

In the context of the Iub interface, there are typically two VPC per Node B. One VPC is used for Operation and Maintenance (O&M) functions while the other is used by the protocol stacks illustrated in Figure 5.1, i.e. used to transfer NBAP, ALCAP and user plane data. Each protocol

```
ERQ - ESTABLISH REQUEST
DSAID-Dest. sign. assoc. ident.:  00 00 00 00
CEID-Connection element ident.
      - Path identifier:        2001 (7D1h)
      - Channel identifier:       20 (14h)
NSEA-Dest. NSAP serv. endpoint addr.
      - Address:  49 00 00 20 01 FF FF FF FF FF FF FF
                  FF FF FF FF FF FF FF FF
LC-Link Charactreristics
      - Maximum forward CPS-SDU bit rate:     82 (52h)
      - Maximum backwards CPS-SDU bit rate:   84 (54h)
      - Average forward CPS-SDU bit rate:     23 (17h)
      - Average backwards CPS-SDU bit rate:   25 (19h)
      - Maximum forward CPS-SDU size:         24 (18h)
      - Maximum backwards CPS-SDU size:       26 (1Ah)
      - Average forward CPS-SDU size:         24 (18h)
- Average backwards CPS-SDU size:            26 (1Ah)
OSAID-Orig. sign. assoc. ident.
      - Signalling association identifier:    45 70 E3 00
SUGR-Served user gen. reference
      - Field:  00 00 00 02
```

Log File 5.1 Example Establish Request ALCAP message

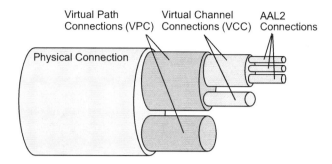

Figure 5.6   General hierarchy of VPC, VCC and AAL2 connections within a physical connection

stack has its own VCC within the VPC. The NBAP protocol stack may use two VCC– one for common NBAP messages and one for dedicated NBAP messages. These Iub VCC are illustrated within Figure 5.7.

The set of VCC presented in Figure 5.7 represents a relatively simple example. It is possible that multiple user plane VCC are configured to differentiate between services. Real time and non-real time services could be provided with separate VCC. Likewise, Section 6.7 introduces the possibility of HSDPA being configured with a separate VCC. Configuring separate VCC for each service type allows the quality of service across the transport network to be matched to the requirements of the service. In addition, Section 5.1.3 explains that it may be necessary to introduce additional user plane VCC if the number of AAL2 connections reaches its maximum limit of 248.

Returning to the ALCAP ERQ message presented within Log File 5.1, the path identifier and channel identifier are followed by a series of downlink (forward) and uplink (backward) bit rates. Both maximum and average bit rates are specified in units of 64 bps, e.g. the figure of 82 corresponds to 5.248 kbps. These bit rates are specified for the AAL2 Common Part Sublayer (CPS) packet payload, i.e. the CPS SDU (described within Section 5.1.3). This means that they exclude the overheads generated by the AAL2 and ATM layers, but include the overhead generated by the Frame Protocol layer. This example is for a 3.4 kbps Signalling Radio Bearer (SRB) and so the bit rates are relatively low. The maximum bit rates can include the impact of any Frame Protocol control frames which may be transferred during the same Transmission Time Interval (TTI) as the Frame Protocol data frames. The average bit rates can include the impact of discontinuous transmission, i.e. SRB are not active during every TTI. The CAC algorithm at the Node B uses these bit rates to determine whether or not the new AAL2 connection should be admitted.

The ALCAP ERQ message also specifies the maximum and average CPS SDU sizes in the uplink and downlink directions. These figures are specified in terms of bytes. In this example, the downlink

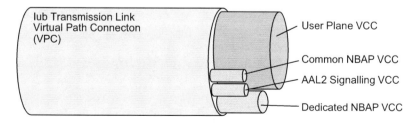

Figure 5.7   Set of VCC belonging to the Iub transport network protocol stacks

```
ECF - ESTABLISH CONFIRM
DSAID-Dest. sign. assoc. ident.:  45 70 E3 00
OSAID-Orig. sign. assoc. ident.
     - Signalling association identifier:  00 00 00 0F
```

Log File 5.2   Example Establish Confirm ALCAP message

CPS SDU size is specified to be 24 bytes (192 bits) whereas the uplink SDU size is specified to be 26 bytes (208 bits). These figures are defined by the SRB transport block size of 148 bits with the addition of the Frame Protocol data frame header, padding and tail bits. The Frame Protocol header for the SRB data frame is 24 bits in both the uplink and downlink directions. 4 bits of padding are included to increase the transport block size to an integral number of bytes. The Frame Protocol data frame includes 16 tail bits in the downlink direction and 32 tail bits in the uplink direction. The number of tail bits is greater in the uplink direction because the CRC check result and a received quality indication are included.

Once the Node B receives the ALCAP ERQ message, it completes its CAC decision and returns an ALCAP Establish Confirm (ECF) message. An example ECF message is presented within Log File 5.2.

The ECF message echoes the Originating Signalling Association Identifier (OSAID) included within the ERQ message as the Destination Signalling Association Identifier (DSAID). The user plane AAL2 connection is established once the RNC has received the ALCAP ECF message. The AAL2 connection can then be used by the user plane protocol stack.

Figure 5.5 also illustrates the procedure used to release an AAL2 connection using the Release Request and Release Confirm messages. An example Release Request message is presented in Log File 5.3.

The connection to be released is identified by its DSAID and a cause is included to provide a reason for the release. In this example, the cause is normal which indicates that the AAL2 connection is no longer required by the higher layers. An example Release Confirm message is presented in Log File 5.4.

Similar to the Release Request message, the Release Confirm message uses the DSAID as a connection identifier.

```
REL - RELEASE REQUEST
DSAID-Dest. sign. assoc. ident.:  00 00 00 0F
CAU-Cause
Cause:
     - Coding Standard: ITU-T standard Q.805 & Q.2610
     - Cause: Normal, unspecified (31 (1Fh))
Diagnostics:
```

Log File 5.3   Example Release Request ALCAP message

```
RLC - RELEASE CONFIRM
DSAID-Dest. sign. assoc. ident.:  45 70 E3 00
```

Log File 5.4   Example Release Confirm ALCAP message

5.1.3 Transport Network User Plane

The user plane VCC is responsible for transferring both the control plane and user plane data belonging to the Radio Interface Protocol (RIP) stack, i.e. all RRC messages and all application data is passed through the user plane VCC. This information is packaged using either common channel or dedicated channel Frame Protocol data frames. The user plane VCC is also responsible for transferring Frame Protocol control frames, e.g. outer loop power control and HSDPA flow control messages.

As illustrated in Figure 5.1, the release 99 and release 4 versions of 3GPP TS 25.426 specify that dedicated channel data streams are transferred across the Iub interface using ATM Adaptation Layer 2 (AAL2). Likewise, the release 99 and release 4 versions of 3GPP TS 25.434 specify that common channel data streams are transferred across the Iub interface using AAL2. The release 5 and release 6 versions of these specifications introduce the option of using IP.

AAL2 is intended for low-rate, delay-sensitive applications which generate short and variable-length packets. It provides the ability to multiplex connections within the ATM cells belonging to a single VCC. These characteristics make AAL2 particularly well suited to the speech service. Assuming that AAL2 is used as the user plane transport protocol, Frame Protocol data and control frames are passed down to the AAL2 layer. The AAL2 layer includes two sublayers: Service Specific Convergence Sublayer (SSCS) and the Common Part Sublayer (CPS). These sublayers are illustrated in Figure 5.8.

In general, the SSCS includes three functions: Service Specific Assured Data Transfer (SSADT), Service Specific Transmission Error Detection (SSTED) and Service Specific Segmentation and Reassembly (SSSAR). 3GPP TS 25.426 specifies that in the case of the Iub transport network, only the SSSAR function is applicable. The SSSAR function is specified within ITU-T recommendation I.366.1 and is responsible for Frame Protocol packet segmentation at the transmitting end, and Frame Protocol packet reassembly at the receiving end. The payload belonging to the CPS sublayer can accommodate a maximum of 45 bytes. Any Frame Protocol packet which has a length of greater than 45 bytes requires segmentation.

Frame Protocol packets belonging to the 12.2 kbps speech service are less than 45 bytes and do not require segmentation. The 12.2 kbps speech service generates 244 bits of user plane data every 20 ms. In the downlink direction, the Frame Protocol layer adds a 40 bit header, 12 bits of padding and a 16 bit tail, resulting in a packet size of 312 bits (39 bytes). The CPS sublayer supports variable length packets and so further padding is not necessary.

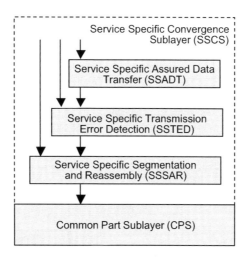

Figure 5.8   Sublayers belonging to AAL2 protocol stack layer

Table 5.2   Frame Protocol packet sizes illustrating the requirement for SSSAR segmentation

Service	Direction	TBS size (bits)	TTI (ms)	Frame protocol Header (bits)	Padding (bits)	Tail (bits)	Total (bytes)	Segmentation necessary
12.2 kbps	UL	244	20	40	12	32	41	No
speech	DL	244	20	40	12	16	39	No
64 kbps	UL	1344	20	24	0	32	175	Yes
PS data	DL	1344	20	24	0	16	173	Yes
128 kbps	UL	1344	20	24	0	32	175	Yes
PS data	DL	2688	10	24	0	16	341	Yes
384 kbps	UL	4032	10	24	0	40	512	Yes
PS data	DL	4032	10	24	0	16	509	Yes

Frame protocol packets belonging to the 384 kbps data service are greater than 45 bytes and so require segmentation. The 384 kbps data service generates 4032 bits of user plane data every 10 ms (based upon 12 transport blocks of 336 bits with 16 bits of RLC header per transport block). In the downlink direction, the Frame Protocol layer adds a 24 bit header and a 16 bit tail resulting in a packet size of 4072 bits (509 bytes).

Table 5.2 presents a set of typical uplink and downlink Frame Protocol packet sizes for a range of services. These figures can be used to deduce whether or not segmentation is necessary.

Segmentation by the SSSAR function does not generate any additional overhead in terms of packet size. Instead, the SSSAR function defines part of the header added by the CPS sublayer. ITU-T recommendation I.363.2 specifies the structure of an AAL2 CPS packet. This structure is illustrated in Figure 5.9.

The SSSAR sublayer defines the content of the User to User Indication (UUI) within the header of the CPS packet. The UUI is used to signal whether or not the payload forms the final section of a Frame Protocol packet. A value of 27 indicates that more data is to follow whereas any value between 0 and 26 indicates the final section of a Frame Protocol packet (values 28 and 29 are reserved for future use whereas values 30 and 31 are used for layer management).

The AAL2 CPS packet header also includes the Connection Identifier (CID). These 8 bits provide a range from 0 to 255. The value 0 is not used for channel identification because the all zero octet is used for a separate padding function. Values 1 and 2 are reserved for layer management peer-to-peer procedures and signalling. Values 3–7 are also reserved. This leaves values 8–255 to identify connections within a specific VCC, i.e. a maximum of 248 AAL2 connections can be established.

A subset of the total CID can be reserved for the common channels when a Node B is switched on. The ALCAP ERQ and ECF handshake illustrated in Figure 5.5 is completed for each common channel, i.e. each common channel has its own AAL2 connection. If each cell at a Node B is configured to use the RACH, PCH, FACH-c and FACH-u transport channels then each cell requires 4 AAL2 connections to support the common channels. A 1+1+1 Node B would require 12 AAL2 connections, whereas a 2+2+2 Node B would require 24 connections. Once CID have been assigned to the AAL2 connections for the common channels, the remaining CID are available for DCH, DSCH and E-DCH connections.

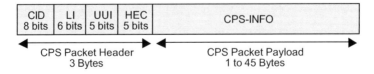

Figure 5.9   Structure of an AAL2 Common Part Sublayer (CPS) Packet

A DCH connection for a standalone SRB uses a single CID. If a user plane DCH is added then a second CID is required, i.e. separate CID are used to address each transport channel. This results in a CS speech call, a CS video call and a PS data connection each requiring two CID when DCH transport channels are used (the number of CID would increase for multi-RAB connections). The requirements for HSDPA and HSUPA are discussed within Chapters 6 and 7, respectively.

Based upon four common channels per cell, a 1+1+1 Node B would have 236 CID available for DCH, DSCH and E-DCH connections. This equates to approximately 78 AAL2 connections per cell. Assuming a traffic profile based upon only release 99 DCH connections means that there could be 39 users per cell. This figure reduces to 18 users per cell for a 2+2+2 Node B. These figures are relatively low and it is likely that as the quantity of traffic increases there will be a requirement for more than 248 CID. In this case, a second VCC must be configured to introduce a further 248 CID. This represents a logical capacity upgrade rather than a physical capacity upgrade because the additional user plane VCC can be configured without necessarily increasing the physical bandwidth of the Iub transport link.

The header belonging to the AAL2 CPS packet illustrated in Figure 5.9 does not specify the VCI nor the VPI. This information is included as part of the ATM cell header. An ATM cell header can be based upon either the User to Network Interface (UNI) format or the Network to Network Interface (NNI) format. The UNI format is used when ATM cells are transferred between a network end point and an ATM switch. The NNI format is used when ATM cells are transferred between ATM switches. The NNI format uses an increased number of bits to represent the VPI. This increases the number of VPI which can be multiplexed onto a single physical interface. The structure of a UNI ATM cell is presented in Figure 5.10.

The payload belonging to an ATM cell has a fixed size of 48 Bytes. Padding must be used if these 48 bytes are not fully occupied. One or more AAL2 CPS packets can be included within the payload of an ATM cell. When ATM cells are used to transfer AAL2 CPS packets, the first byte of the payload is always occupied by an AAL2 CPS PDU start field. This start field includes:

- Offset Field (OSF) of 6 bits which is used to specify the number of bytes between the end of the start field and the first start of an AAL2 CPS packet, or in the absence of a first start of an AAL2 packet, the start of the padding. A value of 47 indicates that there is no start boundary within the AAL2 CPS PDU.
- Sequence Number (SN) of 1 bit which is used to number consecutive AAL2 CPS PDU using modulo 2 arithmetic.
- Parity (P) of 1 bit which is used to detect errors in the start field. The parity bit is set such that the parity across the 8 bit start field is odd.

The processing of Frame Protocol packets by the AAL2 sublayers is summarised in Figure 5.11. In this example, there are two speech users and a single data user. Only the data user requires segmentation by the SSSAR sublayer.

The CPS sublayer adds a 3 byte header to generate a series of AAL2 CPS packets. These packets are segmented, concatenated and padded to generate blocks of 47 bytes. The 1 byte CPS start field is then added to generate the 48 byte AAL2 CPS PDU. These PDU occupy the payload of the ATM cells. In this example, the third AAL2 CPS PDU requires padding because there is insufficient data to occupy the full 48 bytes. In practice, this padding could be replaced by data belonging to another AAL2 connection. Non-consecutive CID have been assigned to illustrate that each user has a second AAL2

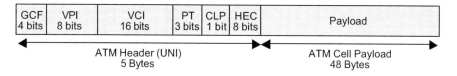

Figure 5.10   Structure of an ATM cell - User to Network Interface (UNI) header

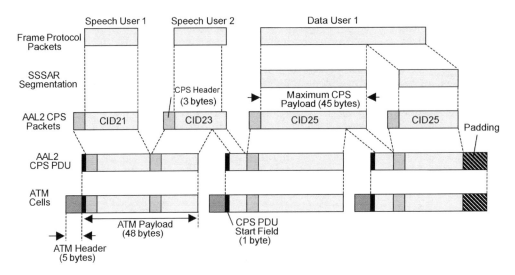

Figure 5.11   AAL2 processing of Frame Protocol packets

connection for their SRB. If it is assumed that this example is based upon a 1+1+1 Node B with 12 AAL2 connections for the common channels then CID 20 would be assigned to the first SRB.

5.2 Architecture

- The Iub interface is often illustrated as a single direct connection between the Node B and RNC. In practice, the Iub is more complex and is typically based upon a network of connections. Switches can be used to aggregate ATM traffic in the uplink direction and distribute ATM traffic in the downlink direction.
- The section of the Iub transport network connecting the Node B to the first ATM switch is known as the last mile. This section has the lowest bandwidth and is usually the section which is most likely to experience congestion. E1 transport connections are typically used for the last mile in countries outside of Japan and North America. JT1 transport connections are more common in Japan, whereas T1 transport connections are more common in North America.
- An E1 transport connection is capable of supporting a maximum bit rate of 2.048 Mbps. This maximum bit rate is generated by 32 channels each of 64 kbps. Two of these channels are used for signalling purposes and the effective available bit rate from a single E1 is 1.92 Mbps. Both JT1 and T1 transport connections are able to support a maximum bit rate of 1.544 Mbps.
- The bandwidth of the last mile connection can be increased by combining the bandwidth offered by multiple E1, JT1 or T1. Inverse Multiplexing for ATM (IMA) can be used to combine a group of individual physical connections. The use of IMA requires the transmission of IMA Control Protocol (ICP) ATM cells. These cells represent an overhead, but are necessary to help the receiver cope with differences in the delay associated with each physical connection.
- The section of the Iub transport network connecting the RNC to a traffic aggregation switch is known as the hub section. The physical connection for the hub section of the Iub could be an STM-1. STM-1 transport connections are widespread outside of North America. OC-3 represents the equivalent transport connection which is more popular in North America. Both connection types make use of optical connections. The hub section of the Iub transport network is able to benefit from a multiplexing gain if the set of Node B do not all peak in activity simultaneously.

The Iub interface is often illustrated as a single direct connection between the Node B and RNC as illustrated in Figure 1.1. In practice, the Iub is more complex and is typically based upon a network of connections. Switches can be used to aggregate ATM traffic in the uplink direction and distribute ATM traffic in the downlink direction. The section of the Iub transport network connecting the Node B to the first ATM switch is known as the last mile. This section has the lowest bandwidth and is usually the section which is most likely to experience congestion. Sections of the Iub transport network between the first ATM switch and the RNC typically have higher bandwidths and are less likely to cause congestion. An example Iub transport network based upon ATM is illustrated in Figure 5.12.

ATM switches are able to provide a cross-connect function to map a set of VPI on one side of the switch to a different set of VPI on the other side of the switch. For example, if every Node B is configured with two VPC (the first for O&M traffic and the second for NBAP, ALCAP and user plane traffic) then those VPC could be addressed across the last mile using VPI 0 and VPI 1. It is possible to re-use the same pair of VPI for every Node B because each Node B has a dedicated physical connection for the last mile. On the RNC side of the first switch, the physical connection is shared between Node B and it is no longer possible to re-use the same VPI numbering, i.e. all VPI sharing the same physical connection must be unique. Renumbering the VPI within the ATM switch allows the numbering to remain unique on both sides of the switch. The switch then becomes responsible for maintaining knowledge of the mapping between the VPI numbering on each side of the switch. This represents a form of ATM multiplexing, i.e. VPC multiplexing.

An example of VPC multiplexing is illustrated in Figure 5.13. In this example, there are three Node B using the same pair of VPI to transfer traffic across the last mile. Uplink traffic is aggregated by an ATM switch prior to forwarding towards the RNC. The ATM switch provides a mapping of VPI $(0,1) \times 3$ on one side of the switch to VPI $(0,1,2,3,4,5)$ on the other side of the switch. In this example, each Node B is configured to transfer traffic using VPI 0 and 1 whereas the RNC is configured to transfer traffic using VPI 0 to 5. The ATM switch has knowledge of both the Node B and RNC VPI configurations and is able to provide an intermediate cross-connect function.

Figure 5.13 indicates that the physical connection providing the last mile of the Iub transport network is an E1. E1 transport connections are widespread outside of Japan and North America. JT1 transport connections are more common in Japan, whereas T1 transport connections are more common in North America. All three interface types are based upon Plesiochronous Digital Hierarchy (PDH)

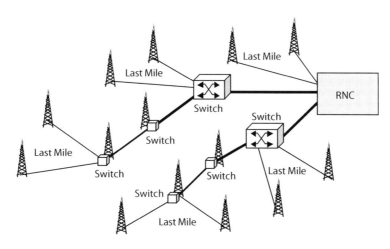

Figure 5.12   Iub transport network based upon ATM

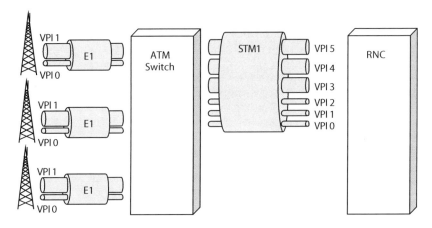

Figure 5.13 Iub transport network aggregation using VPI switching

which means that different sections of the network have similar timing, but are not perfectly synchronised. An E1 transport connection is capable of supporting a maximum bit rate of 2.048 Mbps. This maximum bit rate is generated by 32 channels each of 64 kbps. Two of these channels are used for signalling purposes and the effective available bit rate from a single E1 is 1.92 Mbps. Both JT1 and T1 transport connections are able to support a maximum bit rate of 1.544 Mbps. This maximum bit rate is generated by 24 channels of 64 kbps plus 8 kbps of control information. The control information is used to help the receiver achieve synchronisation and complete demultiplexing.

The bandwidths offered by single E1, JT1 and T1 transport connections are sufficient for relatively low quantities of traffic, but become insufficient as the quantity of traffic increases. Also, these single transport connections are unable to support the high data rates offered by HSDPA and HSUPA. The bandwidth of the last mile connection can be increased by combining the bandwidth offered by multiple E1, JT1 or T1. Inverse Multiplexing for ATM (IMA) can be used to combine a group of individual physical connections. IMA is specified by the ATM Forum within AF-PHY-0086.001. The concept of IMA is illustrated in Figure 5.14.

The set of individual physical connections is known as an IMA group. Each connection within the IMA group is utilised in a round robin fashion and the serial stream of ATM cells at the transmitter becomes a parallel stream across the last mile of the Iub transport network. The receiver extracts ATM cells from each physical connection to reconstruct the original serial stream. The use of IMA requires

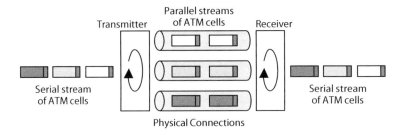

Figure 5.14 Concept of Inverse Multiplexing for ATM (IMA)

the transmission of IMA Control Protocol (ICP) ATM cells. These cells represent an overhead, but are necessary to help the receiver cope with differences in the delay associated with each physical connection. The rate at which ICP cells are sent depends upon the IMA frame duration. The IMA specification states that it is mandatory to support an IMA frame length of 128 cells and optional to support IMA frame lengths of 32, 64 and 256 cells. An IMA frame length of 128 cells represents the default value and means that an ICP cell is sent on each physical connection once every 128 cells, i.e. the overhead generated by the ICP cells is less than 1%. Smaller frame lengths create greater overhead, but allow greater delay differences between the individual physical connections. The IMA specification also states that filler cells are sent when there are no ATM cells to transmit between ICP cells. Filler cells are used to help the receiver maintain synchronisation with each physical connection.

Assuming an IMA frame length of 128 cells means that an IMA group including two E1 connections is able to offer a bandwidth of $3.84 \times 127/128 = 3.810$ Mbps, whereas an IMA group including three E1 connections is able to offer $5.76 \times 127/128 = 5.715$ Mbps. The IMA specifications allow a maximum of 32 physical connections to be included within an IMA group. In practice, the number of physical connections is upgraded to follow the quantity of traffic loading the network. It is common for networks to be deployed using a single E1 and to generate an IMA group of two or three E1 connections as the quantity of traffic increases. Specific last mile connections belonging to hotspot Node B may be configured with five or six E1 connections to provide support for higher quantities of traffic and the higher throughputs offered by HSDPA and HSUPA.

The RNC and Node B databuilds define the allocation of Iub bandwidth to each of the VCC illustrated in Figure 5.7 (recall that Figure 5.7 represents a relatively simple example and it is possible that there are multiple user plane VCC within the VPC). Some example bandwidth allocations for 1, 2 and 3 E1 transmission links are presented in Table 5.3. These allocations are presented in terms of both ATM cells per second and bits per second.

In each case, the majority of the Iub capacity is allocated to the user plane VCC. In the case of 2 and 3 E1 connections, the total throughput figures leave sufficient margin for the ICP ATM cells generated by the IMA protocol. There is also some margin for the O&M VPC. These throughput figures include the overheads generated by the Iub interface, i.e. the user plane figures include the Frame Protocol and AAL2 overheads. These overheads must be subtracted to evaluate the equivalent RLC or application layer throughputs. Section 5.3 quantifies the overheads generated by the Frame Protocol and AAL2 layers.

Figure 5.13 indicates that the physical connection for the hub section of the Iub could be an STM-1 (Synchronous Transport Module 1). STM-1 transport connections are widespread outside of North America. Optical Carrier 3 (OC-3) represents the equivalent transport connection which is more popular in North America. STM-1 connections are based upon Synchronous Digital Hierarchy (SDH) whereas OC-3 connections are based upon Synchronous Optical Networking (SONET). Both

Table 5.3 Example bandwidth allocations for a set of Iub VCC

Number of E1	User plane	AAL2 signalling	Dedicated NBAP	Common NBAP	Total
1	4060 cps	80 cps	160 cps	80 cps	4380 cps
	1.72 Mbps	33.9 kbps	67.8 kbps	33.9 kbps	1.86 Mbps
2	8120 cps	160 cps	320 cps	160 cps	8760 cps
	3.44 Mbps	67.8 kbps	135.7 kbps	67.8 kbps	3.71 Mbps
3	12180 cps	240 cps	480 cps	240 cps	13140cps
	5.16 Mbps	101.8 kbps	203.5 kbps	101.8 kbps	5.57 Mbps

connection types make use of optical connections. SDH provides a frame size of 2430 bytes which is transferred in 125 μs. This corresponds to a bit rate of 155.52 Mbps. The frame of 2430 bytes is organised into nine consecutive subframes each of which includes 9 bytes of control information followed by 261 bytes of payload. This results in a maximum effective throughput for an STM-1 connection of 150.336 Mbps. The use of a single byte within each 125 μs frame corresponds to a 64 kbps connection. SONET uses a frame size of 810 bytes which is transferred in 125 μs. This corresponds to a bit rate of 51.84 Mbps. The frame of 810 bytes is organised into nine consecutive subframes each of which includes 3 bytes of control information followed by 87 bytes of payload. This results in a maximum effective throughput for an OC-1 connection of 50.112 Mbps. An OC-3 connection has triple the bandwidth and so offers an effective throughput of 150.336 Mbps, i.e. the same as STM-1.

The hub section of the Iub transport network is able to benefit from a multiplexing gain if the set of Node B do not all peak in activity simultaneously. For example, if there are 100 Node B connected to the hub section and each Node B is configured with one E1 for the last mile connection then the absolute worst case bandwidth requirement for the hub section would be 100×1.92 Mbps = 192 Mbps. In practise, not all of the last mile connections will be fully loaded, nor will they all peak in activity simultaneously and it is possible that a single STM-1 connection offering 150 Mbps will be sufficient to support the 100 Node B. This reduction in Iub bandwidth allocation is known as overbooking. If an aggressive overbooking strategy is adopted then relatively little Iub bandwidth is allocated and cost savings are achieved, but there is an increased risk of congestion and the subsequent loss of ATM cells. If separate user plane VCC are configured for each service type then different overbooking strategies can be applied to each service. A more aggressive overbooking strategy can be applied to non-real time services which are less sensitive to increased delays.

5.3 Overheads

- It is necessary to quantify the overheads introduced by the Iub interface when dimensioning the Iub capacity requirement. The user plane occupies the majority of the Iub bandwidth and so represents the most important protocol stack when dimensioning.
- The Iub overhead for the 64 kbps and 384 kbps PS data services is typically between 20% and 25% when calculated from the top of the Frame Protocol layer. Referencing the overhead to the top of the RLC layer increases the overhead to between 25% and 30%.
- The instantaneous Iub overhead for the speech service is relatively high during periods of speech activity, i.e. between 50% and 55%. This overhead is reduced when averaging across periods of speech inactivity. Assuming a 50% speech activity factor results in an overhead in the order of −20%, i.e. the average Iub throughput is less than 12.2 kbps.
- Iub overheads can also be calculated for the radio network control plane and transport network control plane protocol stacks. The Iub overheads associated with these protocol stacks can often be relatively large as a result of the messages being small.

It is necessary to quantify the overheads introduced by the Iub interface when dimensioning the Iub capacity requirement. The overheads can be quantified by following the path of a higher layer packet through the appropriate protocol stack. The user plane occupies the majority of the Iub bandwidth and so represents the most important protocol stack when dimensioning the Iub bandwidth requirement. Table 5.4 presents the overheads introduced by each layer of the user plane protocol stack for the 12.2 kbps speech, 64 kbps PS data and 384 kbps PS data services.

Table 5.4 Iub overheads for the 12.2 kbps speech, 64 kbps data and 384 kbps data services

	12.2 kbps speech (downlink)	64 kbps PS data (downlink)	384 kbps PS data (downlink)
Frame protocol layer			
MAC PDU (bits)	81, 103, 60	336	336
# MAC PDU per FP Packet	1, 1, 1	4	12
Frame protocol payload (bits) (including padding)	256	1344	4032
Frame protocol header (bits)	40	24	24
Frame protocol tail (bits)	16	16	16
Frame protocol packet (bits)	312	1384	4072
AAL2 SSSAR layer			
Max. Size of AAL2 SSSAR PDU (bits)	360	360	360
Segmentation required	No	Yes	Yes
# AAL2 SSSAR PDU	1	4	12
AAL2 CPS layer			
AAL2 CPS packet header (bits)	24	24	24
Total AAL2 CPS packet data (bits)	336	1480	4360
ATM layer			
ATM cell header (bits)	40	40	40
ATM cell payload (bits)	384	384	384
AAL2 start field (bits)	8	8	8
Effective ATM cell payload (bits)	376	376	376
# ATM cells	1	4	12
Remaining ATM cell capacity (bits)	40	24	152
Total overhead (bits)	135	325	885
Total payload (bits)	244	1344	4032
Iub protocol overhead (%)	55	24	22
Iub throughput	18.95 kbps	83.45 kbps	491.70 kbps

These examples are based upon the downlink. It is usual for the quantity of downlink traffic to be greater than the quantity of uplink traffic and so dimensioning tends to focus upon the downlink. However, it is good practice to complete the analysis in both directions. Table 5.4 starts by presenting the quantity of data at the top of the Frame Protocol layer, i.e. the bottom of the MAC layer for DCH connections. A set of MAC PDU is known as a transport block set and it transferred across the Iub once per Transmission Time Interval (TTI). The 12.2 kbps speech service makes use of three transport channels with transport block sizes of 81, 103 and 60 bits. These sum to generate 244 bits which defines the 12.2 kbps bit rate when divided by the 20 ms TTI. The 64 kbps PS data service makes use of four transport blocks of 336 bits and a 20 ms TTI. These figures result in a bit rate of 67.2 kbps which includes the overhead introduced by the RLC layer. The 384 kbps PS data service makes use of 12 transport blocks of 336 bits and a 10 ms TTI, i.e. a bit rate of 403.2 kbps.

The Frame Protocol layer adds padding to the speech service transport blocks to ensure that each block is a multiple of 8 bits, i.e. the transport block of 81 bits is padded to 88 bits, the transport block of

103 bits is padded to 104 bits and the transport block of 60 bits is padded to 64 bits. It is not necessary to add any padding for the PS data services because the transport blocks are already a multiple of 8 bits. The size of the header added by the Frame Protocol layer depends upon the number of transport channels. Each additional transport channel increases the size of the header by 8 bits, i.e. a 5 bit Transport Format Indicator (TFI) accompanied by 3 spare bits is required for each transport channel. The PS data services use only a single transport channel and require a 24 bit header. The speech service uses three transport channels and requires a 40 bit header. The size of the Frame Protocol tail is the same for all services, i.e. 16 bits.

The Frame Protocol packet is passed to the AAL2 SSSAR layer for segmentation into blocks no larger than 360 bits. This layer passes the segments to the AAL2 CPS layer without adding any overhead. The AAL2 CPS layer adds 24 bits of header to each segment. These segments are then fed into ATM cells which include a 40 bit ATM header and an 8 bit AAL2 start field. After accounting for the AAL2 start field, the effective ATM cell payload is 376 bits. In general, the quantity of AAL2 CPS data is not a multiple of 376 bits and some ATM cell capacity remains for other AAL2 connections. For example, the speech service occupies 336 of the 376 bits available and there are 40 bits available for other AAL2 connections. This means that the overhead generated by this ATM cell header and 8 bit start field can be partially associated with other AAL2 connections.

The total overhead is calculated by summing the overheads generated by the Frame Protocol, AAL2 and ATM layers whereas the total payload is equal to the transport block set size. The Iub overhead expressed as a percentage is defined as the ratio between the overhead and the payload. This overhead is relative to the bit rate at the top of the Frame Protocol layer rather than the bit rate at the top of the RLC layer. In the case of the speech service, the bit rate at the top of the Frame Protocol layer is equal to the bit rate at the top of the RLC layer and so the overhead does not change. In the case of the data services it can be assumed that each 336 bit transport block includes 16 bits of RLC header. This means that the overheads for the 64 kbps and 384 kbps PS data services increase to 30% and 28%, respectively when measuring the overhead from the top of the RLC layer. The sum of the payload and overhead define the Iub throughput when divided by the TTI.

The Iub overhead for the 12.2 kbps speech service appears very high in Table 5.4. This result represents the overhead during a period of speech activity. The discontinuous nature of the speech service reduces the overhead when averaged over a period of time. When a speech user stops talking the transmitter begins a period of discontinuous transmission (DTX). During periods of DTX the transmitter generates Silence Descriptor (SID) frames once every 8 speech frames, i.e. once every 160 ms. Table 5.5 presents the Iub overheads for a SID frame.

A SID frame uses a single 39 bit transport block which corresponds to a bit rate of 1.95 kbps. The small transport block size means that the relative overheads generated by the Frame Protocol, AAL2 and ATM layers are greater. The resulting Iub overhead is 246% which increases the bit rate to 6.75 kbps. The average speech service throughput on the Iub can be calculated if an assumption is made regarding the percentage of time that the transmitter is in DTX. It is common to assume a figure of 50%, i.e. a speech user talks for half of the time and listens for half of the time. This results in an average Iub throughput of $50\% \times 18.95$ kbps $+ 50\% \times 6.75/8$ kbps $= 9.9$ kbps. The average Iub throughput is less than 12.2 kbps and so the average Iub overhead is negative and has a value of -19%.

Iub overheads can also be calculated for the radio network control plane and transport network control plane protocol stacks, i.e. the protocol stacks using AAL5 rather than AAL2. The calculation for individual messages is less complex for these protocol stacks. However, the radio interface control plane (NBAP) has a large number of different messages and the size of those messages depends upon both the type of message and the content. An NBAP Radio Link Addition Request message is smaller than an NBAP Radio Link Setup Request message and is likely to have a larger Iub overhead. An NBAP Radio Link Setup Request message for an SRB connection is smaller than an NBAP Radio Link

Table 5.5 Iub overheads for SID frames belonging to the 12.2 kbps speech service

	12.2 kbps speech SID frame (downlink)
Frame Protocol layer	
MAC PDU	39, 0, 0 bits
# MAC PDU per FP packet	1, 0, 0
Frame Protocol payload (including padding)	40 bits
Frame Protocol header	40 bits
Frame Protocol tail	16 bits
Frame Protocol packet	96 bits
AAL2 SSSAR layer	
Max. Size of AAL2 SSSAR PDU	360 bits
Segmentation required	No
# AAL2 SSSAR PDU	1
AAL2 CPS layer	
AAL2 CPS packet header	24 bits
Total AAL2 CPS packet data	120 bits
ATM Layer	
ATM cell header	40 bits
ATM cell payload	384 bits
AAL2 start field	8 bits
Effective ATM cell payload	376 bits
# ATM cells	1
Remaining ATM cell capacity	256 bits
Total overhead	96 bits
Total payload	39 bits
Iub protocol overhead	246%
Iub throughput	6.75 kbps

Setup Request message for an SRB plus speech connection. Estimating the Iub throughput requirement for these protocol stacks requires an estimate of the rate at which specific control plane messages are transferred. Table 5.6 presents the Iub overhead calculations for an ALCAP Establish Request message, a common NBAP Radio Link Setup Request message and a dedicated NBAP Radio Link Deletion Request message.

The Iub overheads associated with these messages are relatively large because the messages are relatively small. When AAL5 and ATM are used to transfer IP packets across the Iu–ps interface the overhead is significantly less. For example, the overhead for a 1500 byte IP packet would be 13% which is close to the overhead generated by the ATM layer alone (5 bytes/48 bytes = 10%). The impact of the figures presented in Table 5.6 depends upon the rate at which messages are transferred. This increases as the quantity of traffic loading a Node B increases. The figures presented in Table 5.3 indicate that it is reasonable to assign approximately 10% of the Iub bandwidth to the AAL5 control plane protocol stacks.

Table 5.6 Iub overheads for a set of example control plane messages using AAL5

	ALCAP Establish Request	NBAP Radio Link Setup Request	NBAP Radio Link Deletion Request
Message size (bytes)	66	116	34
AAL5 SSCOP layer			
SSCOP padding (bytes)	2	0	2
SSCOP tail (bytes)	4	4	4
AAL5 CPCS layer			
CPCS padding (bytes)	16	16	0
CPCS tail (bytes)	8	8	8
ATM layer			
ATM cell header (bytes)	5	5	5
ATM cell payload (bytes)	48	48	48
# ATM cells	2	3	1
Total overhead (bytes)	40	43	19
Total payload (bytes)	66	116	34
Iub Protocol overhead (%)	61	37	56

5.4 Service Categories

- The quality of service associated with ATM traffic is managed using service categories. The ATM Forum specifies CBR, rt-VBR, nrt-VBR, UBR, ABR and GFR service categories.
- CBR is intended for services which require a fixed bandwidth to be continuously available for the duration of their connection. A CBR connection is able to transfer cells at the PCR at any time and for any duration. Real time applications such as speech and video services often use CBR VCC to benefit from the high quality of service.
- Speech services can also use rt-VBR to provide a balance between the quality of service and the quantity of resources consumed by each connection. The quality of service offered by a rt-VBR VCC is less than that offered by a CBR VCC but the capacity is greater.
- The quality of service offered by nrt-VBR is only specified in terms of CLR. This means that delays across the ATM network can be more significant and nrt-VBR VCC are more suitable for non-real time applications.
- UBR does not have any quality of service parameters and is limited to providing a best effort capability. This means that UBR is suitable for non-real time applications which are insensitive to delay and packet loss. In practice, the delay and packet loss experienced depends upon the load of the transport network.
- The enhanced version of UBR introduces the optional Minimum Desired Cell Rate parameter. This parameter forms the basis of UBR+ although some implementations of UBR+ re-use of the Minimum Cell Rate parameter normally associated with the ABR and GFR service categories. In both cases, the additional parameter is used to provide information regarding a minimum throughput requirement.
- Key ATM Forum specifications: AF-TM-0121.000 and AF-TM-0150.000

The quality of service associated with ATM traffic is managed using service categories. The ATM Forum specifies six service categories within version 4.1 of its traffic management specification AF-TM-0121.000,

- Constant Bit Rate (CBR)
- Real Time Variable Bit Rate (rt-VBR)
- Non-Real Time Variable Bit Rate (nrt-VBR)
- Unspecified Bit Rate (UBR)
- Available Bit Rate (ABR)
- Guaranteed Frame Rate (GFR)

The GFR service category is the most recently specified and was added in version 4.1 of the specification (March 1999). The UBR service category was enhanced in June 2000 using appendum AF-TM-0150.000. The enhanced version of UBR is often referred to as UBR+.

The set of six service categories allow different quality of service profiles to be applied to different traffic types. VCC are configured to use a specific service category as part of their setup procedure. For example, the Iub transport network could be configured with a CBR VCC for speech traffic, and a UBR VCC for internet browsing traffic. Each service category is defined by a set of parameters which describe both the traffic characteristics and the quality of service requirements. These parameters are presented in Table 5.7.

Traffic belonging to a CBR VCC is characterised using a Peak Cell Rate (PCR) and a Cell Delay Variation Tolerance (CDVT). These parameters are service dependent and so speech service PCR and CDVT values would be different to video service PCR and CDVT values. The PCR defines the maximum throughput generated by the service, whereas the CDVT defines the maximum acceptable jitter in the cell transfer delay. A transport network configured with a CBR VCC is responsible for providing quality of service in terms of peak-to-peak Cell Delay Variation (CDV), maximum Cell Transfer Delay (CTD) and Cell Loss Ratio (CLR). The peak-to-peak CDV represents the difference between the minimum delay across the network and the maximum delay experienced by a high

Table 5.7 Service categories specified by the ATM Forum

	CBR	rt-VBR	nrt-VBR	UBR	ABR	GFR
Traffic characteristics						
Peak cell rate	√	√	√	√	√	√
Cell delay variation tolerance	√	√	√	√	√	√
Sustainable cell rate		√	√			
Maximum burst size		√	√			√
Minimum desired cell rate				Optional		
Minimum cell rate					√	√
Maximum frame size						√
Quality of service requirements						
Peak-to-peak cell delay variation	√	√				
Maximum cell transfer delay	√	√				
Cell loss ratio	√	√	√		Optional	Optional
Feedback information						
Feedback provided					√	

percentage of all ATM cells. The maximum CTD defines the maximum delay and cells which are delivered late or are lost have a CTD which is greater than the maximum. The CLR is defined as the ratio between the number of lost cells relative to the total number of cells transmitted. The CBR service category is intended for services which require a fixed bandwidth to be continuously available for the duration of their connection. A CBR connection is able to transfer cells at the PCR at any time and for any duration. A CBR connection is also able to transfer cells at a rate which is less than the PCR. Real time applications such as speech and video services often use CBR VCC to benefit from the high quality of service. The cost of offering this high quality of service with a CBR VCC is a reduced number of simultaneously active connections, i.e. each connection consumes a relatively large share of the Iub transport network resources.

Traffic belonging to a rt-VBR VCC is characterised using a PCR, CDVT, Sustainable Cell Rate (SCR) and Maximum Burst Size (MBS). The SCR represents an average throughput which quantifies the longer term bandwidth requirement. The MBS provides an indication of the burstiness of the traffic and quantifies the maximum number of ATM cells which can be transferred at the PCR without exceeding the SCR. A transport network configured with a rt-VBR VCC is responsible for providing quality of service in terms of peak-to-peak CDV, maximum CTD and CLR, i.e. the same parameters as used by CBR VCC. Speech services can use rt-VBR VCC to provide a balance between the quality of service offered by the network and the quantity of network resources consumed by each connection. The quality of service offered by a rt-VBR VCC is less than that offered by a CBR VCC but the capacity is greater.

Traffic belonging to a nrt-VBR VCC is characterised using the same parameters as traffic belonging to a rt-VBR VCC, i.e. PCR, CDVT, SCR and MBS. However, the quality of service offered by a nrt-VBR VCC is only specified in terms of CLR. This means that delays across the ATM network can be more significant and nrt-VBR VCC are more suitable for non-real time applications, e.g. an internet browsing service could use an nrt-VBR VCC.

Traffic belonging to a UBR VCC is characterised using a PCR and CDVT. In this case, although PCR and CDVT figures are specified they are not necessarily enforced by the network. UBR VCC do not have any quality of service parameters and are limited to providing a best effort capability, i.e. UBR traffic is queued until capacity becomes available. This means that UBR is suitable for non-real time applications which are insensitive to delay and packet loss. In practice, the delay and packet loss experienced depends upon the load of the transport network. As the load increases there is greater potential for increased delays and packet loss. The benefit of using UBR VCC is that a relatively large number of connections can be admitted and the multiplexing gain is increased. The enhanced version of UBR specified by appendum AF-TM-0150.000 introduces the optional Minimum Desired Cell Rate (MDCR) parameter. This parameter forms the basis of UBR+ although some implementations of UBR+ re-use of the Minimum Cell Rate (MCR) parameter normally associated with the ABR and GFR service categories. In both cases, the additional parameter is used to provide information regarding a minimum throughput requirement. This provides a mechanism to manage the minimum quality of service offered by a UBR VCC.

Traffic belonging to an ABR VCC is characterised using a PCR, CDVT and a Minimum Cell Rate (MCR). Similar to UBR, the PCR and CDVT values may not necessarily be enforced by the transport network. The ABR service category can be viewed as an intelligent version of UBR+. ABR provides a best effort capability operating above a minimum cell rate. However, an ABR VCC makes use of feedback information to control the rate at which ATM cells are sent into the transport network. This flow control mechanism helps to adapt the cell rate to the available bandwidth and helps to avoid cell loss. Specifying a CLR for an ABR VCC is optional. If the flow control mechanism is working well then cell loss should be minimised.

Traffic belonging to a GFR VCC is characterised using a PCR, CDVT, MBS, MCR and a Maximum Frame Size (MFS). The GFR service category was introduced to provide improved support for TCP/IP applications, especially when transferred using AAL5. The use of GFR requires that network elements

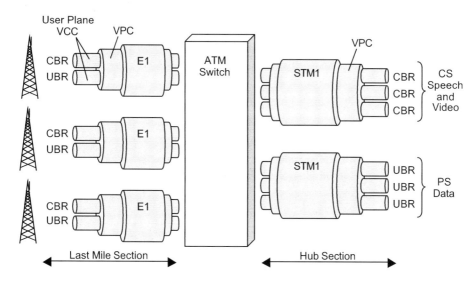

Figure 5.15   Combination of CBR and UBR User Plane VCC for different traffic types

are aware of the AAL5 frame boundaries so that complete AAL5 frames can be discarded when congestion occurs. This is in contrast to discarding individual ATM cells which may belong to different AAL5 frames. The MFS parameter defines the largest AAL5 frame size that is permitted to be sent across the transport network. Similar to UBR and ABR, the GFR service category does not define any quality of service requirements in terms of delay. Similar to ABR, the CLR requirement is optional.

Figure 5.15 presents an example Iub transport network where CS speech and video services are transferred using CBR VCC whereas PS data services are transferred using UBR VCC.

This example illustrates that network aggregation at the hub section can be used to collect all CBR VCC within a first VPC and all UBR VCC within a second VPC. As mentioned in Section 5.2, the two hub section VPC can have different overbooking strategies, i.e. the CBR VPC could have a conservative overbooking strategy while the UBR VPC could have an aggressive overbooking strategy.

6

HSDPA

6.1 Concept

- HSDPA is introduced within the release 5 version of the 3GPP specifications. It provides a solution for achieving increased individual connection throughputs, increased total cell throughputs and reduced round trip times. HSDPA is associated with packet switched services which can be either real time or non-real time.
- HSDPA can be used to transfer both RRC signalling and application data. RRC signalling can be transferred using a parallel DPCH if HSDPA is configured to transfer only application data.
- The RNC buffers HSDPA data until flow control instructions from the Node B allow it to be transferred across the Iub. The Node B scheduler determines when each UE gains access to the common HSDPA air-interface resources. Code multiplexing allows multiple UE to be scheduled simultaneously. Adaptive Modulation and Coding (AMC) within the Node B determines the transport block size and corresponding throughput allocated at any point in time. AMC makes use of channel quality indicators provided by the UE.
- Hybrid ARQ is used as a re-transmission protocol between the Node B and UE. Connections using acknowledged mode RLC can also use RLC re-transmissions between the RNC and UE if Hybrid ARQ re-transmissions fail. TCP re-transmissions between the application server and UE can also be used if the protocol stack includes the TCP layer and both Hybrid ARQ and RLC re-transmissions fail.
- HSDPA connections use a fixed TTI of 2 ms. The modulation scheme can be either QPSK (2 bits per symbol) or 16-QAM (4 bits per symbol). The option of 64-QAM (6 bits per symbol) is introduced within the release 7 version of the 3GPP specifications. A maximum of 15 HS-PDSCH physical channels can be used to support HSDPA within a cell. HSDPA connections do not make use of inner loop power control nor soft handover.
- HSDPA provides a solution which can be applied to downlink user plane data and control plane signalling. The corresponding uplink can remain as a DPCH or it can be configured as an HSUPA connection.

High Speed Downlink Packet Access (HSDPA) is introduced within release 5 of the 3GPP specifications. HSDPA provides a solution for achieving increased individual connection throughputs, increased total cell throughputs and reduced round trip times. The increased individual connection

throughput and reduced round trip time help to improve the quality of service experienced by the
end-user. The increased total cell throughput helps the operator to increase the efficiency with which
air-interface resources are used. HSDPA is associated with packet switched services rather than
circuit switched services. Packet switched services can be either Non-real time (NRT) or Real time
(RT). Example NRT services are email, file transfer and internet browsing. An example RT service is
Voice over IP (VoIP).

An HSDPA connection can be used to transfer both the user plane data originating from an end-user
application and the control plane signalling originating from the RRC layer. Alternatively, an HSDPA
connection can be used to transfer the user plane data while a DPCH is used to transfer the control
plane signalling. These two possibilities are presented in Table 6.1.

The logical channels represent the type of information being transferred and these remain unchanged
irrespective of whether or not an HSDPA connection is used. The set of transport channels depends
upon whether or not the Signalling Radio Bearers (SRB) are transferred using HSDPA. Transferring the
SRB using a DPCH means that a DCH transport channel is used in parallel to an HS-DSCH transport
channel. Transferring the SRB using HSDPA means that the set of transport channels is limited to a
single HS-DSCH. The 3GPP specifications allow a maximum of one HS-DSCH transport channel.
Transferring both the DCCH and DTCH using a single HS-DSCH involves multiplexing within the
Node B section of the MAC layer. The set of physical channels also depends upon whether or not the
SRB are transferred using HSDPA. If the SRB are transferred using a DPCH then a DPDCH is used in
parallel to one or more HS-PDSCH. The 3GPP specifications allow a maximum of 15 HS-PDSCH to be
used at any point in time. Increasing the number of HS-PDSCH increases the rate at which data can be
transferred. The set of physical channels presented in Table 6.1 serves only as an introduction and is not
exhaustive. Additional physical channels used to transfer physical layer control information are
described in Section 6.9.

The introduction of HSDPA requires a significant increase in the functionality of the Node B. This
allows the Node B to become more intelligent and more responsive to changes in the air-interface
conditions. The general concepts associated with HSDPA for a single UE are illustrated in Figure 6.1.
These concepts remain valid when the Node B serves multiple HSDPA UE, but the Node B scheduling
becomes more complex.

The RNC receives downlink user plane data from the packet switched core network in the same way
as for a DPCH connection. Depending upon the application and the RNC implementation, the PDCP
layer may be used to compress the headers generated by the higher layers. The downlink data is then
passed to the RLC layer which segments large packets into smaller packets to ensure that they can be
encapsulated by a set of RLC PDU. If the end-user application is characterised as NRT then it is likely
that the RLC layer will be configured for acknowledged mode and will be responsible for RLC re-
transmissions. RLC PDU are passed to a MAC-d entity within the RNC. The MAC-d entity is
transparent in terms of PDU size (assuming no user plane multiplexing), i.e. MAC-d PDU have the
same size as RLC PDU. The resulting MAC-d PDU are buffered by the RNC until the Node B issues
sufficient flow control credits to allow transfer across the Iub. Frame Protocol data frames are used to
package and transfer MAC-d PDU across the Iub.

Table 6.1 Downlink logical, transport and physical channels for HSDPA

	SRB on DPCH		SRB on HSDPA	
	Control plane signalling	User plane data	Control plane signalling	User plane data
Logical channel	DCCH	DTCH	DCCH	DTCH
Transport channel	DCH	HS-DSCH	HS-DSCH	
Physical channel	DPDCH	HS-PDSCH	HS-PDSCH	

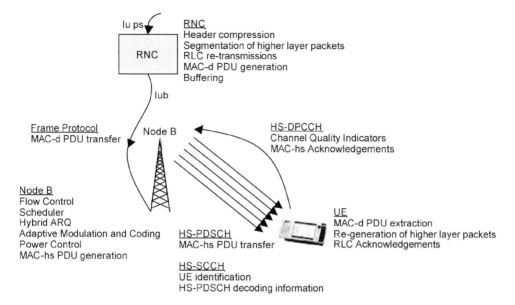

Figure 6.1   General HSDPA concepts for a single HSDPA UE

MAC-d PDU are received by the MAC-hs layer within the Node B. The MAC-hs layer includes the HSDPA scheduler which determines the UE or the group of UE to be served at any point in time. Scheduling can be completed in both the time and code domains, i.e. time multiplexing and code multiplexing. 2 ms subframes are defined for time multiplexing whereas 15 HS-PDSCH channelisation codes are available for code multiplexing. If code multiplexing is not used then only one HSDPA UE is served at any point in time. The identity of the UE being scheduled at any point in time is included within the HS-SCCH physical channel. The HS-SCCH also includes information to help the relevant UE decode the HS-PDSCH data. Hybrid ARQ (HARQ) functionality within the MAC-hs layer allows re-transmissions between the Node B and UE. These re-transmission have a lower delay than those provided by the RLC layer between the RNC and UE. Adaptive Modulation and Coding (AMC) within the MAC-hs layer identifies an appropriate bit rate to allocate during each 2 ms subframe.

HSDPA and DPCH connections have fundamental differences in terms of achieving a specific throughput. Figure 6.2 compares the general relationships between path loss, transmit power and throughput for DPCH and HSDPA connections.

DPCH connections make use of inner loop power control to compensate for changes in the propagation conditions. This helps to ensure that the receiver maintains a relatively constant signal-to-noise ratio and a corresponding constant throughput. HSDPA connections make use of a relatively constant transmit power which does not compensate for changes in the propagation conditions. This results in the receiver experiencing a variable signal-to-noise ratio which allows the transmitter to schedule a variable throughput. DPCH connections benefit from a faster update cycle. DPCH inner loop power control provides 1500 instructions per second whereas the AMC belonging to HSDPA allows a maximum of 500 instructions per second. Nevertheless, the combination of Node B scheduling, HARQ and AMC allows HSDPA to make more efficient use of the air-interface resources. In practice, the behaviour of DPCH and HSDPA connections is more complex than that illustrated in

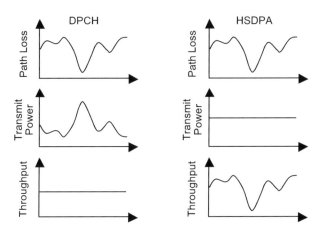

Figure 6.2   Path loss, transmit power and throughput for DPCH and HSDPA connections

Figure 6.2, but the fundamental concepts remain valid. DPCH inner loop power control tracks changes in both the interference conditions and path loss to help ensure that the received signal-to-noise ratio remains relatively constant. DPCH connections can also experience changes in throughput when their bit rates are reconfigured by the RRC layer or when the MAC layer selects a different transport format within the transport format set. The HS-PDSCH transmit power may not be completely constant, but may be updated to compensate for changes in the transmit power used by other physical channels. The HS-PDSCH transmit power may also be reduced when UE are in good coverage and have some margin in their downlink link budget while being scheduled their maximum supported bit rate.

The bit rates allocated by the AMC within the MAC-hs are determined primarily by channel quality indicators provided by the UE. If the UE is experiencing good channel conditions then a high channel quality indicator is signalled to the Node B and a relatively high bit rate is allocated. If the UE is experiencing poor channel conditions then a low channel quality indicator is signalled to the Node B and a relatively low bit rate is allocated. The UE also provides feedback to the Node B in terms of acknowledgements for each MAC-hs PDU. These acknowledgements are used by the HARQ to help determine whether a re-transmission is required or new data can be transmitted. Both the channel quality indicators and the MAC-hs acknowledgements are transferred using the uplink HS-DPCCH physical channel. RLC acknowledgements for downlink user plane data are transferred using the uplink DPDCH, or using the E-DPDCH if HSUPA is available.

HSDPA provides a solution which can be applied to downlink user plane data and control plane signalling. The corresponding uplink can remain as a DPCH or it can be configured as an HSUPA connection. HSUPA is introduced in release 6 of the 3GPP specifications and so the first network and UE implementations of HSDPA support DPCH in the uplink direction. Subsequent implementations support HSUPA in the uplink direction. The throughput offered by the uplink connection can be important, even when a downlink data transfer is completed. For example a downlink file transfer using TCP/IP requires the uplink to return TCP acknowledgements to the application server. If these acknowledgements are not returned at a sufficient rate then the flow control capability of TCP will limit the downlink throughput. The uplink is also required to return RLC acknowledgements when acknowledged mode RLC is used.

Table 6.2 compares the main characteristics of the DPCH and HS-PDSCH. The DPCH is a dedicated physical channel and so every connection has its own DPCH. The HS-PDSCH is a common physical channel and so each connection shares the same set of HS-PDSCH. The HS-PDSCH uses a fixed 2 ms Transmission Time Interval (TTI) and each connection can have only a single transport block

Table 6.2 Comparison of the DPCH and HS-PDSCH

	DPCH	HS-PDSCH
Physical channel type	Dedicated	Common
Transmission time interval (ms)	10, 20, 40, 80	2
Multiple transport blocks per TTI	Yes	No
Modulation scheme	QPSK	QPSK, 16QAM, 64QAM*
Multiple channelisation codes	Yes, but in practice No	Yes
Soft handover	Yes	No
Inner loop power control	Yes	No
Node B scheduling	No	Yes
Node B AMC	No	Yes
Node B flow control	No	Yes
Node B re-transmissions	No	Yes
RNC re-transmissions	Yes	Yes

*64QAM is introduced within the release 7 version of the 3GPP specifications.

transferred during each TTI. The size of the transport block is determined by the Node B AMC functionality. The AMC function also determines whether the HS-PDSCH is modulated using QPSK, 16QAM or 64QAM. 16QAM requires a better signal to noise ratio than QPSK but maps 4 bits rather than 2 bits onto each modulated symbol. Likewise, 64QAM requires a better signal to noise ratio than 16QAM, but maps 6 bits rather than 4 bits onto each modulated symbol. A single HS-PDSCH uses a single downlink channelisation code. Multiple HS-PDSCH are allocated in the vast majority of conditions. Increasing the number of HS-PDSCH allows an increase in the downlink throughput. A single HS-PDSCH is only scheduled when the channel conditions are very poor. The 3GPP specifications allow multiple channelisation codes to be assigned to a single DPCH. However, in practice both network and UE implementations support downlink DPCH using only a single channelisation code. The HS-PDSCH does not support soft handover. This has implications upon the performance at cell edge and means that the serving cell change procedure must be applied when moving between cells. The remaining characteristics in Table 6.2 have already been introduced within this section. The complete set of characteristics is described in greater detail within the following sections of this chapter.

The user plane protocol stack for HSDPA is illustrated in Figure 6.3. This protocol stack was introduced in Section 2.1.2. Use of the PDCP layer is optional and may be limited to specific applications, e.g. VoIP. User plane data belonging to other applications can be processed by the RLC layer directly.

The control plane protocol stack for HSDPA is illustrated in Figure 6.4. This protocol stack is applicable when the SRB are transferred using HSDPA rather than a DPCH. The control plane protocol stack includes the RRC layer to generate and decode signalling messages. The PDCP layer does not apply to the control plane because there are no higher layer headers, e.g. TCP/IP headers to compress.

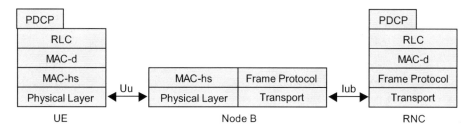

Figure 6.3 User plane HSDPA protocol stack

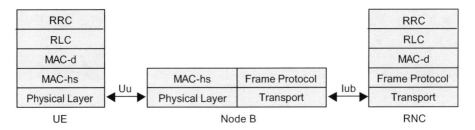

Figure 6.4 Control plane HSDPA protocol stack

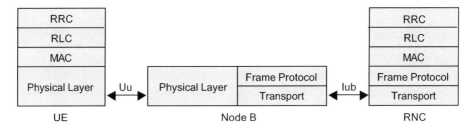

Figure 6.5 Control plane DPCH protocol stack

The control plane protocol stack for a DPCH is illustrated in Figure 6.5. This protocol stack is applicable when the SRB are transferred using a DPCH rather than HSDPA. In this case, the MAC-hs layer is not applied and the dedicated channel Frame Protocol is used rather than the common channel Frame Protocol.

6.2 HSDPA Bit Rates

- HSDPA bit rates depend upon the protocol stack layer from where they are measured. They also depend upon the RF channel conditions used to select an appropriate transport block size. The allocated transport block size determines the modulation scheme, number of HS-PDSCH channelisation codes and coding rate.
- QPSK maps 2 bits onto each symbol whereas 16QAM maps 4 bits onto each symbol. The release 7 version of the 3GPP specifications introduces 64QAM which maps 6 bits onto each symbol. The release 7 version of the specifications also introduces the option to use $2 \times \text{Tx} + 2 \times \text{Rx}$ MIMO for QPSK and 16QAM. This allows the Node B to transmit two transport blocks in parallel and so doubles the throughput.
- The maximum achieved throughput is dependent upon the UE capability which defines the maximum transport block size. The 2 ms TTI is fixed and so the maximum physical layer throughput is directly proportional to the maximum transport block size.
- The UE capability also has an impact upon the selection between chase combining and incremental redundancy for HARQ re-transmissions. UE supporting incremental redundancy are required to buffer larger quantities of soft channel bits after de-spreading.
- The throughput achieved at the higher layers within the protocol stack decreases as the headers and padding added by the lower layers are removed. The throughput is also reduced by re-transmissions at the MAC-hs, RLC and TCP layers. The majority of re-transmissions should

be completed at the MAC-hs layer. The RLC throughput provides a measure of useful data transfer from the RAN perspective.
- HSDPA makes use of common air-interface resources which are shared between all UE with active HSDPA connections. The throughput achieved by an individual UE decreases as the number of active connections increases. The total cell throughput depends upon the scheduler performance and the RF channel conditions experienced by each UE.
- Key 3GPP specifications: TS 25.306, TS 25.212 and TS 25.213

This section describes HSDPA bit rates from a theoretical perspective. It presents the maximum achievable throughputs without accounting for channel conditions. In practice, the maximum bit rates are only achieved when the RF channel conditions are sufficiently good and there are no other capacity bottlenecks throughout the network. The bit rates presented in this section also assume that there is a single active HSDPA connection. HSDPA makes use of common air-interface resources which are shared between all UE with active HSDPA connections. The throughput achieved by an individual UE decreases as the number of active connections increases.

HSDPA bit rates depend upon the protocol stack layer from where they are measured. They also depend upon the allocated modulation scheme, number of HS-PDSCH channelisation codes and coding rate. The throughput capability of HSDPA is often quoted using the figures presented in Table 6.3.

These figures can be derived using the equation below.

$$\text{Throughput} = \frac{\text{Chip rate}}{\text{SF}} \times \text{Bits per symbol} \times \text{Coding rate} \times \text{Number of codes}$$

Where the chip rate is 3.84 Mcps and the HS-PDSCH channelisation code Spreading Factor (SF) is fixed at 16. The ratio of $3.84 \times 10^6/16$ defines the modulated symbol rate before spreading at the Node B transmitter. Multiplying this result by the number of bits per symbol generates the corresponding bit rate. QPSK maps 2 bits onto each symbol, 16QAM maps 4 bits onto each symbol, and 64QAM maps 6 bits onto each symbol. Multiplying by the coding rate references the throughput result to the top of the Physical layer. A low coding rate indicates that the physical layer introduces a large quantity of redundancy to help protect the information bits as they are transferred across the air-interface. A coding rate of 0.5 means that the throughput at the top of the physical layer is half of the throughput at the bottom of the physical layer. The resulting throughput at the top of the Physical layer is applicable to a single HS-PDSCH channelisation code and so the total throughput is obtained by multiplying by the number of codes.

Table 6.3 HSDPA throughput capability

	Coding Rate	5 HS-PDSCH codes (Mbps)	10 HS-PDSCH codes (Mbps)	15 HS-PDSCH codes (Mbps)
QPSK	0.25	0.6	1.2	1.8
	0.50	1.2	2.4	3.6
	0.75	1.8	3.6	5.4
16QAM	0.50	2.4	4.8	7.2
	0.75	3.6	7.2	10.8
	1.00	4.8	9.6	14.4
64QAM*	0.50	3.6	7.2	10.8
	0.75	5.4	10.8	16.2
	1.00	7.2	14.4	21.6

*Introduced by release 7 of the 3GPP specifications.

The throughput figures presented in Table 6.3 do not necessarily reflect the maximum throughput which can be achieved in practice. Not all HSDPA UE are capable of achieving the complete set of bit rates and UE are categorised according to their ability. This UE category information is signalled to the network during connection establishment to ensure that both the RNC and Node B are aware of the UE capability, e.g. to ensure that 16QAM is not used for a UE which only supports QPSK. The set of HSDPA UE categories specified by 3GPP TS 25.306 is presented in Table 6.4.

A category 12 UE is able to support QPSK and 5 codes. According to Table 6.3 this corresponds to a maximum HSDPA throughput of 1.8 Mbps at the top of the physical layer. Table 6.4 specifies that the maximum transport block size is 3630 bits which corresponds to 1.815 Mbps after accounting for the 2 ms TTI, i.e. the two tables are consistent. A category 11 UE is similar to a category 12 UE except that it cannot receive HSDPA data during consecutive TTI. It can only receive HSDPA data during every second TTI. The means that the throughput measured during a single TTI is the same as that for a category 12 UE, but the throughput measured across multiple TTI reduces by a factor of 2.

Category 5 and 6 UE are able to support 16QAM with 5 codes. According to Table 6.3 this corresponds to a maximum HSDPA throughput of 4.8 Mbps at the top of the physical layer. Table 6.4 specifies that the maximum transport block size is 7298 bits which corresponds to 3.649 Mbps after accounting for the 2 ms TTI. This indicates that a UE which supports 16QAM and 5 codes cannot achieve the maximum bit rate which has been calculated based upon a coding rate of 1. The actual maximum bit rate is approximately equal to the 3.6 Mbps which corresponds to a coding rate of 0.75.

A category 10 UE is able to support 16QAM with 15 codes. According to Table 6.3 this corresponds to a maximum HSDPA throughput of 14.4 Mbps at the top of the Physical layer. Table 6.4 specifies that the maximum transport block size is 27952 bits which corresponds to 13.976 Mbps after accounting for the 2 ms TTI. This is approximately equal to the 14.4 Mbps which corresponds to a coding rate of 1. A category 9 UE is also able to support 16QAM with 15 codes, but the maximum transport block size is limited to 20251 bits, i.e. a maximum throughput of 10.126 Mbps which is approximately equal to the throughput figure for a coding rate of 0.75.

Table 6.4 Categories for HSDPA UE

UE category	Modulation schemes	Support for MIMO	Maximum number of codes	Minimum inter-TTI interval	Maximum transport block size	Total soft channel bits
1	QPSK/16QAM	No	5	3	7298	19200
2	QPSK/16QAM	No	5	3	7298	28800
3	QPSK/16QAM	No	5	2	7298	28800
4	QPSK/16QAM	No	5	2	7298	38400
5	QPSK/16QAM	No	5	1	7298	57600
6	QPSK/16QAM	No	5	1	7298	67200
7	QPSK/16QAM	No	10	1	14411	115200
8	QPSK/16QAM	No	10	1	14411	134400
9	QPSK/16QAM	No	15	1	20251	172800
10	QPSK/16QAM	No	15	1	27952	172800
11	QPSK	No	5	2	3630	14400
12	QPSK	No	5	1	3630	28800
13[*]	QPSK/16QAM/64QAM	No	15	1	34800	259200
14[*]	QPSK/16QAM/64QAM	No	15	1	42196	259200
15[*]	QPSK/16QAM	Yes	2×15	1	2×23370	345600
16[*]	QPSK/16QAM	Yes	2×15	1	2×27952	345600

[*]Introduced by release 7 of the 3GPP specifications.

Category 13 and 14 UE are introduced by the release 7 version of the 3GPP specifications. These UE are able to support 64QAM and 15 codes. A category 13 UE has a maximum transport block size of 34800 bits whereas a category 14 UE has a maximum transport block size of 42196 bits. These figures correspond to bit rates of 19.2 and 21.1 Mbps respectively.

The release 7 version of the 3GPP specifications also introduces the possibility of using $2 \times \text{Tx} + 2 \times \text{Rx}$ Multiple Input Multiple Output (MIMO) technology. This allows the Node B to transmit two parallel signals from two antennas. These signals are subsequently received by two UE antennas. When MIMO is not used, HSPDA connections are able to transfer a maximum of one transport block during each TTI. The introduction of MIMO allows two transport blocks to be transmitted in parallel. A category 16 UE has a maximum transport block size of 27952 bits. This is equal to the maximum transport block size for a category 10 UE (16QAM with 15 codes) and corresponds to a maximum throughput of 13.976 Mbps. A category 16 UE can receive two of these transport blocks in parallel and so the maximum throughput increases to 27.952 Mbps.

Table 6.4 also specifies the UE requirement in terms of buffering soft channel bits. Soft channel bits are the values received after de-spreading prior to making a hard decision on whether a 1 or 0 was transmitted. A soft channel bit occupies greater memory than a hard decision bit because it can have more than two values. The figures in Table 6.4 account for the requirement to have parallel HARQ processes and also the requirement for incremental redundancy (both HARQ processes and incremental redundancy are described in section 6.8.4). Table 6.5 presents the dependency of incremental redundancy upon the soft channel bit buffering capability. The maximum number of bits after de-spreading is defined by the modulation symbol rate $\times$ number of bits per symbol $\times$ number of HS-PDSCH codes $\times$ 2 ms TTI. The modulation symbol rate is always 240 ksps because the channelisation code spreading factor has a fixed value of 16.

The ratio of X/Y represents the number of data blocks which can be buffered by the UE. A data block is defined as the set of soft channel bits received during a TTI after de-spreading. For example, a category 3 UE is capable of buffering three data blocks whereas a category 4 UE is capable of buffering four data blocks. Table 6.5 then presents the number of parallel HARQ processes which each UE category must be able to support. This figure is dependent upon the minimum inter-TTI interval presented in Table 6.4. If a UE can receive data during every TTI then it must be able to support six

Table 6.5 Incremental redundancy dependence upon the soft channel bit buffering capability

UE category	Total soft channel bits (X)	Maximum bits after de-spreading (Y)	Ratio X/Y	Parallel HARQ processes	Incremental redundancy
1	19200	9600	2	2	No
2	28800	9600	3	2	Yes
3	28800	9600	3	3	No
4	38400	9600	4	3	Yes
5	57600	9600	6	6	No
6	67200	9600	7	6	Yes
7	115200	19200	6	6	No
8	134400	19200	7	6	Yes
9	172800	28800	6	6	No
10	172800	28800	6	6	No
11	14400	4800	3	3	No
12	28800	4800	6	6	No
13	259200	43200	6	6	No
14	259200	43200	6	6	No
15	345600	57600	6	6	No
16	345600	57600	6	6	No

parallel HARQ processes. If a UE can receive data during every second TTI then it must be able to support three parallel HARQ processes. Each HARQ process must be able to buffer a block of soft channel bits in case re-transmissions are required. A UE is not able to support incremental redundancy if the number of parallel HARQ processes is equal to the number of data blocks which can be buffered. In this case, chase combining must be used rather than incremental redundancy. Chase combining uses the same puncturing pattern at the transmitter for each re-transmission. This means that the data block size at the receiver does not increase after combining. Incremental redundancy uses a different puncturing pattern at the transmitter for each re-transmission and so the data block size increases after combining, i.e. UE using incremental redundancy must be able to buffer blocks of soft channel bits which are larger than the number of bits received during a single TTI.

The bit rates presented in Table 6.3 and those calculated from the transport block sizes in Table 6.4 are applicable to the top of the physical layer. Removing the headers and padding added by each layer results in the throughput decreasing for the higher layers. Table 6.6 presents a set of example maximum throughput figures for a TCP/IP protocol stack assuming a single HSDPA connection in good coverage with 5 HS-PDSCH codes. Results are provided for both QPSK and 16QAM. The impact of re-transmissions at the MAC-hs, RLC and TCP layers has been excluded.

Assuming the UE is in good coverage allows the Adaptive Modulation and Coding (AMC) function to allocate the maximum supported transport block size. In this example, it is assumed that the maximum transport block size of 3440 bits for QPSK and 7168 bits for 16QAM. These figures are less than the UE maximum transport block sizes presented in Table 6.3. In practice, the maximum transport block size is selected by the Node B and is implementation dependent although it should not exceed the capability of the UE. Selecting a smaller maximum transport block size increases the quantity of redundancy which can be introduced by the physical layer. The QPSK transport block size of 3440 bits corresponds to a coding rate of 0.717 whereas the 16QAM transport block size of 7168 bits corresponds to a coding rate of 0.747. The physical layer throughput is calculated by dividing these maximum transport block sizes by the 2 ms TTI.

The throughput at the top of the MAC-hs layer can be derived by removing the MAC-hs header and padding. The MAC-hs header occupies 21 bits if it is assumed that all MAC-d PDU within the MAC-hs PDU have an equal size. This leaves 3419 bits within the QPSK transport block and 7147 bits within the 16QAM transport block. Assuming a MAC-d PDU size of 336 bits allows the QPSK transport block to accommodate 10 PDU and the 16QAM transport block to accommodate 21 PDU. This results in 59 bits of padding within the QPSK transport block and 91 bits of padding within the 16QAM transport block. The MAC-hs throughputs are then defined by $(3440 - 21 - 59)/0.002 = 1.68$ Mbps and $(7168 - 21 - 91)/0.002 = 3.53$ Mbps. The MAC-d throughput equals the MAC-hs throughput unless the MAC-d entity adds a header after multiplexing multiple logical channels onto a single MAC-d flow. Assuming a 16 bit RLC header generates RLC throughputs of $10 \times 320/0.02 = 1.60$ Mbps and $21 \times 320/0.02 = 3.36$ Mbps for QPSK and 16QAM respectively. The RLC layer throughput represents the useful throughput from the perspective of the RAN.

Table 6.6   Example maximum bit rates for a TCP/IP protocol stack

Protocol stack layer	Peak bit rate (QPSK, 5 codes) (Mbps)	Peak bit rate (16QAM, 5 codes) (Mbps)
TCP	1.56	3.27
IP	1.58	3.31
RLC	1.60	3.36
MAC-d	1.68	3.53
MAC-hs	1.68	3.53
Physical layer	1.72	3.58

The IP and TCP throughputs depend upon the size of the Maximum Transmission Unit (MTU) defined at the TCP layer. Larger MTU sizes result in lower TCP/IP overheads, i.e. the ratio of application data to TCP/IP header is greater. The figures in Table 6.6 assume an MTU size of 1420 bytes which is equivalent to the payload associated with 35.5 RLC PDU, i.e. every 35.5 RLC PDU contain one 20 byte TCP header and one 20 byte IP header (IPv4). The set of 10 RLC PDU within the QPSK transport block effectively include 45 bits of TCP header and 45 bits of IP header. The IP throughput is then given by $(320 \times 10 - 45)/0.002 = 1.58$ Mbps whereas the TCP throughput is given by $(320 \times 10 - 90)/0.002 = 1.56$ Mbps. The equivalent throughputs for 16QAM are $(320 \times 21 - 95)/0.002 = 3.31$ Mbps and $(320 \times 21 - 190)/0.002 =$ Mbps.

In practice, the throughput figures presented in Table 6.6 would be reduced by re-transmissions at the MAC-hs, RLC and TCP layers. The majority of re-transmissions should be completed at the MAC-hs layer. The 3GPP specifications target a MAC-hs PDU (transport block) BLER of 10% and so the MAC-hs throughput would be reduced by 10%. Actual implementations may be more aggressive by allowing the AMC function to select relatively large transport block sizes. This would increase the physical layer throughput, but may decrease the MAC-hs throughput as a result of less redundancy and an increased MAC-hs re-transmission requirement. A MAC-hs BLER of less than 10% can be achieved in areas of good coverage where there is margin in the downlink link budget. A benefit of measuring throughput at the RLC layer is that it includes the impact of both MAC-hs and RLC re-transmissions.

If a cell is configured to use 16QAM and a maximum of 5 HS-PDSCH codes then a single HSDPA connection with a coding rate of 0.75 would achieve a physical layer throughput of 3.6 Mbps. The throughput would decrease to 1.8 Mbps if a second HSDPA connection is established and a round robin scheduler is assumed. In this case, each HSDPA connection would be allocated half of the 2 ms TTI and each would have a throughput of 1.8 Mbps. The total HSDPA throughput remains constant at 3.6 Mbps. If the second UE was located in relatively poor coverage then the AMC function would decrease the allocated transport block size to allow a lower coding rate, e.g. the coding rate could be reduced to 0.5. In this case, the first HSDPA connection would continue to achieve 1.8 Mbps while the second HSDPA connection would only achieve 1.2 Mbps. The total HSDPA throughput would decrease from 3.6 to 3 Mbps.

6.3 PDCP Layer

- The PDCP layer can be used as part of the user plane protocol stack. It provides functionality for compressing the headers belonging to the higher layers, e.g. TCP/IP and RTP/UDP/IP headers for IPv4 and IPv6.
- The PDCP layer becomes particularly important for applications which have large overheads generated by the headers belonging to the higher layers, e.g. Voice over IP.
- Key 3GPP specifications: TS 25.323

The Packet Data Convergence Protocol (PDCP) layer can be used by both DPCH and HSDPA connections. In both cases, it operates in the same manner, but its use tends to be more important for some of the applications associated with HSDPA. Its importance depends upon the overhead generated by the headers belonging to the higher layers. Voice over IP (VoIP) represents an example application for which the overhead is large and the PDCP layer is important. In this example, the payload is 244 bits for the 12.2 kbps AMR service while the header is 320 bits when using IPv4 and 480 bits when using IPv6. In this case, header compression by the PDCP layer is essential to reduce the overhead generated by the headers belonging to the higher layers. The benefit of the PDCP layer for VoIP is described and quantified in Section 2.1.2. VoIP is more likely to be provided using HSDPA than a packet switched DPCH because the system delay offered by HSDPA is significantly less. If the system delay becomes too large then the quality of experience decreases for the end-user. The PDCP layer is less important for applications which have relatively small overheads generated by the headers belonging to the higher layers. A TCP/IP file transfer represents an example application for which the overhead is relatively small. In this example, the payload could be 1420 Bytes while the header is 40 bytes when using IPv4 and 60 bytes when using IPv6.

6.4 RLC Layer

- The RLC layer processes data for HSDPA connections in the same way that it processes data for DPCH connections. NRT applications are likely to use acknowledged mode RLC whereas RT applications are likely to use unacknowledged mode RLC.
- The use of HSDPA results in a change to the signalling used to configure the size of an RLC PDU. In the case of an HSDPA connection, the MAC-d PDU size is signalled rather than the RLC PDU size. The MAC-d PDU size is typically 336 bits for a data service. A size of 656 bits may be used for higher HSDPA bit rates. Larger RLC PDU sizes decrease the overhead associated with the RLC header, but also reduce the flexibility available to the MAC-hs layer when packaging MAC-d PDU within a single MAC-hs PDU.
- The RLC layer is responsible for re-transmissions between the RNC and UE when using acknowledged mode. HSDPA helps to reduce the requirement for RLC re-transmissions by allowing re-transmissions from the MAC-hs layer. HSDPA typically generates bursts of RLC re-transmissions rather than individual re-transmissions. These bursts of re-transmissions are caused by packaging groups of RLC PDU within single MAC-hs PDU.
- Key 3GPP specifications: TS 25.322 and TS 25.331.

The general responsibilities of the RLC layer are described in Section 2.3. The RLC layer processes data for HSDPA connections in the same way that it processes data for DPCH connections. NRT applications are likely to be configured with acknowledged mode RLC whereas RT applications are likely to be configured with unacknowledged mode RLC. For example, an internet browsing service would use acknowledged mode RLC whereas a VoIP service would use unacknowledged mode RLC. Figure 6.6 illustrates the set of RLC entities used by the SRB and user plane data when the user plane data is configured with acknowledged mode RLC. This figure remains valid irrespective of whether or not HSDPA is used to transfer the contents of each logical channel.

The RLC layer is responsible for the segmentation, concatenation and padding of higher layer packets to ensure they fit within one or more RLC PDU. Segmentation is used to reduce the size of higher layer

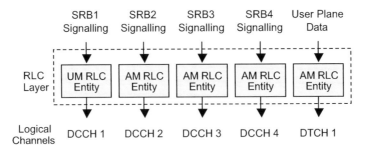

Figure 6.6   RLC entities for control and user planes

packets. Concatenation is used to include both the end of one higher layer packet and the start of another higher layer packet within the same RLC PDU. Padding is used to occupy spare capacity when there are no further higher layer packets. Higher layer packets belonging to the control plane originate from the RRC layer whereas higher layer packets belonging to the user plane originate from the application layer.

The use of HSDPA results in a subtle change to the signalling used to configure the size of an RLC PDU. Log File 6.1 presents an example of the RRC signalling used to specify the size of an SRB RLC PDU when transferred using a DPCH connection. This information could be sent to a UE using either a Radio Bearer Setup message or a Radio Bearer Reconfiguration message.

The RLC PDU size can be derived using the relevant equation from within 3GPP TS 25.331. Table 8.3 in section 8.5.1 summarises these equations. In this example, the RLC PDU size is given by $(8 \times \text{sizeType1}) + 16 = 144$ bits. The equivalent RRC signalling for an HSDPA connection is presented in Log File 6.2. This corresponds to transferring the SRB using an HSDPA connection.

In this case, the MAC-d PDU size is signalled rather than the RLC PDU size and the RLC PDU size must be derived from the MAC-d PDU size. The MAC-d PDU size is signalled using a value of 148 bits. This is equivalent to the RLC PDU size of 144 bits specified in Log File 6.1. The MAC-d entity adds a 4 bit C/T header field when it multiplexes multiple logical channels onto a single transport channel. This header field is used by the receiver to identify which logical channel has been sent at any point in time. There are always at least three DCCH SRB and so the MAC-d entity always has to include these four C/T bits within the MAC-d header.

There is a similar difference between the RRC signalling for a DPCH connection and the RRC signalling for an HSDPA connection when the RLC PDU size is specified for the user plane data. However, it is less common to complete MAC-d multiplexing for user plane data and so in general the C/T header is not necessary and the RLC PDU size equals the MAC-d PDU size (MAC-d multiplexing could be completed for user plane data for multi-RAB connections). Log File 6.3 presents an example of the RRC signalling used to configure the RLC PDU size for a packet switched DPCH data connection.

Similar to the SRB, the RLC PDU size can be derived by applying the appropriate equation from 3GPP TS 25.331. In this example, the RLC PDU size is given by $(32 \times \text{part1}) + 272 = 336$ bits. The equivalent RRC signalling for an HSDPA connection is presented in Log File 6.4.

The MAC-d PDU size is signalled using a value of 336 bits, i.e. equal to the equivalent RLC PDU size for a user plane DPCH connection. The value of 336 bits represents a typical example, but depending upon implementation, the RNC can configure other values. A value of 656 bits is typical for higher HSDPA bit rates. A PDU size of 656 bits allows the payload to double from 320 bits to 640 bits when assuming a 16 bit RLC header. Larger RLC PDU sizes decrease the overhead associated with the RLC header, but also reduce the flexibility available to the MAC-hs layer when packaging MAC-d PDU within a single MAC-hs PDU.

```
tti40 value 2
rlc-Size
octetModeType1
        sizeType1:      16
numberOfTbSizeList
        numberOfTbSizeList value 1:    one
```

Log File 6.1 RRC signalling using to specify the RLC PDU size for a DPCH (SRB)

```
mac-d-PDU-SizeInfo-List
mac-d-PDU-SizeInfo-List value 1
        mac-d-PDU-Size:        148
        mac-d-PDU-Index:         0
```

Log File 6.2 RRC signalling using to specify the MAC-d PDU size for HSDPA (SRB)

```
tti20 value 1
rlc-Size
octetModeType1
        sizeType2
        part1   : 2
numberOfTbSizeList
        numberOfTbSizeList value 1      : zero
        numberOfTbSizeList value 2      : one
        numberOfTbSizeList value 3
        small1  : 4
```

Log File 6.3   RRC signalling using to specify the RLC PDU size for a DPCH (user plane)

```
mac-d-PDU-SizeInfo-List
mac-d-PDU-SizeInfo-List value 1
        mac-d-PDU-Size  : 336
        mac-d-PDU-Index : 0
```

Log File 6.4   RRC signalling using to specify the MAC-d PDU size for HSDPA (user plane)

The RLC layer is responsible for re-transmissions between the RNC and UE. HSDPA helps to reduce the requirement for RLC re-transmissions by allowing re-transmissions from the MAC-hs layer. MAC-hs re-transmissions occur between the Node B and UE. 3GPP TS 25.214 specifies that UE should generate HSDPA Channel Quality Indicators (CQI) based upon the assumption of a 10% block error rate, i.e. the CQI value generated by the UE should correspond to a transport block size which the UE believes it can receive with a 10% probability of error. If three MAC-hs re-transmissions are allowed prior to the first RLC re-transmission, and if it is assumed that each transmission is statistically independent then the probability of error reduces to $0.1^4 - 0.01\%$. In practice, the MAC-hs re-transmissions are not statistically independent because the propagation channel and interference conditions are time correlated. If one MAC-hs transmission fails as a result of poor conditions then there is an increased probability that the subsequent transmission will also fail as a result of poor conditions. This tends to increase the probability of relying upon the RLC layer for re-transmissions. RLC re-transmissions can be required in both good and poor coverage conditions. If the AMC functionality within the MAC-hs layer is too aggressive then the allocated bit rate will be too high and the UE will be more likely to experience difficulties receiving MAC-hs PDU without errors. If the AMC functionality is not sufficiently aggressive then the requirement for re-transmissions will decrease, but the throughput performance will be less.

HSDPA typically generates bursts of RLC re-transmissions rather than individual re-transmissions. These bursts of re-transmissions are caused by packaging groups of RLC PDU within single MAC-hs PDU. The number of RLC PDU within a single MAC-hs PDU depends upon the bit rate allocated by the AMC functionality. If the allocated bit rate is low then a small number of RLC PDU will be included whereas if the allocated bit rate is high then a large number of RLC PDU will be included. For example, if the AMC allocates a MAC-hs PDU size which accommodates 10 RLC PDU, and if the number of MAC-hs PDU re-transmissions has reached the maximum then all 10 of the RLC PDU must be re-transmitted from the RLC layer. Once the re-transmitted RLC PDU reach the MAC-hs layer for the second time they are not necessarily packaged within the same MAC-hs PDU. Their packaging depends upon their position within the queue and also the MAC-hs PDU size allocated by the AMC at that point in time.

Acknowledgements for the downlink RLC PDU are returned to the RNC using the corresponding uplink connection. If the uplink connection is a DPCH then the relevant DCH will be used to return the acknowledgements. If the uplink connection is HSUPA then the relevant E-DCH will be used to return the acknowledgements.

6.5 MAC-d Entity

- Each HSDPA connection has its own MAC-d entity within the RNC. The MAC-d entity generates one or more MAC-d flows. These MAC-d flows can be forwarded across the Iub for subsequent mapping onto the HS-DSCH transport channel within the Node B.
- A maximum of 8 MAC-d PDU sizes can be configured for a single MAC-d flow. The Node B provides instructions regarding the maximum MAC-d PDU size which can be forwarded at any point in time using flow control signalling.
- Priorities can be assigned to MAC-d PDU. These priorities can be used to create a series of priority queues for each MAC-d flow. Both flow control and Hybrid ARQ are able to handle MAC-d PDU based upon their priority.
- Key 3GPP specifications: TS 25.321.

The general responsibilities and architecture of the MAC layer are described in Section 2.4. Each HSDPA connection has its own MAC-d entity within the RNC. The MAC-d entity generates one or more MAC-d flows. These MAC-d flows are either mapped onto the DCH transport channel within the RNC or forwarded across the Iub towards the MAC-hs for subsequent mapping onto the HS-DSCH transport channel within the Node B. Figure 6.7 illustrates the MAC-d flows when the control plane signalling is transferred using a DPCH connection while the user plane data is transferred using an HSDPA connection.

In this case, the MAC-d entity multiplexes the set of SRB DCCH into a single MAC-d flow and maps them onto a DCH transport channel. The user plane data forms a second MAC-d flow which is transferred across the Iub once the Node B has issued the relevant flow control credits.

Figure 6.8 illustrates the MAC-d flows when both the control plane signalling and user plane data are transferred using HSDPA.

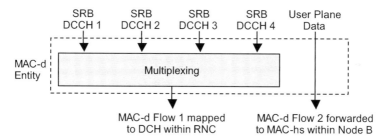

Figure 6.7 MAC-d entity when DCCH is mapped to a DCH

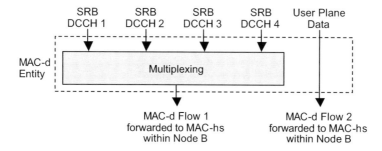

Figure 6.8 MAC-d entity when DCCH is mapped to an HS-DSCH

```
mac-d-PDU-SizeInfo-List
   mac-d-PDU-SizeInfo-List value 1
      mac-d-PDU-Size   : 336
      mac-d-PDU-Index  : 0
   mac-d-PDU-SizeInfo-List value 2
      mac-d-PDU-Size   : 656
      mac-d-PDU-Index  : 1
```

Log File 6.5   Example RRC signalling for multiple MAC-d PDU sizes

In this case, the MAC-d entity generates two separate MAC-d flows for transfer across the Iub. As indicated by the description of the RLC layer these two MAC-d flows are likely to be configured with different MAC-d PDU sizes. The MAC-d flow used by the control plane is likely to have a MAC-d PDU size of 148 bits whereas the MAC-d flow used by the user plane data is likely to have a MAC-d PDU size of 336 or 656 bits. The 3GPP specifications support MAC-d PDU sizes within the range of 1–5000 bits. It is also possible to configure more than a single MAC-d PDU size for each MAC-d flow. For example, the user plane MAC-d flow could be configured with PDU sizes of both 336 and 656 bits. The relevant RRC signalling for this example is presented in Log File 6.5. This signalling could be included as part of a Radio Bearer Setup message or a Radio Bearer Reconfiguration message.

In this example, two MAC-d PDU sizes are configured. The first has an index of 0 whereas the second has an index of 1. A maximum of eight MAC-d PDU sizes can be configured for a single MAC-d flow. The Node B provides instructions regarding the maximum MAC-d PDU size which can be forwarded at any point in time using MAC-hs flow control signalling. Large MAC-d PDU sizes are more appropriate when the HSDPA throughput is high.

Priorities can be assigned to the MAC-d PDU. These priorities can be used to create a series of priority queues for each MAC-d flow. The priority can range from 0 to 15, where 15 corresponds to the highest priority. The priority can be used as part of flow control to determine when specific MAC-d PDU are sent across the Iub. The priority can also be used by the Hybrid ARQ (HARQ) to determine when specific MAC-d PDU are transferred across the air-interface. For example, if the SRB are transferred using HSDPA then the fourth SRB could be assigned a lower priority than the first three SRB. This would be consistent with the approach used when the SRB are transferred using a DPCH. In addition, the priority assigned to the MAC-d flow used to transfer the user plane data could be lower than the priorities assigned to the SRB to ensure that control plane signalling always has priority over user plane data.

6.6 Frame Protocol Layer

- HSDPA uses both data and control frames belonging to the common channel Frame Protocol. Data frames are used to package MAC-d PDU and subsequently transfer them across the Iub. Control frames are used for Iub flow control and congestion control.
- The rate at which the RNC sends HS-DSCH data frames towards the Node B depends upon the instructions provided by the MAC-hs flow control. All MAC-d PDU within a single HS-DSCH data frame have the same length.
- The data frame header specifies the Common Transport Channel Priority Indicator. This informs the Node B of the relative priority of the MAC-d PDU. All of the MAC-d PDU within a single HS-DSCH data frame have the same priority. The data frame header also specifies the RNC buffer occupancy for that priority queue.
- The release 6 version of the protocol includes a Frame Sequence Number (FSN) and Flush indicator within the data frame header. The FSN can be used to detect frame loss and trigger congestion control. The Flush indicator can be used to improve the efficiency of the RLC reset procedure by instructing the Node B to empty a MAC-hs priority queue.

- The release 6 version of the protocol includes a Delay Reference Time within the data frame tail. This allows the Node B to detect increased delay resulting from Iub congestion.
- HS-DSCH Capacity Request and Capacity Allocation control frames are used for Iub flow control purposes on a per MAC-d flow basis.
- The Capacity Allocation control frame is used to provide the RNC with permission to transfer MAC-d PDU belonging to a specific priority queue at a specific maximum rate.
- Key 3GPP specifications: TS 25.435 and TS 25.425.

HSDPA uses both data and control frames belonging to the common channel Frame Protocol. These are specified within 3GPP TS 25.435 for the Iub and within 3GPP TS 25.425 for the Iur. Data frames are used to package a set of MAC-d PDU and subsequently transfer them from the RNC to the Node B. Control frames are used for Iub/Iur flow control and congestion control.

Neither data frames nor control frames include any form of UE identity nor any form of MAC-d flow identity. This means that the receiving Node B (in the case of data frames and downlink control frames) or receiving RNC (in the case of uplink control frames) must rely upon information provided by other layers to associate each Frame Protocol packet with a specific MAC-d flow belonging to a specific HSDPA connection. The lower transport layers can be used for this purpose. When ATM Adaptation Layer 2 (AAL2) is used to transfer data across the Iub each MAC-d flow belonging to an HSDPA connection is assigned an AAL2 Channel Identifier (CID). This CID is unique within a specific ATM Virtual Channel Connection (VCC). The VCC is identified using its Virtual Channel Identifier (VCI). The CID is specified within the header of the AAL2 packet whereas the VCI is specified within the header of the ATM packet (AAL2 packets are encapsulated by ATM packets). These ATM identifiers allow the receiving Node B or receiving RNC to associate each data frame and each control frame with a specific MAC-d flow.

6.6.1 HS-DSCH Data Frame

Figure 6.9 illustrates the structure of the data frame used to package one or more MAC-d PDU for HSDPA. This frame structure was introduced in release 5 of the 3GPP specifications and was subsequently enhanced for release 6. The release 6 version adds the Frame Sequence Number (FSN) and Flush header fields, as well as the Delay Reference Time (DRT) tail field.

The first byte of the HS-DSCH data frame header starts by including a 7 bit Cyclic Redundancy Check (CRC). These CRC bits are used by the Node B to detect whether or not the header information has been received without error. The first byte also includes a 1 bit Frame Type (FT) flag used to indicate whether the subsequent frame is a data frame or a control frame. A value of 0 indicates a data frame whereas a value of 1 indicates a control frame. Figure 6.9 illustrates a data frame and so the FT flag would have a value of 0.

The second byte of the header starts by including the 4 bit Frame Sequence Number (FSN) which was introduced by release 6 of 3GPP TS 25.435. These 4 bits appear as spare within release 5 of 3GPP TS 25.435. In the case of release 6, the FSN is incremented by 1 for each data frame that is transferred to the Node B. This information is used by the Node B to detect frame loss possibly resulting from transport network congestion. If the serving Node B receives non-contiguous FSN then it can inform the RNC and trigger congestion control procedures. The RNC is informed using the Frame Protocol HS-DSCH Capacity Allocation control frame presented in section 6.6.2. The second byte of the header also includes 4 bits to represent the Common Transport Channel Priority Indicator (CmCH-PI). This informs the Node B of the relative priority of the data frame and the associated MAC-d PDU. A value of 0 represents the lowest priority whereas a value of 15 represents the highest priority. All of the MAC-d PDU within a single HS-DSCH data frame have the same priority.

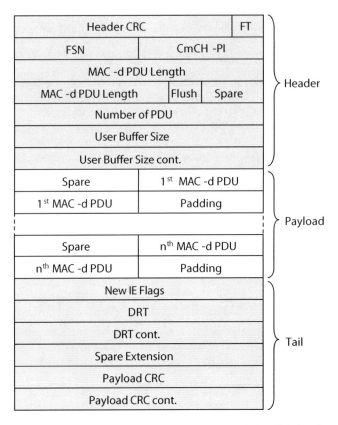

Figure 6.9   Structure of an Iub HS-DSCH Frame Protocol downlink data frame

The third and fourth bytes of the header start by providing 13 bits to specify the MAC-d PDU length. The length is specified in terms of bits. All MAC-d PDU within a single HS-DSCH data frame have the same length. Although the set of 13 bits allows a range from 0 to 8181 bits, the specifications limit the maximum MAC-d PDU length to 5000 bits. As described in Section 6.4, it is typical to use a MAC-d PDU length of 336 bits, or 656 bits if relatively high bit rates are available. The fourth byte of the header also includes the 1 bit Flush flag which was introduced by release 6 of 3GPP TS 25.435. This flag indicates to the Node B whether or not the MAC-hs priority queue should be flushed of all data preceding this data frame. This capability has been introduced to increase the efficiency of the RLC reset procedure. If a UE triggers the RLC reset procedure, the RNC is required to return a reset acknowledgement. The UE discards all data that is received between sending an RLC reset and receiving the acknowledgement. Emptying the priority queue avoids sending data that will be discarded and helps to minimise the delay associated with the UE receiving the acknowledgement.

The fifth byte of the header provides 8 bits to specify the number of MAC-d PDU within the payload of the HS-DSCH data frame. The set of 8 bits allows a maximum of 255 MAC-d PDU within a single data frame. The sixth and seventh bytes provide 16 bits to specify the RNC buffer occupancy for a specific priority queue belonging to the specific HSDPA connection. This represents the number of bytes that the priority queue has buffered at the RNC. The set of 16 bits allows a range from 0 to 65535 bytes. The number of bytes buffered at the RNC depends upon the difference between the rate of incoming data and the rate of outgoing data. If the end-user

```
HS-DSCH DATA FRAME
Header CRC:              33
CmCH-PI:                15
MAC-d PDU Length:        336 bit(s)
NumOfPDU:                28
User Buffer Size:        0 octets
MAC-d PDU
1.   MAC-d PDU 91 50 24 22 30 B4 B5 B4 64 50 F8 B5 E3 87 D1 23 D2 60 39 CA 78
     47 97 2F DD 64 46 E7 36 7F 8C 14 C3 18 47 E6 0A A0 47 FC 9C C7
2.   MAC-d PDU 91 58 0F 99 28 2F 89 A6 31 E0 54 60 C3 D1 F6 A0 8A BB 08 9E 0E
     A4 F4 4B D7 6B 1F A0 D8 70 A0 32 5D E1 C2 7E 1F 02 4B CA 2E 93
3.   MAC-d PDU 91 60 E5 7B 8F 2A 26 6E F4 2E 21 6C 12 DB C2 FF F5 BC EB 60 0A
     8B F1 8A 23 D2 BA 4B BF 59 15 9B 60 31 57 DA 15 2F EE F0 52 38
4.   MAC-d PDU 91 68 96 8B 5C 62 D5 F4 19 D1 5B 4D E9 4B 2C E6 DE 05 4B E4 10
     52 4A 8A 56 37 5B AA 8A 86 EC 6C DC F5 45 7B CB E3 98 4E 23 36
     . . .
     . . .
25.  MAC-d PDU 92 10 35 DA 4B 88 48 55 23 E1 A3 9A C7 6D FF 61 64 93 50 AB 1F
     CF 12 A6 31 5B 13 84 E0 AD DC A2 1A FA 1E 07 26 A5 A0 7E EE 9C
26.  MAC-d PDU 92 18 3F 14 3C 52 22 1D 40 77 BF FF F0 9B 6D 75 E9 9B 2A 5A 36
     9A C2 7D 7F 89 54 1B 62 D6 9C 26 D6 59 8D DD 6F 58 1F FE 0D 51
27.  MAC-d PDU 92 20 CB CA 34 2B F4 DF 2E 84 34 3E 4A BF 15 D7 88 DB 9B 4E 0F
     48 4C 82 3D EE 25 31 32 A2 88 93 4B DB 15 A2 E5 9B B3 D7 A9 12


28.  MAC-d PDU 92 2D 31 FE BC 3A 98 9D B4 8B D8 72 3E 47 1A BF 64 D0 E7 12 EE
     CF 73 34 F9 34 91 45 00 00 00 00 00 00 00 00 00 00 00 00 00 00
Spare Extension
Payload CRC:    51314
```

Log File 6.6 Example HS-DSCH Frame Protocol data frame

application generates a relatively low bit rate then the quantity of data buffered at the RNC is likely to be small because it will be easier for the Node B to transmit all of the data across the air-interface. If the end-user application generates a relatively high bit rate then the quantity of data buffered at the RNC is more likely to be large because it will be less easy for the Node B to transmit all of the data across the air-interface.

The HS-PDSCH is a common physical channel which can be shared between multiple HSDPA connections, i.e. between different UE. As the number of HSDPA connections increases then the bit rate experienced by each individual connection will decrease. This can cause the RNC buffer occupancy to increase. If the end-user application uses TCP and the air-interface is the throughput bottleneck for the connection then the quantity of data buffered at the RNC will approach the TCP congestion window. The application server will continue to send TCP packets at a high rate until the TCP congestion window is reached. This represents a flow control mechanism at the TCP layer and indicates that the RNC buffers should be at least as large as the TCP congestion window.

The tail section of the HS-DSCH data frame includes new Information Element (IE) flags and Delay Reference Time (DRT) sections. These have been introduced for the release 6 version of 3GPP TS 25.435. The new IE flags section has been introduced to allow any subsequent fields to be optional rather than mandatory. Each bit within the new IE flags section indicates whether or not a specific new IE is present. The new IE flags section is only necessary when there is at least one new IE present. The release 6 version of the data frame includes only a single new IE (DRT) and so the flags section is only necessary when that new IE is present. The right most bit within the new IE flags section is used to indicate that the DRT is present. The DRT has been introduced for the purposes of detecting increased delay possibly resulting from Iub congestion.

The rate at which the RNC sends HS-DSCH data frames towards the Node B depends upon the instructions provided by the MAC-hs flow control. For example, flow control may instruct the RNC to forward 100 MAC-d PDU of size 336 bits every 10 ms. This would correspond to an RLC throughput of $320 \times 100/0.01 = 3.2$ Mbps (assuming a 16 bit RLC header). The rate at which MAC-d PDU are transferred across the Iub does not have to equal the 2 ms TTI because the Node B buffers the data and completes its own air-interface scheduling. This is in contrast to DPCH transmissions which are scheduled by the RNC and data is transferred to the Node B once per TTI immediately prior to being transmitted across the air-interface.

An example HS-DSCH data frame is presented in Log File 6.6. A central section of the data frame has been omitted to reduce the length of the log file.

In this example, the set of MAC-d PDU have been assigned the highest priority of 15 and the MAC-d PDU size is 336 bits. The data frame header also specifies that there are 28 MAC-d PDU within the payload and that the RNC buffer is currently empty. This indicates that the RNC is not being provided with data at a sufficient rate to maximise the utilisation of the HS-DSCH. The payload of the data frame includes the 28 MAC-d PDU. The data frame does not include an FSN nor a DRT This implies that the data frame is based upon release 5 of the specifications rather than release 6.

6.6.2 HS-DSCH Control Frames

The HS-DSCH Capacity Request and Capacity Allocation control frames are directly applicable to HSDPA connections. These control frames are used on a per MAC-d flow basis, i.e. each MAC-d flow has its own set of Capacity Request and Capacity Allocation control frames. They are linked to a specific MAC-d flow using lower layer transport addressing, e.g. they could use the same AAL2 CID as the HS-DSCH data frames.

The structure of the HS-DSCH Capacity Request control frame is illustrated in Figure 6.10. This control frame can be sent from the controlling RNC to the serving Node B as part of the Iub flow control procedure.

The Frame Type (FT) flag has a value of 1 to indicate that the subsequent frame is a control frame rather than a data frame. The Control Frame Identifier for a Capacity Request message is 0000 1010. The Capacity Request message provides the Node B with information regarding the RNC buffer occupancy for a specific priority queue belonging to a specific MAC-d flow. This is the same as the information provided within the header of the HS-DSCH data frame. This control frame can be used to initially inform the Node B that the RNC has data to send for a specific priority queue. There is less requirement for this control frame once HS-DSCH data frames are being transferred.

The structure of the HS-DSCH Capacity Allocation control frame is illustrated in Figure 6.11. This control frame is sent from the serving Node B to the controlling RNC as part of the Iub flow control procedure. Its primary purpose is to provide the RNC with permission to transfer MAC-d PDU belonging to a specific MAC-d flow priority queue towards the Node B at a specific maximum rate. Similar to the

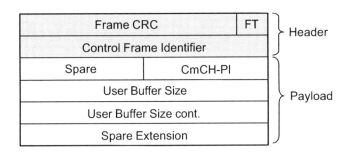

Figure 6.10   Structure of an Iub HS-DSCH Capacity Request control frame

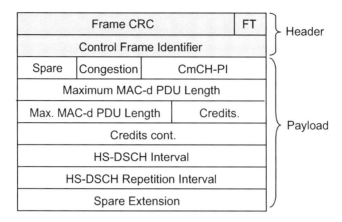

Figure 6.11   Structure of an Iub HS-DSCH Capacity Allocation control frame

Capacity Request message, the FT flag has a value of 1 to indicate that the subsequent frame is a control frame. The Control Frame Identifier for a Capacity Allocation message is 0000 1011.

The first byte of the payload includes 2 bits to report the Iub congestion status to the RNC. These bits have been introduced by the release 6 version of 3GPP TS 25.435. The Node B uses the FSN and DRT fields within the HS-DSCH data frame to detect congestion resulting from either lost or delayed data frames, respectively. A value of 00 indicates that congestion has not been detected whereas a value of 10 indicates that congestion has been detected by an increased delay, and a value of 11 indicates that congestion has been detected by a lost data frame. The first byte of the payload also includes 4 bits to specify the Common Transport Channel Priority Indicator (CmCH-PI). These bits represent the priority queue to which the Capacity Allocation control frame applies. The CmCH-PI value can range from 0 to 15 and should correspond to one of the priority queues used by the RNC.

The second and third bytes of the payload include 13 bits to specify the maximum MAC-d PDU length that the RNC is permitted to forward to the Node B. This information becomes important when a MAC-d flow is configured with multiple MAC-d PDU lengths and the Node B adjusts the maximum length according to the air-interface throughput achieved at any point in time. The third and fourth bytes of the payload include 11 bits to specify the number of flow control credits. These credits define the number of MAC-d PDU which can be transferred within the time interval defined by the HS-DSCH interval. The set of 11 bits provide a range from 0 to 2047. The value of 0 indicates that no MAC-d PDU should be transferred for the relevant priority queue whereas the value of 2047 indicates that there is no limit to the quantity of MAC-d PDU which can be transferred. The HS-DSCH interval is defined in units of 10 ms and so the set of 8 bits allow a maximum value of 2.55 seconds.

The final byte used within the payload specifies the HS-DSCH repetition interval. This value defines the number of consecutive time intervals that the credits can be applied. For example, if the number of credits is 100, the time interval is 10 ms and the repetition period is 20 then the RNC is permitted to transfer 100 MAC-d PDU every 10 ms during the next 200 ms. A repetition interval of 0 is interpreted as unlimited repetition, i.e. for the preceding example, the RNC would be permitted to transfer 100 MAC-d PDU every 10 ms for an indefinite period.

An example HS-DSCH Capacity Allocation control frame is presented in Log File 6.7. In this example, the Node B is permitting the RNC to transfer 49 MAC-d PDU with a maximum size of 336 bits and belonging to priority queue 15, once every 10 ms for an indefinite period. The MAC-hs flow control functionality used to define the contents of this message is described within Section 6.8.1.

```
HS-DSCH CAPACITY ALLOCATION
Control Frame CRC:             22
Control Frame Type:           11
CmCH-PI:                      15
Maximum MAC-d PDU Length:    336 bit(s)
HS-DSCH Credits:              49
HS-DSCH Interval:            10 ms
HS-DSCH Repetition Period:    0 (unlimited repetition period)
```

Log File 6.7   Example HS-DSCH Capacity Allocation control frame

6.7 Iub Transport

- HSDPA is transferred across the Iub using either AAL2/ATM or UDP/IP. The option to use UDP/IP was introduced within the release 5 version of the 3GPP specifications.
- If the AAL2/ATM protocol stack is used then HSDPA MAC-d flows are identified by their AAL2 CID. A total of three CID are required for each HSDPA connection if the SRB are transferred using a DCH. A total of four CID are required for each HSDPA connection if the SRB are transferred using a combination of HSDPA and HSUPA.
- HSDPA can be offered as a best effort service from the perspective of the Iub. Alternatively, a minimum bandwidth can be reserved for HSDPA to help guarantee a minimum HSDPA throughput performance. The priorities assigned to each MAC-d PDU can be used to determine the order in which HSDPA PDU are transferred across the Iub.
- The AAL2/ATM protocol stack generates an overhead of 35% relative to the bit rate at the top of the RLC layer when there are only five MAC-d PDU within each Frame Protocol data frame. This decreases to 29% when there are 50 or 100 PDU within each data frame.
- Packet switched transport networks based upon IP and Ethernet become more attractive as the bandwidth requirements of HSDPA increase. The bandwidth offered by an Ethernet connection can be significantly greater than that offered by a group of E1, T1 or JT1.
- The IETF have specified a solution known as Psuedo Wire Emulation Edge to Edge (PWE3). PWE3 allows ATM cells to be encapsulated within IP packets and subsequently transferred across an Ethernet packet switched network. The PWE3 protocol stack increases the Iub overhead by adding IP and Ethernet, but the availability of Ethernet bandwidth means that this overhead is less significant.
- Key 3GPP specifications: TS 25.430.
- Key IETF documents: RFC 3985, RFC 4717.

HSDPA is transferred across the Iub using either AAL2/ATM or UDP/IP. These two user plane protocol stacks are illustrated in Figure 5.1 and Figure 5.2. The option to use UDP/IP was introduced within the release 5 version of the 3GPP specifications, i.e. the same version that HSDPA was introduced. The majority of current network deployments are based upon the AAL2/ATM protocol stack although it is expected that the UDP/IP protocol stack will increase in importance as the bandwidth requirements increase.

6.7.1 ATM Transport Connections

If the AAL2/ATM protocol stack is used then HSDPA MAC-d flows are identified by their AAL2 Connection Identifier (CID). The number of CID used by an HSDPA connection depends upon whether or not the SRB are transferred using a DCH connection. A total of three CID are required for each

Table 6.7 AAL2 CID requirement for two levels of HSDPA traffic and two Node B configurations

	Low HSDPA traffic		Moderate HSPDA traffic	
	1 + 1 + 1 Node B	2 + 2 + 2 Node B	1 + 1 + 1 Node B	2 + 2 + 2 Node B
Common channels	12	24	12	24
HSDPA (per Node B)	9 (3 users)	9 (3 users)	48 (16 users)	48 (16 users)
DPCH (per cell)	74 (37 users)	34 (17 users)	62 (31 users)	28 (14 users)
Total CID	243	237	246	240

HSDPA connection if the SRB are transferred using a DCH. The first CID is used for the SRB DCH, the second CID is used for the HSDPA MAC-d flow and the third CID is used for the uplink user plane return channel. The uplink return channel could be either a DCH or HSUPA connection. A total of four CID are required for each HSDPA connection if the SRB are transferred using a combination of HSDPA and HSUPA. In this case, the SRB require a first CID for the downlink MAC-d flow and a second CID for the uplink MAC-d flow. The CID requirement would further increase if the HSDPA user plane connection makes use of more than a single MAC-d flow, i.e. each MAC-d flow requires its own CID.

Table 6.7 presents the CID requirements for two example HSDPA traffic scenarios. The first scenario assumes 3 HSDPA connections per Node B whereas the second scenario assumes 16 HSDPA connections per Node B. It is assumed that the SRB are transferred using a DCH connection and so each HSDPA connection requires 3 CID. The results are presented for both 1+1+1 and 2+2+2 Node B configurations. The 2+2+2 Node B configuration requires twice as many CID for the common channels. It is assumed that each cell is configured with AAL2 connections for the PCH, RACH, FACH-c and FACH-u transport channels. These allocations result in 12 common channel CID for a Node B configuration with three cells and 24 common channel CID for a Node B configuration with six cells.

The maximum number of DPCH which can be established in addition to the common channel and HSDPA connections is determined by the limit of 248 CID per AAL2 VCC (Section 5.1.3 explains the limit of 248). The results illustrate that the AAL2 CID belonging to a 2+2+2 Node B configuration are likely to become exhausted with only moderate levels of traffic per cell. Increasing the number of cells at a Node B increases the air-interface capacity, but decreases the average number of CID available per cell. It is necessary to configure a second user plane VCC and introduce a second set of 248 CID once the first set of CID becomes exhausted. This does not necessarily require an increase in the physical bandwidth of the Iub.

HSDPA can be offered as a best effort service from the perspective of the Iub. In this case, the common channels and DCH connections are provided with the greatest priority while the RNC only forwards AAL2 packets associated with HSDPA connections when there is unused bandwidth on the user plane VCC. This scenario is illustrated in Figure 6.12.

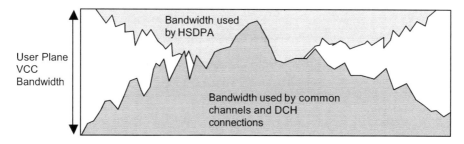

Figure 6.12 HSDPA offered on a best effort basis without any bandwidth reservation

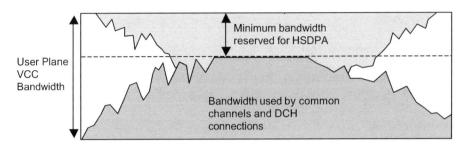

Figure 6.13   HSDPA offered on a best effort basis with a minimum reserved bandwidth

This approach tends to increase the utilisation of the Iub transport resources, but provides a potentially reduced quality of service for the HSDPA connections. If the quantity of DCH traffic becomes high then it is possible that the HSDPA connections will experience congestion and increased delay. Increased delay may not be significant for some applications, but if HSDPA is used to transfer real time data then the reduced quality of service may not be acceptable. Buffering HSDPA data at the Node B allows some variation of the rate at which data is transferred across the Iub. If there is a requirement for HSDPA to support continuously high throughputs then the Iub must have the corresponding bandwidth available.

Alternatively, a minimum bandwidth can be reserved for HSDPA to help guarantee a minimum HSDPA throughput performance. This bandwidth could be reserved during times that HSDPA connections are active. This would allow the common channels and DCH connections to utilise the entire user plane VCC when there are no HSDPA connections active. This scenario is illustrated in Figure 6.13.

This approach has some similarities with the enhanced version of UBR (UBR+) which introduces the optional Minimum Desired Cell Rate (MDCR) parameter to help ensure that at least a minimum target throughput is achieved. The main difference is that UBR+ applies the MDCR requirement to each individual connection so different connections can have different minimum achieved throughputs across the Iub. The solution illustrated in Figure 6.13 is based upon a common bandwidth reservation which is shared by all HSDPA connections. This approach can be compared with the utilisation of air-interface resources which are also common and shared by all HSDPA connections.

The priorities assigned to the PDU belonging to each MAC-d flow can be used to determine the order in which HSDPA PDU are transferred across the Iub. PDU belonging to real time services such as Voice over IP (VoIP) can be assigned a higher priority to help ensure that they are less vulnerable to Iub congestion.

6.7.2 Transport Overheads

It is necessary to quantify the overheads introduced by the Iub interface when dimensioning the Iub capacity requirement. These overheads are dependant upon both the size of the MAC-d PDU and the number of PDU packaged within each Frame Protocol data frame. Table 6.8 presents the overheads when the MAC-d PDU have a size of 336 bits and 5, 50 and 100 PDU are packaged within each data frame. In practice, the number of PDU within each data frame is likely to vary according to the quantity of data buffered at the RNC and the number of credits allocated by the Node B flow control. If Frame Protocol data frames are transferred across the Iub once every 10 ms then five MAC-d PDU per data frame corresponds to an RLC throughput of 160 kbps. Similarly, 50 MAC-d PDU corresponds to an RLC throughput of 1.6 Mbps and 100 MAC-d PDU corresponds to 3.2 Mbps. These figures assume a 16 bit RLC header.

The Frame Protocol layer adds padding to each MAC-d PDU because 3GPP specifies that 4 spare bits precede each PDU and that these spare bits are always positioned at the start of a new byte. The MAC-d PDU are an integer number of bytes and so there are 4 bits of padding following each PDU.

Table 6.8 Iub overheads for an HSDPA connection (ATM transport)

Frame protocol layer			
# MAC-d PDU per FP packet	5	50	100
MAC-d PDU size (bits)	336	336	336
Frame protocol payload (including spare bits and padding) (bits)	1720	16 800	33 600
Frame protocol header (bits)	56	56	56
Frame protocol tail (bits)	16	16	16
Frame Protocol packet (bits)	1792	16 872	33 672
AAL2 SSSAR layer			
Maximum size of AAL2 SSSAR PDU (bits)	360	360	360
Segmentation required	Yes	Yes	Yes
# AAL2 SSSAR PDU	5	47	94
AAL2 CPS layer			
AAL2 CPS packet header (bits)	24	24	24
Total AAL2 CPS packet data (bits)	1912	18 000	35 928
ATM layer			
ATM cell header (bits)	40	40	40
ATM cell payload (bits)	384	384	384
AAL2 start field (bits)	8	8	8
Effective ATM cell payload (bits)	376	376	376
# ATM cells	6	48	96
Remaining ATM cell capacity (bits)	344	48	168
Total overhead (bits)	476	3898	7715
Total payload (bits)	1680	16 800	33 600
Iub protocol overhead (%)	28	23	23
Iub throughput (kbps)	216	2070	4132

This means that the PDU size effectively increases from 336 to 344 bits. The size of the Frame Protocol header is 7 bytes whereas the tail is 2 bytes.

The Frame Protocol packet is passed to the AAL2 SSSAR layer for segmentation into blocks no larger than 360 bits. This layer passes the segments to the AAL2 CPS layer without adding any overhead. The AAL2 CPS layer adds 24 bits of header to each segment. These segments are then fed into ATM cells which include a 40 bit ATM header and an 8 bit AAL2 start field. After accounting for the AAL2 start field, the effective ATM cell payload is 376 bits. In general, the quantity of AAL2 CPS data is not a multiple of 376 bits and some ATM cell capacity remains for other AAL2 connections. The overhead generated by the last ATM cell header and 8 bit start field can be partially associated with other AAL2 connections.

The total overhead is calculated by summing the overheads generated by the Frame Protocol, AAL2 and ATM layers whereas the total payload is equal to the product of the MAC-d PDU size and the number of PDU. The Iub overhead expressed as a percentage is defined as the ratio between the overhead and the payload. This overhead is relative to the bit rate at the top of the Frame Protocol layer rather than the bit rate at the top of the RLC layer. If it is assumed that each MAC-d PDU includes a 16 bit RLC header then the overheads relative to the top of the RLC layer increase to 35% when there are 5 MAC-d PDU within each Frame Protocol packet and 29% when there are either 50 or 100 PDU within each packet. The Iub throughput figures assume that Frame Protocol packets are transferred once every 10 ms.

Packet switched transport networks based upon IP and Ethernet become more attractive as the bandwidth requirements of HSDPA increase. The bandwidth offered by an Ethernet connection can be significantly greater than that offered by a group of TDM network E1, T1 or JT1, e.g. 100 Mbps Ethernet or Gigabit Ethernet. The Internet Engineering Task Force (IETF) document 'draft-ietf-pwe 3-arch-07, RFC 3985' describes a solution which allows ATM cells to be encapsulated within IP packets and subsequently transferred across an Ethernet packet switched network. This approach is known as Psuedo Wire Emulation Edge to Edge (PWE3). The addition of IP and Ethernet to the existing AAL2 and ATM protocol stack increases the overhead generated by the Iub but the availability of Ethernet bandwidth means that this overhead is less significant. Table 6.9 presents the overheads generated by the combined Frame Protocol, AAL2, ATM, IP and Ethernet protocol stack. Similar to the

Table 6.9 Iub overheads for an HSDPA connection (ATM over IP transport)

# ATM cells per IP packet	3	10	25
Total AAL2 CPS packet data (bits)	18000	18000	18000
ATM layer			
ATM cell header (bits)	32	32	32
ATM cell payload (bits)	384	384	384
AAL2 start field (bits)	8	8	8
Effective ATM cell payload (bits)	376	376	376
# ATM cells	48	48	48
Remaining ATM cell capacity (bits)	48	48	48
Control word			
Preferred control word (bits)	32	32	32
Psuedo wire header			
Single MPLS label (bits)	32	32	32
IPv4			
IP header (bits)	160	160	160
# IP packets	16	5	2
IP packet data (bits)	1472	4384	10624
Ethernet			
Ethernet header (bits)	176	176	176
Ethernet CRC tail (bits)	32	32	32
# Ethernet packets	16	5	2
Ethernet packet data (bits)	1680	4592	10832
Total overhead (bits)	10409	5583	4342
Total payload (bits)	16800	16800	16800
Iub protocol overhead (%)	62	33	26
Iub throughput (kbps)	2721	2238	2114

results presented in Table 6.8, these overheads are dependant upon both the size of the MAC-d PDU and the number of PDU packaged within each Frame Protocol data frame. In this case, the overhead is also dependent upon the number of ATM cells encapsulated within each IP packet. Table 6.9 is based upon a MAC-d PDU size of 336 bits, 50 PDU within each Frame Protocol data frame and either 3, 10 or 25 ATM cells within each IP packet. The Frame Protocol and AAL2 sections of the protocol stack remain unchanged and are not presented within Table 6.9.

The IETF document 'draft-ietf-pwe3-atm-encap, RFC 4717' specifies that the Header Error Correction (HEC) bits are excluded from the ATM cell header when one or more ATM cells are packaged within a single IP packet using the 'N-to-one' encapsulation method. This reduces the size of the ATM cell header from 5 bytes to 4 bytes. The 'N-to-one' encapsulation method allows multiple VCC/VPC to be transferred using a single pseudo wire. This is in contrast to the 'one-to-one' encapsulation method which associates a single VCC/VPC to each pseudo wire. The same IETF document specifies that the protocol stack should include a 4 byte control word. This control word can be used to include protocol specific flags and a sequence number to ensure in-sequence delivery.

The protocol stack then includes an MPLS label. In general, MPLS labels are used by routers to identify the next hop when forwarding the packet across the packet switched network. In the case of a pseudo wire connection, the MPLS label is also used to identify the pseudo wire. Figure 6.14 illustrates the structure of an MPLS label.

The label itself occupies the first 20 bits of the MPLS data. The Exp bits can be used to provide visibility of the ATM Cell Loss Priority (CLP) bits at the MPLS layer. The bottom of the stack (S) bit is used to indicate whether or not the label is the last label within the label stack. A value of 1 indicates that it is the last label whereas a value of 0 indicates that it is not the last label. MPLS routers use the first label in the stack for routing purposes, but may add and remove labels across the network. The Time To Live (TTL) field limits the number of hops used to transfer the packet across the MPLS network. The TTL value is decremented by 1 each time a hop is completed. The packet is discarded if the TTL value reaches 0.

A standard IP header is included for the pseudo wire connection. Table 6.9 assumes that IPv4 is used and so the header size is 20 bytes. The Ethernet layer includes a 22 byte header and a 4 byte tail. The header includes an 8 byte preamble, 6 byte source address, 6 byte destination address and a 2 byte Ethertype. The Ethertype field is used to identify the upper layer protocol as IPv4. The size of the Ethernet payload is dependent upon the number of ATM cells within the IP packet.

Table 6.9 illustrates that the total overhead generated by the pseudo wire protocol stack is large when the number of ATM cells included within each IP packet is small. The overhead relative to the bit rate at the top of the Frame Protocol layer is 62% when there are only three ATM cells within the IP packet This overhead decreases to 26% when 25 ATM cells are included within the IP packet. This value is relatively close to the 23% overhead generated by the Frame Protocol, AAL2 and ATM protocol stack presented in Table 6.8. The drawback associated with packaging a large number of ATM cells within each IP packet is an increased delay, i.e. more ATM cells have to be collected before the IP packet can be transferred. System delay can be reduced by including only a small number of ATM cells within each IP packet. This approach is feasible when the bandwidth offered by the Ethernet connection has sufficient margin to accommodate the increased protocol stack overhead.

Figure 6.14   Structure of an MPLS label

6.8 MAC-hs Entity

- MAC-hs entities provide functionality for Iub flow control, scheduling, Adaptive Modulation and Coding (AMC) and Hybrid Automatic Repeat Request (HARQ).
- The objective of flow control is to ensure that the Node B has sufficient data while avoiding buffer overflow. Flow control generates instructions for the RNC in terms of the maximum quantity of MAC-d PDU to be transferred within a specific time interval.
- Flow control makes use of Frame Protocol control frames to signal between the RNC and Node B. Flow control is completed on a per priority queue basis, rather than on a per MAC-d flow basis.
- Flow control should grant a high number of credits if the throughput is high. The number of credits could be adjusted according to the buffer occupancy. If the occupancy is high the number of credits could be decreased.
- The scheduler determines the sequence in which HSDPA UE are assigned access to the 2 ms TTI. The scheduling function is simplified if code multiplexing is disabled so only a single UE can be scheduled during each TTI.
- The simplest form of scheduler is a round robin scheduler which assigns the 2 ms TTI in rotation without accounting for the channel conditions.
- A proportional fair scheduler accounts for the RF channel conditions when allocating the 2 ms TTI. This type of scheduler can increase both the individual connection throughputs and the total cell throughput.
- A proportional fair scheduler attempts to allocate TTI to the UE which is experiencing the best short-term channel conditions relative to its average channel conditions. This approach helps to avoid TTI being scheduled during deep fades.
- A proportional fair scheduler does not offer any benefit over a round robin scheduler if there is only a single HSDPA connection within the cell. The benefit of a proportional fair scheduler increases when UE have relatively low speeds.
- AMC is responsible for selecting appropriate HSDPA throughputs. The throughput can change every 2 ms TTI to account for changes in the RF channel conditions. AMC also selects the modulation scheme and the appropriate number of HS-PDSCH codes.
- UE are responsible for generating Channel Quality Indicator (CQI) values to report the RF channel conditions. CQI values should correspond to the transport block size which the UE believes it can receive with a BLER of 10%.
- AMC may compensate the received CQI reports prior to selecting an appropriate throughput. CQI values may be adjusted to compensate for any difference between the actual HS-PDSCH transmit power and the value assumed by the UE. Compensation may also be completed to account for the connection specific achieved BLER.
- AMC is able to select an appropriate transport block size once a compensated CQI has been derived. Selection typically uses a look-up table which provides a one-to-one mapping between compensated CQI and transport block size. The transport block size determines the throughput.
- HARQ refers to a re-transmission protocol in which the receiver checks for errors in the received data and if an error is detected then the receiver buffers the data and requests a re-transmission from the sender. The receiver is then able to combine the buffered data with the re-transmitted data prior to channel decoding and error detection.
- Chase combining means that the puncturing pattern applied to the re-transmitted bits is the same as the puncturing pattern applied to the original transmission.
- Incremental redundancy means that the puncturing pattern applied to the re-transmitted bits is different to the puncturing pattern applied to the original transmission. The performance of incremental redundancy becomes significantly better than the performance of chase combining when the coding rate is high, i.e. there is increased puncturing.

- The HARQ protocol relies upon receiving acknowledgements from the UE. The round trip time means that acknowledgements are not received instantaneously. Parallel HARQ processes are used to avoid the round trip time affecting the throughput. Six parallel processes ensure a continuous flow of data when there is a single active connection.
- Each transport block includes a Transmission Sequence Number as part of the MAC-hs header. Each priority queue uses its own sequence numbering. Transmit and receive windows are used to avoid sequence number ambiguity. Large window sizes can increase the delay with which data is passed to the higher layers. They also increase the memory requirement of the UE reordering buffer. Small window sizes are more likely to cause the HARQ protocol to stall and force periods of DTX between re-transmissions.
- The total length of the MAC-hs header is equal to $10 + (n \times 11)$, where n is the number of different MAC-d PDU sizes included within the PDU. If only a single size is included then the header length is 21 bits.
- Key 3GPP specifications: TS 25.321, TS 25.211, TS 25.212, TS 25.214.

The MAC-hs entity is the most important part of the protocol stack for HSDPA. It includes the functionality for flow control, scheduling, Adaptive Modulation and Coding (AMC) and Hybrid Automatic Repeat Request (HARQ). Every cell that supports HSDPA has its own MAC-hs entity. The general structure of a MAC-hs entity is illustrated in Figure 6.15.

Frame Protocol data frames arrive at the Node B from the Iub transport connections. Each MAC-d flow has its own transport connection. The example in Figure 6.15 illustrates the MAC-d flows belonging to two HSDPA users. The first user has two MAC-d flows whereas the second user has one MAC-d flow. The Frame Protocol data frames include sets of MAC-d PDU which have associated priority indicators. The MAC-d PDU are extracted from the data frames and placed in the appropriate priority queue belonging to the relevant user. A priority queue represents a Node B buffer which has limited storage capability. Flow control monitors the throughput and occupancy of each buffer to help generate instructions regarding the maximum quantity of additional MAC-d PDU which the RNC should forward. Each HSDPA user has its own HARQ entity which is responsible for serving the priority queues and handling the protocol for MAC-hs re-transmissions. The scheduler determines when each HARQ entity is able to forward data for transfer across the air-interface. If code multiplexing is used then the scheduler can allow more than a single HARQ entity to transfer data during each TTI. Otherwise, only one HARQ entity can transfer data during each TTI. The AMC function determines the transport block size which is allocated at any point in time. This determines the number of MAC-d PDU which are taken from the priority queue when constructing the MAC-hs PDU (transport block). Each HARQ entity can transfer a maximum of one transport block during each TTI unless the $2 \times \text{Tx} + 2 \times \text{Rx}$ Multiple Input Multiple Output (MIMO) introduced by the release 7 version of the 3GPP specifications is used. In this case, the HARQ entity can transfer up to two transport blocks during each TTI. The transport blocks generated by the MAC-hs entity are then forwarded to the physical layer for processing and mapping onto the physical channels.

6.8.1 Flow Control

MAC-hs flow control operates between the Node B and RNC. The objective is to ensure that the Node B has sufficient data to fully utilise the HS-PDSCH, while avoiding overflow of the Node B buffers. If a UE is in poor coverage and the air-interface bit rate is low then it is necessary for the RNC to forward MAC-d PDU at a relatively low rate. If a UE is in good coverage and the air-interface bit rate is high then it is necessary for the RNC to forward MAC-d PDU at a relatively high rate. The MAC-hs entity within the Node B generates instructions for the RNC in terms of the maximum quantity of MAC-d PDU which can be transferred within a specific time interval. Flow control makes use of Frame Protocol control frames to

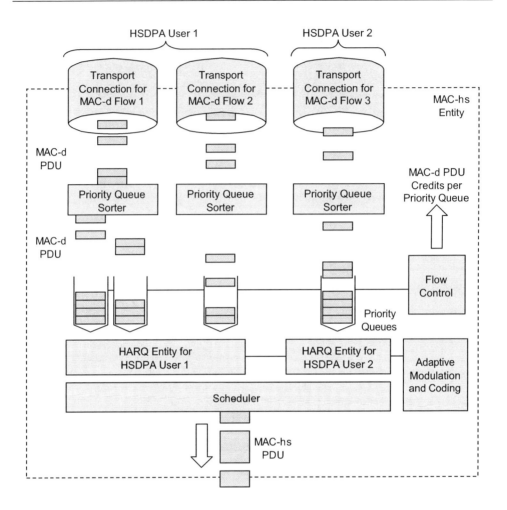

Figure 6.15   General structure of the MAC-hs entity

signal between the RNC and Node B. The downlink Capacity Request control frame and the uplink Capacity Allocation control frame are described in Section 6.6.2. Flow control is completed on a per priority queue basis, rather than on a per MAC-d flow basis, i.e. there is separate flow control signaling for every priority queue belonging to every MAC-d flow. A single HSDPA connection can have one or more MAC-d flows and each MAC-d flow can have one or more priority queues. The simplest scenario is for an HSDPA connection to have a single MAC-d flow with a single priority queue. In this case, flow control can be viewed as being completed on a per HSDPA connection basis.

The general principles of flow control during steady state conditions are illustrated in Figure 6.16. This figure illustrates the example of four HSDPA connections, each with a single priority queue. Steady state conditions are preceded by the RNC using a Frame Protocol Capacity Request message to inform the Node B that a specific priority queue has data to transfer. The Node B then starts to provide instructions using Frame Protocol Capacity Allocation messages. The Node B can also be proactive in terms of returning Capacity Allocation messages, i.e. these messages can be sent prior to receiving a

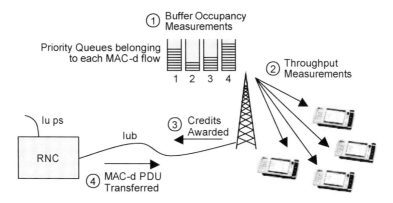

Figure 6.16 Principles of HSDPA flow control after initialisation

Capacity Request message. At this stage, the Node B buffer for the priority queue is empty and the Node B is likely to grant a relatively large number of credits to start filling the buffer.

Flow control can operate in a routine manner once the first set of MAC-d PDU arrive at the Node B and the HARQ entity starts to serve the corresponding priority queue. The detailed implementation of flow control is vendor dependent rather than standardised by 3GPP. A typical flow control algorithm makes use of periodic buffer occupancy and throughput measurements. The throughput measurements quantify the rate at which MAC-d PDU are emptied from the priority queue buffer. Table 6.10 presents an example of the dependency of the allocated flow control credits upon the Node B buffer occupancy and throughput.

If the throughput is high then the Node B should grant a correspondingly high number of credits. If a high number of credits is not granted, the Node B could run out of MAC-d PDU and the Iub flow control would start to limit the end-to-end throughput performance. The precise number of credits granted could be adjusted according to the buffer occupancy. If the buffer occupancy is low then the number of credits would be increased whereas if the buffer occupancy is high then the number of credits would be decreased. The magnitude of the increase or decrease could depend upon how high or how low the buffer occupancy becomes. If the throughput is low then the Node B should grant a correspondingly low number of credits. If a low number of credits is not granted, the Node B buffers could overflow. Similar to the high throughput scenario, the precise number of credits could be adjusted according to the buffer occupancy. The Node B should grant zero credits if the buffer occupancy reaches a maximum threshold at any time.

The flow control algorithm must be sufficiently responsive to cope with rapid changes in throughput. For example, the throughput experienced by a first HSDPA connection can suddenly decrease by a factor of 2 if a second HSDPA connection becomes active. If the number of credits assigned to the priority queue is not decreased rapidly then the Node B buffer could overflow. It's typical for flow control algorithms to use an update cycle in the order of 10 ms, i.e. the buffer occupancy and throughput measurements as well as the number of credits granted are updated every 10 ms. Longer flow control update cycles result in the procedure becoming less responsive and increase the requirement for buffering at the Node B.

Table 6.10 Flow control credits granted as a function of throughput and buffer occupancy

	Low buffer occupancy	High buffer occupancy
Low throughput	Low credits $+\Delta$	Low credits $-\Delta$
High throughput	High credits $+\Delta$	High credits $-\Delta$

6.8.2 Scheduler

The Node B scheduler determines the sequence in which HSDPA UE are assigned access to the 2 ms TTI. Scheduler design is implementation dependent although the 3GPP specifications define the signalling used to inform each UE of when TTI have been assigned, i.e. using an HS-SCCH physical channel. The scheduling function is simplified if code multiplexing is disabled so only a single UE can be scheduled during each TTI. The scheduler is able to schedule more than a single UE during each TTI if code multiplexing is enabled. In this case, a separate HS-SCCH is required for each UE scheduled during the same TTI.

The simplest form of scheduler is a round robin scheduler which assigns the 2 ms TTI in rotation without accounting for the channel conditions experienced by each UE. The concept of a round robin scheduler is illustrated in Figure 6.17.

This example is based upon four simultaneously active UE. Each UE is served in rotation so each is allocated one quarter of the TTI. A round robin scheduler has a linear impact upon the throughput experienced by each UE. Increasing the number of UE by a factor of 2 causes the individual UE throughput to decrease by a factor of 2. This results directly from the UE being allocated half the original number of TTI. An exception to the round robin rotation can occur if the priority queue buffer belonging to one of the UE runs out of MAC-d PDU. This could occur if the data generated by the end-user application is bursty rather than continuous. In this case, the scheduler skips the UE and the next UE is served. The UE without data continues to be skipped until further data arrives at the Node B.

A proportional fair scheduler accounts for the RF channel conditions when allocating the 2 ms TTI. This type of scheduler can increase both the individual connection throughputs and the total cell throughput. A proportional fair scheduler does not offer any benefit over a round robin scheduler if the quantity of HSDPA traffic is relatively low and there is only a single HSDPA connection within the cell, i.e. the single connection is allocated all TTI in both cases. The benefits of a proportional fair scheduler increase as the number of HSDPA UE increases. A proportional fair scheduler attempts to allocate TTI to the UE which is experiencing the best short-term channel conditions relative to its average channel

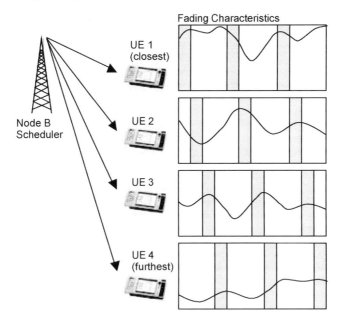

Figure 6.17   Round robin scheduler

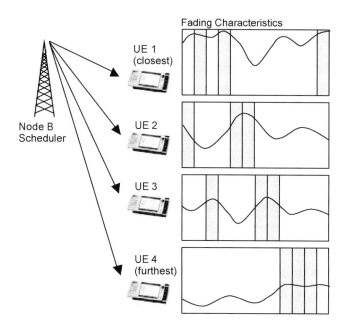

Figure 6.18 Proportional fair scheduler

conditions. This approach helps to avoid TTI being scheduled during deep fades. The concept of a proportional fair scheduler is illustrated in Figure 6.18.

The number of TTI allocated over a longer period of time does not change relative to a round robin scheduler, i.e. each UE is allocated an equal share of the TTI. The benefit of using a proportional fair scheduler results from allocating the TTI at the best possible instants. This requires the scheduler to track the short-term channel conditions experienced by each UE. Locating the scheduler within the Node B rather than within the RNC reduces delay and increases the availability of RF channel condition information. Nevertheless, it remains difficult to track the conditions unless they are changing relatively slowly. The benefit of a proportional fair scheduler increases when UE have relatively low speeds. Higher UE speeds cause the RF channel conditions to change prior to scheduling the next TTI. A proportional fair scheduler can increase HSDPA throughput by between 30% and 50% when UE are moving at 3 km/hr (relative to a round robin scheduler). The gain decreases to less than 5% when UE move at 30 km/hr. These figures assume there are at least five simultaneously active UE. The probability of all UE experiencing a fade at the same time increases as the number of UE decreases. If all UE experience fades at the same time then the scheduler will be forced to allocate one or more TTI during a fade.

Figure 6.19 and Figure 6.20 illustrate an equal throughput scheduler and a maximum throughput scheduler respectively. These scheduler types are less likely to be included within practical network implementations because their performance is less balanced in terms of maximising both end-user experience and total cell throughput. The equal throughput scheduler attempts to ensure that each UE experiences the same average throughput. This approach is fair from the perspective of providing all UE with a similar throughput performance, but can be inefficient from the perspective of the total cell throughput. If there are UE located in poor RF channel conditions then those UE would require significantly more TTI than the UE in good channel conditions. This would decrease the total cell throughput as well as the throughput of the UE in good channel conditions. The maximum throughput scheduler attempts to ensure that the total cell throughput is maximised by always allocating TTI to the UE with the best absolute channel conditions. This approach can cause UE with poor RF channel conditions to become completely excluded and not be scheduled any TTI.

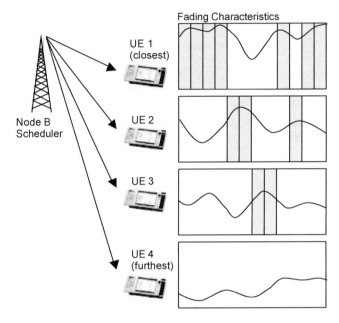

Figure 6.19   Equal throughput scheduler

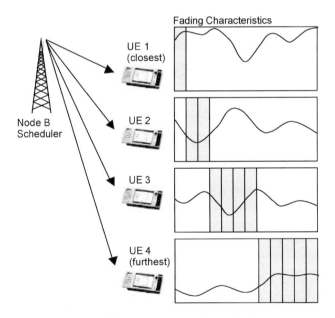

Figure 6.20   Maximum throughput scheduler

6.8.3 Adaptive Modulation and Coding

Adaptive Modulation and Coding (AMC) is responsible for selecting appropriate HSDPA throughputs. A throughput is selected for every TTI that the scheduler allocates to a specific UE. The throughput can vary every 2 ms TTI to account for changes in the RF channel conditions. Each UE reports the RF channel conditions to the Node B using the HS-DPCCH. AMC also selects the modulation scheme and the appropriate number of HS-PDSCH. Each HS-PDSCH is defined by its channelisation code. The general procedure for AMC is illustrated in Figure 6.21.

UE are responsible for generating Channel Quality Indicator (CQI) values when instructed by the RNC. The RNC instructs the UE to start generating and reporting CQI values during HSDPA connection establishment. 3GPP TS 25.214 specifies that UE should generate CQI values which correspond to the transport block size which the UE believes it can receive with a BLER of 10%. TS 25.214 includes a set of look-up tables which define the relationship between CQI value and transport block size. These tables depend upon the HSDPA UE category because different category UE can support different transport block sizes. The relationship between CQI and transport block size for UE categories 1 to 6 is presented in Table 6.11.

The detailed implementation for selecting an appropriate CQI value is not specified by 3GPP although the general procedure is based upon measuring the CPICH Ec/Io. CPICH Ec/Io measurements are scaled to account for the difference between the CPICH transmit power and the HS-PDSCH transmit power. In general, the HS-PDSCH will be transmitted with greater power than the CPICH and so the adjusted Ec/Io value will be greater than the CPICH Ec/Io measurement. The RNC signals a measurement power offset value to the UE to support the scaling of the CPICH Ec/Io measurement. This power offset represents the difference between the CPICH and HS-PDSCH transmit powers. It can be signalled within a Radio Bearer Setup, Radio Bearer Reconfiguration, Transport Channel Reconfiguration or Physical Channel Reconfiguration message. An example of the relevant section from these messages is presented within Log File 6.8.

In this example, the measurement power offset is signalled using a value of 10. The actual value is equal to the signalled value divided by 2 and so the actual measurement power offset is 5 dB. This indicates that the HS-PDSCH are transmitted with a total power of 38 dBm if the CPICH is transmitted with a power of 33 dBm. The actual value of the measurement power offset can range from -6 to

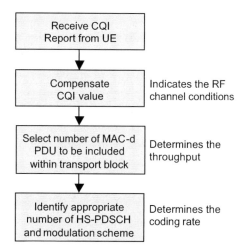

Figure 6.21   General procedure for Adaptive Modulation and Coding (AMC)

Table 6.11 Look-up table used by UE categories 1 to 6 when generating CQI values

CQI	TrBlk Size (bits)	Codes	Modulation	Power Adjustment (dB)	CQI	TrBlk Size (bits)	Codes	Modulation	Power Adjustment (dB)
0	N/A	1	QPSK	0	16	3565	5	16QAM	0
1	137	1	QPSK	0	17	4189	5	16QAM	0
2	173	1	QPSK	0	18	4664	5	16QAM	0
3	233	1	QPSK	0	19	5287	5	16QAM	0
4	317	1	QPSK	0	20	5887	5	16QAM	0
5	377	1	QPSK	0	21	6554	5	16QAM	0
6	461	1	QPSK	0	22	7168	5	16QAM	0
7	650	2	QPSK	0	23	7168	5	16QAM	−1
8	792	2	QPSK	0	24	7168	5	16QAM	−2
9	931	2	QPSK	0	25	7168	5	16QAM	−3
10	1262	3	QPSK	0	26	7168	5	16QAM	−4
11	1483	3	QPSK	0	27	7168	5	16QAM	−5
12	1742	3	QPSK	0	28	7168	5	16QAM	−6
13	2279	4	QPSK	0	29	7168	5	16QAM	−7
14	2583	4	QPSK	0	30	7168	5	16QAM	−8
15	3319	5	QPSK	0					

13 dB so it is possible to signal HS-PDSCH transmit powers which are less than the CPICH transmit power. Log File 6.8 also includes the CQI feedback cycle which defines the rate at which the UE reports CQI information to the Node B. This example has a value of 4 ms which means that the UE is instructed to send a CQI report once every second TTI. Increasing the rate at which CQI reports are sent increases the responsiveness of the AMC function, i.e. AMC has more up-to-date information when selecting an appropriate bit rate. The drawback associated with increasing the rate at which CQI reports are sent is an increased UE transmit power overhead. The CQI repetition factor indicates whether or not the UE should repeat each CQI report. This can be used to improve the reliability of CQI reporting, but also represents an increase in the UE transmit power overhead. This example specifies a repetition factor of 1 which means that repetition is not applied. The delta CQI value defines the amplitude ratio between the DPCCH and HS-DPCCH. This amplitude ratio is described in Section 6.9.3. A value of 4 corresponds to an amplitude ratio of 12/15 which means that the CQI section of the HS-DPCCH is transmitted with 1.9 dB less power than the DPCCH.

The UE must also account for orthogonality when using the CPICH Ec/Io measurement to generate a CQI value. This is necessary because CPICH Ec/Io measurements are based upon RSSI which excludes the impact of downlink orthogonality. Reception of the HS-PDSCH benefits from orthogonality and so the signal-to-noise ratio should be improved. The procedure used by the UE to estimate the orthogonality is not specified by 3GPP, but is likely to be based upon an estimate of the RF propagation channel. Table 6.11 also specifies the number of HS-PDSCH and the modulation scheme associated

```
dl-HSPDSCH-Information
      measurement-feedback-Info
      modeSpecificInfo fdd
            measurementPowerOffset:    10
            feedback-cycle:            fc4
            cqi-RepetitionFactor:      1
            deltaCQI:                  4
```

Log File 6.8 Example of signalling used to configure CQI reporting

with each transport block size. This allows the UE to estimate the signal-to-noise ratio on a per HS-PDSCH basis and to account for when 16QAM is necessary. UE can then select an appropriate transport block size. This selection process is not specified by 3GPP, but is likely to be based upon a UE internal look-up table which defines the relationship between HS-PDSCH signal-to-noise ratio and the transport block size which corresponds to a BLER of 10%.

AMC receives the CQI reports from the UE, but does not necessarily use the received values directly. The received values may be adjusted to compensate for transmit power differences and for variations in the achieved BLER. The HS-PDSCH transmit power assumed by the UE, i.e. the CPICH transmit power plus the measurement power offset, may not be an accurate reflection of the actual HS-PDSCH transmit power. This is likely to be the case if any form of power control is applied to the HS-PDSCH. The measurement power offset is typically signalled to the UE during connection establishment and during the serving cell change procedure (handover procedure). It is not continuously updated and so it can become out-of-date if power control is applied to the HS-PDSCH. AMC within the Node B must account for the difference between the actual HS-PDSCH transmit power and the transmit power assumed by the UE. This requires the Node B to have knowledge of the measurement power offset value signalled to the UE by the RNC. The RNC can use either the NBAP Radio Link Reconfigure Prepare or NBAP Radio Link Setup messages to signal the measurement power offset to the Node B. It can be assumed that there is a 1 dB difference between each of the CQI values. For example, the CQI value can be increased by 5 if the UE is assuming an HS-PDSCH transmit power which is 5 dB too low. Adjusting the received CQI values according to the achieved BLER helps to normalise the impact of differences in the UE implementations. The received CQI can be decreased by the Node B if a specific UE is generating relatively large CQI values and is consequently experiencing a BLER of greater than 10%. Likewise, the received CQI can be increased if a specific UE is generating relatively small CQI values and is experiencing a BLER of less than 10%.

AMC is able to select an appropriate transport block size once a compensated CQI has been derived from the received CQI. This is typically completed using a Node B internal look-up table which provides a one-to-one mapping between compensated CQI and transport block size. This look-up table does not have to be the same as the one used by the UE to generate the CQI values, i.e. the Node B implementation can use a different set of transport block sizes. The set of allowed transport block sizes is specified within 3GPP TS 25.321. Transport block sizes are calculated using an intermediate variable k_t which is equal to the sum of the transport block size index and a variable $k_{o,i}$. The transport block size index can have a value from 0 to 63, whereas the range of values for $k_{o,i}$ is presented within Table 6.12.

Table 6.12 Values specified for the variable $k_{o,i}$ within 3GPP TS 25.321

Number of HS-PDSCH codes	QPSK	16QAM
1	1	40
2	40	79
3	63	102
4	79	118
5	92	131
6	102	141
7	111	150
8	118	157
9	125	164
10	131	169
11	136	175
12	141	180
13	145	184
14	150	188
15	153	192

The transport block size is then calculated using the following equations:

$$\text{If } kt < 40 \text{ then Transport block size} = 125 + 12 \times k_t, \text{ else}$$

$$\text{Transport block size} = \left\lfloor 296 \times (2085/2048)^{k_t} \right\rfloor$$

The release 7 version of the specifications introduce a second set of $k_{0,i}$ values which become applicable when 64QAM is enabled. The equations used to calculate the transport block size are also modified for this scenario. Table 6.13 presents a subset of the transport block sizes which have been calculated for QPSK.

The same transport block size can appear in different columns of the table. For example, if the table is extended downwards then the first column includes the transport block sizes 605, 616 and 627 bits (starting from a transport block size index of 39). These same transport block sizes appear at the top of the second column. This overlap provides the AMC function with some flexibility regarding the coding rate used to transfer a specific transport block size. Selecting the transport block size from the first column will result in a higher coding rate (less redundancy) but will allow all of the HS-PDSCH transmit power to be assigned to a single code. Selecting the same transport block size from the second column will result in a lower coding rate (greater redundancy) but requires the HS-PDSCH transmit power to be shared between two codes.

Selecting transport block sizes from the set of values presented in Table 6.13 depends upon both the MAC-d PDU size and the MAC-hs header size. The transport block size should be selected to ensure that both the set of MAC-d PDU and the MAC-hs header can be accommodated with the least margin, i.e. the quantity of padding is minimised. Table 6.14 presents a set of transport block sizes which can be used to accommodate

Table 6.13 An example subset of transport block sizes (bits) for QPSK

| Transport block size index | Number of HS-PDSCH | | | | |
	1 $k_{0,i} = 1$	2 $k_{0,i} = 40$	3 $k_{0,i} = 63$	4 $k_{0,i} = 79$	5 $k_{0,i} = 92$
0	137	605	914	1217	1537
1	149	616	931	1239	1564
2	161	627	947	1262	1593
3	173	639	964	1285	1621
4	185	650	982	1308	1651
5	197	662	1000	1331	1681
6	209	674	1018	1356	1711
7	221	686	1036	1380	1742
8	233	699	1055	1405	1773
9	245	711	1074	1430	1805
10	257	724	1093	1456	1838
11	269	737	1113	1483	1871
12	281	751	1133	1509	1905
13	293	764	1154	1537	1939
14	305	778	1175	1564	1974
15	317	792	1196	1593	2010
16	329	806	1217	1621	2046
17	341	821	1239	1651	2083
18	353	836	1262	1681	2121
19	365	851	1285	1711	2159
20	377	866	1308	1742	2198

Table 6.14 Transport block sizes selected to accommodate a specific number of MAC-d PDU

Number of MAC-d PDU	MAC-d PDU data (bits)	MAC-hs header (bits)	Total data (bits)	Transport block size (bits)	Padding (bits)
1	336	21	357	365	8
2	672	21	693	699	6
3	1008	21	1029	1036	7
4	1344	21	1365	1380	15
5	1680	21	1701	1711	10
6	2016	21	2037	2046	9
7	2352	21	2373	2404	31
8	2688	21	2709	2726	17
9	3024	21	3045	3090	45
10	3360	21	3381	3440	59

between 1 and 10 MAC-d PDU when assuming a MAC-d PDU size of 336 bits and a MAC-hs header size of 21 bits.

The MAC-hs must identify an appropriate number of HS-PDSCH codes and modulation scheme once the transport block size has been selected. This can be done by considering either the coding rate or the puncturing ratio. The coding rate is defined as the ratio between the transport block size and the number of bits offered by the set of HS-PDSCH, i.e. the coding rate is defined by the following equation:

$$\text{Coding rate} = \text{Transport block size} \left/ \left[\frac{\text{Chip rate}}{\text{SF}} \times \text{Bits per symbol} \times \text{Number of codes} \times \text{TTI} \right] \right.$$

This equation can be simplified to:

$$\text{Coding rate} = \text{Transport block size}/[960 \times \text{Number of codes}] \text{ for QPSK, and}$$
$$\text{Coding rate} = \text{Transport block size}/[1920 \times \text{Number of codes}] \text{ for 16QAM.}$$

Table 6.15 presents the coding rates for the set of transport block sizes identified within Table 6.14 when assuming QPSK. This table could be applied to a category 12 UE which supports a maximum of

Table 6.15 QPSK coding rates for typical transport block sizes

Transport block size (bits)	Number of codes and capacity offered				
	1 code 960 bits	2 codes 1920 bits	3 codes 2880 bits	4 codes 3840 bits	5 codes 4800 bits
365	0.38	0.19	0.13	0.10	0.08
699	0.73	0.36	0.24	0.18	0.15
1036	—	0.54	0.36	0.27	0.22
1380	—	0.72	0.48	0.36	0.29
1711	—	0.89	0.59	0.43	0.36
2046	—	—	0.71	0.53	0.43
2404	—	—	0.83	0.63	0.50
2726	—	—	0.95	0.71	0.57
3090	—	—	—	0.80	0.64
3440	—	—	—	0.90	0.72

five codes and does not support 16QAM. Coding rates greater than 1 have been excluded because they represent cases where the capacity offered by the set of HS-PDSCH is less than the transport block size.

A coding rate of 0.35 represents a typical target. This minimises the requirement for puncturing after the rate 1/3 Turbo coding has been completed. Table 6.15 illustrates that 1 HS-PDSCH can be used to transfer a transport block size of 365 bits (1 MAC-d PDU), 2 HS-PDSCH codes can be used to transfer a transport block size of 699 bits (2 MAC-d PDU), and so forth. Once the maximum number of codes has been reached then it is necessary to start allowing the coding rate to increase. This corresponds to an increase in the puncturing which is necessary after the rate 1/3 Turbo coding. The coding rate is typically allowed to increase to a value of 0.75. This corresponds to a category 12 UE receiving a maximum transport block size of 3440 bits (10 MAC-d PDU and a maximum RLC throughput of 1.6 Mbps).

Table 6.16 presents a similar table for category 5 and 6 UE which support 16QAM and a maximum of 5 HS-PDSCH codes. In this case, a larger set of transport block sizes is possible. The transport block size of 7168 bits corresponds to 21 MAC-d PDU of 336 bits with a 21 bit MAC-hs header and 91 bits of padding. This allows a maximum RLC throughput of 3.36 Mbps. The QPSK modulation scheme is used until the coding rate approaches 0.75. Larger transport block sizes are then transferred using 16QAM and 5 HS-PDSCH codes. The use of 16QAM reduces the coding rate, but requires a higher signal-to-noise ratio. The transport block size is increased with 16QAM until the coding rate reaches 0.75.

A puncturing ratio can also be calculated for each transport block size. This quantifies the percentage of bits which remain after puncturing. Physical layer processing for the HS-DSCH transport channel always uses a 24 bit CRC and rate 1/3 Turbo coding with 12 tail bits. These factors allow the puncturing ratio for QPSK to be expressed using the following equation:

$$\text{Puncturing ratio} = [960 \times \text{Number of codes}]/[(\text{Transport block size} + 24) \times 3 + 12]$$

Table 6.16 QPSK and 16QAM coding rates for typical transport block size

Transport block size (bits)	Modulation, number of codes and capacity offered					
	QPSK 1 code 960 bits	QPSK 2 codes 1920 bits	QPSK 3 codes 2880 bits	QPSK 4 codes 3840 bits	QPSK 5 codes 4800 bits	16QAM 5codes 9600 bits
365	0.38	0.19	0.13	0.10	0.08	0.04
699	0.73	0.36	0.24	0.18	0.15	0.07
1036	—	0.54	0.36	0.27	0.22	0.11
1380	—	0.72	0.48	0.36	0.29	0.14
1711	—	0.89	0.59	0.45	0.36	0.18
2046	—	—	0.71	0.53	0.43	0.21
2404	—	—	0.83	0.63	0.50	0.25
2726	—	—	0.95	0.71	0.57	0.28
3090	—	—	—	0.80	0.64	0.32
3440	—	—	—	0.90	0.72	0.35
4420	—	—	—	—	0.92	0.46
4748	—	—	—	—	0.99	0.49
5480	—	—	—	—	—	0.57
6101	—	—	—	—	—	0.64
6793	—	—	—	—	—	0.71
7168	—	—	—	—	—	0.75

Table 6.17 QPSK puncturing ratios for typical transport block sizes

Transport block size (bits)	Number of codes and capacity offered				
	1 code 960 bits	2 codes 1920 bits	3 codes 2880 bits	4 codes 3840 bits	5 codes 4800 bits
365	0.81	1.63	2.44	3.26	4.07
699	0.44	0.88	1.32	1.76	2.20
1036		0.60	0.90	1.20	1.50
1380		0.45	0.68	0.91	1.14
1711		0.37	0.55	0.74	0.92
2046			0.46	0.62	0.77
2404			0.40	0.53	0.66
2726			0.35	0.47	0.58
3090				0.41	0.51
3440				0.37	0.46

Table 6.17 presents the puncturing ratios corresponding to the coding rates within Table 6.15. The coding rate of 0.35 corresponds to a puncturing ratio of approximately 0.90. Allowing the coding rate to increase to 0.75 causes the puncturing ratio to decrease to approximately 0.45, i.e. only 45% of the channel coded bits remain after puncturing and the majority of the redundancy introduced by Turbo coding is removed.

The preceding discussion on AMC assumes that the Node B always has sufficient MAC-d PDU to fully occupy the maximum transport block size allowed by the channel conditions. In practice, it is likely that there will be times when the Node B does not have sufficient MAC-d PDU. This could be characteristic of an application which generates bursty data. In this case, the AMC can either continue to select the maximum allowed transport block size, or it can select a smaller transport block size based upon the number of MAC-d PDU within the Node B buffer. Continuing to select the maximum allowed transport block size increases the requirement for padding within the transport block. Selecting a smaller transport block size generates some margin within the link budget which provides some scope for reducing the transmit power.

6.8.4 Hybrid Automatic Repeat Request (HARQ)

Automatic Repeat Request (ARQ) refers to a re-transmission protocol in which the receiver checks for errors in the received data and if an error is detected then the receiver discards the data and requests a re-transmission from the sender. Hybrid ARQ (HARQ) refers to a re-transmission protocol in which the receiver checks for errors in the received data and if an error is detected then the receiver buffers the data and requests a re-transmission from the sender. A HARQ receiver is then able to combine the buffered data with the re-transmitted data prior to channel decoding and error detection. The re-transmission and subsequent combining process can be based upon either chase combining or incremental redundancy.

Chase combining means that the puncturing pattern applied to the re-transmitted bits is the same as the puncturing pattern applied to the original transmission. This results in re-transmissions including the same set of bits as the original transmission. The principle of chase combining is presented in Figure 6.22. Turbo coding generates systematic, parity 1 and parity 2 bits. The systematic bits are the same as the original bit sequence prior to Turbo coding. These bits are the most important to the receiver and are provided with the greatest priority, i.e. the parity 1 and parity 2 bits are punctured in preference to the systematic bits. This example is based upon a puncturing ratio of 0.44 which

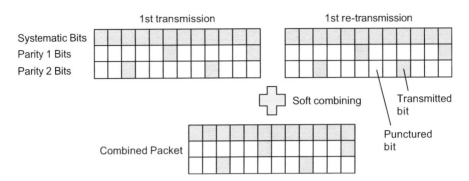

Figure 6.22   Principle of HARQ chase combining

represents a relatively low value. In general, less puncturing would be completed and a larger quantity of parity 1 and parity 2 bits would remain.

In the case of chase combining, the received symbols are combined prior to demodulation. This improves the received signal to noise ratio if it is assumed that the signal power is correlated while the noise power is uncorrelated. The benefits of chase combining relative to incremental redundancy are its simplicity and lower UE memory requirement.

Incremental redundancy means that the puncturing pattern applied to the re-transmitted bits is different to the puncturing pattern applied to the original transmission. This results in re-transmissions including different sets of bits. The principle of incremental redundancy is illustrated in Figure 6.23. The first transmission provides the systematic bits with the greatest priority while subsequent re-transmissions can provide either the systematic or the parity 1 and parity 2 bits with the greatest priority.

In the case of incremental redundancy, the received bit stream is combined prior to Turbo decoding. This allows the re-transmitted data to fill some of the gaps generated by puncturing at the transmitter. The performance of incremental redundancy is similar to the performance of chase combining when the coding rate is low, i.e. there is little puncturing. This applies to the coding rates of 0.35 which were presented in Table 6.15 and Table 6.16. The performance of incremental redundancy becomes significantly better than the performance of chase combining when the coding rate is high, i.e. there is an increased quantity of puncturing. This typically applies to the higher HSDPA throughputs which make use of coding rates greater than 0.6. Incremental redundancy performs better than chase

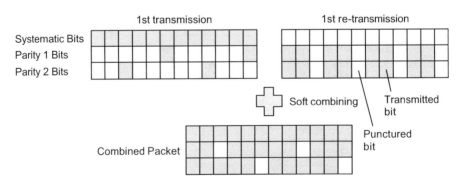

Figure 6.23   Principle of HARQ incremental redundancy

combining when the coding rate is high because channel coding gain is greater than soft combining gain. The drawbacks associated with incremental redundancy are increased complexity and increased UE memory requirements.

The UE is provided with information regarding the use of chase combining and incremental redundancy within the HS-SCCH. This information is received in advance of the HS-PDSCH data. The HS-SCCH is described within Section 6.9.1.

The HARQ protocol relies upon receiving acknowledgements from the UE. The round trip time means that these acknowledgements are not received instantaneously. Parallel HARQ processes are used to avoid the round trip time having an impact upon throughput, i.e. a second HARQ process transmits a transport block while the first HARQ process is waiting for its acknowledgement. These parallel HARQ processes are known as Stop and Wait (SAW) processes because they stop and wait for an acknowledgement prior to transmitting any further data. The concept of parallel HARQ processes is illustrated in Figure 6.24.

The first HARQ process receives data belonging to the first transport block. AMC is responsible for determining the associated number of MAC-d PDU within that transport block. The transport block is transmitted and buffered in case it is necessary to complete a re-transmission. The second HARQ process receives data during the second TTI that the scheduler assigns to the UE. This allows the second transport block to be transmitted and buffered in case of re-transmission. Likewise, the third HARQ process receives data during the third scheduled TTI and the third transport block is transmitted. In this example, the acknowledgement for the first transport block is received after the third transport

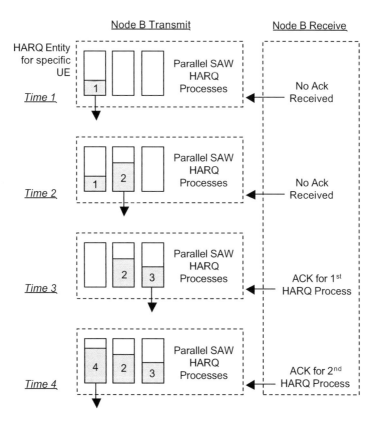

Figure 6.24   Principle of parallel Stop and Wait HARQ processes

block has been transmitted. It is assumed that a positive acknowledgement is received and the first transport block is deleted from the first HARQ process. This allows the first HARQ process to be re-used by the forth transport block. This cycle continues as further acknowledgements are received. A HARQ process completes a re-transmission rather than a new transmission if a negative acknowledgement is received. This reduces the rate at which new data is transferred, i.e. the BLER increases and the higher layer throughput decreases.

This example assumes that three parallel HARQ processes are sufficient to allow a continuous flow of data while waiting for an acknowledgement. It will be shown that this is valid when there are two simultaneously active HSDPA connections and the scheduler allocates half of the TTI to each UE. Six parallel HARQ processes are necessary to ensure a continuous flow of data when there is only a single active connection. Table 6.5 presented the number of parallel HARQ processes associated with each HSDPA UE category. Some UE categories are unable to receive data during every TTI and so they have a reduced requirement for HARQ processes. Category 1 and category 2 UE are limited to receiving data after every third TTI and so it is only necessary for them to use two parallel HARQ processes. Category 3 and category 4 UE are able to receive data after every second TTI and so it is necessary for them to use three HARQ processes. The 3GPP specifications allow a maximum of eight parallel HARQ processes to be configured for an individual UE. Figure 6.25 illustrates the timing associated with receiving an acknowledgement and demonstrates the reason for requiring six parallel HARQ processes to allow a continuous flow of data.

The Node B transmits the HS-SCCH with a timing which is advanced by two time slots relative to the HS-PDSCH. The HS-SCCH is used to signal which UE is scheduled during the subsequent TTI and to provide the information necessary to extract the transport block. The HS-PDSCH is used to transfer the transport block. Both the HS-SCCH and HS-PDSCH are explained in greater detail within Section 6.9. There is a propagation delay associated with the UE receiving the HS-SCCH and HS-PDSCH. This delay is likely to be small compared with the duration of a 2 ms TTI, e.g. a 5 km cell range corresponds to a propagation delay of 17 μs. 3GPP TS 25.211 indicates that the delay between a UE receiving the end of an HS-PDSCH subframe (2 ms TTI) and the UE starting to transmit the HS-DPCCH subframe which includes the corresponding acknowledgement is approximately 7.5 time slots (2.5 subframes). The uplink propagation delay results in the Node B receiving the acknowledgement some time later. This coincides

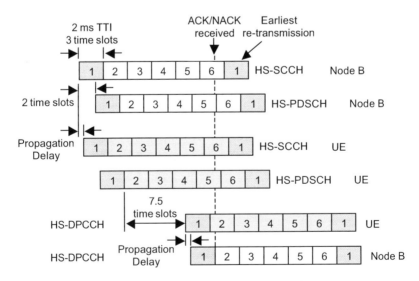

Figure 6.25   Round trip time for a HARQ acknowledgement

with the sixth HS-SCCH subframe and so the earliest time that a re-transmission or new transmission can be scheduled from the same HARQ process is during the seventh subframe.

Each transport block includes a Transmission Sequence Number (TSN) as part of the MAC-hs header. This sequence number ranges from 0 to 63 and is used by the receiver to provide in-sequence delivery to the higher layers. Each priority queue uses its own sequence numbering and so reordering at the receiver is completed on a per priority queue basis. The following descriptions assume a single priority queue and a single set of sequence numbers. Transmit and receive windows are used to avoid sequence number ambiguity. Without the use of transmit and receive windows a UE could confuse transport block y with transport block $y \pm 64$. This type of scenario could happen if the Node B misinterprets an ACK as a NACK. In this case the Node B will re-transmit transport block y while the UE believes it is receiving transport block $y + 64$. The node B uses the transmit window to limit the range of sequence numbers which can be sent. The UE uses the receive window to limit the range of sequence numbers which can be received. The maximum window size which can be used while avoiding TSN ambiguity is half of the sequence number range, i.e. 32. The 3GPP specifications allow window sizes of 4, 6, 8, 12, 16, 24 and 32. Large window sizes can increase the delay with which data is passed to the higher layers. They also increase the memory requirement of the UE reordering buffer. Small window sizes are more likely to cause the HARQ protocol to stall and force periods of DTX between re-transmissions.

Table 6.18 illustrates an example of operating within the TSN range defined by the transmit window. This example is based upon the scheduler assigning all TTI to a single HSDPA connection. The Node B transmits a series of transport blocks using consecutive sequence numbers beginning from 0. The round trip delay illustrated in Figure 6.25 means that the acknowledgement for the first transport block is received during the transmission of the sixth transport block.

The transmit sequence number range is defined as the difference between the most recent TSN and the oldest TSN which has not received a positive acknowledgement. For example, when transport block 11 is transmitted, the most recent TSN is 10 and the oldest TSN which has not yet received a positive acknowledgement is 6. This results in a TSN range of 5. The transmit sequence number range is not permitted to exceed the size of the transmit window. The range increases from 1 while the Node B waits for the first acknowledgement. Once the first acknowledgement has been received the range stabilises at a steady state value which remains constant as long as positive acknowledgements are received.

Table 6.19 illustrates an example where one transport block requires a re-transmission. The transmit TSN range starts to increase after the negative acknowledgement has been received. Transport block 31 with TSN 30 remains buffered by the HARQ process and so the difference between the most recent and oldest TSN increases as new transport blocks are transferred by the other HARQ processes.

Table 6.18 HARQ processes, TSN and transmit TSN range (one active UE)

UE	Transmission	HARQ process	TSN	Tx TSN range	ACK/NACK
1	TrBlk 1	0	0	—	
1	TrBlk 2	1	1	1	
1	TrBlk 3	2	2	2	
1	TrBlk 4	3	3	3	
1	TrBlk 5	4	4	4	
1	TrBlk 6	5	5	5	
1	TrBlk 7	0	6	5	ACK for TSN 0
1	TrBlk 8	1	7	5	ACK for TSN 1
1	TrBlk 9	2	8	5	ACK for TSN 2
1	TrBlk 10	3	9	5	ACK for TSN 3
1	TrBlk 11	4	10	5	ACK for TSN 4
1	TrBlk 12	5	11	5	ACK for TSN 5

Table 6.19 Increase in transmit TSN range caused by a negative acknowledgement

UE	Transmission	HARQ process	TSN	Tx TSN range	ACK/NACK
1	TrBlk 31	0	30	5	ACK for TSN 24
1	TrBlk 32	1	31	5	ACK for TSN 25
1	TrBlk 33	2	32	5	ACK for TSN 26
1	TrBlk 34	3	33	5	ACK for TSN 27
1	TrBlk 35	4	34	5	ACK for TSN 28
1	TrBlk 36	5	35	5	ACK for TSN 29
1	TrBlk 31	0	30	5	NACK for TSN 30
1	TrBlk 37	1	36	6	ACK for TSN 31
1	TrBlk 38	2	37	7	ACK for TSN 32
1	TrBlk 39	3	38	8	ACK for TSN 33
1	TrBlk 40	4	39	9	ACK for TSN 34
1	TrBlk 41	5	40	10	ACK for TSN 35
1	TrBlk 42	0	41	10	ACK for TSN 30
1	TrBlk 43	1	42	5	ACK for TSN 36
1	TrBlk 44	2	43	5	ACK for TSN 37

The TSN range returns to the steady state value of 5 once a positive acknowledgement has been received. This example is based upon a single re-transmission. The TSN range increases further when multiple re-transmissions are required. Multiple re-transmissions may be required if the RF channel conditions suddenly become worse after the AMC has selected a large transport block size. The re-transmitted transport block size is always equal to the original transport block size and cannot be reduced to suit the changing channel conditions.

The example presented in Table 6.19 assumes a transmit window size greater than 10. This prevents the HARQ protocol from stalling while the single re-transmission is completed. If the transmit window size had been configured with a value of 8 then the transmission of the transport blocks with TSN of 38 and above would have been delayed until after the positive acknowledgement had been received for TSN 30. This scenario is illustrated in Table 6.20.

Table 6.20 HARQ protocol stalling when using a transmit window size of 8

UE	Transmission	HARQ process	TSN	Tx TSN range	ACK/NACK
1	TrBlk 31	0	30	5	ACK for TSN 24
1	TrBlk 32	1	31	5	ACK for TSN 25
1	TrBlk 33	2	32	5	ACK for TSN 26
1	TrBlk 34	3	33	5	ACK for TSN 27
1	TrBlk 35	4	34	5	ACK for TSN 28
1	TrBlk 36	5	35	5	ACK for TSN 29
1	TrBlk 31	0	30	5	NACK for TSN 30
1	TrBlk 37	1	36	6	ACK for TSN 31
1	TrBlk 38	2	37	7	ACK for TSN 32
1	DTX (stalled)	—	—	8	ACK for TSN 33
1	DTX (stalled)	—	—	8	ACK for TSN 34
1	DTX (stalled)	—	—	8	ACK for TSN 35
1	TrBlk 39	0	38	2	ACK for TSN 30
1	TrBlk 40	1	39	2	ACK for TSN 36
1	TrBlk 41	2	40	2	ACK for TSN 37
1	TrBlk 42	3	41	3	DTX
1	TrBlk 43	4	42	4	DTX

The HARQ protocol is unable to transfer new transport blocks once the TSN range has reached the transmit window size. The TSN range remains equal to the window size until the positive acknowledgement for TSN 30 has been received. The TSN range then drops to a relatively small value because reception of acknowledgements has continued while the HARQ protocol has been stalled. The TSN range increases back towards its steady state value as new transport blocks are transmitted. This example is based upon a single re-transmission and a relatively small transmit window size. In practice, it is likely that larger transmit windows are used, but HARQ protocol stalling can still occur when more than a single re-transmission is necessary. The HARQ protocol can stall after three re-transmissions when a transmit window size of 16 is used.

The steady state transmit TSN range becomes smaller as the number of simultaneously active HSDPA connections increases. This reduces the potential for HARQ protocol stalling because the TSN range is less likely to reach the transmit window. Table 6.21 and Table 6.22 present examples based upon two and three simultaneously active connections using a round robin scheduler. Table 6.21 illustrates that when there are two simultaneously active connections, three HARQ processes are used by each connection and the TSN range has a steady state value of 2. Table 6.22 illustrates that when there are three simultaneously active connections, two HARQ processes are used by each connection and the TSN range has a steady state value of 1.

Table 6.21 HARQ processes, TSN and window size (two active UE)

UE	Transmission	HARQ process	TSN	Tx TSN range	ACK/NACK
1	TrBlk 18	0	17	2	ACK for TSN 14
2	TrBlk 33	2	32	2	ACK for TSN 29
1	TrBlk 19	1	18	2	ACK for TSN 15
2	TrBlk 34	0	33	2	ACK for TSN 30
1	TrBlk 20	2	19	2	ACK for TSN 16
2	TrBlk 35	1	34	2	ACK for TSN 31
1	TrBlk 21	0	20	2	ACK for TSN 17
2	TrBlk 36	2	35	2	ACK for TSN 32
1	TrBlk 22	1	21	2	ACK for TSN 18
2	TrBlk 37	0	36	2	ACK for TSN 33
1	TrBlk 23	2	22	2	ACK for TSN 19
2	TrBlk 38	1	37	2	ACK for TSN 34

Table 6.22 HARQ processes, TSN and window size (three active UE)

UE	Transmission	HARQ process	TSN	Tx TSN range	ACK/NACK
1	TrBlk 18	0	17	1	ACK for TSN 15
2	TrBlk 32	1	31	1	ACK for TSN 29
3	TrBlk 8	0	7	1	ACK for TSN 5
1	TrBlk 19	1	18	1	ACK for TSN 16
2	TrBlk 33	0	32	1	ACK for TSN 30
3	TrBlk 9	1	8	1	ACK for TSN 6
1	TrBlk 20	0	19	1	ACK for TSN 17
2	TrBlk 34	1	33	1	ACK for TSN 31
3	TrBlk 10	0	9	1	ACK for TSN 7
1	TrBlk 21	1	20	1	ACK for TSN 18
2	TrBlk 35	0	34	1	ACK for TSN 32
3	TrBlk 11	1	10	1	ACK for TSN 8

The preceding examples assume that the Node B receives either a positive or negative acknowledgement for each transport block that has been transmitted. It is also possible that the UE fails to receive the HS-SCCH in the downlink direction and so does not attempt to decode the HS-PDSCH. This would mean that the UE does not return an acknowledgement. In this case, the Node B would experience a period of DTX when expecting to receive an acknowledgement. The Node B treats this period of DTX in the same way as a negative acknowledgement, i.e. a re-transmission is scheduled. These periods of DTX can also be used to trigger an increase in the HS-SCCH transmit power.

Once a UE has successfully received a transport block it is passed to the MAC-hs layer where it is placed within a reordering buffer. The reordering buffer is used to ensure that the RLC layer receives PDU in the correct order, i.e. there is in-sequence delivery. Transport blocks are passed from the reordering buffer to the disassembly unit once all preceding transport blocks have been received. For example, if sequence numbers 2, 3 and 4 have been received while sequence number 1 requires a re-transmission, the MAC-hs layer waits for sequence number 1 prior to passing sequence numbers 1, 2, 3 and 4 to the disassembly unit. The disassembly unit extracts the MAC-d PDU and forwards them to the MAC-d layer for subsequent transfer to the RLC layer. There is a danger that the procedure can stall if the Node B erroneously detects a NACK as an ACK, i.e. the UE will wait for a re-transmission which the Node B will never send. Alternatively, the procedure could stall if the UE fails to receive a transport block after the maximum number of re-transmissions. 3GPP have defined the T1_TSN timer to avoid these scenarios. This timer is started by the UE whenever a transport block is received with a sequence number greater than the next expected sequence number. If T1_TSN expires then all of the transport blocks, from the missing transport block which triggered the start of T1_TSN to the next missing transport block are passed to the disassembly unit. An RLC re-transmission would then be necessary to recover the missing data.

6.8.5 Generation of MAC-hs PDU

A MAC-hs PDU represents an HSDPA transport block. A single HSDPA connection can transfer a maximum of one transport block during each TTI. Code multiplexing allows transport blocks belonging to separate HSDPA connections to be transferred during the same TTI. In this case, a different set of HS-PDSCH channelisation codes are used to transfer each transport block. All of the MAC-d PDU within a single transport block must belong to the same priority queue. The structure of a MAC-hs PDU is illustrated in Figure 6.26.

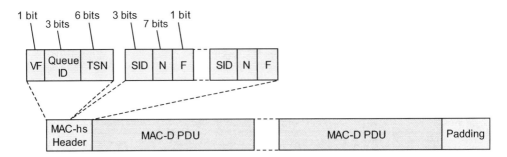

Figure 6.26   Structure of a MAC-hs PDU

The header is divided into a fixed-length section and a variable-length section. The fixed-length section includes the Version Flag (VF), Queue ID, and Transmission Sequence Number (TSN). The variable-length section includes one or more instances of the Size Index Identifier (SID), Number of MAC-d PDU (N) and header end point Flag (F). If all of the MAC-d PDU within the MAC-hs PDU have an equal size then it is sufficient to have a single instance of the SID, N and F header fields. If there are multiple MAC-d PDU sizes then it is necessary to have an instance of the SID, N and F header fields for each MAC-d PDU size.

The VF header field has a length of 1 bit and a fixed value of 0. This header field is not used by the release 5 nor release 6 versions of the 3GPP specifications. It has been included to allow future changes to the format of the MAC-hs header, i.e. if the receiver decodes a value of 0 then it can deduce that the header format illustrated in Figure 6.26 has been used, whereas it can deduce that a different (not yet specified) header format has been used if a value of 1 is decoded. The queue ID header field has a length of 3 bits and is included to allow the receiving MAC-hs entity to organise the received packets into independent reordering queues. These queues correspond to the priority queues at the transmitter. The TSN header has a length of 6 bits and allows the receiving MAC-hs entity to support in-sequence delivery to the higher layers. The number of bits assigned to the TSN allows a sequence number range from 0 to 63. Each priority queue uses its own sequence numbering.

The SID header has a length of 3 bits and is used to specify the length of the associated MAC-d PDU. The size index corresponds to one of the values configured during HSDPA connection establishment. One or more MAC-d PDU sizes can be signalled to the UE within a Radio Bearer Setup or Radio Bearer Reconfiguration message. An example extracted from a Radio Bearer Reconfiguration message is presented in Log File 6.9.

In this example, 2 MAC-d sizes are configured. The first has an index of 0 and a size of 336 bits, whereas the second has an index of 1 and a size of 656 bits. The Node B can be informed of the same information using an NBAP Radio Link Setup Request or Radio Link Reconfiguration Prepare message. It is possible to configure a maximum of eight different MAC-d PDU sizes.

The N header field has a length of 7 bits and is used to specify the number of MAC-d PDU which are included with a length corresponding to the preceding SID. Although 7 bits are assigned to this field it is assumed that the maximum number of MAC-d PDU transmitted within a single TTI is 70. This results a maximum RLC throughput of 11.2 Mbps when the MAC-d PDU size is 336 bits. The F header field has a length of 1 bit and is used as a flag to indicate whether the subsequent bits represent the start of another set of SID, N and F header fields, or the start of the MAC-d PDU. A value of 1 indicates the start of the MAC-d PDU.

The total length of the MAC-hs header is equal to $10 + (n \times 11)$, where n is the number of different MAC-d PDU sizes included within the PDU. If only a single size is included then the header length is 21 bits. The overhead generated by the MAC-hs header depends upon the size of the payload, i.e. the number of MAC-d PDU and their size. Padding is included at the end of the MAC-hs PDU if the accumulated length of the header and MAC-d PDU does not equal the size of the transport block assigned by the AMC function.

```
mac-d-PDU-SizeInfo-List
    mac-d-PDU-SizeInfo-List value 1
        mac-d-PDU-Size        : 336
        mac-d-PDU-Index       : 0
    mac-d-PDU-SizeInfo-List value 2
        mac-d-PDU-Size        : 656
        mac-d-PDU-Index       : 1
```

Log File 6.9 Example MAC-d PDU size indices

6.9 Physical Channels

- The introduction of HSDPA requires the HS-PDSCH and HS-SCCH in the downlink direction combined with the HS-DPCCH in the uplink direction.
- All three physical channels are based upon 2 ms subframes generated from three consecutive time slots, i.e. a 2 ms Transmission Time Interval (TTI).
- The first time slot of the HS-SCCH includes channelisation code set, modulation scheme, and UE identity information. The last two time slots include transport block size index, HARQ process, redundancy and constellation version and new data indicator information.
- The HS-SCCH frame structure precedes the HS-PDSCH frame structure by two time slots to allow UE to extract and act upon information included within the first HS-SCCH time slot prior to the start of the corresponding HS-PDSCH sub-frame.
- A single HS-SCCH supports all HSDPA connections within a cell when code multiplexing is not used. Multiple HS-SCCH are required when code multiplexing is used.
- The HS-SCCH is always spread using a spreading factor of 128.
- Power control can reduce the average transmit power allocation of the HS-SCCH. The transmit power can be changed on a per UE basis every 2 ms.
- The HS-PDSCH is responsible for providing the high data rates associated with HSDPA. The HS-PDSCH is used to transfer application data and can also be used to transfer the downlink Signalling Radio Bearers (SRB).
- A single HS-PDSCH is defined by its channelisation code which always has a spreading factor of 16. Increasing the number of HS-PDSCH increases the maximum achievable throughput. A maximum of 15 HS-PDSCH can be assigned to a single connection. The use of 15 HS-PDSCH is only possible when there is little other traffic using the code tree.
- Physical layer processing for the HS-PDSCH includes CRC attachment, bit scrambling, code block segmentation, channel coding, HARQ functionality, physical channel segmentation, interleaving, constellation re-arrangement and physical channel mapping.
- The transmit power allocated to the set of HS-PDSCH is shared equally between those physical channels. A dynamic transmit power allocation offers the benefit of allowing HSDPA to make use of any unused transmit power capability.
- The HS-DPCCH is responsible for transferring MAC-hs acknowledgements and Channel Quality Indicators (CQI). The MAC-hs acknowledgement occupies the first time slot whereas the CQI occupies the second and third time slots.
- The MAC-hs acknowledgement does not include any information regarding the identity of the relevant HARQ process. The Node B deduces the relevant HARQ process identity from the timing of the acknowledgement.
- The ACK, NACK and CQI can be configured with different transmit powers relative to the DPCCH. Increasing their power improves reliability but increases their overhead.
- A Node B will re-transmit a transport block unnecessarily if it misinterprets an ACK as either a NACK or a period of DTX. The Node B will fail to re-transmit a transport block if it misinterprets a NACK or a period of DTX as an ACK.
- It is useful for a Node B to distinguish between a NACK and a period of DTX when incremental redundancy is used. This can determine whether or not the systematic bits are provided with priority during the rate matching for a re-transmission.
- The release 6 version of the 3GPP specifications introduce an optional change to the ACK/NACK signalling aimed at improving reliability. The change allows UE to send preamble and post-amble sequences within the acknowledgement section of the HS-DPCCH.
- The HS-DPCCH is always spread using a spreading factor of 256.
- Key 3GPP specifications: TS 25.211, TS 25.212, TS 25.213, TS 25.214, TS 25.104.

Table 6.23 Physical channels used by an HSDPA connection

Physical channels	DPCH	HSDPA (SRB on DPDCH)	HSDPA and HSUPA (SRB on DPDCH)	HSDPA and HSUPA (SRB on HS-PDSCH)
Downlink				
HS-PDSCH		√	√	√
HS-SCCH		√	√	√
DPCCH	√	√	√	
DPDCH	√	√	√	
F-DPCH				√
Uplink				
DPCCH	√	√	√	√
DPDCH	√	√	√	
HS-DPCCH		√	√	√
E-DPCCH			√	√
E-DCDCH			√	√

Table 6.23 compares the physical channels used by an HSDPA connection with those used by a DPCH connection. A DPCH connection uses DPCCH and DPDCH physical channels in both the uplink and downlink directions. These physical channels are described within Section 3.3.9. The introduction of HSDPA requires the HS-PDSCH and HS-SCCH in the downlink direction combined with the HS-DPCCH in the uplink direction. These three physical channels are described within this section. The requirement for the DPCCH and DPDCH remains if only the user plane data is transferred using HSDPA while the SRB is transferred using an associated DPCH. The introduction of HSUPA does not change the physical channel types used by HSDPA. The requirement for the DPDCH is removed if the SRB is transferred using HSDPA in the downlink direction and HSUPA in the uplink direction. In this case, the downlink DPCCH can be swapped for the more efficient F-DPCH while the uplink DPCCH remains. The F-DPCH is described within Section 3.3.10.

6.9.1 High Speed Shared Control Channel (HS-SCCH)

The HS-SCCH is introduced within the release 5 version of the 3GPP specifications. It is a common downlink physical channel and is used to transfer HSDPA control information. This control information identifies the UE to which the scheduler has allocated the subsequent HSDPA TTI. This triggers the appropriate UE to decode the HS-PDSCH. The HS-SCCH also includes control information which supports the decoding process. A single HS-SCCH supports all HSDPA connections within a cell when code multiplexing is not used. Multiple HS-SCCH are required when code multiplexing is used. Figure 6.27 illustrates the concept of code multiplexing.

This example is based upon using code multiplexing to simultaneously schedule 3 HSDPA connections. A separate HS-SCCH is required for each UE that is scheduled during the same TTI. This means that three HS-SCCH are required for this example. The 3GPP specifications allow a maximum of four HS-SCCH to be configured within a single cell, i.e. a maximum of four UE can be simultaneously scheduled during the same TTI. Code multiplexing increases the potential to use a larger percentage of the 15 HS-PDSCH. Without code multiplexing, the maximum number of HS-PDSCH is limited by the UE capability. If the population of UE support a maximum of five HS-PDSCH codes then AMC will never be able to allocate more than one-third of the total HS-PDSCH code resource. This could be an appropriate solution when HSDPA shares an RF carrier with the DPCH connections. In this case, both the channelisation code tree and downlink transmit power resources are shared and there is less flexibility

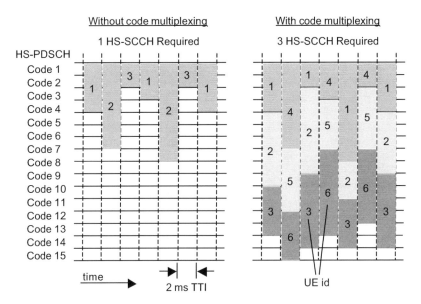

Figure 6.27 Concept of HS-PDSCH code multiplexing

to allocate increased resources to HSDPA. There is greater potential to benefit from code multiplexing if a dedicated RF carrier is allocated. In this case, HSDPA connections can make use of the full channelisation code tree and the total downlink transmit power capability.

The structure of the HS-SCCH is illustrated in Figure 6.28. A single 2 ms HS-SCCH subframe is generated from three consecutive time slots. Physical layer data is mapped onto both in-phase and quadrature modulation branches. 3GPP TS 25.213 specifies that the HS-SCCH is always spread using a spreading factor of 128. This results in a channel symbol rate of 30 ksps and a corresponding channel bit rate of 60 kbps, i.e. 120 bits of physical layer data every 2 ms TTI.

The timing of the HS-SCCH relative to the CPICH and HS-PDSCH is illustrated in Figure 6.29. The HS-SCCH frame structure is time aligned with the CPICH, but precedes the frame structure of the HS-PDSCH by two time slots. The HS-PDSCH frame timing is delayed by two time slots to allow time for UE to extract and act upon information included within the first HS-SCCH time slot prior to the start of the corresponding HS-PDSCH subframe.

The content of the HS-SCCH is presented in Table 6.24. This table also indicates which information is included within the first HS-SCCH time slot, i.e. information which is decoded prior to the start of the corresponding HS-PDSCH subframe.

The channelisation code set information defines the block of HS-PDSCH channelisation codes assigned to a specific HSDPA connection during the subsequent 2 ms TTI. HS-PDSCH channelisation codes are always assigned in a contiguous block, i.e. (SF16,O) to (SF16,$O + P - 1$) where O represents the index belonging to the first channelisation code within the set and P represents the number of channelisation codes. The 7 bits of information are divided into two sections. The first 3 bits define the Code Group Indicator (CGI) whereas the last 4 bits define the Code Offset Indicator (COI). These variables are defined using the equations below.

$$\text{Code group indicator} = \min(P - 1, 15 - P)$$
$$\text{Code offset indicator} = |O - 1 - \lfloor P/8 \rfloor \times 15|$$

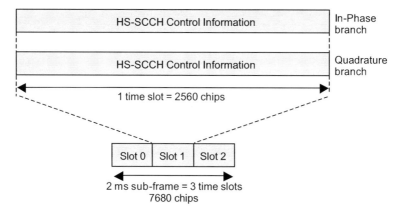

Figure 6.28   Structure of the HS-SCCH

For example, if the HS-PDSCH channelisation code allocation during a specific subframe is (SF16,11) to (SF16,15) then $O = 11$ and $P = 5$ which results in a CGI of 4 and a COI of 10. Both the CGI and COI are required to deduce either O or P. The complete set of all possible CGI and COI values is presented in Table 6.25.

The number of combinations of CGI and COI decreases as the size of the channelisation code allocation increases because there is less flexibility in terms of where the allocation can be placed within the code tree. When 15 HS-PDSCH channelisation codes are allocated there is only a single position which is possible for those 15 codes, i.e. (SF16,1) to (SF16,15).

The modulation scheme information included within the HS-SCCH subframe indicates whether the HS-PDSCH are modulated using QPSK or 16QAM. A value of 0 indicates QPSK whereas a value of 1 indicates 16QAM. The potential to also signal the use of 64QAM is introduced within the release 7 version of 3GPP TS 25.212.

The UE identity corresponds to the HS-DSCH Radio Network Temporary Identity (H-RNTI). The controlling RNC allocates the H-RNTI during connection establishment and potentially during the serving cell change procedure. Its value is unique within a cell so there is no ambiguity regarding the UE being scheduled at any point in time.

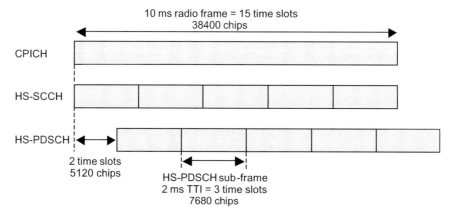

Figure 6.29   Timing of the HS-SCCH relative to the CPICH and HS-PDSCH

Table 6.24 Content of the HS-SCCH

	Number of bits	Subframe location (time slot)
Channelisation code set	7	1st
Modulation scheme	1	1st
UE identity	16	1st
Transport block size index	6	2nd and 3rd
HARQ process	3	2nd and 3rd
Redundancy and constellation version	3	2nd and 3rd
New data indicator	1	2nd and 3rd

The channelisation code set and modulation scheme information (8 bits) has 8 tail bits attached and is then channel coded using rate 1/3 convolutional coding. The resultant 48 channel coded bits are punctured to generate 40 bits. These 40 bits are then masked by the UE identity before being mapped onto the first time slot of the HS-SCCH physical channel.

The transport block size index included within the second and third time slots of the HS-SCCH can be used in combination with the channelisation code set information to calculate the transport block size. This calculation is described in Section 6.8.3. The 6 bit size index has a range from 0 to 63.

The HARQ process information defines the HARQ process responsible for managing the transport block. The 3 bits of information signal a HARQ process identity ranging from 0 to 7. HARQ processes are described within Section 6.8.4.

The redundancy and constellation version information references a row within a look-up table. This look-up table is specified within 3GPP TS 25.212. It defines two information elements if QPSK is being used and three information elements if 16QAM is being used. Table 6.26 presents the content of the look-up table.

The s information element is used to assign greater priority to either the systematic bits or parity bits during rate matching. The Turbo coding procedure generates both systematic and parity bits. The systematic bits are provided with priority if $s = 1$, while the parity 1 and parity 2 bits are provided with priority if $s = 0$. The parity bits are punctured in preference to the systematic bits when the systematic

Table 6.25 Coding for the channelisation code set (code group indicator, code offset indicator)

Number of codes (P)	Starting index for block of channelisation codes (O)														
	1	2	3	4	5	6	7	8	9	10	11	12	13	14	15
1	0,0	0,1	0,2	0,3	0,4	0,5	0,6	0,7	0,8	0,9	0,10	0,11	0,12	0,13	0,14
2	1,0	1,1	1,2	1,3	1,4	1,5	1,6	1,7	1,8	1,9	1,10	1,11	1,12	1,13	
3	2,0	2,1	2,2	2,3	2,4	2,5	2,6	2,7	2,8	2,9	2,10	2,11	2,12		
4	3,0	3,1	3,2	3,3	3,4	3,5	3.6	3,7	3,8	3,9	3,10	3,11			
5	4,0	4,1	4,2	4,3	4,4	4,5	4,6	4,7	4,8	4,9	4,10				
6	5,0	5,1	5,2	5,3	5,4	5,5	5,6	5,7	5,8	5,9					
7	6,0	6,1	6,2	6,3	6,4	6,5	6,6	6,7	6,8						
8	7,15	7,14	7,13	7,12	7,11	7,10	7,9	7,8							
9	6,15	6,14	6,13	6,12	6,11	6,10	6,9								
10	5,15	5,14	5,13	5,12	5,11	5,10									
11	4,15	4.14	4,13	4,12	4,11										
12	3,15	3,14	3,13	3,12											
13	2,15	2,14	2,13												
14	1,15	1,14													
15	0,15														

Table 6.26 Mapping of redundancy and constellation version values

Redundancy and constellation version	QPSK		16QAM		
	s	r	s	r	b
0	1	0	1	0	0
1	0	0	0	0	0
2	1	1	1	1	1
3	0	1	0	1	1
4	1	2	1	0	1
5	0	2	1	0	2
6	1	3	1	0	3
7	0	3	1	1	0

bits have priority. Selecting a redundancy and constellation version from Table 6.26 depends upon whether or not HARQ incremental redundancy and constellation re-arrangement are supported. Constellation re-arrangement is only applicable to 16QAM. Table 6.27 presents examples of the redundancy and constellation versions which could be applied when using chase combining or incremental redundancy with either QPSK or 16QAM. This table assumes three MAC-hs re-transmissions so there are four redundancy and constellation versions associated with each entry.

The redundancy and constellation version must be the same for each MAC-hs transmission if chase combining is used without constellation re-arrangement. Version 0 is selected to ensure that the systematic bits are handled with greater priority during rate matching. If incremental redundancy is supported then the redundancy and constellation versions can be selected to allow both the s and r variables to change. If constellation re-arrangement is supported the versions can be selected to allow the b variable to change. Figure 6.30 illustrates the impact of constellation re-arrangement upon the bits mapped onto each 16QAM symbol.

The mapping between bits and symbols remains unchanged when $b = 0$. The two least significant bits are swapped with the two most significant bits when $b = 1$, whereas the two least significant bits are inverted when $b = 2$. The two most significant bits are inverted and the two least significant bits are swapped with the two most significant bits when $b = 3$. Constellation re-arrangement can improve performance because the probability of error belonging to each of the four bits mapped onto a 16QAM symbol is non-uniform. The probability of error tends to be lower for the two most significant bits because these bits are correctly deduced as long as the correct quadrant is identified. Changing the mapping between the input bit sequence and the 16QAM symbols helps to provide a more uniform probability of error.

The new data indicator within the HS-SCCH informs the UE of whether or not the transport block is a new transmission or a re-transmission. The transmission sequence number within the MAC-hs header cannot be used for this purpose because it is only available after decoding the transport block.

Table 6.27 Example redundancy and constellation versions (3 MAC-hs re-transmissions)

		Chase combining	Incremental redundancy
QPSK	Without constellation re-arrangement	0,0,0,0	0,1,2,3
16QAM	Without constellation re-arrangement	0,0,0,0	0,1,7,0
	With constellation re-arrangement	0,4,5,6	0,1,2,3

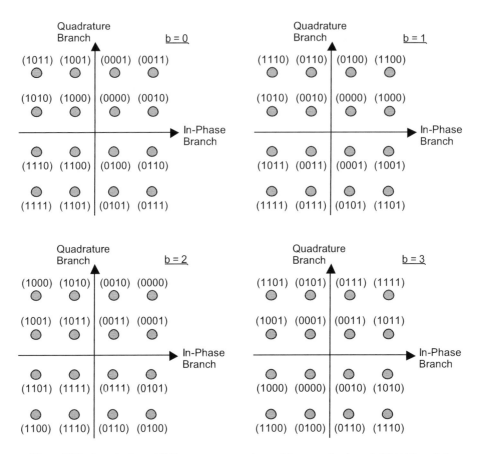

Figure 6.30   Impact of constellation re-arrangement upon bits mapped onto each 16QAM symbol

The transport block size index, HARQ process, redundancy and constellation version and new data indicator information (13 bits) have a 16 bit CRC attached. The resultant 29 bits are masked by the UE identity, extended using 8 tail bits and subsequently channel coded using rate 1/3 convolutional coding. The block of 111 bits is then punctured to generate the 80 bits which are mapped onto the second and third time slots of the HS-SCCH physical channel.

The HS-SCCH physical channel represents an overhead and so the allocated downlink transmit power should be minimised. The transmit power requirement will be increased if code multiplexing is used and multiple HS-SCCH are active. Power control can be used to reduce the transmit power allocation when UE are relatively close to the serving cell because it is not necessary for all UE within the cell to receive the HS-SCCH during every subframe. The transmit power can be changed on a per UE basis every 2 ms. The reported CQI values can be used to identify an appropriate HS-SCCH transmit power for each UE. In addition, uplink reception of either a positive or negative MAC-hs acknowledgement indicates that the UE has successfully received the HS-SCCH and at least attempted to decode the corresponding HS-PDSCH subframe. This can be used as an additional mechanism to adjust the HS-SCCH transmit power although it assumes that the HS-DPCCH coverage used to transfer the acknowledgements is not a limiting factor.

6.9.2 High Speed Physical Downlink Shared Channel (HS-PDSCH)

The HS-PDSCH common physical channel is introduced within the release 5 version of the 3GPP specifications. It is responsible for providing the high data rates associated with HSDPA. Each HSDPA connection is able to generate a maximum of one transport block during each 2 ms TTI. That transport block is processed by the physical layer prior to being mapped onto one or more HS-PDSCH. A single HS-PDSCH is defined by its channelisation code which always has a spreading factor of 16. An HSDPA connection making use of 5 HS-PDSCH requires 5 channelisation codes. Increasing the number of HS-PDSCH increases the maximum achievable throughput. 3GPP TS 25.213 specifies that the HS-PDSCH and HS-SCCH are always assigned channelisation codes from the same code tree. This means that there can be a maximum of 15 HS-PDSCH assigned to a single connection. The 16th channelisation code with spreading factor 16 is always blocked by the HS-SCCH. In general, the HS-PDSCH and HS-SCCH share the same code tree as the common channels, i.e. the code tree belonging to the primary scrambling code, and so the common channels also block the 16th channelisation code. In practice, the use of 15 HS-PDSCH is only possible when there is either no or very little other traffic using the code tree. Figure 6.31 illustrates the code tree occupancy when 15 HS-PDSCH, 3 HS-SCCH and the common channels are allocated.

Table 3.35 within Section 3.3.9 presents the code tree occupancy for a range of HS-PDSCH and HS-SCCH allocations. The scenario in Figure 6.31 excludes all DPCH and F-DPCH allocations, but still generates an occupancy of 99.3%. Disabling the use of code multiplexing and transmitting only a single HS-SCCH would help to decrease the code tree occupancy, but would reduce the flexibility with which the HS-PDSCH can be allocated, i.e. it would only be possible to allocate them to a single UE during each TTI.

The structure of the HS-PDSCH is illustrated in Figure 6.32. A single 2 ms HS-PDSCH subframe is generated from three consecutive time slots. Physical layer data is mapped onto both in-phase and quadrature branches. The spreading factor of 16 results in a modulation symbol rate of 240 ksps. This corresponds to a channel bit rate of 480 kbps when QPSK is used and 960 kbps when 16QAM is used. The channel bit rate increases to 1440 kbps when 64QAM is used. These figures are summarised within Table 6.28 which presents the three slot formats belonging to the three modulation schemes.

These channel bit rates are applicable to a single HS-PDSCH. If an HSDPA connection is using multiple HS-PDSCH then the channel bit rate increases in direct proportion, e.g. the channel bit rate increases to 4800 kbps if 16QAM is used in combination with 5 HS-PDSCH.

The HS-PDSCH is used to transfer downlink application data and can also be used to transfer the downlink Signalling Radio Bearers (SRB). The application data and SRB generate MAC-d PDU which are packaged within an HS-DSCH transport block. AMC determines the number of MAC-d PDU within the transport block. Table 6.29 presents the physical layer processing for three example HS-DSCH transport block sizes. These transport block sizes correspond to RLC throughputs of 800 kbps, 1.6 Mbps and 3.36 Mbps. The first two examples use QPSK whereas the third uses 16QAM.

The physical layer starts by attaching Cyclic Redundancy Check (CRC) bits. These bits are used by the receiver to detect whether or not a transport block includes any errors after the channel decoding process. 3GPP TS 25.212 specifies that HSDPA connections always use a 24 bit CRC. This represents the longest CRC length specified by 3GPP, i.e. the error detection capability is maximised for HSDPA.

The transport block and CRC bits are then scrambled to randomise their order. This process is completed in case the data includes large sections of consecutive ones or consecutive zeros. This type of sequence can have a negative impact upon the demodulation performance of 16QAM. A UE receiving 16QAM must apply an amplitude threshold to deduce whether a received symbol belongs to the inner or outer ring of constellation points. Large sections of consecutive ones or consecutive zeros effectively add a DC offset to the received signal and the average value is no longer at the centre of the modulation constellation. This type of offset makes it difficult for UE to derive an accurate amplitude threshold.

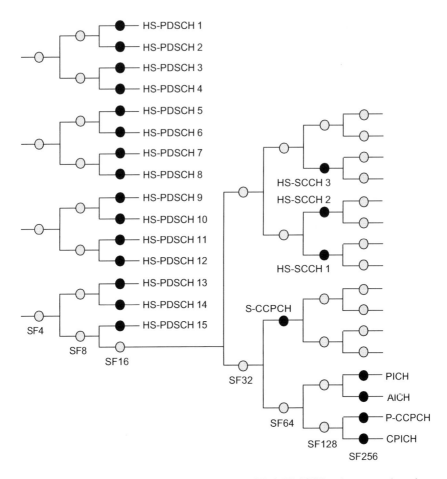

Figure 6.31 Code tree occupancy with 15 HS-PDSCH, 3 HS-SCCH and common channels

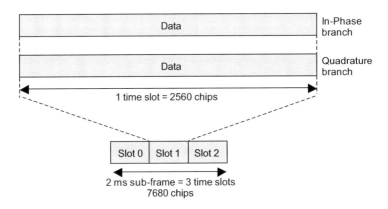

Figure 6.32 Structure of the HS-PDSCH

Table 6.28 HS-PDSCH slot formats

Slot format	Modulation scheme	Spreading factor	Channel symbol rate (ksps)	Channel bit rate (kbps)	Bits per subframe	Bits per time slot
0	QPSK	16	240	480	960	320
1	16QAM	16	240	960	1920	640
2[*]	64QAM	16	240	1440	2880	960

[*]Introduced by release 7 of the 3GPP specifications

The physical layer evaluates the requirement for code block segmentation after bit scrambling. Segmentation is necessary if the resulting code blocks are too large for the subsequent channel coding procedure. The maximum code block size for turbo coding is 5114 bits. Table 6.29 illustrates that code block segmentation is not necessary when the transport block size is 1711 or 3440 bits, but is necessary when the transport block size is 7168 bits. Code block segmentation divides the code block into smaller blocks of equal size. The number of blocks generated is the minimum necessary to ensure that the maximum size of 5114 bits is not exceeded. Padding is added if the original code block size is not a multiple of the number of blocks after segmentation.

3GPP TS 25.212 specifies that HSDPA connections always use 1/3 rate turbo coding. The coding rate of 1/3 means that the number of bits after channel coding is three times the number of bits before channel coding. Turbo coding generates three streams of output data: systematic, parity 1 and parity 2. The systematic bits are equal to the input bits whereas the parity 1 and parity 2 bits are calculated from the input bits to generate redundancy. 12 tail bits are added after the channel coding process. Turbo coding is completed separately for each code block if code block segmentation has been completed.

The Hybrid ARQ (HARQ) functionality is responsible for rate matching. As explained in Section 6.8.4, this can be based upon either chase combining or incremental redundancy. The s and r bits within the 'redundancy and constellation version' section of the HS-SCCH define the puncturing pattern and whether puncturing focuses upon the systematic bits or parity bits. The number of bits after rate matching is exactly equal to the capacity offered by the set of HS-PDSCH. This capacity depends upon the modulation scheme and the number of HS-PDSCH. Table 6.28 presents the number of bits offered by a single HS-PDSCH as a function of the modulation scheme. Table 6.29 illustrates that in this example, rate matching generates an equal number of bits for transport block sizes of 1711 and 3440 bits. Both transport block sizes are assumed to be transferred using QPSK and 5 HS-PDSCH. This

Table 6.29 HS-DSCH transport block processing (three example transport block sizes)

	HSDPA RLC Throughput		
	800 kbps (QPSK)	1.60 Mbps (QPSK)	3.36 Mbps (16QAM)
Transport block size	1711	3440	7168
CRC attachment	$1711 + 24 = 1735$	$3440 + 24 = 3464$	$7168 + 24 = 7192$
Bit scrambling	1735	3464	7192
Code block segmentation	Not necessary	Not necessary	2×3840
Channel coding	$(1735 \times 3) + 12 = 5217$	$(3464 \times 3) + 12 = 10404$	$2 \times [(3840 \times 3) + 12] = 23064$
HARQ functionality	4800	4800	9600
Physical channel segmentation	5×960	5×960	5×1920
HS-DSCH interleaving	5×960	5×960	5×1920
Const. re-arrangement	Not applicable	Not applicable	5×1920
Physical channel mapping	5×960	5×960	5×1920

means that greater puncturing must be applied to the channel coded bits belonging to the transport block size of 3440 bits. Rate matching generates twice as many bits for the transport block size of 7168 bits because it is assumed that 16QAM is used rather than QPSK.

Physical channel segmentation distributes the data across the set of HS-PDSCH. All of the examples presented in Table 6.29 are based upon 5 HS-PDSCH and so the data is divided into five sections. The data would have been divided into three sections if an example had been based upon three HS-PDSCH. All subsequent physical layer processing is completed on a per physical channel basis, i.e. there are five parallel streams of processing when five HS-PDSCH are used.

Interleaving for the HS-DSCH is the same as the second stage of interleaving for the DCH, i.e. block interleaving with column reordering (described in Section 2.6.1). The interleaving array has dimensions of 32 rows by 30 columns and so can accommodate 960 bits. This equals the capacity of a single HS-PDSCH when QPSK is used. Two interleavers are used in parallel for 16QAM and three interleavers are used in parallel for 64QAM.

Interleaving is followed by constellation re-arrangement. This is only applicable to 16QAM and 64QAM. Constellation re-arrangement is described in Section 6.9.1. Finally, the resultant data stream is mapped onto the physical channels ready for spreading and scrambling prior to modulating the RF carrier.

The downlink transmit power allocated to the set of HS-PDSCH is shared equally between those physical channels. For example, the transmit power of a single HS-PDSCH would be $2W$ if the total HS-PDSCH transmit power is $8W$ and four HS-PDSCH are allocated during a specific TTI. This transmit power could increase to $4W$ during the next TTI if AMC allocates only two HS-PDSCH. The total transmit power of the HS-PDSCH may be fixed or may be power controlled by the Node B. If the transmit power is fixed then it is likely that an RNC databuild parameter would be used to configure the transmit power allocation. If the transmit power is dynamic then an RNC databuild parameter could be used to configure the maximum transmit power allocation. A dynamic transmit power allocation offers the benefit of allowing HSDPA to make use of any unused transmit power capability. This helps to maximise the downlink signal-to-noise ratio for HSDPA and allows AMC to allocate increased throughputs. The drawback of allowing HSDPA to use any unused transmit power is that the non-HSDPA connections will tend to experience an increased downlink interference floor. Ideally, this would not be an issue if the network has been planned to cope with Node B transmitting at their maximum power. In practice, there may be areas of poor dominance where an increased interference floor has a negative impact upon performance. The impact would be similar if DPCH connections fully loaded a cell and generated an increased interference floor.

6.9.3 High Speed Dedicated Physical Control Channel (HS-DPCCH)

The HS-DPCCH uplink dedicated physical channel is introduced within the release 5 version of the 3GPP specifications. It is responsible for transferring MAC-hs acknowledgements and Channel Quality Indicators (CQI). MAC-hs acknowledgements are used to inform the Node B of whether or not a transport block belonging to a specific HARQ process has been received successfully. This allows the Node B to decide whether or not a re-transmission is necessary. The CQI values are used to inform the Node B of the channel conditions experienced by the UE. These values are used by AMC when allocating an appropriate HSDPA throughput.

The structure of the HS-DPCCH is illustrated in Figure 6.33. Three time slots are used to generate a 2 ms subframe. The MAC-hs acknowledgement occupies the first time slot whereas the CQI occupies the second and third time slots. The acknowledgement is transmitted first to ensure that the Node B receives this information with minimum delay. The CQI is allocated a larger section of the subframe because it requires the transmission of a numerical value ranging from 0 to 30 whereas the acknowledgement requires only a simple positive or negative indication.

The HS-DPCCH is always spread using a spreading factor of 256. The channelisation code index depends upon the number of uplink DPDCH. The HS-DPCCH is spread using channelisation code (SF256,64) if there is a single uplink DPDCH. This corresponds to the case where HSUPA does not

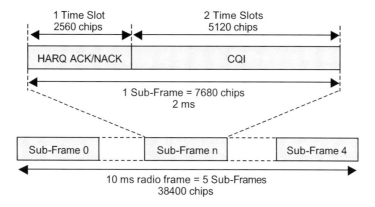

Figure 6.33   Structure of the HS-DPCCH

transfer both the uplink application data and uplink SRB. Section 3.3.9 specified that the uplink DPDCH is spread using channelisation code (SFx,x/4). This means that the HS-DPCCH and DPDCH share the same branch of the code tree and actually share the same code when the DPDCH has a spreading factor of 256. Orthogonality between the two physical channels is maintained by transmitting the DPDCH on the in-phase modulation branch and the HS-DPCCH on the quadrature modulation branch. Figure 6.34 illustrates the HS-DPCCH and DPDCH on separate modulation branches.

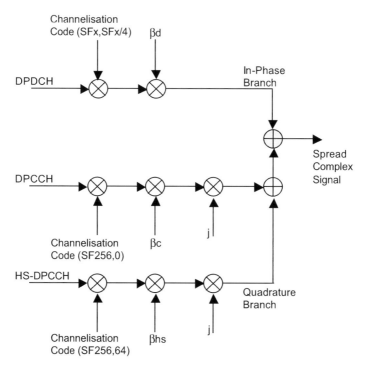

Figure 6.34   Uplink spreading for the HS-DPCCH and DPCCH when there is a single DPDCH

Table 6.30   HS-DPCCH slot format

Slot format	Spreading factor	Channel symbol rate	Channel bit rate	Bits per subframe	Bits per time slot
0	256	15 ksps	15 kbps	30	10

The HS-DPCCH is spread using channelisation code (SF256,33) if there are no uplink DPDCH. This corresponds to the case where HSUPA is used to transfer both the uplink application data and uplink SRB. In this case, the HS-DPCCH is still transmitted using the quadrature branch while the E-DPCCH physical channel is transmitted using the in-phase branch and the E-DPDCH can be transmitted using both the in-phase and quadrature branches. The E-DPCCH and E-DPDCH are described in greater detail within Section 7.7.

The HS-DPCCH spreading factor of 256 results in a channel symbol rate of 15 ksps and a corresponding bit rate of 15 kbps, i.e. 1 bit per modulation symbol. The bit rate of 15 kbps generates 30 bits per subframe. This figure is presented as part of the HS-DPCCH slot format within Table 6.30.

The first 10 bits occupy the first time slot and are used by the MAC-hs acknowledgement. A series of 10 consecutive ones is used to represent a positive acknowledgement (ACK) whereas a series of 10 consecutive zeros is used to represent a negative acknowledgement (NACK). The MAC-hs acknowledgement does not include any information regarding the identity of the relevant HARQ process. The HARQ protocol is synchronous and so the Node B is able to deduce the relevant HARQ process identity from the timing of the acknowledgement, i.e. there is a fixed timing relationship between transmitting a transport block and receiving the corresponding acknowledgement. This timing relationship is illustrated in Figure 6.25. The last 20 bits of the HS-DPCCH occupy the second and third time slots. These bits are used by the CQI which can range from 0 to 30. The CQI value requires 5 bits when expressed in binary format. Channel coding is used to provide redundancy and increase these 5 bits to 20 bits prior to insertion within the HS-DPCCH.

The ACK, NACK and CQI can each be configured with a different transmit power relative to the DPCCH. Table 6.31 presents the set of amplitude ratios specified by 3GPP TS 25.213. The first column defines the values signalled to the UE when establishing or modifying an HSDPA connection. The UE maps these signalled values onto the β_{hs}/β_c amplitude ratios presented in the second column. β_{hs} and β_c

Table 6.31   ACK, NACK and CQI amplitude ratios relative to the DPCCH

$\Delta_{ACK}, \Delta_{NACK}, \Delta_{CQI}$	β_{hs}/β_c	Power relative to the DPCCH (dB)
8	30/15	6.0
7	24/15	4.1
6	19/15	2.1
5	15/15	0.0
4	12/15	−1.9
3	9/15	−4.4
2	8/15	−5.5
1	6/15	−8.0
0	5/15	−9.5

also appear within Figure 6.34. The third column presents the equivalent power difference between the ACK/NACK/CQI and the DPCCH, i.e. $10 \times \log(\beta_{hs}^2/\beta_c^2)$.

The RNC can signal these amplitude ratios to the UE within a Radio Bearer Setup, Radio Bearer Reconfiguration, Transport Channel Reconfiguration or Physical Channel Reconfiguration message. The RNC also informs the Node B of the amplitude ratios to help it take decisions when decoding the HS-DPCCH. An example of the relevant section from a Radio Bearer Reconfiguration message is presented in Log File 6.10.

In this example, the transmit power of the ACK and NACK are both equal to the transmit power of the DPCCH, i.e. the amplitude ratio is 15/15. The transmit power of the CQI is 1.9 dB less than the transmit power of the DPCCH. Increasing the amplitude ratio increases the reliability of successful reception, but generates an increased UE transmit power overhead. The UE can also be configured to repeat the ACK, NACK and CQI. The repetition factors in Log File 6.10 are configured with a value of 1 which indicates that repetition is not used. The drawback associated with repeating the ACK and NACK is a reduced maximum achievable throughput. Repeating the ACK and NACK once means that the maximum rate at which data can be scheduled is once every second TTI. Repeating the ACK and NACK twice further reduces the maximum rate to once every third TTI.

The Node B must decide whether it has received an ACK, NACK or period of DTX when decoding a MAC-hs acknowledgement. Reception of an ACK allows the Node B to discard the successfully received transport block and proceed with a new transmission. Reception of either a NACK or a period of DTX triggers a re-transmission. The Node B will re-transmit a transport block unnecessarily if it misinterprets an ACK transmission as either a NACK or period of DTX. This decreases throughput by occupying a TTI but the impact upon system performance is relatively small. The Node B will fail to

```
ul-DPCH-PowerControlInfo   fdd
    dpcch-PowerOffset: -46
    pc-Preamble:        0
    sRB-delay:          7
    powerControlAlgorithm
      algorithm1:       0
    deltaACK:           5
    deltaNACK:          5
    ack-NACK-repetition-factor: 1
modeSpecificInfo   fdd
    scramblingCodeType: longSC
    scramblingCode:     1000112
    spreadingFactor:    sf16
    tfci-Existence:     true
    puncturingLimit:    pl0-68
modeSpecificPhysChInfo fdd:
dl-HSPDSCH-Information
  hs-scch-Info
    modeSpecificInfo   fdd
  hS-SCCHChannelisationCodeInfo
    hS-SCCHChannelisationCodeInfo value: 4
measurement-feedback-Info
  modeSpecificInfo   fdd
    measurementPowerOffset: 20
    feedback-cycle: fc4
    cqi-RepetitionFactor: 1
    deltaCQI: 4
```

Log File 6.10   Example signalling used to configure the ACK, NACK and CQI amplitude offsets

re-transmit a transport block if it misinterprets a NACK or period of DTX as an ACK. This has a more significant impact upon performance because it becomes necessary for the RNC to recover the data using an RLC re-transmission. 3GPP TS 25.104 specifies that the probability of the Node B misinterpreting a period of DTX as an ACK should be less than the probability of misinterpreting an ACK as either a NACK or period of DTX.

Although both NACKs and periods of DTX trigger a MAC-hs re-transmission, it is useful for the Node B to distinguish between the two when incremental redundancy is used rather than chase combining. If a period of DTX is received then it indicates that the UE has failed to receive the corresponding HS-SCCH subframe and has not even attempted to decode the HS-PDSCH. This means that the UE has not buffered any data for combining with the re-transmission. If a NACK is received then it indicates that the UE has attempted to decode the HS-PDSCH and has buffered the received data ready for combination. Puncturing can prioritise either the systematic bits or parity bits when incremental redundancy is used (the systematic bits are always prioritised when chase combining is used). The systematic bits are the most important and in general, the UE should receive at least one transmission for which the systematic bits have been prioritised. It is not always possible to decode the original data from a single transmission if the parity bits have been prioritised. If a NACK is received for a first transmission which prioritised the systematic bits then the re-transmission can prioritise the parity bits. If a period of DTX is received for a first transmission which prioritised the systematic bits then the re-transmission should also prioritise the systematic bits.

The release 6 version of the 3GPP specifications introduce an optional change to the ACK/NACK signalling aimed at improving reliability [50]. The RNC can enable this change by signalling a HARQ_Preamble_Mode value of 1 to both the UE and Node B. Enabling the HARQ preamble mode allows the UE to send preamble and post-amble sequences within the MAC-hs acknowledgement section of the HS-DPCCH. These sequences are presented within Table 6.32.

The preamble and post-amble sequences allow decisions to be based upon pairs of HS-DPCCH subframes rather than individual subframes. When the HARQ preamble mode is enabled, the UE transmits a preamble sequence during the HS-DPCCH subframe preceding the subframe where the first ACK/NACK is to be transmitted. The Node B knows that the UE has successfully received the HS-SCCH if a preamble is received. It also knows that the following HS-DPCCH subframe must include either an ACK or a NACK rather than a period of DTX. This means that the possible combinations are preamble+ACK, preamble+NACK or DTX+DTX. Once the preamble and first ACK/NACK has been received the Node B knows that all following subframes must include either an ACK or a NACK until either a post-amble or further preamble is received, i.e. it is not necessary for the Node B to detect a period of DTX. A post-amble is sent if there are two consecutive subframes during which the UE is not scheduled. A post-amble is not sent if there is only a single subframe during which the UE is not scheduled because a preamble for the subsequent sub-frame is sent instead. The general sequence of events is illustrated in Figure 6.35. This example is based upon the reception of three transport blocks and assumes that CQI reports are sent once every 4 ms.

Table 6.32   Bit sequences for the MAC-hs acknowledgement section of the HS-DPCCH

ACK	1111111111
NACK	0000000000
PRE*	0010010010
POST*	0100100100

*Introduced as optional within release 6 of the 3GPP specifications

3GPP Release 5

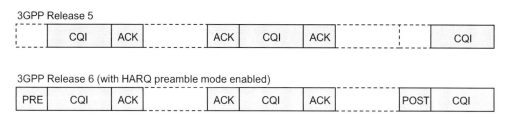

3GPP Release 6 (with HARQ preamble mode enabled)

Figure 6.35 MAC-hs acknowledgement PRE and POST sequences on the HS-DPCCH

6.10 Mobility

- The HS-DSCH serving cell is responsible for transmitting the HS-PDSCH and the corresponding HS-SCCH. It is also responsible for receiving the HS-DPCCH. If a UE is mobile then it is necessary to change the serving cell as the UE crosses cell boundaries.
- The serving cell change procedure represents a hard handover because the HSDPA connection is switched between cells without both cells simultaneously serving the UE.
- Soft handover is not possible for the HS-DSCH because the MAC-hs functionality is located within the Node B. In contrast, the dedicated channels associated with an HSDPA connection can use soft handover.
- The RNC selects the HS-DSCH serving cell when the connection is established. The RNC is then responsible for identifying the requirement for the serving cell change procedure.
- The serving cell change procedure could be triggered using an event 1d (change of best active set cell) measurement report from the UE. Alternatively, it could be triggered using periodic measurement reporting from the UE.
- The power offsets applied to the HS-DPCCH are typically increased when the dedicated channels enter soft handover. This helps to maintain reliability when non-serving cells instruct the UE to decrease its transmit power.
- The MAC-hs entity is reset as part of the serving cell change procedure when the cell change is between two different Node B. The RLC layer becomes responsible for recovering data which was buffered at the original Node B.
- Key 3GPP specifications: TS 25.214 and TS 25.331.

The set of HS-PDSCH and the corresponding HS-SCCH associated with a specific HSDPA connection are always transmitted from a single cell. Similarly, the uplink HS-DPCCH is received by the same single cell. This cell is known as the HS-DSCH serving cell. If a UE is mobile then it becomes necessary to change the serving cell as the UE moves across cell boundaries. This is known as the serving cell change procedure. The serving cell change procedure represents a hard handover because the UE switches its HSDPA connection from one cell to another without both cells simultaneously serving the UE. Soft handover is not possible for the HS-DSCH because the MAC-hs functionality is within the Node B rather than the RNC. Soft handover would require the Node B schedulers to be synchronised in terms of allocating the same TTI to the UE. Soft handover would also require the Node B AMC to be synchronised in terms of allocating the same transport block sizes and including the same set of MAC-d PDU within each transport block. These requirements prevent the use of soft handover for the HS-DSCH.

In contrast, the dedicated channels associated with an HSDPA connection can be transmitted and received by multiple cells, i.e. they can use soft handover. Figure 6.36 illustrates an HSDPA connection provided by the HS-DSCH serving cell while the associated dedicated channels are in soft handover between the serving cell and a neighbouring cell. This example assumes that the SRB are transferred using the DPDCH. If the SRB are transferred using the HS-PDSCH then the downlink DPDCH is no longer required and the downlink DPCCH can be replaced by the F-DPCH. Nevertheless, the same concept remains applicable, i.e. the F-DPCH can be in soft handover while the HSDPA connection is provided by the serving cell.

The RNC selects the HS-DSCH serving cell when the connection is established. If there is only a single cell within the active set then that cell becomes the serving cell. The RNC can make use of UE measurements to select the serving cell if a UE is in soft handover. For example, the RNC could select the serving cell based upon CPICH Ec/Io measurements reported by the UE.

The RNC is responsible for identifying the requirement for the serving cell change procedure after an HSDPA connection has been established. The RNC typically uses UE measurement reports to determine when the serving cell change procedure is necessary. An example of the sequence of events associated with the serving cell change procedure is illustrated in Figure 6.37. In this example, the UE is moving from one cell towards another cell. The UE starts with a single cell in the active set and that cell corresponds to the HS-DSCH serving cell. The RNC instructs the dedicated channels to enter soft handover as the UE moves towards the second cell. This could be triggered by an event 1a measurement report based upon CPICH Ec/Io, i.e. the UE recognises that the CPICH Ec/Io measured from the neighbouring cell has entered the soft handover addition window. Alternatively, the UE could have periodic measurement reporting enabled and the RNC could identify when the neigbouring cell becomes relatively strong without the requirement for a specific measurement reporting event. The dedicated channels are then in soft handover while the original cell continues to act as the HS-DSCH serving cell.

The quality of the second cell continues to improve as the UE moves further towards it. The RNC is responsible for identifying the point at which the serving cell change procedure is triggered. This could be based upon an event 1d measurement report from the UE, i.e. the UE could be instructed to inform the RNC when there has been a change of the best active set cell. Alternatively, it could be based upon periodic measurement reporting from the UE. The HS-PDSCH, HS-SCCH and HS-DPCCH connections are switched from the first cell to the second cell once the serving cell change procedure has been triggered. Finally, the original cell is removed from the dedicated channel active set as the UE continues to move towards the second cell.

An example of the RRC signalling associated with the serving cell change procedure is illustrated in Figure 6.38. In this example, it is assumed that the UE starts with scrambling code 100 within the active set and that the UE is moving towards a cell with scrambling 200. The UE has already received Measurement Control messages to configure the intra-frequency neighbour list and measurement reporting events 1a, 1b and 1d. The overall procedure starts when the UE sends an event 1a

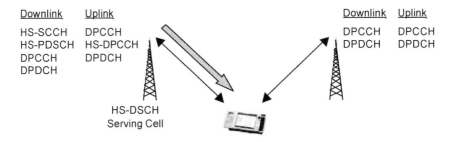

Figure 6.36   HS-DSCH serving cell with associated dedicated channels in soft handover

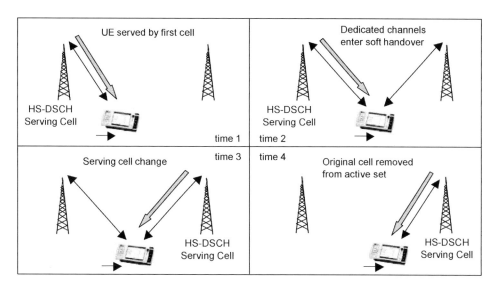

Figure 6.37 Sequence of events for an example serving cell change procedure

Measurement Report to inform the RNC that the cell with scrambling code 200 has entered the addition window. This triggers the RNC to instruct an active set update which adds the new cell to the dedicated channel active set. The active set update is immediately followed by a physical channel reconfiguration. This reconfiguration is used to increase the power offsets applied to the HS-DPCCH. An increased transmit power is required because this physical channel can become less reliable when a UE is in soft handover. The reason for this loss of reliability is explained later in this section. A Measurement Control message may be forwarded to the UE after the physical channel reconfiguration has been completed. This can be used to update the neighbour list using neighbour list information from the new cell within the active set. Alternatively, it may be used to instruct the UE to start periodic reporting of the CPICH Ec/Io. Periodic reporting would allow the RNC to determine when the serving cell should be switched from the original to the new cell. This example is based upon the RNC using an event 1d measurement report to trigger the switch and so periodic reporting is not necessary.

The UE sends an event 1d measurement report once it detects a change of the best cell within the active set. This example assumes that a Measurement Control message was used to configure event 1d during connection establishment, i.e. the event 1d Measurement Control message does not appear within Figure 6.38. An example of the relevant section from an event 1d Measurement Control message is presented in Log File 6.11.

This message instructs the UE to inform the RNC when the CPICH Ec/Io of an active set cell becomes better than the CPICH Ec/Io of the previous best active set cell by a margin of 2.5 dB and for a minimum duration of 640 ms. The margin of 2.5 dB is obtained by dividing the hysteresis value by 2. This division is specified by 3GPP TS 25.331 which defines the following equation for event 1d.

$$\text{Measurement}_{\text{NotBest}} + \text{CIO}_{\text{NotBest}} \geq \text{Measurement}_{\text{Best}} + \text{CIO}_{\text{Best}} + \text{Hysteresis}/2$$

The hysteresis is used to help avoid ping-pong between cells belonging to the active set. The Cell Individual Offset (CIO) values can be configured on a per cell basis to make specific cells appear either more or less attractive as candidates for the best active set cell.

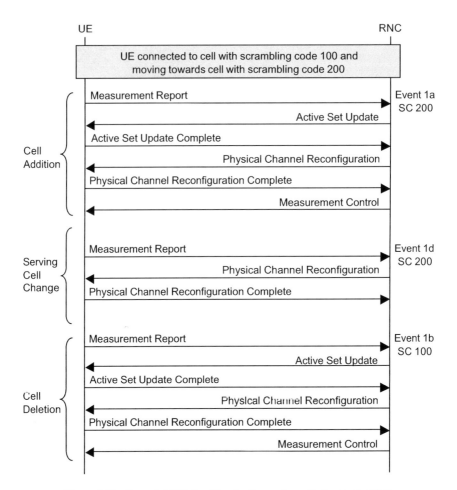

Figure 6.38 Example RRC signalling for the serving cell change procedure

The RNC uses a physical channel reconfiguration procedure to complete the serving cell change once the event 1d measurement report has been received. The relevant content from the Physical Channel Reconfiguration message is presented in Log File 6.12.

The Physical Channel Reconfiguration message instructs the UE to stop using the cell with scrambling code 100 as the serving cell and to start using the cell with scrambling code 200. Alternatively, a Radio Bearer Reconfiguration or Transport Channel Reconfiguration message could be used to inform the UE of the change in serving cell. Log File 6.12 also informs the UE that the MAC-hs entity should be reset as part of the serving cell change. This is necessary when the serving cell change

```
reportCriteria intraFreqReportingCriteria :
eventCriteriaList
  event e1d:
    hysteresis        5,
    timeToTrigger     ttt640,
    reportingCellStatus withinActiveSet: e3
```

Log File 6.11 Example configuration for measurement reporting event 1d

```
modeSpecificInfo fdd : NULL
dl-CommonInformation
modeSpecificInfo fdd :
     mac-hsResetIndicator true
dl-InformationPerRL-List
modeSpecificInfo fdd :
     primaryCPICH-Info
     primaryScramblingCode 200
     servingHSDSCH-RL-indicator TRUE
modeSpecificInfo fdd :
     primaryCPICH-Info
     primaryScramblingCode 100
     servingHSDSCH-RL-indicator FALSE
```

Log File 6.12   Physical Channel Reconfiguration message instructing serving cell change

procedure is between two different Node B. The RLC layer is then responsible for recovering data which was buffered at the original Node B. If the serving cell change is between two cells belonging to the same Node B then it is possible that a MAC-hs reset is not necessary. The data buffered at the Node B can be re-directed towards the new serving cell without requiring the RLC layer to complete re-transmissions.

An event 1b Measurement Report is triggered as the UE continues to move towards the second cell. This causes the RNC to instruct an active set update to delete the original cell from the active set. The active set update is immediately followed by a physical channel reconfiguration. This reconfiguration is used to return the DPCCH power offsets to their original values, i.e. the increased values are only necessary when the UE is in soft handover. Finally, a Measurement Control message is used to update the neighbour list information.

The preceding description referenced a reduction in the reliability of the HS-DPCCH when a UE enters soft handover. This reduction is typically addressed by increasing the relative transmit power allocated to the HS-DPCCH, i.e. the ACK, NACK and CQI amplitude offsets are increased when a UE is in soft handover. This concept is illustrated in Figure 6.39.

The reduction in reliability is caused by the action of inner loop power control. A UE in soft handover receives inner loop power control commands from all active set cells. 3GPP TS 25.214 states that a UE should decrease its transmit power if any of the active set cells instruct a decrease. This creates an issue for the HS-DPCCH because a non-serving cell can instruct the UE to decrease its

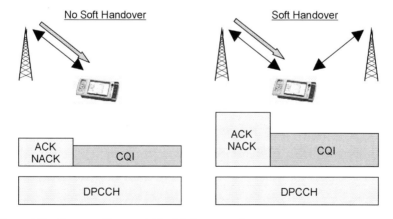

Figure 6.39   Concept of increased HS-DPCCH power offsets when the UE is in soft handover

transmit power while the serving cell is having difficulty receiving the HS-DPCCH. This scenario is illustrated in Figure 6.40.

Increasing the relative transmit power of the HS-DPCCH helps the serving cell to maintain reception even when other active set cells instruct the UE to decrease its transmit power. In general, there should be a limit to the extent that non-serving cells instruct the UE to decrease its transmit power. Requesting a large decrease implies that the link loss to the non-serving cell is less than the link loss to the serving cell and that a serving cell change should be triggered. Table 6.33 presents some example amplitude offsets for the HS-DPCCH.

These offsets typically depend upon the allocated uplink DPDCH bit rate. Smaller offsets can be used when higher uplink bit rates are allocated because the DPCCH tends to have an increased transmit power. Repetition of the ACK, NACK and CQI can also be used to improve reliability when the UE is in soft handover. However, repetition of the MAC-hs acknowledgement reduces the maximum achievable throughput by preventing the scheduler from assigning consecutive TTI to the same UE. The HARQ_Preamble_Mode introduced as part of the release 6 version of the 3GPP specifications can also be used to improve the reliability of the MAC-hs acknowledgement.

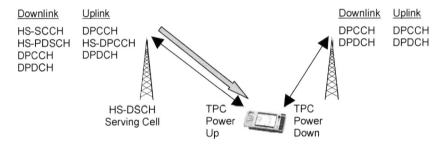

Figure 6.40   Reason for increasing the HS-DPCCH power offsets when UE are in soft handover

Table 6.33   Example amplitude offsets for the HS-DPCCH relative to the DPCCH

	No soft handover	Soft handover
ACK	15/15	24/15
NACK	15/15	24/15
CQI	12/15	24/15

7

HSUPA

7.1 Concept

- HSUPA is introduced within release 6 of the 3GPP specifications. It allows increased individual connection throughputs, increased total cell throughputs and reduced round trip times. HSUPA can support both real time and non-real time packet switched services.
- HSUPA can be used to transfer both RRC signalling and application data. Alternatively, RRC signalling can be transferred using a parallel DPDCH.
- The E-DPDCH is a dedicated physical channel whereas the HS-PDSCH is a common physical channel. Only one HSDPA connection can use a specific HS-PDSCH at any point in time. All HSUPA connections can use their E-DPDCH simultaneously.
- The E-DPDCH can use either a 2 or 10 ms TTI. A 2 ms TTI offers the benefit of reduced system delays and higher potential throughputs. A 10 ms TTI offers the benefit of improved physical layer performance.
- HSUPA allows a maximum E-DPDCH code allocation of $2 \times SF2 + 2 \times SF4$. This corresponds to four E-DPDCH and is only possible when RRC signalling is transferred using HSUPA. Otherwise, the maximum E-DPDCH code allocation is $2 \times SF2$.
- The release 6 version of the 3GPP specifications is limited to supporting BPSK for the E-DPDCH. Release 7 introduces the possibility of using 4 level Pulse Amplitude Modulation (4PAM) to increase the maximum achievable throughput.
- HSUPA is able to benefit from soft handover. The serving E-DCH cell has the main scheduling responsibility while non-serving E-DCH cells maintain sufficient control to protect themselves from increased uplink interference.
- The HSUPA scheduler within the Node B allocates resources in terms of the maximum allowed power offset between the E-DPDCH and the DPCCH. Allocating a large power offset allows the UE to transfer increased quantities of data, but generates an increased contribution to the total uplink interference margin.
- The serving E-DCH cell can use the E-AGCH to change the maximum E-DPDCH power ratio allocated to the UE. Active set cells belonging to the same Node B as the serving cell are able to use the E-RGCH to increase, maintain and decrease the power ratio. Other active set cells are able to use the E-RGCH to maintain or decrease the power ratio.

- The UE selects the E-DCH transport block size during E-TFC selection. This procedure accounts for the E-DPDCH power offset allocated by the Node B scheduler. It also accounts for the UE transmit power and the quantity of data buffered for transmission.
- The HARQ protocol provides support for re-transmissions between the UE and Node B. Each active set cell is responsible for returning MAC-e acknowledgements using the E-HICH. Re-transmissions are avoided if at least one active set cell returns a positive acknowledgement. The RLC protocol provides support for re-transmissions between the UE and RNC when used in acknowledged mode.

High Speed Uplink Packet Access (HSUPA) is introduced within release 6 of the 3GPP specifications. Similar to HSDPA, HSUPA provides a solution for achieving increased individual connection throughputs, increased total cell throughputs and reduced round trip times. The increased individual connection throughput and reduced round trip time help to improve the quality of service experienced by the end-user. The increased total cell throughput helps the operator to increase the efficiency with which air-interface resources are used. HSUPA is associated with packet switched services rather than circuit switched services. Packet switched services can be either non-real time (NRT) or real time (RT). Example NRT services are email, file transfer and internet browsing. An example RT service is voice over IP (VoIP).

An HSUPA connection can be used to transfer both the user plane data originating from an end-user application and the control plane signalling originating from the RRC layer. Alternatively, an HSUPA connection can be used to transfer the user plane data while a DPCH is used to transfer the control plane signalling. These two possibilities are presented in Table 7.1.

The logical channels represent the type of information being transferred and these remain unchanged irrespective of whether or not an HSUPA connection is used. The set of transport channels depends upon whether or not the Signalling Radio Bearers (SRB) are transferred using HSUPA. Transferring the SRB using a DPCH means that a DCH transport channel is used in parallel to an E-DCH transport channel. Transferring the SRB using HSUPA means that the set of transport channels is limited to a single E-DCH. The 3GPP specifications allow a maximum of one E-DCH transport channel per UE. The MAC layer within the UE is then responsible for multiplexing logical channels onto the single E-DCH.

The set of physical channels also depends upon whether or not the SRB are transferred using HSUPA. If the SRB are transferred using a DPCH then a DPDCH is used in parallel to one or more E-DPDCH. Otherwise, the single E-DCH is mapped onto one or more E-DPDCH. The E-DPDCH can use spreading factors ranging from 256 to 2. Decreasing the spreading factor increases the rate at which data can be transferred, i.e. the same concept as for a DPDCH. The 3GPP specifications allow a maximum of 2 E-DPDCH when the SRB are transferred using a DPCH. This increases to a maximum of 4 E-DPDCH when the SRB are transferred using HSUPA, i.e. the maximum HSUPA throughputs can only be achieved when the SRB are transferred using HSUPA. The set of physical channels presented in Table 7.1 serves only as an introduction and is not exhaustive. Additional physical channels used to transfer control information are described in Section 7.7.

Table 7.1 Uplink logical, transport and physical channels for HSUPA

	SRB on DPCH		SRB on HSUPA	
	Control plane signalling	User plane data	Control plane signalling	User plane data
Logical channel	DCCH	DTCH	DCCH	DTCH
Transport channel	DCH	E-DCH	E-DCH	
Physical channel	DPDCH	E-DPDCH	E-DPDCH	

Table 7.2 Comparision of the DPDCH, HS-PDSCH and E-DPDCH

	DPDCH	HS-PDSCH	E-DPDCH
Physical Channel Type	Dedicated	Common	Dedicated
Downlink shared resource	Tx power and codes	Tx power and codes	–
Uplink shared resource	Interference margin	–	Interference margin
TTI	10, 20, 40, 80 ms	2 ms	2, 10 ms
Multiple transport blocks per TTI	Yes	No	No
Modulation scheme	BPSK	QPSK, 16QAM, 64QAM*	BPSK, 4PAM*
Spreading factor	256 to 4**	16	256 to 2
Multiple codes	Not in practice	15×SF16	2×SF2 + 2×SF4
Soft handover	Yes	No	Yes
Inner loop power control	Yes	No	Yes
Node B scheduling	No	Yes	Yes
Node B AMC	No	Yes	No
UE TFC/E-TFC selection	Yes	No	Yes
Node B flow control	No	Yes	No
UE ↔ Node B re-transmission	No	Yes	Yes
UE ↔ RNC Re-transmission	Yes	Yes	Yes

*64QAM for the HS-PDSCH and 4 level Pulse Amplitude Modulation (4PAM) for the E-DPDCH are introduced within the release 7 version of the 3GPP specifications.
**Downlink also allows a spreading factor of 512.

Table 7.2 compares the main characteristics of the DPDCH, HS-PDSCH and E-DPDCH. Both the DPDCH and E-DPDCH are dedicated physical channels. This represents a significant difference between HSDPA and HSUPA. The HS-PDSCH is a common channel shared between all HSDPA connections. Only one HSDPA connection can use a specific HS-PDSCH at any point in time. In contrast, all HSUPA connections can use their E-DPDCH simultaneously without the requirement for time multiplexing. In terms of air-interface resources, DPCH and HSDPA connections share the downlink transmit power capability and the downlink channelisation code tree. The RNC and Node B are responsible for sharing these resources between all downlink connections. DPCH and HSUPA connections share the uplink interference margin at the Node B receiver. The RNC and Node B are responsible for ensuring that the planned maximum increase in uplink interference is not exceeded. Exceeding the planned maximum is likely to have an impact upon uplink coverage as UE towards cell edge will struggle to maintain their uplink C/I requirements. The RNC packet scheduler controls the uplink interference contribution of each DPCH connection, i.e. allocating a 384 kbps uplink DPCH generates a larger contribution than allocating a 64 kbps uplink DPCH. The Node B packet scheduler controls the uplink interference contribution of each HSUPA connection. The HSUPA scheduler within the Node B allocates resources in terms of the maximum allowed power offset between the E-DPDCH and the DPCCH. Allocating a large power offset allows the UE to transfer increased quantities of data, but generates an increased contribution to the total uplink interference margin.

The E-DPDCH can be configured to use either a 2 or 10 ms Transmission Time Interval (TTI). A 2 ms TTI offers the benefit of reduced system delays and higher potential throughputs. A 10 ms TTI offers the benefit of improved physical layer performance. This can be important when channel conditions are relatively poor and the UE is already transmitting at its maximum power, e.g. at cell edge. In general, the limited UE transmit power capability means that a 10 ms TTI is more important for HSUPA than for HSDPA. However, HSUPA is able to benefit from soft handover at locations towards the cell edge. This allows more than a single cell to receive the E-DPDCH. The RNC combines the streams of MAC-es PDU received from each active set Node B prior to forwarding to the higher layers. The serving E-DCH cell has the main scheduling responsibility while non-serving E-DCH cells maintain sufficient control to protect themselves from increased uplink interference.

The maximum HSUPA throughput has been specified to be less for the 10 ms TTI to avoid the code block sizes becoming too large. Similar to HSDPA, an HSUPA connection can transfer a maximum of one transport block during each TTI. The combination of transport block size and TTI determines the connection throughput. The TTI is configured by the RNC during connection establishment. The UE selects the transport block size during the E-TFC selection procedure. This procedure accounts for the E-DPDCH power ratio allocated by the Node B scheduler. A UE is only able to select the larger transport block sizes when the Node B scheduler has granted a relatively large E-DPDCH power ratio. The RNC provides the UE with a look-up table which defines the relationship between a set of reference transport block sizes and a corresponding set of reference power ratios. The UE uses this information to identify a transport block size which corresponds to the allocated power ratio. The E-TFC selection procedure also accounts for the UE transmit power, i.e. whether or not the UE has sufficient power to transmit using the allocated power ratio, and the quantity of data buffered for transmission. E-TFC selection can be viewed as being equivalent to the Node B Adaptive Modulation and Coding (AMC) used for HSDPA. E-TFC selection has an impact upon the physical layer coding rate and the selection of an appropriate set of channelisation codes. It also has an impact upon the modulation scheme in release 7 of the specifications.

The release 6 version of the 3GPP specifications is limited to supporting BPSK for the E-DPDCH. Release 7 introduces the possibility of using 4 level Pulse Amplitude Modulation (4PAM) to increase the maximum achievable throughput. 4PAM requires a better signal to noise ratio than BPSK (and so generates a larger increase in interference at the Node B receiver) but maps 2 bits rather than 1 bit onto each modulated symbol. The use of two E-DPDCH in parallel results in BPSK appearing similar to QPSK and 4PAM appearing similar to 16QAM. HSUPA allows a maximum E-DPDCH code allocation of $2\times SF2 + 2\times SF4$. This corresponds to four E-DPDCH and is only possible when the SRB are transferred using HSUPA. Otherwise, the maximum E-DPDCH code allocation is $2\times SF2$. Higher spreading factors can be used when the Node B scheduler allocates smaller E-DPDCH powers ratios, or when the UE transmit power capability has become exhausted, or when the UE does not have sufficient data buffered to justify the use of the maximum code allocation.

The E-DPDCH makes use of inner loop power control in the same way as the DPDCH. The active set cells transmit uplink Transmit Power Control (TPC) commands on either the DPCCH or F-DPCH. These commands instruct the UE to increase or decrease the transmit power of the DPCCH. If the power offset from the DPCCH remains constant, the DPDCH and E-DPDCH transmit powers increase or decrease by the same step size as the DPCCH.

HSUPA does not require flow control between the RNC and Node B because data transfer is in the uplink direction and all of the data buffered for transmission is within the UE. Similar to HSDPA, HSUPA supports MAC layer re-transmissions between the UE and Node B as well as RLC layer re-transmissions between the UE and RNC. Higher layer re-transmissions between the UE and application server can also be used when necessary, e.g. TCP layer re-transmissions.

Similar to HSDPA, the introduction of HSUPA tends to shift functionality from the RNC to the Node B. This allows HSUPA connections to be more responsive to changes in the air-interface and network load. The general concepts associated with an HSUPA connection are illustrated in Figure 7.1.

An HSUPA connection is established using both RRC signalling between the RNC and UE, and NBAP signalling between the RNC and Node B. An initial serving cell grant is allocated to the UE during connection establishment. This grant is generated by the serving Node B, but is passed to the RNC for subsequent transfer to the UE. The initial serving cell grant defines a maximum E-DPDCH to DPCCH power ratio. The UE is responsible for translating this power ratio into an appropriate transport block size, i.e. selecting a transport block size which can be transferred using the allocated E-DPDCH transmit power. This translation process uses a look-up table provided by the RNC which includes a set of reference transport block sizes and a corresponding set of reference power ratios.

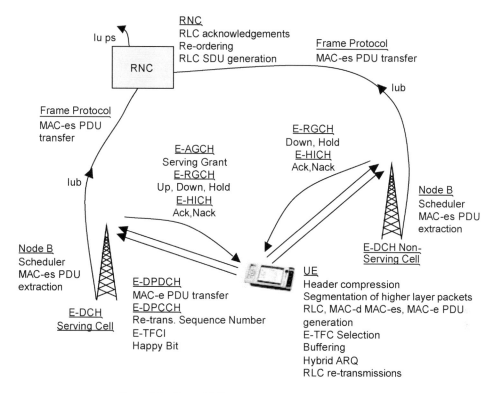

Figure 7.1 General HSUPA concepts for a single HSUPA UE

The UE can start transferring E-DPDCH data using the initial serving cell grant. The PDCP layer within the UE may be used to compress the headers generated by the higher layers. The uplink data is then passed to the RLC layer which segments large packets into smaller packets to ensure that they can be encapsulated by a set of RLC PDU. If the end-user application is characterised as NRT then it is likely that the RLC layer will be configured for acknowledged mode and will be responsible for RLC re-transmissions. RLC PDU are passed to the MAC-d entity within the UE. The MAC-d entity is transparent in terms of PDU size (assuming no user plane multiplexing), i.e. MAC-d PDU have the same size as RLC PDU. The resulting MAC-d PDU are forwarded to the MAC-es/e layer as a MAC-d flow. The UE uses the E-TFC selection procedure to identify an appropriate transport block size. This determines the instantaneous throughput and the number of MAC-d PDU which can be packaged within the MAC-es PDU. Further header information is added to generate a MAC-e PDU. The MAC-e PDU may also include Scheduling Information (SI) to provide the Node B scheduler with information regarding UE transmit power headroom and UE buffer occupancy. The MAC-e PDU represents the E-DCH transport block. The transport block is processed by the physical layer prior to being used to modulate the RF carrier.

The UE transmits the E-DPCCH in parallel to the E-DPDCH. The E-DPCCH informs the Node B of the E-DCH Transport Format Combination Indicator (E-TFCI) corresponding to the transport block size within the E-DPDCH. It also informs the Node B of the Re-transmission Sequence Number (RSN) and the Happy Bit. The RSN can be used by the Node B to determine the redundancy version when the Hybrid ARQ (HARQ) protocol uses incremental redundancy. The RSN is also used to record the number of re-transmissions necessary to successfully receive a transport block. This information is signalled to the RNC to allow the outer loop power control function to calculate the BLER performance and adjust the uplink DPCCH SIR target appropriately. The Happy Bit is used to inform the Node B packet scheduler of whether or not the UE would like an increased power ratio allocation. The scheduler can use this information when updating its allocation of resources.

The serving E-DCH cell can use the E-DCH Absolute Grant Channel (E-AGCH) to rapidly change the maximum E-DPDCH power ratio allocated to the UE. This can be used to either increase or decrease the existing allocation. Both the serving and non-serving E-DCH cells can use the E-DCH Relative Grant Channel (E-RGCH) to adjust the maximum power ratio allocated to the UE. However, only active set cells belonging to the same Node B as the serving cell are able to use the E-RGCH to increase the power ratio allocation. Other active set cells are limited to either maintaining or decreasing the current allocation. This allows those cells to protect themselves from increased interference from UE which they are not serving. Each active set cell is responsible for returning MAC-e acknowledgements using the E-DCH Hybrid ARQ Indicator Channel (E-HICH). These acknowledgements inform the UE of whether or not re-transmissions are necessary. Re-transmissions are avoided if one or more active set cells return a positive acknowledgement.

The user plane protocol stack for HSUPA is illustrated in Figure 7.2. This protocol stack was introduced in Section 2.1.2. Use of the PDCP layer is optional and may be limited to specific applications, e.g. VoIP. User plane data belonging to other applications can be processed by the RLC layer directly.

The control plane protocol stack for HSUPA is illustrated in Figure 7.3. This protocol stack is applicable when the SRB are transferred using HSUPA rather than a DPCH. The control plane protocol stack includes the RRC layer to generate and decode signalling messages. The PDCP layer does not apply to the control plane because there are no higher layer headers, e.g. TCP/IP headers to compress.

The control plane protocol stack for a DPCH is illustrated in Figure 7.4. This protocol stack is applicable when the SRB are transferred using a DPCH rather than HSUPA. In this case, the MAC-es/e layer is not present.

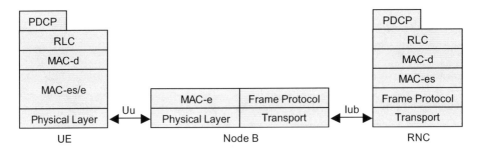

Figure 7.2 User plane HSUPA protocol stack

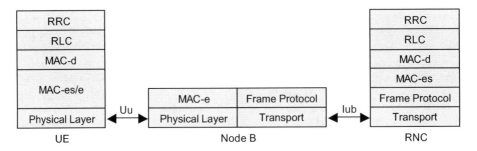

Figure 7.3 Control plane HSUPA protocol stack

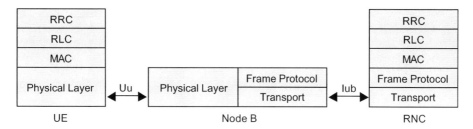

Figure 7.4   Control plane DPCH protocol stack

7.2 HSUPA Bit Rates

- HSUPA bit rates depend upon the protocol stack layer from where they are measured. They also depend upon the spreading factor, number of E-DPDCH channelisation codes, coding rate and modulation scheme.
- The release 6 version of the 3GPP specifications uses BPSK as a modulation scheme. This is sometimes referred to as QPSK because BPSK appears similar to QPSK when two E-DPDCH are used in parallel. BPSK maps 1 bit onto each modulation symbol.
- The release 7 version of the 3GPP specifications introduces 4PAM as a modulation scheme. This is sometimes referred to as 16QAM because 4PAM appears similar to 16QAM when two E-DPDCH are used in parallel. 4PAM maps 2 bits onto each symbol.
- Not all HSUPA UE are capable of achieving the complete set of bit rates. Category 3 UE support 2×SF4 and a 10 ms TTI. Category 5 UE support 2×SF2 and a 10 ms TTI. Category 6 UE support 2×SF4 + 2×SF2 and both 10 ms and 2 ms TTI. Category 7 UE support 2×SF4 + 2×SF2, both 10 ms and 2 ms TTI, and 4PAM.
- Category 3, 5, 6 and 7 UE have maximum physical layer throughputs of 1.45, 2.00, 5.74 and 11.50 Mbps respectively. The maximum throughput figures for the category 6 and 7 UE assume a coding rate of 1.0, i.e. puncturing removes the redundancy added by channel coding. The use of high coding rates increases the probability of error unless the RF channel conditions are particularly good.
- The throughput achieved at the higher layers within the protocol stack decreases as the headers and padding added by the lower layers are removed. The throughput is also reduced by re-transmissions at the MAC-e, RLC and TCP layers. The majority of re-transmissions should be completed at the MAC-e layer. The RLC throughput provides a measure of useful data transfer from the RAN perspective.
- The instantaneous HSUPA throughput is determined during the E-TFC selection procedure. The UE uses this procedure to select a transport block size based upon the E-DPDCH power ratio allocated by the Node B packet scheduler, the UE transmit power limitations and quantity of data buffered for transmission.
- The interference margin at the Node B receiver represents a shared resource. If the allocated E-DPDCH power ratios cause the planned maximum interference margin to be reached then increasing the number of connections will decrease both the allocated power ratio per connection and the corresponding throughput per connection.
- Key 3GPP specifications: TS 25.306, TS 25.321, TS 25.211, TS 25.212 and TS 25.213.

This section describes HSUPA bit rates from a theoretical perspective. It presents the maximum achievable throughputs without considering the limitations imposed by the UE transmit power capability and the E-DPDCH to DPCCH power ratio allocated by the Node B packet scheduler. In practice, the maximum bit rates are only achieved when the UE has sufficient transmit power, i.e. the link loss is not large, and the Node B packet scheduler has allocated a relatively high power ratio. The bit rates presented in this section are applicable to a single HSUPA connection. The E-DPDCH used for HSUPA is a dedicated rather than a common physical channel and so the throughput per connection does not necessarily decrease as the number of connections increase. The total HSUPA cell throughput is obtained by summing the throughputs achieved by each connection. However, the interference margin at the Node B receiver represents a shared resource. If the allocated power ratios cause the planned maximum interference margin to be reached then increasing the number of connections will decrease both the allocated power ratio per connection and the corresponding throughput per connection.

HSUPA bit rates depend upon the protocol stack layer from where they are measured. They also depend upon the spreading factor, number of E-DPDCH channelisation codes, coding rate and modulation scheme. The throughput capability of HSUPA is often quoted using the figures presented in Table 7.3.

An E-DPDCH is defined by its channelisation code so the first throughput column in Table 7.3 is applicable to one E-DPDCH, whereas the second and third columns are applicable to two E-DPDCH and the fourth column is applicable to four E-DPDCH. The BPSK modulation scheme is sometimes referred to as QPSK because BPSK appears similar to QPSK when two E-DPDCH are used in parallel on the in-phase and quadrature branches of the modulation constellation. Similarly, the 4 level Pulse Amplitude Modulation (4PAM) scheme is sometimes referred to as 16QAM because they appear similar when two E-DPDCH are used in parallel. As an example, 3GPP TS 25.213 refers to the modulation schemes as BPSK and 4PAM whereas 3GPP TS 25.306 refers to them as QPSK and 16QAM.

The throughput figures presented in Table 7.3 can be derived using the equation below.

$$\text{Throughput} = \sum_{n=1}^{\#\text{Codes}} \frac{\text{Chip rate}}{\text{SF}_n} \times \text{Bits per symbol} \times \text{Coding rate}$$

Where the chip rate is 3.84 Mcps and the Spreading Factor (SF) ranges from 256 to 2. The ratio of $3.84 \times 10^6 / \text{SF}_n$ defines the modulated symbol rate before spreading at the UE transmitter. Multiplying this result by the number of bits per symbol generates the corresponding bit rate. BPSK maps 1 bit onto each symbol whereas 4PAM maps 2 bits onto each symbol. Multiplying by the coding rate references the throughput result to the top of the physical layer. A low coding rate indicates that the physical layer introduces a large quantity of redundancy to help protect the information bits as they are transferred

Table 7.3 HSUPA throughput capability

	Coding Rate	1×SF4 Code	2×SF4 Codes	2×SF2 Codes	2×SF4 Codes + 2×SF2 Codes
BPSK	0.50	480 kbps	960 kbps	1.92 Mbps	2.88 Mbps
(QPSK)	0.75	720 kbps	1.44 Mbps	2.88 Mbps	4.32 Mbps
	1.00	960 kbps	1.92 Mbps	3.84 Mbps	5.76 Mbps
4PAM[*]	0.50	960 kbps	1.92 Mbps	3.84 Mbps	5.76 Mbps
(16QAM)	0.75	1.44 Mbps	2.88 Mbps	5.76 Mbps	8.64 Mbps
	1.00	1.92 Mbps	3.84 Mbps	7.68 Mbps	11.52 Mbps

[*]Introduced by release 7 of the 3GPP specifications.

across the air-interface. A coding rate of 0.5 means that the throughput at the top of the physical layer is half of the throughput at the bottom of the physical layer. The resulting throughput at the top of the physical layer is applicable to a single E-DPDCH channelisation code so the total throughput is obtained by summing the throughputs generated by each code.

The throughput figures presented in Table 7.3 do not necessarily reflect the maximum throughput which can be achieved in practice. Not all HSUPA UE are capable of achieving the complete set of bit rates and UE are categorised according to their ability. This UE category information is signalled to the network during connection establishment to ensure that both the RNC and Node B are aware of the UE capability, e.g. to ensure that the resources corresponding to $2 \times SF4 + 2 \times SF2$ are not allocated to a UE which only supports $2 \times SF4$. The set of UE categories specified by 3GPP TS 25.306 is presented in Table 7.4.

A category 1 UE has the lowest capability. It is able to support a maximum of one E-DPDCH using a spreading factor of 4. This provides a maximum E-DPDCH capacity of 9600 bits per 10 ms radio frame. The maximum transport block size of 7110 bits allows a maximum physical layer throughput of 711 kbps and a corresponding coding rate of $7110/9600 = 0.74$.

A category 3 UE is able to support a maximum of two E-DPDCH using spreading factors of 4. This provides a maximum E-DPDCH capacity of 19200 bits per 10 ms radio frame. The maximum transport block size of 14484 bits allows a maximum physical layer throughput of 1.45 Mbps and a corresponding coding rate of $14484/19200 = 0.75$.

A category 5 UE is able to support a maximum of two E-DPDCH using spreading factors of 4 plus two E-DPDCH using spreading factors of 2. This provides a maximum E-DPDCH capacity of 57600 bits per 10 ms radio frame. The maximum transport block size of 20000 bits allows a maximum physical layer throughput of 2.00 Mbps and a corresponding coding rate of $20000/57600 = 0.35$.

Table 7.5 summarises the maximum physical layer throughputs for each UE category. It also includes the coding rates corresponding to the maximum transport block size and the maximum E-DPDCH capacity. The maximum transport block size for a category 7 UE using a 2 ms TTI requires the use of 16QAM. QPSK is unable to accommodate the transport block size of 22996 bits within a 2 ms TTI. Typical target maximum coding rates are of the order of 0.75. Coding rates greater than 0.75 require relatively aggressive puncturing after channel coding. The maximum physical layer through-puts associated with category 6 and 7 UE require coding rates of 1. This indicates that puncturing removes all of the redundancy added by the channel coding process, i.e. all of the parity 1 and parity 2 bits generated by Turbo coding would be removed. The use of high coding rates increases the probability of error during the decoding process unless the RF channel conditions are particularly good.

The bit rates presented in Table 7.3 and those calculated from the transport block sizes in Table 7.5 are applicable to the top of the physical layer. These figures include the overheads generated by the MAC and

Table 7.4 Categories for HSUPA UE

UE category	Modulation schemes	Maximum number of codes	TTI (ms)	Maximum transport block size (bits)	
				10 ms TTI	2 ms TTI
1	QPSK	$1 \times SF4$	10	7110	–
2	QPSK	$2 \times SF4$	10 and 2	14484	2798
3	QPSK	$2 \times SF4$	10	14484	–
4	QPSK	$2 \times SF2$	10 and 2	20000	5772
5	QPSK	$2 \times SF2$	10	20000	–
6	QPSK	$2 \times SF4 + 2 \times SF2$	10 and 2	20000	11484
7[*]	QPSK/16QAM	$2 \times SF4 + 2 \times SF2$	10 and 2	20000	22996

[*]Introduced by release 7 of the 3GPP specifications.

Table 7.5 Maximum physical layer throughputs for each UE category

UE category	Maximum transport block size (bits)		Physical layer throughput (Mbps)		Coding rate	
	10 ms TTI	2 ms TTI	10 ms TTI	2 ms TTI	10 ms TTI	2 ms TTI
1	7110		0.71		0.74	
2	14484	2798	1.45	1.40	0.75	0.73
3	14484		1.45		0.75	
4	20000	5772	2.00	2.89	0.52	0.75
5	20000		2.00		0.52	
6	20000	11484	2.00	5.74	0.35	1.00
7	20000	22996	2.00	11.50	0.35	1.00*

*Assumes that 16QAM is used rather than QPSK.

RLC layers. The RLC layer throughput represents the useful throughput from the perspective of the radio access network. The RLC throughput can be quantified by identifying the number of MAC-d PDU which can be accommodated by each transport block size, and subsequently dividing the total RLC payload by the TTI. If it is assumed that the MAC-d PDU size is 336 bits and the acknowledged mode RLC header size is 16 bits then the RLC payload per MAC-d PDU is 320 bits. In practice, the size of the RLC header will sometimes be greater than 16 bits, e.g. when length indicators are included to signal the end of a higher layer packet. Nevertheless, an RLC header size of 16 bits represents a reasonable approximation. The MAC-es and MAC-e header sizes should be subtracted from the transport block size prior to calculating the number of MAC-d PDU which can be accommodated. The MAC-es header size is 6 bits and the MAC-e header size is 12 bits if it is assumed that there is a single MAC-es PDU within each MAC-e PDU. Table 7.6 presents the resultant maximum RLC throughputs for each UE category.

A MAC-d PDU size of 656 bits has been assumed for the category 7 UE when using a 2 ms TTI because a maximum of 63 MAC-d PDU can be concatenated within a single MAC-es PDU. This limitation originates from the length of the header field used to inform the RNC of the number of MAC-d PDU within a specific MAC-es PDU, i.e. the header field length of 6 bits supports a range from 0 to 63 PDU. Assuming an acknowledged mode RLC header size of 16 bits results in an RLC payload of 640 bits.

In practice, the maximum transport block size may differ slightly from the figures presented within the preceding tables. The release 6 version of 3GPP TS 25.321 specifies two transport block size tables for the 2 ms TTI and two transport block size tables for the 10 ms TTI. The release 7 version of TS 25.321 specifies an additional two tables for the 2 ms TTI. These additional tables define the transport block sizes

Table 7.6 Maximum RLC layer throughputs for each UE category

UE category	Maximum transport block size (bits)		Maximum number of MAC-d PDU		RLC throughput (Mbps)	
	10 ms TTI	2 ms TTI	10 ms TTI	2 ms TTI	10 ms TTI	2 ms TTI
1	7110		21		0.67	
2	14484	2798	43	8	1.38	1.28
3	14484		43		1.38	
4	20000	5772	59	17	1.89	2.72
5	20000		59		1.89	
6	20000	11484	59	34	1.89	5.44
7	20000	22996	59	35*	1.89	11.20

*Assumes a MAC-d PDU size of 656 bits.

Table 7.7 Maximum transport block sizes from the E-TFC selection transport block size tables

Table	Maximum transport block size (bits)	
	10 ms TTI	2 ms TTI
0	20000	11484
1	19950	11478
2*		22995
3*		22996

*Introduced by release 7 of the 3GPP specifications.

used for 16QAM. The RNC instructs the UE and Node B to use a specific transport block size table during connection establishment. The UE uses this table to select appropriate transport block sizes during the E-TFC selection procedure. This procedure is described in greater detail within Section 7.6.1. The Node B uses the same table when identifying the received transport block size from the E-TFCI within the E-DPCCH. Table 7.7 presents the maximum transport block sizes within each of these tables.

The maximum transport block size is slightly smaller when using table index 1 rather than table index 0. However, the difference is relatively small and does not affect the maximum number of MAC-d PDU which can be accommodated, e.g. the transport block sizes of 20000 and 19950 bits can both accommodate a maximum of 59 MAC-d PDU when the PDU size is 336 bits.

Table 7.8 presents a set of example maximum throughput figures for a TCP/IP protocol stack assuming UE categories 3, 5, 6 and 7. The throughput figures for the category 6 and 7 UE are relatively optimistic because they assume a coding rate of 1. The impact of re-transmissions at the MAC-e, RLC and TCP layers has been excluded.

The physical layer throughputs are the same as those presented within Table 7.5 whereas the RLC layer throughputs are the same as those presented within Table 7.6. The MAC-d throughput is obtained by adding the overhead generated by the RLC layer, i.e. 16 bits for every acknowledged mode RLC PDU. This represents a 5% overhead for an RLC payload of 320 bits and a 2.5% overhead for an RLC payload of 640 bits. The MAC-es throughput is equal to the MAC-d throughput because the MAC-d layer is transparent in terms of PDU size. The MAC-e layer throughput is slightly greater than the MAC-es layer throughput because the MAC-es layer adds a 6 bit sequence number.

The IP and TCP throughputs depend upon the size of the Maximum Transmission Unit (MTU) defined at the TCP layer. Larger MTU sizes result in lower TCP/IP overheads, i.e. the ratio of application data to TCP/IP header is greater. The figures in Table 7.8 assume an MTU size of 1420

Table 7.8 Example maximum bit rates for a TCP/IP protocol stack

Protocol stack layer	Category 3 UE 10 ms TTI QPSK 2×SF4 (Mbps)	Category 5 UE 10 ms TTI QPSK 2×SF2 (Mbps)	Category 6 UE 2 ms TTI QPSK 2×SF4+2×SF2 (Mbps)	Category 7 UE 2 ms TTI 16QAM 2×SF4+2×SF2 (Mbps)
TCP	1.34	1.84	5.29	10.89
IP	1.36	1.86	5.36	11.04
RLC	1.38	1.89	5.44	11.20
MAC-d	1.44	1.98	5.71	11.48
MAC-es	1.44	1.98	5.71	11.48
MAC-e	1.45	1.98	5.72	11.48
Physical layer	1.45	2.00	5.74*	11.50*

*Requires a coding rate of 1.0.

bytes (11360 bits) which includes 160 bits of IP header and 160 bits of TCP header. These figures mean that the IP layer throughput is equal to the RLC layer throughput × 0.986 and the TCP layer throughput is equal to the RLC layer throughput × 0.972.

In practice, the throughput figures presented in Table 7.8 would be reduced by re-transmissions at the MAC-e, RLC and TCP layers. The majority of re-transmissions should be completed at the MAC-e layer. A typical target BLER for the MAC-e PDU (transport blocks) is 10% and in this case, the MAC-e throughput would be reduced by 10%. Implementations may be more aggressive by defining relatively large transport block sizes for each reference E-DPDCH power ratio. This would increase the physical layer throughput, but may decrease the MAC-e throughput as a result of less redundancy and an increased MAC-e re-transmission requirement.

The HUSPA total cell throughput can be estimated using the load equation. If the radio network has been planned using link budgets which include a 4 dB interference margin the corresponding maximum uplink cell load is 60%. This can be derived using the expression below.

$$\text{Interference margin} = -10 \times \text{LOG}(1 - \text{Maximum cell load})$$

The uplink load generated by a single connection is given by:

$$\text{Own cell load} = 1 \left/ \left(1 + \frac{\text{Processing gain}}{\text{Eb/No}} \right) \right.$$

This expression requires assumptions regarding the throughput of the HSUPA connection and the associated uplink Eb/No requirement. The Eb/No requirement depends upon the propagation channel, target BLER and Node B receiver performance. It also depends upon the overheads generated by the HS-DPCCH, E-DPCCH, DPCCH and DPDCH physical channels. Table 7.9 presents some example figures for a 3 km/hr Pedestrian A propagation channel when assuming a 10% transport block target BLER and a RAKE receiver with dual branch receive diversity

These figures indicate that a 1.28 Mbps HSUPA connection generates an own cell load of 30%. The own cell load is increased by inter-cell interference generated by UE connected to neighbouring cells. The quantity of inter-cell interference depends upon the geometry of the radio network plan and the activity at the neighbouring cells. A figure of 65% is typically assumed for a macrocell network with an even distribution of traffic. This can decrease to 25% for a microcell and potentially to less than 20% for an indoor solution. An inter-cell interference ratio of 65% would increase the 30% own cell load to a total cell load of 50%. The corresponding HSUPA total cell throughput is approximately 1.5 Mbps when assuming a maximum total cell load of 60%. Assuming an inter-cell interference ratio of 25% increases the HSUPA total cell throughput to more than 2 Mbps. These figures increase if the maximum total cell load has been planned to be greater than 60%. This could be the case for macrocells which provide capacity for areas of high traffic, microcells and indoor solutions. In practice, cells will also be loaded by DPCH connections and these connections will use a share of the total uplink cell load. The Eb/No requirements would increase by approximately 3 dB for cells which are not configured with dual branch receive diversity. This could be applicable to indoor solutions and microcells. The total cell throughput can be increased by using advanced receivers to decrease the uplink Eb/No requirements.

Table 7.9 Example own cell loads generated by an HSUPA connection

RLC throughput	Eb/No (dB)	Processing gain (dB)	Own cell load (%)
320 kbps	0.6	10.8	9
640 kbps	0.0	7.8	14
1.28 Mbps	1.1	4.8	30

7.3 PDCP Layer

- The PDCP layer can be used as part of the user plane protocol stack. It provides functionality for compressing the headers belonging to the higher layers, e.g. TCP/IP and RTP/UDP/IP headers for IPv4 and IPv6.
- The PDCP layer becomes particularly important for applications which have large overheads generated by the headers belonging to the higher layers, e.g. Voice over IP.
- Key 3GPP specifications: TS 25.323

Section 6.3 describes the Packet Data Convergence Protocol (PDCP) layer for HSDPA. Section 6.3 is also applicable to HSUPA.

7.4 RLC Layer

- The RLC layer processes data for HSUPA connections in the same way that it processes data for DPCH and HSDPA connections. NRT applications are likely to use acknowledged mode whereas RT applications are likely to use unacknowledged mode.
- The use of HSUPA results in a change to the signalling used to configure the size of an RLC PDU. In the case of an HSUPA connection, the RLC PDU size is signalled explicitly. This is in contrast to a DPCH connection which requires an equation to translate the signalled information into a PDU size. It is also in contrast to an HSDPA connection in which case the MAC-d PDU size is signalled rather than the RLC PDU size.
- The RLC layer is responsible for re-transmissions between the UE and RNC when using acknowledged mode. HSUPA helps to reduce the requirement for RLC re-transmissions by allowing re-transmissions from the MAC-e layer. HSUPA typically generates bursts of RLC re-transmissions rather than individual re-transmissions. These bursts of re-transmissions are caused by packaging groups of RLC PDU within single MAC-es PDU.
- Key 3GPP specifications: TS 25.322 and TS 25.331.

The general responsibilities of the RLC layer are described in Section 2.3. The RLC layer processes data for HSUPA connections in the same way that it processes data for DPCH connections. NRT applications are likely to be configured with acknowledged mode RLC whereas RT applications are likely to be configured with unacknowledged mode RLC. For example, an internet browsing service would use acknowledged mode RLC whereas a VoIP service would use unacknowledged mode RLC.

HSUPA results in a change to the signalling used to configure the size of an RLC PDU. Log File 7.1 presents an example of the RRC signalling used to specify the size of an SRB RLC PDU when

```
ul-TrCH-Type
e-dch
        logicalChannelIdentity:    1
        e-DCH-MAC-d-FlowIdentity: 0
        ddi: 0
        rlc-PDU-SizeList
        rlc-PDU-SizeList value 1
            RLC-PDU-Size: 144
        includeInSchedulingInfo: TRUE
```

Log File 7.1 RRC signalling using to specify the RLC PDU size for HSUPA (SRB)

```
mac-d-PDU-SizeInfo-List
mac-d-PDU-SizeInfo-List value 1
      mac-d-PDU-Size:    148
      mac-d-PDU-Index:   0
```

Log File 7.2   RRC signalling using to specify the MAC-d PDU size for HSDPA (SRB)

transferred using HSUPA. This information could be sent to the UE using a Radio Bearer Setup message or a Radio Bearer Reconfiguration message.

The RLC PDU size is signalled explicitly using a value of 144 bits. This is in contrast to the equivalent DPCH signalling shown in Section 6.4 which requires an equation to derive the RLC PDU size from the signalled information. The equivalent RRC signalling for an HSDPA connection is shown in Log File 7.2. This corresponds to transferring the SRB using an HSDPA connection.

In this case, the MAC-d PDU size is signalled rather than the RLC PDU size and the RLC PDU size must be derived from the MAC-d PDU size. The MAC-d PDU size is signalled using a value of 148 bits. This is equivalent to the RLC PDU size of 144 bits specified in Log File 7.1. The MAC-d entity adds a 4 bit C/T header field when it multiplexes multiple logical channels onto a single transport channel. This header field is used by the receiver to identify which logical channel has been sent at any point in time. There are always at least three DCCH SRB and so the MAC-d entity always has to include these 4 C/T bits within the MAC-d header.

There is a similar difference between the RRC signalling for an HSUPA connection and HSDPA connection when the RLC PDU size is specified for the user plane data. However, it is less common to complete MAC-d multiplexing for user plane data and so in general the C/T header is not necessary and the RLC PDU size equals the MAC-d PDU size (MAC-d multiplexing could be completed for user plane data for multi-RAB connections). Log File 7.3 presents an example of the RRC signalling used to configure the RLC PDU size for an HSUPA user plane data connection. Note that Log File 7.1 and Log File 7.3 are not extracted from the same UE log file otherwise they would have different MAC-d flow identities and different Data Description Indicator (DDI) values. DDI values are described in greater detail within Section 7.6.3.

The RLC PDU size is signalled explicitly using a value of 336 bits. The equivalent RRC signalling for an HSDPA connection is presented in Log File 7.4. The HSDPA MAC-d PDU size is signalled using a value of 336 bits, i.e. equal to the equivalent RLC PDU size for an HSUPA connection.

The RLC layer is responsible for re-transmissions between the UE and RNC. HSUPA helps to reduce the requirement for RLC re-transmissions by allowing re-transmissions from the MAC-e layer. MAC-e re-transmissions occur between the UE and Node B. The RNC outer loop power control function adjusts the uplink DPCCH SIR target to help ensure that the target MAC-e PDU BLER performance is achieved. Similar to HSDPA, a typical target BLER performance is 10%.

HSUPA typically generates bursts of RLC re-transmissions rather than individual re-transmissions. These bursts of re-transmissions are caused by packaging groups of RLC PDU within single MAC-es PDU. The number of RLC PDU within a single MAC-es PDU depends upon the transport block size

```
ul-TrCH-Type
e-dch
      logicalChannelIdentity:    5
      e-DCH-MAC-d-FlowIdentity:  0
      ddi: 0
      rlc-PDU-SizeList
      rlc-PDU-SizeList value 1
          RLC-PDU-Size:    336
      includeInSchedulingInfo:   TRUE
```

Log File 7.3   RRC signalling using to specify the RLC PDU size for HSUPA (user plane)

```
mac-d-PDU-SizeInfo-List
mac-d-PDU-SizeInfo-List value 1
      mac-d-PDU-Size:    336
      mac-d-PDU-Index:   0
```

Log File 7.4   RRC signalling using to specify the MAC-d PDU size for HSDPA (user plane)

selected by the UE during the E-TFC selection procedure. If the UE selects a transport block size which accommodates 10 RLC PDU, and if the number of MAC-e PDU re-transmissions has reached the maximum then all 10 of the RLC PDU must be re-transmitted from the RLC layer. Once the re-transmitted RLC PDU reach the MAC-es layer for the second time they are not necessarily packaged within the same MAC-es PDU. Their packaging depends upon their position within the queue and also the MAC-es PDU size selected by the UE at that point in time.

Acknowledgements for the uplink RLC PDU are returned to the UE using the corresponding downlink connection. If the downlink connection is HSDPA then the HS-DSCH will be used to return the acknowledgements.

7.5 MAC-d Entity

- The MAC-d entity generates one or more MAC-d flows. These MAC-d flows can be mapped directly onto a DCH transport channel or they can be forwarded to the MAC-es/e layer for subsequent mapping onto the E-DCH transport channel. It is common to have a first MAC-d flow for the control plane SRB and a second MAC-d flow for the user plane data.
- Key 3GPP specifications: TS 25.321.

The general responsibilities and architecture of the MAC layer are described in Section 2.4. The MAC-d entity within the UE generates one or more MAC-d flows. These MAC-d flows are either mapped directly onto a DCH transport channel or forwarded to the MAC-es/e layer within the UE for subsequent mapping onto the E-DCH transport channel. Figure 7.5 illustrates the MAC-d flows when the control plane signalling is transferred using a DPCH connection while the user plane data is transferred using an HSUPA connection.

In this case, the MAC-d entity multiplexes the set of SRB DCCH into a single MAC-d flow and maps them onto a DCH transport channel. The user plane data forms a second MAC-d flow which is forwarded to the MAC-es/e layer within the UE.

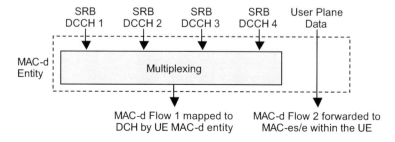

Figure 7.5   MAC-d entity when DCCH is mapped to a DCH

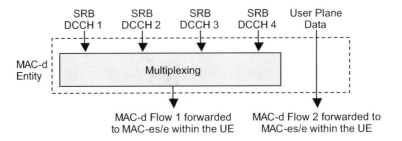

Figure 7.6 MAC-d entity when DCCH is mapped to an E-DCH

Figure 7.6 illustrates the MAC-d flows when both the control plane signalling and user plane data are transferred using HSUPA.

In this case, the MAC-d entity generates two separate MAC-d flows for transfer towards the MAC-es/e layer. As indicated by the description of the RLC layer these two MAC-d flows are likely to be configured with different MAC-d PDU sizes. The MAC-d flow used by the control plane is likely to have a MAC-d PDU size of 148 bits whereas the MAC-d flow used by the user plane data is likely to have a MAC-d PDU size of 336 or 656 bits.

7.6 MAC-es/e Entity (UE)

- The MAC-es/e entity within the UE is responsible for E-TFC selection, HARQ and generating both MAC-es and MAC-e PDU. The peer MAC-e layer is within the Node B whereas the peer MAC-es layer is within the RNC.
- E-TFC selection involves the UE selecting a single transport block size from a transport block size table. The release 6 version of 3GPP TS 25.321 defines two tables for the 2 ms TTI and two tables for the 10 ms TTI. The set of transport block sizes within each table has been optimised to minimise the requirement for padding.
- The number of transport block sizes within each table is limited by the number of bits used to inform the Node B of the E-TFCI. The E-DPCCH includes 7 bits to represent the E-TFCI. This allows a maximum of 128 transport block sizes within each table.
- The first transport block size within each table has only 18 bits. This size is used when the UE is required to transfer Scheduling Information without transferring any MAC-d PDU.
- The UE selects a transport block size using the E-TFC selection procedure once every TTI. This procedure accounts for the UE maximum transport block size capability, the UE transmit power capability, the scheduled grant allocated by the Node B packet scheduler and the quantity of MAC-d PDU buffered within the UE.
- The UE calculates the E-DPDCH to DPCCH power ratio required to transfer each transport block size using a set of reference power ratios signalled by the RNC. The RNC is able to signal a maximum of eight reference power ratios during connection establishment.
- The RNC is able to configure a minimum set of E-TFCI which cannot be blocked by the UE transmit power capability. This allows the UE to continue attempting data transfer at a low bit rate, even when its uplink transmit power capability has become exhausted.
- Scheduled grant values represent the total E-DPDCH to DPCCH power ratio. Scheduled grant values which include ×2, ×4 and ×6 factors are intended for scheduling resources for multiple physical channels. A single SF2 E-DPDCH is counted as two physical channels, i.e. values which

include a $\times 2$, $\times 4$ and $\times 6$ factor are intended for the $2 \times SF4$, $2 \times SF2$ and $2 \times SF4 + 2 \times SF2$ configurations respectively.

- The HSUPA HARQ protocol is based upon parallel Stop and Wait (SAW) HARQ processes. An HSUPA connection using a 10 ms TTI always uses four parallel processes whereas a connection using a 2 ms TTI always uses eight parallel processes.

- It is not necessary to inform the Node B of which HARQ process is being used at any point in time. This can be avoided because the HSUPA protocol is synchronous and the timing is deterministic.

- The maximum number of HARQ re-transmissions can be configured independently for each MAC-d flow. A maximum of eight re-transmissions are always used when Scheduling Information is transferred alone, i.e. without any MAC-d flow data.

- The HARQ protocol can use chase combining or incremental redundancy. Incremental redundancy does not require the transmission of any redundancy version parameters because they are deterministic and can be calculated by both the Node B and UE.

- A MAC-es PDU represents a collection of equal sized MAC-d PDU belonging to the same MAC-d flow and the same logical channel. The header of a MAC-es PDU includes a 6 bit Transmission Sequence Number (TSN). The RNC uses the TSN when re-ordering MAC-es PDU. Each logical channel has its own sequence numbering.

- The MAC-es layer also generates a Data Description Indicator (DDI) and a Number of MAC-d PDU (N) indicator for every MAC-es PDU. This information is not attached to the MAC-es PDU but is forwarded to the MAC-e layer for subsequent inclusion within the MAC-e PDU. The DDI value is used to identify the MAC-d flow, logical channel identity and MAC-d PDU size.

- A MAC-e PDU represents an HSUPA transport block. Each HSUPA connection can transfer a maximum of one transport block during each TTI. A transport block can include zero, one or more MAC-es PDU. It can also include Scheduling Information.

- Scheduling Information is included to help support the operation of the Node B packet scheduler. It provides a measure of both the resource requirement and the ability of the UE to make use of additional resources.

- The MAC-e and MAC-es headers sum to a total of ($n \times 18$) bits, where n is the number of MAC-es PDU within the MAC-e PDU. Scheduling Information adds a further 18 bits whenever it is included.

- Key 3GPP specifications: TS 25.321, TS 25.133, TS 25.213 and TS 25.214.

7.6.1 E-TFC Selection

The E-DCH Transport Format Combination (E-TFC) selection procedure involves the UE selecting a single transport block size from a transport block size table. The release 6 version of 3GPP TS 25.321 defines two transport block size tables for the 2 ms TTI and two transport block size tables for the 10 ms TTI. The release 7 version of the TS 25.321 introduces an additional two tables for the 2 ms TTI. These additional tables include larger transport block sizes to support the increased throughputs provided by the 16QAM modulation scheme. The set of transport block sizes within each table has been optimised to minimise the requirement for padding. This has been done assuming either an arbitrary MAC-d PDU size or a fixed MAC-d PDU size of 168 or 336 bits. Table 7.10 shows which transport block size tables have been optimised for an arbitrary MAC-d PDU size and which have been optimised for a fixed PDU size.

An exponential increase of transport block size has been used to optimise the transport block size tables for an arbitrary MAC-d PDU size, i.e. plotting the transport block sizes generates an exponential curve. This means that the difference between consecutive transport block sizes is relatively small at the bottom of the table whereas the difference is relatively large at the top of the table. This approach

Table 7.10 MAC-d PDU sizes for which the transport block size tables have been optimised

Table	10 ms TTI	2 ms TTI
0	Arbitrary MAC-d PDU sizes	Arbitrary MAC-d PDU sizes
1	MAC-d PDU sizes of 168 and 336 bits	MAC-d PDU sizes of 168 and 336 bits
2*		Arbitrary MAC-d PDU sizes
3*		MAC-d PDU sizes of 168 and 336 bits

*Introduced by release 7 of the 3GPP specifications for the 16QAM modulation scheme

helps to minimise the quantity of padding when expressed as a percentage of the transport block size. Optimising the set of transport block sizes for fixed MAC-d PDU sizes of 168 and 336 bits is based upon defining transport block sizes which are a multiple of either 168 or 336 bits after accounting for the MAC-es header, the MAC-e header and the Scheduling Information (SI).

The number of transport block sizes within each table is limited by the number of bits used to inform the Node B of the E-DCH Transport Format Combination Indicator (E-TFCI). The E-DPCCH includes 7 bits which are dedicated to signalling the E-TFCI value. This allows a range from 0 to 127 and means there can be a maximum of 128 transport block sizes within each table. The RNC informs both the UE and Node B of the table to be used during connection establishment. An example section from a message which instructs the UE to use table index 1 is illustrated within Log File 7.5. The TTI is

```
E-DPDCH-Info
        e-TFCI-TableIndex:              1
        e-DCH-MinimumSet-E-TFCI:        4
        reference-E-TFCIs
        E-DPDCH-Reference-E-TFCI value 1
                reference-E-TFCI.                    3
                reference-E-TFCI-PO:               11
        E-DPDCH-Reference-E-TFCI value 2
                reference-E-TFCI:                   7
                reference-E-TFCI-PO:               13
        E-DPDCH-Reference-E-TFCI value 3
                reference-E-TFCI:                  11
                reference-E-TFCI-PO:               15
        E-DPDCH-Reference-E-TFCI value 4
                reference-E-TFCI:                  19
                reference-E-TFCI-PO:               18
        E-DPDCH-Reference-E-TFCI value 5
                reference-E-TFCI:                  39
                reference-E-TFCI-PO:               20
        E-DPDCH-Reference-E-TFCI value 6
                reference-E-TFCI:                  59
                reference-E-TFCI-PO:               22
        E-DPDCH-Reference-E-TFCI value 7
                reference-E-TFCI:                  79
                reference-E-TFCI-PO:               24
        maxChannelisationCodes: sf4x2
pl-NonMax:      17
schedulingInfoConfiguration
        periodicityOfSchedInfo-NoGrant:    ms200
        periodicityOfSchedInfo-Grant:      ms200
        powerOffsetForSchedInfo:           3
threeIndexStepThreshold:    6
twoIndexStepThreshold:     15
```

Log File 7.5 RRC signalling used to provide the UE with E-DCH transport block size information signalled elsewhere within the same message. This section of RRC signalling could belong to either a Radio Bearer Setup or a Radio Bearer Reconfiguration message.

Table 7.11 presents the content of transport block size table 1 for a 10 ms TTI. This table has been optimised for fixed MAC-d PDU sizes of 168 and 336 bits. The first transport block size within the table is significantly smaller than 168 bits, i.e. 18 bits. This size is used when the UE is required to transfer SI without transferring any MAC-d PDU, e.g. when there is no application data to transfer. The content of the SI is described within Section 7.6.4. It occupies exactly 18 bits. The second transport block size is equal to $168 + 18 = 186$ bits. This assumes a single MAC-d PDU of 168 bits with the addition of an 18 bit MAC-es/e header. The third transport block size is equal to $168 + 18 + 18 = 204$ bits. This assumes a single MAC-d PDU of 168 bits with the addition of 18 bits of MAC-es/e header and 18 bits of SI. All subsequent transport block sizes are generated from a multiple of either 168 or 336 bits plus a multiple of 18 bits. Some of the larger transport block sizes are based upon $(n \times 336) + (7 \times 18)$ bits. These transport block sizes assume that there are either 6 MAC-es PDU within each MAC-e PDU and that SI is included, or that there are 7 MAC-es PDU within each MAC-e PDU and that SI is excluded. For example, the maximum transport block size is 19950 bits which is generated from $(59 \times 336) + (7 \times 18)$.

The optimised set of transport block sizes does not always avoid the requirement for padding. However, the padding represents a relatively small overhead when it is required. Table 7.12 presents the padding requirement for a range of example payloads. These figures assume a single MAC-es PDU within each MAC-e PDU and so the total MAC-es/e header size is always 18 bits. Examples with and without SI are included. Padding is not necessary for the smaller transport block sizes, but as the transport block sizes increase, padding is required for the examples which exclude SI. The quantity of padding is significantly greater for the example based upon 50 MAC-d PDU. This results from the top end of the table including transport block sizes based upon every alternate number of MAC-d PDU

Table 7.11   HSUPA transport block size table 1 for a 10 ms TTI

TB index	TB size (bits)	TB index	TB size (bits)	TB index	TB size (bits)	TB index	TB size (bits)	TB index	TB size (bits)	TB index	TB size (bits)
0	18	21	1866	42	5094	63	8772	84	12186	105	15828
1	186	22	1884	43	5412	64	8808	85	12468	106	15900
2	204	23	2034	44	5430	65	9108	86	12522	107	16164
3	354	24	2052	45	5748	66	9144	87	12804	108	16236
4	372	25	2370	46	5766	67	9444	88	12858	109	16500
5	522	26	2388	47	6084	68	9480	89	13140	110	16572
6	540	27	2706	48	6102	69	9780	90	13194	111	17172
7	690	28	2724	49	6420	70	9816	91	13476	112	17244
8	708	29	3042	50	6438	71	10116	92	13530	113	17844
9	858	30	3060	51	6756	72	10152	93	13812	114	17916
10	876	31	3378	52	6774	73	10452	94	13866	115	18516
11	1026	32	3396	53	7092	74	10488	95	14148	116	18606
12	1044	33	3732	54	7110	75	10788	96	14202	117	19188
13	1194	34	3750	55	7428	76	10824	97	14484	118	19278
14	1212	35	4068	56	7464	77	11124	98	14556	119	19860
15	1362	36	4086	57	7764	78	11178	99	14820	120	19950
16	1380	37	4404	58	7800	79	11460	100	14892		
17	1530	38	4422	59	8100	80	11514	101	15156		
18	1548	39	4740	60	8136	81	11796	102	15228		
19	1698	40	4758	61	8436	82	11850	103	15492		
20	1716	41	5076	62	8472	83	12132	104	15564		

Table 7.12   Padding requirement for a range of MAC-d PDU payloads (PDU size of 336 bits)

Number of MAC-d PDU	RLC throughput	MAC-d PDU data (bits)	MAC-es/e header (bits)	SI (bits)	TB Size (bits)	Padding (bits)
0	0 kbps	0	0	18	18	0
1	32 kbps	336	18	0	354	0
1	32 kbps	336	18	18	372	0
5	160 kbps	1680	18	0	1698	0
5	160 kbps	1680	18	18	1716	0
10	320 kbps	3360	18	0	3378	0
10	320 kbps	3360	18	18	3396	0
20	640 kbps	6720	18	0	6756	18
20	640 kbps	6720	18	18	6756	0
30	960 kbps	10080	18	0	10116	18
30	960 kbps	10080	18	18	10116	0
40	1.28 Mbps	13440	18	0	13476	18
40	1.28 Mbps	13440	18	18	13476	0
50	1.60 Mbps	16800	18	0	17172	354
50	1.60 Mbps	16800	18	18	17172	336
59	1.89 Mbps	19824	18	0	19860	18
59	1.89 Mbps	19824	18	18	19860	0

rather than every consecutive number of MAC-d PDU, i.e. the top end of the table is based upon 46, 47, 48 49, 51, 53, 55, 57 and 59 MAC-d PDU. In practice, if the UE selects a transport block size of 17172 bits then it will include 51 MAC-d PDU rather than 50 MAC-d PDU. The UE would only limit itself to including 50 PDU if a 51st PDU is not available within its buffers.

The UE selects a transport block size using the E-TFC selection procedure once every TTI. This procedure accounts for the UE maximum transport block size capability, the UE transmit power capability, the scheduled grant allocated by the Node B packet scheduler and the quantity of MAC-d PDU buffered within the UE.

The UE maximum transport block size capability can be used to truncate the top end of the transport block size table. Table 7.4 illustrates that a category 3 UE is able to support a maximum transport block size of 14484 bits. This means that only transport block size indices 0–97 are applicable to category 3 UE. A category 5 UE is able to support a maximum transport block size of 19950 bits and so the complete set of transport block size indices are applicable. The Node B packet scheduler is aware of the UE capability and should not allocate resources which would allow the UE to exceed its capability. This could result in resources being reserved unnecessarily.

The UE is responsible for identifying which transport block sizes cannot be transferred due to insufficient transmit power. If a DPDCH is configured in parallel to the E-DPDCH then TFC selection always has priority over E-TFC selection, i.e. data sent on the E-DPDCH is only able to use the transmit power remaining after subtracting the DPDCH power. 3GPP TS 25.133 specifies the calculation used to evaluate the maximum available E-DPDCH to DPCCH power ratio. The principle of this calculation is illustrated by the equation below.

$$\text{Maximum available power ratio} = \frac{P_{\text{Max}} - P_{\text{DPCCH}} - P_{\text{DPDCH}} - P_{\text{HS-DPCCH}} - P_{\text{E-DPDCH}}}{P_{\text{DPCCH}}}$$

The maximum available E-DPDCH transmit power is defined as the difference between the UE maximum transmit power capability and the sum of the transmit powers allocated to the other uplink physical channels. The UE compares the resultant maximum available power ratio with the power ratio required to transfer each of the transport block sizes within the transport block size table. A transport block size is blocked from the subsequent selection procedure if its power ratio requirement exceeds the maximum available power ratio.

The UE calculates the power ratio required to transfer each transport block size using a set of reference power ratios signalled by the RNC. The RNC is able to signal a maximum of eight reference power ratios during HSUPA connection establishment. These reference power ratios are likely to depend upon the UE category, i.e. they should span an appropriate range of transport block sizes. Log File 7.5 includes an example set of reference power ratios for a category 3 UE. Each reference power ratio is paired with a reference E-TFCI which points towards a specific transport block size. The reference power ratios themselves are signalled using a pointer towards an E-DPDCH to DPCCH amplitude ratio. These pointers and their corresponding amplitude ratios are presented in Table 7.13. Each amplitude ratio is applicable to a single E-DPDCH. This is in contrast to the maximum available power ratio calculated from the UE transmit power capability which is based upon the total E-DPDCH power.

The content of Table 7.13 has been extracted from the release 6 version of 3GPP TS 25.213. The release 7 version includes an additional table with larger amplitude offsets. These larger offsets are intended for use with 16QAM.

Table 7.14 summarises the set of reference E-TFCI and reference power offsets shown in Log File 7.5.

Table 7.13 Signalled E-DPDCH to DPCCH amplitude ratios

Signalled power offset	Amplitude ratio, β_{ed}/β_c	Signalled power offset	Amplitude ratio, β_{ed}/β_c	Signalled power offset	Amplitude ratio, β_{ed}/β_c
29	168/15	19	53/15	9	17/15
28	150/15	18	47/15	8	15/15
27	134/15	17	42/15	7	13/15
26	119/15	16	38/15	6	12/15
25	106/15	15	34/15	5	11/15
24	95/15	14	30/15	4	9/15
23	84/15	13	27/15	3	8/15
22	75/15	12	24/15	2	7/15
21	67/15	11	21/15	1	6/15
20	60/15	10	19/15	0	5/15

Table 7.14 Example reference E-TFCI with corresponding reference power ratios

Reference E-TFCI	Transport block size	Reference power offset	E-DPDCH to DPCCH amplitude ratio	Equivalent power ratio (dB)
3	354	11	21/15	2.9
7	690	13	27/15	5.1
11	1026	15	34/15	7.1
19	1698	18	47/15	9.9
39	4740	20	60/15	12.0
59	8100	22	75/15	14.0
79	11460	24	95/15	16.0

The UE uses this reference information to calculate the power ratio required for each transport block size. The procedure for this calculation is specified within 3GPP TS 25.214. The calculation is based upon the following equation.

$$\beta_{ed,j,harq} = \beta_c \times \text{Amplitude ratio}_{ref} \times \sqrt{\frac{\#\text{PhysCh}_{ref}}{\#\text{PhysCh}_j}} \times \sqrt{\frac{\text{TrBlkSize}_j}{\text{TrBlkSize}_{ref}}} \times 10^{\frac{\Delta_{harq}}{20}}$$

Where $\beta_{ed,j,harq}$ represents a temporary E-DPDCH variable for the jth transport block size and a specific HARQ profile. β_c is the DPCCH amplitude ratio which can range from 1/15 to 15/15. β_c and β_d are used to configure the relative powers of the DPCCH and DPDCH respectively. β_c, β_d, β_{ed}, β_{ec} and β_{hs} are illustrated in Figure 7.17. If a DPDCH has not been configured then β_c is assumed to have a value of 1. Amplitude ratio$_{ref}$ is the amplitude ratio associated with the appropriate reference transport block size. If the jth transport block size is equal to or larger than the largest reference size then the largest reference size and its associated amplitude ratio are used. If the jth transport block size is smaller than the smallest reference size then the smallest reference size and its associated amplitude ratio are used. Otherwise, the reference size and its associated amplitude ratio are identified by rounding down to the nearest reference size.

The UE then has to calculate the number of E-DPDCH physical channels required to transfer the jth transport block size and the number of E-DPDCH physical channels required to transfer the reference transport block size. This calculation is based upon the requirement for puncturing and is described within Section 7.7.2. In this context, a single SF2 E-DPDCH is counted as $2 \times$ SF4 E-DPDCH, i.e. it is counted as two physical channels. This leads to the $2 \times$ SF4 configuration having 2 physical channels, the $2 \times$ SF2 configuration having four physical channels and the $2 \times$ SF4 + $2 \times$ SF2 configuration having six physical channels.

The next part of the equation is based upon the ratio of the jth transport block size and the reference transport block size. The final part of the equation accounts for an additional power offset, Δ_{harq} which can be configured by the RNC during connection establishment. This power offset can be configured independently for each MAC-d flow. Configuring a large value for a specific MAC-d flow causes the set of transport block sizes to have an increased power ratio requirement, i.e. they will be transmitted with increased E-DPDCH transmit power and will benefit from a reduced BLER performance. This allows the quality of service to be adjusted on a per MAC-d flow basis. The value of Δ_{harq} can range from 0 to 6 dB. A separate value which is independent of any MAC-d flow is configured for the transport block size of 18. This represents a power offset which is applied when the transport block includes only SI. An example of this power offset is included within Log File 7.5, i.e. 'power-OffsetForSchedInfo = 3 dB'.

Once the temporary variable $\beta_{ed,j,harq}$ has been calculated, an unquantised gain factor is generated for each E-DPDCH. This gain factor is defined by the equations below.

$$\beta_{ed,k,j,unquantised} = \sqrt{2} \times \beta_{ed,j,harq}$$

if the kth E-DPDCH has a spreading factor equal to 2,

$$\beta_{ed,k,j,unquantised} = \beta_{ed,j,harq}$$

if the kth E-DPDCH has a spreading factor greater than 2.

The unquantised value for each E-DPDCH is then mapped onto a quantised value from Table 7.13. If $\beta_{ed,k,j,unquantised} / \beta_c$ is less than the smallest value within Table 7.13 then the quantised value of $\beta_{ed,k} / \beta_c$ is set equal to the smallest value within the table. Otherwise, $\beta_{ed,k} / \beta_c$ is set to the largest value for which $\beta_{ed,k} \leq \beta_{ed,k,j,unquantised}$.

Finally, the total E-DPDCH power ratio requirement is obtained by summing the power ratios associated with each E-DPDCH, i.e.

$$\text{Total E-DPDCH to DPCCH power ratio requirement} = \sum_{j=1}^{\#\text{E-DPDCH}} \left(\frac{\beta_{ed,j}}{\beta_c}\right)^2$$

This result is then compared with the maximum available power ratio calculated from the UE transmit power capability. A transport block size is blocked if its power ratio requirement is greater than the maximum available power ratio. A set of example total power ratio requirements is presented in Table 7.15. These examples are based upon the reference power offsets within Log File 7.5. They assume that $\beta_c = 1$ and $\Delta_{harq} = 0$ dB. In these examples, the number of physical channels required for the jth transport block size is always equal to the number of physical channels required for the reference transport block size. The figures are applicable to a category 3 UE and so the maximum transport block size is 14484 bits and the maximum number of physical channels is $2\times$SF4.

Based upon these figures, if the UE determines that its maximum available power ratio is 17 dB then transport block sizes of 11460, 12858 and 14484 (plus other intermediate sizes not included within Table 7.15) will be blocked from the subsequent selection procedure.

The RNC is able to configure a minimum set of E-TFCI which cannot be placed in the blocked stated while the DPDCH is inactive. This allows the UE to continue attempting data transfer at a low bit rate, even when its uplink transmit power capability has become exhausted. An example of the signalling used to configure the minimum set of E-TFCI is included within Log File 7.5. In this example, the minimum set is configured with an index of 4 which corresponds to a transport block size of 372 bits. This allows the UE to continue attempting to transfer a single MAC-d PDU plus SI during each TTI.

The UE considers the resources allocated by the Node B packet scheduler after it has identified which E-TFC are in the supported state. The packet scheduler manages the uplink interference margin at the Node B receiver. It ensures that the maximum allowed interference margin is not exceeded by limiting the E-DPDCH power ratio allocated to each UE. Resources can be allocated using either the E-DCH Absolute Grant Channel (E-AGCH) or the E-DCH Relative Grant Channel (E-RGCH). Resources can also be allocated using a combination of NBAP and RRC signalling, i.e. the Node B

Table 7.15 Example calculations of the total E-DPDCH to DPCCH power ratio requirement

Transport block size (bits)	Reference transport block size (bits)	Reference amplitude ratio	Number of physical channels	Unquantised amplitude ratio	Quantised amplitude ratio	Total E-DPDCH to DPCCH Power ratio required (dB)
354	354	21/15	1	1.40	21/15	2.9
522	354	21/15	1	1.70	24/15	4.1
690	690	27/15	1	1.80	27/15	5.1
858	690	27/15	1	2.01	30/15	6.0
1026	1026	34/15	1	2.27	34/15	7.1
1362	1026	34/15	1	2.61	38/15	8.1
1698	1698	47/15	1	3.13	47/15	9.9
3042	1698	47/15	1	4.19	60/15	12.0
4740	4740	60/15	2	4.00	60/15	15.1
6420	4740	60/15	2	4.66	67/15	16.0
8100	8100	75/15	2	5.00	75/15	17.0
9780	8100	75/15	2	5.49	75/15	17.0
11460	11460	95/15	2	6.33	95/15	19.0
12858	11460	95/15	2	6.71	95/15	19.0
14484	11460	95/15	2	7.12	106/15	20.0

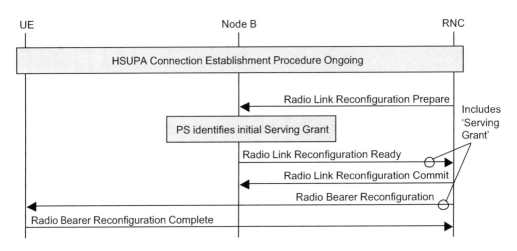

UE   Node B   RNC

HSUPA Connection Establishment Procedure Ongoing

Radio Link Reconfiguration Prepare

Includes 'Serving Grant'

PS identifies initial Serving Grant

Radio Link Reconfiguration Ready

Radio Link Reconfiguration Commit

Radio Bearer Reconfiguration

Radio Bearer Reconfiguration Complete

Figure 7.7 Allocating initial resources using Radio Link and Radio Bearer reconfigurations

informs the RNC of the allocation using NBAP signalling and the RNC subsequently informs the UE of the allocation using RRC signalling.

The E-DPDCH power ratio allocated by the Node B can be used to truncate the top end of the transport block size table. This could involve truncating a section which is already in the blocked state after the evaluation of the UE transmit power capability. It could also involve truncating a section which is in the supported state, i.e. the packet scheduler limits the power ratio resources to a greater extent than the UE transmit power capability. Ideally, the packet scheduler would never allocate a section of the transport block size table which is in the blocked state because it results in resources being reserved unnecessarily.

An initial resource allocation can be granted by the Node B packet scheduler during connection establishment. This resource allocation is signalled to the RNC as part of the NBAP Radio Link Setup or NBAP Radio Link Reconfiguration procedures. An example based upon the Radio Link Reconfiguration procedure is illustrated in Figure 7.7.

The initial resource allocation is signalled using a Serving Grant value which can range from 0 to 38. The set of values between 0 and 37 allocate an E-DPDCH power ratio whereas the value of 38 indicates a zero allocation. The set of power ratios associated with each Serving Grant is presented within Table 7.16. The term Serving Grant is generalised to Scheduled Grant within this table because the set

Table 7.16 Scheduled Grant values

Index	Scheduled grant	Index	Scheduled grant	Index	Scheduled grant	Index	Scheduled grant
37	$(168/15)^2 \times 6$	27	$(134/15)^2$	17	$(42/15)^2$	7	$(13/15)^2$
36	$(150/15)^2 \times 6$	26	$(119/15)^2$	16	$(38/15)^2$	6	$(12/15)^2$
35	$(168/15)^2 \times 4$	25	$(106/15)^2$	15	$(34/15)^2$	5	$(11/15)^2$
34	$(150/15)^2 \times 4$	24	$(95/15)^2$	14	$(30/15)^2$	4	$(9/15)^2$
33	$(134/15)^2 \times 4$	23	$(84/15)^2$	13	$(27/15)^2$	3	$(8/15)^2$
32	$(119/15)^2 \times 4$	22	$(75/15)^2$	12	$(24/15)^2$	2	$(7/15)^2$
31	$(150/15)^2 \times 2$	21	$(67/15)^2$	11	$(21/15)^2$	1	$(6/15)^2$
30	$(95/15)^2 \times 4$	20	$(60/15)^2$	10	$(19/15)^2$	0	$(5/15)^2$
29	$(168/15)^2$	19	$(53/15)^2$	9	$(17/15)^2$	38	0
28	$(150/15)^2$	18	$(47/15)^2$	8	$(15/15)^2$		

of values are also used for the more general scheduling procedure. The values within Table 7.16 have been extracted from the release 6 version of 3GPP TS 25.321. The release 7 version includes a second set of values which are applicable when the 16QAM modulation scheme is used. This second set of values ranges from $(11/15)^2$ to $(376/15)^2 \times 4$, i.e. larger power ratios are included.

Scheduled grant values represent the total E-DPDCH to DPCCH power ratio. This is in contrast to the amplitude ratios within Table 7.13 which are applicable to individual E-DPDCH. The scheduled grant values which include $\times 2$, $\times 4$ and $\times 6$ factors are intended for scheduling resources for multiple E-DPDCH physical channels. Similar to during the evaluation of the UE transmit power capability, a single SF2 E-DPDCH is counted as 2 physical channels, i.e. values which include a $\times 2$, $\times 4$ and $\times 6$ factor are intended for the $2 \times SF4$, $2 \times SF2$ and $2 \times SF4 + 2 \times SF2$ configurations, respectively.

The UE compares the scheduled grant with the power ratio requirement associated with each transport block size. The power ratio requirements have already been calculated from the set of reference power ratios while evaluating the UE transmit power capability. A specific transport block size is categorised as scheduled if its power ratio requirement is less than or equal to the scheduled grant power ratio. The UE can then identify the maximum scheduled transport block size. The same procedure is completed when subsequent allocations are made using the E-AGCH and E-RGCH. Similar to the Serving Grant value, these physical channels indicate a position within Table 7.16. The E-AGCH and E-RGCH are described in greater detail within Section 7.7.

The final stage of E-TFC selection considers the quantity of MAC-d PDU buffered within the UE. The UE should not select a large transport block size if there are only a few MAC-d PDU ready for transmission. This would generate increased quantities of padding within the transport block. It would be preferable to select a smaller transport block size which accommodates the set of MAC-d PDU, but which reduces the quantity of padding. Using a smaller transport block size helps to reduce the

TB	TB Size	TB	TB Size	TB	TB Size	TB	TB Size	TB	TB Size	TB	TB Size
0	(hatched)	21	1866	42	5094	63	8772	84	12186	105	15828
1	Minimum Set	22	1884	43	5412	64	8808	85	Not allocated by Node B PS	106	15900
2		23	2034	44	5430	65	9108	86		107	16164
3	(hatched)	24	2052	45	5748	66	(hatched)	87		108	16236
4	(hatched)	25	2370	46	5766	67	9444	88		109	16500
5	522	26	2388	47	6084	68	9480	89	13140	110	16572
6	540	27	2706	48	6102	69	9780	90	13194	111	Excluded by UE category
7	UE selects maximum allowed transport block size if it has sufficient MAC-d PDU	28		49		70	9816	91	13476	112	
8		29		50		71	10116	92	Blocked by UE Transmit Power	113	
9		30		51		72	10152	93		114	17916
10	876	31	3378	52	6774	73	10452	94		115	18516
11	1026	32	3396	53	7092	74	Not allocated by Node B PS	95		116	18606
12	UE selects smaller transport block size if it has a smaller number of buffered MAC-d PDU	33		54	'10	75		96	14202	117	19188
13		34		55	128	76		97	14484	118	19278
14		35		56	164	77		98	14556	119	19860
15	1362	36	4086	57	7764	78	11178	99	14820	120	19950
16	1380	37	4404	58	7800	79	11460	100	Excluded by UE category		
17	1530	38	4422	59	8100	80	11514	101			
18	(hatched)	39	4740	60	8136	81	11796	102			
19	1698	40	4758	61	8436	82	11850	103	15492		
20	1716	41	5076	62	8472	83	12132	104	15564		

Figure 7.8   Example of E-TFC selection from transport block size table

requirement for puncturing within the physical layer and also helps to allow the selection of larger spreading factors. In contrast, the UE should select the largest allowed transport block size if it has sufficient MAC-d PDU ready for transmission.

The complete set of principles used for E-TFC selection are summarised within Figure 7.8. This figure is based upon a category 3 UE which supports a maximum transport block size of 14484 bits. Evaluation of the UE transmit power capability has blocked transport block sizes from 13194 to 14484 bit, while the Node B packet scheduler has allocated resources for a maximum transport block size of 9144 bits. The UE selects the transport block size of 9144 bits if it has at least 27 MAC-d PDU ready for transmission (assuming a MAC-d PDU size of 336 bits). Otherwise it selects a smaller transport block size which accommodates the complete set of PDU while minimising the requirement for padding. In this example, it is not necessary to use the minimum set of transport block sizes because the UE transmit power capability supports a relatively large set of transport block sizes.

The E-TFC selection procedure can also be influenced by the logical channel priority. If multiple logical channels are multiplexed onto the E-DCH then the E-TFC selection procedure is responsible for ensuring that transport block sizes are selected to allow the highest priority logical channel data to be transferred first. The choice of logical channel can have an impact upon the transport block size because each logical channel can use a different MAC-d PDU size. In addition, multiplexing multiple logical channels within the same transport block requires multiple MAC-es PDU and so the total number of MAC-es/e header bits increases.

7.6.2 Hybrid Automatic Repeat Request (HARQ)

Hybrid Automatic Repeat Request (HARQ) refers to a re-transmission protocol in which the receiver checks for errors in the received data and if an error is detected then the receiver buffers the data and requests a re-transmission from the sender. A HARQ receiver is then able to combine the buffered data with the re-transmitted data prior to channel decoding and error detection. The HARQ protocol used for HSUPA shares many common principles with the HARQ protocol used for HSDPA.

Both HSUPA and HSDPA HARQ protocols are based upon parallel HARQ processes. These HARQ processes are categorised as Stop and Wait (SAW) because they stop and wait for an acknowledgement prior to transmitting any further data. The round trip time means that acknowledgements are not received instantaneously. The use of parallel HARQ processes avoids the round trip time having an impact upon throughput, i.e. a second HARQ process transmits a transport block while the first HARQ process is waiting for its acknowledgement. In the case of HSDPA, the number of HARQ processes used at any point in time is dynamic and changes according to the rate at which TTI are scheduled. If the number of TTI allocated to a UE decreases then the requirement for parallel HARQ processes also decreases. An HSDPA connection can use a maximum of 8 parallel HARQ processes although six are sufficient to ensure a continuous flow of data when a UE is allocated every consecutive TTI. In the case of HSUPA, the number of HARQ processes is fixed. An HSUPA connection using a 10 ms TTI always uses 4 parallel HARQ processes whereas a connection using a 2 ms TTI always uses 8 parallel HARQ processes. In the case of HSDPA, the HS-SCCH is used to inform the UE of the HARQ process which is being used at any point in time. In contrast, it is not necessary to inform the Node B of which HSUPA HARQ process is being used at any point in time. This can be avoided because the HSUPA protocol is synchronous and the timing is deterministic. The HARQ processes are allocated to consecutive TTI in a cyclic fashion.

Figure 7.9 illustrates an example of the HARQ round trip timing for an HSUPA connection using a 10 ms TTI. In this example it is assumed that the downlink DPCH frame timing follows the CPICH frame timing by two time slots (20×256 chips). This means the corresponding E-HICH frame timing is four time slots in advance of the CPICH. Section 7.7.3 describes the relationship between the downlink DPCH and E-HICH timing. 3GPP TS 25.214 specifies that when a 10 ms TTI is used, an E-HICH acknowledgement during SFN y is applicable to the E-DPDCH transmission during SFN $y - 3$.

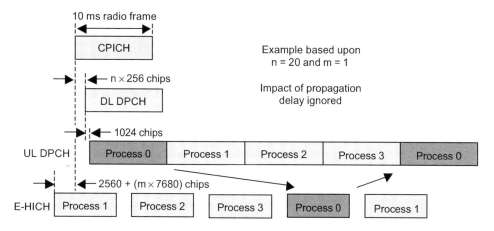

Figure 7.9 HARQ round trip timing when using a 10 ms TTI

Reception of the E-HICH acknowledgement allows the UE to re-use the HARQ process during the next available TTI. Three consecutive TTI pass before the UE is able to re-use the HARQ process and so a total of 4 HARQ processes are required to ensure a continuous flow of data. E-HICH acknowledgements will be sent from each active set cell if a UE is in soft handover. Those cells will have different CPICH and E-HICH timings so there will be some variance in the timing of the received acknowledgements.

Figure 7.10 illustrates an example of the HARQ round trip timing for an HSUPA connection using a 2 ms TTI. Similar to the previous example, it is assumed that the downlink DPCH frame timing follows the CPICH frame timing by two time slots. This means the corresponding E-HICH frame timing follows the CPICH frame timing by eight time slots.

3GPP TS 25.214 specifies the rules which link a specific E-HICH acknowledgement to a specific E-DPDCH transmission. The content of Table 7.17 demonstrates these rules for a 2 ms TTI. For example, an E-HICH acknowledgement received during SFN y, subframe 0 is applicable to the E-DPDCH transmission during SFN $y - 1$, subframe 2. This relationship introduces a three subframe delay which adds to the delay between the start of the E-DPDCH radio frame and the start of the E-HICH radio frame.

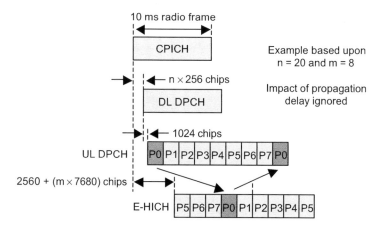

Figure 7.10 HARQ round trip timing when using a 2 ms TTI

Table 7.17 Link between E-HICH acknowledgements and E-DPDCH transmissions (2 ms TTI)

E-HICH SFN	y	y	y	y	y	$y+1$	$y+1$	$y+1$	$y+1$	$y+1$	$y+2$	$y+2$	$y+2$
E-HICH subframe	0	1	2	3	4	0	1	2	3	4	0	1	2
E-DPDCH SFN	$y-1$	$y-1$	$y-1$	y	y	y	y	y	$y+1$	$y+1$	$y+1$	$y+1$	$y+1$
E-DPDCH subframe	2	3	4	0	1	2	3	4	0	1	2	3	4

The use of eight parallel HARQ processes for a 2 ms TTI provides some margin for variance in receiving the acknowledgements from multiple active set cells. Nevertheless, the HARQ round trip time for a 2 ms TTI is significantly less than that for a 10 ms TTI, i.e. HARQ processes are re-used every 16 ms compared with every 40 ms for the 10 ms TTI. This allows re-transmissions to be completed more rapidly when using the 2 ms TTI.

The maximum number of re-transmissions can be configured independently for each MAC-d flow. An example of this is included within Log File 7.6. In this example, the MAC-d flow with identity 0 is configured to use a maximum of 3 re-transmissions. The 3GPP specifications support a maximum of 15 re-transmissions. 3GPP TS 25.321 specifies that a maximum of 8 re-transmissions are always used when Scheduling Information is transferred alone, i.e. without any MAC-d flow data.

The HSUPA HARQ protocol can use either chase combining or incremental redundancy. The principles of these re-transmission combining techniques are described within Section 6.8.4. In the case of HSDPA, redundancy version bits are included within the HS-SCCH to inform the UE of which puncturing pattern has been used at the transmitter, and so which combining technique should be used at the receiver, i.e. chase combining is used if re-transmissions use the same puncturing pattern as the original transmission. In the case of HSUPA, the RNC uses RRC signaling during connection establishment to configure the use of either chase combining or incremental redundancy. An example of the relevant signaling is included within Log File 7.6. In this example, HARQ information is signaled using a value of 'rvtable' rather than 'rv0'. This indicates that incremental redundancy should be used rather than chase combining. Signaling a value of 'rv0' indicates that chase combining should be used. The use of incremental redundancy with HSUPA does not require the transmission of any redundancy version parameters during the data transfer. This is avoided by using redundancy version parameters which are deterministic and can be calculated independently by both the Node B and UE. The first transmission always assigns priority to the systematic bits (the most important bits generated by the Turbo encoder). Subsequent re-transmissions can assign priority to either the systematic bits or the parity bits. If the number of systematic bits divided by the total number of bits offered by the

```
UL-AddReconfTransChInformation-r6
e-dch
   tti:                     tti10
   harq-Info:               rvtable
   addReconf-MAC-d-FlowList
   E-DCH-AddReconf-MAC-d-FlowList
      E-DCH-AddReconf-MAC-d-Flow
         mac-d-FlowIdentity:       0
         mac-d-FlowPowerOffset:    0
         mac-d-FlowMaxRetrans:     3
         transmissionGrantType
            scheduledTransmissionGrantInfo
```

Log File 7.6 HARQ information within MAC-d flow configuration

E-DPDCH physical channel(s) is less than 0.5, the systematic bits are assigned priority for all re-transmissions. A result of less than 0.5 corresponds to a puncturing ratio greater than 0.67. In this case, the quantity of puncturing is only moderate and it is not necessary to take the priority away from the systematic bits because there is sufficient capacity to transfer a reasonable number of parity bits in addition to the systematic bits. Incremental redundancy is achieved between transmissions by changing the puncturing pattern within the parity bits. The performance of incremental redundancy is likely to be similar to the performance of chase combining for this scenario. If the number of systematic bits divided by the total number of bits offered by the E-DPDCH physical channel(s) is greater than 0.5, the parity bits can also be assigned priority during subsequent re-transmissions. In this case, the quantity of puncturing is more significant and it is necessary to puncture the systematic bits to ensure that the Node B receives a reasonable number of parity bits. The performance of incremental redundancy is likely to be greater than the performance of chase combining for this scenario.

HSDPA uses a 1 bit new data indicator within the HS-SCCH to allow the UE to differentiate between a new transmission and a re-transmission. HSUPA uses a 2 bit Re-transmission Sequence Number (RSN) within the E-DPCCH. The Node B can use this information to differentiate between a new transmission and a re-transmission. This avoids the Node B receiver combining data which belongs to different transport blocks. MAC-es PDU include sequence numbers, but these are only visible after the transport block has been successfully decoded and are only intended for RNC re-ordering purposes. The RSN within the E-DPCCH is also used to calculate the puncturing pattern for incremental redundancy. In addition, the Node B uses this information to inform the RNC of the number of re-transmissions required to successfully receive a transport block.

Access to individual HARQ processes can be controlled for HSUPA connections using the 2 ms TTI. In this case, the RNC can configure each MAC-d flow to have access to a specific group of HARQ processes. MAC-d flows with higher priority can be assigned access to an increased number of HARQ processes to allow greater throughputs. By default, all MAC-d flows have access to all HARQ processes. Individual HARQ processes belonging to connections using the 2 ms TTI can also be activated and deactivated using the E-DCH Absolute Grant Channel (E-AGCH).

7.6.3 Generation of MAC-es PDU

A MAC-es PDU represents a collection of equal-sized MAC-d PDU belonging to the same MAC-d flow and the same logical channel. The structure of a MAC-es PDU is illustrated in Figure 7.11. The header of a MAC-es PDU includes a 6 bit Transmission Sequence Number (TSN) which ranges from 0 to 63. The RNC uses this TSN when re-ordering MAC-es PDU prior to extracting the MAC-d PDU and delivering them to the higher layers. Each logical channel has its own sequence numbering to allow the RNC to re-order the data belonging to each logical channel independently.

The MAC-es layer also generates a Data Description Indicator (DDI) and a Number of MAC-d PDU (N) indicator for every MAC-es PDU. This information is not attached to the MAC-es PDU, but is forwarded to the MAC-e layer for subsequent inclusion within the MAC-e PDU. The DDI value is used

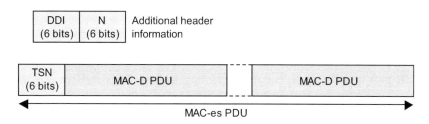

Figure 7.11   Structure of a MAC-es PDU

to identify the MAC-d flow, logical channel identity and MAC-d PDU size. This information is not explicitly included within the DDI but instead the DDI identifies a set of values which were configured during connection establishment. Examples of associating a DDI with a specific MAC-d flow, logical channel and MAC-d PDU size are shown in Log File 7.1 and Log File 7.3. These DDI configurations could be included within a Radio Bearer Setup or Radio Bearer Reconfiguration message. The Node B is informed of the same information using NBAP signalling. A single DDI value can be used if an HSUPA connection has been configured with a single MAC-d flow, a single logical channel and a single MAC-d PDU size. The N value which defines the number of MAC-d PDU within the MAC-es PDU has a range from 0 to 63. This means there can be a maximum of 63 MAC-d PDU within a single MAC-es PDU.

The total length of the MAC-es header is equal to 6 bits. The additional 12 bits generated by the DDI and N information is forwarded to the MAC-e layer for inclusion within the MAC-e header. MAC-es PDU can have a variable length and padding is not necessary.

7.6.4 Generation of MAC-e PDU

A MAC-e PDU represents an HSUPA transport block. A single HSUPA connection can transfer a maximum of one transport block during each TTI. The size of the transport block is determined by the E-TFC selection procedure. A transport block can accommodate zero, one or more MAC-es PDU. It can also accommodate Scheduling Information (SI). The structure of a transport block is illustrated in Figure 7.12.

The header is generated by concatenating the DDI/N pairs received from the MAC-es layer. A single DDI/N pair accompanies every MAC-es PDU. The payload is generated by concatenating the corresponding MAC-es PDU. The order of the DDI/N pairs corresponds to the order of the MAC-es PDU, i.e. the nth DDI/N pair belongs to the nth MAC-es PDU. If an HSUPA connection has been configured with a single DDI (single MAC-d flow, single logical channel and single MAC-d PDU size) there will be a single DDI/N pair and a single MAC-es PDU within each transport block that includes data rather than just SI. The smallest transport block size of 18 bits can be used to transfer SI without any MAC-es PDU.

SI is included to help support the operation of the Node B packet scheduler. It provides a measure of both the resource requirement and the ability of the UE to make use of additional resources. The structure of the SI is illustrated in Figure 7.13.

The UE power headroom (UPH) provides a measure of the ratio between the UE maximum transmit power capability and the current transmit power of the DPCCH. It does not account for the DPDCH, HS-DPCCH, E-DPCCH nor E-DPDCH. These physical channels have been excluded to help improve the measurement accuracy. The UE power headroom allows the Node B to evaluate whether or not there is any benefit associated with allocating an increased E-DPDCH power ratio. A

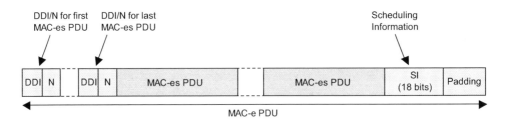

Figure 7.12   Structure of a MAC-e PDU (transport block)

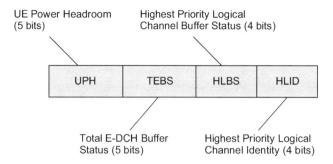

UE Power Headroom
(5 bits)

Highest Priority Logical
Channel Buffer Status (4 bits)

| UPH | TEBS | HLBS | HLID |

Total E-DCH Buffer
Status (5 bits)

Highest Priority Logical
Channel Identity (4 bits)

Figure 7.13 Structure of the Scheduling Information (SI)

UE with a small power headroom will be less able to benefit from an increased allocation. This implies that the Node B packet scheduler should avoid allocating increased resources to UE with small power headrooms.

The Total E-DCH Buffer Status (TEBS) provides a measure of the total quantity of data ready for transmission and re-transmission at the RLC layer. This helps to indicate whether or not the UE requires additional resources to transfer its data. It can also indicate when the UE buffers are empty and there is potential to reduce the allocated resources. The TEBS value is represented using 5 bits which provide a range from 0 to 31. A value of 0 indicates that the UE buffers are empty. The values 1 to 30 are used to represent a range of exponentially increasing buffer occupancies. An exponential scale has been defined to help minimise the quantisation error when expressed as a percentage of the buffer occupancy. The value of 31 corresponds to a buffer occupancy which is greater than 37 642 bytes. The range of values is intended to be sufficiently large to ensure that the UE does not transfer all of the reported data prior to the Node B having a chance to increase the resource allocation.

The Highest Priority Logical Channel Identity (HLID) informs the Node B of the identity of the highest priority logical channel which has data buffered at the RLC layer. If multiple logical channels have the highest priority and have data to transfer then the logical channel with the largest buffer occupancy is selected. The HLID is signalled with a value of 0 if the TEBS value indicates that all buffers are empty.

The Highest Priority Logical Channel Buffer Status (HLBS) quantifies the percentage of data that the highest priority logical channel (identified by the HLID) contributes towards the total buffer occupancy reported by the TEBS. A large result indicates that the majority of the data buffered at the RLC layer belongs to the highest priority logical channel. This can increase the incentive for the Node B packet scheduler to allocate additional resources. If the reported TEBS value is equal to 31 (buffer occupancy > 37642 bytes) the HLBS result is calculated relative to 50000 bytes. The HLBS is signalled with a value of 0 if the TEBS value indicates that all buffers are empty.

It is not always necessary to include all logical channels when generating SI reports. The RNC can instruct the UE to exclude specific logical channels. The RNC indicates whether or not each logical channel should be included during connection establishment. Log File 7.1 and Log File 7.3 provide examples of the RNC instructing the UE to include specific logical channels when generating SI reports. UE also exclude logical channels which have been configured to transfer data using the 'non-scheduled grant' mode of operation. This mode is described in greater detail within Section 7.8.1.

The UE includes SI within a transport block whenever there is space to do so. In addition, its inclusion can be triggered periodically or by specific events. Log File 7.5 shows two periodic timers which can be configured by the RNC during connection establishment. The first timer is applicable when the UE has a zero grant or if all HARQ processes are deactivated. The second timer is applicable when the UE has a non-zero grant and at least one active HARQ process. In the example of Log File 7.5, both timers are configured with values of 200 ms. The default value for these timers instructs the UE not to send periodic SI reports. The maximum rate at which periodic SI reports can be transferred is once per TTI whereas the minimum rate is once per second. These periodic timers are not applicable if the TEBS value is zero, i.e. the UE does not have any data to transfer. In addition to the periodic triggering mechanisms, a UE which has a zero grant or all HARQ processes deactivated also triggers an SI report as soon as the TEBS value becomes greater than zero, and when data arrives which has higher priority than the data already in the transmission buffer. A UE which has a non-zero grant and at least one active HARQ process also triggers an SI report after a serving cell change when the new serving cell does not belong to the same radio link set as the previous serving cell. This triggering mechanism ensures that the Node B associated with the new serving cell has an SI report.

The MAC-e and MAC-es headers sum to a total of $(n \times 18)$ bits, where n is the number of MAC-es PDU within the MAC-e PDU. If the HSUPA connection has a single DDI (single MAC-d flow, single logical channel and single MAC-d PDU size) then the total header length will be 18 bits. The SI report adds a further 18 bits whenever it is included.

7.7 Physical Channels

- HSUPA introduces the E-DPCCH and E-EPDCH physical channels in the uplink direction, and the E-HICH, E-AGCH and E-RGCH physical channels in the downlink direction. An HSUPA UE has to be capable of simultaneously transmitting up to five physical channels: DPCCH, DPDCH, E-DPCCH, E-DPDCH and HS-DPCCH.
- The E-DPCCH provides information to help the Node B decode the E-DPDCH. It also provides information to help the Node B packet scheduler allocate resources. It includes a re-transmission sequence number, an E-TFCI and a happy bit.
- The E-DPCCH content occupies a single 2 ms subframe if a 2 ms TTI is used. Otherwise, the content is repeated five times to occupy the five subframes belonging to a 10 ms TTI.
- The E-DPDCH is responsible for providing the high data rates associated with HSUPA. An HSUPA connection is able to transfer a maximum of one transport block during each 2 or 10 ms TTI. A single E-DPDCH is defined by its channelisation code which can have a spreading factor between 256 and 2.
- 3GPP specifies the following sequence of E-DPDCH configurations: {SF256, SF128, SF64, SF32, SF16, SF8, SF4, 2×SF4, 2×SF2, 2×SF4+2×SF2 (BPSK), 2×SF4+2×SF2 (4PAM)}. The release 7 version of the specifications introduce the 4PAM configuration.
- The 2×SF4+2×SF2 configuration cannot be used when a DPDCH is configured. This results from the DPDCH blocking the section of the code tree where the E-DPDCH SF4 codes are allocated.
- Physical layer processing for the E-DPDCH includes CRC attachment, code block segmentation, channel coding, HARQ functionality (rate matching), physical channel segmentation, interleaving and physical channel mapping. The HARQ functionality is responsible for selecting an appropriate E-DPDCH spreading factor configuration.
- The E-HICH is used by the MAC-e layer within the Node B to acknowledge uplink transport blocks transferred on the E-DPDCH. A re-transmission is triggered if a UE does not receive a positive acknowledgement from at least one active set cell.
- The content of the E-HICH occupies a single 2 ms subframe if a 2 ms TTI is used. Otherwise, the content is repeated across four consecutive subframes if a 10 ms TTI is used.
- The radio frame timing of the E-HICH is linked to the radio frame timing of the downlink DPCH. The E-HICH frame timing is quantised into 6 steps rather than the 150 steps used by the downlink DPCH.

- The E-RGCH is used to adjust the E-DPDCH to DPCCH power ratio allocated to one or more HSUPA connections. Cells belonging to the serving radio link set are permitted to use the E-RGCH to issue 'up' commands. All cells within the E-DCH active set are able to issue 'hold' and 'down' commands.
- The serving radio link set E-RGCH occupies a single 2 ms sub-frame if a 2 ms TTI is used and four consecutive subframes if a 10 ms TTI is used. The non-serving radio link set E-RGCH occupies five consecutive subframes irrespective of the TTI.
- The E-HICH and E-RGCH share the same channelisation code and the same set of 40 signature sequence hopping patterns. A single channelisation code can be used to allocate 20 E-HICH hopping patterns and 20 E-RGCH hopping patterns.
- The E-AGCH is only applicable to the E-DCH serving cell. The Node B packet scheduler can use the E-AGCH to instruct a UE to change its scheduled grant. The E-AGCH can also be used to activate and deactivate individual HARQ processes if a 2 ms TTI is used.
- The content of the E-AGCH occupies a single 2 ms subframe if a 2 ms TTI is used. Otherwise, the content is repeated across five consecutive subframes if a 10 ms TTI is used.
- The E-AGCH is addressed to specific UE by masking the CRC bits with an E-RNTI. The serving Node B can allocate both primary and secondary E-RNTI to each UE. If both a primary and secondary E-RNTI are allocated then the primary E-RNTI can be common (used to address a group of UE) while the secondary E-RNTI can be dedicated.
- Key 3GPP specifications: TS 25.211, TS 25.212, TS 25.213, TS 25.215 and TS 25.321.

Table 7.18 compares the physical channels used by an HSUPA connection with those used by HSDPA and DPCH connections. A DPCH connection uses DPCCH and DPDCH physical channels in both the uplink and downlink directions. HSDPA introduces the HS-PDSCH, HS-SCCH and HS-DPCCH physical channels. HSUPA introduces the E-DPCCH, E-EPDCH, E-HICH, E-AGCH and E-RGCH physical channels. HSDPA and HSUPA connections continue to require the DPDCH if the SRB is transferred using a DCH rather than the HS-DSCH and E-DCH. The DPDCH can also be required for multi-service connections, e.g. establishing a speech call in parallel to an ongoing HSDPA/HSUPA data transfer. The F-DPCH can replace the downlink DPCCH if the downlink DPDCH is not used.

Table 7.18 Physical channels used by an HSUPA connection

	Physical channels	DPCH	HSDPA (SRB on DPDCH)	HSDPA & HSUPA (SRB on DPDCH)	HSDPA & HSUPA (SRB on E-DCH)
Downlink	HS-PDSCH		√	√	√
	HS-SCCH		√	√	√
	DPCCH	√	√	√	
	DPDCH	√	√	√	
	F-DPCH				√
	E-HICH			√	√
	E-AGCH			√	√
	E-RGCH			√	√
Uplink	DPCCH	√	√	√	√
	DPDCH	√	√	√	
	HS-DPCCH		√	√	√
	E-DPCCH			√	√
	E-DCDCH			√	√

7.7.1 E-DCH Dedicated Physical Control Channel (E-DPCCH)

The E-DCH Dedicated Physical Control Channel (E-DPCCH) is introduced within the release 6 version of the 3GPP specifications. It is responsible for providing information to help the Node B decode the E-DPDCH. It also provides information to help the Node B packet scheduler allocate resources.

The E-DPCCH is always spread using the second channelisation code with spreading factor 256, i.e. (SF256,1). The location of this channelisation code within the code tree is illustrated in Figure 7.15. The spreading factor of 256 generates a BPSK symbol rate of 15 ksps. There is a single bit per symbol and so the corresponding bit rate is 15 kbps. The E-DPCCH slot format is summarised within Table 7.19.

The frame timing of the E-DPCCH is aligned with the frame timing of the uplink DPCCH. The frame and time slot structure for both the E-DPCCH and E-DPDCH is illustrated in Figure 7.14. Each 10 ms radio frame is divided into five subframes. Each subframe has a duration of 2 ms and includes three time slots. Each time slot includes 2560 chips. The E-DPCCH is always mapped onto the in-phase branch of the modulation constellation. This is illustrated in Figure 7.17.

The content of the E-DPCCH is shown in Table 7.20. The physical layer applies channel coding to generate 30 bits from the initial set of 10 bits. These 30 bits occupy a single 2 ms subframe if a 2 ms TTI is being used. Otherwise, they are repeated five times to occupy the five subframes belonging to a 10 ms TTI.

Table 7.19 E-DPCCH slot format

Slot format	Spreading factor	Channel symbol rate	Channel bit rate	Bits per subframe	Bits per time slot
0	256	15 ksps	15 kbps	30	10

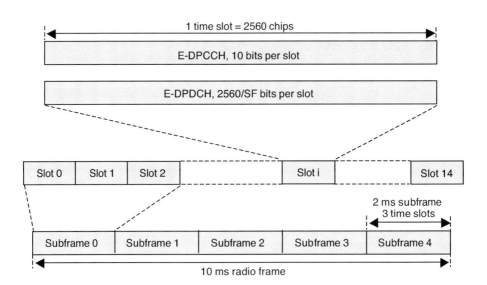

Figure 7.14 Frame and time slot structure for the E-DPCCH and E-DPDCH

Table 7.20 E-DPCCH content

Re-transmission Sequence Number (RSN)	2 bits
E-DCH Transport Format Combination Indicator (E-TFCI)	7 bits
Happy Bit	1 bit

The Re-transmission Sequence Number (RSN) informs the Node B of whether the transport block included within the E-DPDCH is a new transmission or a re-transmission. This helps the Node B to avoid combining buffered data belonging to an old transport block with new data belonging to a new transport block. The RSN is also used to calculate the puncturing pattern for incremental redundancy. In addition, the Node B can use the RSN to inform the RNC of how many re-transmissions were necessary to successfully receive a transport block. The RNC can then use this information to calculate the uplink BLER when outer loop power control adjusts the uplink DPCCH SIR target. The RNC can also use the number of re-transmissions to help deduce the timing of the original transmission. This can be useful for re-ordering purposes. The RSN is represented using only 2 bits so has a relatively limited range of 0 to 3. The vast majority of transport blocks should be successfully received without requiring a fourth re-transmission. However, if more than three re-transmissions are necessary the RSN value is maintained at 3.

The E-DCH Transport Format Combination Indicator (E-TFCI) informs the Node B of the transport block size included within the E-DPDCH. The E-TFCI represents an index within the relevant transport block size table. An example transport block size table is shown in Table 7.11. The RNC instructs the UE and Node B to use a specific transport block size table during connection establishment. The UE uses this table during the E-TFC selection procedure. The E-TFCI is represented using 7 bits which allows a range from 0 to 127. This limits the maximum number of transport block sizes to 128.

The happy bit is used by the Node B packet scheduler when identifying which UE would like increased resource allocations. A UE signals that it is unhappy if all three of the following criteria are true:

1. The UE is transmitting as much data as allowed by the current serving grant.
2. The UE has sufficient transmit power to increase its data rate.
3. The total E-DCH buffer occupancy requires longer than 'Happy Bit Delay Condition' when data is transferred at the rate defined by the current serving grant.

A UE signals that it is happy if any of the criteria above are false. In the case of a 2 ms TTI, the time required to transfer the data using the current serving grant is scaled by the ratio of the number of active HARQ processes to the total number of HARQ processes. This scaling is applied prior to the comparison with the 'Happy Bit Delay Condition' to ensure that both delays are based upon having all HARQ processes active. The 'Happy Bit Delay Condition' is configured during connection establishment. An example of the signalling used to inform the UE of its value is presented in Log File 7.7.

In this example, the delay condition is configured with a value of 200 ms. The maximum value which can be configured is 1 s whereas the minimum value is 2 ms. Configuring a small value will increase the number of UE reporting that they are unhappy. Log File 7.7 also configures the transmit power of the E-DPCCH relative to the DPCCH. In this example, a value of 5 is signalled to the UE. This value is mapped onto an amplitude ratio using the relationship presented in Table 7.21.

```
E-DPCCH-Info
    e-DPCCH-DPCCH-PowerOffset: 5
    happyBit-DelayCondition: ms200
```

Log File 7.7 E-DPCCH configuration information

Table 7.21 Signalled E-DPCCH to DPCCH amplitude ratios

Signalled ΔE-DPCCH	$A_{ec} = \beta_{ec}/\beta_c$	Signalled ΔE-DPCCH	$A_{ec} = \beta_{ec}/\beta_c$
8	30/15	3	9/15
7	24/15	2	8/15
6	19/15	1	6/15
5	15/15	0	5/15
4	12/15		

The amplitude ratio defined by A_{ec} is multiplied by the actual value of β_c prior to applying β_{ec} as an amplitude scaling factor during the modulation process. β_c is the DPCCH amplitude ratio which can range from 1/15 to 15/15. β_c and β_d are used to configure the relative powers of the DPCCH and DPDCH respectively. If a DPDCH has not been configured then β_c is assumed to have a value of 1. β_c, β_d and β_{ec} are illustrated in Figure 7.17.

7.7.2 E-DCH Dedicated Physical Data Channel (E-DPDCH)

The E-DCH Dedicated Physical Data Channel (E-DPDCH) is introduced within the release 6 version of the 3GPP specifications. It is responsible for providing the high data rates associated with HSUPA. An HSUPA connection is able to generate a maximum of one transport block during each 2 or 10 ms TTI. That transport block is processed by the Physical layer prior to being mapped onto one or more E-DPDCH. A single E-DPDCH is defined by its channelisation code which can have a spreading factor between 256 and 2. Decreasing the spreading factor and increasing the number of E-DPDCH increases the maximum achievable throughput. The release 7 version of the specifications also allows the modulation scheme to be changed to further increase the throughput.

The set of E-DPDCH slot formats is presented in Table 7.22. Slot formats 0 to 7 are based upon BPSK modulation (1 bit per symbol) whereas slot formats 8 and 9 are based upon 4PAM modulation (2 bits per symbol). In the context of HSUPA, the BPSK modulation scheme is often referred to as QPSK while the 4PAM modulation scheme is often referred to as 16QAM, e.g. within TS 25.306.

3GPP specifies the following sequence of E-DPDCH configurations: {SF256, SF128, SF64, SF32, SF16, SF8, SF4, 2×SF4, 2×SF2, 2×SF4+2×SF2 (BPSK), 2×SF4+2×SF2 (4PAM)}. A single E-DPDCH with a spreading factor of 256 represents the lowest throughput configuration whereas

Table 7.22 E-DPDCH slot formats

Slot format	Spreading factor	Channel symbol rate (ksps)	Channel bit rate (kbps)	Bits per 2 ms TTI	Bits per 10 ms TTI
0	256	15	15	30	150
1	128	30	30	60	300
2	64	60	60	120	600
3	32	120	120	240	1200
4	16	240	240	480	2400
5	8	480	480	960	4800
6	4	960	960	1920	9600
7	2	1920	1920	3840	19200
8*	4	960	1920	3840	19200
9*	2	1920	3840	7680	38400

*Introduced by release 7 of the 3GPP specifications.

the $2\times$SF4$+2\times$SF2 E-DPDCH configuration using 4PAM represents the highest throughput configuration. The sequence of E-DPDCH configurations illustrates that the 4PAM modulation scheme is only applicable to the $2\times$SF4$+2\times$SF2 configuration. The sequence also illustrates that a spreading factor of 2 can only be used when there is more than a single E-DPDCH.

Table 7.23 presents the set of aggregate channel symbol rates and channel bit rates when multiple E-DPDCH are used. UE avoid the use of multiple E-DPDCH unless the puncturing requirement associated with a single E-DPDCH becomes too great.

HSUPA can have a significant impact upon the utilisation of the uplink channelisation code tree. 3GPP have coordinated the allocation of channelisation codes with the allocation of the in-phase and quadrature branches of the modulation constellation. In some instances, two physical channels are allocated the same channelisation code or channelisation code tree branch. In these cases, the two physical channels are allocated different branches of the modulation constellation. Orthogonality is preserved because the in-phase and quadrature branches are orthogonal to one another in the domain of the modulation constellation. This is similar to using a single channelisation code to spread both the in-phase and quadrature branches of the downlink DPCH or HS-PDSCH.

Table 7.24 presents the uplink channelisation codes and modulation branches allocated to the DPCCH, DPDCH, HS-DPCCH and E-DPCCH. The first section of the table is applicable when a single DPDCH has been configured whereas the second section is applicable when a DPDCH has not been configured. Excluding the DPDCH does not change the allocations for the DPCCH and E-DPCCH, but moves the HS-DPCCH to a lower position in the code tree.

Table 7.25 presents the corresponding set of E-DPDCH allocations. This table indicates that the $2\times$SF4$+2\times$SF2 configuration cannot be used when a DPDCH is configured. This results from the DPDCH blocking the section of the code tree where the E-DPDCH SF4 codes are allocated. When a DPDCH is configured, the E-DPDCH are limited to the top half of the code tree, i.e. code tree branches originating from SF2,1 rather than SF2,0. The HS-DPCCH also blocks the same section of the code tree when a DPDCH is configured. However, the HS-DPCCH is moved to a lower section within the code tree when a DPDCH has not been configured.

Table 7.23 E-DPDCH combinations

Spreading factor	Aggregate channel symbol rate (ksps)	Aggregate channel bit rate (kbps)	Aggregate bits per 2 ms TTI	Aggregate bits per 10 ms TTI
$2\times$SF4	1920	1920	3840	19200
$2\times$SF2	3840	3840	7680	38400
$2\times$SF4 $+ 2\times$SF2 (BPSK)	5760	5760	11520	57600
$2\times$SF4 $+ 2\times$SF2 (4PAM)[*]	5760	11520	23040	115200

[*]Introduced by release 7 of the 3GPP specification.

Table 7.24 Uplink modulation branch and channelisation code allocations (excluding E-DPDCH)

Physical Channel	1 DPDCH configured		0 DPDCH configured	
	Modulation branch	Channelisation code	Modulation branch	Channelisation code
DPCCH	Quadrature	SF256,0	Quadrature	SF256,0
DPDCH	In-phase	SFx,x/4		
HS-DPCCH	Quadrature	SF256,64	Quadrature	SF256,33
E-DPCCH	In-phase	SF256,1	In-phase	SF256,1

Table 7.25   E-DPDCH channelisation code allocations

	1 DPDCH Configured		0 DPDCH Configured	
$1\times$SFy	In-phase	SFy,y/2	In-phase	SFy,y/4
$2\times$SF4	In-phase+quadrature	$2\times$SF4,2	In-phase+quadrature	$2\times$SF4,1
$2\times$SF2	In-phase+quadrature	$2\times$SF2,1	In-phase+quadrature	$2\times$SF2,1
$2\times$SF4+$2\times$SF2			$2\times$In-phase+$2\times$quadrature	$2\times$SF2,1+$2\times$SF4,1

Figure 7.15 illustrates the channelisation code allocations when a DPDCH is configured and the HSUPA connection is using its maximum configuration of $2\times$SF2 E-DPDCH. In this example, it is assumed that the DPDCH is used to transfer Signalling Radio Bearer (SRB) information so the DPDCH is configured with a high spreading factor. This results in the DPDCH and HS-DPCCH sharing the same channelisation code. Table 7.24 indicates that orthogonality is maintained because the DPDCH uses the in-phase branch while the HS-DPCCH uses the quadrature branch. The bit rate

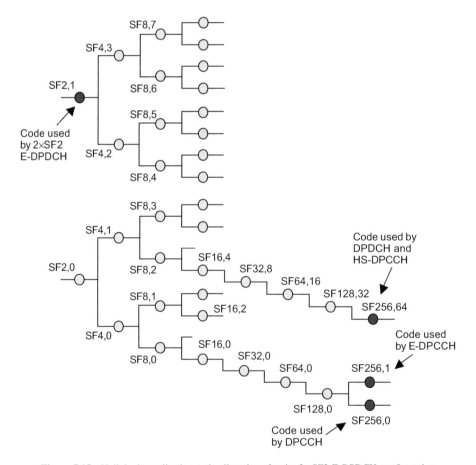

Figure 7.15   Uplink channelisation code allocations for the $2\times$SF2 E-DPDCH configuration

requirement of the DPDCH would increase if a circuit switched service were to be established in parallel to the HSUPA packet switched service. This would cause the DPDCH to use a channelisation code further up the same branch of the code tree. The DPDCH and HS-DPCCH would continue to prevent the E-DPDCH from using the SF4,1 section of the code tree.

Figure 7.16 illustrates the maximum E-DPDCH channelisation code allocations when a DPDCH is not configured, i.e. the 2×SF4+2×SF2 configuration. These channelisation code allocations are applicable to both the BPSK and 4PAM modulation schemes. As indicated by Table 7.24, the HS-DPCCH code allocation is moved into the lower quarter of the code tree when the DPDCH is not present. This creates sufficient space for the HSUPA connection to use the SF4,1 channelisation codes. This figure illustrates that the 2×SF4+2×SF2 configuration uses a total of two E-DPDCH channelisation codes, both of which are re-used by the in-phase and quadrature modulation branches.

Figure 7.17 illustrates the allocation of the in-phase and quadrature branches when the DPDCH is present. When two E-DPDCH are used, there are three physical channels allocated to the in-phase branch and three physical channels allocated to the quadrature branch. The two E-DPDCH are distributed between the in-phase and quadrature branches. The transmit power of each E-DPDCH will

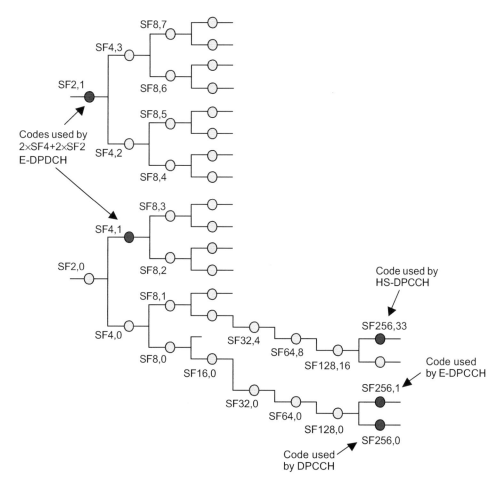

Figure 7.16 Uplink channelisation code allocations for the 2×SF4+2×SF2 E-DPDCH configuration

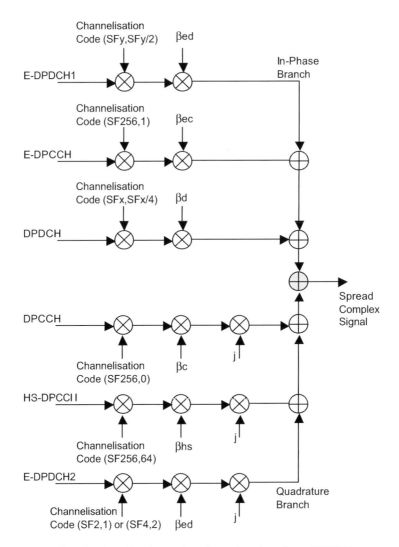

Figure 7.17 Allocation of in-phase and quadrature branches when a DPDCH is present

be greater than the transmit power of the other physical channels when two or more E-DPDCH are used. This results from the E-DPDCH being allocated a relatively high power ratio and used to transfer high bit rates when multiple E-DPDCH are present. When only a single E-DPDCH is present then it is allocated to the in-phase branch of the modulation constellation. In this case, the power of the E-DPDCH can be more comparable to the power of the other physical channels, i.e. when a lower power ratio is allocated and the throughput is decreased.

Table 7.26 presents the physical layer processing for three example E-DCH transport block sizes extracted from transport block size table 1 for a 10 ms TTI. These transport block sizes correspond to RLC throughputs of 384 kbps, 1.38 Mbps and 1.89 Mbps (assuming a MAC-d PDU size of 336 bits with a payload of 320 bits). The transport block size of 14484 bits corresponds to the maximum for a

Table 7.26 E-DCH transport block processing for 10 ms TTI (three example transport block sizes)

	HSUPA RLC throughput			
	384 kbps (SF4) Category 3 UE	1.38 Mbps (2×SF4) Category 3 UE	1.38 Mbps (2×SF2) Category 5 UE	1.89 Mbps (2×SF2) Category 5 UE
Transport block Size	4068 bits	14484 bits	14484 bits	19860 bits
CRC attachment	4068+24=4092	14484+24=14508	14484+24=14508	19860+24=19884
Code block Segmentation	Not necessary	3×4836	3×4836	4×4971
Channel coding	(4092×3)+12 = 12288	3×[(4836×3)+12] = 43560	3×[(4836×3)+12] = 43560	4×[(4971×3)+12] = 59700
HARQ functionality	9600	19200	38400	38400
Physical channel segmentation	1×9600	2×9600	2×19200	2×19200
E-DCH Interleaving	1×9600	2×9600	2×19200	2×19200
Physical channel mapping	1×9600	2×9600	2×19200	2×19200

category 3 UE. The Physical layer processing for this transport block size is shown for both category 3 and category 5 UE. Both UE categories have been included to demonstrate that different UE categories can select different E-DPDCH configurations for the same transport block size. The transport block size of 19860 bits does not correspond to the absolute maximum for a category 5 UE, but represents the maximum size necessary when a single MAC-e PDU is included within each MAC-es PDU, i.e. using the absolute maximum transport block size of 19950 bits increases the requirement for padding without accommodating any additional payload.

The physical layer starts by attaching Cyclic Redundancy Check (CRC) bits. These bits are used by the receiver to detect whether or not a transport block includes any errors after the channel decoding process. 3GPP TS 25.212 specifies that HSUPA connections always use a 24 bit CRC. This represents the longest CRC length specified by 3GPP, i.e. the error detection capability is maximised for HSUPA.

The physical layer then evaluates the requirement for code block segmentation. Segmentation is necessary if the code blocks are too large for the subsequent channel coding procedure. The maximum code block size for turbo coding is 5114 bits. Table 7.26 illustrates that code block segmentation is not necessary when the transport block size is 4068 bits, but is necessary when the transport block size is either 14484 or 19860 bits. Code block segmentation divides the code block into smaller blocks of equal size. The number of blocks generated is the minimum necessary to ensure that the maximum size of 5114 bits is not exceeded. Padding is added if the original code block size is not a multiple of the number of blocks after segmentation.

3GPP TS 25.212 specifies that HSUPA connections always use 1/3 rate turbo coding. The coding rate of 1/3 means that the number of bits after channel coding is three times the number of bits before channel coding. Turbo coding generates three streams of output data: systematic, parity 1 and parity 2. The systematic bits are equal to the input bits whereas the parity 1 and parity 2 bits are calculated from the input bits to generate redundancy. 12 tail bits are added after the channel coding process. Turbo coding is completed separately for each code block if code block segmentation has been completed.

The Hybrid ARQ (HARQ) functionality is responsible for rate matching. As explained in Section 7.6.2, this can be based upon either chase combining or incremental redundancy. The number of bits after rate matching is exactly equal to the capacity offered by the set of E-DPDCH. This capacity depends upon the spreading factor, the number of E-DPDCH and the modulation scheme. Table 7.22 and Table 7.23 present the number of bits offered by each E-DPDCH configuration. The UE is responsible for selecting the appropriate configuration after identifying the transport block size. 3GPP have fully specified this selection procedure to ensure that it is deterministic and to allow the receiving Node B to complete the same procedure after reading the E-TFCI from the E-DPCCH. This allows the Node B to identify the E-DPDCH configuration used by the UE prior to attempting any decoding.

Before starting the selection procedure, a list of the capacities associated with each of the supported E-DPDCH configurations is generated. This list is known as SET 0. Examples of SET 0 for category 3 and 5 UE are shown below.

SET 0 for category 3 UE $= \{N_{SF256}, N_{SF128}, N_{SF64}, N_{SF32}, N_{SF16}, N_{SF8}, N_{SF4}, N_{2 \times SF4}\}$
$= \{150, 300, 600, 1200, 2400, 4800, 9600, 19200\}$ bits
SET 0 for category 5 UE $= \{N_{SF256}, N_{SF128}, N_{SF64}, N_{SF32}, N_{SF16}, N_{SF8}, N_{SF4}, N_{2 \times SF4}, N_{2 \times SF2}\}$
$= \{150, 300, 600, 1200, 2400, 4800, 9600, 19200, 38400\}$ bits

---

The subsequent selection procedure is based upon completing up to three steps.

Step 1. The first step attempts to completely avoid puncturing while using a single E-DPDCH. If this is possible, the maximum spreading factor which avoids puncturing is selected, i.e. a single E-DPDCH is selected and the spreading gain is maximised. This step is specified in terms of defining SET 1.

SET 1 $= \{N_{SFx}$ within SET 0 such that $N_{SFx} - N_{coded}$ is non-negative$\}$
where N_{coded} is the number of bits generated by the physical layer after channel coding.
If SET 1 is not empty and the smallest element within SET 1 requires a single E-DPDCH then the smallest element within SET 1 is selected. Else continue to step 2.

This first step fails to identify an E-DPDCH configuration for the examples presented in Table 7.26. The smallest transport block size shown in Table 7.26 generates 12288 bits after channel coding whereas the largest capacity associated with a single E-DPDCH is 9600 bits. This indicates that the channel coded data cannot be accommodated by a single E-DPDCH without puncturing. As an example, a transport block size of 2706 bits generates 8202 bits after channel coding. In this case, a single SF4 E-DPDCH would be able to accommodate the channel coded data without puncturing.

Step 2. The second step attempts to use a single E-DPDCH while keeping the quantity of puncturing below $(1 - PL_{non-max})$. If this is not possible then the second step attempts to use the smallest multiple E-DPDCH configuration while keeping the quantity of puncturing below $(1 - PL_{non-max})$. If this is not possible then the selection procedure moves to the third step. The puncturing limit, $PL_{non-max}$ is configured by the RNC during connection establishment. An example of the relevant signalling is shown in Log File 7.5. In this case, $PL_{non-max}$ is signalled using a value of 17. The actual value is obtained by multiplying the signalled value by 0.04, i.e. the actual value is 0.68. The second step is specified in terms of defining SET 2.

SET 2 $= \{N_{SFx}$ within SET 0 such that $N_{SFx} - PL_{non-max} \times N_{coded}$ is non-negative$\}$
If SET 2 is not empty and the smallest element within SET 2 requires multiple E-DPDCH then the smallest element within SET 2 is selected. If the smallest element within SET 2 requires a single E-DPDCH then the largest element within SET 2 which requires a single E-DPDCH is selected. If SET 2 is empty then continue to step 3.

The second step succeeds in finding an E-DPDCH configuration for the transport block size of 4068 bits and also the transport block size of 14484 bits when used by a category 5 UE. The transport block size of 4068 bits generates 12288 bits after channel coding. The puncturing limit of 0.68 allows this to be reduced to 8356 bits which can be accommodated by a single SF4 E-DPDCH. The transport block size of 14484 bits generates 43560 bits after channel coding. The puncturing limit of 0.68 allows this to be reduced to 29621 bits which can be accommodated by $2 \times$ SF2 E-DPDCH belonging to a category 5 UE. The second step fails to find an E-DPDCH configuration for the transport block size of 14484 bits when used by a category 3 UE and also for the transport block size of 19860 bits. The third step must be used to identify the appropriate configuration for these examples.

Step 3. The third step attempts to minimise the quantity of puncturing beyond $(1-\text{PL}_{\text{non-max}})$. The maximum quantity of puncturing allowed by the third step is defined by $(1-\text{PL}_{\text{max}})$. PL_{max} is specified by 3GPP TS 25.212 to have a value of 0.44 except when the $2 \times$ SF4 $+ 2 \times$ SF2 configuration is used in which case it has a value of 0.33. These values of 0.33 and 0.44 are relatively low and correspond to a large quantity of puncturing. A value of 0.33 means that only one-third of the channel coded bits remain after puncturing, i.e. the redundancy added by the rate 1/3 Turbo coding is removed and there will be a high probability of requiring re-transmissions unless the RF channel conditions are particularly good. The third step allows E-DPDCH configurations to be selected for the transport block size of 14484 bits when used by a category 3 UE and also for the transport block size of 19860 bits. The transport block size of 14484 bits generates 43560 bits after channel coding. The maximum puncturing limit of 0.44 allows this to be reduced to 19166 bits which can be accommodated by $2 \times$ SF4 E-DPDCH. The transport block size of 19860 bits generates 59700 bits after channel coding. The maximum puncturing limit of 0.44 allows this to be reduced to 26268 bits which can be accommodated by $2 \times$ SF2 E-DPDCH.

Physical channel segmentation is completed after selecting an E-DPDCH configuration and completing the appropriate puncturing or repetition. Physical channel segmentation distributes the data across the set of E-DPDCH. The first example within Table 7.26 is based upon a single E-DPDCH and so physical channel segmentation is not necessary. The remaining examples are based upon two E-DPDCH and so the data is divided into two equal sections. All subsequent physical layer processing is completed on a per physical channel basis, i.e. there are two parallel streams of processing when two E-DPDCH are used.

Interleaving for the E-DCH is the same as the second stage of interleaving for the DCH, i.e. block interleaving with column re-ordering (described in section 2.6.1). The interleaving array has dimensions of y rows by 30 columns where the number of rows is dependent upon the quantity of data associated with each physical channel. Finally, the resultant data stream is mapped onto the physical channels ready for spreading and scrambling prior to modulating the RF carrier.

Table 7.27 presents the physical layer processing for three example E-DCH transport block sizes extracted from transport block size table 1 for a 2 ms TTI, and one example E-DCH transport block size extracted from transport block size table 3 for a 2 ms TTI. The example from table 3 is included to illustrate the processing for the 4PAM modulation scheme. The transport block sizes in Table 7.27 correspond to RLC throughputs of 3.2, 4.16, 5.44 and 11.20 Mbps (assuming a MAC-d PDU size of 336 bits with a payload of 320 bits for the first three examples, and a MAC-d PDU size of 636 bits with a payload of 640 bits for the fourth example). The transport block size of 11460 bits does not correspond to the absolute maximum for a category 6 UE, but represents the maximum size necessary when a single MAC-e PDU is included within each MAC-es PDU, i.e. using the absolute maximum transport block size of 11478 bits increases the requirement for padding without accommodating any additional payload. The transport block size of 22996 bits corresponds to the maximum for a category 7 UE when using table 3 for a 2 ms TTI.

The physical layer processing for these examples is the same as that for the examples within Table 7.26 until the selection of an appropriate spreading factor. The SET 0 values for category 6 and 7 UE are shown below. The category 7 UE includes an additional E-DPDCH configuration based upon the 4PAM modulation scheme, i.e. $M_{2 \times \text{SF4} + 2 \times \text{SF2}}$.

Table 7.27 E-DCH transport block processing for 2 ms TTI (four example transport block sizes)

	HSUPA RLC throughput			
	3.20 Mbps (2×SF4+2×SF2) BPSK Category 6 UE	4.16 Mbps (2×SF4+2×SF2) BPSK Category 6 UE	5.44 Mbps (2×SF4+2×SF2) BPSK Category 6 UE	11.20 Mbps (2×SF4+2×SF2) 4PAM Category 7 UE
Transport block size	6756	8772	11460	22996
CRC attachment	6756+24=6780	8772+24=8796	11460+24=11484	22996+24=23020
Code block segmentation	2×3390	2×4398	3×3828	5×4604
Channel coding	2×[(3390×3)+12] = 20364	2×[(4398×3)+12] = 26412	3×[(3828×3)+12] = 34488	5×[(4604×3)+12] = 69120
HARQ functionality	11520	11520	11520	23040
Physical channel segmentation	2×1920 + 2×3840	2×1920 + 2×3840	2×1920 + 2×3840	2×3840 + 2×7680
E-DCH Interleaving	2×1920 + 2×3840	2×1920 + 2×3840	2×1920 + 2×3840	2×3840 + 2×7680
Physical channel Mapping	2×1920 + 2×3840	2×1920 + 2×3840	2×1920 + 2×3840	2×3840 + 2×7680

SET 0 for category 6 UE = $\{N_{SF256}, N_{SF128}, N_{SF64}, N_{SF32}, N_{SF16}, N_{SF8}, N_{SF4}, N_{2\times SF4}, N_{2\times SF2},$
$N_{2\times SF4+2\times SF2}\}$
= {30, 60, 120, 240, 480, 960, 1920, 3840, 7680, 11520} bits
SET 0 for category 7 UE = $\{N_{SF256}, N_{SF128}, N_{SF64}, N_{SF32}, N_{SF16}, N_{SF8}, N_{SF4}, N_{2\times SF4}, N_{2\times SF2},$
$N_{2\times SF4+2\times SF2}, M_{2\times SF4+2\times SF2}\}$
= {30, 60, 120, 240, 480, 960, 1920, 3840, 7680, 11520, 23040} bits

---

The smallest transport block size within Table 7.27 generates 20364 bits after channel coding. The first step of selecting an E-DPDCH configuration fails because puncturing is required for all single E-DPDCH configurations. Likewise, the second step fails because the assumed puncturing limit of 0.68 generates 13848 bits which cannot be supported by any of the category 6 UE configurations. The third step succeeds because the PL_{max} value of 0.33 is able to reduce the 20364 bits to 6720 bits. This pattern is the same for the other two category 6 UE examples. All three of these examples use the 2×SF4+2×SF2 configuration although the quantity of puncturing increases as the RLC throughput increases. The first example has a puncturing ratio of 0.57. This decreases to 0.44 for the second example and to 0.33 for the third example. The example based upon the category 7 UE also relies upon the third step of the selection procedure to identify an appropriate E-DPDCH configuration. In this case there are 69120 bits after channel coding. The PL_{max} value of 0.33 is able to decrease this to 22810 bits. The introduction of the 4PAM modulation scheme within the release 7 version of the specifications generates an addition to the second step of the procedure used to select an E-DPDCH configuration. This addition does not influence the examples presented in Table 7.27 but has an impact if SET 2 includes only the $N_{2\times SF4+2\times SF2}$ and $M_{2\times SF4+2\times SF2}$ configurations. In this case, an additional puncturing threshold is used to select between the two configurations. The additional puncturing threshold is called $PL_{mod,switch}$ and is specified to have a value of 0.468. The 4PAM 2×SF4+2×SF2 configuration is selected if $N_{coded} \times 0.468$ is greater than the number of bits which can be accommodated by the BPSK 2×SF4+2×SF2 configuration.

The interleaving stage is also modified for the 4PAM modulation scheme. Two interleavers are used in parallel for each physical channel when 4PAM is used. Otherwise, the remainder of the physical layer processing is the same and the resultant data stream is mapped onto the physical channels ready for spreading and scrambling.

The transmit power of the E-DPDCH is defined by a combination of the E-DPDCH to DPCCH power ratio identified during E-TFC selection, and the inner loop power control which determines the transmit power of the DPCCH. Inner loop power control causes the transmit power of the E-DPDCH to increase and decrease in the same way as the DPCCH. Changes in the E-DPDCH to DPCCH power ratio are superimposed upon these changes. Outer loop power control within the RNC also has an impact upon the E-DPDCH transmit power because it defines the uplink SIR target used by inner loop power control. If Outer loop power control increases the DPCCH SIR target the absolute transmit power of the E-DPDCH will increase by the same amount as the DPCCH.

7.7.3 E-DCH Hybrid ARQ Indicator Channel (E-HICH)

The E-DCH Hybrid ARQ Indicator Channel (E-HICH) is a downlink dedicated physical channel introduced within the release 6 version of the 3GPP specifications. The MAC-e layer within the Node B uses the E-HICH to acknowledge the reception of uplink transport blocks transferred on the E-DPDCH. A transport block re-transmission is triggered if a UE does not receive a positive acknowledgement from at least one E-DCH active set cell.

The E-HICH is always spread using a channelisation code with a spreading factor of 128. This generates a modulation symbol rate of 30 ksps and a corresponding channel bit rate of 60 kbps. These figures result in 120 bits per 2 ms subframe and 40 bits per time slot. 3GPP TS 25.211 specifies 40 E-HICH signature sequences. Each signature sequence has a length of 40 bits and occupies a single time slot. The 40 signature sequences are orthogonal to allow a group of HSUPA connections to share the same E-HICH channelisation code. A single HSUPA connection does not use the same signature sequence during every time slot. TS 25.211 specifies 40 signature sequence hopping patterns. Each hopping pattern defines a series of three different signature sequences to occupy the three time slots associated with a single 2 ms subframe. If an HSUPA connection is using a 2 ms TTI, the E-HICH acknowledgement is transferred using a single subframe (three time slots, single repetition of the hopping pattern).

If an HSUPA connection is using a 10 ms TTI, the E-HICH acknowledgement is transferred using four consecutive subframes (12 time slots, 4 repetitions of the hopping pattern). The four subframes are followed by a 2 ms period of DTX to generate an overall cycle duration of 10 ms, i.e. to match the TTI.

The RNC configures the E-HICH channelisation code and signature sequence hopping pattern during connection establishment. The allocated channelisation code and hopping pattern are signalled to both the Node B and UE. An example of the relevant section from a Radio Bearer Reconfiguration message sent to a UE is presented in Log File 7.8. In this example, the UE is allocated an E-HICH channelisation code index of 5 and a signature sequence hopping pattern index of 4.

The set of 40 hopping patterns belonging to an E-HICH channelisation code are shared with the E-DCH Relative Grant Channel (E-RGCH). Sharing the set of hopping patterns means that the RNC could use a single channelisation code to allocate 20 E-HICH hopping patterns and 20 E-RGCH hopping patterns, i.e. 20 HSUPA connections. The number of hopping patterns and connections can be increased by allocating more than a single channelisation code for the E-HICH and E-RGCH. The

```
E-HICH-Information
    channelisationCode:    5
    signatureSequence:     4
```

Log File 7.8 E-HICH configuration information

E-HICH and E-RGCH hopping patterns belonging to a specific UE and transmitted by a specific active set cell always use the same channelisation code.

A UE only triggers a MAC-e transport block re-transmission if it does not receive any positive acknowledgements from the active set cells. Receiving a single positive acknowledgement is sufficient for the UE to discard the relevant transport block and proceed to transferring the next. All E-DCH active set cells generate positive acknowledgements by multiplying the signature sequence by $+1$. Cells belonging to the same radio link set (generally the same Node B) as the serving E-DCH cell generate negative acknowledgements by multiplying the signature sequence by -1. Other cells within the active set generate a period of DTX rather than explicitly signalling a negative acknowledgment. These rules are summarised within Table 7.28.

This approach of combining BPSK type signalling from the serving radio link set with on–off type signalling from the non-serving radio link set is intended to reduce the downlink transmit power requirement for the cells belonging to non-serving radio link sets. The BPSK type signalling is more reliable than the on–off type signalling because there is a greater difference between the ACK and the NACK signals. This increases the potential to misinterpret a NACK as an ACK. This type of misinterpretation has no impact if another active set cell has sent a genuine ACK. However, if all other active set cells have sent negative acknowledgements the UE should complete a re-transmission but instead will proceed to transfer the next transport block. The RLC layer will then become responsible for recovering the data using RLC re-transmissions. RLC re-transmissions are relatively slow and have a more significant impact upon connection performance. If a UE misinterprets an ACK as a NACK without receiving an ACK from any other active set cell, the UE will trigger a re-transmission unnecessarily. This has a relatively minor impact upon throughput by occupying a TTI.

The radio frame timing of the E-HICH is linked to the radio frame timing of the downlink DPCH. The timing of the downlink DPCH is offset from the CPICH by ($n \times 256$) chips, where n can range from 0 to 149. If n is small, the DPCH frame timing follows the CPICH frame timing with relatively little delay. This delay increases as the value of n increases. The E-HICH frame timing follows the same pattern, i.e. UE which have a DPCH frame timing which starts relatively early also use an E-HICH frame timing which starts relatively early. However, the E-HICH frame timing is quantised into 6 steps rather than 150 steps. The timing of these 6 steps is specified within 3GPP TS 25.211 and is summarised within Table 7.29.

These E-HICH frame timings are shown in Figure 7.18 and Figure 7.19 for a 10 and 2 ms TTI, respectively. The corresponding DPCH frame timings are also shown. For example, a UE whose DPCH frame timing starts (20×256) chips after the CPICH and uses a 10 ms TTI for the E-DCH has an E-HICH frame timing which starts four time slots in advance of the CPICH. The six E-HICH starting times are separated by three time slots. This allows the signature sequence hopping patterns to remain time aligned and there is no danger of different connections using the same signature sequence during the same time slot. In the case of the 10 ms TTI, the E-HICH starting time of -7 time slots can be viewed as being the same as the E-HICH starting time of +8 time slots (based upon the cycle of 15 time slots within a radio frame). However these two timings are differentiated

Table 7.28 Multiplication factors to generate positive and negative E-HICH acknowledgements

	Radio link set including the serving E-DCH cell	Radio link set not including the serving E-DCH cell
ACK	1	1
NACK	-1	0

Table 7.29 E-HICH frame timing relative to CPICH

DPCH timing $n \times 256$ chips	10 ms TTI	2 ms TTI
0–9	−7 time slots	5 time slots
10–39	−4 time slots	8 time slots
40–69	−1 time slots	11 time slots
70–99	2 time slots	14 time slots
100–129	5 time slots	17 time slots
130–149	8 time slots	20 time slots

by their cell System Frame Numbers (SFN). When the CPICH has an SFN of z, the E-HICH frame which starts 7 time slots in advance has an SFN of z and likewise the E-HICH frame which starts with 8 time slots delay also has an SFN of z. This means that the E-HICH frames which coincide have SFN of z and $z - 1$. The same argument is applicable to the E-HICH frame timings of $+5$ and $+20$ time slots for the 2 ms TTI. In the case of the 10 ms TTI, the distribution of E-HICH frame timings has the benefit of distributing the periods of DTX which helps to reduce the peak downlink transmit power requirement.

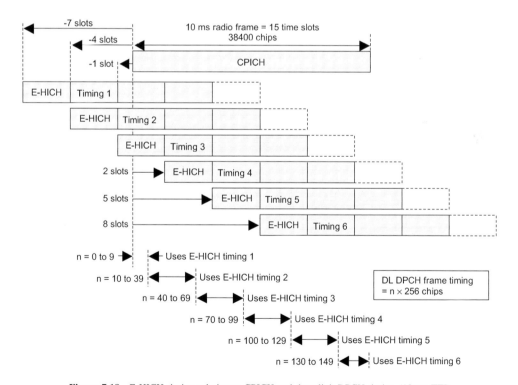

Figure 7.18 E-HICH timing relative to CPICH and downlink DPCH timing (10 ms TTI)

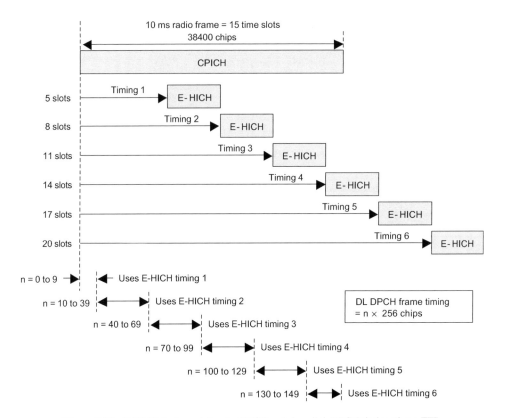

Figure 7.19 E-HICH timing relative to CPICH and downlink DPCH timing (2 ms TTI)

7.7.4 E-DCH Relative Grant Channel (E-RGCH)

The E-DCH Relative Grant Channel (E-RGCH) is a downlink dedicated physical channel introduced within the release 6 version of the 3GPP specifications. It is used by the E-DCH active set cells to adjust the E-DPDCH to DPCCH power ratio allocated to one or more HSUPA connections. Adjustments to the allocated power ratio are made by stepping up or stepping down within the table of scheduled grant values.

The E-RGCH is always spread using a channelisation code with a spreading factor of 128. This provides 40 bits per time slot which are occupied by one of 40 signature sequences. Signature sequence hopping patterns define combinations of three signature sequences to occupy three consecutive time slots. As described in Section 7.7.3, the E-RGCH shares channelisation codes and signature sequence hopping patterns with the E-HICH. The RNC configures the E-RGCH signature sequence hopping pattern during connection establishment. The allocated hopping pattern is signalled to both the Node B and UE. It is not necessary for the RNC to configure an E-RGCH channelisation code because it is always the same as the code allocated for the E-HICH. An example of the signalling used to inform the UE of the relevant signature sequence hopping pattern is presented in Log File 7.9. In this example, the UE is allocated a hopping pattern index of 5. The RNC also includes a combination index to inform the UE of which cells belong to the same radio link set, i.e. all cells belonging to the same radio link set

```
e-RGCH-Info
    E-RGCH-Information
        signatureSequence:      5
        rg-CombinationIndex:    0
```

Log File 7.9 E-RGCH configuration information

are allocated the same index value. Cells belonging to the serving radio link set always generate equal E-RGCH values. This allows the UE to soft combine these signals prior to decoding the content. The E-RGCH signals from other cells in the active set cannot be soft combined because their content can be different.

Only cells belonging to the serving radio link set are permitted to use the E-RGCH to issue 'up' commands. An 'up' command allows the UE to increase its E-DPDCH to DPCCH power ratio. All cells within the E-DCH active set are able to issue 'hold' and 'down' commands. This allows all cells within the active set to protect themselves from increased quantities of uplink power. A 'hold' command corresponds to generating a period of DTX on the E-RGCH. An 'up' command is generated by multiplying the signature sequences by +1 whereas a 'down' command is generated by multiplying the signature sequences by −1. These rules are summarised within Table 7.30.

An E-RGCH 'up' or 'down' command instructs the UE to change its scheduled grant, i.e. its E-DPDCH to DPCCH power ratio. This is done using the scheduled grant table presented within Table 7.16. A 'down' command always instructs the UE to reduce its scheduled grant by moving down 1 row within the scheduled grant table, i.e. the index decreases by 1. The interpretation of an 'up' command depends upon the current location within the table and the configuration of the 'SG2IndexStepThreshold' and 'SG3IndexStepThreshold' parameters. The RNC configures these two parameters during connection establishment. An example of them being sent to a UE within an RRC message is included within Log File 7.5. A UE increases its scheduled grant by moving up three rows within the scheduled grant table if it receives an 'up' command while operating below 'SG3IndexStepThreshold'. A UE increases its grant by moving up two rows if it receives an 'up' command while operating below 'SG2IndexStepThreshold'. If a UE receives an 'up' command while operating above 'SG2IndexStepThreshold' then it increases its scheduled grant by moving up one row. This design allows the serving radio link set to increase the allocated grant more rapidly when the existing grant is relatively low. These concepts are illustrated within Figure 7.20.

The timing of the E-RGCH for cells belonging to the serving radio link set is the same as that used by the E-HICH. This timing is illustrated within Figure 7.18 and Figure 7.19. The E-RGCH belonging to cells within the serving radio link set occupies 12 time slots when a 10 ms TTI is used and 3 time slots when a 2 ms TTI is used. The E-RGCH belonging to other cells in the active set occupies 15 time slots irrespective of the TTI. These 15 time slots occupy 5 subframes and accommodate 5 repetitions of the signature sequence hopping pattern. The quantity of repetition is increased for cells outside the serving radio link set to increase the reliability of their commands. This reflects the importance of those cells being able to protect themselves from increased quantities of uplink power. A UE starts a guard timer if it receives a 'down' command from a cell outside the serving radio link set. This guard timer prevents the UE from subsequently increasing its scheduling grant during a period of 16 ms for a 2 ms

Table 7.30 Multiplication factors to generate up, 'down' and 'hold' E-RGCH commands

	Radio link set including the serving E-DCH cell	Radio link set not including the serving E-DCH cell
UP	1	Not allowed
HOLD	0	0
DOWN	−1	−1

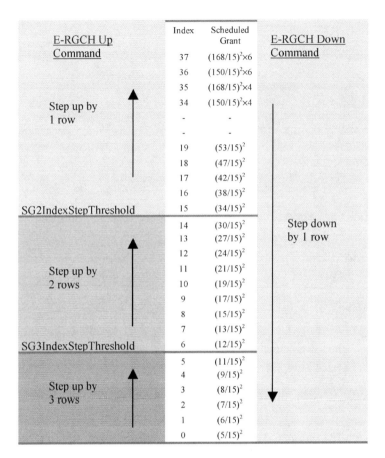

Figure 7.20   Interpretation of E-RGCH 'up' and 'down' commands

TTI connection, or during a period of 40 ms for a 10 ms TTI connection. The E-RGCH frame timing for cells outside the serving radio link set is fixed irrespective of the associated DPCH frame timing. The set of 15 time slots always has a frame timing which is 2 time slots behind the frame timing of the CPICH.

The E-RGCH signature sequence hopping patterns belonging to cells outside the serving radio link set can be allocated to groups of UE rather than individual UE. This helps to reduce the requirement for additional hopping patterns and allows non-serving cells to simultaneously instruct groups of HSUPA connections to reduce their allocated power ratios. The drawback of this approach is that it reduces the level of control because UE cannot be instructed to make changes individually.

7.7.5 E-DCH Absolute Grant Channel (E-AGCH)

The E-DCH Absolute Grant Channel (E-AGCH) is a common physical channel introduced within the release 6 version of the 3GPP specifications. It is only applicable to the E-DCH serving cell. The Node B packet scheduler belonging to the E-DCH serving cell can use the E-AGCH to instruct a UE to change its scheduled grant from one value to another without stepping through the intermediate values. The E-AGCH can also be used to activate and deactivate individual HARQ processes if a 2 ms TTI is used.

The E-AGCH is always spread using a channelisation code with a spreading factor of 256. The RNC is responsible for allocating the specific channelisation code during connection establishment. The channelisation code index is signalled to both the UE and Node B. An example of the relevant RRC signalling is presented in Log File 7.10.

The spreading factor of 256 generates a modulation symbol rate of 15 ksps and a corresponding channel bit rate of 30 kbps. These figures result in 60 bits per 2 ms subframe. The content of the E-AGCH is shown in Table 7.31. Concatenating the absolute grant value with the absolute grant scope generates a total of 6 bits. The physical layer adds a 16 bit CRC and 8 tail bits prior to completing rate 1/3 convolutional coding. This increases the total number of bits to 90 before puncturing is used to remove 30 bits. The resultant 60 bits occupy a single 2 ms subframe if a 2 ms TTI is used. Otherwise, they are repeated across five consecutive subframes if a 10 ms TTI is used.

A single E-AGCH channelisation code can be allocated to a relatively large group of HSUPA connections. The content of the E-AGCH does not explicitly include a UE identity to link each absolute grant to a specific HSUPA connection. Instead, the physical layer uses an E-RNTI to mask the CRC bits which are attached prior to channel coding. The group of HSUPA connections are then responsible for matching their E-RNTI with the value used to mask the CRC bits. The serving Node B can allocate both primary and secondary E-RNTI to each HSUPA connection. If only a primary E-RNTI is allocated then the value will be unique across all connections having the same serving cell. In this case, the serving cell is only able to send absolute grants to one connection at a time. If both a primary and secondary E-RNTI are allocated then the primary E-RNTI can be common while the secondary E-RNTI is dedicated. This approach allows the serving cell to use the primary E-RNTI to address groups of connections and the secondary E-RNTI to address individual connections.

The 5 bits allocated to the absolute grant value provide a range from 0 to 31. 3GPP TS 25.212 defines the mapping between the signalled value and the actual absolute grant value. The release 6 version of TS 25.212 defines the mapping presented in Table 7.32. The release 7 version includes an additional table which becomes applicable when the 4PAM modulation scheme is used. The 4PAM version of the table includes scheduled grant values which extend as high as $(377/15)^2 \times 4$.

These absolute grant values are a subset of the scheduled grant values presented in Table 7.16. Each value represents a total E-DPDCH power to DPCCH power ratio. The scheduled grant values which have been excluded from the set of absolute grant values are all from the lower end of the table, i.e. power ratios of $(21/15)^2$ and below.

The absolute grant value of 'Inactive' represents a special value which can be used to activate and deactivate individual HARQ processes belonging to connections using a 2 ms TTI. This value of 'Inactive' is used in combination with the value of the absolute grant scope. If the value 'Inactive' is received while the absolute grant scope has a value of 1 (per HARQ process) then the current HARQ process is deactivated and is no longer able to transfer data. The HARQ process can be reactivated subsequently if an absolute grant is received with the appropriate timing, i.e. during the subframe which corresponds to that HARQ process.

```
e-AGCH-Information
    e-AGCH-ChannelisationCode: 16
```

Log File 7.10   E-AGCH configuration information

Table 7.31   E-AGCH content

Absolute grant value	5 bits
Absolute grant scope	1 bit

Table 7.32 Absolute grant values

Index	Scheduled grant	Index	Scheduled grant	Index	Scheduled grant
31	$(168/15)^2 \times 6$	20	$(119/15)^2$	9	$(34/15)^2$
30	$(150/15)^2 \times 6$	19	$(106/15)^2$	8	$(30/15)^2$
29	$(168/15)^2 \times 4$	18	$(95/15)^2$	7	$(27/15)^2$
28	$(150/15)^2 \times 4$	17	$(84/15)^2$	6	$(24/15)^2$
27	$(134/15)^2 \times 4$	16	$(75/15)^2$	5	$(19/15)^2$
26	$(119/15)^2 \times 4$	15	$(67/15)^2$	4	$(15/15)^2$
25	$(150/15)^2 \times 2$	14	$(60/15)^2$	3	$(11/15)^2$
24	$(95/15)^2 \times 4$	13	$(53/15)^2$	2	$(7/15)^2$
23	$(168/15)^2$	12	$(47/15)^2$	1	Zero
22	$(150/15)^2$	11	$(42/15)^2$	0	Inactive
21	$(134/15)^2$	10	$(38/15)^2$		

If the absolute grant has a value of 'Inactive' while the absolute grant scope has a value of 0 (all HARQ processes) then the UE has to check whether or not a secondary E-RNTI was configured during connection establishment. This case is applicable to connections using both 2 and 10 ms TTI. The absolute grant scope is always sent using a value of 'all HARQ processes' for connections using a 10 ms TTI. If a secondary E-RNTI was configured then the UE interprets this scenario as a trigger to start using the secondary E-RNTI. The UE then switches from monitoring for only the primary E-RNTI to monitoring for both the primary and secondary E-RNTI. Absolute grants are accepted using the secondary E-RNTI until a subsequent grant is received using the primary E-RNTI. This triggers the UE to revert back to using only the primary E-RNTI.

If a UE using a 2 ms TTI has not had a secondary E-RNTI configured when receiving an absolute grant value of 'inactive' and an absolute grant scope of 'all HARQ processes' then the UE deactivates all HARQ processes. HARQ processes can then be reactivated either individually or all together if the UE subsequently receives an absolute grant with a scope of 'per HARQ process' or 'all HARQ processes' respectively.

3GPP TS 25.211 specifies that the frame timing of the E-AGCH follows the frame timing of the CPICH by two time slots (5120 chips). A maximum of one absolute grant can be transferred per 10 ms if a 10 ms TTI is used. This increases to a maximum of five absolute grants per 10 ms if a 2 ms TTI is used. If necessary, a cell can be configured to transmit more than a single E-AGCH by allocating more than a single channelisation code.

7.8 MAC-e Entity (Node B)

- The MAC-e layer supports the HARQ protocol from the receiver perspective. It also supports de-multiplexing of MAC-e PDU and provides HSUPA packet scheduler functionality.
- The HSUPA packet scheduler within the Node B operates in combination with the DPCH packet scheduler within the RNC. The total interference margin is shared between the HSUPA and DPCH connections.
- The RNC packet scheduler typically operates using its own target for uplink interference power. This target effectively splits the total interference margin into a share which can be used for DPCH connections and a share which can be used for HSUPA connections.
- If the RNC recognises that the Node B packet scheduler is not making full use of its HSUPA resources then the RNC may increase its own target interference margin to avoid limiting the performance of DPCH connections unnecessarily.

- The RNC can use NBAP signalling to instruct the Node B to start applying a specific target uplink interference margin. The RNC can also instruct the Node B to limit the interference power contribution of HSUPA connections served by neighbouring cells.
- The Node B packet scheduler may trigger load control actions if the RNC reports that congestion has been detected on the Iub as a result of either lost or delayed Frame Protocol packets.
- If the Node B is operating below its target interference margin it has scope to upgrade existing connections. Candidates can be identified using the happy bit, the Scheduling Information and by identifying connections relatively small existing allocations.
- If the Node B is operating above its target interference margin then it becomes necessary to identify connections to downgrade. Connections with low utilisation can be selected. Otherwise, connections with relatively high resource allocations can be selected.
- In general, the Node B packet scheduler can use either the E-RGCH or the E-AGCH to increase or decrease existing allocations. Downgrades triggered as a result of a cell receiving too much uplink power from non-serving UE must use the E-RGCH.
- The RNC is also able to configure HSUPA connections to operate in a non-scheduled mode. This allows the UE to transfer HSUPA data at any time without receiving any scheduling information from the Node B.
- Key 3GPP specifications: TS 25.321 and TS 25.433.

The MAC-e layer within the Node B provides peer functionality for some of the MAC-e/es capability within the UE. The Node B supports the requirements of the HARQ re-transmission protocol from the receiver perspective. This allows the uplink receiver to complete either chase combining or incremental redundancy. It also allows the receiver to generate acknowledgements for subsequent transfer on the E-HICH physical channel. The Node B also supports de-multiplexing of MAC-e PDU (transport blocks). This capability allows MAC-es PDU to be extracted and forwarded across the Iub using Frame Protocol.

The MAC-e layer within the Node B is also responsible for providing the HSUPA packet scheduling function. This function defines the algorithms used to allocate resources to the population of HSUPA connections. These resources are allocated using a combination of RRC signalling via the RNC, and Physical layer signalling using the E-RGCH and E-AGCH.

7.8.1 Packet Scheduler

The basic air-interface resource in the uplink direction is the interference margin. This represents the rise over thermal noise assumed during link budget analysis and radio network planning. Large interference margins correspond to high uplink capacities and small cell ranges. The interference margin assumed during radio network planning is likely to depend upon the environment type. Rural areas may be planned with relatively low interference margins to allow the cell range to be maximised while placing less emphasis upon the air-interface capacity. Urban areas are more likely to be planned with higher interference margins to provide greater capacity at the cost of an increased site density.

A single uplink connection generates an increase in the uplink interference floor. The increase will be greater if the uplink throughput is high. Table 7.9 illustrates this trend for an HSUPA connection. As the number of uplink connections increases then so too does the uplink interference floor. The uplink capacity is defined by the uplink throughput which generates an interference floor increase equal to the increase assumed during radio network planning. This throughput is likely to depend upon the traffic profile, i.e. the percentage of connections using a particular service type. The uplink throughput should be maximised if all connections are using HSUPA. If the increase in uplink interference exceeds the value assumed during radio network planning then the performance of connections towards the cell

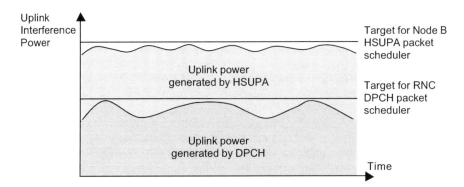

Figure 7.21   Sharing uplink air-interface resources between HSUPA and DPCH connections

edge may start to decrease. In this case, uplink transmit powers become insufficient to allow UE to achieve their uplink C/I requirements.

The Node B packet scheduler operates in combination with the RNC packet scheduler. The total interference margin must be shared between the HSUPA and DPCH connections. The Node B packet scheduler allocates resources to HSUPA connections whereas the RNC packet scheduler allocates resources to DPCH connections. The RNC maintains overall control, but allows the Node B packet scheduler to operate relatively autonomously after defining the target maximum uplink interference power. The general concept of the RNC and Node B packet schedulers sharing the total uplink interference margin is illustrated in Figure 7.21.

The RNC packet scheduler typically operates using its own target for uplink interference power. This target effectively splits the total uplink interference margin into a section which can be used for DPCH connections and a section which can be used by HSUPA connections. There will be increased margin for HSUPA connections if the RNC operates with a relatively low target power. However, this approach could have a negative impact upon the DPCH performance to which end-users have already become accustomed (assuming that DPCH connections were supported by the network prior to the introduction of HSUPA and prior to the requirement to share the total uplink interference margin between the two connection types). The RNC packet scheduler will typically allocate uplink DPCH resources until its target interference margin is reached. Subsequent connection requests may trigger a downgrade of the bit rate allocated to existing connections or may be denied. If the RNC recognises that the Node B packet scheduler is not making full use of its HSUPA resources then the RNC may increase its own target interference margin to avoid limiting the performance of DPCH connections unnecessarily.

The Node B packet scheduler typically operates using a target uplink interference margin provided by the RNC. The RNC is able to use the NBAP signalling specified within 3GPP TS 25.433 to instruct the Node B to start applying a specific target. The relevant NBAP information elements are presented in Table 7.33.

The 'Maximum Target Received Total Wideband Power' defines the target uplink interference power to be used by the Node B packet scheduler. This information element defines an absolute power in dBm. In practice, measurements of increases in received power can be more accurate than measure-

Table 7.33   NBAP information elements for the target uplink interference margin

Maximum target received total wideband power	Indicates the maximum target uplink interference power including received wideband power from all sources.
Reference received total wideband power	Indicates the reference uplink interference power. This corresponds to the interference floor measured when the network is unloaded.

ments of absolute power. This means that the Node B packet scheduler may be designed to target a specific increase in uplink interference floor rather than to target a specific absolute power. The 'Reference Received Total Wideband Power' defines the unloaded uplink interference floor and allows the Node B to calculate the interference floor increase corresponding to the absolute value of the target uplink interference power.

The target power used by the Node B packet scheduler should be greater than the target power used by the RNC packet scheduler. The target used by the Node B represents the total uplink interference power whereas the RNC target represents only part of the total power. The Node B packet scheduler can allocate an increased share of the total interference margin if the RNC is not fully utilising its target power. This allows increased HSUPA throughputs during times when the DPCH traffic is relatively low. If a network has used a specific uplink interference power target for DPCH packet scheduling prior to the introduction of HSUPA then it should be possible to increase this target for the purposes of Node B packet scheduling without having a negative impact upon uplink coverage. This results from the increased responsiveness of the Node B packet scheduler relative to the RNC packet scheduler, i.e. it takes the RNC longer to respond after air-interface conditions have changed. The slower response rate of RNC can generate an increased variance of the uplink power around the target value. The more responsive Node B packet scheduler reduces this variance and can increase its mean operating point without allowing increased levels of uplink interference.

In addition to instructing the Node B to use a specific target uplink interference power, the RNC can also instruct the Node B to limit the contribution of HSUPA connections served by neighbouring cells. The receiving cell will have limited resources to serve its own HSUPA connections if the contribution from connections served by neighbouring cells becomes too great. The general concept of dividing the total HSUPA uplink interference power into served and non-served contributions is illustrated in Figure 7.22.

The RNC can use NBAP signalling to instruct the Node B to start applying a specific target for the maximum ratio of non-serving HSUPA connection power to total HSUPA connection power. The relevant NBAP information element is presented in Table 7.34.

If the Node B detects that the contribution from HSUPA connections served by neighbouring cells becomes greater than the target, and if the receiving cell is fully utilising its interference margin then it can use the E-RGCH to instruct the non-serving connections to reduce their E-DPDCH to DPCCH power ratios. The Node B should not apply any load control actions to the non-serving UE if the receiving cell is not fully utilising its interference margin because this would reduce the throughput of the non-serving connections unnecessarily.

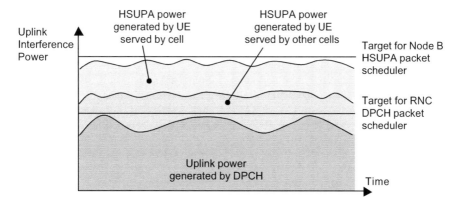

Figure 7.22   Sharing resources between UE served by receiving cell and UE served by other cells

Table 7.34   NBAP information element for the target ratio of non-serving to total E-DCH power

Target non-serving E-DCH to total E-DCH Power Ratio	Indicates the target ratio of the received E-DCH power from non-serving UE to the total E-DCH power.

The Node B packet scheduler may have some responsibilities in terms of coping with congestion across the Iub transport network. 3GPP has defined mechanisms which allow the RNC to detect either lost or delayed Frame Protocol packets. The 'Tunnel Congestion Indication' Frame Protocol control frame allows the RNC to subsequently report any congestion to the Node B. This control frame is described in greater detail within Section 7.9. Its reception can be used to trigger the Node B to initiate load control actions. The severity of these actions should depend upon the cause value within the congestion indication, i.e. lost frames are more significant than delayed frames. This form of Iub congestion is detected and reported on a per HSUPA connection basis. This allows the Node B to target overload actions towards the specific connection for which the congestion was detected.

The Node B packet scheduler can monitor for connections which are not making full use of their allocated E-DPDCH to DPCCH power ratios. This can be achieved by comparing the allocated power ratio with the power ratio required by the transport block size being sent. The RNC sends both the UE and Node B the same set of reference power ratios and reference transport block sizes. This allows the Node B to duplicate the mapping used by the UE to relate the power ratio to the transport block size. If a connection is detected to have a low utilisation then its allocated resources can be downgraded or released to make them available for another connection.

If the Node B is operating below its target interference margin it has scope to upgrade existing connections. Candidates for upgrading can be identified using the happy bit within the E-DPCCH, the Scheduling Information within the E-DPDCH and by identifying connections which currently have relatively small allocations. The happy bit provides an indication of whether or not a UE believes that it requires a greater allocation of resources based upon its current buffer occupancies and the configuration of the 'Happy Bit Delay Condition' parameter. UE which report themselves as unhappy are more likely to be targeted by the packet scheduler for an upgrade. The E-DPDCH Scheduling Information includes a measure of the UE transmit power headroom. If the headroom reported by a UE is small then the benefit of allocating increased resources is likely to be limited. This results from the UE not having sufficient transmit power capability to take advantage of an increased resource allocation. The Scheduling Information also includes RLC buffer occupancy measurements - both total and for the highest priority logical channel. UE which report high buffer occupancies for high-priority logical channels are more likely to be targeted for increased resource allocations. The Node B packet scheduler is also likely to consider the size of existing resource allocations when identifying candidates for upgrades. If all connections have been configured with an equal quality of service then the packet scheduler should attempt to allocate equal resources across all connections.

If the Node B is operating above its target interference margin then it becomes necessary to identify specific connections for downgrading. Connections with low utilisation can be selected as described above. Otherwise, connections with relatively high resource allocations can be selected. This selection procedure can account for the quality of service profile associated with each connection, e.g. connections based upon the background traffic class can be downgraded before connections based upon the interactive traffic class.

In general, the Node B packet scheduler can use either the E-RGCH or the E-AGCH to increase or decrease existing allocations. An exception is when downgrades are triggered as a result of a cell receiving too much interference power from non-serving UE. In that case the E-RGCH must be used because the E-AGCH can only be sent from the serving cell. The Node B packet scheduler can use RRC signalling to allocate the initial E-DPDCH to DPCCH power ratio. This initial allocation is transferred to the RNC using NBAP signalling and subsequently to the UE using RRC signalling. This sequence is illustrated in Figure 7.7.

The RNC is also able to configure HSUPA connections to operate in a non-scheduled mode. This allows the UE to transfer HSUPA data at any time without receiving any scheduling information from the Node B. This approach is intended to reduce both scheduling delay and the signalling requirement. MAC-d flows used to transfer the Signalling Radio Bearers may be configured to use non-scheduled transmissions. The RNC configures a non-scheduled transmission for a specific MAC-d flow by allocating a maximum MAC-e PDU size, i.e. a maximum transport block size. This limits the quantity of resource which the non-scheduled transmission can use. In addition, specific HARQ processes can be reserved for non-scheduled and scheduled transmissions if a 2 ms TTI is used.

7.8.2 De-multiplexing

The Node B MAC-e layer receives MAC-e PDU from the physical layer after they have been successfully received. The de-multiplexing function is responsible for extracting the MAC-es PDU from the payload and the DDI/N pairs from the header. Both the MAC-es PDU and the DDI/N pairs are forwarded to the Frame Protocol layer for packaging within a data frame.

7.9 Frame Protocol Layer

- The Node B uses Frame Protocol E-DCH data frames to package MAC-es PDU prior to transferring them across the Iub. The RNC can use Tunnel Congestion Indication control frames to inform the Node B of transport network congestion or subsequent recovery.
- Frame Protocol packets do not include any form of UE identity. The receiving RNC or Node B use the identity of the lower layer transport connection to associate packets with specific HSUPA connections.
- The header of the E-DCH data frame includes a Frame Sequence Number to allow the RNC to detect frame loss. It also includes a Connection Frame Number to help detect increased delays. The number of HARQ re-transmissions is included to support MAC-es PDU re-ordering and to allow outer loop power control to calculate the BLER.
- The Tunnel Congestion Indication control frame indicates either no congestion, congestion detected from frame loss or congestion detected from increased delay. The Node B can use this information to initiate congestion control actions.
- Key 3GPP specifications: TS 25.427.

HSUPA uses E-DCH data frames to package MAC-es PDU prior to transferring them across the Iub transport network. These data frames belong to the dedicated channel Frame Protocol specified within 3GPP TS 25.427. HSUPA can also use Tunnel Congestion Indication control frames to allow the RNC to inform the Node B if it detects transport network congestion or subsequently detects a recovery from congestion. Outer Loop Power Control frames are used in the same way as for a DCH connection. These control frames also belong to the dedicated channel Frame Protocol.

Frame Protocol packets do not include any form of UE identity. The receiving RNC (in the case of E-DCH data frames) or receiving Node B (in the case of downlink control frames) must rely upon information provided by other layers to associate each Frame Protocol packet with a specific HSUPA connection. The lower transport layers can be used for this purpose. When ATM Adaptation Layer 2 (AAL2) is used to transfer data across the Iub each HSUPA connection is assigned an AAL2 Channel Identifier (CID). This CID is unique within a specific ATM Virtual Channel Connection (VCC). The VCC is identified using its Virtual Channel Identifier (VCI). These ATM identifiers allow the receiving

Node B or receiving RNC to associate each data frame and each control frame with a specific HSUPA connection.

7.9.1 E-DCH Data Frame

Figure 7.23 illustrates the structure of the E-DCH data frame used to package MAC-es PDU prior to transferring them across the Iub transport network. The header can be divided into a general section followed by a series of one or more subframe specific sections.

The general section of the header starts with an 11 bit Cyclic Redundancy Check (CRC). These CRC bits are used by the RNC to detect whether or not the header information has been received correctly. A 1 bit Frame Type (FT) flag is used to indicate whether the subsequent frame is a data frame or a control frame. A value of 0 indicates a data frame whereas a value of 1 indicates a control frame. Figure 7.23 illustrates a data frame and so the FT flag has a value of 0. The 4 bit Frame Sequence Number (FSN) is incremented by 1 for every Frame Protocol data frame sent to the serving RNC. This information is

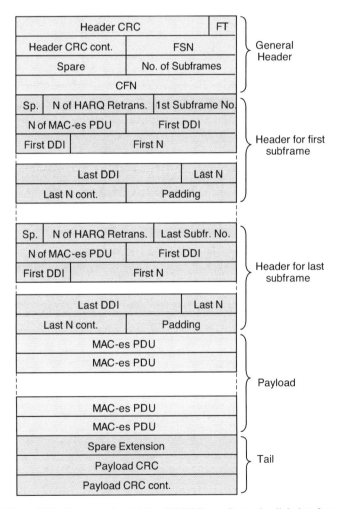

Figure 7.23 Structure of an Iub/Iur E-DCH Frame Protocol uplink data frame

used by the serving RNC to detect frame loss possibly resulting from transport network congestion. If the serving RNC receives non-contiguous FSN then it can instruct the Node B to initiate transport network congestion control procedures. The general section of the header then includes a 4 bit indication of how many subframes are included within the Frame Protocol data frame. If the HSUPA TTI is 10 ms and the E-DCH data frames are transferred to the serving RNC once every 10 ms then each data frame would include one subframe. If the HSUPA TTI is 2 ms and the E-DCH data frames are transferred to the serving RNC once every 10 ms then each data frame would include five subframes. The signalled value for the number of subframes is equal to the actual value minus 1, i.e. a signalled value of 0 corresponds to an actual value of 1. The final part of the general section of the header is an 8 bit Connection Frame Number (CFN) used to indicate the CFN during which the Node B successfully decoded the MAC-es PDU. This information can be used by the serving RNC to help detect increased delays possibly resulting from transport network congestion. Similar to identifying a missing FSN, the RNC can use this information to instruct the Node B to initiate transport network congestion control procedures.

The subframe specific section of the header starts with a 4 bit indication of the number of HARQ re-transmissions required to successfully decode the corresponding MAC-e PDU. A value of 15 means that the number of re-transmissions is not known. Values of 13 and 14 are reserved and ignored whereas a value of 12 means that 12 or more re-transmissions were required. The Node B can use the Re-transmission Sequence Number (RSN) from within the E-DPCCH to help identify the number of re-transmissions. The synchronous nature of the HARQ protocol makes it easier for the Node B to keep track of the number of re-transmissions when the Node B fails to receive data during specific subframes, i.e. if the Node B can deduce the subframe during which the first transmission occurred then it can calculate the number of re-transmissions. The number of re-transmissions can be used by the MAC-es layer within the RNC to help re-order MAC-es PDU. It can also be used by the RNC outer loop power control functionality when calculating the BLER performance.

If a 2 ms TTI is used, the 3 bit subframe number indicates the subframe during which the transport block was received. The subframe number can range from 0 to 4. The RNC can use it for re-ordering purposes and also to help calculate transport network delays. The subframe number is always set to 0 if a 10 ms TTI is used. The next part of the header uses 4 bits to indicate the number of MAC-es PDU within the payload of the relevant transport block, i.e. the transport block received during the specified subframe. The remainder of the header includes the set of DDI/N pairs which the Node B has extracted from the header of the MAC-e PDU. The number of DDI/N pairs equals the number of MAC-es PDU within the payload.

The payload of the E-DCH data frame includes the set of MAC-es PDU. The payload will include only a single MAC-es PDU if data frames are transferred every 10 ms, while a 10 ms TTI is used and each MAC-e PDU includes a single MAC-es PDU. The final part of the E-DCH data frame includes an optional CRC field for the contents of the payload. These CRC bits can be used by the RNC to detect whether or not the payload information has been received correctly.

7.9.2 Tunnel Congestion Indication Control Frame

The structure of the Tunnel Congestion Indication control frame is illustrated in Figure 7.24. This control frame is only used in the downlink direction. The content is included within the 2 bit congestion

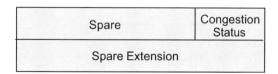

Figure 7.24   Structure of a Tunnel Congestion Indication control frame

status section. The congestion status can be signalled using a value of 0 to represent no congestion, a value of 2 to represent congestion detected by increased delay and a value of 3 to represent congestion detected by frame loss.

The Tunnel Congestion Indication control frame is not used under normal operating conditions. However, if the RNC detects transport network congestion from either an increased delay or a lost frame, it can use this control frame to inform the Node B. The Node B may then initiate congestion control actions to help alleviate the network load. The congestion is detected and reported on a per HSUPA connection basis. The control frame itself does not include any form of identity, but the lower layer transport connection identifies the relevant HSUPA connection. The Node B may target the individual connection which has experienced congestion or it may target a broader group of connections. Congestion control is likely to involve reducing the E-DPDCH to DPCCH power ratio to reduce the transport block sizes received by the Node B. Frame loss is more significant than an increased delay and so the congestion control actions may be more severe if frame loss is reported. The Node B remains in a load control state until the RNC subsequently sends another control frame indicating no congestion. The Node B can then return to normal operation.

7.10 MAC-es Entity (RNC)

> - The MAC-es layer provides MAC-es PDU re-ordering and macro diversity combining. It also extracts MAC-d PDU before forwarding them to the MAC-d layer.
> - Key 3GPP specifications: TS 25.321.

The MAC-es layer within the RNC provides a re-ordering capability to ensure that in-sequence delivery can be provided to the higher layers, i.e. MAC-d PDU are delivered to the MAC-d layer in the correct order. The re-ordering function makes use of the Transmission Sequence Number (TSN) included within the MAC-es PDU header. It can also use the Connection Frame Number (CFN) and number of HARQ re-transmissions specified within the Frame Protocol data frame header. This information allows the RNC to identify the TTI during which the MAC-es PDU was first transmitted. Each logical channel has its own series of sequence numbers so re-ordering is completed on a per logical channel basis. MAC-es PDU are delivered to the disassembly function once consecutive TSN have been received. If specific TSN remain missing, the RNC proceeds to deliver the next MAC-es PDU to the disassembly function and subsequently relies upon RLC re-transmissions to recover the data.

The MAC-es layer also provides macro diversity combining for HSUPA connections which are in soft handover. Macro diversity combining at the RNC is not necessary for softer handover because combining is completed within the Node B receiver. Macro diversity combining involves selecting individual MAC-es PDU from each soft handover connection to generate the final stream of data used by the re-ordering function. Duplicate MAC-es PDU are discarded as part of the combining process.

The disassembly function is responsible for extracting the set of MAC-d PDU from the series of MAC-es PDU and subsequently forwarding them to the MAC-d layer.

7.11 Mobility

> - HSUPA connections use a combination of the soft handover and serving cell change procedures.
> - It is not necessary for the E-DCH active set to be the same as the DCH active set. If cells are included within the DCH active set, but not within the E-DCH active set, there is a danger that those cells can experience increased quantities of uplink interference.

• The support for soft handover means that the responsiveness of the serving cell change procedure is less critical for HSUPA. The HSUPA serving cell change procedure can use the same triggering mechanisms as the HSDPA serving cell change procedure.
• Key 3GPP specifications: TS 25.331.

HSUPA connections use a combination of soft handover and serving cell change procedures. In contrast to the HS-PDSCH used for HSDPA, the E-DPDCH used for HSUPA can take advantage of soft handover. This helps to improve the performance of HSUPA towards the cell edge, but generates an increased overhead in terms of Node B baseband processing and Iub transport capacity. Similar to HSDPA, it is necessary for an HSUPA connection to change its serving cell as the UE moves throughout the network.

When an HSUPA connection is in soft handover, it is not necessary for the E-DCH active set to be the same as the DCH active set. An example of a UE with different E-DCH and DCH active sets is illustrated in Figure 7.25. In this example, the UE has an E-DCH active set size of 2 while the DCH active set size is 3. Soft handover radio link addition can be triggered in the usual manner when an HSUPA connection is active, e.g. using measurement reporting event 1a. The subsequent radio link addition procedure includes an admission control decision for the E-DCH. If the new cell does not support HSUPA or if the cell is already supporting its maximum number of HSUPA connections then the new radio link can be added to the DCH active set without being added to the E-DCH active set. The content of the Active Set Update message sent from the RNC to the UE indicates that it should include the new radio link. If conditions change, it is possible that a subsequent Active Set Update message is used to add a cell which is already in the DCH active set to the E-DCH active set.

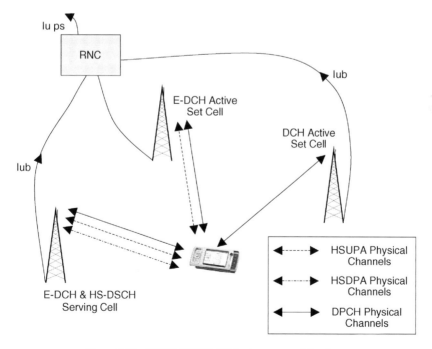

Figure 7.25   HSUPA/HSDPA UE in three-way soft handover

If cells are included within the DCH active set, but not within the E-DCH active set, there is a danger that those cells can experience increased quantities of uplink interference. Including a cell within the DCH active set means that inner loop power control can be used to limit the received power of the DPCCH based upon the SIR target defined by the outer loop power control function within the RNC. However, inner loop power control cannot limit the power ratio between the E-DPDCH and DPCCH. The serving HSUPA cell may allocate relatively large power ratios to increase the E-DPDCH transmit power without having knowledge of the impact upon the cells which are only in the DCH active set. It is likely that the RNC will release the HSUPA connection and replace it with a DCH connection if the UE moves too close to the cell which cannot be added to the E-DCH active set.

The UE illustrated in Figure 7.25 is configured with both HSDPA and HSUPA connections. HSDPA data is only transferred from the HS-DSCH serving cell. In this example, it is assumed that the HS-DSCH serving cell is the same as the E-DCH serving cell. The support for soft handover means that the responsiveness of the serving cell change procedure is less critical for HSUPA. The performance of an HSDPA connection relies upon the serving cell always corresponding to the best cell. The HSUPA serving cell change procedure can use the same triggering mechanisms as the HSDPA serving cell change procedure, e.g. downlink CPICH Ec/Io. This approach allows the two serving cell responsibilities to move together and removes the requirement to have separate triggering mechanisms and additional signalling.

8

Signalling Procedures

8.1 RRC Connection Establishment

- RRC connection establishment is used to move a UE from RRC Idle mode to RRC Connected mode. The transition can be to CELL_DCH or CELL_FACH. The transition to CELL_FACH requires less signalling, but may provide a reduced quality of service.
- A UE can be directed towards a different RF carrier as part of the RRC connection establishment procedure. This could be triggered by network congestion.
- RRC connection establishment involves a combination of RRC, NBAP, ALCAP and Frame Protocol signalling. It also involves synchronisation on the air-interface and Iub.
- A UE uses SRB 0 to send the RRC Connection Request on the RACH. An RNC uses SRB 0 to send the RRC Connection Setup on the FACH. A UE uses SRB 2 to send the RRC Connection Setup Complete on a DCH (CELL_DCH) or on a FACH (CELL_FACH).
- The release 4 and newer versions of the RRC Connection Request message include the release version of the UE. This allows the RNC to respond with the appropriate version of the RRC Connection Setup message.
- The release 6 and never versions of the RRC Connection Request message include the HSDPA and HSUPA capability of the UE. This allows the RNC to direct the UE to an appropriate RF carrier.
- RRC Connection Setup messages are addressed to specific UE by including the UE Identity received within the relevant RRC Connection Request. The setup message configures SRB 1, 2 and 3 as a minimum. SRB 4 may also be configured.
- The RRC Connection Request and RRC Connection Setup messages can be re-transmitted from the RRC layer, i.e. by re-transmitting the entire message. Individual transport blocks of the RRC Connection Setup Complete message can be re-transmitted by the RLC layer.
- If the UE is moved into CELL_DCH, a common NBAP radio link setup procedure is used to configure the Node B; an ALCAP handshake is used to establish an AAL2 connection on the Iub; and Frame Protocol control frames are used to achieve Iub synchronisation.
- If the UE is moved into CELL_DCH, the Node B starts transmitting the DPCCH after receiving the Radio Link Setup Request message. The UE synchronises with the downlink DPCCH after receiving the RRC Connection Setup message. The UE then starts to transmit the uplink DPCCH allowing the Node B to achieve synchronisation.

Radio Access Networks for UMTS   Chris Johnson
© 2008 John Wiley & Sons, Ltd

- The RRC Connection Setup Complete message provides the RNC with a specification of the UE radio access capability. This allows the RNC to identify the maximum supported bit rates. It can also include HSDPA and HSUPA capability information.
- Key 3GPP specifications: TS 25.331, TS 24.008, TS 25.133, TS 25.306.

RRC connection establishment is used to move a UE from RRC Idle mode to RRC Connected mode. It is necessary for a UE to be in RRC Connected mode before it is able to start signalling or transferring data to and from the network. The transition from RRC Idle mode can move the UE into either CELL_DCH or CELL_FACH. RRC connection establishment is always initiated by the UE, but can be triggered by either the UE or the network. For example, a UE triggers RRC connection establishment when it is in RRC Idle mode and wishes to establish a mobile originated speech call or a mobile originated data connection. The network uses the paging procedure to trigger the UE to initiate RRC connection establishment. For example, a UE can be paged for a mobile terminated speech call or to receive an incoming SMS. Specific signalling procedures can also trigger the establishment of an RRC connection. For example, an RRC connection is established to complete a location area update if a UE is registered with the CS core network and crosses a location area boundary while in RRC Idle mode. A UE can be directed towards a different RF carrier as part of the RRC connection establishment procedure. The network may decide to direct a UE towards a different RF carrier if the current carrier is relatively congested. Alternatively, if an HSDPA/HSUPA UE is establishing a data connection then the network may direct the UE towards an RF carrier which supports HSDPA and HSUPA connections.

The signalling associated with RRC connection establishment is illustrated in Figure 8.1. This figure is based upon the transition from RRC Idle mode to CELL_DCH. From the UE perspective, RRC

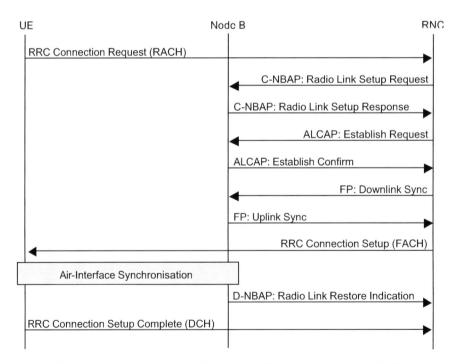

Figure 8.1 RRC connection establishment signalling (RRC Idle to CELL_DCH)

connection establishment to CELL_DCH involves three RRC messages combined with air-interface synchronisation. From the network perspective, RRC connection establishment to CELL_DCH involves RRC, NBAP, ALCAP and Frame Protocol signalling combined with both air-interface and Iub synchronisation.

A UE initiates the RRC connection establishment procedure by sending an RRC Connection Request message to the RNC. This message uses Signalling Radio Bearer (SRB) 0 on the CCCH logical channel, RACH transport channel and PRACH physical channel. These channel types require the UE to complete the random access procedure. This involves sending one or more PRACH power control preambles before receiving a Node B acknowledgement on the AICH physical channel. The UE is only able to send the RRC Connection Request message after receiving a positive acknowledgement on the AICH. If the UE is located within an area of poor coverage, the Node B may not acknowledge a preamble, or the Node B may acknowledge a preamble, but the UE may not receive the acknowledgement. If the UE has not received an acknowledgement after completing a cycle of PRACH preambles the UE MAC layer is able to request further cycles of PRACH preambles. The delay introduced by the UE sending a relatively large number of PRACH preambles is dependent upon the AICH transmission timing. The UE is able to send one preamble every six time slots (4.0 ms) if the transmission timing is set to 0. The UE is able to send one preamble every eight time slots (5.3 ms) if the transmission timing is set to 1. The number of preambles required to obtain an acknowledgement can be influenced by the PRACH open loop power control calculation. A UE is more likely to receive a rapid acknowledgement if the open loop power control calculation generates a relatively high transmit power result. However, generating a high transmit power result also increases the probability of generating excessive uplink interference. A UE typically sends less than five preambles to receive an acknowledgement so the delay associated with the PRACH procedure is relatively small. If the UE is in poor coverage and has to send multiple preamble cycles then the delay could exceed 200 ms. The RRC connection establishment procedure fails if the UE has not received an acknowledgement after the maximum allowed number of PRACH preamble cycles.

The UE is able to send the RRC Connection Request message after receiving a positive acknowledgement on the AICH. The message is relatively small and requires only a single transport block. The transport block size is typically 168 bits while the TTI is typically 20 ms. Transparent mode RLC is always used to transfer the RRC Connection Request message. An example RRC Connection Request message is presented in Log File 8.1.

```
UL-CCCH-Message
rrcConnectionRequest
    initialUE-Identity
        tmsi-and-LAI
            tmsi
                Bin: 01 13 E9 27
            lai
                plmn-Identity
                    mcc value: 234
                    mnc value: 99
                lac
                    Bin: 27 83
    establishmentCause: originatingConversationalCall
    protocolErrorIndicator:    noError
    measuredResultsOnRACH
        currentCell
            modeSpecificInfo fdd
                measurementQuantity
                cpich-Ec-N0: 41   (-4.0 to -3.5 dB)
```

Log File 8.1 RRC Connection Request message

In this example, the UE identifies itself to the RNC using its Temporary Mobile Subscriber Identity (TMSI). TMSI are allocated to be unique within individual location areas, but may be repeated across different locations areas and different PLMN. The UE specifies the Location Area Identity (LAI) in combination with the TMSI to ensure that its identity is unique globally. In general, UE are located within the location area where the TMSI was assigned. However, if a UE has just crossed a location area boundary then it will establish an RRC connection to complete a location area update and refresh its TMSI. In that case, the LAI included within the RRC Connection Request message corresponds to the old LAI within which the old TMSI was allocated. 3GPP TS 25.331 specifies that UE should use the TMSI and LAI as the initial UE identity whenever possible. If a TMSI is not available then the Packet-TMSI (P-TMSI) should be used in combination with the routing area identity (RAI). If neither the TMSI nor P-TMSI are available then the IMSI should be used, and finally if the IMSI is not available then the IMEI should be used. A TMSI may not be available if a UE has not previously registered with the network or if a previous location update request has been rejected. 3GPP TS 24.008 specifies that a UE should delete its stored TMSI and LAI after a rejected location update request.

The RRC Connection Request message includes an establishment cause after the UE identity. This provides the network with an initial indication of why the UE is requesting an RRC connection. The release 6 version of 3GPP TS 25.331 specifies a total of 22 establishment causes. The example shown in Log File 8.1 indicates an establishment cause of mobile originating conversational call. It is likely that this corresponds to either a speech or video call. The relationship between the higher layer establishment cause and the RRC connection establishment cause is defined within 3GPP TS 24.008. Examples for the CS domain are presented within Table 8.1. The RNC may use the establishment cause for pre-emption purposes, e.g. an emergency call could pre-empt an ongoing PS data connection if the network is experiencing congestion.

The RRC connection establishment cause is followed by a protocol error indicator. In general this indicator should not show an error. The indicator can be set to show an error if the UE has previously received an invalid RRC Connection Setup message or an invalid RRC Connection Reject message. An RRC Connection Setup message can be categorised as invalid if it attempts to configure an unsupported configuration.

The final section of the RRC Connection Request message provides the RACH measurements which the UE has been instructed to include by SIB 11. In this example, SIB 11 has instructed the UE to provide the RNC with a CPICH Ec/Io measurement recorded from the current serving cell. The CPICH Ec/Io measurement requires a mapping to derive the actual value from the signalled value. A look-up table generated from the mapping specified within 3GPP TS 25.133 is presented in Table 8.5. A signalled value of 41 corresponds to an actual value of between -4.0 and -3.5 dB. The network can use this CPICH Ec/Io measurement for its downlink DPCH open loop power control calculation. The

Table 8.1 Mapping between higher layer and RRC connection establishment causes

CS Non-Access Stratum Procedure	RRC Establishment Cause
Originating CS speech call	Originating conversational call
Originating CS data call	Originating conversational call
CS emergency call	Emergency call
Call re-establishment	Call re-establishment
Location update	Registration
IMSI detach	Detach
MO SMS via CS domain	Originating low-priority signalling
Supplementary services	Originating high-priority signalling
Answer to circuit switched paging	Equal to cause value within paging message
Location services	Originating high priority signalling

measurement is also useful for performance monitoring purposes. It allows CPICH Ec/Io coverage data to be collected and linked to the RRC connection establishment success rate.

The RRC Connection Request message in Log File 8.1 has been recorded from a UE using the release 99 version of the 3GPP specifications. The release 4 version of the specifications introduces an additional mandatory information element to indicate which release of the specifications a UE is using. An example of this access stratum release indicator is presented within Log File 8.2. The RNC can use this information to select the appropriate version of any subsequent RRC messages. It can also be used to redirect a UE towards a specific RF carrier. A release 5 UE may be HSDPA capable and could be redirected to an HSDPA RF carrier if requesting a service supported by HSDPA.

The release 6 version of the specifications introduces an optional information element which can be used to inform the RNC of whether a UE supports HSDPA, or both HSDPA and HSUPA. An example of this UE capability indicator is presented in Log File 8.3. The RNC can assume that a release 6 UE does not support HSDPA nor HSUPA if this information element is absent from the RRC Connection Request message. Similar to the access stratum release indicator, the HSDPA and HSUPA capability information can be used to redirect a UE towards a specific RF carrier.

If the Node B successfully receives the RACH transport block which includes the RRC Connection Request message, the Node B packages it within a common channel Frame Protocol data frame. This data frame is then forwarded across the Iub to the controlling RNC. An example of a common channel Frame Protocol data frame is presented in Log File 2.44 within Section 2.5.3. If the UE is in an area of poor uplink coverage then it is possible that the Node B does not receive the RRC Connection Request message. The UE will trigger a re-transmission if the timer T300 expires while the counter V300 is less than or equal to N300. Both T300 and V300 are broadcast within SIB 1. The principle of these re-transmissions from the RRC layer is illustrated in Figure 8.2.

T300 is re-started and V300 is incremented after each transmission. The re-transmission procedure is relatively slow and can have a significant impact upon the RRC connection establishment delay. Successful RRC connection establishment typically requires less than 700 ms when a response is received from the RNC after the first transmission of the RRC Connection Request message. Typical values for T300 and N300 are 2 s and 3, respectively. This means that the delay associated with the establishment procedure can potentially increase by a factor of 10. Decreasing the value of T300

```
rrcConnectionRequest-v4b0ext
    accessStratumReleaseIndicator: rel-5
```

Log File 8.2 Release indicator within RRC Connection Request message

```
RRCConnectionRequest-v690ext-IEs
    ueCapabilityIndication: hsdch-edch
```

Log File 8.3 HSDPA and HSUPA capability within RRC Connection Request message

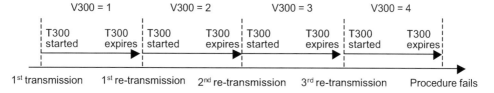

Figure 8.2 Re-transmission of the RRC Connection Request message (N300 = 3)

reduces the delay associated with the re-transmissions, but increases the probability of consecutive attempts failing. If one attempt has failed as a result of poor radio propagation conditions then a subsequent attempt is also likely to fail if the radio conditions have not had sufficient time to change. The overall RRC connection establishment procedure fails if the UE has not received a response from the RNC when T300 expires and V300 is greater than N300.

If the RNC successfully receives the RRC Connection Request message then admission control is used to determine whether or not the request can be accepted. In general, checks are completed in terms of downlink transmit power, uplink interference, channelisation code availability and Iub transport capacity. The RNC uses periodic NBAP measurement reports from the Node B to gain visibility of the both the downlink transmit power and uplink interference. The RNC is responsible for allocating channelisation codes and so has knowledge of whether or not any codes are available. Likewise the RNC should be aware of the Iub transport network load. If admission control denies the connection request then an RRC Connection Reject message is returned to the UE. This message uses Signalling Radio Bearer (SRB) 0 on the CCCH logical channel, FACH transport channel and S-CCPCH physical channel. Unacknowledged mode RLC is always used to transfer the RRC Connection Reject message. An example of the message is shown in Log File 8.4.

The UE starts to decode RRC messages from the FACH after sending the RRC Connection Request message. The FACH is a common transport channel shared between all UE within the same cell. This means that the UE may receive RRC messages addressed to other UE while waiting for a response to its RRC Connection Request. The UE has to check the initial UE identity within any RRC Connection Reject or RRC Connection Setup messages to determine whether or not the message should be processed or discarded. Messages are only processed if the initial UE identity matches the identity within the RRC Connection Request message. RRC Connection Reject messages include a cause value which can have one of two values: congestion or unspecified. The example shown in Log File 8.4 indicates that the RRC Connection Request has been rejected as a result of congestion. The reject message also includes a wait time which defines how long the UE must wait prior to re-attempting RRC connection establishment. A UE can continue to send RRC Connection Request messages after the wait time has expired. The value of V300 is not reset and so the number of further request messages depends upon how many were sent prior to receiving the reject message. The wait time is likely to depend upon the RRC connection establishment cause. A wait time of 1 s could be applied for the registration cause while a wait time of 5 s could be applied for the background cause. The overall RRC connection establishment procedure fails if a wait time of 0 s is included within the reject message. The RRC Connection Reject message can optionally include redirection information. This information can be used to re-direct the UE to either a different RF carrier or a different radio access technology.

```
DL-CCCH-Message
RRCConnectionReject
rrcConnectionReject-r3
initialUE-Identity
    tmsi-and-LAI
        tmsi
            Bin: 01 13 E9 27
        lai
            plmn-Identity
                    mcc value: 234
                    mnc value: 99
            lac
                Bin: 27 83
rrc-TransactionIdentifier: 0
rejectionCause: congestion
waitTime: 3
```

Log File 8.4   RRC Connection Reject message

If admission control accepts the request for an RRC connection, the RNC initiates the NBAP radio link setup procedure. This procedure is initiated by the RNC sending an NBAP Radio Link Setup Request message to the Node B (shown within Figure 8.1). This message provides sufficient information for the Node B to start transmitting and receiving. The first part of an example NBAP Radio Link Setup Request message is presented in Log File 8.5.

The message starts by specifying the controlling RNC communication context identity. The Node B uses this identity to identify the connection within subsequent NBAP messages. The message then specifies the uplink scrambling code and the uplink spreading factor. These have now been allocated by the controlling RNC. It is not necessary to signal the uplink channelisation code indices because the DPCCH is always spread using channelisation code (SF256,0) whereas the DPDCH is always spread using (SFx,x/4), i.e. in this example the uplink DPDCH is spread using (SF256,64). The uplink puncturing limit is also signalled to the Node B. A mapping is necessary to derive the actual value from the signalled value. The actual value = (signalled value $\times$ 4) + 40, i.e. a signalled value of 7 translates to an actual value of 68%.

```
RADIO LINK SETUP REQUEST
NBAP-PDU
RadioLinkSetupRequestFDD
CRNC-CommunicationContextID: 1039
UL-DPCH-Information-RL-SetupRqstFDD
        ul-ScramblingCode
                - uL-ScramblingCodeNumber: 1000015
                - uL-ScramblingCodeLength: long
        - minUL-ChannelisationCodeLength: v256
        - ul-PunctureLimit: 7
        tFCS    tFCSvalues
                no-Split-in-TFCI
                cTFC - ctfc2bit: 0
                        tFC-Beta - computedGainFactors: 0
                cTFC - ctfc2bit: 1
                        tFC-Beta - signalledGainFactors
                                gainFactor fdd
                                - betaC: 15
                                - betaD: 11
                                - refTFCNumber: 0
        - ul-DPCCH-SlotFormat: 0
        - ul-SIR-Target: 46
        - diversityMode: none
DL-DPCH-Information-RL-SetupRqstFDD
        tFCS    tFCSvalues
                no-Split-in-TFCI
                cTFC - ctfc2bit: 0
                cTFC - ctfc2bit: 1
        - dl-DPCH-SlotFormat: 5
        tFCI-SignallingMode
                - tFCI-SignallingOption: normal
        - multiplexingPosition: flexible
        powerOffsetInformation
                - pO1-ForTFCI-Bits: 0
                - pO2-ForTPC-Bits: 12
                - pO3-ForPilotBits: 0
        - fdd-TPC-DownlinkStepSize: step-size1
        - limitedPowerIncrease: not-used
        - innerLoopDLPCStatus: active
```

Log File 8.5 NBAP Radio Link Setup Request message (part 1)

The Radio Link Setup Request message then proceeds to inform the Node B of the uplink Transport Format Combination Set (TFCS). The TFCS is relatively simple because the RNC is only configuring a standalone SRB which uses a single transport channel with two transport formats. The TFCS includes two Transport Format Combinations (TFC) which are specified using calculated Transport Format Combinations (cTFC) of 0 and 1. The DPCCH and DPDCH amplitude gain factors, β_c and β_d are specified for the second TFC. The Node B is informed that these gain factors are to be used as a reference to calculate the gain factors for the first TFC. This part of the Radio Link Setup Request message does not provide any information regarding the bit rates of the transport format combinations. The Node B is informed that the uplink DPCCH slot format is 0. This allows the Node B to deduce that each time slot of the uplink DPCCH includes 6 pilot bits, 2 Transmit Power Control (TPC) bits, 2 Transport Format Combination Indicator (TFCI) bits and 0 transmit diversity Feedback Indication (FBI) bits. It is not necessary to signal a slot format for the uplink DPDCH because the uplink spreading factor has already been specified and there is a one-to-one mapping between uplink DPDCH slot format and spreading factor. An uplink SIR target is included to initialise uplink inner loop power control at the Node B. A mapping is necessary to derive the actual value from the signalled value. The actual value is equal to the signalled value / 10, i.e. a signalled value of 46 translates to an actual value of 4.6 dB. The Node B is also informed that downlink transmit diversity is not to be used.

The next part of the Radio Link Setup Request message provides similar information for the downlink direction. The downlink TFCS includes two TFC which are defined using cTFC values of 0 and 1. In the downlink direction, a single slot format defines both the DPCCH and DPDCH. Slot format 5 indicates a downlink spreading factor of 256 as well as 12 DPDCH and 8 DPCCH bits per time slot. The 8 DPCCH bits correspond to 2 TPC bits, 2 TFCI bits and 4 pilot bits. The downlink multiplexing position is specified to be flexible rather than fixed. This does not have any implications because the radio link is configured with only a single transport channel. The downlink DPCCH power offsets are specified as PO1, PO2 and PO3. PO1 is applicable to the TFCI bits, PO2 is applicable to the TPC bits and PO3 is applicable to the pilot bits. All three of these values are signalled using units of 0.25 dB, i.e. the actual value = signalled value / 4. In this example, the TPC bits are transmitted using 3 dB more power than the DPDCH, whereas the TFCI and pilot bits are transmitted using a power which is equal to the power of the DPDCH. Finally, the downlink inner loop power control step size is specified to equal 1 dB and the limited power increase algorithm is disabled. Downlink inner loop power control itself is enabled.

The second part of the same NBAP Radio Link Setup Request message is presented within Log File 8.6. This section of the message starts by defining the dedicated channel Frame Protocol parameters. The Frame Protocol layer within the Node B is instructed to include a payload CRC as part of each uplink data frame. The Node B is also instructed to use silent mode. This means that the Node B only transfers data frames when DPDCH data has been received. The Node B is also provided with time of arrival window start (toAWS) and end (toAWE) points. These parameters define the time window during which the Node B must receive downlink data from the RNC in order to have sufficient time to complete its Physical layer processing and transmit the resulting data during the appropriate TTI. The toAWS point is defined in ms relative to the toAWE point, i.e. in this example the receive window has a width of 15 ms. The toAWE point is defined in ms relative to the latest possible time that the Node B can receive downlink data in order to have sufficient time to complete physical layer processing. This parameter is intended to allow some margin so that when data is received late, i.e. after the end point, the Node B has sufficient time to correct the downlink timing between itself and the RNC using the DCH frame protocol timing adjustment procedure.

The Radio Link Setup Request message then proceeds to configure a single transport channel. Only a single transport channel is necessary because the RNC is configuring a standalone SRB. In this example, the transport channel is assigned an identity of 24. The uplink and downlink TFS include two Transport Formats (TF). The first TF corresponds to no data transfer. In the uplink direction this is indicated by the transport format having no transport blocks whereas in the downlink direction this is

```
DCH-FDD-Information
DCH-FDD-InformationItem
        - payloadCRC-Presence: included
        - ul-FP-Mode: silent
        - toAWS: 15
        - toAWE: 5
dCH-SpecificInformationList
DCH-Specific-FDD-Item
        - dCH-ID: 24
        ul-TransportFormatSet
                dynamicParts
                        - nrOfTransportBlocks: 0
                        - nrOfTransportBlocks: 1
                        - transportBlockSize: 148
                semi-staticPart
                        - transmissionTimeInterval: msec-40
                        - channelCoding: convolutional-coding
                        - codingRate: third
                        - rateMatchingAttribute: 1
                        - cRC-Size: v16
        dl-TransportFormatSet
                dynamicParts
                        - nrOfTransportBlocks: 1
                        - transportBlockSize: 0
                        - nrOfTransportBlocks: 1
                        - transportBlockSize: 148
                semi-staticPart
                        - transmissionTimeInterval: msec-40
                        - channelCoding: convolutional-coding
                        - codingRate: third
                        - rateMatchingAttribute: 1
                        - cRC-Size: v16
        allocationRetentionPriority
                - priorityLevel: 14
                - pre-emptionCapability: shall-not-trigger-pre-emption
                - pre-emptionVulnerability: pre-emptable
                - frameHandlingPriority: 14
                - qE-Selector: selected
RL-InformationList-RL-SetupRqstFDD
        RL-InformationItem-RL-SetupRqstFDD
        - rL-ID: 1
        - c-ID: 121
        - firstRLS-indicator: first-RLS
        - frameOffset: 5
        - chipOffset: 8192
        - propagationDelay: 3
        dl-CodeInformation
        FDD-DL-CodeInformationItem
        - dl-ScramblingCode: 0
        - fdd-DL-ChannelisationCodeNumber: 10
        - initialDL-transmissionPower: -148
        - maximumDL-power: -51
        - minimumDL-power: -180
```

Log File 8.6   NBAP Radio Link Setup Request message (part 2)

indicated by the transport block size being 0 bits. The second TF in both the uplink and downlink directions is based upon a single transport block with a size of 148 bits. The TTI is 40 ms and so the corresponding bit rate at the top of the Physical layer is 3.7 kbps. The definition of the uplink and downlink transport format sets is completed by indicating that rate 1/3 convolutional coding should be applied and a 16 bit CRC should be included. A rate matching attribute is also specified, but the value has no meaning because only a single transport channel is being configured. Rate matching attributes are used to define the share of the DPDCH that each transport channel is allocated when multiple transport channels are multiplexed by the physical layer.

The Radio Link Setup Request message then goes on to define the priority of the dedicated channel. In this example, the DCH is not able to trigger pre-emption in case the Node B resources are congested, and the DCH will be susceptible to pre-emption from other higher-priority DCH. A priority level of 14 is assigned which represents the lowest signalled priority. A priority level of 1 represents the highest priority and a value of 15 indicates that the DCH is not assigned a priority. The frame handling priority is also configured with a value of 14. In this case, a value of 14 represents a high priority. A value of 0 represents the lowest priority whereas a value of 15 represents the highest priority. This priority value is used in case the Node B becomes overloaded (in contrast to the preceding priority which is applicable to pre-emption when new capacity requests are received). A high priority is assigned in this example because the DCH is established for signalling purposes rather than user plane data purposes. A connection is more likely to fail completely if the signalling connection is not maintained. The quality estimate selector instructs the Node B to include either an estimate of the uplink transport channel BER or the uplink physical channel BER within the payload of each uplink dedicated channel Frame Protocol data frame. The Node B includes an estimate of the transport channel BER whenever possible if a value of 'selected' is received. In that case, the Node B only includes the physical channel BER if the transport channel BER is not available. The Node B includes an estimate of the physical channel BER if a value of 'non-selected' is received.

The final section of the NBAP Radio Link Setup Request message defines some further characteristics of the radio link. The radio link identifier is used to identify the radio link at the Node B in case more than a single radio link is established for the same UE. The cell identifier is used to identify the cell at which the new radio link is to be established. The Node B is also informed that this radio link belongs to the first radio link set for the UE, i.e. the UE is not in soft handover with another Node B. The frame offset defines the time offset between the downlink DCH connection frame number (CFN) and the cell System Frame Number (SFN). The chip offset defines the radio link timing within the resolution of a single radio frame. The chip offset defines the timing of the DPCH frames relative to the timing of the CPICH frames. The propagation delay reflects the value reported by the Node B within the header of the RACH common channel Frame Protocol data frame. The units are in steps of three chips so the actual value is obtained by multiplying the signalled value by three, i.e. the propagation delay is nine chips in this example.

The downlink scrambling code value of 0 indicates that the radio link should use the primary scrambling code belonging to the cell. Each primary scrambling code has 15 secondary scrambling codes. Values 1 to 15 can be signalled to indicate that one of the secondary scrambling codes should be used. The downlink channelisation code index is 10. The spreading factor has already been signalled by specifying downlink slot format 5. This indicates that the downlink channelisation code to be used is (SF256,10). The result of the DPCH open loop power control calculation is signalled using a value of -148. This figure has units of 0.1 dB and so the actual value is -14.8 dB. This value represents the initial DPDCH transmit power relative to the CPICH transmit power. If the CPICH transmit power is 33 dBm then the initial DPDCH transmit power will be 18.2 dBm. The maximum and minimum downlink transmit powers for this radio link are signalled in the same way, i.e. with units of 0.1 dB and relative to the CPICH transmit power. The values signalled in this example indicate a maximum DPDCH transmit power of 27.9 dBm and a minimum DPDCH transmit power of 15 dBm (assuming a CPICH transmit power of 33 dBm). The maximum downlink transmit power allocation is relatively low due to the low bit rate of the DCH.

The Node B starts to transmit the downlink DPCCH once it has read and accepted the content of the NBAP Radio Link Setup Request message. At this point there is no data to transfer so the DPDCH remains in DTX. Also, the UE is not yet attempting to receive the downlink DPCCH nor is the UE transmitting the uplink DPCCH. This means that the Node B does not receive any inner loop power control commands and so transmits using a fixed power. Figure 8.1 illustrates that the Node B responds to the Radio Link Setup Request message using a Radio Link Setup Response message. An example of this message is presented in Log File 8.7. The Radio Link Setup Response message uses the controlling RNC communication context to identify the relevant connection. This allows the RNC to associate the response message with the appropriate request message. The Node B also provides the RNC with a Node B communication context. The RNC uses this identity to identify the connection within subsequent NBAP messages, i.e. the CRNC communication context is used for addressing in the uplink direction whereas the Node B communication context is used for addressing in the downlink direction.

A communication control port is also allocated for any subsequent signalling procedures, i.e. dedicated NBAP procedures are directed towards a specific control port at the Node B. If a Node B is configured to use multiple ATM VCC for dedicated NBAP signalling then the control port identifies which VCC should be used. The Radio Link Setup Response message then specifies the relevant radio link and radio link set identities. It also includes an uplink interference power measurement. The signalled value of 72 corresponds to a measurement result between -104.9 and -104.8 dBm. This mapping is specified within 3GPP TS 25.133. The lower bound is equal to $-112 + (\text{signalled value} - 1) \times 0.1$, whereas the upper bound is equal to $-112 + \text{signalled value} \times 0.1$. The Node B then assigns a binding identity and transport layer address for the dedicated channel connection. Both of these are used during the subsequent establishment of the Iub transport connection. The binding identity identifies the DCH during Iub transport connection establishment. The transport layer address defines the address at which the Iub transport connection should be established. If a Node B is configured to use multiple ATM VCC for user plane data transfer then the transport layer address identifies which VCC should be used. Finally, the message specifies that Site Selection Diversity Transmit power control (SSDT) is not supported. SSDT provides functionality to limit downlink transmissions to a single active set cell when a UE is in soft handover. This reduces the soft handover overhead in terms of downlink load upon the air-interface. The UE provides feedback to the network using the DPCCH FBI bits to indicate which active set cell should transmit at any point in time.

The RNC initiates the establishment of an Iub transport connection after receiving the information within the NBAP Radio Link Setup Response. The RNC initiates this procedure by sending an ALCAP

```
RADIO LINK SETUP RESPONSE
NBAP-PDU
RadioLinkSetupResponseFDD
CRNC-CommunicationContextID: 1039
NodeB-CommunicationContextID: 160001
CommunicationControlPortID: 2
RL-InformationResponseList-RL-SetupRspFDD
    RL-InformationResponseItem-RL-SetupRspFDD
    - rL-ID: 1
    - rL-Set-ID: 1
    - received-total-wide-band-power: 72
    diversityIndication
      nonCombiningOrFirstRL
      dCH-InformationResponse
          DCH-InformationResponseItem
            - dCH-ID: 24
            - bindingID: '00000003'H
            - transportLayerAddress: '49 00 00 10 20 FF FF FF FF
            FF FF FF FF FF FF FF FF FF FF FF'H
    - sSDT-SupportIndicator: sSDT-not-supported
```

Log File 8.7   NBAP Radio Link Setup Response message

```
ERQ - ESTABLISH REQUEST
DSAID-Dest. sign. assoc. ident.:  00 00 00 00
CEID-Connection element ident.
- Path identifier: 10001
- Channel identifier: 20
NSEA-Dest. NSAP serv. endpoint addr.
- Address:  49 00 00 10 20 FF FF FF FF FF FF FF FF FF FF FF
FF FF FF FF
LC-Link Charactreristics
- Maximum forward CPS-SDU bit rate: 82
- Maximum backwards CPS-SDU bit rate: 84
- Average forward CPS-SDU bit rate: 75
- Average backwards CPS-SDU bit rate: 82
- Maximum forward CPS-SDU size: 24
- Maximum backwards CPS-SDU size: 26
- Average forward CPS-SDU size: 24
- Average backwards CPS-SDU size: 26
OSAID-Orig. sign. assoc. ident.
- Signalling association identifier:  45 09 5C 00
SUGR-Served user gen. reference
- Field:  00 00 00 03
```

Log File 8.8   ALCAP Establish Request message

Establish Request message to the Node B. An example Establish Request message is presented within Log File 8.8.

The Destination Signalling Association Identifier (DSAID) is initialised with value of 0. The response to the Establishment Request message provides a value for this field which can be used during any subsequent signalling procedures. It allows signalling procedures to be linked with one another. For example, the DSAID is used to identify the connection during the release procedure. The connection element identifier specifies the ATM identities to be associated with the new connection. The path identifier is used to identify the appropriate user plane VCC whereas the Channel Identifier (CID) is used to identify the individual AAL2 connection within the user plane VCC. The concepts of VCC and CID are described within Chapter 5. The destination Network Service Access Point (NSAP) address matches the transport layer address provided by the Node B within the NBAP Radio Link Setup Response message.

The Establish Request message then includes a set of characteristics describing the new connection in terms of maximum and average bit rates and SDU sizes. The Node B can use this information to complete its own admission control procedure. The bit rates are specified using units of 64 bps, e.g. a signalled value of 82 corresponds to 5.248 kbps. These bit rates are specified for the AAL2 Common Part Sublayer (CPS) packet payload, i.e. the CPS SDU (described within Section 5.1.3). This means they exclude the overheads generated by the AAL2 and ATM layers, but include the overhead generated by the Frame Protocol layer. This example is for a 3.4 kbps Signalling Radio Bearer (SRB) so the bit rates are relatively low. The average bit rates can include the impact of discontinuous transmission, i.e. SRB are not active during every TTI. The maximum and average SDU sizes are specified in terms of bytes. In this example, the downlink SDU size is specified to be 24 bytes (192 bits) whereas the uplink SDU size is specified to be 26 bytes (208 bits). These figures are defined by the SRB transport block size of 148 bits with the addition of the Frame Protocol data frame header, padding and tail bits. The Frame Protocol header for the SRB data frame is 24 bits in both the uplink and downlink directions. 4 bits of padding are included to increase the transport block size to an integral number of bytes. The Frame Protocol data frame includes 16 tail bits in the downlink direction and 32 tail bits in the uplink direction. The number of tail bits is greater in the uplink direction because the CRC check result and a received quality indication are included.

The final section of the Establish Request message specifies the Originating Signalling Association Identifier (OSAID) and the Served User Generated Reference (SUGR). The OSAID is similar to the DSAID except that it links signalling procedures from the perspective of the sender. The OSAID becomes the DSAID within any messages sent in the opposite direction. The SUGR matches the binding identity provided by the Node B within the NBAP Radio Link Setup Response. This allows the Node B to link the new Iub transport connection with the appropriate radio link.

The Node B returns an ALCAP Establish Confirm message if its admission control accepts the request for a new Iub transport connection. An example of this message is presented within Log File 8.9.

The DSAID mirrors the OSAID within the Establish Request message. This allows the RNC to associate the Establish Confirm message with the appropriate Establish Request message. The Establish Confirm message also specifies an OSAID which is used by the RNC as a DSAID for any subsequent ALCAP signalling procedures.

The RNC achieves synchronisation across the new Iub transport connection after receiving the ALCAP Establish Confirm message. This helps to ensure that the RNC is able to send Frame Protocol data frames such that they arrive at the Node B within the time window defined by toAWS and toAWE. The Node B has already been informed of the toAWS and toAWE parameters within the NBAP Radio Link Setup Request message. Synchronisation is achieved using dedicated channel Frame Protocol control frames. The RNC starts by sending a Downlink Synchronisation frame to the Node B. This control frame includes a target CFN value. After receiving the downlink control frame, the Node B responds with an Uplink Synchronisation control frame which echoes the target CFN and adds a time of arrival (toA). An example pair of Downlink and Uplink Synchronisation frames are presented within Log File 8.10

The RNC can use the time of arrival value to compensate its timing and help ensure that the next transmission is received between toAWS and toAWE. Figure 8.1 illustrates a single pair of Downlink and Uplink Synchronisation frames although it is common to achieve synchronisation using two pairs of frames. The first pair provides the information necessary to achieve synchronisation while the second pair compensates and confirms that synchronisation has been achieved.

The RNC then proceeds to transfer the RRC Connection Setup message to the UE. Similar to the RRC Connection Request message, this message uses Signalling Radio Bearer (SRB) 0 on the CCCH logical channel. In this case, the FACH transport channel and S-CCPCH physical channel are used. The

```
ECF - ESTABLISH CONFIRM
DSAID-Dest. sign. assoc. ident.:  45 09 5C 00
OSAID-Orig. sign. assoc. ident.
- Signalling association identifier:  00 00 00 14
```

Log File 8.9   ALCAP Establish Confirm message

```
DOWNLINK SYNC
Control Frame CRC: 27
Control Frame Type: 3
Conn. Frame number: 164

UPLINK SYNC
Control Frame CRC: 66
Control Frame Type: 4
Conn. Frame number: 164
Time of Arrival: 9.875 ms
```

Log File 8.10   Downlink and Uplink Synchronisation control frames

RRC Connection Setup message does not use the new Iub transport connection nor does it use the new radio link. The RNC packages the transport blocks belonging to the RRC Connection Setup message within a series of common channel Frame Protocol data frames. These data frames are transferred across the Iub using a common channel transport connection. The RRC Connection Setup message is relatively large and typically occupies seven transport blocks. An example of these seven transport blocks is presented within Log File 2.23 in section 2.3.2. In this example, each transport block has a size of 168 bits and an associated TTI of 10 ms. An example of a common channel Frame Protocol data frame which includes two transport blocks belonging to an RRC Connection Setup message is presented within Log File 2.45 in section 2.5.3. This illustrates that two transport blocks are transferred across the Iub and subsequently across the air-interface during each 10 ms TTI. The complete message occupies four TTI although the fourth TTI transfers only a single transport block. SRB 0 uses unacknowledged mode RLC and so the complete set of seven transport blocks must be re-transmitted from the RRC layer if any one transport block is received in error. The parameters defining re-transmission of the RRC Connection Setup message are not specified by 3GPP. The RNC databuild may include implementation dependent parameters defining the period between re-transmissions and the maximum allowed number of re-transmissions.

The first part of an example RRC Connection Setup message is presented in Log File 8.11. The UE may receive RRC messages on the FACH addressed to other UE while waiting for a response to its RRC Connection Request. The UE has to check the initial UE identity within any RRC Connection Setup messages to determine whether or not the message should be processed or discarded. Messages are only processed if the initial UE identity matches the identity within the RRC Connection Request message.

The RRC Connection Setup message allocates a UTRAN Radio Network Temporary Identity (U-RNTI). The U-RNTI is a concatenation of the SRNC identity with the S-RNTI. The U-RNTI provides a unique identifier for the RRC connection within the PLMN. The RRC Connection Setup

```
DL-CCCH-Message
rrcConnectionSetup
rrcConnectionSetup-r3
initialUE-Identity
    tmsi-and-LAI
        tmsi
            Bin: 01 13 E9 27
        lai
            plmn-Identity
                    mcc value: 234
                    mnc value: 99
            lac
                    Bin: 27 83
rrc-TransactionIdentifier: 0
new-U-RNTI
    srnc-Identity
        Bin: 03 F
    s-RNTI
        Bin: 12 48 C
rrc-StateIndicator: cell-DCH
utran-DRX-CycleLengthCoeff: 5
capabilityUpdateRequirement
    ue-RadioCapabilityFDDUpdateRequirement: true
    ue-RadioCapabilityTDDUpdateRequirement: false
    systemSpecificCapUpdateReqList
        systemSpecificCapUpdateReqList value 1: gsm
```

Log File 8.11   RRC Connection Setup message (part 1)

message then proceeds to instruct the UE to move into CELL_DCH and provides a UTRAN DRX cycle length coefficient. The DRX cycle length is calculated as 2^k, where k is the cycle length coefficient. In this example, the coefficient has a value of 5 and so the UTRAN DRX cycle length is 32 radio frames. When a UE is in RRC Connected mode, the DRX cycle length is the minimum of the UTRAN DRX cycle length and the DRX cycle length belonging to any core network domains with which the UE is registered, but does not have a signalling connection. The core network DRX cycle lengths can be read from SIB 1. The RNC then uses the RRC Connection Setup message to request specific UE capability information. The UE reports this information within the RRC Connection Setup Complete message. In this example, the UE is requested to provide information regarding its UMTS FDD and GSM capabilities.

The second section of the RRC Connection Setup message is presented in Log File 8.12. This section configures the set of SRB. It is mandatory to configure at least SRB 1 to 3, and optional to configure SRB 4. SRB 1 and 2 are used for signalling which either originates or terminates at the RNC. SRB 3 and 4 are used for signalling which either originates or terminates the core network (non-access stratum signalling). SRB 1 is allocated a radio bearer identity of 1 and is configured to use unacknowledged mode RLC in both the uplink and downlink directions. SRB 1 is then mapped onto two different transport channels. The first mapping is applicable when the UE is in CELL_DCH whereas the second mapping is applicable when the UE is in CELL_FACH. The first mapping links SRB 1 to the DCH transport channel which has identity 24. The UE is instructed to use the uplink RLC PDU sizes configured later in the message. The uplink is also allocated a logical channel priority of 1. This priority can range from 1 to 8, where 1 represents the highest priority. The RLC PDU size and priority are not configured for the downlink direction because they are important to the sender rather than the receiver, i.e. the RNC is configured with the equivalent downlink information. The second mapping

```
srb-InformationSetupList value 1                timerPollProhibit: tpp120
rb-Identity: 1                                  timerPoll: tp300
ul-RLC-Mode: ul-UM-RLC-Mode                     poll-SDU: sdu1
dl-RLC-Mode: dl-UM-RLC-Mode                     lastTransPDU-Poll: false
rb-MappingInfo value 1                          lastRetransPDU-Poll: true
  ul-LogicalChannelMappings                     pollWindow: pw70
    ul-TransportChType dch: 24               dl-RLC-Mode: dl-AM-RLC-Mode
    logicalChannelIdentity: 1                   inSequenceDelivery: true
  rlc-SizeList: configured                      receivingWindowSize: rw64
    mac-LogicalChannelPriority: 1               dl-RLC-StatusInfo
  dl-LogChMappingList value 1                   missingPDU-Indicator: true
    dl-TransportChType dch: 24             rb-MappingInfo value 1
    logicalChannelIdentity: 1                 ul-LogicalChannelMappings
rb-MappingInfo value 2                          ul-TransportChType dch: 24
  ul-LogicalChannelMappings                     logicalChannelIdentity: 2
    ul-TransportChannelType: rach             rlc-SizeList: configured
    logicalChannelIdentity: 1                 mac-LogicalChannelPriority: 1
  rlc-SizeList explicList value 1             dl-LogicalChMappingList value 1
    rlc-SizeIndex: 1                             dl-TransportChType dch: 24
  mac-LogicalChannelPriority: 1                 logicalChannelIdentity: 2
  dl-LogicalChMappingList value 1         rb-MappingInfo value 2
    dl-TransportChannelType: fach             ul-LogicalChannelMappings
    logicalChannelIdentity: 1                   ul-TransportChannelType: rach
srb-InformationSetupList value 2                logicalChannelIdentity: 2
rb-Identity: 2                                rlc-SizeList explicList value 1
ul-RLC-Mode: ul-AM-RLC-Mode                     rlc-SizeIndex: 1
  transmissionRLC-Discard                     mac-LogicalChannelPriority: 1
  noDiscard: dat20                            dl-LogicalChMappingList value 1
  transWindowSize: tw64                         dl-TransportChannelType: fach
  timerRST: tr300                               logicalChannelIdentity  : 2
  max-RST: rst1
```

Log File 8.12   RRC Connection Setup message (part 2)

```
srb-InformationSetupList value 3          srb-InformationSetupList value 4
rb-Identity: 3                            rb-Identity: 4
ul-RLC-Mode: ul-AM-RLC-Mode               ul-RLC-Mode: ul-AM-RLC-Mode
  transmissionRLC-Discard                   transmissionRLC-Discard
  noDiscard: dat20                          noDiscard: dat20
  transWindowSize: tw64                     transWindowSize: tw32
  timerRST: tr300                           timerRST: tr300
  max-RST: rst1                             max-RST: rst1
  timerPollProhibit: tpp120                 timerPollProhibit: tpp120
  timerPoll: tp300                          timerPoll: tp300
  poll-SDU: sdu1                            poll-SDU: sdu1
  lastTransPDU-Poll: false                  lastTransPDU-Poll: false
  lastRetransPDU-Poll: true                 lastRetransPDU-Poll: true
  pollWindow: pw70                          pollWindow: pw70
dl-RLC-Mode: dl-AM-RLC-Mode               dl-RLC-Mode: dl-AM-RLC-Mode
  inSequenceDelivery: true                  inSequenceDelivery: true
  receivingWindowSize: rw64                 receivingWindowSize: rw32
  dl-RLC-StatusInfo                         dl-RLC-StatusInfo
  missingPDU-Indicator: true                missingPDU-Indicator: true
rb-MappingInfo value 1                    rb-MappingInfo value 1
  ul-LogicalChannelMappings                 ul-LogicalChannelMappings
    ul-TransportChType dch: 24                ul-TransportChType dch: 24
    logicalChannelIdentity: 3                 logicalChannelIdentity: 4
  rlc-SizeList: configured                  rlc-SizeList: configured
  mac-LogicalChannelPriority: 1             mac-LogicalChannelPriority: 5
  dl-LogicalChMappingList value 1           dl-LogicalChMappingList value 1
    dl-TransportChType dch: 24                dl-TransportChType dch: 24
    logicalChannelIdentity: 3                 logicalChannelIdentity: 4
rb-MappingInfo value 2                    rb-MappingInfo value 2
  ul-LogicalChannelMappings                 ul-LogicalChannelMappings
    ul-TransportChType: rach                  ul-TransportChannelType: rach
    logicalChannelIdentity: 3                 logicalChannelIdentity: 4
  rlc-SizeList explicList value 1           rlc-SizeList explicList value 1
    rlc-SizeIndex: 1                          rlc-SizeIndex: 1
  mac-LogicalChannelPriority: 1             mac-LogicalChannelPriority: 5
  dl-LogicalChMappingList value 1           dl-LogicalChMappingList value 1
    dl-TransportChType: fach                  dl-TransportChannelType: fach
    logicalChannelIdentity: 3                 logicalChannelIdentity: 4
```

Log File 8.12   *(Continued)*

links SRB 1 to the RACH in the uplink direction and the FACH in the downlink direction. In this case
the UE is informed explicitly that the uplink RLC PDU size should be based upon transport format 1
from within the RACH transport format set. This defines the uplink bit rate of SRB 1 when the UE is in
CELL_FACH.

SRB 2 is allocated a radio bearer identity of 2 and is configured to use acknowledged mode RLC in
both the uplink and downlink directions. The set of acknowledged mode RLC parameters are described
in greater detail within Section 2.3.3. Similar to SRB 1, SRB 2 is mapped onto two different transport
channels, i.e. the DCH with an identity of 24 for when the UE is in CELL_DCH and the RACH/FACH
combination for when the UE is in CELL_FACH. SRB 2 is allocated the same logical channel priority
as SRB 1.

SRB 3 is allocated a radio bearer identity of 3 and is also configured to use acknowledged mode
RLC. The configuration information for SRB 3 is the same as that for SRB 2. Finally, SRB 4 is
allocated a radio bearer identity of 4 and is configured to use acknowledged mode RLC. In this case,
the acknowledged mode transmit and receive windows are configured with smaller values, i.e. 32 RLC
PDU rather than 64. In addition, the logical channel priority is configured with a value of 5 rather
than 1. These differences reflect the lower priority of SRB 4 relative to SRB 1, 2 and 3. The logical
channel priority is used for scheduling purposes so data belonging to SRB 1, 2 and 3 is transferred prior
to data belonging to SRB 4.

The third section of the RRC Connection Setup message is presented in Log File 8.13. This section configures the DCH Transport Format Sets (TFS). These TFS are the same as those signalled to the Node B within the NBAP Radio Link Setup Request. In this case, RLC PDU sizes are signalled rather than transport block sizes. The RLC PDU size must be derived from the size type information. 3GPP TS 25.331 specifies the set of equations presented in Table 8.3 (Section 8.5.1). In this case, the appropriate equation to calculate the RLC PDU size is: $(8 \times \text{sizeType1}) + 16$. This equation generates a result of 144 bits. These 144 bits correspond to the transport block size of 148 bits signalled within the NBAP Radio Link Setup Request. The transport block size is applicable to the top of the physical layer whereas the RLC PDU size is applicable to the top of the MAC layer, i.e. the RLC PDU size excludes the overhead generated by the MAC layer. The MAC layer adds a header of 4 bits to generate the transport block size of 148 bits.

This section of the RRC Connection Setup message also defines power offsets between the acknowledged PRACH preamble and the PRACH message. These power offsets are only applicable when the UE is in CELL_FACH and using the RACH transport channel. The downlink quality target is also specified. The signalled value is equal to $10 \times \log(\text{BLER})$ and so in this example, the BLER target is equal to 1%.

The final section of the RRC Connection Setup message is presented in Log File 8.14. This section focuses upon providing physical layer information. The UE is instructed to use a maximum transmit

```
ul-CommonTransChInfo                         semistaticTF-Information
modeSpecificInfo fdd                           channelCodingType
ul-TFCS                                           convolutional: third
normalTFCI-Signalling                             rateMatchingAttribute: 1
complete                                          crc-Size: crc16
ctfcSize                                     dl-CommonTransChInfo
  ctfc2Bit value 1                           modeSpecificInfo fdd
    ctfc2: 0                                 dl-Parameters: sameAsUL
    powerOffsetInformation                   dl-AddReconfTransChInfoList
      gainFactorInformation                    dl-AddReconfTrChInfoList value 1
        computedGainFactors: 0                   dl-TransportChType: dch
      powerOffsetPp-m: 4                         dl-transportChIdentity: 24
  ctfc2Bit value 2                               tfs-SignallingMode
    ctfc2: 1                                        explicit-config
    powerOffsetInformation                           dedicatedTransChTFS
      gainFactorInformation                            tti40
        signalledGainFactors                           tti40 value 1
          modeSpecificInfo fdd                           rlc-Size
            gainFactorBetaC: 15                            bitMode
            gainFactorBetaD: 11                            sizeType1: 0
            referenceTFC-ID: 0                         numberOfTbSizeList
      powerOffsetPp-m: 5                                numberOfTbSizeLst 1: one
ul-AddReconfTransChInfoList                          logicalChList: allSizes
  ul-AddReconfTrChInfoList value 1                 tti40 value 2
    ul-TransportChannelType: dch                     rlc-Size
    transportChannelIdentity: 24                       octetModeType1
    transportFormatSet                                 sizeType1: 16
    dedicatedTransChTFS                            numberOfTbSizeList
      tti40                                        numberOfTbSizeLst 1: one
      tti40 value 1                                logicalChList: allSizes
        rlc-Size                                 semistaticTF-Information
          octetModeType1                           channelCodingType
          sizeType1: 16                              convolutional: third
        numberOfTbSizeList                           rateMatchingAttribute: 1
          numberOfTbSizeList 1: zero                 crc-Size: crc16
          numberOfTbSizeList 2: one          dch-QualityTarget
        logicalChannelList: allSizes           bler-QualityValue: -20
```

Log File 8.13 RRC Connection Setup message (part 3)

power of 24 dBm. The UE is then provided with information to support its initial uplink transmission. The DPCCH power offset is used within the uplink DPCH open loop power control calculation. The power control preamble defines the number of DPCCH radio frames that the UE must transmit prior to sending any DPDCH data. This provides the Node B with some time to achieve uplink synchronisation prior to receiving any data. The SRB delay provides additional time for the Node B to synchronise when the DPDCH data belongs to an SRB, i.e. in this example, SRB data cannot be sent by the UE until it has transmitted a total of 14 DPCCH radio frames. This corresponds to 140 ms and represents a significant contribution to the overall RRC connection establishment delay. In practice, the power control preamble is often configured with a value of 0 to allow the UE to send SRB data after 70 ms rather than 140 ms. The UE is also instructed to use uplink inner loop power control algorithm 1. This means that the UE changes its transmit power every time slot rather than every five time slots. The power control step size is not explicitly labeled, but is the value which follows the power control algorithm, i.e. the value of 0 in this example. The signalled value for the power control step size is equal to the actual value minus 1, i.e. in this example, the power control step size is 1 dB.

The UE is informed of the uplink scrambling code allocated by the RNC and is instructed to use a minimum uplink spreading factor of 256. The UE is also instructed to include TFCI bits within the DPCCH and to apply a puncturing limit of 68%. The RRC Connection Setup message then provides information regarding the downlink physical channel. The UE is informed of the power offset between the DPDCH and DPCCH pilot bits (PO3). This information helps the UE to identify an appropriate SIR target for inner loop power control. The BLER target presented within Log File 8.13 is applicable to the DPDCH and allows the UE to estimate an appropriate DPDCH SIR target. Inner loop power control is based upon SIR measurements from the DPCCH pilot bits. Signalling the value of PO3 allows the UE to translate the DPDCH SIR target to an equivalent DPCCH SIR target.

The UE is also informed that the downlink spreading factor is 256, the DPCCH includes 4 pilot bits and that downlink TFCI bits should be included. This information allows the UE to deduce that downlink DPCH slot format 5 should be expected. The UE is then informed of the default DPCH offset using a value 391. The signalled value is mapped to the actual value by multiplying by 512 chips, i.e. the actual value is 200192 chips. The actual value of the default DPCH offset provided in the RRC Connection Setup message is equal to the combination of the frame offset and chip offset provided to the Node B in

```
maxAllowedUL-TX-Power: 24              spreadingFactorAndPilot
ul-ChannelRequirement                    sfd256: pb4
  ul-DPCH-Info                           positionFixedOrFlex: flexible
  ul-DPCH-PowerControlInfo fdd           tfci-Existence: true
    dpcch-PowerOffset: -49             modeSpecificInfo fdd
    pc-Preamble: 7                       defaultDPCH-OffsetValue: 391
    sRB-delay: 7                       dl-InformationPerRL-List
    powerControlAlgorithm             dl-InformationPerRL-List value 1
      algorithm1: 0                     modeSpecificInfo fdd
  modeSpecificInfo fdd                    primaryCPICH-Info
    scramblingCodeType: longSC            primaryScramblingCode: 296
    scramblingCode: 1000015           dl-DPCH-InfoPerRL fdd
    spreadingFactor: sf256            pCPICH-UseForChEst: mayBeUsed
    tfci-Existence: true              dpch-FrameOffset: 32
    puncturingLimit: pl0-68           dl-ChannelisationCodeList
dl-CommonInformation                  dl-ChannelisationCodeList 1
dl-DPCH-InfoCommon                      sf-AndCodeNumber
  cfnHandling: maintain                   sf256: 10
  modeSpecificInfo fdd                tpc-CombinationIndex: 0
    powerOffsetPilot-pdpdch: 0
```

Log File 8.14   RRC Connection Setup message (part 4)

the NBAP Radio Link Setup message. The example in Log File 8.6 has a frame offset of 5 and a chip offset of 8192. These figures combine to generate an offset of $(5 \times 38\,400) + 8192 = 200192$ chips. The UE is then able to derive the CFN timing from the SFN timing in the same way as the Node B. The UE reads the SFN from the BCCH information broadcast on the BCH transport channel and P-CCPCH physical channel. Downlink DPCH timing is described in greater detail within Section 3.3.9.2.

The RRC Connection Setup message then informs the UE that the new radio link should be based upon primary scrambling code 296 and that the UE can use the primary CPICH for channel estimation purposes. This implies that beamforming is not used and that the DPCH will be transmitted using the same antenna gain pattern as the primary CPICH. The optional information element specifying a secondary scrambling code is excluded which means that the new radio link will use the primary scrambling code. The UE is then provided with a radio link specific frame offset. In this example, the signalled value is equal to 32. The signalled value is mapped to the actual value by multiplying by 256 chips, i.e. the actual value is 8192 chips which equals the chip offset provided to the Node B within the NBAP Radio Link Setup message. In the case of the RRC Connection Setup message this radio link specific DPCH frame offset does not add any further information because the value of 8192 chips can be derived from the default DPCH offset by applying modulo 38400 arithmetic. The radio link specific DPCH frame offset information element is more important when new radio links are subsequently added to the active set. In this case, the default DPCH offset is not signalled because it has already been used to initialise the CFN timing. The new radio link specific DPCH frame offset specifies the timing of the new radio link relative to its CPICH. This is illustrated within Figure 3.34.

Finally, the RRC Connection Setup message informs the UE that the downlink spreading factor is 256 and that channelisation code number 10 should be used. The Transmit Power Control (TPC) combination index is used to identify groups of radio links belonging to the same radio link set, i.e. all radio links which have the same combination index belong to the same radio link set. TPC commands received from different radio links within the same radio link set are always equal because they have been generated by the same Node B receiver. This information allows the UE to improve the reliability of detection by combining TPC commands prior to taking a decision upon the transmitted value.

The UE has sufficient information to attempt air-interface synchronisation once it has received the RRC Connection Setup message. The UE must achieve downlink synchronisation before starting to transmit, i.e. the UE achieves synchronisation prior to the Node B. The UE uses the timing information within the RRC Connection Setup message to determine when to expect the downlink DPCCH pilot bits. If the downlink open loop power control calculation resulted in a relatively low power then the UE will experience greater difficulty when attempting to synchronise. The rate at which a UE achieves air-interface synchronisation also depends upon the 3GPP specified parameters T312 and N312. The Physical layer of the UE must generate N312 in-sync primitives within a time period defined by T312. T312 is started when a UE initiates the procedure for dedicated channel establishment. The procedure fails if N312 in-sync primitives are not received within the time interval defined by T312. The UE evaluates whether or not an in-sync primitive should be generated every 10 ms. The values of N312 and T312 are read from SIB 1 prior to attempting RRC connection establishment. Typical values for N312 and T312 are 10 and 6 s respectively.

The UE starts transmitting the uplink DPCCH after generating N312 in-sync primitives. This allows the Node B to achieve air-interface synchronisation. Uplink synchronisation at the Node B is based upon the N_INSYNC_IND parameter. The RNC signals this parameter to the Node B when the Node B is switched on, i.e. within the NBAP Cell Setup Request message. N_INSYNC_IND represents the uplink equivalent of N312 and the Node B has to generate N_INSYNC_IND successive in-sync primitives to achieve uplink synchronisation. The Node B must achieve synchronisation within the time interval defined by the sum of the power control preamble and SRB delay otherwise the UE can start transmitting DPDCH data prior to the Node B being able to receive. The Node B informs the RNC that it has achieved uplink synchronisation using the dedicated NBAP Radio Link Restore message. An example of this message is presented within Log File 8.15. The Controlling RNC communication context is used to identify the connection. The Radio Link Restore Indication then specifies the identity of the radio link which has achieved uplink synchronisation.

The UE transmits the RRC Connection Setup Complete message after transmitting the DPCCH during the number of frames defined by the power control preamble and SRB delay. This message represents the first DPDCH data transferred across the new dedicated channel connection. The RRC Connection Setup Complete message typically requires two 148 bit transport blocks. The TTI depends upon the bit rate of the dedicated channel. It is common to use either a 10 or 40 ms TTI which leads to a bit rate of either 14.8 or 3.7 kbps (a single transport block is transmitted during each TTI). The 14.8 kbps bit rate allows the transport blocks to be transmitted at four times the rate and so reduces the delay associated with any subsequent signalling procedure. However, the 14.8 kbps bit rate also has a greater C/I requirement and a greater peak Iub transport resource requirement. The larger C/I requirement has an impact upon coverage. The impact upon air-interface capacity should be minimal because the increased C/I requirement is offset by a reduced activity factor, i.e. the increased bit rate allows signalling procedures to be completed more rapidly so the activity factor decreases. In fact, the 14.8 kbps bit rate may increase capacity because the air-interface typically becomes more efficient for higher bit rates. The peak Iub transport resource requirement is greater for the 14.8 kbps bit rate, but the average resource requirement remains unchanged, i.e. the same quantity of data is transferred irrespective of whether the connection is configured to use a bit rate of 3.7 or 14.8 kbps. If the SRB use a 14.8 kbps DCH during the establishment of a user plane connection, e.g. speech connection, the bit rate is typically reduced to 3.7 kbps once the user plane connection has been established, i.e. the 14.8 kbps bit rate is only allocated to standalone SRB.

The RRC Connection Setup Complete message is transferred to the RNC using acknowledged mode RLC. This means that it uses the logical channel configured for SRB 2. An example RRC Connection Setup Complete message is presented within Log File 8.16. The message starts by specifying Start values for both the CS and PS core network domains. These values are used for ciphering and integrity protection. They have a length of 20 bits and are used to initialise the 20 Most Significant Bits (MSB) of the RRC Hyper Frame Number (HFN), RLC HFN and MAC-d HFN. Initialising the RRC HFN is applicable to integrity protection whereas initialising the RLC and MAC-d HFN is applicable to ciphering.

The RRC Connection Setup Complete message provides the capability update requested within the RRC Connection Setup message. The UE starts by specifying its Packet Data Convergence Protocol (PDCP) capability. This has implications upon both lossless SRNS relocation and header compression. Lossless SRNS relocation means that packets are not lost as the Iu connection is moved from one RNC to another. This requires PDCP layer support from both the UE and the RNC. The release 99 version of 3GPP TS 25.323 specifies the use of IETF RFC 2507 for header compression. The release 4 version introduces the use of IETF RFC 3095 which is also known as Robust Header Compression (ROHC). The RRC Connection Setup Complete message within Log File 8.16 indicates that neither lossless

```
RADIO LINK RESTORE INDICATION [FDD]
NBAP-PDU
RadioLinkRestoreIndicationFDD
CRNC-CommunicationContextID: 1039
rL-InformationList-RL-RestoreIndFDD
      RL-InformationItemIE-RL-RestoreIndFDD
          - rL-ID: 1
```

Log File 8.15   NBAP Radio Link Restore message

```
UL-DCCH-Message                              fddPhysChCapability
rrcConnectionSetupComplete                   downlinkPhysChCapability
rrc-TransactionIdentifier: 0                   maxNoDPCH-PDSCH-Codes: 1
startList                                      maxNoPhysChBitsReceiv: b9600
  cn-DomainIdentity: cs-domain,                supportForSF-512: FALSE
    start-Value: '00 01 0'H                    supportOfPDSCH: FALSE
  cn-DomainIdentity ps-domain,                 simultSCCPCH-DPCH: notSupp
    start-Value: '00 00 D'B                  uplinkPhysChCapability
ue-RadioAccessCapability                       maxNoDPDCH-BitsTrans: b4800
  pdcp-Capability                              supportOfPCPCH: FALSE
    losslessSRNS-RelocSupp: FALSE          ue-MultiModeRAT-Capability
    suppForRfc2507: notSupported             multiRAT-CapabilityList
  rlc-Capability                               supportOfGSM: TRUE
    totalRLC-AM-BufferSize: kb50               supportOfMulticarrier: FALSE
    maximumRLC-WindowSize: mws2047           multiModeCapability fdd
    maximumAM-EntityNumber: am6            securityCapability
  transportChannelCapability                 cipheringAlgCap: uea1, uea0
    dl-TransChCapability                       integrityProtectAlgCap: uia1
      maxNoBitsReceived: b8960            ue-positioning-Capability
      maxConvCodeBitsReceiv: b1280           standaloneLocMethodsSup: FALSE
      turboDecodingSupport: b8960            ue-BasedOTDOA-Supported: FALSE
      maxSimultTransChs: e8                  nwAssistGPS-Supp: NoNwAssGPS
      maxSimultCCTrCH-Count: 1               suppForUE-GPS-TimCellFr: FALSE
      maxReceivedTrBlocks: tb32              supportForIPDL: FALSE
      maxNumOfTFC-InTFCS: tfc128          measurementCapability
      maxNumberOfTF: tf64                    downlinkCompressedMode
    ul-TransChCapability                       fdd-Measurements: TRUE
      maxNoBitsTransmitted: b3840            gsm-Measurements
      maxConvCodeBitsTrans: b1280              gsm900: TRUE
      turboDecodingSupport: b3840             dcs1800: TRUE
      maxSimultaneousTransChs: e8             gsm1900: TRUE
      modeSpecificInfo fdd:                   multiCarrierMeas: FALSE
      maxTransmittedBlocks: tb32            uplinkCompressedMode
      maxNumberOfTFC-InTFCS: tfc64           fdd-Measurements: TRUE
      maxNumberOfTF: tf32                    gsm-Measurements
  rf-Capability                                gsm900: TRUE
    fddRF-Capability                           dcs1800: TRUE
      ue-PowerClass 4                          gsm1900: TRUE
      txRxFreqSeparation: mhz190              multiCarrierMeas: FALSE
physicalChannelCapability
```

Log File 8.16   RRC Connection Setup Complete message

SRNS relocation nor RFC 2507 are supported. Support for RFC 3095 is not specified because the message has been generated by a release 99 UE.

The UE then reports its RLC capability in terms of total AM buffer size, maximum RLC window size and the maximum number of AM entities. The total AM buffer size is quantified in kbytes and includes both the transmitter and receiver buffers. This information can be used by the RNC when identifying an appropriate RLC window size. Larger window sizes require greater buffering capabilities. The UE also specifies its maximum supported RLC window size. This can be signalled using a value of either 2047 or 4095. The final section of the RLC capability specifies the maximum number of acknowledged mode entities. Three entities are used by SRB 2, 3 and 4. The remaining entities can be used by user plane connections. SRB 4 is optional and can be released if there is a requirement for an additional entity.

The transport channel capability information starts by specifying the maximum number of bits which the UE is able to receive at any point in time. This information can be used by the RNC to limit the maximum allocated downlink bit rate. Similarly, the UE specifies the maximum number of bits which the UE is able to transmit at any point in time. The RNC can use this information to limit the maximum allocated uplink bit rate. These maximum bit rates are referenced to the top of the physical layer, i.e. they represent the maximum number of bits belonging to a specific Transport Format Combination (TFC). For example, a 384 kbps connection multiplexed with a set of SRB requires a maximum of at least $12 \times 336 + 1 \times 148 = 4180$ bits. The example presented in Log File 8.16 indicates that 384 kbps is supported in the downlink direction, but not in the uplink direction. There are similar maxima for the subset of transport channels which use convolutional coding and the subset of transport channels which use turbo coding. The example in Log File 8.16 indicates that the sum of the transport blocks which are to be convolutional coded cannot exceed 1280 bits. This is sufficient to support the speech service which requires at least $1 \times 244 + 1 \times 148 = 392$ bits.

The UE also specifies the maximum number of uplink and downlink transport channels, and the maximum number of downlink Coded Composite Transport Channels (CCTrCh). If a UE does not support simultaneous reception of the DCH and FACH transport channels then it is usual for the maximum number of CCTrCh to be 1. If a connection is using multiple DCH then those DCH are multiplexed within the physical layer to generate a single CCTrCh. The transport channel section of the message then specifies the maximum number of downlink transport blocks which end within the same 10 ms time interval, and the maximum number of uplink transport blocks which start at the same time. The downlink definition is based upon a 10 ms time interval because not all transport channels are synchronised, e.g. FACH, PCH and DCH use different downlink timing. The UE also specifies the maximum number of TFC within the uplink and downlink Transport Format Combination Set (TFCS). Likewise, the UE specifies the maximum number of Transport Formats (TF). This represents a limit for the total number of TF belonging to all transport channels rather than a limit for an individual transport channel.

The RF capability information specifies the UE transmit power class and the frequency separation between the uplink and downlink carriers. The example shown in Log File 8.16 specifies a UE power class of 4 which corresponds to a maximum transmit power capability of 21 dBm. The duplex spacing has a value of 190 MHz which represents the default for operating band I. The release 7 and earlier versions of 3GPP TS 25.331 only allow the default duplex spacing to be signalled.

The physical channel capability information starts by specifying the maximum number of DPCH codes which the UE is able to receive in parallel. In practice, multi-code reception is not supported and the maximum number of codes is always equal to 1. The title of this information element refers to the Physical Downlink Shared Channel (PDSCH). This physical channel has been removed from the 3GPP specifications and is no longer applicable. It should not be confused with the HS-PDSCH used for HSDPA. The capability information then specifies the maximum number of physical channel bits which can be received within a 10 ms time interval. This information defines the minimum downlink spreading factor. A value of 9600 bits corresponds to a minimum downlink spreading factor of 8, i.e. 3.84×10^6 chips per second $\times 2$ bits per symbol $\times 0.01$ seconds $/ 8 = 9600$ bits. This maximum number of physical channel bits includes data received on both the S-CCPCH and the DPCH. If the UE is capable of receiving the S-CCPCH and DPCH simultaneously then the number of bits received on the S-CCPCH must also be taken into account. In this example, the UE specifies that it does not support simultaneous reception of the S-CCPCH and DPCH. It also specifies that it does not support the downlink spreading factor of 512, nor does it support the PDSCH.

The equivalent uplink physical channel information specifies the maximum number of bits which can be transmitted at any point in time. Similar to the downlink information, this can be used to deduce the minimum supported spreading factor. In this example, the maximum number of physical layer bits which can be transmitted is 4800. This corresponds to a minimum uplink spreading factor of 8, i.e. 3.84×10^6 chips per second $\times 1$ bit per symbol $\times 0.01$ seconds $/ 8 = 4800$ bits. This result is consistent with the previous deduction that the UE does not support 384 kbps in the uplink direction. An uplink 384 kbps connection typically uses a spreading factor of 4. The UE also specifies that it does not support the Physical Common Packet Channel (PCPCH). This information is obsolete because the PCPCH has been removed from the specifications.

The multi-mode and multi-RAT capability information specifies whether or not the UE supports the FDD and TDD variants of UMTS. It also specifies whether or not the UE supports GSM and the multi-carrier 3G technology. In this example, the UE supports both GSM and the FDD variant of UMTS. The security capability information specifies which ciphering and integrity protection algorithms are supported. Ciphering algorithms include UMTS Encryption Algorithms (UEA) 0 and 1. UEA0 corresponds to no encryption whereas UEA1 corresponds to Kasumi encryption. Only a single integrity protection algorithm can be specified - UMTS Integrity protection Algorithm (UIA) 1. This is also based upon Kasumi.

The UE positioning capability information specifies the availability of methods to identify the geographic location of the UE. The standalone method indicates that the UE is able to determine its location without any support from UMTS, e.g. using a standalone GPS receiver. The Observed Time Difference Of Arrival (OTDOA) method is based upon the UE measuring SFN-SFN timing differences between neighbouring cells. This method relies upon the UE being able to record measurements from at least three cells. In practice, radio network planning aims to maximise cell dominance and there is likely to be many locations where a UE is unable to record measurements from three cells. The Idle Period Downlink (IPDL) method is intended to help improve the 'hearability' of neighbouring cells by decreasing the transmit power of specific cells for short periods of time. The network assisted GPS method allows the network to provide support to a GPS receiver within the UE. In this case, the network uses its own set of GPS receivers to record satellite information. These network GPS receivers should have visibility of the same satellite constellation as the UE. The network can then provide information to help the UE acquire its own satellite information when necessary. The UE positioning capability information also specifies whether or not the UE is able to measure the GPS timing of cell frames. This corresponds to measuring the time difference between the start of a specific radio frame and the GPS Time of Week.

The UE measurement capability information specifies which measurements require the use of compressed mode. In general, all measurements require the use of compressed mode unless the UE has more than a single receiver, or has a variable duplex spacing. A UE can record measurements without compressed mode if it has two receivers and there is sufficient isolation between the UE transmitter and the second receiver. The isolation will be relatively small if the measurements are recorded from a frequency band which is close to the UE transmit frequency. In this case, compressed mode could be applied in the uplink direction to avoid the UE transmissions interfering with its measurements. A UE can potentially record measurements without uplink compressed mode if it has a variable duplex spacing, i.e. the receiver could move to a different RF carrier while the transmitter remains at its original frequency. This approach also relies upon having sufficient isolation between the UE transmitter and receiver. The example presented in Log File 8.16 illustrates that both uplink and downlink compressed mode is necessary for inter-frequency FDD measurements and all GSM measurements. The UE signals that compressed mode is not necessary for measurements from the multi-carrier 3G technology, but the UE has already informed the RNC that the multi-carrier system is not supported and so this compressed mode information is ignored.

The section of the RRC Connection Setup message presented in Log File 8.11 also instructs the UE to provide system specific information for GSM. The example RRC Connection Setup Complete message shown in Log File 8.16 does not include this information (an example without this information has been presented to help reduce the length of the log file). The GSM system specific information provides the RNC with the MS classmark 2 and MS classmark 3. An example of an MS classmark 2 is shown in Log File 8.19. This information can be used by the RNC during an inter-system handover.

The example RRC Connection Setup Complete message presented in Log File 8.16 is based upon a release 99 UE. Release 5 UE are able to signal their HSDPA capability whereas release 6 UE are able to signal their HSDPA and HSUPA capabilities. An example section of an RRC Connection Setup Complete message from a release 5 UE is presented in Log File 8.17.

```
ue-RadioAccessCapability-v590ext
   dl-CapabilityWithSimultaneousHS-DSCHConfig: kbps64
   rlc-Capability-r5-ext:
   physicalChannelCapability
      fdd-hspdsch
         supported
         hsdsch-physical-layer-category: 12
```

Log File 8.17   HSDPA capability within an RRC Connection Setup Complete message

In this example, the UE signals that it is HS-DSCH category 12, i.e. it supports QPSK and a maximum of five HS-PDSCH codes. The UE also signals that it can support a downlink DCH in parallel to an HS-DSCH. This capability can be used to support multi-RAB connections, e.g. a DCH speech connection in parallel to an HS-DSCH data transfer. The specifications allow the UE to signal parallel DCH capabilities of 32, 64, 128 and 384 kbps. The detailed interpretation of these values is specified within 3GPP TS 25.306 which defines a look-up table to associate a set of transport channel and physical channel capabilities with each of the four bit rates. A subset of this look-up table is presented within Table 8.2. The characteristics within this table are similar to those defined for the non-HSDPA radio access capability.

The preceding description of RRC connection establishment assumes that the RNC instructs the UE to move from RRC Idle mode to CELL_DCH. The RNC can also instruct the UE to move from RRC Idle mode to CELL_FACH. This transition is relatively simple because it does not require a dedicated radio link, nor does it require a dedicated Iub transport connection. This reduces both the signalling requirement and corresponding RRC connection establishment delay. The signalling used for RRC connection establishment to CELL_FACH is presented within Figure 8.3.

Table 8.2   Interpretation of the simultaneous DCH and HS-DSCH capability

	32 kbps	64 kbps	128 kbps	384 kbps
Maximum number of received TrBlk bits	640	3840	3840	6400
Maximum number of convolutional coded TrBlk bits	640	640	640	640
Support for turbo decoding	No	Yes	Yes	Yes
Maximum number of turbo coded TrBlk bits	-	3840	3840	6400
Maximum number of transport channels	8	8	8	8
Maximum nunber of TFC	32	48	96	128
Maximum number if TF	32	64	64	64
Maximum received DPCH and S-CCPCH PhyCh bits	1200	2400	4800	19200

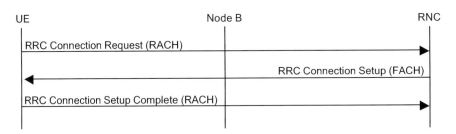

Figure 8.3   RRC connection establishment signalling (RRC Idle to CELL_FACH)

In this case, it is not necessary to use any NBAP nor ALCAP signalling. The RRC Connection Setup Complete message is returned to the RNC using the RACH transport channel rather than a DCH transport channel. The DCCH bit rates offered by the RACH and FACH tend to be greater than those offered by a DCH. This helps to increase the rate at which signalling procedures are completed. A UE can then be moved from CELL_FACH to CELL_DCH if a DCH, E-DCH or HS-DSCH is required for an end-user application. The drawback associated with establishing an RRC connection into CELL_FACH is a potentially reduced quality of service. The RACH and FACH are common transport channels which may experience congestion if relatively large numbers of UE are using them simultaneously.

8.2 Speech Call Connection Establishment

- An end-user initiates a mobile originated speech call by pressing the 'call' button after entering the relevant telephone number. The core network initiates a mobile terminated speech call by paging a UE after receiving a connection request.
- The RRC Connection Request message specifies 'originating conversational call' or 'terminating conversational call'. This does not differentiate between a speech call and a video call. The release 6 version of the RRC Connection Request specifies a call type.
- After moving into RRC connected mode, the UE sends an Initial Direct Transfer message to the RNC. This contains a mobility management CM Service Request for an originating connection, or radio resource management Paging Response for a terminating connection.
- The RNC packages the NAS message from the Initial Direct Transfer within a RANAP Initial UE Message. This is then packaged within an SCCP Connection Request message. SCCP signalling is used to establish an Iu–cs signalling connection.
- The security mode procedure is used to enable integrity protection and ciphering. This helps to secure all subsequent signalling between the UE and RNC.
- The originating procedure uses an uplink Setup message to inform the core network of the target telephone number and call type. It's acknowledged by a Call Proceeding message. The terminating procedure uses a downlink Setup message to inform the UE of the source telephone number and call type. It's acknowledged by a Call Confirmed message.
- RAB establishment is the same for both mobile originating and terminating connections. A RANAP RAB Assignment Request specifies the QoS requirements and Iu user plane mode. It also provides information to help establish an Iu user plane transport connection.
- RAB establishment includes an NBAP synchronised radio link reconfiguration procedure. This adds the speech service bearers to the existing SRB. The activation time accounts for the delay associated with sending the equivalent information to the UE within a Radio Bearer Setup message. The UE and Node B apply the new configuration simultaneously.
- RAB establishment includes ALCAP signalling across both the Iub and Iu–cs. This is used to establish AAL2 transport connections for the user plane speech data.
- The speech service requires a single RAB with three RAB subflows. These subflows are used to transfer the class A, B and C bits generated by the speech codec. Each RAB subflow is allocated its own logical channel and its own transport channel. The transport channels are multiplexed with the SRB transport channel within the physical layer.
- The speech service transport format combination set includes six transport format combinations. These correspond to the transmission of speech data, the transmission of comfort noise data, and speech service DTX, each with and without SRB activity.
- The terminating procedure uses an uplink Alerting message to inform the core network that the telephone is ringing. The originating procedure uses a downlink Alerting message to inform the originating UE that the target telephone is ringing.

- The terminating procedure uses an uplink Connect message to inform the core network that the telephone has been answered. The originating procedure uses a downlink Connect message to inform the originating UE that the target telephone has been answered. In both cases, the Connect messages are acknowledged using Connect Acknowledge messages.
- Key 3GPP specifications: TS 25.331, TS 25.413, TS 25.415, TS 24.008 and TS 44.018.

8.2.1 Mobile Originated

An end-user initiates a mobile originated speech call by pressing the 'call' button after having entered or selected the relevant telephone number. The UE triggers RRC connection establishment if it is in RRC Idle mode when the end-user presses the 'call' button. This procedure is described in detail within Section 8.1. The RRC Connection Request message specifies a cause value of 'originating conversational call'. This does not differentiate between different conversational call types, e.g. between a speech call and a video call. The release 6 version of 3GPP TS 25.331 specifies that the RRC Connection Request message also includes a 'call type' information element when an RRC connection is established for a CS domain application. This information element allows the UE to inform the RNC that the RRC connection is being requested for a speech call, a video call or another CS domain service type. Providing the RNC with this information during RRC connection establishment allows the RNC to configure the appropriate user plane resources at the same time as the configuring the SRB resources. This reduces the connection establishment delay by avoiding the subsequent reconfiguration procedure normally required to add the user plane resources to the existing SRB resources. The signalling procedure described in this section assumes that the RNC does not take advantage of the 'call type' information element within the RRC Connection Request message.

The first phase of signalling used to establish a mobile originated speech call is presented in Figure 8.4. After moving into RRC connected mode, the UE sends an Initial Direct Transfer message to the RNC. This RRC message is used to package a Non-Access Stratum (NAS) message addressed to the CS core

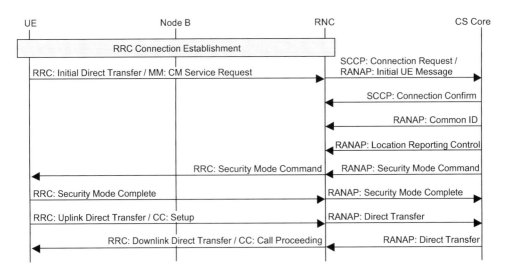

Figure 8.4 Signalling for mobile originated speech call (part 1)

network. It is transferred to the RNC using SRB 3, i.e. the high priority SRB used to transfer NAS messages with AM RLC.

An example Initial Direct Transfer message is presented within Log File 8.18. This message starts by instructing the RNC to forward the NAS message to the CS core network domain. An intra-domain routing parameter is then specified. This parameter can be used if the RNC is connected to more than a single MSC. In a simple scenario, the RNC is connected to only a single MSC and this routing parameter is not required. If the RNC is connected to multiple MSC, the routing parameter identifies the MSC to which the NAS message is directed. This is also the MSC which subsequently establishes a signalling connection with the UE. The UE generates the routing parameter from the TMSI which was allocated by the core network during registration. This allows the core network to maintain control over the routing parameter generated by each UE, i.e. the core network can control which MSC handles each UE. The routing parameter is derived from the TMSI by extracting bits 14 to 23, i.e. a section of 10 bits. The routing parameter can also be derived from the IMSI or IMEI if a TMSI is not available. The example in Log File 8.18 is based upon the TMSI. The message includes a label which indicates that the TMSI was allocated within the current location area, i.e. it is a local TMSI. This label uses the term P-TMSI because it is a general label used for both CS and PS domain messages. It is also possible for the label to indicate a TMSI from a different location area within the same PLMN, a TMSI from a different PLMN, an IMSI or an IMEI.

The Initial Direct Transfer message then includes the NAS message itself. The format of the NAS message is specified within 3GPP TS 24.008. Figure 8.5 illustrates the structure of the NAS message from Log File 8.18. The first byte always includes 4 bits to signal the protocol to which the message belongs. A value of 5 indicates that the message belongs to the mobility management protocol. Mobility management messages also include a skip indicator which represents a section of padding. The second byte of the NAS message specifies the message type belonging to the appropriate protocol. A value of 24 for the mobility management protocol corresponds to a CM Service Request message.

The third byte of the CM Service Request specifies the ciphering key sequence number and the CM service type. The ciphering key sequence number is used by the core network to identify the ciphering and integrity keys stored within the UE without having to trigger the authentication procedure. The CM service type provides an initial indication of the service which will be requested by the UE. A value of 1 corresponds to a general mobile originating call establishment. Other values include emergency call, short message service, supplementary service activation and voice group call establishment.

```
INITIAL_DIRECT_TRANSFER
UL-DCCH-Message
initialDirectTransfer
cn-DomainIdentity: cs-domain
intraDomainNasNodeSelector
  version: release99
    cn-Type: gsm-Map-IDNNS
    routingbasis
      localPTMSI
        routingparameter
          Bin: 3F C
nas-Message
  Hex: 05 24 11 03 57 18 A1 05 F4 B0 3F EF 67
v3a0NonCriticalExtensions
initialDirectTransfer-v3a0ext
  start-Value
    Bin: 00 04 9
```

Log File 8.18 Initial Direct Transfer message for a mobile originated speech call

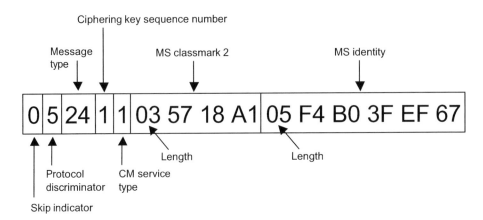

Figure 8.5   Structure of a CM Service Request message

The next section within the CM Service Request represents MS classmark 2 information. This provides the core network with some general information regarding the UE. The first byte specifies the length of the remaining content. The structure of the MS classmark 2 information is specified within 3GPP TS 24.008. The example illustrated in Figure 8.5 can be decoded to generate the information shown in Log File 8.19.

The final section of the CM Service Request message specifies the UE identity. Similar to the MS classmark 2 information, the first byte specifies the length of the remaining content. The first 3 bits of the second byte define the type of UE identity. In this example, the first 3 bits of the second byte are 100 which indicates that a TMSI is included. The remaining bytes within the message identify the TMSI as B0 3F EF 67. This TMSI is consistent with the value of the intra-domain routing parameter included within the Initial Direct Transfer message, i.e. extracting bits 14 to 23 from the TMSI generates the intra-domain routing parameter.

The RNC extracts the NAS message from the Initial Direct Transfer message and re-packages it within a Radio Access Network Application Part (RANAP) Initial UE Message. RANAP is specified

```
Mobile Station Classmark 2 (03 57 18 A1)
Revision level: 2
A5/1 algorithm available
"Controlled Early Classmark Sending" is implemented in MS
RF power capability: RF Power capability is irrelevant
PS cabability: not present
SS screening indicator: ellipsis notation and phase 2 error handling
SM capability: MS supports mobile terminated point to point SMS
no VBS capability or no notifications wanted
no VGCS capability or no notifications wanted
FC frequency capability: MS does not support E-GSM or R-GSM band
CM3: MS does support options indicated in Classmark 3
LCS value added location request notification capability supported
UCS2: the ME has a preference for the default alphabet over UCS2
SoLSA: The ME does not support
"NW initiated MO CM connection request" supported for at least 1 CM protocol
A5/2 algorithm available
A5/3 algorithm not available
```

Log File 8.19   Mobile Station Classmark 2 information

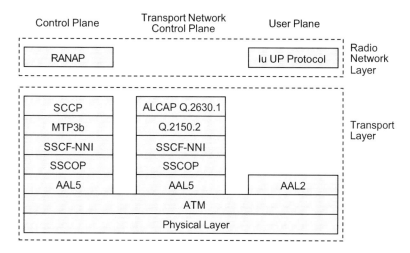

Figure 8.6 Iu–cs transport network protocol stacks (3GPP Releases 99 and 4)

within 3GPP TS 25.413 and represents the equivalent of NBAP for the Iu interface. Figure 8.6 illustrates RANAP within the set of the Iu–cs protocol stacks. Similar to the Iub, the user plane section of the protocol stack is based upon a Frame Protocol layer combined with AAL2 and ATM. Also similar to the Iub, the transport network control plane is responsible for establishing and releasing the AAL2 connections belonging to the user plane. This figure is applicable to the release 99 and release 4 versions of the specifications. The release 5 version introduces the use of IP on the Iu–cs.

An example RANAP Initial UE Message is shown in Log File 8.20. The header starts by confirming that the message is addressed to the CS core network domain. It also specifies the Location Area Identity (LAI) and Service Area Identity (SAI). These information elements define the location of the UE. A service area represents a group of one or more cells belonging to the same location area. It is common to use the cell identity to define the Service Area Code (SAC). This approach generates a unique SAI for every cell within the network. The emergency services can use SAI information to locate an end-user when making an emergency call.

The RANAP Initial UE Message then includes the NAS PDU which represents the CM Service Request. This remains unchanged from the NAS PDU within the Initial Direct Transfer message

```
RANAP-PDU
InitialUE-Message
CN-DomainIndicator: cs-domain
LAI
   - pLMNidentity: '32F499'H
   - lAC: '0056'H
SAI
   - pLMNidentity: '32F499'H
   - lAC: '0056'H
   - sAC: 'E9D7'H
NAS-PDU: '05 24 11 03 57 18 A1 05 F4 B0 3F EF 67'H
IuSignallingConnectionIdentifier: '0D 5B 01'H
GlobalRNC-ID
   - pLMNidentity: '32F499'H
   - rNC-ID: 6
```

Log File 8.20 RANAP Initial UE Message used to transfer the CM Service Request

```
CONNECTION REQUEST
Source Local Reference
    - 1A5145h
Protocol Class
    - 2h, connection oriented
Called Party Address
    - no global title present
    - routing based on SSN and MTP routing label
    - subsystem: RANAP
    Calling Party Address
        - no global title present
        - routing based on SSN and MTP routing label
        - point code: 3088
        - subsystem: RANAP
    SCCP User Data
        - data: 00 13 40 40 00 00 06 00 03 40 01 00 00 0F 40 06 00 32
          F4 01 75 3B 00 3A 40 08 00 32 F4 01 75 3B 4E 1B 00 10 40 0E
          0D 05 24 61 03 4F 18 80 05 F4 01 00 A3 BD 00 4F 40 03 07 13
          01 00 56 40 05 32 F4 01 00 7F
End of Optional Parameters
```

Log File 8.21   SCCP Connection Request message

generated by the UE. The NAS PDU is followed by an Iu signalling connection identifier. This identifier can be used to reference specific signalling connections in case there is a requirement to release one of more connections using the RANAP reset resource procedure. Finally, the Initial UE Message includes the identity of the RNC from which the message is originating.

The Signalling Connection Control Part (SCCP) layer packages the RANAP Initial UE Message within a Connection Request message. An SCCP Connection Request message is used because the RNC recognises that the UE does not yet have an Iu signalling connection. The Connection Request message is used to both transfer the RANAP message and initiate the establishment of an Iu signalling connection for any subsequent signalling procedures. An example SCCP Connection Request message is shown in Log File 8.21. The message starts by specifying a source local reference which becomes the destination local reference for any subsequent downlink messages. This allows the SCCP layer to identify downlink messages which belong to the same signalling connection. The section of user data within the Connection Request message represents the coded version of the RANAP message.

The CS core network responds to the SCCP Connection Request message using an SCCP Connection Confirm message. An example Connection Confirm message is presented within Log File 8.22. This message includes a destination local reference which reflects the source local reference within the Connection Request message. It also includes a source local reference which becomes the destination local reference for any subsequent uplink messages.

The Iu signalling connection is established once the RNC receives the Connection Confirm message. The core network then proceeds to send a RANAP Common Identity message. This message is used to inform the RNC of the permanent UE identity, i.e. the IMSI. An example Common Identity message is

```
CONNECTION CONFIRM
Destination Local Reference
    - 1A5145h
Source Local Reference
    - 000110h
Protocol Class
    - 2h, connection oriented
```

Log File 8.22   SCCP Connection Confirm message

```
RANAP-PDU
CommonID
PermanentNAS-UE-ID
   - iMSI: '32199059000054F8'H
```

Log File 8.23   RANAP Common Identity message

presented in Log File 8.23. The RNC can use the IMSI to help process incoming paging messages and also to support specific radio resource management functions, e.g. IMSI-based handover. IMSI-based handover allows the RNC to define different neighbour lists and different handover paths for specific groups of UE. This can be useful when RAN sharing principles are applied across part of the network. RAN sharing allows two or more operators to share a group of cells rather than each operator deploying its own dedicated cells across the entire network. As UE move from an area of shared cells to an area of dedicated cells, their IMSI can be used to direct them towards the appropriate layer, i.e. towards the cells belonging to their home PLMN.

The RANAP Common Identity message is packaged within an SCCP Data Form 1 message before being transferred across the Iu–cs. Log File 8.24 illustrates the Data Form 1 message containing the Common Identity message. The Data Form 1 message uses the newly established Iu–cs signalling connection. The destination local reference within the SCCP message reflects the value allocated during the signalling connection establishment procedure.

The CS core network then proceeds to send a RANAP Location Reporting Control message to the RNC. This message is used to instruct the RNC to inform the core network whenever the UE moves from one service area to another. This allows the core network to track the location of the UE. An example Location Reporting Control message is presented in Log File 8.25.

The core network then initiates the security mode procedure. This procedure is used to enable integrity protection and ciphering. The RANAP Security Mode Command message provides the RNC with a list of the integrity protection and ciphering algorithms permitted by the core network. 3GPP have only specified a single integrity protection algorithm (UIA1), and that algorithm is always specified as permitted. 3GPP have specified two encryption algorithms (UEA0 and UEA1). UEA0 corresponds to not using encryption whereas UEA1 corresponds to using encryption. The RNC is responsible for selecting an appropriate pair of algorithms (either UIA1 with UEA0, or UIA1 with UEA1) based upon both its own capability and the UE capability. The UE capability has already been

```
DT1 - DATA FORM 1
Destination Local Reference
- 1A5145h
Segmenting/Reassembling
- no more data
- the Segment./Reas. octet :  00
SCCP User Data
- data :  00 0F 40 10 00 00 01 00 17 40 09 50 32 19 90 59 00 00 54 F8
```

Log File 8.24   SCCP message used to package the RANAP Common Identity message

```
RANAP-PDU
LocationReportingControl
RequestType
    - event: change-of-servicearea
    - reportArea: service-area
```

Log File 8.25   RANAP Location Reporting Control message

signalled to the RNC within the RRC Connection Setup Complete message. The pair of selected algorithms are signalled to the UE within an RRC Security Mode Command message. This message is transferred using SRB 2 because it uses AM RLC, but does not encapsulate any NAS messages. The RRC Security Mode Command also specifies the RLC sequence numbers corresponding to the time at which ciphering will start in the downlink direction. A series of four sequence numbers will be provided if four SRB have been configured during RRC connection establishment. In general, ciphering and integrity protection starts immediately after the security mode procedure. The UE responds to the Security Mode Command using a Security Mode Complete message. This message acknowledges the Security Mode Command and specifies the RLC sequence numbers corresponding to the time at which ciphering will start in the uplink direction. Reception of this message allows the RNC to respond to the core network using a RANAP Security Mode Complete message. This message informs the core network of which integrity protection and ciphering algorithms have been selected.

The UE then initiates call control signalling. This signalling is encrypted if ciphering has been enabled during the preceding security mode procedure. The UE generates a call control Setup message to provide the core network with details of the requested connection. The structure of an example Setup message is illustrated in Figure 8.7. Similar to the mobility management CM Service Request message, the first byte always includes 4 bits to signal the protocol to which the message belongs. A value of 3 indicates that the message belongs to the call control protocol. Call control messages also include a transaction identifier to allow the UE and core network to distinguish between multiple message flows belonging to the same protocol. This allows parallel sets of transactions to be completed without confusing which messages belong to which transaction. All messages belonging to the same transaction are assigned the same transaction indicator. The second byte of the Setup message specifies the message type belonging to the call control protocol. A value of 45 corresponds to a Setup message.

The bearer capability information element defines the requested connection type and provides information regarding the UE capability. This information element is optional within the Setup message so it is necessary to include an Information Element Identifier (IEI). This is in contrast to the mandatory information elements within the CM Service Request which do not require an IEI. The IEI is followed by a length indicator to specify the length of the remaining content. The information content itself can be decoded using the rules specified within 3GPP TS 24.008. The information content from this example is presented in Log File 8.26. The information transfer capability shows that the UE is requesting a speech connection. The UE also signals its capability in terms of speech codecs.

The next part of the Setup message specifies the telephone number called by the originating UE. The third byte within this information element defines the type of number and numbering plan. In this example, the type of number is unknown while the numbering plan is identified as ISDN or telephony. The destination telephone number then starts in the forth byte using Binary Coded Decimal (BCD).

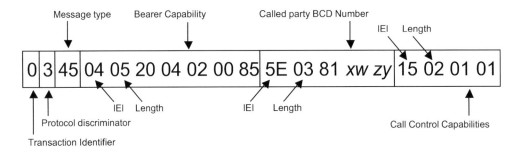

Figure 8.7 Structure of a mobile originated Setup message (the actual BCD telephone number has been replaced by *xwzy*. In practice, this section of the message would include the target telephone number)

```
Radio channel requiment MS to NW direction: MS supports at least full rate
speech version 1, but does not support half rate speech version 1
Coding standard: GSM standardized coding
Transfer mode: circuit mode
Information transfer capability: speech
CTM text telephony: is not supported
Speech version 1: GSM full rate speech version 3 - FR AMR
Speech version 2: GSM full rate speech version 2 - GSM Enhanced FR
Speech version 3: GSM full rate speech version 1 - GSM FR
Speech version 4: GSM half rate speech version 3 - HR AMR
```

Log File 8.26   Bearer Capability information

The pair of digits belonging to each byte are read from right to left so the destination telephone number is *wxyz* in this example.

The final part of the Setup message shown in Figure 8.7 specifies a set of call control capabilities. In this example, the UE signals that it supports Dual Tone Multi-frequency (DTMF) signalling, that it supports a maximum of one speech bearer and that it does not support the Prolonged Clearing Procedure (PCP).

The UE packages the Setup message within an Uplink Direct Transfer message before sending it to the RNC. The Uplink Direct Transfer message is sent using SRB 3 because it contains a NAS message and is treated with high priority. The resultant message is presented in Log File 8.27.

The RNC receives the Uplink Direct Transfer message, extracts the NAS PDU and packages it within a RANAP Direct Transfer message. This RANAP message is then forwarded to the CS core network using the Iu–cs signalling connection. The core network responds by returning a Call Proceeding message. The Call Proceeding message serves as an acknowledgement which indicates that no further connection establishment information will be accepted until the ongoing request has been processed, i.e. the core network is acknowledging that it has now received all of the information necessary to establish the connection. An example Call Proceeding message is illustrated in Figure 8.8. The protocol discriminator indicates a call control message while the transaction identifier indicates that the message belongs to the same signalling procedure as the preceding Setup message. The Setup message had a transaction identifier of 0 rather than 8. However, the most significant bit of the transaction identifier represents a flag to indicate whether the message is being sent from the side which originated the transaction (flag = 0), or is being sent to the side which originated the transaction (flag = 1). In this case, the transaction is mobile originated so downlink messages belonging to the same transaction as uplink messages with an identifier of 0 have an identifier of 8.

This example of a Call Proceeding message is relatively simplistic because it is sufficient to send only header information without any payload. Payload information can be included when necessary. For example, if the network supports the enhanced Multi-Level Precedence and Pre-emption (eMLPP) service then the Call Proceeding message includes a priority level, or if the network supports the multi-call service then the payload can include network call control capability information. The Call

```
UPLINK_DIRECT_TRANSFER
UL-DCCH-Message
integrityCheckInfo
messageAuthenticationCode
  Bin: 67 D9 E7 70
rrc-MessageSequenceNumber: 1
uplinkDirectTransfer
  cn-DomainIdentity: cs-domain
  nas-Message
    Hex: 03450405 20040200 855E0381 xwzy1502 0101
```

Log File 8.27   Uplink Direct Transfer message containing a call control Setup message

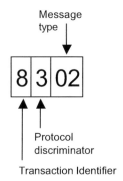

Figure 8.8   Structure of a Call Proceeding message

Proceeding message is transferred from the core network to the RNC using a RANAP Direct Transfer message, and subsequently from the RNC to the UE using an RRC Downlink Direct Transfer message.

The second phase of signalling used to establish a mobile originated speech call is presented in Figure 8.9. This phase uses RAB establishment to generate a user plane connection between the core network and UE. The existing radio link used by the set of SRB is reconfigured to accommodate user plane data. In addition, a new Iub transport connection is added in parallel to the existing SRB

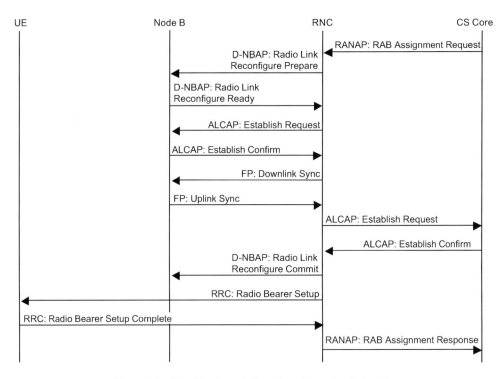

Figure 8.9   Signalling for mobile originated speech call (part 2)

connection, and an Iu–cs transport connection is added to allow user plane data transfer between the RNC and core network. A Radio Bearer Setup procedure informs the UE of the new logical and transport channels. It also instructs the UE to reconfigure its physical channels.

RAB establishment starts when the core network forwards a RANAP RAB Assignment Request message to the RNC. This message allows the core network to request a specific Quality of Service (QoS) profile. An example RAB Assignment Request message is presented in Log File 8.28. The speech service requires a single RAB with three RAB subflows. These subflows are used to transfer the class A, B and C bits generated by the speech codec. The traffic class requested by the core network is conversational rather than streaming, interactive or background. This reflects the real time require-ments of the speech service. The RAB is specified as bidirectional and symmetric. This means that the maximum and guaranteed bit rate requirements apply equally to both the uplink and downlink directions. Both of these bit rate requirements are specified to be 12.2 kbps. This represents the total throughput generated by the sum of the three RAB subflows. The RAB Assignment Request specifies that the maximum SDU size is 244 bits. This also represents a total across the three RAB subflows. The combination of maximum bit rate and maximum SDU size indicates that higher-layer packets are transferred once every 20 ms.

The RAB Assignment Request then specifies a set of parameters to associate with each RAB subflow. The first RAB subflow corresponds to the class A speech bits. These represent the most important bits generated by the speech codec. An SDU error ratio requirement of 7×10^{-3} is specified, i.e. no more than 0.7% of the higher layer packets should be received in error. The RLC layer does not complete segmentation for the speech service, nor does it complete re-transmissions. This means that the SDU error ratio is equal to the transport block error ratio and that the corresponding BLER target for outer loop power control is 0.7%. The Node B and UE use the physical layer CRC bits to determine the percentage of transport blocks received in error. Outer loop power control adjusts the DPCCH SIR target to help ensure that the target BLER requirement is achieved. The RAB Assignment Request also specifies that the residual bit error ratio requirement is 1×10^{-6}. This represents the Bit Error Ratio (BER) requirement after physical layer processing, i.e. after taking advantage of the error protection capability of convolutional coding. The RNC is then informed that erroneous SDU should be delivered to the higher layers rather than being discarded. Forwarding erroneous SDU is likely to decrease speech quality and the corresponding end-user experience, but is preferable to discarding them and consequently generating periods of silence. SDU sizes of 81, 39 and 0 bits are defined for the class A subflow. The SDU size of 81 bits is used during periods of speech activity. This corresponds to a subflow throughput of 4.05 kbps. This throughput decreases during periods of inactivity while the speech codec generates only comfort noise parameters. Comfort noise parameters are used to generate audio background noise at the receiver rather than allowing total silence during periods of speech inactivity. The SDU size of 39 bits corresponds to the transmission of comfort noise parameters. These parameters are transmitted once every eight speech frames during periods of speech inactivity resulting in an instantaneous bit rate of 1.95 kbps and an average bit rate of 244 bps.

The second and third RAB subflows correspond to the speech service class B and C bits respectively. These subflows do not have an SDU error ratio requirement because they are specified to be configured without any consideration of error detection. This means that the physical layer does not need to include any CRC bits nor does it need to identify which transport blocks are received in error. The residual BER for the class B subflow is 1×10^{-3} whereas the residual BER for the class C subflow is 5×10^{-3}. These figures reflect the relative importance of the class A, B and C subflows. Residual BER requirements can be used to help determine the physical layer rate matching attributes which determine the share of DPDCH bits allocated to each subflow. The class B and C subflows use SDU sizes of 103 and 60 bits respectively. These figures correspond to bit rates of 5.15 and 3 kbps. The class B and C subflows do not any transfer data during periods of speech inactivity so it's not necessary to include any

```
    RANAP-PDU
    RAB-AssignmentRequest
    RAB-SetupOrModifyList
    firstValue
      - rAB-ID: '00000001'B
      rAB-Parameters
        trafficClass: conversational
        rAB-AsymmetryIndicator: symmetric-bidirectional
        maxBitrate
          - MaxBitrate: 12200
        guaranteedBitRate
          - GuaranteedBitrate: 12200
        deliveryOrder: delivery-order-requested
  maxSDU-Size: 244
  sDU-Parameters

    sDU-ErrorRatio
      - mantissa: 7
      - exponent: 3
    residualBitErrorRatio
      - mantissa: 1
      - exponent: 6
    deliveryOfErroneousSDU: yes
    sDU-FormatInformationParameters
      - subflowSDU-Size: 81
      - subflowSDU-Size: 39
      - subflowSDU-Size: 0

    residualBitErrorRatio
      - mantissa: 1
      - exponent: 3
    deliveryOfErroneousSDU: no-error-detection-consideration
    sDU-FormatInformationParameters
      - subflowSDU-Size: 103
      - subflowSDU-Size: 0
      - subflowSDU-Size: 0

    residualBitErrorRatio
      - mantissa: 5
      - exponent: 3
    deliveryOfErroneousSDU: no-error-detection-consideration
    sDU-FormatInformationParameters
      - subflowSDU-Size: 60
      - subflowSDU-Size: 0
      - subflowSDU-Size: 0

  transferDelay: 100
  allocationOrRetentionPriority
    - priorityLevel: 3
    - pre-emptionCapability: may-trigger-pre-emption
    - pre-emptionVulnerability: not-pre-emptable
    - queuingAllowed: queueing-allowed
  sourceStatisticsDescriptor: speech
userPlaneInformation
  - userPlaneMode: support-mode-for-predefined-SDU-sizes
  - uP-ModeVersions: '0000000000000001'B
transportLayerInformation
  - transportLayerAddress: '47 00 00 11 11 11 11 11 11 11 11
  4B C6 00 20 48 0D 01 02 00'H
iuTransportAssociation
  - bindingID: '00000744'H
secondValue
```

Log File 8.28   RANAP RAB Assignment Request for a CS speech call

additional SDU sizes. Summing the bit rates of 4.05, 5.15 and 3 kbps generates the total speech service bit rate of 12.2 kbps.

The RAB Assignment Request then proceeds to specify the transfer delay requirement. The example presented in Log File 8.28 defines a requirement of 100 ms. This means that at least 95% of all SDU should have a transfer delay of less than 100 ms. A transfer delay of 100 ms is relatively small, but this reflects the requirement of the real time speech service. Allocation and retention parameters are then specified in terms of a priority level, pre-emption capability, pre-emption vulnerability and permission to queue. These parameters become significant during periods of network congestion in which case it is necessary to distinguish between high and low priority connections. The priority level can range from 1 to 15 where 1 represents the highest priority, 14 represents the lowest priority and 15 represents no priority. The speech service RAB in Log File 8.28 has been allocated a priority of 3. This corresponds to allocating a relatively high priority to speech connections while leaving some margin to allocate even higher priorities to other connection types, e.g. emergency calls. In this example, the speech service RAB request is able to pre-empt existing RAB but once established, the speech service RAB cannot be pre-empted by other RAB. The RAB Assignment Request message also specifies that the RAB request can be queued at the RNC if necessary and that the source data is speech.

The RAB Assignment Request then provides some information regarding the transfer of user plane data across the Iu interface. User plane data can be transferred across the Iu using either a transparent mode or a support mode. Transparent mode is applicable to connections which only require a data transfer service from the Iu user plane protocol stack. Support mode is applicable to connections which require additional services from the protocol stack. 3GPP TS 25.415 specifies a single support mode known as 'support mode for predefined SDU sizes'. Speech connections use this support mode because the SDU sizes are pre-defined and speech connections require other services from the Iu user plane protocol stack, e.g. support for rate adaptation when more than a single AMR codec bit rate is used. The RAB Assignment Request also specifies the version of the data transfer mode. The transparent mode of operation is limited to version 1. The release 99 version of the specifications define version 1 of the support mode for predefined SDU sizes, whereas the release 4 and newer versions define version 2.

The final part of the RAB Assignment Request provides Iu transport network information similar to that provided for the Iub within the NBAP Radio Link Setup Response (described for RRC Connection Establishment in Section 8.1 and shown in Log File 8.7). The transport layer address defines the address at which the Iu transport connection should be established. The binding identity provides a link to the RAB during Iu transport connection establishment. This ensures that the correct RAB is associated with the new transport connection. The binding identity appears as the 'served user-generated reference' within the subsequent Iu ALCAP Establish Request message.

The RNC uses Admission Control to determine whether or not the RAB Assignment Request should be accepted. In general, checks are completed in terms of downlink transmit power, uplink interference, channelisation code availability and Iub transport capacity. The RNC uses periodic NBAP measurement reports from the Node B to gain visibility of the both the downlink transmit power and uplink interference. The RNC is responsible for allocating channelisation codes and so has knowledge of whether or not any codes available. Likewise, the RNC should be aware of the Iub transport network load. If admission control denies the RAB request, a RAB Assignment Response message is used to inform the core network (the same message type is used when RAB establishment is successful, but the message content is different).

If admission control accepts the RAB request, the RNC initiates the NBAP radio link reconfiguration procedure. This procedure reconfigures the radio link at the Node B to prepare for the transfer of user plane data. The procedure is initiated by the RNC sending an NBAP Radio Link Reconfiguration Prepare message to the Node B (shown within Figure 8.9). The first part of an example NBAP Radio Link Reconfiguration Prepare message is shown in Log File 8.29.

The message starts by specifying the Node B communication context. This identity was allocated by the Node B during RRC connection establishment, i.e. within the NBAP Radio Link Setup Response.

```
NBAP-PDU
RadioLinkReconfigurationPrepare
NodeB-CommContextID: 480158
UL-DPCH-Information-RL-ReconfPrep        DL-DPCH-Information-RL-ReconfPrep
ul-scramblingCode                        tFCS
- ul-ScramblingCodeNumber: 10685         - ctfc6bit: 0
- ul-ScramblingCodeLength: long          - ctfc6bit: 1
- minUL-ChCodeLength: len64              - ctfc6bit: 11
- punctureLimit: 7                       - ctfc6bit: 12
tFCS                                     - ctfc6bit: 13
- ctfc6bit: 0                            - ctfc6bit: 23
- ctfc6bit: 1                            dL-DPCH-SlotFormat: 8
- ctfc6bit: 11                           - multiplexPosition: fixed-posit
- ctfc6bit: 12                           - tPC-DL-StepSize: one
- ctfc6bit: 13                           - limitedPowerInc: not-used
- ctfc6bit: 23
ul-DPCCH-SlotFormat: 0
```

Log File 8.29   NBAP Radio Link Reconfiguration Prepare message (part 1)

The message then specifies the uplink scrambling code which remains unchanged. If the standalone SRB have been using a bit rate of 3.4 kbps, the minimum uplink spreading factor decreases from 256 to 64. Otherwise, if the standalone SRB have been using a bit rate of 13.6 kbps, the minimum uplink spreading factor remains unchanged from 64. In this case, it is not necessary to reduce the spreading factor because the SRB bit rate is reduced to 3.4 kbps at the same time as introducing the 12.2 kbps speech connection. An uplink spreading factor of 64 is sufficient to transfer both a 3.4 kbps SRB and a 12.2 kbps speech connection. It is not necessary to signal the uplink channelisation code indices because the DPCCH is always spread using channelisation code (SF256,0) whereas the DPDCH is always spread using (SFx,x/4). The uplink puncturing limit is also signalled to the Node B. The actual value = (signalled value × 4) + 40, i.e. a signalled value of 7 translates to an actual value of 68%.

The Radio Link Reconfiguration Prepare message then proceeds to inform the Node B of the new uplink Transport Format Combination Set (TFCS). The introduction of the speech connection increases the number of Transport Format Combinations (TFC) from two to six. These six TFC are signalled using calculated TFC (cTFC) values of 0, 1, 11, 12, 13 and 23. The Node B can use these cTFC values to identify the allowed TFC. The translation between a specific cTFC value and its corresponding TFC is based upon the intermediate variables L and P. These variables are defined for each transport channel after ordering them according to their transport channel identity. L is equal to the number of transport formats within the transport format set, whereas P is equal to the product of the L values belonging to the preceding transport channels. By default, the P value for the first transport channel is equal to 1. The L and P variables belonging to a speech connection are illustrated in Figure 8.10.

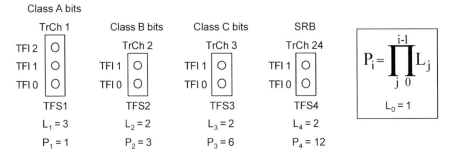

Figure 8.10   Variables used to calculate the cTFC for the speech service

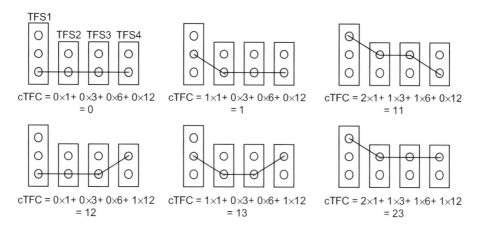

Figure 8.11   Calculation of cTFC values for the speech service

The cTFC values are then generated by summing the products of the appropriate Transport Format Indicator (TFI) and the corresponding value of P for each transport channel. This calculation is illustrated in Figure 8.11. There is always a one-to-one relationship between a specific cTFC value and the corresponding TFC. The first part of the NBAP Radio Link Reconfiguration Prepare message has not specified the new transport channel identities nor the number of transport formats within each transport format set. However, this information is provided within subsequent sections of the message.

The NBAP Radio Link Reconfiguration message then informs the Node B that the uplink DPCCH slot format is 0. In general, this remains unchanged from the slot format used for the standalone SRB. It is not necessary to signal a slot format for the uplink DPDCH because the uplink spreading factor has already been specified and there is a one-to-one mapping between uplink DPDCH slot format and spreading factor.

The rules for generating the downlink cTFC values are the same as those for the uplink. The uplink and downlink TFCS are the same for a speech connection so the cTFC values are equal. In the downlink direction, a single slot format defines both the DPCCH and DPDCH. Slot format 8 indicates a downlink spreading factor of 128 as well as 34 DPDCH and 6 DPCCH bits per time slot. The 6 DPCCH bits correspond to 2 TPC bits and 4 pilot bits. There are no TFCI bits to inform the UE of which TFC is being sent at any point in time. This means that the UE must complete blind detection of the TFC. Blind detection requires the physical layer to use fixed transport channel locations within the DPDCH. The UE can then deduce the TFC by identifying which sections of the DPDCH include data and which sections are using DTX. Finally, the first part of the NBAP Radio Link Reconfiguration Prepare message specifies that the downlink inner loop power control step size is equal 1 dB and the limited power increase algorithm is disabled.

The second part of the NBAP Radio Link Reconfiguration Prepare message is shown in Log File 8.30. This section can be used to define changes to the existing SRB transport channel. The information is similar to the content of the Radio Link Setup Request message sent during RRC connection establishment. In this case, there is an additional information element to specify whether or not it is necessary to establish a new Iub transport connection for the SRB. The example in Log File 8.30 indicates that it is not necessary. The SRB transport block size remains unchanged. Likewise, the number of transport blocks belonging to each transport format remains unchanged. If the standalone SRB have been using a bit rate of 3.4 kbps, the TTI remains unchanged from 40 ms. If the standalone SRB have been using a bit rate of 13.6 kbps, the TTI increases from 10 to 40 ms. This corresponds to decreasing the SRB bit rate from 13.6 to 3.4 kbps.

```
DCH-Modify-Inf-RL-ReconfPrep          - codingRate: third
DCH-Modify-InfItem-RL-ReconfPrep      - rateMatchingAttribute: 210
- ul-FP-Mode: silent                  cRC-Size: 16
- aAL2-ChRealloc: not-be-realloc      ul-TransportFormatSet
dCH-Info-RL-ReconfPrep                dynamicTransportFormatInfo
DCH-Modify-Info-Item-RL-ReconfPrep    - nrOfTransportBlocks: 0
- dCH-ID: 24                          - nrOfTransportBlocks: 1
dL-TransportFormatSet                 - transportBlockSize: 148
dynamicTransportFormatInfo            semiStaticTransportFormatInfo
- nrOfTransportBlocks: 1              - transTimeInterval: msec-40
- transportBlockSize: 0               - channelCodingType: convolutional
- nrOfTransportBlocks: 1              - codingRate: third
- transportBlockSize: 148             - rateMatchingAttribute: 210
semiStaticTransportFormatInfo         cRC-Size: 16
- transTimeInterval: msec-40          - frameHandlingPriority: 14
- channelCodingType: convolutional
```

Log File 8.30   NBAP Radio Link Reconfiguration Prepare message (part 2)

The second part of the Radio Link Reconfiguration Prepare message also defines the uplink and downlink rate matching attributes for the SRB transport channel. Rate matching attributes determine the share of the DPDCH allocated to each transport channel when multiple transport channels are multiplexed within the physical layer. Previously, only a single transport channel was mapped onto the DPDCH so rate matching attributes had no impact. Section 2.6.1 describes the use of rate matching attributes in greater detail.

The third part of the NBAP Radio Link Reconfiguration Prepare message is shown in Log File 8.31. This section introduces the set of three transport channels used to transfer the class A, B and C speech data. The SRB already have their own Iub transport connection and are already configured to use Frame Protocol data frames. The set of three transport channels belonging to the speech service share a second connection across the Iub and use their own set of Frame Protocol data Frames. Log File 8.31 starts by configuring the Frame Protocol data frames for the speech service transport channels. These parameters are the same as those specified for the SRB during RRC connection establishment (described in Section 8.1 and shown in Log File 8.6).

The next part of the message defines the uplink and downlink transport format sets for each of the three new transport channels. Transport channel identities 1, 2 and 3 are allocated which confirms the order illustrated in Figure 8.10. The transport format sets also match those illustrated in Figure 8.10. The transport block sizes are equal to the SDU sizes defined within the RANAP RAB Assignment Request message. This indicates that the RLC layer does not provide any segmentation nor concatenation function, and that neither the RLC layer nor the MAC layer add any overhead. The TTI is configured with a value of 20 ms for all three transport channels. Only the first transport channel is configured to include CRC bits as part of the physical layer processing. This reflects the requirement within the RAB Assignment Request for the class A subflow to have both an error detection capability and an associated SDU error ratio. The RAB Assignment Request specified that the data belonging to the class B and class C subflows can be forwarded without error detection. The first transport channel is selected to provide the quality estimate appearing within the Frame Protocol data frames. This means that the Node B will include an estimate of BER experienced by the class A speech data whenever possible. Otherwise, the Node B will include an estimate of the physical channel BER.

The fourth and final part of the NBAP Radio Link Reconfiguration Prepare message is shown in Log File 8.32. This section allows a change to the downlink channelisation index which is likely to be necessary if the downlink spreading factor has changed. It is not necessary if the downlink spreading factor has remained unchanged. Log File 8.32 also defines an initial Node B transmit power to be applied when the new radio link configuration becomes active. This result can be calculated by scaling

```
DCH-Add-Inf-RL-ReconfPrep
DCH-Add-InfItem-RL-ReconfPrep
- payloadCRC-Presence: included
- ul-FP-Mode: silent
- toAWS: 15
- toAWE: 5

dCH-Add-Info-RL-ReconfPrep
DCH-Add-Info-Item-RL-ReconfPrep
- dCH-ID: 1
dL-TransportFormatSet
dynamicTransportFormatInfo
- nrOfTransportBlocks: 1
- transportBlockSize: 0
- nrOfTransportBlocks: 1
- transportBlockSize: 39
- nrOfTransportBlocks: 1
- transportBlockSize: 81
semiStaticTransportFormatInfo
- transTimeInterval: msec-20
- channelCodingType: convolutional
- codingRate: third
- rateMatchingAttribute: 196
cRC-Size: 12
ul-TransportFormatSet
dynamicTransportFormatInfo
- nrOfTransportBlocks: 1
- transportBlockSize: 0
- nrOfTransportBlocks: 1
- transportBlockSize: 39
- nrOfTransportBlocks: 1
- transportBlockSize: 81
semiStaticTransportFormatInfo
- transTimeInterval: msec-20
- channelCodingType: convolutional
- codingRate: third
- rateMatchingAttribute: 196
cRC-Size: 12
- frameHandlingPriority: 13
- qE-Selector: selected-DCH

DCH-Add-Info-Item-RL-ReconfPrep
- dCH-ID: 2
dL-TransportFormatSet
dynamicTransportFormatInfo
- nrOfTransportBlocks: 0
- nrOfTransportBlocks: 1
```

```
- transportBlockSize: 103
semiStaticTransportFormatInfo
- transTimeInterval: msec-20
- channelCodingType: convolutional
- codingRate: third
- rateMatchingAttribute: 202
cRC-Size: 0
ul-TransportFormatSet
dynamicTransportFormatInfo
- nrOfTransportBlocks: 0
- nrOfTransportBlocks: 1
- transportBlockSize: 103
semiStaticTransportFormatInfo
- transTimeInterval: msec-20
- channelCodingType: convolutional
- codingRate: third
- rateMatchingAttribute: 202
cRC-Size: 0
- frameHandlingPriority: 13
- qE-Selector: non-selected-DCH

DCH-Add-Info-Item-RL-ReconfPrep
- dCH-ID: 3
dL-TransportFormatSet
dynamicTransportFormatInfo
- nrOfTransportBlocks: 0
- nrOfTransportBlocks: 1
- transportBlockSize: 60
semiStaticTransportFormatInfo
- transTimeInterval: msec-20
- channelCodingType: convolutional
- codingRate: half
- rateMatchingAttribute: 256
cRC-Size: 0
ul-TransportFormatSet
dynamicTransportFormatInf
- nrOfTransportBlocks: 0
- nrOfTransportBlocks: 1
- transportBlockSize: 60
semiStaticTransportFormatInfo
- transTimeInterval: msec-20
- channelCodingType: convolutional
- codingRate: half
- rateMatchingAttribute: 256
cRC-Size: 0
- frameHandlingPriority: 13
- qE-Selector: non-selected-DCH
```

Log File 8.31   NBAP Radio Link Reconfiguration Prepare message (part 3)

```
RLModifyInforList-RL-ReconfPrep
RLModifyInfItem-RL-ReconfPrep
- rL-ID: 1
dL-ChCodeInf-RL-ReconfPrep
DL-ChCodeInfItem-RL-ReconfPrep
- dL-ChannelisationCodeNumber: 5
- dL-TransmissionPower: -116
- maxDL-Power: -8
- minDL-Power: -158
```

Log File 8.32   NBAP Radio Link Reconfiguration Prepare message (part 4)

the existing downlink transmit power. An appropriate scaling factor can be generated by comparing the C/I requirement of the existing radio link with the C/I requirement of the new radio link. The signalled value of −116 has units of 0.1 dB and so the actual value is −11.6 dB. This value represents a transmit power relative to the CPICH transmit power. If the CPICH transmit power is 33 dBm then the initial DPDCH transmit power for the reconfigured radio link will be 21.4 dBm. The maximum and minimum downlink transmit powers for the reconfigured radio link are signalled in the same way, i.e. with units of 0.1 dB and relative to the CPICH transmit power. The values signalled in this example indicate a maximum DPDCH transmit power of 32.2 dBm and a minimum DPDCH transmit power of 17.2 dBm (assuming a CPICH transmit power of 33 dBm).

The Node B does not apply the new radio link configuration immediately after receiving the Radio Link Reconfiguration Prepare message. The new configuration is evaluated to ensure that it is acceptable, but is then stored until the RNC specifies the Connection Frame Number (CFN) during which it should become active. This approach corresponds to the synchronous radio link reconfiguration procedure. The Node B acknowledges that it has accepted the new configuration by returning an NBAP Radio Link Reconfiguration Ready message. An example of this message is presented in Log File 8.33. The controlling RNC communication context is used to identify the relevant connection, whereas the radio link identity is used to identify the relevant radio link belonging to that connection. The reconfiguration involves the addition of transport channels which require a new connection across the Iub, i.e. the speech service will use a new Iub transport connection which will run in parallel to the existing SRB transport connection. This triggers the Node B to include transport network information within the Radio Link Reconfiguration Ready message. This information is used by the RNC when establishing the new Iub transport connection.

The RNC proceeds to establish the speech service Iub transport connection by sending an ALCAP Establish Request message to the Node B. The Node B responds with an ALCAP Establish Confirm message. This procedure is the same as that described for RRC connection establishment in Section 8.1. The Establish Request message includes the AAL2 Channel Identifier (CID) for the new transport connection. The destination Network Service Access Point (NSAP) address matches the transport layer address provided by the Node B within the Radio Link Reconfiguration Ready message. Similarly, the Served User Generated Reference (SUGR) matches the binding identity provided by the Node B. The SUGR allows the Node B to link the new Iub transport connection with the appropriate transport channels. The maximum and average bit rates within the Establish Request message are based upon the requirements of the speech service. These bit rates exclude the overheads generated by the AAL2 and ATM layers, but include the overhead generated by the Frame Protocol layer. The average bit rates can be less than the maximum bit rates to account for the impact of speech inactivity. Similarly, the maximum and average SDU sizes are based upon the requirements of the speech service.

The RNC achieves synchronisation across the new Iub transport connection after receiving the ALCAP Establish Confirm message. This helps to ensure that the RNC is able to send Frame Protocol

```
NBAP-PDU
RadioLinkReconfigurationReady
CRNC-CommunicationContextID: 3617
RadioLinkInformationList-RL-ReconfReady
RadioLinkInformationItem-RL-ReconfReady
- rL-ID: 1
dCH-Information-RL-ReconfReady
DCH-List-RL-ReconfReady
DCH-List-Item-RL-ReconfReady
- dCH-ID: 1
- bindingID: '00000002'H
transportLayerAddress: '4900003001FFFFFFFFFFFFFFFFFFFFFFFFFFFFFF'H
```

Log File 8.33   NBAP Radio Link Reconfiguration Ready message

data frames such that they arrive at the Node B within the time window defined by toAWS and toAWE. The Node B has already been informed of the toAWS and toAWE parameters within the NBAP Radio Link Reconfiguration Prepare message. Synchronisation is achieved using dedicated channel Frame Protocol control frames. This procedure is the same as that described for RRC connection establishment in Section 8.1.

The RNC also proceeds to establish an equivalent AAL2 transport connection across the Iu–cs. For the end-user, this represents the first AAL2 connection across the Iu–cs. The existing Iu–cs signalling connection uses the RANAP protocol stack and AAL5. For example, RANAP signalling across AAL5 was used when the call control Setup message was transferred from the RNC to the core network. This is in contrast to the Iub which uses AAL2 transport connections for both the SRB and speech data (although NBAP signalling uses AAL5). The RNC sends an ALCAP Establish Request message to the core network, and the core network responds with an ALCAP Establish Confirm. The Establish Request includes the AAL2 CID for the new Iu–cs transport connection. The destination NSAP matches the transport layer address provided by the core network within the RAB Assignment Request. Similarly, the SUGR matches the binding identity within the RAB Assignment Request. The bit rates specified within the Iu–cs Establish Request are smaller than those specified within the equivalent Iub Establish Request. This reflects the relatively small overhead generated by the Iu–cs user plane protocol stack when compared with the overhead generated by the Iub DCH Frame Protocol. The data frames belonging to the Iu–cs 'support mode for predefined SDU sizes' protocol include 3 bytes of header information when a payload CRC is included. This generates a total SDU size of 35 bytes after adding 0.5 bytes of padding to the 31.5 bytes (244 bits) of speech data. This data frame size is applicable to both the uplink and downlink directions. It generates a corresponding maximum throughput of $35 \times 8/0.02 = 14$ kbps. The DCH Frame Protocol on the Iub generates larger overheads with downlink data frame sizes reaching 39 bytes and uplink data frame sizes reaching 41 bytes.

The RNC then proceeds to provide the Node B with the CFN during which it should start to apply the new radio link configuration. This information is sent using an NBAP Radio Link Reconfiguration Commit message. An example of this message is presented in Log File 8.34.

The RNC calculates the CFN by considering that the UE must be ready to apply the new radio link configuration at the same time as the Node B. The UE has not yet been informed of the new configuration so the CFN must allow sufficient time for the Radio Bearer Setup message to be transferred to the UE. In general, some margin is included to allow for one or more RLC retransmissions which may or may not be necessary. If the RNC specifies an activation CFN which passes after only a short delay there is a danger that the UE does not receive the Radio Bearer Setup message before the Node B applies the new configuration. This would result in a connection establishment failure. If the activation CFN is specified to pass after a relatively long delay the UE is more likely to receive the Radio Bearer Setup message prior to the Node B applying the new configuration but the connection setup delay is increased.

The first part of an example Radio Bearer Setup message is shown in Log File 8.35. This section starts by providing the UE with the same activation CFN as already signalled to the Node B within the Radio Link Reconfiguration Commit message. The UE is then instructed to remain in CELL_DCH when the new configuration becomes active. The UE is also informed of the new CS domain RAB identity and instructed to use T314 rather than T315 as a re-establishment timer. The re-establishment timer becomes applicable if the UE experiences radio link failure, i.e. air-interface synchronisation is

```
NBAP-PDU
RadioLinkReconfigurationCommit
NodeB-CommunicationContextID: 480158
CFN: 172
```

Log File 8.34 NBAP Radio Link Reconfiguration Commit message

```
DL-DCCH-Message                                  rb-InformationSetupList value 2
  integrityCheckInfo                               rb-Identity: 6
  messageAuthenticationCode                        rlc-Info
    Bin: B9 2C F8 FC                                 ul-RLC-Mode
  rrc-MessageSequenceNumber: 4                         ul-TM-RLC-Mode
RadioBearerSetup                                        segmentationIndication: false
rrc-TransactionIdentifier: 0                         dl-RLC-Mode
activationTime: 172                                    dl-TM-RLC-Mode
rrc-StateIndicator: cell-DCH                           segmentationIndication: false
rab-InformationSetupList                           rb-MappingInfo
  rab-InformationSetupList value 1                 rb-MappingInfo value 1
    rab-Info                                         ul-LogicalChannelMappings
    rab-Identity                                     ul-TransportChannelType dch: 2
    gsm-MAP-RAB-Identity: 01                         rlc-SizeList: configured
    cn-DomainIdentity: cs-domain                     mac-LogicalChannelPriority: 2
    re-EstablishmentTimer: useT314                 dl-LogicalChannelMappingList
rb-InformationSetupList                            dl-LogicalChMappingList value 1
                                                     dl-TransportChType dch: 2
rb-InformationSetupList value 1
  rb-Identity: 5                                 rb-InformationSetupList value 3
  rlc-Info                                         rb-Identity: 7
    ul-RLC-Mode                                    rlc-Info
      ul-TM-RLC-Mode                                 ul-RLC-Mode
      segmentationIndication: false                    ul-TM-RLC-Mode
    dl-RLC-Mode                                        segmentationIndication: false
      dl-TM-RLC-Mode                               dl-RLC-Mode
      segmentationIndication: false                  dl-TM-RLC-Mode
  rb-MappingInfo                                     segmentationIndication: false
  rb-MappingInfo value 1                           rb-MappingInfo
    ul-LogicalChannelMappings                      rb-MappingInfo value 1
    ul-TransportChannelType: dch: 1                  ul-LogicalChannelMappings
    rlc-SizeList: configured                         ul-TransportChannelType dch: 3
    mac-LogicalChannelPriority: 2                    rlc-SizeList: configured
  dl-LogicalChannelMappingList                      mac-LogicalChannelPriority: 2
  dl-LogicalChMappingList value 1                  dl-LogicalChannelMappingList
    dl-TransportChType: dch: 1                     dl-LogicalChMappingList value 1
                                                     dl-TransportChType dch: 3
```

Log File 8.35   Radio Bearer Setup message (part 1)

lost. In this case, the UE has the time interval defined by T314 to re-establish synchronisation and continue with the connection. T314 is often used for real time services while T315 is used for non-real time services. The UE reads the values of T314 and T315 from SIB 1 while in RRC Idle mode. T314 may be configured with a value of 0 s to allow UE with real time connections to drop immediately to RRC Idle mode as soon as radio link failure occurs.

The first part of the Radio Bearer Setup message then provides information regarding each of the new radio bearers. Three new radio bearers are established to transfer the speech data belonging to the class A, B and C subflows. These radio bearers are allocated identities of 5, 6 and 7. These identities follow on from the identities already allocated to the four SRB. All three of the new radio bearers are configured with Transparent Mode (TM) RLC in both the uplink and downlink directions. In each case, the UE is instructed that the RLC layer is not required to complete segmentation of the higher layer packets. The set of three logical channels are mapped onto a set of three Dedicated Channels (DCH) which have identities of 1, 2 and 3. These transport channel identities match those within the NBAP Radio Link Reconfiguration Prepare message. The Node B has not been informed of the radio bearer information because the peer RLC and MAC layers on the network side are within the RNC.

The second part of the Radio Bearer Setup message is shown in Log File 8.36. This section is used to configure the uplink Transport Format Combination Set (TFCS) as well as the individual uplink

```
ul-CommonTransChInfo                    transportFormatSet
modeSpecificInfo fdd                    dedicatedTransChTFS
ul-TFCS                                   tti20 value 1
  normalTFCI-Signalling                     rlc-Size bitMode
  ctfcSize                                    sizeType1: 81
  ctfc6Bit                                  numberOfTbSizeList
    ctfc6Bit value 1                        numberOfTbSizeList 1: zero
      ctfc6: 0                              logicalChannelList: allSizes
      powerOffsetInformation              tti20 value 2
        gainFactorInformation               rlc-Size
        computedGainFactors: 0              bitMode
    ctfc6Bit value 2                          sizeType1: 39
      ctfc6: 1                              numberOfTbSizeList
      powerOffsetInformation              numberOfTbSizeList 1: one
        gainFactorInformation               logicalChannelList: allSizes
        computedGainFactors: 0            tti20 value 3
    ctfc6Bit value 3                        rlc-Size
      ctfc6: 11                             bitMode
      powerOffsetInformation                  sizeType1: 81
        gainFactorInformation             numberOfTbSizeList
        computedGainFactors: 0            numberOfTbSizeList 1: one
    ctfc6Bit value 4                        logicalChannelList: allSizes
      ctfc6: 12                         semistaticTF-Information
        powerOffsetInformation            channelCodingType
        gainFactorInformation               convolutional: third
        computedGainFactors: 0            rateMatchingAttribute: 196
    ctfc6Bit value 5                        crc-Size: crc12
      ctfc6: 13
        powerOffsetInformation        ul-AddReconfTransChInfoList value 3
        gainFactorInformation             ul-TransportChannelType: dch
        computedGainFactors: 0          transportChannelIdentity: 2
    ctfc6Bit value 6                    transportFormatSet
      ctfc6: 23                         dedicatedTransChTFS
        powerOffsetInformation            tti20 value 1
        gainFactorInformation               rlc-Size bitMode
        signalledGainFactors                  sizeType1: 103
          modeSpecificInfo fdd            numberOfTbSizeList
          gainFactorBetaC: 10             numberOfTbSizeList 1: zero
          gainFactorBetaD: 15             numberOfTbSizeList 2: one
          referenceTFC-ID: 0              logicalChannelList: allSizes
ul-AddReconfTransChInfoList           semistaticTF-Information
                                          channelCodingType
ul-AddReconfTransChInfoList value 1       convolutional    : third
  ul-TransportChannelType: dch            rateMatchingAttribute: 202
  transportChannelIdentity: 24            crc-Size: crc0
  transportFormatSet
  dedicatedTransChTFS               ul-AddReconfTransChInfoList value 4
    tti40 value 1                        ul-TransportChannelType: dch
      rlc-Size octetModeType1          transportChannelIdentity: 3
        sizeType1: 16                   transportFormatSet
      numberOfTbSizeList               dedicatedTransChTFS
      numberOfTbSizeList 1: zero          tti20 value 1
      numberOfTbSizeList 2: one             rlc-Size bitMode
      logicalChannelList: allSizes          sizeType1: 60
  semistaticTF-Information             numberOfTbSizeList
    channelCodingType                  numberOfTbSizeList 1: zero
      convolutional: third             numberOfTbSizeList 2: one
    rateMatchingAttribute: 210           logicalChannelList: allSizes
    crc-Size: crc16                 semistaticTF-Information
                                        channelCodingType
ul-AddReconfTransChInfoList value 2       convolutional: half
  ul-TransportChannelType: dch            rateMatchingAttribute: 256
  transportChannelIdentity: 1             crc-Size: crc0
```

Log File 8.36 Radio Bearer Setup message (part 2)

Transport Format Sets (TFS). The information signalled to the UE is similar to the information already signalled to the Node B. The same uplink cTFC values are signalled, but in this case they are accompanied by information to define the DPDCH and DPCCH gain factors. The largest TFC is defined as a reference TFC with gain factors which are signalled explicitly. The UE is then responsible for using the reference gain factors to calculate the gain factors for the remainingTFC.

The Radio Bearer Setup message then proceeds to provide information regarding the transport format sets belonging to both the SRB and the speech connection, i.e. a total of four transport format sets. RLC PDU sizes are signalled rather than transport block sizes. The RLC PDU size must be derived from the size type information. 3GPP TS 25.331 specifies the set of equations presented in Table 8.3 (Section 8.5.1). In the case of the SRB on DCH 24, the appropriate equation to calculate the RLC PDU size is: $(8 \times \text{sizeType1}) + 16$. This equation generates a result of 144 bits. These 144 bits correspond to a transport block size of 148 bits. The transport block size is applicable to the top of the physical layer whereas the RLC PDU size is applicable to the top of the MAC layer, i.e. the RLC PDU size excludes the overhead generated by the MAC layer. The MAC layer adds a header of 4 bits to generate the transport block size of 148 bits. In the case of the speech service transport channels, the RLC PDU sizes are signalled using bit mode size type 1 information. This means that the signalled sizeType1 value is equal to the RLC PDU size. The speech service RLC PDU sizes are equal to the corresponding transport block sizes because the MAC layer does not add any overhead. The TTI is specified to be 40 ms for the SRB transport channel, and 20 ms for each of the speech service transport channels. The channel coding type, rate matching attributes and the number of CRC bits are also specified for each transport channel.

The third part of the Radio Bearer Setup message is shown in Log File 8.37. This section represents the equivalent of the second section, but for the downlink direction. An information element is provided to signal that the downlink cTFC values are equal to the uplink cTFC values. Each of the four downlink transport format sets are specified and BLER targets are defined for the SRB and speech service class A transport channels. These correspond to the only two transport channels which use CRC bits within the physical layer. The UE applies these BLER target requirements as part of the downlink outer loop power control procedure. The signalled values are equal to $10 \times \log(\text{BLER})$, i.e. the signalled values of -20 correspond to BLER targets of 1%.

The fourth and final part of the Radio Bearer Setup message is shown in Log File 8.38. This section provides the UE with information regarding the uplink and downlink physical channels. This information matches the information provided to the Node B within the Radio Link Reconfiguration Prepare message. The UE is informed of the power offset between the downlink DPDCH and the downlink DPCCH pilot bits (PO3) to help identify an appropriate SIR target for inner loop power control. The BLER target requirements appearing within Log File 8.37 are applicable to the DPDCH. These allow the UE to estimate an appropriate DPDCH SIR requirement. Inner loop power control is based upon SIR measurements from the DPCCH pilot bits. Signalling the value of PO3 allows the UE to translate the DPDCH SIR requirement to an equivalent DPCCH SIR target.

After receiving the Radio Bearer Setup message, the UE waits until the activation CFN occurs. It then starts to apply the new configuration and returns a Radio Bearer Setup Complete message to the RNC. If the standalone SRB had previously been using a bit rate of 13.6 kbps, the Radio Bearer Setup Complete represents the first message transferred with the reconfigured bit rate of 3.4 kbps. An example Radio Bearer Setup Complete message is presented in Log File 8.39. This message only includes a COUNT-C activation time and a START value for the purposes of ciphering.

Once the RNC has received the Radio Bearer Setup Complete message, it forwards a RANAP RAB Assignment Response to the core network. This message is used to report that RAB establishment has been successful and call control signalling can proceed. An example RAB Assignment Response is presented in Log File 8.40. This message simply specifies the identity of the RAB which has been established.

```
dl-CommonTransChInfo                              numberOfTbSizeList 1: one
modeSpecificInfo fdd                              logicalChannelList: allSizes
dl-Parameters: sameAsUL                         tti20 value 2
dl-AddReconfTransChInfoList                       rlc-Size
                                                  bitMode
dl-AddReconfTransChInfoList 1                       sizeType1: 39
  dl-TransportChannelType: dch                    numberOfTbSizeList
  dl-transportChannelIdentity: 24                   numberOfTbSizeList 1: one
  tfs-SignallingMode                              logicalChannelList: allSizes
    explicit-config                             tti20 value 3
  dedicatedTransChTFS                             rlc-Size
  tti40 value 1                                   bitMode
    rlc-Size                                        sizeType1: 81
    bitMode                                       numberOfTbSizeList
      sizeType1: 0                                  numberOfTbSizeList 1: one
    numberOfTbSizeList                            logicalChannelList: allSizes
      numberOfTbSizeList 1: one               semistaticTF-Information
    logicalChannelList: allSizes                channelCodingType
  tti40 value 2                                     convolutional: third
    rlc-Size                                      rateMatchingAttribute: 196
    octetModeType1                                crc-Size: crc12
      sizeType1: 16                           dch-QualityTarget
    numberOfTbSizeList                            bler-QualityValue: -20
      numberOfTbSizeList 1: one
    logicalChannelList: allSizes          dl-AddReconfTransChInfoList value 3
  semistaticTF-Information                    dl-TransportChannelType: dch
    channelCodingType                         dl-transportChannelIdentity: 2
      convolutional: third                    tfs-SignallingMode
    rateMatchingAttribute: 210                  sameAsULTrCH
    crc-Size: crc16                           ul-TransportChannelType: dch
  dch-QualityTarget                           ul-TransportChannelIdentity: 2
    bler-QualityValue: -20                    dch-QualityTarget
                                                bler-QualityValue: 0
dl-AddReconfTransChInfoList 2
  dl-TransportChannelType: dch            dl-AddReconfTransChInfoList value 4
  dl-transportChannelIdentity: 1            dl-TransportChannelType: dch
  tfs-SignallingMode                        dl-transportChannelIdentity: 3
    explicit-config                         tfs-SignallingMode
  dedicatedTransChTFS                         sameAsULTrCH
  tti20 value 1                             ul-TransportChannelType: dch
    rlc-Size                                ul-TransportChannelIdentity: 3
    bitMode                                 dch-QualityTarget
      sizeType1: 0                            bler-QualityValue: 0
    numberOfTbSizeList
```

Log File 8.37   Radio Bearer Setup message (part 3)

At this stage, the radio access network and the Iu transport connection are both ready to start transferring speech data. However, the originating end-user remains unconnected to the destination telephone number. Figure 8.12 illustrates the final phase of signalling used to connect the originating end-user. This phase uses RRC and RANAP messages to transfer call control signalling between the UE and core network.

The core network forwards a call control Alerting message to the originating UE after the destination telephone number has been contacted. The Alerting message triggers the originating UE to play a ring tone to indicate that the remote terminal is ringing and waiting to be answered. If the destination telephone number is a second UE then the Alerting message is usually sent after the RAB have been established for both the originating and terminating UE (shown in Figure 8.23 for a video call). Some core networks may send the Alerting message prior to establishing the RAB at the terminating UE to

```
ul-ChannelRequirement                    positionFixedOrFlexible: fixed
ul-DPCH-Info                              tfci-Existence: false
  modeSpecificInfo fdd                  dl-InformationPerRL-List
    scramblingCodeType: longSC          dl-InformationPerRL-List value 1
    scramblingCode: 10685                 modeSpecificInfo fdd
    spreadingFactor: sf64                 primaryCPICH-Info
    tfci-Existence: true                  primaryScramblingCode: 106
    puncturingLimit: p10-68             dl-DPCH-InfoPerRL fdd
modeSpecificPhysChInfo fdd:               pCPICH-UseForChEst: mayBeUsed
dl-CommonInformation                      dpch-FrameOffset: 136
dl-DPCH-InfoCommon                        dl-ChannelisationCodeList
  cfnHandling: maintain                   dl-ChannelisationCodeList 1
  modeSpecificInfo fdd                      sf-AndCodeNumber
    powerOffsetPilot-pdpdch: 0              sf128: 7
    spreadingFactorAndPilot             tpc-CombinationIndex: 0
      sfd128: pb4
```

Log File 8.38   Radio Bearer Setup message (part 4)

```
UL-DCCH-Message
integrityCheckInfo
messageAuthenticationCode: 1C 15 87 8C
rrc-MessageSequenceNumber: 2
radioBearerSetupComplete
rrc-TransactionIdentifier: 0
  start-Value: 00 00 4
  count-C-ActivationTime: 32
```

Log File 8.39   Radio Bearer Setup Complete message

```
RANAP-PDU
RAB-AssignmentResponse
RAB-SetupOrModifiedList
  RAB-SetupOrModifiedItem
   - rAB-ID: '00000001'B
```

Log File 8.40   RANAP RAB Assignment Response message

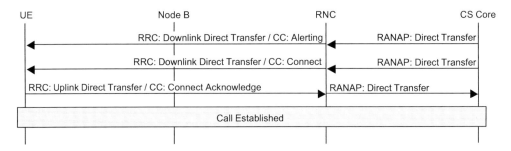

Figure 8.12   Signalling for mobile originated speech call (part 3)

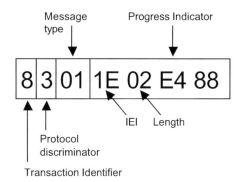

Figure 8.13 Structure of a downlink Alerting message

generate the impression that connection establishment is progressing more rapidly. The structure of an example Alerting message is illustrated in Figure 8.13.

The transaction identifier matches the values within the Setup and Call Proceeding messages (after accounting for the direction flag occupying the most significant bit). This indicates that all of these messages belong to the same signalling sequence. The Alerting message includes an optional progress indicator to provide a status update. The example shown in Figure 8.13 indicates that the remote user is served by a public network and that 'in-band' information or an appropriate pattern is now available. The core network can use the newly established RAB to transfer 'in-band' information to the UE. This signalling mechanism can be used to provide the ring tone played by the originating UE. Otherwise, the originating UE itself generates the ring tone.

The originating UE then continues to play the ring tone until either the call is answered at the remote end or a supervision timer expires. Supervision timers are used within both the UE and network to avoid playing the ring tone for prolonged periods. Once these timers expire, the originating UE may be directed towards an answering machine or the connection establishment attempt may be cleared. If the call is answered at the remote end, the core network can proceed to send a Connect message to the UE. The structure of an example Connect message is illustrated in Figure 8.14.

Similar to the Alerting message, this example of a Connect message includes an optional progress indicator. In this case, it specifies that the local user is served by a public network and that the call is end-to-end across the PLMN/ISDN. The originating UE then completes the call establishment procedure by returning a Connect Acknowledge message to the core network. The structure of a

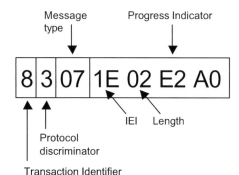

Figure 8.14 Structure of a downlink Connect message

Message
type

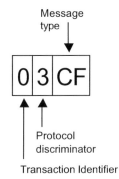

Protocol
discriminator

Transaction Identifier

Figure 8.15   Structure of an uplink Connect Acknowledge message

Connect Acknowledge message is shown in Figure 8.15. This message never includes any content other than the transaction identifier, protocol discriminator and message type.

The Connect Acknowledge message completes the connection establishment procedure. Speech data can then be transferred both to and from the originating UE.

8.2.2 Mobile Terminated

A mobile terminated speech call is initiated when the core network receives a connection request for a UE registered within one of its location areas. This triggers the core network to generate a RANAP Paging message and to forward that message to all RNC which have cells belonging to the relevant location area. The Paging message is forwarded to a single RNC if the strategy for location area planning has been based upon a single RNC per location area. Allowing location areas to span multiple RNC reduces the requirement for location area updates resulting from mobility, but increases the network paging load. Figure 8.16 illustrates the core network transferring a RANAP paging message to a single RNC as part of the first phase of signalling for a mobile terminated speech call.

An example RANAP Paging message is presented in Log File 8.41. It starts by specifying the core network domain in which the message was generated. It then addresses the UE using a mandatory permanent identity and an optional temporary identity. The permanent identity is always an IMSI whereas the temporary identity can be either a TMSI (CS connections) or P-TMSI (PS connections). The RNC uses the IMSI to identity whether or not the UE is already in RRC Connected mode with a signalling connection to another core network domain. For example, a paging message could be received from the CS core network domain while the UE is in CELL_PCH with a signalling connection to the PS core network. The IMSI can also be used to address the UE within the subsequent RRC Paging Type 1 message. However, it is preferable to use a temporary identity within the RRC message to help maintain user confidentiality. The RANAP Paging message also specifies the location area within which the UE is registered. This information is optional and excluding it allows the RNC to deduce that the paging message should be broadcast across the whole RNC area (unless the IMSI has already identified the location of the UE). The final section of the Paging message specifies a cause value. In this example, the cause value indicates a terminating conversational call. At this stage there is no distinction between a speech call and a video call.

The RNC reads the information from within the RANAP Paging message and uses it to generate an appropriate RRC paging message. The RNC generates a Paging Type 2 message if the IMSI has shown that the UE is already in CELL_DCH or CELL_FACH with a signalling connection to the PS core network domain. The Paging Type 2 message is transferred directly to the UE because the RNC has

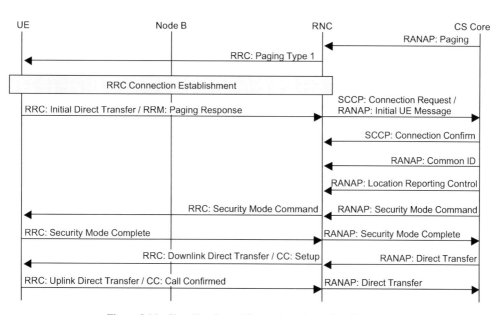

Figure 8.16 Signalling for mobile terminated speech call (part 1)

knowledge of either its active set cells in CELL_DCH, or the cell upon which it is camped in CELL_FACH. The RNC generates a Paging Type 1 message if it has identified that the UE is in CELL_PCH or URA_PCH, or if it believes that the UE is in RRC Idle mode. If the RANAP Paging message has been sent to multiple RNC because the location area spans more than a single RNC, the UE can be in RRC Connected mode with one RNC without the other RNC having any knowledge of its RRC state. In this case, the RNC to which the UE is connected can transfer a paging message directly towards the UE while the other RNC have to broadcast paging messages from all cells belonging to the relevant location area. The example illustrated in Figure 8.16 assumes that the UE is in RRC Idle mode when the RNC receives the RANAP Paging message. An example Paging Type 1 message is shown in Log File 8.42.

This example Paging Type 1 message includes two paging records. 3GPP TS 25.331 specifies that a single Paging Type 1 message can include up to eight paging records, depending upon the capacity (transport block size) of the PCH transport channel. The RANAP Paging message is only able to

```
RANAP-PDU
Paging
CN-DomainIndicator: cs-domain
PermanentNAS-UE-ID
 - iMSI: '32199059000066F1'H
TemporaryUE-ID
 - tMSI: '1400001F'H
PagingAreaID
 lAI
  - pLMNidentity: '32F499'H
  - lAC: '7779'H
PagingCause: terminating-conversational-call
```

Log File 8.41 RANAP Paging message

```
PCCH-Message
pagingType1
pagingRecordList

pagingRecordList value 1
  cn-Identity
  pagingCause: terminatingConversationalCall
  cn-DomainIdentity: cs-domain
  cn-pagedUE-Identity
    tmsi-GSM-MAP: 14 00 00 1F

pagingRecordList value 2
  cn-Identity
  pagingCause: terminatingInteractiveCall
  cn-DomainIdentity: ps-domain
  cn-pagedUE-Identity
    p-TMSI-GSM-MAP: 36 00 FF 00
```

Log File 8.42   Paging Type 1 message

accommodate a single paging request so the RNC becomes responsible for grouping paging messages to transmit within each Paging Type 1 message. This grouping must account for the timing of the DRX cycles used by each UE. In this example, the first paging record corresponds to the RANAP Paging message shown in Log File 8.41. The information within this record reflects the content of the RANAP message except that the IMSI and location area identity information has been removed. If the UE had been in CELL_PCH or URA_PCH the TMSI would have been replaced by the U-RNTI. The second paging record shows the PS core network paging a UE to establish an interactive data connection. This could correspond to a push-email application transferring an email header to a UE registered with the PS core network.

Reception of a Paging Type 1 message triggers the UE to initiate the RRC connection establishment procedure. This procedure is described in detail within section 8.1. The RRC Connection Request message specifies a cause value of 'terminating conversational call', i.e. it reflects the value within the paging message. The UE sends a radio resource management Paging Response to the CS core network as soon as it has established the RRC connection. This is in contrast to the mobility management CM Service Request which is sent as part of the signalling for a mobile originated connection. Mobility management messages are specified within 3GPP TS 24.008 whereas radio resource management messages are specified within 3GPP TS 44.018. SRB 3 is used to transfer the Paging Response to the RNC within an RRC Initial Direct Transfer message. The RNC then uses a RANAP Initial UE message to forward the Paging Response to the core network. An example Paging Response is shown in Figure 8.17.

The content of this message is similar to the content of the CM Service Request shown in Figure 8.5. The 0.5 bytes of spare capacity within the Paging Response replaces the CM service type within the CM Service Request. Reception of the Initial Direct Transfer message containing the Paging Response triggers the RNC to establish an Iu–cs signalling connection. This procedure is the same as that for a mobile originated connection and is described in Section 8.2.1. Likewise, the RANAP Common Identity and Location Reporting Control messages, and the security mode procedure are the same as for a mobile originated connection.

The core network then proceeds to forward a call control Setup message to the UE. This is in contrast to the mobile originated scenario in which case the UE is responsible for sending the Setup message. An example mobile terminated Setup message is presented in Figure 8.18.

The mobile terminated Setup message uses an identifier of 05 rather than the corresponding mobile originated identifier of 45. The two most significant bits of the mobile originated identifier represent a

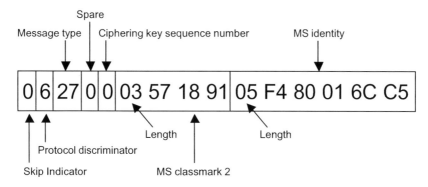

Figure 8.17 Structure of a Paging Response message

send sequence number. The core network can use this sequence number to identify a duplicate message if it receives more than a single copy of the same message type. The bearer capability informs the UE of the requested connection type. This allows the UE to recognise that a speech call has been requested rather than a video call. The information content from this example is presented in Log File 8.43.

The last part of the Setup message specifies the telephone number belonging to the originator of the speech call. This information allows the terminating end-user to recognise the caller prior to answering the phone. The third byte within this information element defines the type of number and numbering plan. The telephone number itself starts in the fourth byte using Binary Coded Decimal (BCD). The pair of digits belonging to each byte are read from right to left so in this example the telephone number is *pq rs tu vw xy z* (in practice these characters would be replaced by decimal digits).

The UE acknowledges the Setup message by sending a Call Confirmed message. This represents the uplink equivalent of the Call Proceeding message. An example Call Confirmed message is presented

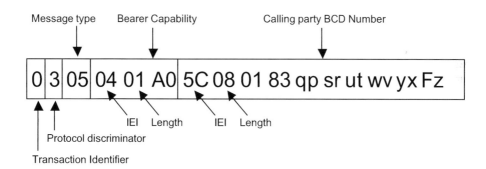

Figure 8.18 Structure of a mobile terminated Setup message (the actual BCD telephone number has been replaced by 'qpsrutwvyxz', in practice, this section of the message would include the originating telephone number)

```
Radio  channel  requiment  MS  to  NW  direction:  Full  rate  support  only  MS/full
rate speech version 1 supported
Coding standard: GSM standardized coding
Transfer mode: circuit mode
Information transfer capability: speech
```

Log File 8.43 Bearer Capability information

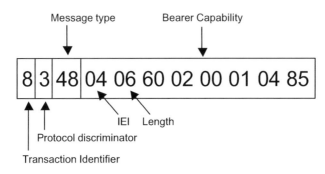

Figure 8.19   Structure of a Call Confirmed message

within Figure 8.19. The bearer capability information element is included to inform the core network of the UE capability, e.g. which speech codecs are supported.

The core network then proceeds to initiate RAB establishment. This phase of establishing a mobile terminated connection is identical to the equivalent phase for establishing a mobile originated connection. RAB establishment is described in section 8.2.2 for a mobile originated connection and is illustrated in Figure 8.9.

Once RAB establishment has been completed, the radio access network and the Iu transport connection are both ready to start transferring speech data. However, the originating and terminating end-users remain unconnected. Figure 8.20 illustrates the call control signalling used to connect the two users.

The terminating UE forwards a call control Alerting message to the core network. This indicates that the terminating UE is notifying the end-user of the incoming call, e.g. by playing its ringing tone or by vibrating. An example uplink Alerting message is shown in Figure 8.21. Similar to the Setup message, the uplink version of the Alerting message includes a send sequence number within the two most significant bits of the message type identifier.

The terminating UE continues to notify the end-user of the incoming call until the end-user either accepts or rejects the call, or a supervision timer expires. Supervision timers are used within both the UE and network to avoid the terminating UE attempting to notify the end-user for prolonged periods of time. Once these timers expire, the terminating UE records a missed call. If the end-user accepts the incoming call, the UE can proceed to send a Connect message to the core network. The core network subsequently responds using a Connect Acknowledge message. Example Connect and Connect Acknowledge messages are shown in Figure 8.22.

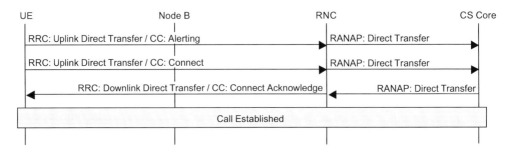

Figure 8.20   Signalling for mobile terminated speech call (part 3)

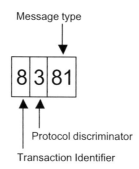

Figure 8.21   Structure of an uplink Alerting message

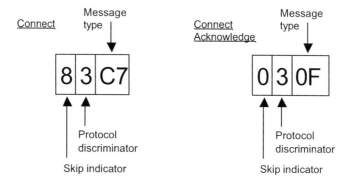

Figure 8.22   Structure of uplink Connect and downlink Connect Acknowledge messages

The Connect Acknowledge message completes the connection establishment procedure. Speech data can then be transferred both to and from the terminating UE.

8.3 Video Call Connection Establishment

- Video call connections are usually established between two mobiles and connection establishment is mobile-to-mobile. This requires the mobile originated and mobile terminated connection establishment procedures to be completed in parallel.
- The message types used to establish a video call connection are the same as those used to establish a speech call, i.e. the message flow for a mobile-to-mobile video call is the same as that for a mobile-to-mobile speech call. However, the content of the messages differ.
- The Setup message from the originating UE specifies the requirement for a 64 kbps Unrestricted Digital Information (UDI) connection using H.223 and H.245. The Setup message also includes the telephone number of the terminating UE.
- UDI is an ISDN term used to describe a data connection which supports the transmission of any bit pattern. The terms UDI and video call are often used synonymously within the context of UMTS. The ITU-T Recommendation H.223 specifies the multiplexing of video, audio and control data whereas, H.245 specifies the control plane signalling for connection establishment at the higher layers.

- The terminating UE is paged after the core network has received the Setup message from the originating UE. The core network subsequently forwards a Setup message to the terminating UE. This message requests a 64 kbps UDI connection using H.223 and H.245.
- RAB are established for both the originating and terminating UE after the terminating UE has acknowledged the downlink Setup message using a Call Confirmed message. RAB establishment is the same for both the originating and terminating connections.
- RAB establishment includes an NBAP synchronised radio link reconfiguration procedure. It also includes ALCAP signalling across both the Iub and Iu–cs to establish AAL2 transport connections for the user plane video call data.
- The video call service does not require RAB subflows, i.e. there is only a single flow of data. This is in contrast to the speech service which requires three RAB subflows. The video call data uses a single logical channel and a single transport channel.
- The video call service transport format combination set includes four transport format combinations. These correspond to video call activity and video call inactivity, each with and without SRB activity.
- Alerting, Connect and Connect Acknowledge messages are used in the same way as for the speech service, i.e. to indicate when the terminating UE is ringing and when the terminating UE has been answered.
- Video content is not transferred immediately after the Connect Acknowledge messages. The higher layers have to complete their own signalling procedures prior to transferring any video content, e.g. the H.245 protocol has to negotiate the video codec.
- Key 3GPP specifications: TS 25.331, TS 25.413, TS 25.415, TS 24.008 and TS 44.018.
- Key ITU Recommendations: ITU-T Rec H.223, Rec H.245, Rec H.324.

8.3.1 Mobile Originated and Mobile Terminated

Video call connections are usually established between two mobiles and connection establishment is mobile-to-mobile rather than mobile-to-fixed line. In this case, the mobile originated and mobile terminated connection establishment procedures are completed in parallel. The interaction of these procedures is illustrated in Figure 8.23. This figure can also be used to represent the equivalent signalling procedure for a mobile-to-mobile speech call. Only the message content differentiates a speech call establishment procedure from a video call establishment procedure.

The originating UE initiates the call by establishing an RRC connection and moving into RRC Connected mode. This procedure is described in detail within Section 8.1. The RRC Connection Request message specifies a cause value of 'originating conversational call', i.e. the same value as for a speech connection. The originating UE then proceeds to forward a mobility management CM Service Request to the CS core network. The content of this message is the same as that used for a speech connection. The RNC establishes an Iu–cs signalling connection when forwarding the CM Service Request to the core network. The SCCP messages used to establish the signalling connection are shown in Section 8.2.1. The core network then sends Common Identity and Location Reporting Control messages to the RNC. These messages have been excluded from Figure 8.23 to help simplify the diagram. The Common Identity message provides the RNC with the permanent identity of the UE whereas the Location Reporting Control message instructs the RNC to inform the core network whenever the UE changes service area. The core network then proceeds to enable integrity protection and ciphering by completing the security mode procedure.

The originating UE sends a call control Setup message once the security mode procedure has been completed. This is the first message that indicates a video connection is being requested rather than a speech connection. Figure 8.24 illustrates an example of a Setup message generated by the originating

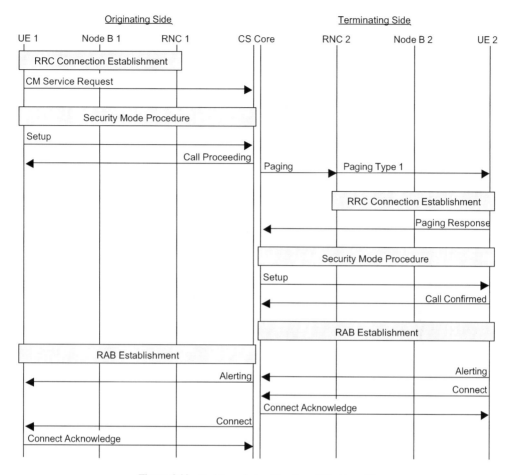

Figure 8.23   End-to-end signalling for a CS video call

UE of a video call. The rules for defining the transaction identifier, protocol discriminator and message type are the same as those for a speech connection.

The bearer capability belonging to a video call can be compared to the equivalent capability for a speech call. Log File 8.26 presents the speech call version whereas Log File 8.44 presents the video call version. The information transfer capability is specified as 'speech' for the speech call and 'Unrestricted Digital Information' (UDI) for the video call. UDI is an ISDN term used to describe a data connection which supports the transmission of any bit pattern. This is in contrast to Restricted Digital Information (RDI) connections which do not allow the transmission of all bit patterns, i.e. the 8 bit zero bit pattern is not permitted. In general, UDI describes the connection type rather than the application. However, UMTS uses the UDI connection type to transfer video call data so the terms UDI and video call are often used synonymously within the context of UMTS. A UDI bit rate of 64 kbps is specified lower down the capability information. The video call bearer capability information also specifies the use of ITU-T Recommendations H.223 and H.245. These recommendations are introduced by Recommendation H.324 which describes a terminal for low bit rate multimedia communication. H.223 specifies the multiplexing of video, audio and control data whereas H.245 specifies the control

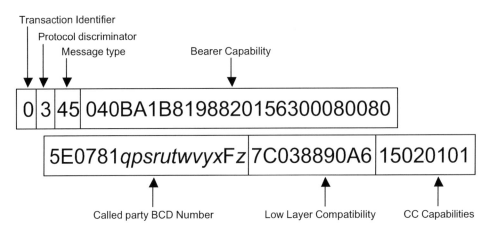

Figure 8.24   Structure of a mobile originated Setup message for a video connection

```
Radio channel requiment MS to NW direction: Full rate support only MS
Coding standard: GSM standardized coding
Transfer mode: circuit mode
Information transfer capability: unrestricted digital information
Data Compression (MS to NW): not allowed
Structure: unstructured
Dublex mode: full duplex
Configuration: point-to-point
Negotiation of intermediate rate requested: No meaning
Establishment: demand
Access identity: octet identifier
Rate adaption: Other rate adaption
Signalling access protocol: I.440/450
Other ITC: value ignored
Other rate adaption: H.223 & H.245
Layer 1 identity: octet identifier
User information layer 1 protocol: default layer 1 protocol
Synchronous
Stop bits: 1 bit (this value is also used in the case of synchronous mode)
Negotiation: in-band negotiation not possible
Data bits: 8 bits (this value is also used for bit oriented protocols)
User rate: 9.6 kbit/s Recommendation X.1 and V.110
Intermediate rate: 16 kbit/s
Network independent clock (NIC) on transmission doesn't require with NIC
Network independent clock (NIC) on reception cannot accept data with NIC
Parity information: none
Connection element: transparent
Modem type: none
Other modem type: no other modem type specified in this field
Fixed network user rate: 64.0 kbit/s bit transparent
Accept TCH/F14.4 channel coding (MS to NW): not acceptable
Accept TCH/F9.6 channel coding (MS to NW): not acceptable
Accept TCH/F4.8 channel coding (MS to NW): not acceptable
Maximum number of TCH (MS to NW): 1
User initiated modification indication: not allowed/required
Wanted air interface user rate: not applicable/no meaning
```

Log File 8.44   Bearer Capability information

```
Coding Standard: ITU-T recommendation and TTC standard
Information transfer capability: Unrestricted digital information
Transfer mode: circuit mode
Information transfer rate 64 kbit/s
Layer 1 identifier 1
User information layer 1 protocol Reserved
```

Log File 8.45   Low layer compatibility information

plane signalling for connection establishment at the higher layers. The bearer capability information does not specify the video codec because this is negotiated by the higher layers once the connection has been established.

The destination telephone number follows the bearer capability within the Setup message. This is coded in the same way as for a speech call. The video call Setup message then includes the low layer compatibility information presented in Log File 8.45. This information is subsequently forwarded to the remote terminal (within the mobile terminating Setup message) to specify compatibility requirements for the video call connection. The content is similar to sections of the bearer capability information, i.e. the remote terminal must support a 64 kbps UDI connection.

The core network acknowledges the mobile originated Setup message using a Call Proceeding message. An example is shown in Figure 8.25. The bearer capability information included within this message reflects the content of the bearer capability within the preceding Setup message.

The core network then proceeds to contact the terminating UE. A paging message is broadcast using a cause value of 'terminating conversational call'. This triggers the terminating UE to establish an RRC connection and forward a Paging Response message to the core network. The content of the Paging Response is the same as that for a mobile terminated speech call. The RNC establishes an Iu–cs signalling connection at the same time as forwarding the Paging Response within a RANAP Initial UE message. The core network sends Common Identity and Location Reporting Control messages to the RNC before using the security mode procedure to enable integrity protection and ciphering.

Once the security mode procedure has been completed, the core network sends a call control Setup message to the terminating UE. This is the first message received by the terminating UE that indicates a video connection is being requested rather than a speech connection. An example mobile terminated Setup message for a video connection is shown in Figure 8.26. The bearer capability information reflects the capability requested within the mobile originated Setup message, i.e. a 64 kbps UDI connection using the H.245 and H223 protocols is requested. The mobile terminated Setup message also includes the telephone number of the caller and the low layer compatibility information forwarded from the originating UE.

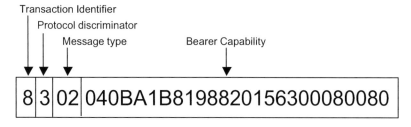

Figure 8.25   Structure of a Call Proceeding message for a video connection

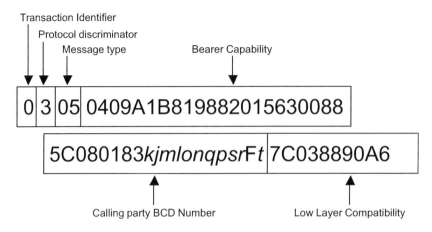

Figure 8.26   Structure of a mobile terminated Setup message for a video connection

The terminating UE acknowledges the Setup message by returning a call control Call Confirmed message. An example of this message is presented in Figure 8.27. This is the forth message to include bearer capability information. In this case, the terminating UE acknowledges that it can support the requested 64 kbps UDI bearer. The content is identical to that included within the Call Proceeding message. The Call Confirmed message also includes a section of Call Control (CC) Capability information, e.g. support for DTMF. The originating UE has already signalled its CC Capability information within the original Setup message.

The core network then proceeds to establish RAB for each of the originating and terminating UE. The RAB establishment procedure is identical for each UE and the message types are the same as those used to establish a speech call. These are illustrated in Figure 8.9. The RAB Assignment Request specifies the Quality of Service (QoS) requirements. An example for a CS video call is presented in Log File 8.46. This can be compared with the equivalent speech message in Log File 8.28. The core network requests maximum and guaranteed bit rates of 64 kbps. This corresponds to the UDI bit rate specified within the call control Setup messages. The maximum SDU size is 640 bits and the lower layers are requested to deliver SDU in the correct order. The RAB Assignment Request includes only a single set of SDU parameters. It's not necessary for the video call message to include multiple sets of SDU parameters because the application generates only a single flow of SDU. This is in contrast to the

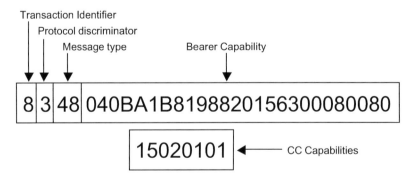

Figure 8.27   Structure of a Call Confirmed message for a video connection

```
RANAP-PDU
RAB-AssignmentRequest
RAB-SetupOrModifyList
firstValue
  - rAB-ID: '00000001'B
  rAB-Parameters
    trafficClass: conversational
    rAB-AsymmetryIndicator: symmetric-bidirectional
    maxBitrate
      - MaxBitrate: 64000
    guaranteedBitRate
      - GuaranteedBitrate: 64000
    deliveryOrder: delivery-order-requested
    maxSDU-Size: 640
    sDU-Parameters
      residualBitErrorRatio
        - mantissa: 1
        - exponent: 4
      deliveryOfErroneousSDU: no-error-detection-consideration
    transferDelay: 160
    allocationOrRetentionPriority
      - priorityLevel: lowest
      - pre-emptionCapability: shall-not-trigger-pre-emption
      - pre-emptionVulnerability: not-pre-emptable
      - queuingAllowed: queueing-not-allowed
    sourceStatisticsDescriptor: unknown
  userPlaneInformation
    - userPlaneMode: transparent-mode
    - uP-ModeVersions: '0000000000000001'B
  transportLayerInformation
    - transportLayerAddress: '45 00 04 47 80 23 01 50 1F 00 00
      00 00 00 00 00 00 00 00 00'H
  iuTransportAssociation
    - bindingID: '0060FA12'H
secondValue
```

Log File 8.46   RANAP RAB Assignment Request for a CS video call

speech service which generates three subflows of SDU. An SDU error ratio is not included because the RAB Assignment Request specifies that error detection is not necessary. However, the residual BER requirement is specified to be 1×10^{-4}, i.e. 0.01%. The transfer delay requirement is specified to be 160 ms rather than the 100 ms specified for a speech call within Log File 8.28. The source statistics descriptor is included whenever the traffic class is either conversation or streaming. 3GPP TS 25.413 specifies two values for this information element, 'speech' and 'unknown'. Video calls use of the 'unknown' value.

The RNC initiates its admission control procedure after receiving the RAB Assignment Request. Similar to a speech call, this typically involves completing checks in terms of downlink transmit power, uplink interference, channelisation code availability and Iub transport capacity. If admission control denies the RAB request, a RAB Assignment Response is used to inform the core network (the same message type is used when RAB establishment is successful, but the message content is different). If admission control accepts the RAB request, the RNC initiates the NBAP synchronised radio link reconfiguration procedure. This procedure reconfigures the radio link at the Node B to prepare for the transfer of user plane data.

The RNC starts by sending an NBAP Radio Link Reconfiguration Prepare message to the Node B (shown in Figure 8.9). This message informs the Node B of the new uplink and downlink Transport Format Combination Sets (TFCS). Introducing the video call connection increases the number of

```
DL-DPCH-Information-RL-ReconfPrepFDD
tFCS
- ctfc2bit: 0
- ctfc2bit: 1
- ctfc2bit: 2
- ctfc2bit: 3
dl-DPCH-SlotFormat: 13
- multiplexingPosition: flexible
- limitedPowerIncrease: not-used
```

Log File 8.47   Video call TFCS within a Radio Link Reconfiguration Prepare message

transport channels from 1 to 2, and increases the number of Transport Format Combinations (TFC) from 2 to 4. These TFC are signalled using calculated TFC (cTFC) values of 0, 1, 2 and 3. Log File 8.47 illustrates the section of the Radio Link Reconfiguration Prepare message which specifies these values for the downlink direction.

The Node B can use these cTFC values to identify the allowed TFC. The translation between a specific cTFC value and its corresponding TFC is based upon the intermediate variables L and P. These variables are defined for each transport channel after ordering them according to their transport channel identity. L is equal to the number of transport formats within the transport format set, whereas P is equal to the product of the L values belonging to the preceding transport channels. By default, the P value for the first transport channel is equal to 1. The L and P variables belonging to a video call connection are illustrated in Figure 8.28.

The cTFC values are then generated by summing the products of the appropriate Transport Format Indicator (TFI) and the corresponding value of P for each transport channel. This calculation is illustrated in Figure 8.29. There is always a one-to-one relationship between a specific cTFC value and the corresponding TFC.

The transport channel identities and Transport Format Sets (TFS) are specified within a subsequent section of the Radio Link Reconfiguration Prepare message. Log File 8.48 presents the section which

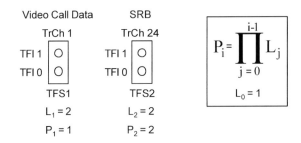

Figure 8.28   Variables used to calculate the cTFC for the video call service

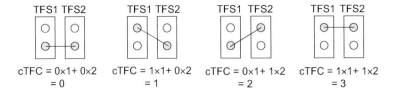

Figure 8.29   Calculation of cTFC values for the video call service

```
ul-TransportFormatSet                          dl-TransportFormatSet
dynamicParts                                   dynamicParts
- nrOfTransportBlocks: 0                        - nrOfTransportBlocks: 0
- nrOfTransportBlocks: 2                        - nrOfTransportBlocks: 2
- transportBlockSize: 640                       - transportBlockSize: 640
semi-staticPart                                semi-staticPart
- transTimeInterval: msec-20                    - transTimeInterval: msec-20
- channelCoding: turbo-coding                   - channelCoding: turbo-coding
- codingRate: third                             - codingRate: third
- rateMatchingAttribute: 182                    - rateMatchingAttribute: 182
- cRC-Size: v16                                 - cRC-Size: v16
```

Log File 8.48   Video call TFS within a Radio Link Reconfiguration Prepare message

defines the uplink and downlink TFS for the user plane transport channel, i.e. the transport channel responsible for transferring the video call application data.

The 64 kbps UDI connection throughput is achieved by transferring two 640 bit transport blocks every 20 ms. The transport block size is equal to the maximum SDU size specified within the RAB Assignment Request. This implies that transparent mode RLC is used and that neither the RLC nor MAC layers add any overhead. Rate 1/3 turbo coding is specified in combination with a rate matching attribute of 182. The rate matching attribute must be interpreted in combination with the rate matching attribute for the SRB transport channel. Although not shown in Log File 8.48, the SRB is configured with a rate matching attribute of 256. This means that the SRB transport channel would be allocated a larger share of the DPDCH if both transport channels generated an equal number of channel coded bits per 10 ms radio frame. In practice, the SRB transport channel generates fewer bits (transport block size of 148 bits and a TTI of 40 ms) so the SRB share of the DPDCH is scaled down. The use of rate matching attributes is described in Section 2.6.1. Log File 8.48 also indicates that CRC bits are included as part of the physical layer processing. These are included despite the RAB Assignment Request stating that error detection is not necessary. The RNC and UE can use these CRC bits to measure the BLER for outer loop power control. An outer loop power control BLER target can be derived from the residual BER requirement specified within the RAB Assignment Request.

The Radio Link Reconfiguration Prepare message also specifies an uplink spreading factor of 16 and a downlink slot format of 13. Downlink slot format 13 corresponds to a spreading factor of 32. It also allows downlink TFCI bits to be included as part of the DPCCH. The inclusion of downlink TFCI bits means that the UE is not required to complete blind detection of the transport format combination. This is consistent with the use of flexible transport channel locations indicated within Log File 8.47, i.e. blind detection requires fixed transport channel locations.

The Node B responds to the Radio Link Reconfiguration Prepare using a Radio Link Reconfiguration Ready message. This message includes the transport layer address and binding identity for the new Iub transport connection. The RNC uses this information when sending the subsequent ALCAP Establish Request to the Node B. The Establish Request specifies bit rates and SDU sizes applicable to a video call connection. The Node B responds using an Establish Confirm, and Frame Protocol control frames are then used to achieve synchronisation. The RNC proceeds to use the transport layer address and binding identity within the RAB Assignment Request to establish an Iu–cs transport connection for the user plane data. Once this has been completed, the end-user has two AAL2 connections across the Iub (SRB + user plane data) and a single AAL2 connection across the Iu–cs (user plane data).

The RNC then forwards an NBAP Radio Link Reconfiguration Commit message to the Node B. This message informs the Node B of the CFN during which the new configuration should become active. The RNC calculates the CFN by considering that the UE must be ready to apply the new radio link configuration at the same time as the Node B. The UE has not yet been informed of the new configuration so the CFN must allow sufficient time for the Radio Bearer Setup message to be transferred to the UE. The first part of the Radio Bearer Setup specifies the activation CFN already signalled to the Node B. The general format of the message is the same as that used for a speech connection. However, the radio bearer setup procedure for a video call configures only a single user plane logical channel and a single user plane transport channel. The equivalent procedure for a speech call configures three user plane logical channels and three user plane transport channels. Log File 8.49 presents the section of the Radio Bearer Setup message which configures the user plane logical and transport channels.

The user plane data is mapped onto radio bearer 5 which is configured to use transparent mode RLC in both the uplink and downlink directions. It is not necessary for the RLC layer to complete segmentation because the higher layer packets have the same size as the RLC PDU. Radio bearer 5 is mapped onto transport channel 1 and its uplink is allocated a priority of 2. This priority can range from 1 to 8, where 1 represents the highest priority. SRB are typically allocated a priority of 1 during RRC connection establishment. This helps to ensure that signalling information always has priority over user plane data. The priority is not configured for the downlink direction because it is important to the sender rather than the receiver, i.e. the RNC is configured with the equivalent downlink information. The RLC PDU size for transport channel 1 is configured using octet mode 1 and sizeType2 information. 3GPP TS 25.331 specifies that the corresponding PDU size is given by $(\text{part}1 \times 32) + 272 + (\text{part}2 \times 8) = 352 + 272 + 16 = 640$ bits. This is consistent with the transport block size specified within the Radio Link Reconfiguration Prepare message and confirms that the RLC and MAC layers do not add any overhead. The associated Transport Format Set (TFS) includes a first

```
rb-InformationSetupList                          numberOfTbSizeList
rb-InformationSetupList 1                         numberOfTbSizeList 1: zero
  rb-Identity: 5                                  numberOfTbSizeList 2: one
  rlc-Info                                        logicalChannelList: allSizes
    ul-RLC-Mode                                   semistaticTF-Information
      ul-TM-RLC-Mode                                channelCodingType
        segmentationIndication: false               convolutional: third
    dl-RLC-Mode                                      rateMatchingAttribute: 256
      dl-TM-RLC-Mode                                crc-Size: crc16
        segmentationIndication: false
  rb-MappingInfo                               ul-AddReconfTransChInfoList value 2
  rb-MappingInfo value 1                         ul-TransportChannelType: dch
    ul-LogicalChannelMappings                    transportChannelIdentity: 1
      ul-TransportChType dch: 1                  transportFormatSet
      rlc-SizeList: configured                   dedicatedTransChTFS
      mac-LogicalChannelPriority: 2               tti20 value 1
    dl-LogicalChannelMappingList                  rlc-Size
    dl-LogicalChMappingList 1                       octetModeType1
      dl-TransportChType dch: 1                     sizeType2
ul-AddReconfTransChInfoList                         part1: 11
                                                   part2: 2
ul-AddReconfTransChInfoList 1                   numberOfTbSizeList
  ul-TransportChannelType: dch                  numberOfTbSizeList 1: zero
  transportChannelIdentity: 24                  numberOfTbSizeList 2
  transportFormatSet                             small1: 2
  dedicatedTransChTFS                           logicalChannelList: allSizes
    tti40 value 1                               semistaticTF-Information
    rlc-Size                                      channelCodingType: turbo
      octetModeType1                              rateMatchingAttribute: 182
        sizeType1: 16                             crc-Size: crc16
```

Log File 8.49 Logical and transport channel information from a Radio Bearer Setup message

Transport Format (TF) based upon no transport blocks and a second TF based upon two transport blocks. These correspond to 0 and 64 kbps when considering the 20 ms TTI. In general, video call connections have an activity factor of 100% so the second TF is used continuously. Log File 8.49 also includes TFS information for the SRB. This could be reconfigured at this stage if the SRB had previously been using a bit rate of 13.6 kbps. The RLC PDU size for the SRB on transport channel 24 is specified using octet mode 1 and sizeType1 information. In this case, the RLC PDU size is given by $(8 \times \text{sizeType1}) + 16 = 128 + 16 = 144$ bits. The TFS allows a maximum of 1 transport block to be transferred during each 40 ms TTI. This generates a corresponding bit rate of 3.6 kbps at the top of the MAC layer and a bit rate of 3.4 kbps at the top of the RLC layer when assuming an 8 bit RLC header.

The RAB establishment procedure is completed when the RNC returns a RAB Assignment Response to the core network. This message reports that RAB establishment has been successful and that further call control signalling can proceed.

The final phase of call control signalling can be completed once RAB establishment has been completed for both the originating and terminating UE. The messages used during this phase are the same as those used for a speech call. The terminating UE forwards an Alerting message to the core network. This indicates that the terminating UE is notifying its end-user of the incoming call, e.g. by playing its ringing tone or by vibrating. This message is relayed to the originating UE as a downlink Alerting message. The originating UE then starts to play a ringing tone. The terminating UE continues to notify its end-user of the incoming call until the end-user either accepts or rejects the call, or a supervision timer expires. Supervision timers are used within both the UE and network to avoid the terminating UE attempting to notify the end-user for prolonged periods of time. Once these timers expire, the terminating UE records a missed call while the originating UE is either directed towards an answering machine or the connection establishment attempt is cleared.

If the terminating end-user accepts the incoming call, the terminating UE can proceed to send an uplink Connect message to the core network. The core network responds with a downlink Connect Acknowledge message, while forwarding a Connect message to the originating UE. Finally, the originating UE returns an uplink Connect Acknowledge message. At this point, the end-to-end connection has been established, but video content is not transferred between the two UE. The higher layers have to complete their own signalling procedures prior to transferring any video content, e.g. the H.245 protocol has to negotiate the video codec. The connection is established from the perspective of the end-users after the higher layers have completed their signalling procedures and video content is transferred between the two UE.

8.4 Short Message Service (SMS)

- An end-user initiates the mobile originated short message service by pressing the 'send' button after having compiled a message and selecting the relevant telephone number. If the UE starts from RRC Idle mode, an RRC Connection Request is sent using a cause value of 'Originating Low Priority Signalling'.
- The transfer of an SMS text message does not require any user plane radio bearers nor does it require an AAL2 connection across the Iu–cs, i.e. a text message is transferred to the RNC using the SRB and is transferred across the Iu–cs using the signalling connection.
- The SMS text message is sent as a NAS message within an Uplink Direct Transfer after the security mode procedure has been completed.
- The Short Message Application Layer (SM-AL) within the UE passes the text message to the Short Message Transfer Layer (SM-TL). The SM-TL codes the text message into a binary string and packages it within an SMS-Submit TPDU. Header information specifies the coding method, the target telephone number and the requirement for a status report.

- The SM-TL within the UE passes the SMS-Submit TPDU to the Short Message Relay Layer (SM-RL). The SM-RL generates an RP-Data packet by adding further header information. This includes the telephone number of the SMS message centre.
- The SM-RL within the UE passes the RP-Data packet to the Connection Management (CM) sublayer which generates a CP-Data packet by adding further header information. The CP-Data packet corresponds to the NAS PDU within the Uplink Direct Transfer.
- The originating UE receives two acknowledgements after sending the text message. The first is generated by the CM sublayer within the MSC to confirm that the CP-Data packet has been received successfully. The second is generated by the SM-TL within the SMS message centre to confirm that the SMS-Submit TPDU has been received successfully.
- An SMS message centre initiates the mobile terminated short message service after receiving an SMS-Submit TPDU. The message centre triggers an MSC to page the relevant UE.
- The SM-TL within the message centre packages the coded text message within an SMS-Deliver TPDU. Header information specifies the coding method, the originating telephone number and the requirement for a status report. The SM-RL within the message centre packages the SMS-Deliver TPDU within an RP-Data packet and forwards to the MSC.
- The CM sublayer within the MSC packages the RP-Data packet within a CP-Data packet. This corresponds to the NAS PDU within the Downlink Direct Transfer.
- The MSC receives an acknowledgement from the UE CM sublayer to confirm successful reception of the CP-Data packet. The message centre receives an acknowledgement from the UE SM-TL to confirm successful reception of the SMS-Deliver TPDU.
- Key 3GPP specifications: TS 24.008, TS 23.038, TS 23.040, TS 24.011.

8.4.1 Mobile Originated

An end-user initiates the mobile originated Short Message Service (SMS) by pressing the 'send' button after having compiled a message and selecting the relevant telephone number. RRC connection establishment is triggered if the UE is in RRC Idle mode. This procedure is described in detail within Section 8.1. 3GPP TS 24.008 specifies that the RRC Connection Request message should specify a cause value of 'Originating Low Priority Signalling'. The remainder of the signalling used to transfer a mobile originated SMS is presented in Figure 8.30. In contrast to the speech and video call signalling procedures, the transfer of an SMS does not require any user plane radio bearers nor does it require an AAL2 connection across the Iu–cs, i.e. the SMS is transferred to the RNC using the SRB and is transferred across the Iu–cs using the signalling connection.

The UE uses SRB 3 to send an Initial Direct Transfer message to the RNC. This message includes a mobility management CM Service Request. The content of the CM Service Request is the same as that described for a speech call within Section 8.2.1. The RNC extracts the CM Service Request and packages it within a RANAP Initial UE Message before forwarding it to the CS core network. The RNC also establishes an Iu–cs signalling connection using the SCCP signalling described in Section 8.2.1. The core network returns a RANAP Common Identity message and a RANAP Location Reporting Control message. These messages are also described in Section 8.2.1. The core network then initiates the security mode procedure to enable integrity protection and ciphering.

The UE is able to transfer the SMS data as soon as the security mode procedure has been completed. The text message is forwarded to the RNC as a NAS PDU within an Uplink Direct Transfer message. A series of protocol stack layers within the UE generate the NAS PDU from the original text message. The Short Message Application Layer (SM-AL) provides the text message to the Short Message Transfer Layer (SM-TL). This layer is responsible for generating a Transfer Protocol User Data

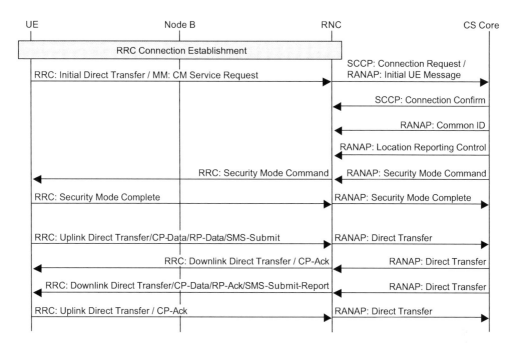

Figure 8.30   Signalling for mobile originated SMS (part 1)

(TP-User-Data) packet by coding the original text message into a binary string. 3GPP TS 23.038 specifies the set of coding methods which can be used for this translation. The SM-TL layer packages the coded text message into an SMS-Submit Transfer Protocol Data Unit (TPDU). The format of the SMS-Submit TPDU is specified within 3GPP TS 23.040. The header information specifies the coding method applied to the original text message. It also specifies the destination telephone number and whether or not an SMS status report is required. The peer SM-TL layer is within the SMS message centre, i.e. the SMS message centre uses this header information to direct the text message towards the appropriate UE and when requested, prompt that UE to provide a status report acknowledging reception of the message.

The SM-TL layer passes the SMS-Submit TPDU to the Short Message Relay Layer (SM-RL). The SM-RL layer generates a Relay Protocol Data (RP-Data) packet by adding further header information. The format of an RP-Data packet is specified within 3GPP TS 24.011. The header includes the telephone number of the SMS message centre. The peer SM-RL layer is within the MSC, i.e. the MSC uses the header information within the RP-Data packet to forward the text message towards the SMS message centre.

The SM-RL layer passes the RP-Data packet to the Connection Management (CM) sublayer. The CM sublayer generates a Control Protocol Data (CP-Data) packet by adding further header information. The CP-Data packet corresponds to the NAS PDU included within the Uplink Direct Transfer message. The header information added by the CM sublayer includes the usual NAS PDU information elements, i.e. protocol discriminator, transaction identifier and message type. The format of the CP-Data packet is specified within 3GPP TS 24.011.

The processing of an SMS text message between the SM-AL and RRC layers is summarised in Figure 8.31. The start of an example CP-Data packet is shown in Figure 8.32. This represents the NAS

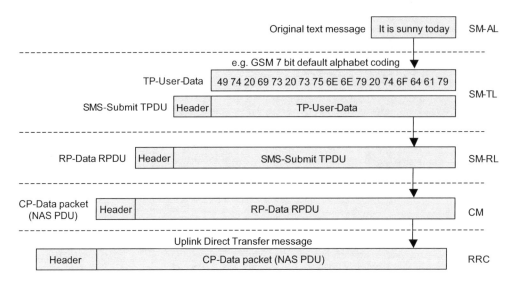

Figure 8.31 Coding and packaging of an SMS text message

PDU included within the Uplink Direct Transfer message. It encapsulates an RP-Data RPDU which then includes an SMS-Submit TPDU. The RNC extracts the CP-Data packet (NAS PDU) from within the Uplink Direct Transfer and forwards it to the core network using a RANAP Direct Transfer message. The CM sublayer within the MSC is able to read the header information to identify that it has received a CP-Data packet. The protocol discriminator value of 9 indicates SMS, whereas message type 1 indicates CP-Data.

The CM sublayer within the MSC extracts the RP-Data RPDU shown in Figure 8.33. The SM-RL layer within the MSC is then able to read the RP-Data header information to identify the telephone number of the relevant SMS message centre. The originator address within an uplink RP-Data packet is always signalled as zero. The SM-RL layer forwards the SMS-Submit TPDU towards the SMS message centre.

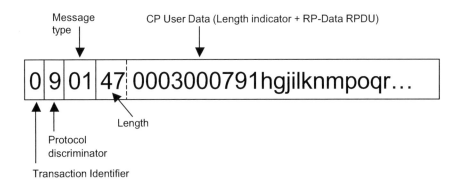

Figure 8.32 Example CP-Data / RP-Data / SMS-Submit packet (NAS PDU) (telephone number information has been replaced by a string of characters)

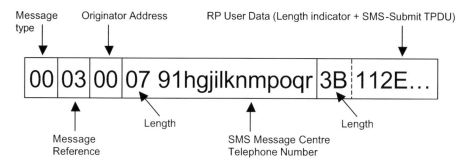

Figure 8.33   Example RP-Data / SMS-Submit packet

Figure 8.30 illustrates that the UE receives two acknowledgements after sending the CP-Data/ RP-Data/SMS-Submit packet. The first acknowledgement is generated by the CM sublayer within the MSC to confirm that the CP-Data packet has been successfully received. This acknowledgement is sent as a CP-Ack packet in contrast to a CP-Data packet. An example CP-Ack packet is illustrated in Figure 8.34. The protocol discriminator value of 9 indicates SMS, whereas message type 4 indicates CP-Ack.

The second acknowledgement is generated by the SM-TL layer within the SMS message centre. This acknowledgement is sent within a downlink CP-Data packet and is used to confirm that the SMS-Submit TPDU has been received successfully. In this case, an RP-Ack packet is used rather than an RP-Data packet. An example CP-Data packet which encapsulates an RP-Ack is shown in Figure 8.35.

The CM sublayer within the UE extracts the RP-Ack shown in Figure 8.36. The message type value of 3 indicates an RP-Ack. In this case, it is optional to include the RP User Data so an Information Element Identifier (IEI) is included. The RP User Data represents an SMS-Submit-Report. This acknowledgement does not represent the SMS status report which can be requested from the terminating UE, but represents an acknowledgement to confirm that the SMS message centre has received the SMS-Submit. The payload SMS-Submit-Report includes a time stamp recording the time at which the SMS-Submit was received by the SMS message centre.

The CM sublayer within the UE completes the mobile originating SMS signalling procedure by generating a CP-Ack to acknowledge reception of the preceding CP-Data packet. This acknowledgement has the same structure as the downlink CP-Ack shown in Figure 8.34. In this case, the CP-Ack is sent to the MSC.

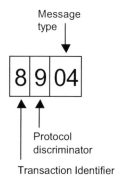

Figure 8.34   Example CP-Ack packet

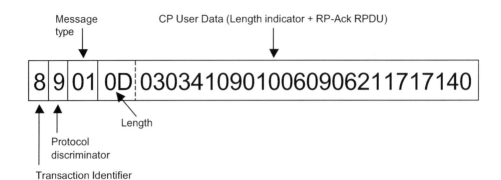

Figure 8.35   Example CP-Data / RP-Ack packet (NAS PDU)

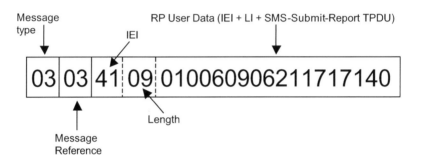

Figure 8.36   Example RP-Ack packet

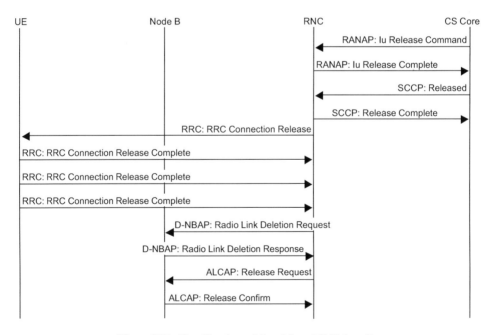

Figure 8.37   Signalling for mobile originated SMS (part 2)

The core network is able to initiate the connection release procedure after receiving the CP-Ack. This procedure is illustrated in Figure 8.37. The core network sends a RANAP Iu Release Command to the RNC. This message instructs the RNC to release all resources associated with the relevant Iu connection. The core network also sends an SCCP Released message to clear the Iu–cs signalling connection. The RNC acknowledges these messages using RANAP Iu Release Complete and SCCP Release Complete messages respectively.

The RNC proceeds to forward an RRC Connection Release message to the UE. This message specifies a value for the constant N308. The UE responds by returning a series of RRC Connection Release Complete messages. The number of messages returned is equal to N308+1, i.e. N308 defines the number of re-transmissions. The UE returns to RRC Idle mode once the RRC Connection Release Complete messages have been sent. The RNC proceeds to clear the resources at the Node B by sending an NBAP Radio Link Deletion Request. The RNC also clears the Iub transport connection for the SRB by sending an ALCAP Release Request. In the case of an SMS data transfer, neither the Iub nor Iu–cs have any user plane AAL2 connections to release.

8.4.2 Mobile Terminated

An SMS message centre initiates the mobile terminated Short Message Service (SMS) after receiving an SMS-Submit TPDU as part of the mobile originated SMS signalling procedure. The first phase of the mobile terminated procedure is illustrated in Figure 8.38. Similar to the mobile originated procedure, the SMS data is transferred using a combination of the SRB and Iu–cs signalling connection, i.e. user plane connections are not necessary. The core network pages the UE using a cause value of 'Terminating Low Priority Signalling'. This triggers the UE to establish an RRC connection using the same cause value. The RRC connection establishment procedure is described in detail within Section 8.1.

After establishing an RRC connection, the UE uses SRB 3 to send an Initial Direct Transfer message to the RNC. This message includes a radio resource management Paging Response. The content of the Paging Response is the same as that described for a speech call within Section 8.2.2. The RNC extracts the Paging Response and packages it within a RANAP Initial UE Message before forwarding it to the CS core network. The RNC also establishes an Iu–cs signalling connection using the SCCP signalling described in Section 8.2.1. The core network returns a RANAP Common Identity message and a RANAP Location Reporting Control message. These messages are also described in Section 8.2.1. The core network then initiates the security mode procedure to enable integrity protection and ciphering.

The core network is able to transfer the SMS data as soon as the security mode procedure has been completed. The SM-TL layer within the SMS message centre extracts the binary string representing the text message from within the SMS-Submit TPDU. The binary string is then packaged within an SMS-Deliver TPDU. The format of the SMS-Deliver TPDU is specified within 3GPP TS 23.040. The header information specifies the coding method applied to the original text message. It also specifies the originating telephone number, whether or not an SMS status report is required and the time stamp of when the SMS message centre received the message.

The SM-TL layer passes the SMS-Deliver TPDU to the SM-RL layer which generates an RP-Data packet by adding further header information. The format of an RP-Data packet is specified within 3GPP TS 24.011. The header includes the telephone number of the SMS message centre from where the SMS-Deliver TPDU is being sent. The SMS message centre then forwards the RP-Data packet to the relevant MSC. The CM sublayer within the MSC generates a Control Protocol Data (CP-Data) packet by adding further header information. The CP-Data packet corresponds to the NAS PDU included within the RANAP Direct Transfer message. The header information added by the CM sublayer includes the usual NAS PDU information elements, i.e. protocol discriminator,

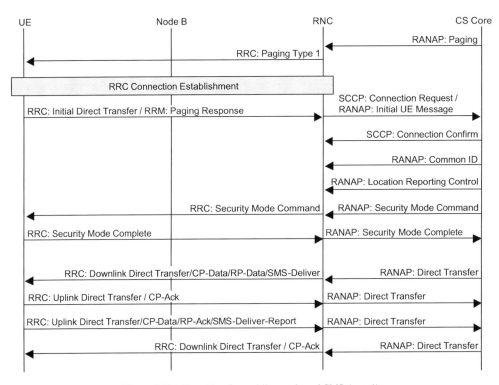

Figure 8.38   Signalling for mobile terminated SMS (part 1)

transaction identifier and message type. The format of the CP-Data packet is specified within 3GPP TS 24.011.

The NAS PDU is packaged within a RANAP Direct Transfer message and forwarded to the RNC using the Iu–cs signalling connection. The RNC extracts the NAS PDU and packages it within a Downlink Direct Transfer message before forwarding to the UE. The CM sublayer within the UE extracts the RP-Data packet from within the CP-Data packet (NAS PDU). The SM-RL layer extracts the SMS-Deliver TPDU while the SM-TL layer extracts the binary coded text message. The original text message is then decoded and passed to the SM-AL layer before being displayed to the end-user.

Figure 8.38 illustrates that the core network receives two acknowledgements after sending the CP-Data / RP-Data / SMS-Deliver packet. The first acknowledgement is generated by the CM sublayer within the UE to confirm that the CP-Data packet has been successfully received. This acknowledgement is sent as a CP-Ack packet. The second acknowledgement is generated by the SM-TL layer within the UE to confirm that the SMS-Deliver TPDU has been successfully received. This acknowledgement is sent within an uplink CP-Data / RP-Ack packet. The payload of the RP-Ack packet includes an SMS-Deliver-Report TPDU.

The CM sublayer within the MSC completes the mobile terminating SMS signalling procedure by generating a CP-Ack to acknowledge reception of the preceding CP-Data packet. The core network is then able to initiate the connection release procedure. This represents the second phase of signalling for a mobile terminating SMS. The release procedure is illustrated in Figure 8.37, i.e. it is the same as the release procedure for a mobile originating SMS.

8.5 PS Data Connection Establishment

- An end-user establishes a mobile originated PS data connection when using an application which requires data transfer through the PS core network, e.g. internet browsing or email.
- The RRC Connection Request cause value depends upon the initial state of the UE. The cause value will be registration if the UE has not already registered with the PS core network. Otherwise, the cause value can be 'originating interactive call', 'originating background call' or 'originating high-priority signalling'.
- The GPRS attach procedure must be completed if the UE has not already registered with the PS core network. This involves a GMM Attach Request/Attach Accept handshake.
- If the UE is already registered with the PS core network, the Attach Request message is replaced by a GMM Service Request message.
- The UE proceeds to forward an SM Activate PDP Context Request to the PS core network if a PDP context is not already active. A PDP context defines the QoS attributes associated with the data connection. The UE can request a specific QoS profile or can indicate that the core network should allocate the QoS profile to which the end-user has subscribed.
- The Activate PDP Context Request includes PDP address information which may correspond to an IP address. The UE can request an IP address from the core network by indicating that dynamic addressing is used.
- The Activate PDP Context Request also includes the Access Point Name (APN) and may include protocol configuration information for external network protocols, e.g. PAP information can be used to specify a user name and password for the APN.
- The SGSN interrogates the HLR to obtain the QoS profile to which the end-user has subscribed. The SGSN then forwards a RAB Assignment Request to the RNC. This includes QoS profile and user plane GPRS Tunnelling Protocol (GTP-U) information.
- The RNC and SGSN use GTP-U to provide user plane connectivity across the Iu–ps. The GTP-U protocol stack layer typically runs above UDP, IP, AAL5 and ATM. AAL5 avoids the requirement to use ALCAP signalling to establish and release transport connections.
- The GTP-U header includes a connection identity (GTP Tunnel Endpoint Identifier) which allows the RNC and SGSN to separate the data belonging to different connections.
- The RNC uses Admission Control to determine whether or not the RAB Assignment Request should be accepted. The RNC may allocate a finite bit rate and select between allocating a DPCH or HSDPA/HSUPA connection. Alternatively, the RNC can postpone the allocation of a finite bit rate until after RAB establishment.
- Postponing the allocation of a finite bit rate means that RAB establishment does not involve any NBAP nor ALCAP signalling but includes a radio bearer setup procedure to configure a 0/0 kbps DCH for the user plane radio bearer.
- After receiving the RAB Assignment Response, the SGSN completes a handshake with the GGSN and forwards an Activate PDP Context Accept to the UE. This message confirms the allocated QoS profile and specifies an IP address for the UE when necessary.
- If a finite bit rate has not been allocated during RAB establishment, the RNC can use traffic volume measurements to allocate an appropriate bit rate. The UE can report uplink RLC buffer occupancy while the RNC can monitor downlink RLC buffer occupancy.
- The initial capacity request often originates from the uplink irrespective of the direction of the subsequent application data transfer. This is typical of TCP applications which send an uplink SYN packet prior to transferring any application data.
- After receiving a capacity request and selecting the channel types, the RNC forwards an NBAP Radio Link Reconfiguration Prepare message to the Node B. The reconfiguration message includes the relevant extensions if HSDPA or HSUPA have been selected.

- The Node B does not apply the new radio link configuration immediately. The new configuration is stored until the RNC specifies the Connection Frame Number (CFN) during which it should become active. The Node B acknowledges the new configuration by returning an NBAP Radio Link Reconfiguration Ready message.
- The RNC initiates the establishment of the Iub transport connections by sending one or more ALCAP Establish Request messages to the Node B. Multiple connections are necessary if HSDPA or HSUPA is used.
- The RNC uses an NBAP Radio Link Reconfiguration Commit message to provide the Node B with the CFN during which the new configuration should be applied. The CFN is calculated by considering that the UE must apply the new configuration at the same time.
- The Radio Bearer Reconfiguration message provides the UE with the same activation CFN. This message also includes configuration information similar to what was provided to the Node B within the Radio Link Reconfiguration Prepare message.
- After receiving the Radio Bearer Reconfiguration message, the UE waits until the activation CFN occurs. It then applies the new configuration and returns a Radio Bearer Reconfiguration Complete message.
- Key 3GPP specifications: TS 24.008, TS 25.331, TS 23.107 and TS 25.211.

8.5.1 Mobile Originated

An end-user establishes a mobile originated PS data connection when using an application which requires data transfer through the PS core network, e.g. internet browsing, email transfer, instant messenger, file download and video streaming. The UE triggers RRC connection establishment if starting from RRC Idle mode. This procedure is described in detail within Section 8.1. The cause value within the RRC Connection Request message depends upon the initial state of the UE. If the UE has not already registered with the PS core network domain, i.e. the UE is not GPRS attached, then the cause value will be 'registration'. In this case, the GPRS attach signalling procedure has to be completed prior to any PS data transfer. If the UE is already registered with the PS core network, the cause value within the RRC Connection Request message can be 'originating interactive call', 'originating background call' or 'originating high-priority signalling'. 3GPP TS 24.008 states that the interactive and background causes should be used when requesting RAB establishment or PDP context activation, whereas the high-priority signalling cause should be used when modifying an existing PDP context.

The first phase of signalling used to establish a mobile originated PS data connection is presented in Figure 8.39. This figure assumes that the UE has not already registered with the PS core network so the GPRS attach procedure is included. After moving into RRC connected mode, the UE sends an Initial Direct Transfer message to the RNC. This RRC message is used to package a Non-Access Stratum (NAS) message addressed to the PS core network. It is transferred to the RNC using SRB 3, i.e. the high priority SRB used to transfer NAS messages with AM RLC.

The NAS PDU represents a GPRS Mobility Management (GMM) Attach Request message. An example is shown in Figure 8.40. The format of this message is specified within 3GPP TS 24.008. The first byte always includes 4 bits to signal the protocol to which the message belongs. A value of 8 indicates that the message belongs to the GMM protocol. GMM messages include a skip indicator which represents a section of padding. The second byte of the NAS message specifies the message type. A value of 1 for the GMM protocol corresponds to an Attach Request.

The Attach Request includes a section which specifies the MS network capability. This provides information regarding the GPRS capability of the UE. The message also includes a ciphering key sequence number and an indication of the attach type. The core network uses the ciphering key sequence number to identify the ciphering and integrity keys stored within the UE. The attach type identifies the signalling procedure as either a GPRS attach or a joint GRPS/IMSI attach. The attach

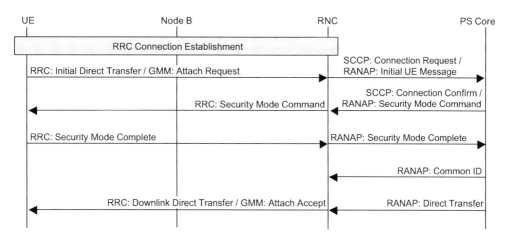

Figure 8.39 Signalling for mobile originated PS data connection (part 1 - option 1)

type also specifies whether or not the UE has data to transfer after the registration procedure has been completed. This avoids the core network from releasing the Iu connection while the UE wishes to transfer data. Decoding the attach type shown in Figure 8.40 generates the information in presented in Log File 8.50.

The Attach Request also includes a DRX parameter to indicate whether or not the UE supports DRX mode on GPRS. The UE identity is included after the DRX parameter using either a P-TMSI or IMSI. The UE includes a P-TMSI whenever it has one. Otherwise an IMSI is included. The example shown in Figure 8.40 includes a P-TMSI. This can be deduced from the three least significant bits of the second

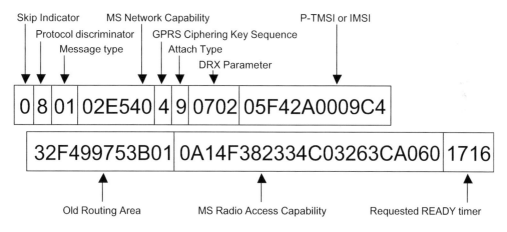

Figure 8.40 Structure of a GPRS mobility management Attach Request message

```
Attach Type (9 = 1001)
- GPRS attach (001)
- Follow-on request pending (1)
```

Log File 8.50 Attach type information from within an Attach Request message byte, i.e. a value of 100 indicates a P-TMSI whereas a value of 001 indicates an IMSI. The P-TMSI shown in Figure 8.40 has a value of 2A0009C4.

The UE also specifies the routing area that it has recorded. The Routing Area Identity (RAI) includes the Mobile Country Code (MCC), Mobile Network Code (MNC), Location Area Code (LAC) and Routing Area Code (RAC). If the UE does not have an RAI in memory it signals the LAC using a decimal value of 65534 (FFFE in hexadecimal). Figure 8.40 includes an MCC = 234, MNC = 99, LAC = 753B and RAC = 1. The MS Radio Access Capability information within the Attach Request provides further capability information, e.g. GPRS multi-slot class, HSCSD multi-slot class, support for UMTS FDD, support for UMTS TDD and support for CDMA2000. The final section of the Attach Request message specifies the GPRS READY timer. This timer is not applicable to UMTS.

The RNC forwards the NAS PDU representing the Attach Request to the PS core network using a RANAP Initial UE message. The RANAP message is packaged within an SCCP Connection Request message which is used to initiate the establishment of an Iu–ps signalling connection. An example SCCP Connection Request is shown in Log File 8.21. The core network extracts the NAS PDU from within the RANAP message, and responds to the SCCP Connection Request using an SCCP Connection Confirm. The Connection Confirm message includes a RANAP Security Mode Command. This allows the core network to initiate the security mode procedure at the same time as completing the Iu–ps signalling connection establishment. Log File 8.51 illustrates an example SCCP Connection Confirm which includes a Security Mode Command. Log File 8.22 illustrates an example which does not include a RANAP message.

The RNC forwards an RRC version of the Security Mode Command after receiving the RANAP version. The UE responds by returning an RRC Security Mode Complete which triggers the RNC to return a RANAP version to the core network. Completion of the security mode procedure allows integrity protection and ciphering to become active. The core network then proceeds to provide the RNC with a RANAP Common Identity message. This message informs the RNC of the permanent identity of the UE, i.e. the IMSI. An example Common Identity message is presented in Log File 8.23. The RNC can use the IMSI to help process incoming paging messages and also to support specific radio resource management functions, e.g. IMSI-based handover.

The core network completes the registration procedure by returning a GMM Attach Accept message. This is sent as a NAS PDU within a RANAP Direct Transfer message. An example Attach Accept is shown in Figure 8.41. In this case, the GMM message type is signalled using a value of 2 rather than 1. The force to standby information is not relevant to UMTS and should always indicate a value of 0.

The 'attach result' represents the equivalent of the 'attach type' within the Attach Request message. The attach result confirms that the UE is either GPRS attached, or both GPRS and IMSI attached. It

```
CC - CONNECTION CONFIRM
Destination Local Reference
- 551045h
Source Local Reference
- 000108h
Protocol Class
- 2h, connection oriented
SCCP User Data
- length: 56
- data :
00 06 00 34 00 00 03 00 0C 00 12 00 00 99 55 1A 62 E5 67 71 EC
6D A3 55 07 8B B1 E3 B3 00 0B 40 12 08 80 7D E3 85 C7 EC 81 89
EC 49 7F 21 3D 42 D2 0B EF 00 4B 00 01 00
```

Log File 8.51 SCCP Connection Confirm message (including a RANAP message)

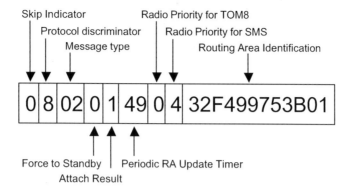

Figure 8.41 Structure of a GPRS mobility management Attach Accept message

also specifies whether or not the UE can proceed to establish a connection for data transfer. Decoding the attach result shown in Figure 8.41 generates the information in presented in Log File 8.52. The UE is GPRS attached and is able to proceed to establish a connection for data transfer.

The Attach Accept also specifies a timer for periodic routing area updates. This corresponds to timer T3312, which is equivalent to timer T3212 for periodic location area updates. T3212 is broadcast as part of the CS domain NAS message within SIB 1. The three most significant bits of the periodic routing area update timer define the units whereas the five least significant bits define the timer itself. A value of 49 indicates a timer of 9 decihours (54 minutes). Priorities are included for mobile originated Tunnelling of Messages 8 (TOM8) and SMS messages at the lower layers. Both priorities can range from 1 to 4, where 4 represents the lowest priority. A value of 0 is also interpreted as a value of 4 so the example in Figure 8.41 allocates a priority of 4 to both TOM8 and SMS data. The final section of the Attach Accept message specifies the current Routing Area Identifier (RAI). This may be the same as the RAI already stored by the UE. Otherwise, the UE uses this value to update its memory.

The UE becomes registered with the PS core network domain after receiving the GPRS Attach Accept message. It is not necessary for the UE to complete the registration procedure if the UE is already registered when the end-user initiates the PS data connection. The majority of UE allow the end-user to chose whether the GPRS attach procedure is completed by default when the UE is switched on, or is only completed when necessary, i.e. whenever the end-user wishes to transfer PS data. This may be shown within the user interface of the UE menu system as having a packet data connection 'when needed' or 'when available'. Figure 8.42 illustrates the first phase of signalling used to establish a mobile originated PS data connection when the UE is already registered with the PS core network domain.

In this case, the Initial Direct Transfer includes a NAS PDU which represents a GMM Service Request message rather than an Attach Request message. The structure of an example Service Request is shown in Figure 8.43. The protocol discriminator indicates GPRS mobility management while the message type indicates a Service Request.

```
           Attach Result
           - GPRS only attached
           - Follow-on proceed
```

Log File 8.52 Attach result information from within an Attach Accept message

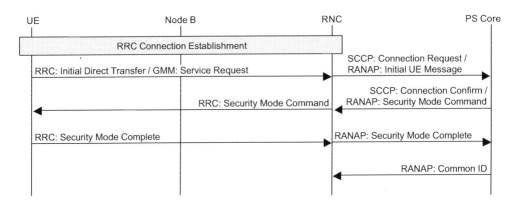

Figure 8.42   Signalling for mobile originated PS data connection (part 1, option 2)

The service type information indicates the reason for sending the Service Request. Cause values of signalling (0), data (1), paging response (2), MBMS Multicast Service Reception (4) and MBMS Broadcast Service Reception (4) can be included. The example shown in Figure 8.43 indicates a cause value of signalling. Signalling is used rather than data because the UE still has to complete the PDP context activation procedure prior to transferring any data. The Service Request also includes a ciphering key sequence number and a P-TMSI. The UE should always have a P-TMSI when sending the Service Request message because it has already registered with the PS core network.

The final part of the Service Request provides optional PDP context information. The first byte (32) represents the Information Element Identifier (IEI) whereas the second byte (02) represents the length indicator. The third and fourth bytes (0000) represent the Session Management (SM) state of specific PDP contexts. Each of the 16 bits corresponds to an individual Network layer Service Access Point Identifier (NSAPI). NSAPI are used to address specific PDP contexts. An individual UE can activate a maximum of 11 PDP contexts (either primary or secondary) and so the first five NSAPI are not used. Signalling a value of 1 for any of the remaining NSAPI indicates that it is already used and that the corresponding PDP context does not have an SM state of inactive. The value of 0000 indicates that none of the NSAPI are used and all PDP contexts are inactive.

The subsequent signalling within Figure 8.42 is the same as that within Figure 8.39 except that it is not necessary for the core network to respond with an Attach Accept message.

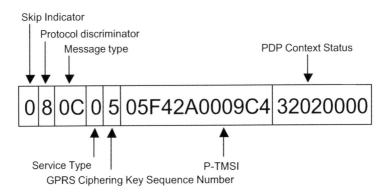

Figure 8.43   Structure of a GPRS mobility management Service Request message

The UE then proceeds to activate a PDP context to define the Quality of Service (QoS) attributes associated with the PS data connection. The signalling used to activate a PDP context is shown in Figure 8.44. This signalling procedure includes RAB establishment. It is possible that the UE is already GPRS attached and already has an active PDP context when the end-user initiates the PS data connection. In that case, the signalling within Figure 8.44 reduces to only the RAB Assignment and Radio Bearer Setup procedures, i.e. the Activate PDP Context messages are not necessary.

The UE uses an Uplink Direct Transfer to forward a session management Activate PDP Context Request to the PS core network. The content of an example Activate PDP Context Request is shown in Log File 8.53. The UE allocates NSAPI 5 to the PDP context. NSAPI numbering starts from 0 so this represents the sixth NSAPI and the first available NSAPI (the first five NSAPI are not used). This NSAPI identifies the PDP context within the UE and the SGSN. The request message also allocates a SAPI for the Logical Link Control (LLC) layer. This SAPI can be allocated a value of 3, 5, 9 or 11 and is used to identify the PDP context within the LLC layer.

The Activate PDP Context Request proceeds to specify the QoS profile requested by the UE. In this example, the UE does not request any particular QoS attributes, but instead indicates that the core network should allocate the QoS profile to which the end-user has subscribed. This information is stored within the core network Home Location Register (HLR). The Activate PDP Context Request also includes PDP address information. The example in Log File 8.53 does not include a specific address, but specifies that dynamic addressing should be used with IPv4, i.e. the UE requests the network to assign it with an IP address. The Access Point Name (APN) is specified using a hexadecimal string based upon a combination of length indicators and ASCII coded characters. The first byte (03) represents the length of the first section of the APN (74 6F 70). The first byte following the first section (04) represents the length of the second section (75 6D 74 73), and so on. Decoding the ASCII in Log File 8.53 generates an APN of 'top.umts.base3'.

The final section of the Activate PDP Context Request message provides protocol configuration information for external network protocols. In this example, information is signalled to the Password Authentication Protocol (PAP) and to the IP Control Protocol (IPCP). PAP information provides a user name and password for the APN. This information is ASCII coded within the hexadecimal string. The RRC message used to transfer the Activate PDP Context Request is ciphered, but there is no further encryption once the NAS PDU has been extracted, i.e. the user name and password can be extracted by inspection. The PAP message represents an Authenticate Request which has a 4 byte header. The next byte (04) represents a length indicator for the user name. The user name (75 6D 74 73) is followed by a length indicator for the password (08). Decoding the ASCII in Log File 8.53 generates the user name and password combination of 'umts' and 'balloons'.

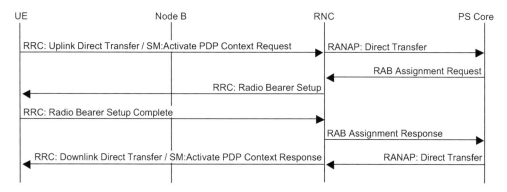

Figure 8.44   Signalling for mobile originated PS data connection (part 2)

```
ACTIVATE PDP CONTEXT REQUEST
- Transaction identifier: 0
Network Serv. Access Point Id
- NSAPI 5
LLC Serv. Access point Id
- SAPI 3
Quality of Service
- Reliab. class:                    Subscribed
- Delay class:                      Subscribed
- Precedence class:                 Subscribed
- Peak throughput:                  Subscribed
- Mean throughput:                  Best effort
- Traffic class:                    Subscribed
- Delivery order:                   Subscribed
- Delivery of erroneous SDUs:       Subscribed
- Maximum SDU size:                 Subscribed
- Maximum uplink bit rate:          Subscribed
- Maximum downlink bit rate:        Subscribed
- Residual BER:                     Subscribed
- SDU error ratio:                  Subscribed
- Transfer delay:                   Subscribed
- Traffic handling priority:        Subscribed
- Guaranteed uplink bit rate:       Subscribed
- Guaranteed downlink bit rate: Subscribed
Packet Data Protocol Address
- Dynamic PDP addressing is applied
- PDP type organisation: IETF allocated address
- PDP type number: IPv4
Access Point Name: 03 74 6F 70 04 75 6D 74 73 05 62 61 73 65 33
Protocol Conf. Options
- Configuration protocol: IP PDP type
- Password Auth. Protocol:
    01E90012 04 756D7473 08 62616C6C6F6F6E73
- IP Control Protocol: 01020010810600000000830600000000
```

Log File 8.53   Activate PDP Context Request message

Once the SGSN has received the Activate PDP Context Request, the HLR is interrogated to obtain the QoS profile to which the end-user has subscribed. This interrogation is completed within the core network and is not shown in Figure 8.44. The SGSN is then able to generate a RANAP RAB Assignment Request. An example RAB Assignment Request is presented in Log File 8.54. The RAB is allocated an identity which equals the NSAPI allocated by the UE.

The RAB parameters reflect the QoS profile received from the HLR. The traffic class is interactive rather than conversational, streaming or background. The interactive traffic class is intended for non-real time applications which have a higher priority than those using the background traffic class. The RAB is specified as bidirectional and asymmetric. This means that the maximum bit rate has to be specified separately for the uplink and downlink. In this example, the maximum downlink bit rate is 384 kbps whereas the maximum uplink bit rate is 64 kbps. The RNC should account for these maximum bit rates when configuring the radio bearer between itself and the UE. The interactive traffic class does not have a guaranteed bit rate. This is in contrast to the conversational traffic class used for the CS speech service (shown in Log File 8.28). The inclusion of a guaranteed bit rate for the conversational traffic class reflects its potential to have a relatively high QoS. 3GPP TS 23.107 specifies the set of QoS attributes associated with each traffic class.

The RAB Assignment Request also informs the RNC that the higher layers do not require in-order delivery. It is assumed that the higher layers have their own mechanism for reordering packets, e.g. the header belonging to a TCP packet includes a sequence number which can be used for re-ordering. This

```
RANAP-PDU
RAB-AssignmentRequest
RAB-SetupOrModifyList
firstValue
  - rAB-ID: '00000101'B
  rAB-Parameters
    trafficClass: interactive
    rAB-AsymmetryIndicator: asymmetric-bidirectional
    maxBitrate
      - MaxBitrate: 384000
      - MaxBitrate: 64000
    deliveryOrder: delivery-order-not-requested
    maxSDU-Size: 12000
    sDU-Parameters
      sDU-ErrorRatio
        - mantissa: 1
        - exponent: 4
      residualBitErrorRatio
        - mantissa: 1
        - exponent: 5
      deliveryOfErroneousSDU: no
    trafficHandlingPriority: 2
    allocationOrRetentionPriority
      - priorityLevel: 5
      - pre-emptionCapability: shall-not-trigger-pre-emption
      - pre-emptionVulnerability: pre-emptable
      - queuingAllowed: queueing-allowed
    relocationRequirement: none
  userPlaneInformation
    - userPlaneMode: transparent-mode
    - uP-ModeVersions: '0000000000000001'B
  transportLayerInformation
    - transportLayerAddress: '0A2A0631'H
    - iuTransportAssociation: gTP-TEI: 'F4500999'H
secondValue
  pDP-TypeInformation
    - PDP-Type: ipv4
    - dataVolumeReportingIndication: do-report
```

Log File 8.54   RAB Assignment Request message (PS data connection)

is in contrast to the CS speech service which requires higher layer packets to be delivered in order. Log File 8.54 indicates that the maximum SDU size is configured with a value of 12000 bits (1500 bytes). This represents a typical value for a PS data connection (based upon an Ethernet packet size) and is significantly greater than the speech service maximum SDU size of 244 bits. The RLC layer is responsible for segmenting the relatively large SDU into sections of typically 320 bits.

The SDU parameters define an SDU error ratio of 1×10^{-4} and a residual BER of 1×10^{-5}. In addition, the RNC is informed that SDU which include bit errors should be discarded rather than delivered to the higher layers. This assumes that the higher layers have their own mechanism for recovering lost data, e.g. TCP re-transmissions. The SDU error ratio of 0.01% is significantly less than the typical 1% transport block BLER target used by outer loop power control. The relationship between the SDU error ratio and the transport block BLER after RLC re-transmissions can be derived from the equality below:

$$(1 - \text{SDU\_Error\_Ratio})^{\text{N\_TB}} = (1 - \text{TB\_BLER})^{\text{N\_SDU}}$$

Where N_TB is the number of bits within a transport block and N_SDU is the number of bits within an SDU. This equality states that the probability of transferring (N_TB × N_SDU) bits without error in

N_TB SDU is equal to the probability of transferring the same number of bits without error in N_SDU transport blocks. Rearranging this equality generates the expression below:

$$TB\_BLER = 1 - (1 - SDU\_Error\_Ratio)^{\frac{N\_TB}{N\_SDU}}$$

Assuming a transport block size of 336 bits, an SDU size of 12000 bits and an SDU error ratio of 1×10^{-4} generates a transport block BLER of 3×10^{-6}. This corresponds to the residual transport block BLER after RLC re-transmissions. If the BLER for each transmission is 1% and re-transmissions are assumed to be independent then the BLER reduces to 1×10^{-6} after two re-transmissions. This figure is reasonably well aligned with the residual BLER.

The interactive traffic class does not include a transfer delay requirement, but includes a Traffic Handling Priority (THP). Only the interactive traffic class uses the THP QoS attribute. The THP can be allocated a value of 1, 2 or 3 to prioritise between different interactive connections. If the network experiences congestion, a connection with a THP of 1 should be served prior to a connection with a THP of 2 or 3. Likewise, all streaming connections should be served prior to all interactive connections and all conversational connections should be served prior to all streaming connections.

Allocation and retention parameters are specified in terms of a priority level, pre-emption capability, pre-emption vulnerability and permission to queue. These parameters become significant during periods of network congestion in which case it is necessary to distinguish between high and low priority connections. The priority level can range from 1 to 15 where 1 represents the highest priority, 14 represents the lowest priority and 15 represents no priority. The PS data RAB in Log File 8.54 has been allocated a priority of 5. This represents a lower priority then the allocation of 2 for the CS speech RAB shown in Log File 8.28. In this example, the PS data RAB request is not able to pre-empt existing RAB and once established, the PS data RAB can be pre-empted by other RAB. The RAB Assignment Request message also specifies that the RAB request can be queued at the RNC if necessary.

The relocation requirement information within the RAB Assignment Request message was originally introduced for connections towards the PS core network domain. However, this information element is no longer used and should always be configured with a value of 'none'.

The RAB Assignment Request then provides information regarding the transfer of user plane data across the Iu–ps. Figure 8.45 illustrates both the user plane and control plane sections of the Iu–ps

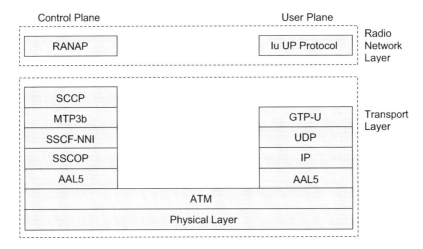

Figure 8.45   Iu–ps transport network protocol stacks (3GPP Releases 99 and 4)

protocol stack. This figure is applicable to the release 99 and release 4 versions of the specifications. The release 5 version allows IP to replace the AAL5 and ATM layers. The user plane section of the Iu–ps protocol stack is based upon a Frame Protocol layer operating above the user plane GPRS Tunnelling Protocol (GTP-U). The user plane Frame Protocol layer can operate using either transparent mode or support mode. Transparent mode is applicable to connections which only require a data transfer service. Support mode is applicable to connections which require additional services. In contrast to speech service connections, PS data connections can use transparent mode. The RAB Assignment Request shown in Log File 8.54 specifies that version 1 of transparent mode should be used. The 3GPP specifications define only a single version of transparent mode.

The GTP-U layer provides connectivity between the SGSN and RNC, i.e. the GTP-U header includes a connection identity which allows the RNC and SGSN to separate the data belonging to each connection. A GTP layer also provides connectivity between the SGSN and GGSN within the core network. The Iu–ps GTP-U runs over UDP, IP, AAL5 and ATM. The use of AAL5 rather than AAL2 differentiates the Iu–ps from the Iu–cs. The use of AAL5 means that it is not necessary to include the ALCAP transport network control plane protocol stack. User plane connections across the Iu–ps rely upon the GTP-U layer to identify the data belonging to each connection (user plane connections across the Iu–cs rely upon the AAL2 CID). The RAB Assignment Request specifies the transport layer address to which uplink user plane data should be addressed. It also specifies the GTP Tunnel Endpoint Identifier (TEI) which represents the connection identifier within the GTP-U header.

The final part of the RAB Assignment Request message includes some PDP context information. The RNC is informed that the PDP context is based upon IPv4 and that it should report the quantity of unsuccessfully transmitted data.

The RNC uses Admission Control to determine whether or not the RAB Assignment Request should be accepted. There are two main implementation options at this stage. The first option allows the RNC to allocate a finite bit rate during RAB establishment. In this case, the RNC can also use the RAB establishment procedure to select between allocating an HSDPA/HSUPA connection or a DPCH connection. The second option postpones the allocation of a finite bit rate until after RAB establishment. In this case, the choice between allocating an HSDPA/HSUPA or DPCH connection is also postponed. The first option reduces the connection establishment delay whereas the second option allows the allocated bit rate to be based upon traffic volume measurement reports. If the first option is used, admission control has to run checks in terms of downlink transmit power, uplink interference, channelisation code availability and Iub transport capacity. If the second option is used, admission control is able to accept the RAB request without checking the availability of resources at this stage.

The signalling procedure described in this section assumes that the RNC postpones the allocation of a finite bit rate until after RAB establishment. Figure 8.44 indicates that RAB establishment for a PS data connection does not involve any NBAP signalling to reconfigure the radio link at the Node B, nor does it include any ALCAP signalling to establish a user plane Iub connection. This pattern is consistent with postponing the allocation of a finite bit rate. These procedures would be included if a non-zero bit rate was allocated. An example Radio Bearer Setup message allocating 0 kbps is shown in Log File 8.55. In contrast to the Radio Bearer Setup messages used for the speech and video call services, this message does not specify an activation time. The UE interprets this as meaning that the new configuration should applied as soon as the message is received. An activation time is not necessary because the radio bearer setup is not part of a synchronised radio link reconfiguration procedure.

The Radio Bearer Setup message informs the UE that the 0 kbps radio bearer belongs to RAB identity 5, i.e. the RAB identity specified within the RANAP RAB Assignment Request. The UE is also informed that the RAB belongs to the PS core network domain and that T315 should be used as a re-establishment timer rather than T314. The optional PDCP information states that lossless SRNS relocation is not supported and that a PDCP header will not be included, i.e. the PDCP

```
DL-DCCH-Message                                    dl-LogicalChannelMappingList
  integrityCheckInfo                               dl-LogicalChMappingList 1
  messageAuthenticationCode                          dl-TransportChType: dch: 1
    Bin: 3C 51 E3 B6                             ul-CommonTransChInfo
  rrc-MessageSequenceNumber: 2                   ul-TFCS
RadioBearerSetup                                 normalTFCI-Signalling
rrc-TransactionIdentifier: 0                     ctfcSize
rrc-StateIndicator: cell-DCH                       ctfc2Bit
rab-InformationSetupList                          ctfc2Bit value 1
  rab-InformationSetupList value 1                  ctfc2: 0
    rab-Info                                        gainFactorInformation
    rab-Identity                                    computedGainFactors: 0
    gsm-MAP-RAB-Identity: 05                       ctfc2Bit value 2
    cn-DomainIdentity: ps-domain                    ctfc2: 1
    re-EstablishmentTimer: useT315                  gainFactorInformation
    rb-InformationSetupList                         signalledGainFactors
    rb-InformationSetupList value 1                  gainFactorBetaC: 15
      rb-Identity: 5                                 gainFactorBetaD: 15
      pdcp-Info                                      referenceTFC-ID: 0
        losslessSRNS-Reloc: notSupp            ul-AddReconfTransChInfoList
        pdcp-PDU-Header: absent                ul-AddReconfTransChInfoList 1
      rlc-Info                                 ul-TransportChannelType: dch
      ul-RLC-Mode                              transportChannelIdentity: 1
        ul-AM-RLC-Mode                         transportFormatSet
        transmissionRLC-Discard                dedicatedTransChTFS
        maxDAT-Retransmissions                 tti20 value 1
        maxDAT: dat8                             rlc-Size
        timerMRW: te300                          octetModeType1
        maxMRW: mm12                             sizeType2
        transmissionWindowSize: tw128            part1: 2
        timerRST: tr250                        numberOfTbSizeList
        max-RST: rst12                         numberOfTbSizeList 1: zero
        pollingInfo                            logicalChannelList: allSizes
          timerPollProhibit: tpp120            semistaticTF-Information
          timerPoll: tp240                     channelCodingType
          poll-SDU: sdu1                         convolutional: third
          lastTransPDU-Poll: false             rateMatchingAttribute: 256
          lastRetransPDU-Poll: true            crc-Size: crc16
          pollWindow: pw70                     dl-CommonTransChInfo
      dl-RLC-Mode                              dl-Parameters: sameAsUL
        dl-AM-RLC-Mode                         dl-AddReconfTransChInfoList
        inSequenceDelivery: true               dl-AddReconfTransChInfoList 1
        receivingWindowSize: rw256             dl-TransportChannelType: dch
        dl-RLC-StatusInfo                      dl-transportChannelIdentity: 1
        missingPDU-Indicator: false            tfs-SignallingMode
      rb-MappingInfo                             sameAsULTrCH
      rb-MappingInfo value 1
        ul-LogicalChannelMappings              ul-TransportChannelType: dch
          ul-TransportChType: dch: 1           ul-TransportChannelIdentity: 1
          rlc-SizeList: configured             dch-QualityTarget
          mac-LogicalChPriority: 4             bler-QualityValue: -20
```

Log File 8.55 Radio Bearer Setup message allocating a 0 kbps connection

layer is effectively transparent. The radio bearer is configured to use Acknowledged Mode (AM) RLC in both the uplink and downlink directions. The set of AM RLC parameters are described in Section 2.3.3. However, the set of AM parameters may be changed when the connection is allocated a finite bit rate. This is necessary because some of the parameters depend upon the allocated bit rate.

The radio bearer is then mapped onto a dedicated transport channel with an identity of 1. The uplink is allocated a logical channel priority of 4. This priority can range from 1 to 8, where 1 represents the highest priority. The UE is not informed of a priority for the downlink because the priority is important to the sender rather than the receiver, i.e. the RNC is configured with the equivalent downlink information. The Transport Format Combination Set (TFCS) information within this message is not important because the Transport Format Set (TFS) belonging to the new transport channel is configured with only a single Transport Format (TF), and that TF does not include any transport blocks. This information will be reconfigured when a finite bit rate is allocated to the new DCH transport channel. The final part of the message specifies that $10 \times \log(\text{BLER}) = -20$, i.e. the transport block BLER for a single transmission should be 1%. The Radio Bearer Setup message does not change the spreading factor of the uplink and downlink physical channels, nor does it change the DPCH slot formats, i.e. the UE and Node B continue to transmit and receive in the same way as before the radio bearer setup.

The UE acknowledges the Radio Bearer Setup message using a Radio Bearer Setup Complete message. An example speech connection Radio Bearer Setup Complete is shown in Log File 8.39. The PS data version has the same format except that the COUNT-C activation time is excluded. This parameter is only applicable to Transparent Mode (TM) RLC connections.

The RNC proceeds to forward a RAB Assignment Response to the SGSN. An example is shown in Log File 8.56. This message confirms that the RAB with an identity of 5 has been established. The SGSN is provided with the transport layer address to which the downlink application data should be sent. The RNC also provides the SGSN with a GTP-TEI to include within the downlink GTP-U headers. The SGSN has already provided the RNC with the uplink GTP-TEI within the RAB Assignment Request. These identifiers allow the RNC and SGSN to associate individual packets with specific connections. The uplink and downlink identifiers do not have to be equal, i.e. the identifiers within Log File 8.54 and Log File 8.56 are different.

After receiving the RAB Assignment Response, the SGSN forwards a PDP Context Request to the GGSN. This signalling is within the PS core network and is not shown in Figure 8.44. It is used to provide the GGSN with the QoS requirements and also to establish a GTP tunnel between the SGSN and GGSN. After receiving a positive response from the GGSN, the SGSN completes the PDP context activation procedure by sending an Activate PDP Context Accept message to the UE. An example accept message is shown in Log File 8.57. The main purpose of this message is to confirm the final set of QoS attributes and to allocate an IP address to the UE.

The quality of service section of the Activate PDP Context Accept message has a fixed format for all traffic classes. This means that all QoS attributes are present for all traffic classes irrespective of whether or not they are applicable. The message in Log File 8.57 specifies an interactive traffic class and a transfer delay requirement of 200 ms. The UE knows that the transfer delay requirement does not apply to the interactive traffic class and so ignores the figure of 200 ms. Likewise, the guaranteed bit rate requirements are ignored.

The IP address allocated to the UE is 10.3.255.250. This IP address is used by the higher end-to-end IP layer. It is not used by any lower IP layers, e.g. the IP layer below the Iu–ps GTP-U layer. The accept

```
RANAP-PDU
RAB-AssignmentResponse
RAB-SetupOrModifiedList
RAB-SetupOrModifiedItem
  - rAB-ID: '00000101'B
  - transportLayerAddress:
    - '00001010001010100000011000011010'B
  - iuTransportAssociation
    - gTP-TEI: '00000000'H
```

Log File 8.56 RAB Assignment Response message

```
ACT. PDP CONTEXT ACCEPT
- Transaction identifier: 08h
LLC Serv. Access point Id
- SAPI 3
Quality of Service
- Reliab. class: Unack. GTP & LLC; Ack. RLC; Protected data
- Delay class:                        Delay class 2
- Precedence class:                   Normal priority
- Peak throughput:                    Up to 32 000 octet/s
- Mean throughput:                    Best effort
- Traffic class:                      Interactive class
- Delivery order:                     Without delivery order
- Delivery of erroneous SDUs:         not delivered
- Maximum SDU size:                   1500 Octets
- Maximum bit rate for uplink:        64 kbps
- Maximum bit rate for downlink:      384 kbps
- Residual BER:                       1*10-5
- SDU error ratio:                    1*10-4
- Transfer delay:                     4000 ms
- Traffic handling priority:          level 2
- Guaranteed bit rate for uplink:     0 kbps
- Guaranteed bit rate for downlink:   0 kbps
Radio Priority
- Priority level 2
Packet Data Protocol Address
- PDP type organisation: IETF allocated address
- PDP type number: IPv4
- Address: 10.3.255.250
Protocol Conf. Options
- Configuration protocol: IP PDP type
Password Authentic. Protocol
Contents:   02 E9 0005 00
Internet Protocol Control Protocol
Contents:   04 02 00 10 81 06 00 00 00 00 83 06 00 00 00 00
```

Log File 8.57 Activate PDP Context Accept message

message also includes responses from the external network protocols included within the Activate PDP Context Request message. The PAP protocol has included an Authenticate Acknowledge message to signal its acceptance of the user name and password included within the Authenticate Request.

The final phase of establishing a PS data connection is illustrated in Figure 8.46. This phase is not necessary if a finite bit rate has already been allocated during RAB establishment, i.e. the NBAP, ALCAP and Frame Protocol signalling is completed during RAB establishment rather than during this phase.

The RNC uses a Measurement Control message to configure uplink traffic volume measurement reporting at the UE. Reporting is based upon event 4a which is triggered when the uplink RLC buffer payload exceeds a specific threshold. The RNC is responsible for completing the equivalent downlink measurements. The downlink measurements do not require any external signalling because they are completed and reported within the RNC. Log File 2.19 presents an example Measurement Control message used to configure reporting event 4a. In this example, the reporting event is configured with a measurement identity of 16 and the UE is instructed to apply the measurements to the DCH transport channel with identity 1, i.e. the user plane transport channel which is currently configured with 0 kbps in both the uplink and downlink directions. The Measurement Control proceeds to specify that the UE should report the RLC buffer payload when event 4a is triggered. This information helps the RNC allocate an appropriate uplink bit rate. If the UE reports a high buffer payload then the RNC can attempt to allocate a high uplink bit rate. The final section of the Measurement Control message defines

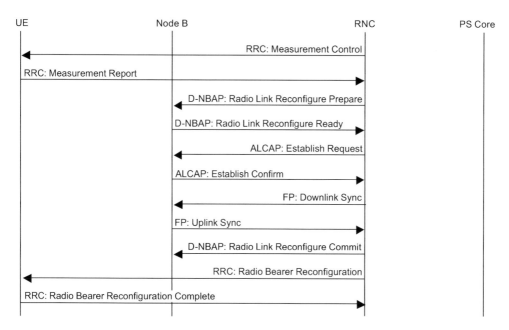

Figure 8.46   Signalling for mobile originated PS data connection (part 3)

the conditions which trigger event 4a. A reporting threshold of 8 bytes is configured with a time-to-trigger of 240 ms. This means that if the RLC buffer occupancy for DCH transport channel 1 exceeds 8 bytes for 240 ms then the UE will provide the RNC with a Measurement Report including the relevant RLC buffer occupancy measurement.

A low triggering threshold is used to ensure that the UE informs the RNC as soon as it has data to send. Using a high threshold at this stage could result in the connection establishment procedure failing. TCP applications establish their connection at the TCP layer using the three-way handshake shown in Figure 8.47. Application data is not transferred until this handshake has been completed. The handshake starts with the TCP layer within the UE (client) sending a SYN packet to the application server. A SYN packet is a 20 byte TCP header which includes a positive SYN flag, an initial sequence number and a target port number. A 20 byte IP header is added if IPv4 is used under the TCP layer. This results in a total packet size of 40 bytes. If the RLC buffer threshold used to trigger event 4a is greater than 40 bytes then the UE would not inform the RNC that it has data to transfer and the RNC

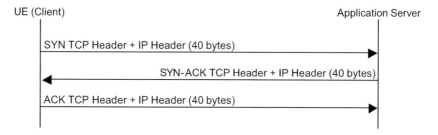

Figure 8.47   Connection establishment at the TCP layer

would not allocate a finite bit rate. This would result in the UE being unable to transfer the SYN packet and so unable to initiate the TCP connection establishment.

Figure 8.46 illustrates the UE sending an event 4a Measurement Report prior to the RNC allocating a finite bit rate. This indicates that the uplink RLC buffer within the UE has exceeded its threshold prior to the downlink RLC buffer within the RNC. This is typical of a TCP application which sends the uplink SYN packet prior to transferring any data irrespective of whether the subsequent application data is to be transferred in the uplink or downlink direction. Log File 2.19 presents an example Measurement Report which has been triggered by an event 4a. The Measurement Report includes an identity which matches the identity within the corresponding Measurement Control message. The main part of the Measurement Report specifies that radio bearer 5 which belongs to DCH transport channel 1 has triggered event 4a with an RLC buffer occupancy of 1024 bytes (1 kbyte). This represents an uplink capacity request which triggers the RNC to allocate a non-zero bit rate.

Having received the uplink capacity request, the RNC has to select the appropriate channel types for the uplink and downlink of the user plane connection. The uplink/downlink combinations could be DCH/DCH, DCH/HSDPA or HSUPA/HSDPA. HSDPA and HSUPA are typically allocated in preference to a DCH if they are supported by both the network and UE. The RNC forwards an NBAP Radio Link Reconfiguration Prepare message to the Node B after selecting the appropriate channel types (and appropriate bit rates in the case of a DCH connection). An example Radio Link Reconfiguration Prepare message which allocates DCH to both the uplink and downlink is presented in Log File 8.58.

The SRB DCH is referenced using its identity of 24. It is listed as a DCH to be modified because it was added to the radio link during RRC connection establishment. The SRB Transmission Time Interval (TTI) is reconfigured from 10 to 40 ms if a 10 ms TTI was allocated during RRC connection establishment. This corresponds to decreasing the SRB bit rate from 13.6 to 3.4 kbps. The rate matching attributes can also be reconfigured at this stage. These attributes had no impact while there was only a single transport channel associated with the radio link. Adding the user plane DCH means that the rate matching attributes start to have an impact.

The user plane DCH is referenced using its identity of 1 and is listed as a DCH to be added. The identity matches that of the 0 kbps DCH signalled to the UE within the preceding Radio Bearer Setup message. The user plane DCH includes three Transport Formats (TF) within its uplink Transport Format Set (TFS) and five TF within its downlink TFS. The uplink TF correspond to bit rates of 0, 16 and 64 kbps. The downlink TF correspond to bit rates of 0, 32, 128, 256 and 384 kbps. These bit rates are generated by assuming that the transport block size of 336 bits includes 320 bits of RLC payload, i.e. the overhead generated by the RLC layer is 16 bits. The uplink TTI is 20 ms whereas the downlink TTI is 10 ms.

The downlink Transport Format Combination Set (TFCS) is configured with ten calculated Transport Format Combinations (cTFC). These are based upon all possible combinations of the TF belonging to the SRB and user plane DCH. The Node B uses the cTFC values to identify the allowed TFC. The translation between a specific cTFC value and its corresponding TFC is based upon the intermediate variables L and P. These variables are defined for each transport channel after ordering them according to their transport channel identity. L is equal to the number of transport formats within the transport format set, whereas P is equal to the product of the L values belonging to the preceding transport channels. By default, the P value for the first transport channel is equal to 1. The L and P variables belonging to the downlink of the PS data connection are illustrated in Figure 8.48.

The cTFC values are generated by summing the products of the appropriate Transport Format Indicator (TFI) and the corresponding value of P for each transport channel. This calculation is illustrated in Figure 8.49. There is always a one-to-one relationship between a specific cTFC value and the corresponding TFC. The uplink cTFC values are calculated in the same manner although the user plane DCH has only three TF within its uplink TFS.

```
NBAP-PDU                                      semiStaticTransportFormatInfo
RadioLinkReconfigurationPrepare                - transTimeInterval: msec-40
NodeB-CommContextID: 160053                    - channelCodingType: conv
UL-DPCH-Inf-RL-ReconfPrep                      - codingRate: third
ul-scramblingCode                              - rateMatchingAttribute: 256
  - ul-ScramblingCodeNumber: 10872             - cRC-Size: 16
  - ul-ScramblingCodeLength: long            frameHandlingPriority: 14
  - minUL-ChannelCodeLength: len16
  - punctureLimit: 7                         DCH-Add-Item-RL-ReconfPrep
tFCS                                           - payldCRC-PresInd: crc-included
  - ctfc4bit: 0                                - ul-FP-Mode: silent
  - ctfc4bit: 1                                - toAWS: 25
  - ctfc4bit: 2                                - toAWE: 10
  - ctfc4bit: 3                              DCH-Add-Spec-Item-RL-ReconfPrep
  - ctfc4bit: 4                              - dCH-ID: 1
  - ctfc4bit: 5                              dL-TransportFormatSet
ul-DPCCH-SlotFormat: 0                        dynamicTransportFormatInfo
DL-DPCH-Information-RL-ReconfPrep              - nrOfTransportBlocks: 0
tFCS                                           - nrOfTransportBlocks: 1
  - ctfc4bit: 0                                - transportBlockSize: 336
  - ctfc4bit: 1                                - nrOfTransportBlocks: 4
  - ctfc4bit: 2                                - transportBlockSize: 336
  - ctfc4bit: 3                                - nrOfTransportBlocks: 8
  - ctfc4bit: 4                                - transportBlockSize: 336
  - ctfc4bit: 5                                - nrOfTransportBlocks: 12
  - ctfc4bit: 6                                - transportBlockSize: 336
  - ctfc4bit: 7                              semiStaticTransportFormatInfo
  - ctfc4bit: 8                                - transTimeInterval: msec-10
  - ctfc4bit: 9                                - channelCodingType: turbo
dL-DPCH-SlotFormat: 15                         - codingRate: third
  - tFCI-Presence: present                     - rateMatchingAttribute: 176
  - multiplexPosition: flex-pos                - cRC-Size: 16
  - tPC-DL-StepSize: one                     ul-TransportFormatSet
  - limitedPowerInc: not-used               dynamicTransportFormatInfo
                                               - nrOfTransportBlocks: 0
DCH-Modify-Item-RL-ReconfPrep                  - nrOfTransportBlocks: 1
  - ul-FP-Mode: silent                         - transportBlockSize: 336
  - aAL2-ChRealloc: not-be-realloc             - nrOfTransportBlocks: 4
DCH-Modify-Spec-Item-RL-ReconfPrep             - transportBlockSize: 336
- dCH-ID: 24                                 semiStaticTransportFormatInfo
dL-TransportFormatSet                          - transTimeInterval: msec-20
dynamicTransportFormatInfo                     - channelCodingType: turbo
  - nrOfTransportBlocks: 1                      - codingRate: third
  - transportBlockSize: 0                       - rateMatchingAttribute: 178
  - nrOfTransportBlocks: 1                      - cRC-Size: 16
  - transportBlockSize: 148                  frameHandlingPriority: 11
semiStaticTransportFormatInfo                qE-Selector: selected-DCH
  - transTimeInterval: msec-40
  - channelCodingType: conv                  RadioLinkModInfList-RL-ReconfPrep
  - codingRate: third                        RadioLinkModInfItem-RL-ReconfPrep
  - rateMatchingAttribute: 256               - rL-ID: 1
  - cRC-Size: 16                             dL-ChCodeInf-RL-ReconfPrep
ul-TransportFormatSet                        DL-ChCodeInfItem-RL-ReconfPrep
dynamicTransportFormatInfo                     - dL-ChannelisationCodeNumber: 1
  - nrOfTransportBlocks: 0                      - dL-TransmissionPower: 2
  - nrOfTransportBlocks: 1                      - maxDL-Power: 70
  - transportBlockSize: 148                    - minDL-Power: -80
```

Log File 8.58   NBAP Radio Link Reconfiguration Prepare message

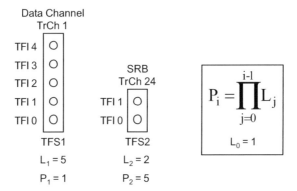

Figure 8.48 Variables used to calculate the cTFC for an example PS data connection

The Radio Link Reconfiguration Prepare message specifies a minimum uplink spreading factor (channelisation code length) of 16 and a downlink slot format of 15. 3GPP TS 25.211 associates slot format 15 with a spreading factor of 8. The UE uses the minimum uplink spreading factor when transmitting the larger TFC within its TFCS. The uplink spreading factor can be increased when transmitting smaller TFC. The algorithm used to select the appropriate spreading factor is described in Section 2.6.1. This algorithm makes use of the uplink puncturing limit. The Node B is provided with the uplink puncturing limit so it can apply the same algorithm as the UE to determine which uplink spreading factor to associate with each uplink TFC. The actual value of the puncturing limit is equal to $0.4 + (0.04 \times \text{signalled value})$, i.e. the signalled value of 7 corresponds to an actual value of 0.68.

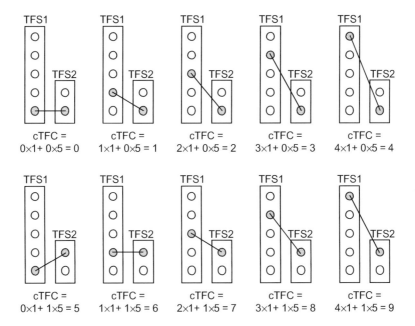

Figure 8.49 Calculation of cTFC values for an example PS data service

The Radio Link Reconfiguration Prepare message also specifies the time of arrival window start (toAWS) and end (toAWE) points for the user plane DCH. These parameters define the time window during which the Node B must receive downlink user plane data from the RNC in order to have sufficient time to complete its physical layer processing and transmit the resulting data during the appropriate TTI. These parameters are applicable to the Iub transport connection which is established after the Node B has responded to the Radio Link Reconfiguration Prepare message.

The Node B is also informed of changes to the maximum and minimum downlink transmit powers. The maximum transmit power is signalled using a value of 70 whereas the minimum transmit power is signalled using a value of -80. These values have units of 0.1 dB and are relative to the CPICH transmit power. Assuming a CPICH transmit power of 33 dBm results in a maximum transmit power of 40 dBm and a minimum transmit power of 25 dBm.

If the RNC had decided to allocate HSDPA to transfer the downlink user plane data the corresponding DCH TFS would have been left empty. The Radio Link Reconfiguration Prepare message would also include an extension to specify the characteristics of the HSDPA connection. An example HSDPA extension is shown in Log File 8.59. This example is based upon configuring a single MAC-d flow for the user plane data. The SRB continues to be transferred using its allocated DCH.

The MAC-d flow is allocated an identity of 0. The allocation and retention parameters associated with the RAB are forwarded to the Node B. A single priority queue is configured with a scheduling priority of 15, a T1 timer of 120 ms and a maximum transmit window size of 16. The relevance of these parameters is described in Chapter 6. The MAC-d PDU size of 336 bits is associated with a Size Index Identifier (SID) of 0. The SID is subsequently included within the MAC-hs header to

```
protocolExtensions
extensionValue
hSDSCH-MACdFlows-Information
hSDSCH-MACdFlow-Specific-Info
HSDSCH-MACdFlow-Specific-InfoItem
   - hsDSCH-MACdFlow-ID: 0
   allocationRetentionPriority
      - priorityLevel: 1
      - pre-emptionCapability: shall-not-trigger-pre-emption
      - pre-emptionVulnerability: pre-emptable
priorityQueue-Info
PriorityQueue-InfoItem
   - priorityQueueId: 0
   - associatedHSDSCH-MACdFlow: 0
   - schedulingPriorityIndicator: 15
   - t1: v120
   - mAC-hsWindowSize: v16
MACdPDU-Size-IndexItem
   - sID: 0
   - macdPDU-Size: 336
   - rLC-Mode: rLC-AM
ueCapability-Info
   - hSDSCH-Physical-Layer-Category: 6
mAChs-Reordering-Buffer-Size-for-RLC-UM: 0

cqiFeedback-CycleK: v4
cqiRepetitionFactor: 1
ackNackRepetitionFactor: 1
cqiPowerOffset: 4
ackPowerOffset: 5
nackPowerOffset: 5
measurement-Power-Offset: 9
```

Log File 8.59 HSDPA extension to an NBAP Radio Link Reconfiguration Prepare message

inform the UE of the transport block size being sent at any point in time. This example is relatively simplistic because only a single transport block size is configured. The Node B is also informed of the UE HS-DSCH category. A category 6 UE supports 16QAM and a maximum of five HS-PDSCH codes. Other HSDPA parameters are configured, including the rate at which the UE will be instructed to send CQI reports, repetition factors for both the CQI and MAC-hs acknowledgements, power offsets for the HS-DPCCH and the CPICH to HS-PDSCH measurement power offset which will be signalled to the UE.

If the RNC had decided to allocate HSUPA to transfer the uplink user plane data the corresponding DCH TFS would have been left empty while the Radio Link Reconfiguration Prepare message would include an extension to specify the characteristics of the HSUPA connection.

The Node B does not apply the new radio link configuration immediately after receiving the Radio Link Reconfiguration Prepare message. The new configuration is evaluated to ensure that it is acceptable but is then stored until the RNC specifies the Connection Frame Number (CFN) during which it should become active. The Node B acknowledges that it has accepted the new configuration by returning an NBAP Radio Link Reconfiguration Ready message. An example of this message for a speech connection is presented in Log File 8.33. The same format is used for a PS data connection. The key content of the message is the binding identity and transport layer address to be used when establishing the Iub transport connection. If the RNC is configuring HSDPA then the Radio Link Reconfiguration Ready message will include two binding identities and two transport layer addresses. One pair will be used for the HSDPA Iub connection while the other pair will be used for the uplink user plane DCH, or alternatively for HSUPA.

The RNC proceeds to establish one or more Iub transport connections by sending one or more ALCAP Establish Request messages to the Node B. Figure 8.46 illustrates a single Iub transport connection being established. This is applicable to a DCH/DCH connection. The Node B responds using an ALCAP Establish Confirm message. This procedure is the same as that described for RRC connection establishment in Section 8.1. The Establish Request message includes the AAL2 Channel Identifier (CID) for the new transport connection. The destination Network Service Access Point (NSAP) address matches the transport layer address provided by the Node B. Similarly, the Served User Generated Reference (SUGR) matches the binding identity provided by the Node B. The SUGR allows the Node B to link the new Iub transport connection with the appropriate transport channels.

The RNC achieves synchronisation across the Iub after receiving the ALCAP Establish Confirm message. Synchronisation is only necessary for Iub transport connections allocated to DCH. Transport connections for HSDPA and HSUPA do not require synchronisation because the Iub timing requirements are more relaxed. In general, HSDPA data is buffered at the Node B rather than immediately processed and scheduled into a specific TTI. The requirement for Iub synchronisation is indicated by the Radio Link Reconfiguration Prepare message including toAWS and toAWE information for a DCH connection, but not for an HSDPA nor HSUPA connection. Iub synchronisation for a DCH helps to ensure that the RNC is able to send Frame Protocol data frames such that they arrive at the Node B within the time window defined by toAWS and toAWE. Synchronisation is achieved using dedicated channel Frame Protocol control frames. This procedure is the same as that described for RRC connection establishment in Section 8.1.

Figure 8.46 does not include any signalling to establish a connection across the Iu–ps. This is in contrast to a speech connection which uses ALCAP signalling to establish a user plane connection across the Iu–cs. The Iu–ps is based upon AAL5 rather than AAL2 and the protocol stack shown in Figure 8.45 does not include an ALCAP transport network control plane. The user plane connection across the Iu–ps is based upon the GTP-U TEI which are exchanged between the RNC and SGSN during RAB establishment.

The RNC proceeds to provide the Node B with the CFN during which it should start to apply the new radio link configuration. This information is sent using an NBAP Radio Link Reconfiguration Commit message. An example of this message for a speech connection is presented in Log File 8.34. The same format is used for a PS data connection. The RNC calculates the CFN by considering that the UE must be ready to apply the new radio link configuration at the same time as the Node B. The UE has not yet been informed of the new configuration so the CFN must allow sufficient time for the Radio Bearer Reconfiguration message to be transferred to the UE. In general, some margin is included to allow for one or more RLC re-transmissions which may or may not be necessary.

Figure 8.46 illustrates that connection establishment is completed using a Radio Bearer Reconfiguration procedure. The Radio Bearer Reconfiguration message starts by providing the UE with the same activation CFN as already signalled to the Node B within the Radio Link Reconfiguration Commit message. The UE is then provided with information which is similar to what was provided to the Node B within the Radio Link Reconfiguration Prepare message. The same cTFC values and the same TFS are specified. The section of the Radio Bearer Reconfiguration message which specifies the TFS is shown in Log File 8.60.

```
ul-AddReconfTransChInfoList
ul-AddReconfTransChInfoList value 1
ul-TransportChannelType: dch
transportChannelIdentity: 24               dl-AddReconfTransChInfoList
  transportFormatSet                       dl-AddReconfTransChInfoList value 1
  dedicatedTransChTFS                      dl-TransportChannelType: dch
    tti40                                  transportChannelIdentity: 24
    tti40 value 1                          tfs-SignallingMode
      rlc-Size                               sameAsULTrCH
      octetModeType1                       ul-TransportChannelType: dch
        sizeType1: 16                      ul-TransportChannelIdentity: 24
    numberOfTbSizeList                       qualityTarget
      numberOfTbSizeList 1: zero             bler-QualityValue: -29
      numberOfTbSizeList 2: one          dl-AddReconfTransChInfoList value 2
    logicalChannelList: allSizes         dl-TransportChannelType: dch
    semistaticTF-Information             transportChannelIdentity: 1
      channelCodingType                   tfs-SignallingMode
        convolutional: third                explicit-config
      rateMatchingAttribute: 256          dedicatedTransChTFS
      crc-Size: crc16                       tti10
ul-AddReconfTransChInfoList value 2         tti10 value 1
ul-TransportChannelType: dch                  rlc-Size
transportChannelIdentity: 1                   octetModeType1
  transportFormatSet                            sizeType2
  dedicatedTransChTFS                             part1: 2
    tti20                                 numberOfTbSizeList
    tti20 value 1                           numberOfTbSizeList 1: zero
      rlc-Size                              numberOfTbSizeList 2: one
      octetModeType1                        numberOfTbSizeList 3
        sizeType2                             small: 4
          part1: 2                          numberOfTbSizeList 4
    numberOfTbSizeList                        small: 8
      numberOfTbSizeList 1: zero            numberOfTbSizeList 5
      numberOfTbSizeList 2: one              small: 12
      numberOfTbSizeList 3               logicalChannelList: allSizes
        small: 4                         semistaticTF-Information
    logicalChannelList: allSizes           channelCodingType: turbo
    semistaticTF-Information               rateMatchingAttribute: 176
      channelCodingType: turbo            crc-Size: crc16
      rateMatchingAttribute: 142        qualityTarget
      crc-Size: crc16                     bler-QualityValue: -20
```

Log File 8.60   Content from a Radio Bearer Reconfiguration message (example section 1)

Uplink and downlink TFS are specified for both the SRB (transport channel 24) and the user plane radio bearer (transport channel 1). The TFS for the SRB are symmetric, so a TFS is only specified for the uplink direction rather than specifying the same information twice. The TTI is configured with a value of 40 ms and the TFS includes TF with either 0 or 1 transport blocks. The transport block size can be derived from the RLC PDU size information. 3GPP TS 25.331 specifies the set of equations presented in Table 8.3.

There are three main groups of equations categorised by the RLC size information mode. Each equation within a group is referenced by its size type. Log File 8.60 indicates that the RLC PDU size for the SRB is signalled using octet mode RLC size information type 1. This indicates that the second set of equations are applicable. Log File 8.60 also indicates that size type 1 is used. This indicates that the first equation from the second set of equations is applicable. The size type 1 information element is signalled using a value of 16 so it is possible to deduce that the RLC PDU size is $(8 \times 16) + 16 = 144$ bits. This has not changed since the SRB were first configured during RRC connection establishment. The RLC PDU size of 144 bits corresponds to a transport block size of 148 bits after the MAC header has been included. The MAC header is necessary to identify which SRB is transferred at any point in time.

The TFS for the user plane data are asymmetric so separate uplink and downlink TFS are specified. The uplink is configured with a 20 ms TTI and a TFS which includes TF with either 0, 1 or 4 transport blocks. The RLC PDU size is specified using octet mode RLC size information type 1, i.e. the second set of equations within Table 8.3. Size type 2 is specified so the second equation from the second set of equations is applicable, i.e. the RLC PDU size is given by $(2 \times 32) + 272 = 336$ bits. The user plane data does not require multiplexing within the MAC layer so the MAC layer does not add a header and the RLC PDU size equals the transport block size. The uplink TFS defines a maximum uplink bit rate of $(4 \times 336)/0.02 = 67.2$ kbps at the top of the physical layer. This corresponds to a maximum RLC throughput of 64 kbps when assuming a 16 bit acknowledged mode RLC header. The downlink is configured with a 10 ms TTI and a TFS which includes TF with either 0, 1, 4, 8 or 12 transport blocks. The downlink RLC PDU size is specified in the same way as the uplink. This results in a maximum downlink Physical layer bit rate of $(12 \times 336)/0.01 = 403.2$ kbps, and a corresponding maximum RLC throughput of 384 kbps.

The Radio Bearer Reconfiguration message shown in Figure 8.46 also specifies physical channel information. This section of the message is shown in Log File 8.61. The uplink puncturing limit matches the value signalled to the Node B within the Radio Link Reconfiguration Prepare message.

The minimum uplink spreading factor is 16 whereas the downlink spreading factor is 8. These figures are applicable to the maximum uplink and downlink bit rates of 64 and 384 kbps, respectively.

Table 8.3 Equations used to calculate RLC PDU sizes

RLC Size Information Mode	Size Type	Equation	Comments
BitModeRLC-SizeInfo	sizeType1	sizeType1	
	sizeType2	$(part1 \times 8) + 128 + part2$	part2 is optional
	sizeType3	$(part1 \times 16) + 256 + part2$	part2 is optional
	sizeType4	$(part1 \times 64) + 1024 + part2$	part2 is optional
OctetModeRLC-SizeInfoType1	sizeType1	$(8 \times sizeType1) + 16$	
	sizeType2	$(part1 \times 32) + 272 + (part2 \times 8)$	part2 is optional
	sizeType3	$(part1 \times 64) + 1040 + (part2 \times 8)$	part2 is optional
OctetModeRLC-SizeInfoType2	sizeType1	$(8 \times sizeType1) + 48$	
	sizeType2	$(16 \times sizeType2) + 312$	
	sizeType3	$(64 \times sizeType3) + 1384$	

```
ul-ChannelRequirement                positionFixedOrFlexible: flexible
ul-DPCH-Info                           tfci-Existence: true
modeSpecificInfo fdd                 dl-InformationPerRL-List
  scramblingCodeType: longSC         dl-InformationPerRL-List value 1
  scramblingCode: 1007629            modeSpecificInfo fdd
  spreadingFactor: sf16                primaryCPICH-Info
  tfci-Existence: true                 primaryScramblingCode: 9
  puncturingLimit: p10-68            dl-DPCH-InfoPerRL fdd
  modeSpecificPhysChInfo fdd           pCPICH-UseForChEst: mayBeUsed
dl-CommonInformation                   dpch-FrameOffset: 48
dl-DPCH-InfoCommon                   dl-ChannelisationCodeList
cfnHandling: maintain                  dl-ChannelisationCodeList 1
modeSpecificInfo fdd                     sf-AndCodeNumber sf8: 1
  powerOffsetPilot-pdpdch: 0         tpc-CombinationIndex: 0
  spreadingFactorAndPilot: sfd8
```

Log File 8.61   Content from a Radio Bearer Reconfiguration message (example section 2)

Table 8.4 presents a set of typical spreading factors for a range of bit rates. In general there is a factor of two difference between the uplink and downlink spreading factors. For a specific bit rate, the downlink requires a spreading factor which is twice as large as the uplink because the data is transferred using both the in-phase and quadrature branches of the modulation scheme. Transferring data on both the in-phase and quadrature branches means that the symbol rate is halved and twice as much spreading is necessary to reach the chip rate.

The 3.4 kbps SRB has equal spreading factors in the uplink and downlink directions because the downlink spreading factor of 512 offers a maximum DPDCH bit rate of 6 kbps. This is insufficient to support an SRB bit rate of 3.4 kbps once the redundancy introduced by channel coding has been taken into account. The relationship between downlink spreading factor and supported bit rate becomes less linear for high spreading factors because the DPCCH bits become more significant. The DPCCH bits do not affect the linearity of the relationship between bit rate and spreading factor in the uplink direction because they are transferred using a separate branch of the modulation scheme.

The Radio Bearer Reconfiguration message would include HSDPA information if the RNC had decided to allocate HSDPA to transfer the downlink user plane data. Log File 8.62 illustrates a section from a Radio Bearer Reconfiguration message which maps the user plane radio bearer onto an HS-DSCH transport channel.

A Radio Bearer Reconfiguration message used to allocate an HSDPA connection would also specify a set of HSDPA parameters. Log File 8.63 presents the section which specifies these parameters. In this

Table 8.4   Relationship between spreading factor and maximum allocated bit rate

Bit rate	Uplink spreading factor	Downlink spreading factor
3.4 kbps SRB	256	256
13.6 kbps SRB	64	128
12.2 kbps Speech + 3.4 kbps SRB	64	128
8 kbps Data + 3.4 kbps SRB	128	256
16 kbps Data + 3.4 kbps SRB	64	128
32 kbps Data + 3.4 kbps SRB	32	64
64 kbps Data + 3.4 kbps SRB	16	32
128 kbps Data + 3.4 kbps SRB	8	16
256 kbps Data + 3.4 kbps SRB	4	8
384 kbps Data + 3.4 kbps SRB	4	8

```
        rb-MappingInfo
        rb-MappingInfo value 1
          ul-LogicalChannelMappings
          ul-TransportChannelType: dch: 1
          rlc-SizeList: configured
          mac-LogicalChannelPriority: 8
          dl-LogicalChannelMappingList
          dl-LogicalChannelMappingList value 1
            dl-TransportChannelType
              hsdsch: 0
```

Log File 8.62 Radio Bearer Reconfiguration message mapping a radio bearer onto HSDPA

example, the HSDPA connection is configured with six HARQ processes and a single MAC-d flow with a single priority queue. The MAC-d PDU size is specified to be 336 bits. The UE is also informed of the HS-SCCH channelisation code and the set of HS-DPCCH power offsets and repetition factors. These parameters are explained in Chapter 6.

If the RNC had decided to allocate HSUPA, the Radio Bearer Reconfiguration message would map the uplink user plane radio bearer onto an E-DCH. The message would then also include a set of HSUPA parameters. It is also possible for the Radio Bearer Reconfiguration message to map the SRB onto HSDPA and HSUPA.

After receiving the Radio Bearer Reconfiguration message, the UE waits until the activation CFN occurs. It then starts to apply the new configuration and returns a Radio Bearer Reconfiguration Complete message. This message acknowledges that the new configuration has been successfully applied and completes the PS data connection establishment procedure.

```
dl-AddReconfTransChInfoList value 2          sRB-delay: 7
dl-TransportChannelType: hsdsch              powerControlAlgorithm
tfs-SignallingMode                             algorithm1: 0
hsdsch                                       deltaACK: 5
  harqInfo                                   deltaNACK: 5
    numberOfProcesses: 6                     ack-NACK-repetition-factor: 1
    memoryPartitioning: implicit           modeSpecificInfo fdd
  addOrReconfMAC-dFlow                        scramblingCodeType: longSC
  mac-hs-AddReconfQueue-List                  scramblingCode: 1154291
  mac-hs-AddReconfQueue-List 1                spreadingFactor: sf8
    mac-hsQueueId: 0                          tfci-Existence: true
    mac-dFlowId: 0                            puncturingLimit: pl0-68
    reorderingReleaseTimer: rt120          modeSpecificPhysChInfo fdd:
    mac-hsWindowSize: mws16                dl-HSPDSCH-Information
    mac-d-PDU-SizeInfo-List                  hs-scch-Info
    mac-d-PDU-SizeInfo-List 1                modeSpecificInfo fdd
      mac-d-PDU-Size: 336                    hS-SCCHChannelisationCodeInfo
      mac-d-PDU-Index: 0                       hS-SCCHChannelCodeInfo: 4
dch-QualityTarget                            measurement-feedback-Info
  bler-QualityValue: -20                     modeSpecificInfo fdd
ul-ChannelRequirement                          measurementPowerOffset: 6
ul-DPCH-Info                                   feedback-cycle: fc4
ul-DPCH-PowerControlInfo fdd                   cqi-RepetitionFactor: 1
  dpcch-PowerOffset: -48                       deltaCQI: 4
  pc-Preamble: 0
```

Log File 8.63 Radio Bearer Reconfiguration message mapping a radio bearer onto HSDPA

8.6 Soft Handover

- Soft handover allows the UE to simultaneously connect to multiple Node B. It helps to provide seamless mobility and improve the RF conditions at cell edge where signal strengths are generally low and cell dominance is poor.
- Soft handover is applicable to DPCH and HSUPA connections. The soft handover procedure uses intra-frequency reporting event 1a for radio link addition, event 1b for radio link deletion and event 1c for radio link replacement.
- UE can read an initial set of parameters which characterise reporting events 1a, 1b and 1c from SIB 11 or SIB 12. UE are typically sent dedicated Measurement Control messages to update the soft handover parameter set after moving into CELL_DCH.
- The measurement quantity for events 1a, 1b and 1c can be either path loss, CPICH RSCP or CPICH Ec/Io. CPICH Ec/Io is typically selected because the corresponding UE measurements are more accurate.
- The filter coefficient defines the memory of the filtering applied by the RRC layer. Filtering reduces short-term variations of the physical layer measurements. The release 5 and newer versions of the specifications state that filtering is to be completed in log units.
- The UE is instructed which quantities to include when sending a Measurement Report. They are specified independently for the active and monitored sets. They can also be specified for detected set measurements which can be used for neighbour list optimisation.
- The RNC completes its admission control procedure after receiving an event 1a Measurement Report. Admission control is used to ensure that there are sufficient resources to introduce the new radio link.
- In the case of soft handover, the RNC forwards a Common NBAP Radio Link Setup Request to the Node B. In the case of softer handover, the RNC forwards a Dedicated NBAP Radio Link Addition Request to the Node B.
- In the case of soft handover, the Node B attempts to achieve uplink synchronisation across the air-interface after receiving the Radio Link Setup Request. In the case of softer handover, the Node B is already synchronised as a result of having an existing radio link.
- In the case of soft handover, the RNC uses the transport layer information within the NBAP Radio Link Setup Response to establish the necessary Iub transport connections. In the case of softer handover, it is not necessary to establish new Iub transport connections.
- The RNC uses an Active Set Update message to instruct the UE to add the new cell to the active set. The same message type is used when deleting and replacing an active set cell. The UE acknowledges the Active Set Update with an Active Set Update Complete.
- A Measurement Control message can be used to configure reporting events 6f and 6g after the UE has entered soft or softer handover. A Measurement Control can also be used to update the neighbour list information.
- Key 3GPP specifications: TS 25.331, TS 25.133 and TS 25.433.

8.6.1 Inter-Node B

Soft handover allows the UE to simultaneously connect to multiple Node B. This is in contrast to hard handover in which case the connection to the first Node B is broken before the connection to the second Node B is established. Soft handover helps to provide seamless mobility as UE move throughout the network. It also helps to improve the RF conditions at cell edge where signal strengths are generally low and cell dominance is poor. Soft handover is only applicable to UE which are in CELL_DCH. In addition, it is only applicable to DPCH and HSUPA connections. The soft handover procedure uses intra-frequency reporting events 1a, 1b and 1c. These events are specified by 3GPP TS 25.331. Event 1a corresponds to radio link addition, event 1b corresponds to radio link deletion and event 1c corresponds to radio link replacement.

A UE in RRC Idle mode, CELL_FACH, CELL_PCH or URA_PCH can read the parameters which characterise reporting events 1a, 1b and 1c from SIB 11 or SIB 12. These parameters become applicable as soon as the UE moves into CELL_DCH. UE are typically sent dedicated Measurement Control messages to update the soft handover parameter set after moving into CELL_DCH. Figure 8.50 illustrates the RNC sending a Measurement Control message prior to the soft handover active set cell addition procedure. This Measurement Control message can also be used to update the neighbour list.

Log File 8.64 presents the section of the Measurement Control message which is applicable to configuring reporting events 1a, 1b and 1c. The Measurement Control starts by specifying that the measurement quantity is CPICH Ec/Io and that measurements should be filtered at the RRC layer using a filter coefficient of 4. 3GPP TS 25.331 specifies that the measurement quantity can be either path loss, CPICH RSCP or CPICH Ec/Io. CPICH Ec/Io is typically selected because the corresponding UE measurements are more accurate. CPICH Ec/Io measurements are more accurate because Ec/Io is calculated from the ratio of RSCP to RSSI. Measurement errors in the RSCP tend to be correlated with measurement errors in the RSSI. This means that the measurement errors cancel to some extent and the accuracy of the final result improves. This is reflected by the measurement accuracy requirements specified within 3GPP TS 25.133. CPICH Ec/Io measurements are specified to have an absolute accuracy of between ±1.5 and ±3.0 dB, depending upon the signal strength conditions. The equivalent

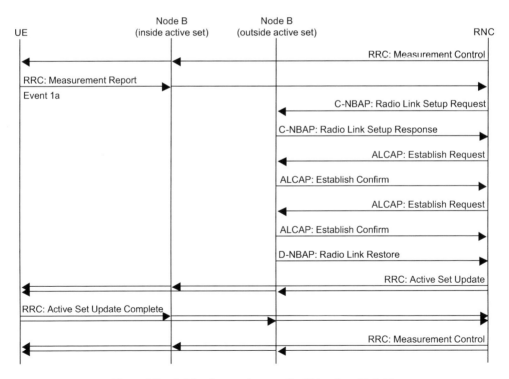

Figure 8.50   Soft handover active set cell addition (inter-Node B)

```
intraFreqMeasQuantity
filterCoefficient: fc4
intraFreqMeasQuantity: cpich-Ec-N0

intraFreqReportingQuantity
activeSetReportingQuantities
cellIdentity-reportingInd: false
cellSyncInfoReportingInd: true
cpich-Ec-N0-reportingInd: true
cpich-RSCP-reportingInd: false
pathloss-reportingInd: false

monitoredSetReportingQuantities
cellIdentity-reportingInd: false
cellSyncInfoReportingInd: true
cpich-Ec-N0-reportingInd: true
cpich-RSCP-reportingInd: false
pathloss-reportingInd: false

reportCriteria
intraFreqReportingCriteria
eventCriteriaList
eventCriteriaList 1
event e1a
triggeringCondition:
monitoredSetCellsOnly
reportingRange: 8
w: 0
reportDeactivationThreshold: t2
reportingAmount: ra-Infinity
reportingInterval: ri0-5
hysteresis: 0
```

```
timeToTrigger: ttt100
reportingCellStatus
allActiveplusMonitoredSet:
    viactCellsPlus2

eventCriteriaList 2
event e1b
triggeringCondition:
activeSetCellsOnly
reportingRange: 12
w: 0
hysteresis: 0
timeToTrigger: ttt640
reportingCellStatus
withinActiveSet: e3

eventCriteriaList 3
event e1c
replacementActThreshold: t3
reportingAmount: ra-Infinity
reportingInterval: ri0-5
hysteresis: 4
timeToTrigger: ttt100
reportingCellStatus
allActiveplusMonitoredSet:
    viactCellsPlus2

measurementReportingMode
measurementReportTransferMode:
    acknowledgedModeRLC
periodicalOrEventTrigger:
    eventTrigger
```

Log File 8.64   Measurement Control content used to configure events 1a, 1b and 1c

absolute accuracy requirements for the CPICH RSCP are between ±6 and ±8 dB. The path loss measurement quantity is calculated by subtracting the CPICH RSCP from the CPICH transmit power. In this case, the UE is informed of the CPICH transmit power as part of the neighbour list information.

The filter coefficient defines the memory of the filtering applied by the RRC layer. The physical layer completes its measurements over a relatively short period of time. This means they are prone to short term variations. Filtering within the RRC layer can be used to reduce these short-term variations. Soft handover events are less likely to be triggered by short-term variations when the filtering has an increased memory. An increased memory also results in the soft handover procedures becoming less responsive. Scenarios which require rapid soft handover, e.g. motorways and high-speed trains, should generally be configured with smaller filter coefficients. 3GPP TS 25.331 specifies the following equation to define the filtering completed by the RRC layer.

$$\text{Filtered Measurement}_n = (1 - a) \times \text{Filtered measurment}_{n-1} + a \times \text{Measurement}_n$$

where $a = 0.5^{\text{Filter coefficient}/2}$

Larger filter coefficients generate smaller values for the variable a. This causes the filtered measurement result to have an increased dependency upon the previous filtered measurement result and a decreased dependency upon the new physical layer measurement. Configuring the filter coefficient with a value of 0 causes the variable a to have a value of 1, which consequently disables filtering at the RRC layer, i.e. each filtered measurement result equals the corresponding physical layer measurement. 3GPP TS 25.133 specifies that the interval between consecutive measurements is 200 ms, i.e. the filter within the RRC layer expects an input every 200 ms and generates an output every 200 ms. Figure 8.51 illustrates the impulse response of the RRC filter for a range of filter coefficients. The impulse response represents the output generated by the filter after inputting a single '1'. This figure illustrates that the output decays more rapidly when the filter coefficient is small. This corresponds to a reduced filter memory.

Measurement reporting events 1a, 1b and 1c have time-to-trigger values associated with them, i.e. time intervals during which specific criteria must be satisfied. The 3GPP specifications allow these time-to-trigger values to be configured with values as small as 10 ms. If RRC filtering generates a result once every 200 ms then all time-to-trigger values between 10 and 200 ms will generate the same behaviour. In practice, the physical layer can provide the RRC layer with measurements more frequently than once every 200 ms. The precise rate is implementation dependant and is likely to vary between UE vendors. If a particular UE generates filtered results once every 50 ms then time-to-trigger values of 10, 20 and 40 ms would generate the same behaviour, but values of 60, 80 and 100 ms would generate a different behaviour. The rate at which the physical layer generates measurements may depend upon the number of cells being measured, i.e. if fewer cells are being measured then those cells may be reported more frequently. It is necessary to up-sample the RRC filter characteristic when the physical layer generates measurements more frequently than once every 200 ms. This ensures that the impulse response of the filter is maintained rather than compressed in time. The release 99 and release 4 versions of TS 25.331 do not specify whether the RRC layer filtering should be completed in linear or logarithmic units. The release 5 and newer versions specify that filtering should be completed in logarithmic units. This approach is intended to make the soft handover reporting events more responsive.

The measurement control information shown in Log File 8.64 instructs the UE which quantities to include when sending a Measurement Report. These reporting quantities are specified independently for cells belonging to the active set and cells belonging to the monitored set. It is also possible to specify a set of reporting quantities for cells belonging to the detected set. Monitored set cells do not belong to the active set, but are included within the neighbour list. Detected set cells are not included within the neighbour list. Detected set cell measurements can be used to help optimise neighbour list information, i.e. identify missing neighbours. The example information presented in Log File 8.64

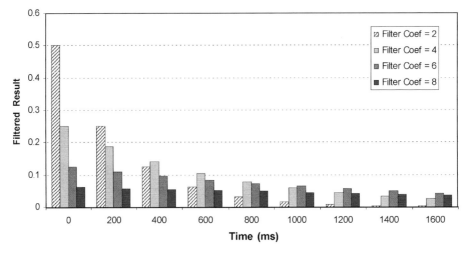

Figure 8.51   Impulse response of RRC filter for a range of filter coefficients

instructs the UE to report the CPICH Ec/Io and timing information for each of the reported active set cells and monitored set cells. The timing information quantifies the time difference between the SFN timing of the cell being measured and the CFN timing of the connection.

The measurement control information shown in Log File 8.64 proceeds to configure the set of parameters directly associated with reporting events 1a, 1b and 1c. Reporting event 1a is used to add a neighbour to the active set. It is only applicable when the active set size is less than its maximum. The maximum active set size can be limited to help avoid excessive soft handover overheads and to place the emphasis upon planning areas of good dominance. The maximum active set size is indicated within Log File 8.64 using the reporting deactivation threshold. This parameter represents the maximum active set size for which event 1a can still be triggered. In this example, the deactivation threshold has a value of 2 so event 1a cannot triggered once the active set size reaches 3. Log File 8.64 also specifies that event 1a should only be applied to monitored set cells, i.e. detected set cells are not allowed to trigger event 1a. 3GPP TS 25.331 specifies that event 1a is triggered using the equation below. This equation is evaluated after the RRC layer has completed its filtering. Event 1a is triggered if the equation remains true during a period of time defined by the time-to-trigger. The time-to-trigger is configured with a value of 100 ms in this example.

$$10 \log[M_{new}] + CIO_{new} \geq W \times 10 \log \left[\sum_{i=1}^{N} M_i \right] + (1 - W) \times 10 \log[M_{best}] - (R_{1a} - H/2)$$

M_{new} is the CPICH Ec/Io measurement belonging to the cell entering the reporting range (addition window). CIO_{new} is the Cell Individual Offset which can be configured for each neighbouring cell. A positive offset increases the probability of the cell triggering event 1a. This can be useful in high-mobility scenarios in which case it is desirable to trigger event 1a earlier than usual. W is a weighting coefficient. N is the number of cells within the active set which are not forbidden from affecting the reporting range. A cell can be forbidden from affecting the reporting range if its CPICH Ec/Io is found to have a large variance. The RNC is responsible for informing the UE of any cells which are forbidden from affecting the reporting range. M_i is the CPICH Ec/Io measured from the ith active set cell which is not forbidden from affecting the reporting range. M_{best} is the highest CPICH Ec/Io measured from the active set cells which are not forbidden from affecting the reporting range. R_{1a} is the reporting range (addition window) for event 1a, and H is a hysteresis parameter.

The equation below represents the equation used to trigger event 1a after inserting the parameter values shown in Log File 8.64. Configuring a weighting coefficient of 0 means that the addition window is relative to the highest CPICH Ec/Io measured from the active set cells. It can be beneficial to configure a weighting coefficient of greater than 0 when the CPICH Ec/Io measurements from the active set cells are relatively unstable.

$$10 \log[M_{new}] + CIO_{new} \geq 10 \log[M_{best}] - 4$$

The reporting range signalled in Log File 8.64 has a value of 8, but the actual value is equal to the signalled value divided by 2. The hysteresis parameter is configured with a value of 0 so does not have an impact. The hysteresis parameter can be used to differentiate between the threshold used to start periodic event 1a reporting and the threshold used to stop periodic event 1a reporting. The UE starts periodic reporting using the window size defined by $(R_{1a} - H/2)$. The time interval between periodic reports is defined by the reporting interval. The reporting interval is configured with a value of 0.5 s in this example. Periodic reporting allows the RNC to receive updated measurement results in case it is unable to add the candidate cell into the active set, e.g. as a result of a negative admission control decision. If the periodic reports indicate that the candidate cell is becoming excessively strong then it may be necessary for the RNC to release the RRC connection to avoid the candidate cell from experiencing increased levels of uplink interference. Otherwise, periodic reporting continues until either the RNC responds with an active set update, the CPICH Ec/Io falls below the window defined by $(R_{1a} + H/2)$ or the maximum number of periodic reports is reached. In this example, the reporting amount is set to infinity so the UE continues to report until the RNC responds or the CPICH Ec/Io falls below the window. Log File 8.64 also indicates that event 1a Measurement Reports should include information regarding the active set cells plus up to two monitored set cells.

Reporting event 1b is used to delete a cell from the active set. It is only applicable when the active set size is greater than 1. 3GPP TS 25.331 specifies that event 1b is triggered using the equation below. This equation is evaluated after the RRC layer has completed its filtering. Event 1b is triggered if the equation remains true during a period of time defined by the time-to-trigger. The time-to-trigger is configured with a value of 640 ms in this example.

$$10\log[M_{old}] + CIO_{old} \leq W \times 10\log\left[\sum_{i=1}^{N} M_i\right] + (1 - W) \times 10\log[M_{best}] - (R_{1b} + H/2)$$

M_{old} is the CPICH Ec/Io measurement belonging to the cell leaving the reporting range (deletion window). CIO_{old} is the Cell Individual Offset which can be configured for each neighbouring cell. A negative offset increases the probability of the cell triggering event 1b (it also reduces the probability of triggering event 1a and being included within the active set in the first place). W is a weighting coefficient. N is the number of cells within the active set which are not forbidden from affecting the reporting range. M_i is the CPICH Ec/Io measured from the ith active set cell which is not forbidden from affecting the reporting range. M_{best} is the highest CPICH Ec/Io measured from the active set cells which are not forbidden from affecting the reporting range. R_{1b} is the reporting range (deletion window) for event 1b, and H is a hysteresis parameter.

The equation below represents the equation used to trigger event 1b after inserting the parameter values shown in Log File 8.64. Configuring a weighting coefficient of 0 means that the deletion window is relative to the highest CPICH Ec/Io measured from the active set cells.

$$10\log[M_{old}] + CIO_{old} \leq 10\log[M_{best}] - 6$$

The reporting range signalled in Log File 8.64 has a value of 12, but the actual value is equal to the signalled value divided by 2. The hysteresis parameter is configured with a value of 0 so does not have an impact. The hysteresis parameter can have an impact upon the UE re-triggering an event 1b for a specific cell. The release 99 and release 4 versions of the specifications do not support periodic reporting for event 1b, i.e. the RNC is expected to react to the first event 1b Measurement Report. In this case, the UE could only send a second event 1b Measurement Report if the CPICH Ec/Io increased to a value within the window defined by $(R_{1b} - H/2)$ and then returned to a value outside the window defined by $(R_{1b} + H/2)$. The release 5 and newer versions of the specifications support periodic reporting for event 1b. In this case, the hysteresis parameter can be used to differentiate between the threshold used to start periodic event 1b reporting and the threshold used to stop periodic event 1b reporting. The example shown in Log File 8.64 does not configure periodic reporting for event 1b. The log file indicates that event 1b Measurement Reports should include information regarding up to three active set cells.

Reporting event 1c is used to replace an active set cell with a higher quality neighbouring cell. In general, it only becomes applicable after the active set size has reached its maximum. Event 1a and cell addition are applicable prior to the active set size reaching its maximum. The active set size at which event 1c becomes applicable is signalled using the replacement activation threshold. In this example, event 1c becomes applicable when the active set size reaches 3, i.e. the maximum active set size. 3GPP

TS 25.331 specifies that event 1c is triggered using the equation below. This equation is evaluated after the RRC layer has completed its filtering. Event 1c is triggered if the equation remains true during a period of time defined by the time-to-trigger. The time-to-trigger is configured with a value of 100 ms in this example.

$$10 \log[M_{new}] + CIO_{new} \geq 10 \log[M_{AS}] + CIO_{AS} + H/2$$

M_{new} is the CPICH Ec/Io measurement belonging to the cell entering the reporting range (replacement window). CIO_{new} is the Cell Individual Offset for the cell entering the reporting range. A positive offset increases the probability of the cell triggering event 1c. M_{AS} is the lowest CPICH Ec/Io measured from the active set cells while CIO_{AS} is the corresponding Cell Individual Offset. H is the hysteresis parameter for event 1c which acts as the reporting range (replacement window). Early versions of the 3GPP specifications did not specify the use of the Cell Individual Offset for event 1c. The specifications were subsequently updated to include the Cell Individual Offset when evaluating the criteria for event 1c.

The equation below represents the equation used to trigger event 1c after inserting the hysteresis value. The hysteresis value signalled in Log File 8.64 has a value of 4, but the actual value is equal to the signalled value divided by 2. The equation itself includes a further division by 2 so the actual replacement window is one quarter of the signalled value.

$$10 \log[M_{new}] + CIO_{new} \geq 10 \log[M_{AS}] + CIO_{AS} + 1$$

Event 1c uses periodic reporting similar to event 1a. The hysteresis parameter differentiates between the threshold used to start periodic event 1c reporting and the threshold used to stop periodic event 1c reporting. The UE starts periodic reporting using the window size defined by $+H/2$. The time interval between periodic reports is defined by the reporting interval. The reporting interval is configured with a value of 0.5 s in this example. Periodic reporting continues until either the RNC responds with an active set update, the CPICH Ec/Io falls below the window defined by $-H/2$ or the maximum number of periodic reports is reached. In this example, the reporting amount is set to infinity so the UE continues to report until the RNC responds or the CPICH Ec/Io falls below the window. Log File 8.64 also indicates that event 1c measurement reports should include information regarding the active set cells plus up to two monitored set cells.

Log File 8.64 specifies that Measurement Reports should be sent using acknowledged mode RLC. In addition, it states that Measurement Reports should be sent only when triggered (event triggered) rather than periodically. An alternative implementation would be to instruct the UE to send Measurement Reports periodically irrespective of whether or not events 1a, 1b or 1c have been triggered. In this case, it is not necessary to configure the parameters associated with events 1a, 1b or 1c. The UE periodically reports its measurements while the RNC evaluates the criteria for active set updates. This approach increases the quantity of information available to the RNC, but has the drawback of an increased signalling load. In addition, this approach does not allow the use of small time-to-trigger values, e.g. if the RNC is sent a Measurement Report once every 500 ms then it cannot become more responsive than when using a time-to-trigger of 500 ms.

Figure 8.50 illustrates the UE sending an event 1a Measurement Report. The content of an example event 1a Measurement Report is shown in Log File 8.65. The initial active set for this example includes a single cell (scrambling code 104). The UE provides measurement results for this active set cell and two monitored set cells (scrambling codes 8 and 48). This is consistent with the instructions provided within the Measurement Control message.

The UE reports the time offset between the P-CCPCH of the cell being measured and its existing DPCH connection. This represents the System Frame Number (SFN) to Connection Frame Number (CFN) observed time difference. The time difference is defined by $OFF \times 38400 + T_m$, where OFF can

```
UL-DCCH-Message                              off: 56
integrityCheckInfo                           tm: 1396
messageAuthenticationCode                    modeSpecificInfo fdd
Bin: 93 7E B3 8E                             primaryCPICH-Info
rrc-MessageSequenceNumber: 7                 primaryScramblingCode: 8
measurementReport                            cpich-Ec-N0: 30 (-9.5 to -9.0 dB)
measurementIdentity: 1
                                             intraFreqMeasuredResultsList 3
intraFreqMeasuredResultsList                 cellSynchronisationInfo
intraFreqMeasuredResultsList 1               modeSpecificInfo fdd
cellSynchronisationInfo                      countC-SFN-Frame-difference
modeSpecificInfo fdd                         countC-SFN-High: 0
countC-SFN-Frame-difference                  off: 252
countC-SFN-High: 0                           tm: 21804
off: 1                                       modeSpecificInfo fdd
tm: 34304                                    primaryCPICH-Info
modeSpecificInfo fdd                         primaryScramblingCode  : 48
primaryCPICH-Info                            cpich-Ec-N0: 1 (-24.0 to -23.5 dB)
primaryScramblingCode: 104
cpich-Ec-N0: 38 (-5.5 to -5.0 dB)            eventResults
                                             intraFreqEventResults
intraFreqMeasuredResultsList 2               eventID : e1a
cellSynchronisationInfo                      cellMeasurementEventResults
modeSpecificInfo fdd                         fdd value 1
countC-SFN-Frame-difference                  primaryScramblingCode: 8
countC-SFN-High: 0
```

Log File 8.65 Measurement Report triggered by event 1a

range from 0 to 255 radio frames and T_m can range from 0 to 38399 chips. In the case of a cell being added to the active set, this information allows the RNC to provide the Node B with appropriate instructions regarding the timing of the new radio link. The UE also reports the CPICH Ec/Io measured from each cell. The mapping between the signalled and actual values of the CPICH Ec/Io is presented in Table 8.5.

Log File 8.65 indicates that the CPICH Ec/Io of the neighbouring cell with scrambling code 8 is 4 dB less than the CPICH Ec/Io of the only active set cell, i.e. the neighbouring cell appears to be only just within the addition window. The CPICH Ec/Io of the neighbouring cell with scrambling code 48 is significantly less than the other two cells. This indicates that the UE is in an area of relatively good

Table 8.5 Mapping between the signalled and actual values of CPICH Ec/Io (result in dB)

0	$x < -24.0$	13	$-18.0 \le x < -17.5$	26	$-11.5 \le x < -11.0$	39	$-5.0 \le x < -4.5$
1	$-24.0 \le x < -23.5$	14	$-17.5 \le x < -17.0$	27	$-11.0 \le x < -10.5$	40	$-4.5 \le x < -4.0$
2	$-23.5 \le x < -23.0$	15	$-17.0 \le x < -16.5$	28	$-10.5 \le x < -10.0$	41	$-4.0 \le x < -3.5$
3	$-23.0 \le x < -22.5$	16	$-16.5 \le x < -16.0$	29	$-10.0 \le x < -9.5$	42	$-3.5 \le x < -3.0$
4	$-22.5 \le x < -22.0$	17	$-16.0 \le x < -15.5$	30	$-9.5 \le x < -9.0$	43	$-3.0 \le x < -2.5$
5	$-22.0 \le x < -21.5$	18	$-15.5 \le x < -15.0$	31	$-9.0 \le x < -8.5$	44	$-2.5 \le x < -2.0$
6	$-21.5 \le x < -21.0$	19	$-15.0 \le x < -14.5$	32	$-8.5 \le x < -8.0$	45	$-2.0 \le x < -1.5$
7	$-21.0 \le x < -20.5$	20	$-14.5 \le x < -14.0$	33	$-8.0 \le x < -7.5$	46	$-1.5 \le x < -1.0$
8	$-20.5 \le x < -20.0$	21	$-14.0 \le x < -13.5$	34	$-7.5 \le x < -7.0$	47	$-1.0 \le x < -0.5$
9	$-20.0 \le x < -19.5$	22	$-13.5 \le x < -13.0$	35	$-7.0 \le x < -6.5$	48	$-0.5 \le x < 0.0$
10	$-19.5 \le x < -19.0$	23	$-13.0 \le x < -12.5$	36	$-6.5 \le x < -6.0$	49	$-0.0 \le x$
11	$-19.0 \le x < -18.5$	24	$-12.5 \le x < -12.0$	37	$-6.0 \le x < -5.5$		
12	$-18.5 \le x < -18.0$	25	$-12.0 \le x < -11.5$	38	$-5.5 \le x < -5.0$		

dominance. The final section of the Measurement Report confirms that the cell with scrambling code 8 was responsible for triggering the event 1a.

The RNC completes its admission control procedure after receiving the event 1a Measurement Report. Admission control is used to ensure that there are sufficient resources to introduce the new radio link. In general, admission control checks the availability of downlink transmit power, Node B baseband processing, Iub transport capacity and downlink channelisation code resources. It is not necessary to check the uplink interference margin because the candidate cell is already receiving uplink interference from the UE. Including the cell within the active set should help to reduce the level of uplink interference because the cell will be able to start power controlling the UE.

After completing its admission control procedure, the RNC forwards a Common NBAP Radio Link Setup Request to the Node B. This message type is applicable because it is assumed that the Node B does not already have a radio link for the UE, i.e. the event 1a is for a soft handover connection rather than a softer handover connection. Section 8.6.2 illustrates that a Dedicated NBAP Radio Link Addition Request is used when the Node B already has at least one existing radio link for the UE. The Radio Link Setup Request configures the new radio link at the Node B. If the UE has triggered the event 1a while having only a standalone SRB then the Node B is only provided with the SRB configuration. Otherwise, if the UE has triggered the event 1a while having both SRB and user plane connections then the Node B is provided with the relevant configuration information for both connection types. The content of the message is similar to the Radio Link Setup Request described in Section 8.1 and the Radio Link Reconfiguration Prepare messages described in Sections 8.2, 8.3 and 8.5. The Node B attempts to achieve uplink air-interface synchronisation after receiving the Radio Link Setup Request. The Node B also acknowledges the setup request using a Radio Link Setup Response message.

The Radio Link Setup Response includes the transport network information necessary for the RNC to establish the Iub transport connections, i.e. the transport layer address and binding identities are included. Figure 8.50 illustrates the RNC using ALCAP signalling to establish two Iub transport connections. This assumes that the UE has both SRB and user plane DCH connections. The RNC would establish only a single Iub transport connection if the UE had only a standalone SRB. The Frame Protocol layer achieves synchronisation across the new transport connections after they have been established. Downlink and Uplink Synchronisation control frames are used for this purpose.

The Node B uses a Dedicated NBAP Radio Link Restore message to inform the RNC that it has successfully achieved uplink synchronisation across the air-interface. Figure 8.50 illustrates the Node B sending this message after the Iub transport connections have been established. However, this message does not rely upon the Iub transport connections and can be sent as soon as synchronisation has been achieved, e.g. it could be sent during the ALCAP signalling procedures. The Node B starts to transmit at its initial transmit power after achieving uplink synchronisation. The Node B can then start to obey the inner loop power control commands sent by the UE. At this stage, the UE does not receive the new downlink transmission.

The RNC proceeds to instruct the UE to add the new cell to the active set. This is done using an Active Set Update message. Both the original and new active set cells transmit this message although the UE will only receive it from the original active set cell. An example Active Set Update message is shown in Log File 8.66.

The Active Set Update message instructs the UE to add a new radio link to the active set. The UE is informed that the new radio link has a scrambling code of 8, a downlink spreading factor of 128 and a channelisation code index of 7. The UE is also informed of the time offset between the new downlink DPCH radio frames and the P-CCPCH radio frames belonging to the new cell. This time offset is illustrated in Figure 3.34. The actual value of the time offset is equal to the signalled frame offset $\times$ 256 chips, i.e. $74 \times 256 = 18944$ chips in this example. The UE uses this information to help achieve downlink synchronisation across the air-interface. The Active Set Update message also specifies a Transmit Power Control (TPC) combination index. The UE uses this index to identify which

```
DL-DCCH-Message
integrityCheckInfo
messageAuthenticationCode
Bin: 55 3B E0 28
rrc-MessageSequenceNumber: 6
activeSetUpdate-r3
  rrc-TransactionIdentifier: 2
  maxAllowedUL-TX-Power: 24
  rl-AdditionInformationList
  rl-AdditionInformationList value 1
    primaryCPICH-Info
      primaryScramblingCode: 8
    dl-DPCH-InfoPerRL
      pCPICH-UsageForChannelEst: mayBeUsed
      dpch-FrameOffset: 74
      dl-ChannelisationCodeList
      dl-ChannelisationCodeList value 1
        sf-AndCodeNumber: sf128: 7
      tpc-CombinationIndex: 1
```

Log File 8.66   Active Set Update message used to add a cell to the active set

radio links belong to the same radio link set. The UE can combine TPC commands originating from the same radio link set because they are always guaranteed to match. In the case of soft handover, the new radio link belongs to a different radio link set so the combination index will differ from the value assigned to the original active set cell. This indicates that the UE should not combine the TPC commands.

The UE acknowledges the Active Set Update message using an Active Set Update Complete message. This message can be sent before the UE has achieved downlink synchronisation with the new active set cell although it should be received and forwarded to the RNC by both active set cells. An example Active Set Update Complete message is shown in Log File 8.67. The transaction identifier within this message matches the value within the corresponding Active Set Update message. This allows the RNC to link the two messages to the same transaction.

Figure 8.50 illustrates that the RNC returns a Measurement Control message after receiving the Active Set Update Complete. This Measurement Control can be used to configure reporting events 6f and 6g. These reporting events are illustrated in Figure 2.34 and Figure 2.35. They are only applicable when the UE has an active set size greater than 1. The UE sends an event 6f or 6g Measurement Report if the downlink timing of an active set cell drifts outside a specific time window. The RNC may also forward a further Measurement Control message to update the neighbour list information now that a new cell has joined the active set.

Figure 8.52 illustrates the signalling associated with deleting a cell from the active set. This example is based upon using an event 1b Measurement Report to trigger the procedure. Alternatively, the procedure could be triggered using one or more periodic Measurement Reports if periodic rather than event driven reporting has been enabled.

```
UL-DCCH-Message
integrityCheckInfo
messageAuthenticationCode
Bin: D1 E4 BA 9B
rrc-MessageSequenceNumber: 8
activeSetUpdateComplete
  rrc-TransactionIdentifier: 2
```

Log File 8.67   Active Set Update Complete message

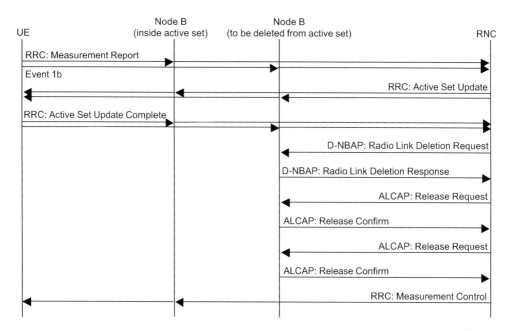

Figure 8.52   Soft handover active set cell deletion (inter-Node B)

An example event 1b Measurement Report is shown in Log File 8.68. The content is similar to that of an event 1a Measurement Report. The original Measurement Control message shown in Log File 8.64 provided instructions for the UE to only report active set cells. In this example, timing and CPICH Ec/Io measurements are reported for two cells, which indicates that the active set size is equal to 2. There is a 7.5 dB difference between the two CPICH Ec/Io measurements.

```
UL-DCCH-Message
integrityCheckInfo
messageAuthenticationCode           intraFreqMeasuredResultsList 2
Bin: 82 3F 7A 07                    cellSynchronisationInfo
rrc-MessageSequenceNumber: 3        modeSpecificInfo fdd
measurementReport                   countC-SFN-Frame-difference
measurementIdentity: 1              countC-SFN-High: 6
                                    off: 4
                                    tm: 3072
intraFreqMeasuredResultsList        modeSpecificInfo fdd
intraFreqMeasuredResultsList 1      primaryCPICH-Info
cellSynchronisationInfo             primaryScramblingCode: 8
modeSpecificInfo fdd                cpich-Ec-N0: 25 (-12 to -11.5 dB)
countC-SFN-Frame-difference
countC-SFN-High: 6                  eventResults
off: 4                              intraFreqEventResults
tm: 3844                            eventID : e1b
modeSpecificInfo fdd                cellMeasurementEventResults fdd
primaryCPICH-Info                   fdd value 1
primaryScramblingCode: 104          primaryScramblingCode  : 8
cpich-Ec-N0: 40 (-4.5 to -4.0 dB)
```

Log File 8.68   Measurement Report triggered by event 1b

```
DL-DCCH-Message
integrityCheckInfo
messageAuthenticationCode
Bin: 2D 07 21 DA
rrc-MessageSequenceNumber: 7
activeSetUpdate-r3
  rrc-TransactionIdentifier: 3
  maxAllowedUL-TX-Power: 24
  rl-RemovalInformationList
  rl-RemovalInformationList value 1
    primaryScramblingCode: 8
```

Log File 8.69 Active Set Update message used to delete a cell from the active set

This exceeds the 6 dB deletion window which explains why the UE has triggered an event 1b Measurement Report.

The RNC responds to the Measurement Report using an Active Set Update message. This message instructs the UE to remove the active set cell whose CPICH Ec/Io has fallen below the deletion window. An example of an Active Set Update message being used to delete an active set cell is shown in Log File 8.69. In this example, the cell using a scrambling code of 8 is being removed from the active set.

The UE responds to the RNC using an Active Set Update Complete message. At this stage, the cell being removed from the active set continues to transmit and receive while the UE stops receiving its downlink signal. The RNC then proceeds to use NBAP signalling to delete the radio link from the Node B. The Node B then stops transmitting and receiving for the relevant UE. The RNC uses ALCAP signalling to release the Iub transport resources. The example shown in Figure 8.52 assumes that the UE has been configured with both SRB and user plane DCH. This means that there are two Iub transport connections to release. Finally, the RNC may forward a new Measurement Control message to update the neighbour list information.

8.6.2 Intra-Node B

The signalling procedures for softer handover (intra-Node B) are similar to those for soft handover (inter-Node B). Figure 8.53 illustrates the procedure used to add cell 2 to the active set when cell 2 belongs to a Node B which already has a cell 1 within the active set. The Measurement Control message is the same as that shown in Figure 8.50, i.e. the Measurement Control message does not differentiate between intra-Node B and inter-Node B soft handover procedures. Similarly, the format of the event 1a Measurement Report is the same as that for an inter-Node B scenario.

However, the RNC uses a Dedicated NBAP Radio Link Addition Request rather than a Common NBAP Radio Link Setup Request. A Radio Link Addition Request is significantly shorter than a Radio Link Setup Request because the Node B already has knowledge of the radio link configuration, i.e. it is not necessary to re-inform the Node B of the transport format sets, the transport format combination sets and the physical channel characteristics. An example Radio Link Addition Request is shown in Log File 8.70. The Node B communication context is used to identify the relevant connection, while the cell identity is used to identify the target cell. The new radio link is allocated an identity which must differ from that of the existing radio link at the same Node B. The Node B is also provided with timing information which has been calculated by the RNC based upon the timing measurements reported by the UE within the preceding Measurement Report. The scrambling code index of 0 indicates that the cell should transmit using its primary scrambling code rather than a secondary scrambling code. The RNC also allocates a channelisation code to the new radio link. The index of the allocated channelisation code

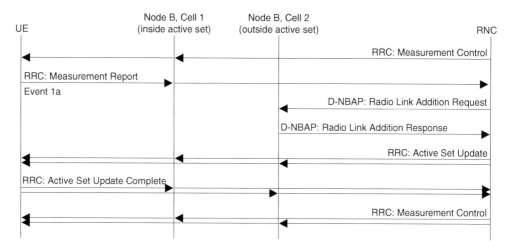

Figure 8.53 Soft handover active set cell addition (intra-Node B)

is signalled without the spreading factor because the Node B already has knowledge of the appropriate spreading factor.

After receiving the Radio Link Addition Response, the RNC proceeds directly with the active set update procedure. It is not necessary to establish any new Iub transport connections because the Node B already has connections for the existing active set cell. This illustrates that softer handover does not generate an overhead in terms of the Iub transport resources. Similarly, softer handover typically does not generate an overhead in terms of Node B baseband processing. This is in contrast to soft handover which generates both Iub transport and Node B baseband processing overheads. The format of the Active Set Update and Active Set Update Complete messages are the same as those for an inter-Node B scenario. In this case, the TPC combination index within the Active Set Update message will match the value already allocated to the existing active set cell.

Figure 8.54 illustrates the procedure used to delete cell 2 from the active set when the active set includes both cell 1 and cell 2 which belong to the same Node B. This procedure is identical to the inter-Node B scenario with the exception that the Iub transport and Node B baseband processing resources are retained. These resources are required by the remaining radio link. The Node B is able to

```
NBAP-PDU
RadioLinkAdditionRequestFDD
NodeB-CommunicationContextID: 160001
RL-InformationList-RL-AdditionRqstFDD
RL-InformationItem-RL-AdditionRqstFDD
- rL-ID: 2
- c-ID: 141
- frameOffset: 4
- chipOffset: 9985
- diversityControlField: may
dl-CodeInformation
FDD-DL-CodeInformationItem
- dl-ScramblingCode: 0
- fdd-DL-ChannelisationCodeNumber: 10
```

Log File 8.70 NBAP Radio Link Addition Request message

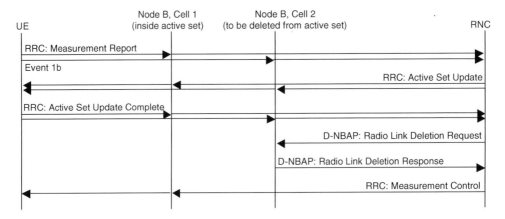

Figure 8.54 Soft handover active set cell deletion (intra-Node B)

deduce that it should not release its baseband processing resources by recognising that it still has a radio link associated with the relevant Node B communication context. The Iub transport resources are not released because the RNC does not send any ALCAP Release Request messages.

8.7 Inter-System Handover

- Dual-mode terminals are capable of inter-system handover between UMTS and GSM. Inter-system handovers are typically triggered by coverage, network load or service type.
- Coverage-based inter-system handovers can be triggered by UE measurements of CPICH RSCP, CPICH Ec/Io or uplink transmit power. Alternatively, they can be triggered by Node B measurements of downlink transmit power.
- Triggering thresholds should be low enough to avoid triggering unnecessary inter-system handover attempts, but should be high enough to allow sufficient time for the inter-system handover procedure to be completed prior to the UE moving completely out of coverage.
- A Physical Channel Reconfiguration Prepare message can be sent to the Node B after the RNC has decided to initiate an inter-system handover attempt. This message is used to configure the parameters associated with compressed mode.
- Compressed mode is used to generate transmission gaps which allow UE to record measurements from GSM. Speech service inter-system handover requires one compressed mode TGPS for RSSI measurements and another for BSIC verification.
- A single TGPS can include a maximum of two transmission gaps. The first gap starts during the time slot specified by the TGSN. The TGL defines the duration of the gap. The TGD defines the number of time slots between the start of the first gap and the start of the second gap. The TGPL specifies the number of radio frames belonging to the gap pattern.
- Compressed mode can use either SF/2 or Higher Layer Scheduling (HLS). The release 99 and release 4 versions of the specifications also define puncturing as a downlink method. Inter-system handover for a speech connection typically uses the SF/2 method.
- The SF/2 method involves halving the spreading factor to double the DPDCH bit rate. The increased bit rate compensates for the transmission gap and the higher layer throughput remains relatively unchanged.

- The HLS method involves the MAC layer selecting a lower transport format combination. This reduces the quantity of data transferred to the physical layer which consequently allows a transmission gap to be generated. The higher layer throughput is reduced.
- The RNC provides the Node B with a compressed mode activation CFN within the Radio Link Reconfiguration Commit message. This CFN must allow sufficient time for the compressed mode configuration to be signalled to the UE.
- The compressed mode configuration is signalled to the UE using a Physical Channel Reconfiguration message. This includes the measurement purpose associated with each TGPS, i.e. RSSI measurements and BSIC verification. The Reconfiguration message also includes the activation CFN.
- The UE is provided with the inter-system neighbour list within a Measurement Control message. This message also includes instructions for reporting the UE measurements to the RNC. The UE then proceeds to measure the RSSI belonging to each GSM neighbour.
- The UE records an RSSI measurement from every RF carrier in the neighbour list before providing any measurement results. This allows the UE to identify the strongest RF carriers and report them in order of their strength.
- The RNC initiates BSIC verification by forwarding a Compressed Mode Command to the Node B and a further Measurement Control to the UE. The length of the neighbour list can be reduced to avoid the UE attempting BSIC verification for weak neighbouring cells.
- BSIC verification allows the RNC to confirm the identity of the target neighbouring cell and so ensures that the GSM network is requested to reserve resources at the correct cell.
- The RNC starts to involve the GSM network once BSIC verification has been completed. The RNC sends a RANAP Relocation Required message to the 3G MSC. This message specifies the target for the inter-system handover using a GSM Cell Global Identity (CGI).
- The 3G MSC forwards a Prepare Handover request to the 2G MSC which subsequently generates a Handover Request for the BSC.
- The Handover Request triggers the BSC to reserve a set resources for the incoming speech connection and to return a Handover Request Acknowledge. The Handover Request Acknowledge includes the Handover Command which is to be provided to the UE.
- The Handover Request is forwarded from the 2G MSC to the 3G MSC and then onto the RNC which packages it within a Handover from UTRAN Command. This message provides the UE with sufficient information to continue its speech connection on GSM.
- The UE moves onto the GSM network and provides the BSC with a Handover Complete message to indicate that inter-system handover has been successful. The 3G MSC starts to release the UMTS resources after receiving notification of the successful handover.
- Key 3GPP specifications: TS 25.211, TS 25.212, TS 25.214, TS 25.215, TS 25.331, TS 25.413, TS 25.433, TS 05.05, TS 08.08, TS 09.02 and TS 04.18.

8.7.1 Speech

Dual-mode terminals are capable of inter-system handovers between the UMTS and GSM networks. Inter-system handovers are typically triggered by coverage, network load or service type. Coverage-based handovers are useful when GSM coverage is more extensive than UMTS coverage. A UE with an active speech connection and moving out of UMTS coverage can trigger an inter-system handover to avoid its connection dropping. Load based handovers are useful when the UMTS network starts to experience congestion. Inter-system handovers can help to reduce congestion by moving speech connections onto the GSM network. Load based handovers typically target speech connections rather than data connections because the impact upon the end-user quality of service is minimal. Data connections typically experience a reduction in throughput when moved onto the GPRS system. Service-based handovers are applicable if the operator wishes to associate specific services with specific network layers. Inter-system handovers can be used to move all speech traffic onto the GSM network while leaving data traffic on the UMTS network.

The first phase of signalling associated with a UMTS to GSM speech connection inter-system handover is shown in Figure 8.55. The RNC starts by forwarding a series of Measurement Control messages. These are used to configure the set of coverage-based triggering mechanisms which can be reported by the UE. They are typically based upon CPICH Ec/Io, CPICH RSCP and UE transmit power. Measurement reporting event 1f can be used to instruct the UE to inform the RNC if the CPICH RSCP or CPICH Ec/Io falls below a specific absolute threshold. Measurement reporting event 6a can be used to instruct the UE to inform the RNC if the UE transmit power exceeds a specific absolute threshold. Events 1e and 6b can be used to instruct the UE to inform the RNC if conditions recover

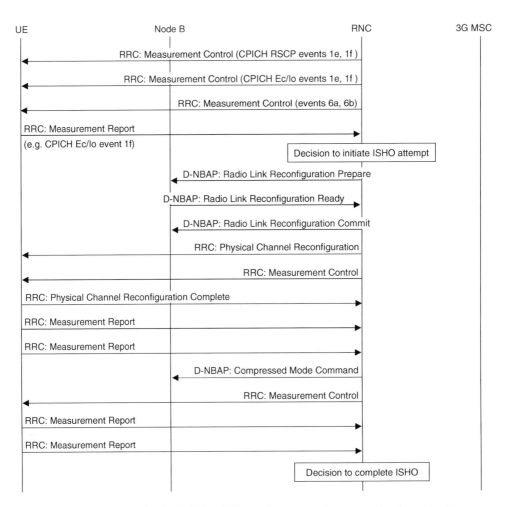

Figure 8.55 Signalling for UMTS to GSM speech connection inter-system handover (part 1)

before the inter-system handover has been completed, i.e. the inter-system handover attempt can be cancelled. The RNC may also have its own coverage-based inter-system handover triggering mechanisms which do not rely upon Measurement Reports from the UE. The RNC could trigger inter-system handover based upon the downlink radio link transmit power exceeding an absolute threshold. This mechanism relies upon the Node B reporting the downlink transmit power used by each individual radio link. The triggering mechanisms based upon CPICH RSCP, CPICH Ec/Io, UE transmit power and downlink transmit power are all associated with coverage, i.e. at least one of these mechanisms will be triggered if the UE moves towards the edge of coverage. The first mechanism to be triggered depends upon whether coverage is uplink service limited, downlink service limited or CPICH limited. Load- and service-based handovers do not require measurement reporting from the UE so Measurement Control messages are not necessary. Figure 8.55 illustrates the Measurement Control messages immediately before an inter-system handover attempt is initiated. In practice, the Measurement Control messages are likely to be sent immediately after connection establishment whereas the inter-system handover could be triggered at any time afterwards.

Log File 8.71 presents an example Measurement Control message used to configure reporting events 1e and 1f for CPICH RSCP. The reporting event is allocated an identity of 3. This identity allows the UE and RNC to link any subsequent Measurement Reports. The intra-frequency measurement quantity is specified to be CPICH RSCP and an RRC layer filter coefficient of 3 is specified. CPICH Ec/Io and path loss can also be specified as intra-frequency measurement quantities. The filter coefficient defines the filtering applied to the Physical layer measurements prior to evaluating whether or not an event 1e or event 1f has been triggered. The filtering method is the same as that described for soft handover within Section 8.6.1. The Measurement Control message proceeds to inform the UE of which measurements to report when either event 1e or 1f is triggered. In this example, the UE is instructed to report timing information, CPICH RSCP and CPICH Ec/Io for the active set cells. The timing information quantifies the time

```
DL-DCCH-Message
integrityCheckInfo
messageAuthenticationCode              reportCriteria
Bin: 83 5C 70 E2                       intraFreqReportingCriteria
rrc-MessageSequenceNumber: 0           eventCriteriaList
measurementControl                     eventCriteriaList value 1
rrc-TransactionIdentifier: 0           event e1e
measurementControl-r4                    trigCond: activeSetCellsOnly
measurementIdentity: 3                   thresholdUsedFrequency: -105
measurementCommand setup                 hysteresis: 0
intraFrequencyMeasurement                timeToTrigger: ttt1280
  intraFreqMeasQuantity                  reportingCellStatus
  filterCoefficient: fc3                   withinActiveSet: e3
  modeSpecificInfo fdd                 eventCriteriaList value 2
  intraFreqMeasQuantity: cpich-RSCP    event e1f
intraFreqReportingQuantity               trigCond: activeSetCellsOnly
activeSetReportingQuantities             thresholdUsedFrequency: -108
  cellIdentity-reportingInd: false       hysteresis: 0
  cellSynchInfoReportingInd: true        timeToTrigger: ttt640
  cpich-Ec-N0-reportingInd: true         reportingCellStatus
  cpich-RSCP-reportingInd: true            withinActiveSet: e3
  pathloss-reportingInd: false
monitoredSetReportingQuantities        measurementReportingMode
  cellIdentity-reportingInd: false     measReportTransferMode: ackModeRLC
  cellSynchInfoReportingInd: false     periodicalOrEventTrig: eventTrigger
  cpich-Ec-N0-reportingInd: false
  cpich-RSCP-reportingInd: false
  pathloss-reportingInd: false
```

Log File 8.71 Measurement Control message used to configure events 1e and 1f for CPICH RSCP difference between the SFN timing of the cell being measured and the CFN timing of the connection. The UE is also instructed not to report anything for the monitored set cells.

The Measurement Control message proceeds to specify the criteria for triggering events 1e and 1f. Event 1e is configured to be triggered if the CPICH RSCP remains greater than $-105\,$dBm for a period of 1280 ms. Event 1f is configured to be triggered if the CPICH RSCP remains less than $-108\,$dBm for a period of 640 ms. Both events are configured to be triggered only by active set cells. In addition, the UE is instructed to provide measurement information for a maximum of three active set cells.

A similar Measurement Control message can be used to configure events 1e and 1f for CPICH Ec/Io. A separate measurement identity is allocated to allow any subsequent Measurement Reports to be differentiated. Typical triggering thresholds for CPICH Ec/Io are $-14\,$dB for event 1f and $-12\,$dB for event 1e. Triggering thresholds should be low enough to avoid triggering unnecessary inter-system handover attempts, but they should be high enough to allow sufficient time for the inter-system handover procedure to be completed prior to the UE moving completely out of coverage.

Log File 8.72 presents an example Measurement Control message used to configure reporting events 6a and 6b. The reporting event is allocated an identity of 5. The UE internal measurement quantity is specified to be UE transmit power and an RRC layer filter coefficient of 8 is specified. Downlink RSSI and Rx−Tx time difference can also be specified as UE internal measurement quantities. The filter coefficient defines the filtering applied to the physical layer measurements prior to evaluating whether or not an event 6a or event 6b has been triggered. The filtering method is the same as that described for soft handover within Section 8.6.1 except that filtering is based upon receiving a physical layer measurement once every time slot (15 times per 10 ms radio frame) rather than once every 200 ms. The increased rate at which physical layer measurements are generated means that larger filter coefficients are typically used for events 6a and 6b. The Measurement Control message proceeds to inform the UE of which measurements to report when either event 6a or 6b is triggered. In this example, the UE is instructed to report its transmit power, but not its Rx−Tx time difference.

The Measurement Control message proceeds to specify the criteria for triggering events 6a and 6b. Event 6a is triggered if the UE transmit power remains greater than 20 dBm for a period of 320 ms. Event 6b is triggered if the UE transmit power remains less than 18 dBm for a period of 320 ms.

```
DL-DCCH-Message                             reportCriteria
integrityCheckInfo                          ue-InternalReportingCriteria
messageAuthenticationCode                   ue-InternalEventParamList
Bin: 12 3D 33 5A                            ue-InternalEventParamList value 1
rrc-MessageSequenceNumber: 1                  event6a
measurementControl                            timeToTrigger: tt320
rrc-TransactionIdentifier: 1                  transmittedPowerThreshold: 20
measurementControl-r4                       ue-InternalEventParamList value 2
measurementIdentity: 5                        event6b
measurementCommand setup                      timeToTrigger: tt320
ue-InternalMeasurement                        transmittedPowerThreshold: 18
  ue-InternalMeasQuantity
  measQuantity: ue-TransPower               measurementReportingMode
  filterCoefficient: fc8                    measReportTransferMode: ackModeRLC
ue-InternalReportingQuantity                periodicalOrEventTrig: eventTrigger
  ue-TransmittedPower: true
  ue-RX-TX-TimeDifference: false
```

Log File 8.72   Measurement Control message used to configure events 6a and 6b

Figure 8.55 illustrates an inter-system handover attempt initiated after an event 1f Measurement Report. It is assumed that the event 1f has been triggered by CPICH Ec/Io rather than CPICH RSCP. Log File 8.73 illustrates an example event 1f Measurement Report triggered by CPICH Ec/Io. The Measurement Report provides results for two cells because there were two cells in the active set when the event 1f was triggered. The active set cells are identified by their scrambling codes. Measurements are provided in terms of timing information and CPICH Ec/Io. In this example, the UE has not been requested to report CPICH RSCP when an event 1f is triggered by CPICH Ec/Io. The final section of the Measurement Report specifies that the event 1f has been triggered by the active set cell with scrambling code 34.

Log File 8.73 reports that one of the two active set cells has triggered event 1f. It is typical to require that all active set cells trigger an event 1f prior to initiating an inter-system handover attempt. This avoids inter-system handovers resulting from individual active set cells with poor coverage while there are other active set cells with good coverage. An inter-system handover attempt is typically initiated if all active set cells have triggered an event 1f without subsequently triggering an event 1e. This requirement is shown in Figure 8.56. The set of event 1f Measurement Reports which trigger an inter-system handover attempt should be associated with the same measurement, i.e. they should all be associated with CPICH Ec/Io or all be associated with CPICH RSCP.

Figure 8.55 illustrates that a dedicated NBAP Physical Channel Reconfiguration Prepare message is sent to the Node B after the RNC has decided to initiate an inter-system handover attempt. This message is used to configure the parameters associated with compressed mode. The precise signalling used to configure these parameters depends upon the RNC implementation. Some implementations may configure the compressed mode parameters during connection establishment. This approach increases the size of the messages during connection establishment for all UE irrespective of whether or not inter-system handover is triggered. The description within this section assumes that the compressed mode parameters are only configured when necessary, i.e. after inter-system handover has been triggered.

Compressed mode is used to generate transmission gaps which allow the UE to record measurements from other RF carriers. Within the context of inter-system handover, the other RF carriers belong to different Radio Access Technologies (RAT). Compressed mode can also be used to measure other UMTS RF carriers for the purposes of inter-frequency handover. The majority of UE

```
UL-DCCH-Message
integrityCheckInfo
messageAuthenticationCode                 intraFreqMeasResultsList value 2
Bin: AC 3C AD 37                            cellSynchronisationInfo
rrc-MessageSequenceNumber: 2                  countC-SFN-Frame-difference
measurementReport                               countC-SFN-High: 6
measurementIdentity: 5                        off: 4
measuredResults                               tm: 3072
intraFreqMeasuredResultsList                modeSpecificInfo fdd
                                            primaryCPICH-Info
intraFreqMeasResultsList value 1              primaryScramblingCode: 34
  cellSynchronisationInfo                   cpich-Ec-N0: 20 (-14.5 to -14 dB)
    countC-SFN-Frame-difference
      countC-SFN-High: 6                  eventResults
    off: 4                                intraFreqEventResults
    tm: 3844                                eventID: e1f
  modeSpecificInfo fdd                      cellMeasurementEventResults
  primaryCPICH-Info                         fdd value 1
    primaryScramblingCode: 33               primaryScramblingCode: 34
  cpich-Ec-N0: 28 (-10.5 to -10 dB)
```

Log File 8.73   Measurement Report triggered by event 1f

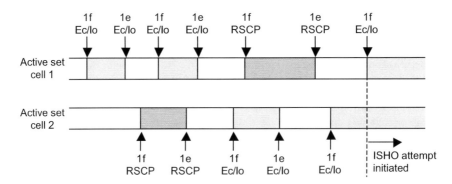

Figure 8.56   Initiating an inter-system handover after all active set cells have triggered an event 1f

require compressed mode when recording measurements on other RF carriers. Compressed mode is necessary in the downlink if a UE has a single receiver. Compressed mode will also be necessary in the uplink if the UE has fixed duplex spacings or if the uplink transmissions are likely to interfere with the downlink measurements. A UE with two receivers may not require compressed mode although this can depend upon the frequency separation between the uplink band within which the UE is transmitting and the downlink band within which the UE is required to measure. UE inform the RNC of which measurements require compressed mode within the UE radio access capability section of the RRC Setup Complete message. An example of this information is shown in Log File 8.16 within Section 8.1. An example Radio Link Reconfiguration Prepare message is shown in Log File 8.74.

This message identifies the relevant connection using the Node B communication context. It also specifies that the reconfiguration is applicable to radio link 1. The downlink scrambling code index of 0

```
NBAP-PDU                                   - UL-Comp-Mode-Method: sFdiv2
RadioLinkReconfigurationPrepare            - dL-FrameType: typeA
protocolIEs                                - deltaSIR1: 15
NodeB-CommContextID: 160004                - deltaSIRafter1: 7
RadioLinkModInfList-RL-ReconfPrep          - deltaSIR2: 0
RadioLinkModInfItem-RL-ReconfPrep          - deltaSIRafter2: 0
- rL-ID: 1
dl-CodeInformation                         - tGPSI: 3
FDD-DL-CodeInformationItem                 - tGSN: 4
- dl-ScramblingCode: 0                     - tGL1: 7
- fdd-DL-ChanCodeNumber: 5                 - tGD: 0
- transGapPatCodeInf: code-change          - tGPL1: 4
                                           - uL-DL-mode: both-ul-and-dl
Trans-Gap-Pattern-Sequence-Inf             - DL-Comp-Mode-Method: sFdiv2
tGPSI: 2                                    - UL-Comp-Mode-Method: sFdiv2
- tGSN: 4                                  - dL-FrameType: typeA
- tGL1: 7                                  - deltaSIR1: 15
- tGD: 0                                   - deltaSIRafter1: 7
- tGPL1: 4                                 - deltaSIR2: 0
- uL-DL-mode: both-ul-and-dl               - deltaSIRafter2: 0
- DL-Comp-Mode-Method: sFdiv2
```

Log File 8.74   Radio Link Reconfiguration Prepare used to configure compressed mode

indicates that the Node B should use the primary scrambling code rather than a secondary scrambling code. The Node B is also instructed to use the channelisation code with an index of 5. The downlink spreading factor is not specified within this message, but has already been configured as 128. The final part of the downlink code information instructs the UE to change its scrambling code when applying the SF/2 compressed mode method. This corresponds to using an alternative scrambling code. Alternative scrambling codes are described in Section 3.3.9.2. The UE always uses the parent channelisation code when applying the SF/2 compressed mode method. In this example, the UE switches from channelisation code (SF128,5) to (SF64,2), and at the same time switches to the alternative scrambling code associated with the combination of primary scrambling code and channelisation code.

The Physical Channel Reconfiguration Prepare message specifies the parameters associated with two Transmission Gap Pattern Sequences (TGPS). These TGPS are identified using their TGPS identities (TGPSI). Different measurement purposes are associated with each TGPS. For example, TGPS 1 can be configured for UMTS inter-frequency measurements, TGPS 2 can be configured for GSM RSSI measurements and TGPS 3 can be configured for GSM BSIC verification. This example is based upon inter-system handover for a CS speech connection which requires GSM RSSI measurements and BSIC verification, i.e. it is necessary to configure two TGPS. The Node B is not informed of the measurement purpose associated with each TGPS. This information is only provided to the UE which becomes responsible for completing the measurements. Log File 8.74 indicates that in this example the TGPS for GSM RSSI measurements is configured using the same parameter set as the TGPS for BSIC verification. The timing parameters associated with a specific TGPS are illustrated in Figure 8.57.

A single TGPS can include a maximum of two transmission gaps. The first transmission gap starts during the time slot specified by the Transmission Gap Starting Slot Number (TGSN). Log File 8.74 indicates that the TGSN is configured with a value of 4. Time slot numbering starts from zero so this represents the fifth time slot belonging to the radio frame. The Transmission Gap Length (TGL) specifies the duration of the transmission gap. Log File 8.74 indicates that the first transmission gap has a duration of seven time slots. The Transmission Gap Distance (TGD) specifies the number of time slots between the start of the first transmission gap and the start of the second transmission gap. In this example, the TGD is configured with a value of 0 which indicates that the TGPS includes only a single transmission gap. This is consistent with the message excluding a second TGL. The Transmission Gap Pattern Length (TGPL) specifies the number of radio frames belonging to the transmission gap pattern. In this example, the TGPL has a value of 4 radio frames. This indicates that there is one transmission gap of 4.67 ms every 40 ms. This TGPS configuration is illustrated in the top half of Figure 8.58.

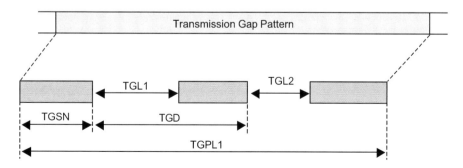

Figure 8.57 Timing parameters associated with a specific TGPS

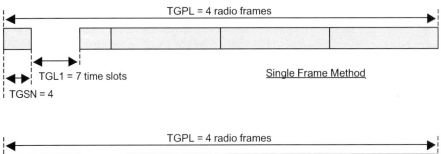

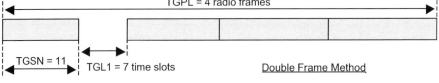

Figure 8.58   Compressed mode transmission gaps illustrating the single and double frame methods

Transmission gaps can be timed such that they start and finish within a single radio frame or such that they span two consecutive radio frames. The former is known as the single frame method whereas the latter is known as the double frame method. In both cases, the transmission gap must allow a minimum of eight time slots to be transmitted within each radio frame. Table 8.6 presents the range of values associated with each of the TGPS timing parameters. TGL values greater than 7 are only applicable to the double frame method.

The release 99 and release 4 versions of 3GPP TS 25.215 allow each TGPS to be configured with two transmission gap patterns (allowing up to four transmission gaps). 3GPP subsequently agreed that this capability was unnecessary so the release 5 and newer versions of the specifications allow only a single transmission gap pattern to be associated with each TGPS.

Log File 8.74 proceeds to specify that compressed mode should be used in both the uplink and downlink directions. It also specifies that the spreading factor division by 2 (SF/2) compressed mode method should be used. 3GPP TS 25.212 specifies that uplink compressed mode can be based upon either SF/2 or Higher Layer Scheduling (HLS). The SF/2 method involves halving the spreading factor to double the bit rate of the DPDCH during the one or two radio frames within which the transmission gap is located. The increased bit rate compensates for the transmission gap and the higher layer throughput remains relatively unchanged. The SF/2 method is not applicable to DPCH connections using a spreading factor of 4 because a DPCH spreading factor of 2 is not supported. The time slots transmitted with SF/2 require approximately 3 dB greater UE transmit power. In addition, compressed mode tends to cause an increase in transmit power immediately after a transmission gap during the inner loop power control recovery period. Power control commands cannot be sent during the

Table 8.6   Range of values associated with the TGPS timing parameters

Parameter		Range
TGPSI	Transmission Gap Pattern Sequence Identifier	1–6
TGSN	Transmission Gap Starting Slot Number	0–14
TGL1	Transmission Gap Length 1	1–14 slots
TGL2	Transmission Gap Length 2	1–14 slots
TGD	Transmission Gap Start Distance	0, 15–269 slots
TGPL1	Transmission Gap Pattern Length 1	1–144 frames

transmission gap so inner loop power control can lose track of the power required at the receiver. UE are usually instructed to over-compensate after the transmission gap to help regain inner loop power control tracking as rapidly as possible. These short term increases in UE transmit power mean that the SF/2 method is less suitable for scenarios in which the UE is transmitting close to its maximum power.

The HLS method involves the MAC layer selecting a lower Transport Format Combination (TFC). The lower TFC reduces the quantity of data transferred to the physical layer which consequently allows the compressed mode transmission gap to be generated. In this case, the higher layer throughput is reduced during compressed mode. This means that in general, HLS is less appropriate for connections supporting real time services. The HLS method can be applied to the AMR speech service if lower AMR bit rates are supported. 3GPP TS 25.212 specifies that HLS can only be used for transport channels which have flexible positions within the radio frame. Speech, video call and PS data connections typically use flexible positions in the uplink direction so this requirement is not a limitation. Similar to the SF/2 method, the HLS method can generate an increase in the UE transmit power during the inner loop power control recovery period following a transmission gap.

The release 99 and release 4 versions of 3GPP TS 25.212 define three compressed mode methods for the downlink direction. These methods are SF/2, HLS and puncturing. 3GPP subsequently agreed to remove the puncturing method so the release 5 and newer versions of the specifications only define the SF/2 and HLS methods. It is common to use the same compressed mode method in both the uplink and downlink directions although it is not mandatory. Similar to the uplink, the SF/2 method involves increasing the bit rate of the DPDCH during the radio frames which include the transmission gap. The availability of the SF/2 channelisation code is more important in the downlink direction because the code tree is shared. An alternative scrambling code can be used when the SF/2 channelisation code is blocked by another connection. Alternative scrambling codes are not orthogonal so they tend to increase the level of downlink interference. Similar to the uplink, the slots transmitted with SF/2 require approximately 3 dB greater transmit power and compressed mode tends to cause an increase in transmit power immediately after a transmission gap. Also similar to the uplink, HLS involves the MAC layer selecting a lower TFC. The requirement to use flexible transport channel positions for HLS can be more significant in the downlink direction. The speech service is often configured with fixed positions to allow blind detection which avoids the requirement to transmit downlink DPCCH TFCI bits.

3GPP TS 25.212 specifies a single uplink compressed mode frame structure and two downlink compressed mode frame structures. In the downlink direction, frame structure A is intended to maximise the transmission gap length whereas frame structure B is intended to minimise the impact upon inner loop power control. Frame structure B allows the Node B to transmit the TPC command belonging to the first time slot within the transmission gap, i.e. effectively reducing the duration of the transmission gap. The example shown in Log File 8.74 indicates that downlink frame type A should be used. The downlink frame structure can be configured independently to the downlink slot format. 3GPP TS 25.211 specifies downlink slot formats A and B for use in compressed mode. Slot format A is applicable to HLS whereas slot format B is applicable to SF/2. The number of bits assigned to the DPDCH and DPCCH tend to double for slot format B as a result of the increased bit rate. In the uplink direction, the DPDCH slot formats are not changed by compressed mode. However, 3GPP TS 25.211 specifies DPCCH slot formats A and B for use during uplink compressed mode. These slot formats are applicable when TFCI bits are included within the DPCCH. The choice between slot format A and slot format B is determined by the number of time slots transmitted during the radio frame rather than by the compressed mode method. Slot format A is used if 10 to 14 slots are transmitted whereas slot format B is used if 8 to 9 slots are transmitted.

3GPP TS 25.214 specifies changes to the operation of inner loop power control when compressed mode is active. The objective of these changes is to help ensure that inner loop power control can recover as rapidly as possible after a compressed mode transmission gap. In the case of uplink inner loop power control, the first changes are applicable to the SIR target used when generating the uplink TPC commands at the Node B. The active set Node B modify the SIR target both before and after the compressed mode measurement gaps. DeltaSIR1 defines a SIR target increase to be applied during the radio frame which includes the start of the first transmission gap in the transmission gap pattern. DeltaSIRafter1 defines a SIR target increase to be applied one frame after the frame containing the start of the first transmission gap in the transmission gap pattern. These parameters are signalled in Log File 8.74 using units of 0.1 dB, i.e. the signalled values of 15 and 7 correspond to 1.5 dB and 0.7 dB. DeltaSIR2 and DeltaSIRafter2 can also be configured if there are two transmission gaps within the transmission gap pattern. The uplink SIR target is also scaled by the ratio between the number of DPCCH pilot bits in a normal time slot and the number of DPCCH pilot bits in a compressed time slot.

Figure 8.55 illustrates the Node B responding to the Radio Link Reconfiguration Prepare using a Radio Link Reconfiguration Ready message. An example of this message is shown in Log File 8.75. The Node B uses the RNC communication context to identify the connection.

At this point, the Node B has accepted the compressed mode configuration, but has not started to apply it. The RNC finalises the procedure by forwarding a Radio Link Reconfiguration Commit message. An example commit message is shown in Log File 8.76. The connection is identified using the Node B communication context. The Connection Frame Number (CFN) defines the radio frame during which the Node B deactivates any TGPS from preceding compressed mode measurements.

3GPP TS 25.433 specifies that the compressed mode configuration change CFN is ignored when the active pattern sequence information is included within a Radio Link Reconfiguration Commit message. The active pattern sequence information specifies that the transmission gap pattern with TGPSI 2 should become active during CFN 224. The activation CFN must allow sufficient time for the compressed mode configuration to be signalled to the UE. This allows both the Node B and UE to start compressed mode simultaneously. The transmission gap pattern is repeated according to the value of the Transmission Gap Pattern Repetition Count (TGPRC). The value of 0 indicates that the Node B should continue applying the transmission gap pattern until the RNC instructs otherwise.

The compressed mode configuration is signalled to the UE using a Physical Channel Reconfiguration message. An example of this message is presented in Log File 8.77. The content is similar to the information already provided to the Node B. The UE is informed of the measurement purpose associated with each TGPS. In this example, TGPS 2 is associated with GSM RSSI measurements

```
NBAP-PDU
RadioLinkReconfigurationReady
protocolIEs
CRNC-CommunicationContextID: 2493
```

Log File 8.75 Radio Link Reconfiguration Ready acknowledging the compressed mode parameters

```
NBAP-PDU
RadioLinkReconfigurationCommit
protocolIEs
NodeB-CommunicationContextID: 160004
CFN: 200
Active-Pattern-Sequence-Information
- cMConfigurationChangeCFN: 224
- tGPSI: 2
- tGPRC: 0
- tGCFN: 224
```

Log File 8.76 Radio Link Reconfiguration Commit instructing the start of compressed mode

```
DL-DCCH-Message                          tgps-ConfigurationParams
integrityCheckInfo                       tgmp: gsm-initialBSICIdent
messageAuthenticationCode                tgprc: 0
Bin: C3 AF D7 D3                         tgsn: 4
rrc-MessageSequenceNumber: 10            tgl1: 7
physicalChannelReconfiguration           tgd: 270
rrc-TransactionIdentifier: 0             tgpl1: 4
physicalChannelReconfiguration-r5        rpp: mode0
  activationTime: 200                    itp: mode0
  rrc-StateIndicator: cell-DCH           ul-DL-Mode
  dl-CommonInformation                     ul-and-dl
  dl-DPCH-InfoCommon                         ul: sf-2
    cfnHandling: maintain                    dl: sf-2
    powerOffsetPilot-pdpdch: 0           dl-FrameType: dl-FrameTypeA
    spreadingFactorAndPilot              deltaSIR1: 15
    sfd128: pb4                          deltaSIRAfter1: 7
    positionFixedOrFlexible: fixed       deltaSIR2: 0
    tfci-Existence: false                deltaSIRAfter2: 0
                                         nidentifyAbort: 128
dpch-CompressedModeInfo
tgp-SequenceList                         dl-InformationPerRL-List
tgp-SequenceList value 1                 dl-InformationPerRL-List value 1
  tgpsi: 2                                 primaryCPICH-Info
  tgps-Status                                primaryScramblingCode: 33
  activate tgcfn: 224                        servingHSDSCH-RL-indicator: false
  tgps-ConfigurationParams                 dl-DPCH-InfoPerRL
  tgmp: gsm-CarrierRSSIMeasurement           pCPICH-UsageForChEst: mayBeUsed
  tgprc: 0                                   dpch-FrameOffset: 116
  tgsn: 4                                    dl-ChannelisationCodeList
  tgl1: 7                                    dl-ChannelisationCodeList 1
  tgd: 270                                     sf-AndCodeNumber
  tgpl1: 4                                       sf128: 5
  rpp: mode0                                   scrCodeChange: codeChange
  itp: mode0                                 tpc-CombinationIndex: 0
  ul-DL-Mode                             dl-InformationPerRL-List value 2
    ul-and-dl                              primaryCPICH-Info
      ul: sf-2                                 primaryScramblingCode: 34
      dl: sf-2                                 servingHSDSCH-RL-indicator: false
  dl-FrameType: dl-FrameTypeA              dl-DPCH-InfoPerRL
  deltaSIR1: 15                              pCPICH-UsageForChEst: mayBeUsed
  deltaSIRAfter1: 7                          dpch-FrameOffset: 122
  deltaSIR2: 0                               dl-ChannelisationCodeList
  deltaSIRAfter2: 0                          dl-ChannelisationCodeList 1
                                               sf-AndCodeNumber
tgp-SequenceList value 2                       sf128: 7
tgpsi: 3                                      scrCodeChange: noCodeChange
  tgps-Status: deactivate                  tpc-CombinationIndex: 0
```

Log File 8.77 Physical Channel Reconfiguration used to configure compressed mode

whereas TGPS 3 is associated with BSIC verification. The TGD value differs from the value signalled to the Node B. The Node B was signalled a value of 0 to indicate that the transmission gap pattern includes only a single transmission gap. The UE is signalled a value of 270 to indicate that the transmission gap pattern includes only a single transmission gap.

The UE is also informed of the Recovery Period Power control (RPP) mode. The RPP mode can be configured with a value of either 0 or 1. A value of 0 means that inner loop power control is not affected by the RPP mode. A value of 1 means that if uplink power control algorithm (PCA) 1 is used then the power control step size is changed to the minimum of 3 dB and 2 × the normal power control step size.

If PCA 2 is used then the power control step size is 1 dB during the recovery period. The recovery period during which the RPP mode is applicable is defined to have a duration equal to the minimum of the transmission gap length and seven slots. In addition, the Initial Transmit Power (ITP) mode can be used to modify the initial UE transmit power following a transmission gap. The ITP mode can be configured with a value of either 0 or 1. A value of 0 indicates that the UE should apply the TPC command sent by the Node B during the first time slot of the uplink transmission gap. If the Node B did not send a TPC command during that time slot i.e. downlink compressed mode Frame Format A was used, then the ITP mode does not impact the UE transmit power. Configuring the ITP mode with a value of 1 indicates that the UE should apply a filtered version of the TPC commands received up to and including the first time slot of the transmission gap. The filtering applied to the TCP commands is specified within 3GPP TS 25.214.

Log File 8.77 illustrates that the UE is provided with DeltaSIR and DeltaSIRafter parameters. These represent the downlink equivalent of the parameters signalled to the Node B, i.e. they are used to modify the downlink SIR target rather than the uplink SIR target.

3GPP TS 25.331 specifies that the number of TGPS repetitions should be limited when a UE is completing BSIC verification. The maximum number of repetitions is defined by the NidentifyAbort variable. The example in Log File 8.77 has a value of 128 which corresponds to 5.12 s when the TGPL is 4 radio frames.

The final section of the Physical Channel Reconfiguration message provides parameters which are relevant to individual radio links. The example shown in Log File 8.77 includes information regarding two radio links because the UE has two cells in the active set. Each radio link is identified by its scrambling code. The key information from the perspective of compressed mode is the scrambling code change information. In this example, the UE is instructed to change the downlink scrambling code to an alternative scrambling code for the first radio link, but to continue using the usual scrambling code for the second radio link.

The UE is provided with an activate CFN for TGPS 2 which matches the value signalled to the Node B within the Radio Link Reconfiguration Commit message. TGPS 3 is left inactive at this stage. The UE is now ready to start compressed mode, but has not been provided with an inter-system neighbour list nor the configuration used to report its measurements. An example Measurement Control message used to provide this information is shown in Log File 8.78. This example includes only two inter-system neighbours whereas in practice the number of inter-system neighbours is likely to be of the order of 15 to 20.

The Measurement Control message specifies that each inter-system neighbour belongs to the GSM Radio Access Technology (RAT). Each neighbour is identified using a combination of its Absolute Radio Frequency Channel Number (ARFCN) and its Base Station Identity Code (BSIC). The BSIC is a concatenation of the Network Colour Code (NCC) and the Base station Colour Code (BCC). The relationship between the signalled ARFCN and the actual RF carrier frequency is specified within 3GPP TS 05.05. The UE is instructed to measure and report the GSM RSSI without verifying the BSIC. Periodic reporting is configured using a reporting interval of 0.5 s. The reporting cell status indicates that the UE can include the results from up to six GSM cells within each Measurement Report.

Figure 8.55 illustrates that the UE responds to the Physical Channel Reconfiguration using a Physical Channel Reconfiguration Complete message. This message is sent as soon as the activation CFN is reached. The UE then proceeds to measure the RSSI belonging to each GSM neighbour. The UE also proceeds to send Measurement Reports at a rate defined by the reporting interval. The UE records an RSSI measurement from every RF carrier before providing any measurement results. This allows the UE to identify the strongest RF carriers and report them in order of their strength. If the number of RF carriers is relatively large then the UE may not have finished measuring them when the first Measurement Report is sent. In this case, the Measurement Report is sent without any content. Once the UE has recorded measurements from the complete set of RF carriers, it is able to select the

```
DL-DCCH-Message                              newInterRATCellList value 2
integrityCheckInfo                             interRATCellID: 1
messageAuthenticationCode                      technologySpecificInfo gsm
Bin: DF 28 0E C8                               interRATCellIndOffset: 0
rrc-MessageSequenceNumber: 11                  bsic
measurementControl                               ncc: 1
rrc-TransactionIdentifier: 2                     bcc: 2
measurementControl-r4                          freq-band: dcs1800BandUsed
measurementIdentity: 7                         bcch-ARFCN: 55
measurementCommand
setup                                        interRATMeasQuantity
interRATMeasurement                            ratSpecificInfo gsm
interRATCellInfoList                             measQuantity: gsm-CarrierRSSI
  removeIRATCells: AllIRATCells                  bsic-VerifRequired: notRequired
  newInterRATCellList                        interRATReportingQuantity
                                               utran-EstimatedQuality: false
  newInterRATCellList value 1                  ratSpecificInfo gsm
    interRATCellID: 0                            gsm-Carrier-RSSI: true
      technologySpecificInfo gsm             reportCriteria
      interRATCellIndOffset: 0                 periodicalReportingCriteria
      bsic                                       reportingInterval: ril0-5
        ncc: 1                                   reportingCellStatus
        bcc: 3                                   ActSetOrVirtActSetIRATcells: e6
      freq-band: dcs1800BandUsed           measurementReportingMode
      bcch-ARFCN: 45                           measReportTransMode: unackModeRLC
                                               periodicalOrEventTrig: periodical
```

Log File 8.78 Measurement Control used to configure GSM RSSI measurements and reporting

strongest and report them within the subsequent Measurement Report. Log File 8.79 presents an example Measurement Report which includes a series of 6 GSM RSSI measurements (recorded using a full neighbour list rather than the reduced neighbour list shown in Log File 8.78). The measurement results are reported in order of their signal strength, i.e. the strongest RF carrier is reported first. The BSIC have not been verified so each measurement is referenced by its ARFCN rather than by its BSIC.

```
UL-DCCH-Message                              gsm value 3
integrityCheckInfo                             gsm-CarrierRSSI
messageAuthenticationCode                        rssi: -87 dBm
Bin: 0D 12 A1 F9                               bsicReported
rrc-MessageSequenceNumber: 2                     nonVerifiedBSIC: 59
measurementReport                            gsm value 4
measurementIdentity: 7                         gsm-CarrierRSSI
measuredResults                                  rssi: -91 dBm
interRATMeasuredResultsList                    bsicReported
interRATMeasuredResultsList value 1              nonVerifiedBSIC: 35
  gsm value 1                                gsm value 5
    gsm-CarrierRSSI                            gsm-CarrierRSSI
      rssi: -74 dBm                              rssi: -92 dBm
    bsicReported                               bsicReported
      nonVerifiedBSIC: 55                        nonVerifiedBSIC: 115
  gsm value 2                                gsm value 6
    gsm-CarrierRSSI                            gsm-CarrierRSSI
      rssi: -79 dBm                              rssi: -93 dBm
    bsicReported                               bsicReported
      nonVerifiedBSIC: 31                        nonVerifiedBSIC: 45
```

Log File 8.79 Measurement Report providing GSM RSSI measurements

```
NBAP-PDU
CompressedModeCommand
NodeB-CommunicationContextID: 160004
Active-Pattern-Sequence-Information
- cMConfigurationChangeCFN: 11
transmission-Gap-Pattern-Sequence-Status
- tGPSI: 3
- tGPRC: 0
- tGCFN: 12
```

Log File 8.80   Compressed Mode Command used to switch from one TGPS to another

The UE continues to report GSM RSSI measurements until the RNC instructs otherwise. The RNC may be configured to initiate BSIC verification after receiving the first set of measurements. Alternatively, the RNC may be configured to initiate BSIC verification after averaging a specific number of measurements. Averaging the measurements tends to increase their reliability, but also increases the inter-system handover delay. The RNC may select the best RF carrier or the best group of RF carriers for BSIC verification. Selecting a group of RF carriers increases the potential for successful BSIC verification. Figure 8.55 illustrates the UE sending two Measurement Reports prior to the RNC initiating BSIC verification. In practice, the UE may send more than two Measurement Reports.

The RNC initiates BSIC verification by forwarding a Dedicated NBAP Compressed Mode Command to the Node B and a further Measurement Control message to the UE. An example Compressed Mode Command is shown in Log File 8.80. The compressed mode configuration change CFN defines the radio frame during which the Node B stops applying the active TGPS. The TGCFN defines the radio frame during which the Node B starts applying the TGPS identified by the TGPSI. The TGPRC defines the number of repetitions of the TGPS. The value of 0 indicates that the Node B should continue applying the TGPS until instructed otherwise.

An example of the corresponding Measurement Control message is presented in Log File 8.81. In this example, all but one of the inter-system neighbours are removed. This indicates that the strongest

```
DL-DCCH-Message                        interRATReportingQuantity
integrityCheckInfo                       utran-EstimatedQuality: false
messageAuthenticationCode                ratSpecificInfo gsm
Bin: DF 97 12 E3                         gsm-Carrier-RSSI: true
rrc-MessageSequenceNumber: 12          reportCriteria
measurementControl                     periodicalReportingCriteria
rrc-TransactionIdentifier: 3             reportingInterval: ril0-5
measurementControl-r4                    reportingCellStatus
measurementIdentity: 7                   ActSetOrVirtActSet-RATcells: e6
measurementCommand                     measurementReportingMode
modify                                   measReportTransMode: unackModeRLC
measurementType                          periodOrEventTrigger: periodical
interRATMeasurement                    dpch-CompressedModeStatusInfo
interRATCellInfoList                     tgps-Reconfig-CFN: 11
  removedInterRATCellList                tgp-SequenceShortList
    removeSomeInterRATCells              tgp-SequenceShortList value 1
    removeSomeInterRATCells value:         tgpsi: 2
    0,1,2,4,5,6,7,8,9,10,11,12,13,         tgps-Status: deactivate
    14,15                                tgp-SequenceShortList value 2
interRATMeasQuantity                       tgpsi: 3
  ratSpecificInfo gsm                      tgps-Status: activate
  measQuantity: gsm-CarrierRSSI            tgcfn: 12
  bsic-VerifRequired: required
```

Log File 8.81   Measurement Control used to switch from RSSI reporting to BSIC verification

RF carrier is used by only a single neighbour. The UE is instructed to measure the RSSI and verify the BSIC of the remaining neighbour. It is not mandatory for the RNC to remove any neighbours at this stage and BSIC verification can be completed with a larger group of neighbours. The UE is provided with the same pair of CFN as the Node B, i.e. TGPS 2 stops during CFN 11 while TGPS 3 starts during CFN 12.

The UE proceeds to send Measurement Reports at a rate defined by the reporting interval. The UE may not have completed BSIC verification prior to sending the first Measurement Report. In this case, the UE sends a Measurement Report but indicates that the BSIC has not been verified. Log File 8.82 shows a Measurement Report which has been generated after the UE has successfully completed BSIC verification. The integer following the verified BSIC indication represents the inter-RAT neighbour cell identity. In this example, neighbour cell 3 is the only neighbour which has not been removed from the neighbour list. BSIC verification allows the RNC to confirm the identity of the neighbouring cell and ensures that the GSM network is requested to reserve the correct resources, i.e. at the correct cell.

The RNC starts to involve the GSM network once BSIC verification has been completed. The second phase of signalling associated with a UMTS to GSM inter-system handover for a CS speech connection is shown in Figure 8.59. This phase involves reserving resources on the GSM network, instructing the UE to complete an inter-system handover and clearing the resources from the UMTS network. The RNC sends a RANAP Relocation Required message to the 3G MSC. This message specifies the target for the relocation using the appropriate GSM Cell Global Identity (CGI). The CGI is generated from a concatenation of the PLMN identity, LAC and Cell Identity (CI). The RNC is able to compile this information from its database of GSM neighbours. The Relocation Required message also identifies the source of the relocation using the PLMN identity, RNC identity and Service Area Identity (SAI). The SAI is a concatenation of the PLMN identity, LAC and Service Area Code (SAC). The UE capability is included using its MS Classmark 2 and MS Classmark 3 information elements. This information has been stored by the RNC since connection establishment, i.e. the UE can be requested to provide this information within the RRC Connection Setup Complete message.

The 3G MSC generates a Mobile Application Part (MAP) Prepare Handover request after receiving the Relocation Required message. The 3G MSC forwards this request to the 2G MSC which subsequently generates a Base Station System Mobile Application Part (BSSMAP) Handover Request for the BSC. This Handover Request triggers the BSC to reserve a set resources for the incoming speech connection and to return a Handover Request Acknowledge. The Handover Request Acknowledge includes the Handover Command which is to be provided to the UE currently connected to the

```
UL-DCCH-Message
integrityCheckInfo
messageAuthenticationCode
Bin: 75 37 26 B3
rrc-MessageSequenceNumber: 6
measurementReport
  measurementIdentity: 7
  measuredResults
    interRATMeasuredResultsList
    interRATMeasuredResultsList value 1
      gsm value 1
        gsm-CarrierRSSI
          rssi: -84 dBm
        bsicReported
          verifiedBSIC: 3
```

Log File 8.82   Measurement Report providing BSIC verification

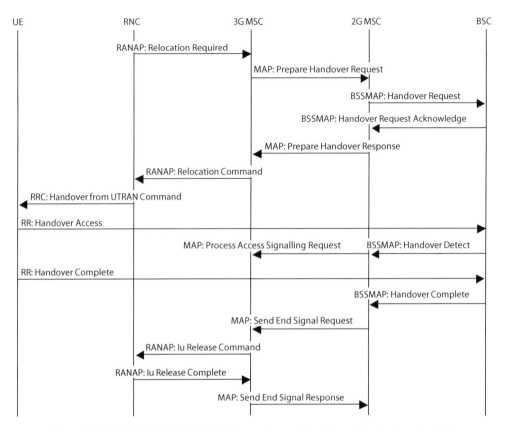

Figure 8.59   Signalling for UMTS to GSM speech connection inter-system handover (part 2)

UMTS network. The 2G MSC forwards the Handover Command to the 3G MSC using a Prepare Handover response message.

The 3G MSC packages the Handover Command within a RANAP Relocation Command. This message is forwarded to the RNC which then extracts the Handover Command and forwards it to the UE within a Handover from UTRAN Command. An example Handover from UTRAN Command is shown in Log File 8.83. The Handover Command originating from the BSC is included as the GSM message.

The Handover Command provides the UE with sufficient information to continue its speech connection on the GSM network. A section from the decoded Handover Command is shown in Log File 8.84. The NCC and BCC match the BSIC verified by the UE. Likewise, the ARFCN matches the RF carrier from which the BSIC was verified. The Handover Command allocates a GSM time slot and specifies the relevant channel types. The UE is also provided with a handover reference value and informed that it is mandatory to send a Handover Access burst when moving onto the GSM network.

The UE proceeds to move onto the GSM network and subsequently send the requested Handover Access burst. This burst includes the handover reference value from within the Handover Command. The BSC uses this burst to trigger sending a Handover Detect message to the 2G MSC. The UE proceeds to send a Handover Complete message to indicate that it has successfully completed the inter-system handover. The BSC informs the 2G MSC which then informs the 3G MSC. The 3G MSC

```
DL-DCCH-Message
integrityCheckInfo
messageAuthenticationCode
Bin: 7A C3 AF 72
rrc-MessageSequenceNumber: 13
handoverFromUTRANCommand-GSM
handoverFromUTRANCommand-GSM-r3
rrc-TransactionIdentifier: 0
toHandoverRAB-Info
  rab-Identity
    gsm-MAP-RAB-Identity
      Bin: 01
  cn-DomainIdentity: cs-domain
  nas-Synchronisation-Indicator
    Bin: 6
  re-EstablishmentTimer: useT314

  frequency-band: dcs1800BandUsed
  gsm-MessageList
  gsm-Messages value 1 (Handover Command)
    Bin: 06 2B 1C 37 0D 90 6E E0 05 D0 62 00 00 00 80 00 00 00
         00 00 40 00 00 00 00 00 00 63 41 72 01 03 91 03 06 20 95 08
         23 89 62
```

Log File 8.83   Handover from UTRAN Command

initiates the release of resources on the UMTS network. The release procedure is initiated using an Iu Release Command. This triggers the RNC to delete the radio links at the Node B which were previously within the active set. The Iu and Iub transport connections are also released. The speech connection is then continued on the GSM network unless an inter-system handover back to UMTS is triggered.

```
Cell Description
  NCC: 3
  BCC: 4
  BCCH carriers absolute RF channel: 55
Description of the first channel, after time
  Timeslot: 5
  Channel type: TCH/F + FACCH/F and SACCH/F
  Training sequence code: 4
  Hopping channel: RF hopping channel
  MAIO: 1
  HSN: 46
Handover Reference
  Handover reference value: 224
Power Command
  ATC: Sending of Handover access is mandatory
  EPC_mode: Channel(s) not in EPC mode
  FPC_EPC: FPC not in use
  Power level: 5
```

Log File 8.84   Section from the Handover Command generated by the BSC

9

Planning

9.1 Link Budgets

- Link budgets are used during both system dimensioning and radio network planning. Link budgets are typically generated for a range of services and a range of environment types. Both uplink and downlink link budgets should be completed.
- It is common to limit uplink DPCH link budgets to the 12.2 kbps speech and the 64 kbps CS and PS data services. Bit rates greater than 64 kbps may be supported, but excluding them from the link budget analysis helps to limit the resultant site density.
- UE transmit powers are typically either 24 or 21 dBm. A figure of 21 dBm represents a relatively worst case assumption, but ensures that the link budget results remain valid for both 21 and 24 dBm UE.
- Maximum downlink transmit powers typically depend upon both the network implementation and the RNC databuild.
- Eb/No requirements reflect the performance of the receiver. They depend upon the air-interface propagation conditions, the physical channel configuration, the target Block Error Rate and the use of receive diversity.
- The target uplink load is likely to depend upon the environment type. The uplink load provides a reflection of the uplink air-interface capacity. High loads correspond to high capacity, but also correspond to high interference margins and reduced cell ranges.
- The fast fading margin is dependent upon the UE speed. Inner loop power control is able to track fast fading at relatively low UE speeds. The fast fading margin allows the peak transmit power to be greater than the average transmit power.
- In general, there are three types of soft handover gain: a reduced Eb/No requirement as a result of diversity combining; a reduced fast fading margin as a result of not having to track all of the fast fades; and a reduced slow fading margin as a result of not having to track all of the slow fades.
- Building penetration loss usually represents the link budget assumption with the greatest uncertainty. This is accounted for by calculating the slow fading margin from an indoor standard deviation which encorporates the variance of the building penetration loss.
- A comparison of the uplink and downlink maximum allowed path loss results requires an offset to account for the frequency difference between the uplink and downlink frequency bands.

Radio Access Networks for UMTS   Chris Johnson
© 2008 John Wiley & Sons, Ltd

- The introduction of HSDPA has an impact upon both the uplink and downlink link budgets. The requirement for the UE to transmit the HS-DPCCH physical channel in addition to the DPDCH and DPCCH decreases the uplink maximum allowed path loss.
- HSDPA and HSUPA link budgets can be completed in the forward direction to determine the maximum allowed path loss for a specific cell edge bit rate. Alternatively, link budgets can be completed in the reverse direction to identify the cell edge bit rate for a specific maximum allowed path loss.
- The HSDPA transmit power depends upon whether a shared or dedicated RF carrier has been allocated. The transmit power available to HSDPA is greater if a dedicated RF carrier has been allocated.
- HSDPA connections have variable Eb/No requirements because they have variable bit rates. Link level simulations can be used to generate a look-up table which associates a range of Eb/No requirements with a range of bit rates.
- HSDPA does not use inner loop power control so it is not necessary to include a fast fading margin. HSDPA uses the serving cell change procedure rather than soft handover. The serving cell change gain represents a reduction in the slow fading margin as a result of the UE being able to select the best serving cell.
- A UE with an HSUPA connection has to transmit up to five physical channels, i.e. the DPCCH, E-DPCCH, E-DPDCH, HS-DPCCH and DPDCH. The Eb/No requirement within an HSUPA link budget should reflect the complete set of physical channels.
- Key 3GPP specifications: TS 25.101, TS 25.104, TS 25.141 and TS 25.213.

Link budgets are used during both system dimensioning and radio network planning. The dimensioning process provides an estimate of the number of network elements required to achieve a specific coverage and capacity performance. Link budgets are used during dimensioning to estimate the maximum allowed path loss and the corresponding cell range. The results from a dimensioning exercise can be used as an input for a business case analysis. The number of network elements and their associated configuration allow the cost of the network to be quantified. The path loss based approach to radio network planning (Section 9.2.1) makes use of link budget results to define maximum allowed path loss thresholds. These thresholds are typically generated for a range of services and a range of environment types. The 3G simulation based approach to radio network planning (Section 9.2.2) makes use of link budget assumptions to define some of the simulation inputs. The interaction between system dimensioning, business case analysis and network planning is illustrated in Figure 9.1.

Link budgets are typically generated for a range of physical channel and service type combinations, e.g. the DPCH 12.2 kbps speech service, the DPCH 64 kbps CS data service, the DPCH 384 kbps PS

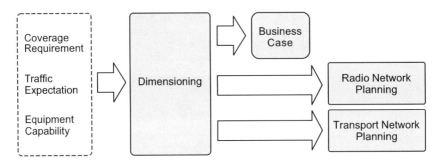

Figure 9.1   Interaction between dimensioning, business case analysis and network planning

data service, the HSDPA 1.6 Mbps PS data service and the HSUPA 1.4 Mbps PS data service. The link budget generating the smallest maximum allowed path loss defines the maximum cell range.

9.1.1 DPCH

9.1.1.1 Uplink

A set of example uplink DPCH link budgets is presented in Table 9.1. It is common to limit uplink DPCH link budgets to the 12.2 kbps speech and the 64 kbps CS and PS data services. Bit rates greater than 64 kbps may be supported, but excluding them from the link budget analysis helps to limit the site density and the associated network cost. In this case, higher bit rate services may not be available at cell edge, but could be offered in areas of relatively good coverage. An additional link budget analysis could be completed to estimate the percentage of the cell area over which the higher bit rate services are supported.

UE transmit powers are typically either 24 or 21 dBm, i.e. power class 3 or power class 4 respectively. These power classes are specified within 3GPP TS 25.101. A figure of 21 dBm represents a relatively worst case assumption, but ensures that the link budget remains valid for both 21 and 24 dBm UE. In addition, a figure of 21 dBm can be adopted for a 24 dBm terminal because 3GPP TS 25.101 specifies a $+1\,dB/-3\,dB$ tolerance for the maximum transmit power of a class 3 UE. The terminal antenna gain can vary from one UE model to another. Datacards may have a higher antenna gain than handheld devices. The majority of UE typically have an antenna gain in the order of 0 dBi. Body loss is dependent upon the relative positions of the UE, the end-user and the active set cells. A figure of 3 dB represents a typical assumption when the UE is held to one side of the end-user's head.

Table 9.1   Uplink DPCH link budgets

Service type	Speech	CS data	PS data
Uplink bit rate (kbps)	12.2	64	64
Maximum transmit power (dBm)	21.0	21.0	21.0
Terminal antenna gain (dBi)	0.0	0.0	0.0
Body loss (dB)	3.0	3.0	3.0
Transmit EIRP (dBm)	18.0	18.0	18.0
Chip rate (Mcps)	3.84	3.84	3.84
Processing gain (dB)	25.0	17.8	17.8
Required Eb/No (dB)	4.4	2.0	2.0
Target uplink load (%)	50	50	50
Rise over thermal noise (dB)	3.0	3.0	3.0
Thermal noise power (dBm)	−108.0	−108.0	−108.0
Receiver noise figure (dB)	2.0	2.0	2.0
Interference floor (dBm)	−103.0	−103.0	−103.0
Receiver sensitivity (dBm)	−123.6	−118.8	−118.8
Node B antenna gain (dBi)	18	18	18
Cable loss (dB)	0.0	0.0	0.0
Fast fading margin (dB)	1.8	1.8	1.8
Soft handover gain (dB)	2.0	2.0	2.0
Building penetration loss (dB)	12.0	12.0	12.0
Indoor location probability (%)	90	90	90
Indoor standard deviation (dB)	10	10	10
Slow fading margin (dB)	7.8	7.8	7.8
Isotropic power required (dBm)	−122.0	−117.2	−117.2
Maximum allowed path loss (dB)	140.0	135.2	135.2

The first main result from the uplink DPCH link budget is the transmitter Effective Isotropic Radiated Power (EIRP). This is defined using the expression below.

$$\text{Transmit EIRP} = \text{Maximum transmit power} + \text{Antenna gain} - \text{Body loss}$$

3GPP TS 25.213 specifies a fixed chip rate of 3.84 Mcps. The processing gain is calculated as the ratio between the chip rate and bit rate. Large processing gains correspond to low bit rates and reduced C/I requirements.

$$\text{Processing gain} = 10 \times \log \frac{\text{Chip rate}}{\text{Bit rate}}$$

Uplink Eb/No requirements reflect the performance of the Node B receiver. They also depend upon the air-interface propagation conditions, the physical channel configuration, the target Block Error Rate (BLER) and the use of uplink receive diversity. RAKE receivers are able to take advantage of the delay spread generated by multi-path propagation. The high chip rate and corresponding short chip period (260 ns) allow relatively small delay spreads to be resolved. A RAKE receiver combines the strongest delay spread components to generate a multi-path diversity gain which is maximised if the delay spread components fade independently and have equal strength. Oversampling within the receiver (more than one digital sample per chip) also helps the RAKE receiver to maximise the multi-path diversity gain. 3GPP TS 25.104 and TS 25.141 specify the four propagation channels presented in Table 9.2.

Different UE speeds have been specified for each operating band so the Doppler frequency remains approximately constant. For example, operating band I has an uplink frequency range from 1920 to 1980 MHz whereas operating band V has an uplink frequency range from 824 to 849 MHz. The 3 km/hr Doppler frequency generated for band I at 1920 MHz is 5.3 Hz whereas the 7 km/hr Doppler frequency generated for band V at 824 MHz is also 5.3 Hz (Doppler frequency is defined as UE speed × operating frequency/speed of light). This approach allows an easier comparison of performance although in practice the requirement for end-user mobility is likely to be similar for all operating bands.

The four propagation channels are intended to test different aspects of the Node B receiver. The first channel includes two delay spread components which are fading independently and are spaced in time by 3.75 chips. The second delay spread component has an average power which is 10 dB less than the first delay spread component. This propagation channel tests the receiver's ability to capture a relatively weak delay spread component which is delayed by a non-integral number of chip periods. The receiver requires oversampling to maximise the benefit of the delay spread components when

Table 9.2   3GPP multi-path propagation channels used to test the Node B receiver

Frequency band	Case 1		Case 2		Case 3		Case 4	
I, II, III, IV, IX	3 km/hr		3 km/hr		120 km/hr		250 km/hr	
V, VI, VIII	7 km/hr		7 km/hr		280 km/hr		583 km/hr	
VII	2.3 km/hr		2.3 km/hr		92 km/hr		192 km/hr	
	Relative delay (ns)	Average power (dB)	Relative delay (ns)	Average power (dB)	Relative delay (ns)	Average power (dB)	Relative delay (ns)	Average power (dB)
	0	0	0	0	0	0	0	0
	976	−10	976	0	260	−3	260	−3
			20000	0	521	−6	521	−6
					781	−9	781	−9

they are spaced by a non-integer number of chip periods. Larger oversampling rates provide greater time resolution but increase the UE processing requirement.

The second channel includes three delay spread components with equal average powers. The third delay spread component arrives 20 µs after the first. This propagation channel tests the receiver's ability to buffer and combine delay spread components with relatively large time separation. 20 µs corresponds to 77 chips and an equivalent path difference of 6 km.

The third and fourth channels include four delay spread components which are spaced by 1, 2 and 3 chips. The UE speed is increased to 120 km/hr for the third channel and 250 km/hr for the fourth channel (operating bands I, II, III, IV and IX). These UE speeds increase the Doppler frequency to approximately 220 and 450 Hz, respectively. These propagation channels test the receiver's ability to cope with high Doppler frequencies. Doppler frequencies have a similar impact to offsets between the UE transmit frequency and the Node B receive frequency, e.g. the UE could transmit at 1922.6 MHz + 25 Hz while the Node B receives at 1922.6 MHz − 25 Hz. These frequency offsets cause the received modulation constellation to rotate over time. If these rotations are not compensated there is an increased risk of the receiver deducing an incorrect modulation constellation point. The receiver requires some form of Automatic Frequency Control (AFC) to track and compensate the phase rotations.

3GPP also specifies a moving propagation channel and a birth–death propagation channel. The moving propagation channel includes two delay spread components of equal strength, but with a variable time delay. The time delay between the two delay spread components varies sinusoidally between 1 and 6 µs. The birth–death propagation channel is based upon two equal strength delay spread components which alternately vanish and reappear every 191 ms. When reappearing, a random time difference is selected from the range −5 to +5 µs while ensuring that the two delay spread components do not coincide.

3GPP specifies Eb/No requirements for these propagation conditions with power control disabled and a continuous DCCH Signaling Radio Bearer (SRB) activity. These requirements are presented in Table 9.3.

These figures indicate that static propagation conditions (no delay spread components and no fading) result in the minimum Eb/No requirement. Comparing case 1 with case 2 indicates that receivers perform better when the received power is equally distributed between multiple delay spread components. The multi-path diversity gain is relatively small when there are only two delay spread components and one is much weaker than the other, i.e. when the strong delay spread component is experiencing a fade the receiver has to rely upon the remaining relatively weak delay spread component.

Comparing case 2 with case 3 indicates that performance is improved when the UE speed is increased to 120 km/hr and the delay spread components are separated by an integer number of chips. Increasing UE speed can improve the performance of the physical layer when channel coding is combined with interleaving. Propagation channels which exhibit fading tend to generate bursts of errors during fades. Channel de-coding at the receiver performs best when errors are randomly

Table 9.3   3GPP Eb/No requirements with power control disabled (1% BLER)

Propagation Conditions	12.2 kbps Speech (dB)	64 kbps Data (dB)	384 kbps Data (dB)
Static	5.1	1.7	1.0
Case 1	11.9	9.2	8.8
Case 2	9.0	6.4	6.1
Case 3	7.2	3.8	3.6
Case 4	10.2	6.8	6.6
Moving	5.7	2.2	
Birth-Death	7.7	4.2	

distributed in time. The de-interleaving process is used to randomise the bursts of errors and so improves the performance of the channel de-coding. Fades have relatively long durations when UE are moving slowly and relatively few, but wide bursts of errors are experienced. Wide bursts of errors are more difficult for the de-interleaving process to randomise and so the performance of channel de-coding decreases. Fades have relatively short durations when UE are moving quickly and relatively many, but short bursts of errors are experienced. The de-interleaving process can cope with short bursts of errors and the receiver performance improves. The Transmission Time Interval (TTI) also has an impact upon the performance of interleaving and channel decoding. Large TTI increase system delay, but allow more effective interleaving by increasing the time window across which data is re-ordered.

Comparing case 3 with case 4 indicates that increasing the UE speed from 120 to 250 km/hr can be expected to increase the Eb/No requirement by 3 dB. This increase illustrates that receivers find it more difficult to track and compensate phase changes for relatively high Doppler frequencies. The signal to noise ratio has to be improved to maintain an equal probability of error.

Disabling power control while measuring the Eb/No requirements in Table 9.3 means that the results effectively include the fast fading margin. This has greatest impact upon the Eb/No requirements at UE speeds less than 50 km/hr. Inner loop power control is unable to track the fast fading at UE speeds greater than 50 km/hr and the Eb/No requirements effectively encorporate the fast fading margin irrespective of whether or not inner loop power control is enabled, i.e. the fast fading margin becomes 0 dB at high speeds. It is important to separate the fast fading margin from the Eb/No at lower speeds because the Eb/No influences both coverage and capacity whereas the fast fading margin influences only coverage.

The Eb/No requirements in Table 9.3 illustrate that requirements decrease for higher bit rate connections. This decrease results from a reduction of the overheads generated by the DPCCH physical channel and the SRB DCCH logical channel. Although Eb/No requirements decrease for higher bit rates the C/I requirements increase. The C/I requirement is calculated by subtracting the processing gain from the Eb/No requirement.

$$C/I \text{ requirement} = Eb/No - \text{Processing gain}$$

This relationship can be derived from the definition of Eb/No, i.e. in linear units,

$$Eb/No = \frac{(\text{Wanted signal power/Bit rate})}{(\text{Interfering signal power/Chip rate})} = PG \times \frac{\text{Wanted signal power}}{\text{Interfering signal power}} = PG \times C/I$$

Table 9.4 presents the C/I requirements when assuming the 3GPP Eb/No requirements corresponding to a static propagation environment. These figures illustrate a significant increase in the C/I requirement as the bit rate is increased from 12.2 to 384 kbps. This increase in C/I requirement has a direct impact upon sensitivity and the maximum allowed path loss.

The BLER target also has an impact upon the Eb/No requirement. High BLER targets allow reduced Eb/No requirements and a corresponding increase in the maximum allowed path loss. Reduced Eb/No

Table 9.4   C/I requirements based upon the 3GPP Eb/No requirements (static propagation channel)

	12.2 kbps Speech	64 kbps Data	384 kbps Data
Eb/No Requirement (dB)	5.1	1.7	1.0
Processing Gain (dB)	25.0	17.8	10.0
C/I Requirement (dB)	−19.9	−16.1	−9

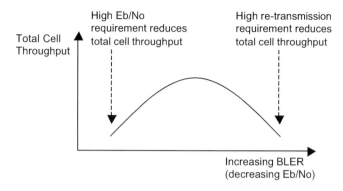

Figure 9.2   Relationship between total cell throughput and BLER target

requirements also increase the maximum number of simultaneously active connections. However, high BLER targets generate an increased quantity of re-transmissions. These re-transmissions do not represent useful throughput so the throughput performance of the end-user decreases. The total cell throughput can also decrease despite the increased number of simultaneously active connections. Low BLER targets require increased Eb/No requirements. In this case, the number of re-transmissions is relatively low so the throughput performance of the individual end-user improves. However, the increased Eb/No requirement reduces the maximum number of simultaneously active connections causing the total cell throughput to decrease. The general relationship between BLER target and total cell throughput is illustrated in Figure 9.2.

The target uplink load appearing within the link budgets shown in Table 9.1 is likely to depend upon the environment type. The uplink load provides a reflection of the uplink air-interface capacity. High loads correspond to high capacity, but also correspond to high interference margins and reduced cell ranges. Rural areas may be planned using a relatively low load to maximise the cell range while placing less emphasis upon the air-interface capacity. Urban areas may be planned using a higher uplink load to provide greater capacity at the cost of an increased site density. Typical assumptions for the target uplink load are 30%, 50% and 75% for rural, suburban and urban environments, respectively. Applying different assumptions for each environment type means that separate link budgets must be generated for each environment.

The equation below allows the rise over thermal noise to be calculated directly from the target load. This equation generates the characteristic exponential relationship between load and interference power.

$$\text{Rise over thermal noise} = -10 \times \log(1 - \text{Target load})$$

The thermal noise power is defined by the assumed temperature of the Node B receiver, Boltzmann's constant and the receiver bandwidth:

$$\text{Thermal noise} = 10 \times \log(kTB) = 10 \times \log(1.4 \times 10^{-23} \times 290 \times 3.84 \times 10^6) = -138\,\text{dBW}$$

The result of this equation has units of dBW whereas the link budget has units of dBm, i.e. power relative to 1 mW rather than power relative to 1 W. Powers quoted in dBW can be translated to dBm by adding $10 \times \log(1000) = 30\,\text{dB}$. This means that a power of $-138\,\text{dBW}$ corresponds to $-108\,\text{dBm}$.

The receiver noise figure assumption reflects the performance of the Node B receiver subsystem. The noise figure belonging to the Node B cabinet should be used if the receiver subsystem does not include a Mast Head Amplifier (MHA). If an MHA is included, the noise figure should be the composite noise figure of the MHA, cable/connectors and Node B cabinet. The composite noise figure can be calculated using Friis' equation,

$$\text{Composite noise figure} = 10 \times \log\left(\text{NF}_{\text{MHA}} + \frac{\text{NF}_{\text{Cable}} - 1}{\text{Gain}_{\text{MHA}}} + \frac{\text{NF}_{\text{NodeB}} - 1}{\text{Gain}_{\text{MHA}} \times \text{Gain}_{\text{Cable}}}\right)$$

With the exception of the composite noise figure result, all of the variables within the preceding equation have linear units. The noise figure of the cable and connectors is equal to their loss. For example, the noise figure is 2 dB in log units and 1.6 in linear units if the cable and connector loss is 2 dB. The gain of the cable and connectors is equal to $-1 \times$ their loss, i.e. -2 dB in log units and 0.6 in linear units. Friis' equation illustrates that when the MHA has a high gain, the noise figure of the receiver subsystem is dominated by the noise figure of the MHA. This emphasises the importance of having a low noise, high gain amplifier for the MHA.

The link budget interference floor is defined as the sum of the thermal noise power, the receiver noise figure and the rise over thermal noise. A high interference floor requires an increased wanted signal strength to achieve the uplink C/I requirement. The second main result from the uplink DPCH link budget is the receiver sensitivity calculated from the C/I requirement and interference floor,

$$\text{Sensitivity} = \text{Interference floor} + \text{Eb/No} - \text{Processing gain}$$
$$= \text{Interference floor} + \text{C/I requirement}$$

Figure 9.3 provides a graphical representation of the sensitivity calculation.

The final section of the link budget includes a series of gains, losses and margins. The Node B antenna gain should be representative of the antenna type planned for network deployment. In practice, networks may include a range of antenna types. Antenna gains tend to decrease as the horizontal and vertical beamwidths increase and the antenna becomes less directional. System dimensioning and radio network planning should account for any significant differences between the antenna gains belonging to different antenna types. The antenna gain figure can encorporate a polarisation loss of approximately

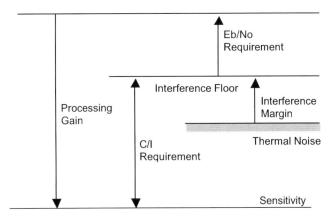

Figure 9.3   Relationship between sensitivity, interference floor, Eb/No and processing gain

0.5 dB. Antenna gains are typically quoted from measurements which have been recorded using a receiving antenna element which has exactly the same polarisation as the transmitting antenna element. In practice, the two antenna elements have different polarisations and a polarisation loss is experienced. Reflections change the polarision of a radio signal and this helps to reduce the loss because many signals with different polarisations can reach the receiving antenna. Cross polar antennas also reduce the potential for polarisation loss because the maximum angular offset is 45° compared with 90° for a vertically polarised antenna.

The cable loss variable within the uplink link budget is only applicable if it has not already been included as part of the receiver noise figure. The uplink cable loss is included within the composite noise figure if an MHA has been assumed. Otherwise, the cable loss should equal all cable and connector losses between the Node B cabinet and antenna.

The fast fading margin is dependent upon the UE speed. Uplink inner loop power control is able to track fast fading at relatively low UE speeds. This results in the requirement for a fast fading margin because the peak UE transmit power requirement is greater than the average UE transmit power requirement. At speeds greater than 50 km/hr, the fades are experienced too rapidly for inner loop power control to track. This increases average UE transmit power and decreases the requirement for a fast fading margin. Assumptions for the fast fading margin are typically derived from link level simulations and are not easy to validate in the field. A figure of 1.8 dB is typical for a UE moving at less than 30 km/hr.

Soft handover gain assumptions are often derived from a combination of link level and system level simulations. In general, there are three types of soft handover gain: a reduction in the Eb/No requirement as a result of macro diversity combining gain at the RNC and potentially greater diversity within the RAKE receiver (softer handover); a reduction in the fast fading margin as a result of not having to track all of the fast fades; and a reduction in the slow fading margin as a result of not having to track all of the slow fades. The reduction in the Eb/No requirement is often assumed to be 0 dB in the uplink direction when dual branch receive diversity is used. The quantity of diversity within the RAKE receiver is already relatively high if dual branch receive diversity is used. This means the incremental impact of soft handover is relatively small. Soft handover has greater impact upon the Eb/No requirement if the Node B is configured with a single antenna element. The soft handover gain appearing within Table 9.1 represents the sum of the fast and slow fading gains. This can differ from the soft handover gain specified as an input to a 3G simulation because radio network planning tools typically model slow fading explicitly and do not require an input parameter to define the slow fading gain, i.e. they only require input parameters for the potential Eb/No reduction and fast fading gain.

Including the building penetration loss as part of the link budget generates an outdoor maximum allowed path loss result which includes sufficient margin to allow UE at the cell edge to establish and maintain connections from within buildings. The building penetration loss may be replaced by a vehicle penetration loss if link budgets are generated for a section of motorway or a rural area. The assumptions for building penetration loss typically depend upon the environment type, e.g. building penetration could be greater within an urban environment than within a suburban environment. The building penetration loss usually represents the link budget assumption with the greatest uncertainty. Eb/No figures and soft handover gains could have an uncertainty of ±1 dB while building penetration loss could have an uncertainty of ±5 dB. This uncertainty is included within the link budget result by calculating the slow fading margin from an indoor standard deviation which encorporates the variance of the building penetration loss. Building penetration loss assumptions are relatively difficult to validate in the field due to the large variance between different buildings and the geometry of those buildings with respect to the radio network plan.

The indoor location probability represents the average probability of experiencing indoor coverage across the cell area. This figure is translated to an equivalent cell edge indoor coverage probability before combining with the standard deviation to generate the slow fading margin. The indoor location probability at cell edge will be less than the average as a result of the higher UE transmit power requirement. The indoor standard deviation represents a combination of the outdoor standard deviation associated with slow fading, and the standard deviation generated by the variance of the building penetration loss.

The third main result from the link budget is the isotropic power required at the Node B antenna. This power is calculated by combining the receiver sensitivity with each of the gains, losses and margins.

$$\text{Isotropic power} = \text{Sensitivity} - \text{Antenna gain} + \text{Cable loss} + \text{Fast fading margin} - \text{SHO gain} + \text{Building penetration loss} + \text{Slow fading margin}$$

The maximum allowed path loss is then calculated as the difference between the transmit EIRP and the isotropic power requirement.

$$\text{Maximum allowed path loss} = \text{Transmit EIRP} - \text{Isotropic power}$$

This maximum allowed path loss result can be compared with the equivalent downlink result to determine whether coverage is uplink or downlink limited. This comparison requires an offset to account for the frequency difference between the uplink and downlink frequency bands. The downlink RF carrier has a frequency which is 190 MHz greater than the uplink RF carrier if operating band I is assumed. Higher frequencies tend to experience greater path loss so coverage will tend to be downlink limited if both the uplink and downlink link budgets generate equal maximum allowed path loss results. The frequency-dependent term within a typical Okumura–Hata path loss model is given by the equation below.

$$\text{Frequency-dependent loss} = 33.9 \times \log(\text{Frequency}(\text{MHz}))$$

This equation indicates that for a fixed cell range, the uplink propagation loss at 1922.6 MHz is 1.4 dB less than the downlink propagation loss at 2112.6 MHz, i.e. the uplink maximum allowed path loss can be 1.4 dB less than the downlink maximum allowed path loss before the uplink starts to determine the cell range.

9.1.1.2 Downlink

A set of example downlink DPCH link budgets is presented in Table 9.5. Downlink DPCH link budgets typically include a broader range of bit rates than the uplink. This reflects the requirement to achieve higher throughputs in the downlink direction. The higher transmit power capability in the downlink direction increases the scope for achieving these throughputs towards the cell edge. Table 9.5 includes link budgets for bit rates up to 384 kbps. The higher downlink bit rates may be excluded when defining the maximum allowed path loss used to plan the network. This helps to reduce the site density and corresponding network cost, but means that the higher bit rates may not be offered at locations around the cell edge.

The maximum downlink transmit powers typically depend upon both the network implementation and the RNC databuild. The RNC databuild may define the maximum transmit powers directly. Alternatively, the RNC may calculate the maximum powers from other parameters within the RNC databuild. The equation below represents an example of how the RNC could calculate the maximum downlink transmit powers using parameters within the RNC databuild.

$$\text{Maximum transmit power} = \text{CPICH transmit power} - \text{Offset} + 10 \times \log\left(\frac{C/I_{\text{SRB}} + C/I_{\text{Service}}}{C/I_{\text{Reference}}}\right)$$

Table 9.5 Downlink DPCH link budgets

Service type	Speech	CS data	PS data	PS data	PS data
Downlink bit rate (kbps)	12.2	64	64	128	384
Maximum transmit power (dBm)	34.2	37.2	37.2	40.0	40.0
Cable loss (dB)	2.0	2.0	2.0	2.0	2.0
MHA insertion loss (dB)	0.5	0.5	0.5	0.5	0.5
Node B antenna gain (dBi)	18	18	18	18	18
Transmit EIRP (dBm)	49.7	52.7	52.7	55.5	55.5
Chip rate (Mcps)	3.84	3.84	3.84	3.84	3.84
Processing gain (dB)	25.0	17.8	17.8	14.8	10.0
Required Eb/No (dB)	7.9	5.3	5.0	4.7	4.8
Target downlink load (%)	80	80	80	80	80
Rise over thermal noise (dB)	7.0	7.0	7.0	7.0	7.0
Thermal noise power (dBm)	−108.0	−108.0	−108.0	−108.0	−108.0
Receiver noise figure (dB)	8.0	8.0	8.0	8.0	8.0
Interference floor (dBm)	−93.0	−93.0	−93.0	−93.0	−93.0
Receiver sensitivity (dBm)	−110.1	−105.5	−105.8	−103.1	−98.2
Terminal antenna gain (dBi)	0.0	0.0	0.0	0.0	0.0
Body loss (dB)	3.0	3.0	3.0	3.0	3.0
Fast fading margin (dB)	1.8	1.8	1.8	1.8	1.8
Soft handover gain (dB)	2.0	2.0	2.0	2.0	2.0
MDC gain (dB)	1.2	1.2	1.2	1.2	1.2
Building penetration loss (dB)	12.0	12.0	12.0	12.0	12.0
Indoor location probability (%)	90	90	90	90	90
Indoor standard deviation (dB)	10	10	10	10	10
Slow fading margin (dB)	7.8	7.8	7.8	7.8	7.8
Isotropic power required (dBm)	−88.7	−84.1	−84.4	−81.7	−76.8
Maximum allowed path loss (dB)	138.4	136.8	137.1	137.2	132.3

This example calculation scales the maximum downlink transmit power according to the C/I requirement of each service. The C/I requirement of the signalling radio bearer is also taken into account. The signalling radio bearer has relatively little impact upon the calculation for high bit rate services, but has a greater impact upon low bit rate services. It is assumed that the maximum downlink transmit power requirement for a reference service is equal to the CPICH transmit power minus an offset. If the reference service is assumed to be the speech service then the offset can be configured with a value of 0 dB, i.e. the maximum downlink transmit power requirement for the speech service is approximately equal the CPICH transmit power. This type of approach based upon C/I requirements aims to provide similar downlink coverage for each service, i.e. if a service has a greater C/I requirement then a correspondingly greater maximum downlink transmit power is allocated.

The calculation for the higher bit rate services may result in maximum downlink transmit powers which are relatively high, e.g. more than 50% of the total downlink transmit power capability. In this case, the maximum downlink transmit power can be rounded down. The set of example link budgets presented in Table 9.5 assume that no more than 10 W can be allocated to a single individual connection. This assumption has a relatively large impact upon the 384 kbps PS data service which consequently has a relatively low maximum allowed propagation loss.

The cable loss and Node B antenna gain equal the values assumed for the uplink link budget. An MHA does not reduce the impact of the cable loss in the downlink direction. Instead, it introduces an additional insertion loss. The first main result from the downlink DPCH link budget is the transmitter EIRP which is defined using the expression below.

$$\text{Transmit EIRP} = \text{Maximum Tx power} - \text{Cable loss} - \text{MHA insertion loss} + \text{Antenna gain}$$

The second part of the downlink link budget is similar to the second part of the uplink link budget. The downlink receiver sensitivity is calculated from the processing gain, Eb/No requirement and downlink interference floor. In this case, the Eb/No requirements reflect the performance of the UE and are likely to vary between UE models. Typical Eb/No figures representative of the UE population should be selected. Downlink Eb/No figures can be validated in the field if SIR measurements are recorded and subsequently translated to Eb/No requirements. The Eb/No requirements for UE capable of downlink receive diversity are reduced by approximately 3 dB. The target downlink load is typically greater than the target uplink load. The downlink load benefits from orthogonality, but includes the soft handover overhead. The downlink inter-cell interference ratio is also likely to be greater than the corresponding uplink figure at cell edge. An 80% downlink load is a typical assumption and results in a 7 dB increase in interference floor. The downlink receiver noise figure assumption is also dependent upon the performance of the UE and is likely to vary between UE models. A noise figure of 8 dB represents a typical assumption.

The final part of the downlink link budget is similar to the final part of the uplink link budget, i.e. all remaining gains, losses and margins are included. The UE antenna gain and body loss figures should be the same as those assumed for the uplink. The downlink fast fade margin represents the difference between the average downlink transmit power and the short-term increases required to compensate for fast fading. In general, downlink inner loop power control has less tracking to complete than uplink inner loop power control. This is because the own-cell contribution to the interference floor experiences the same fading as the wanted signal, i.e. the signal to noise ratio conditions remain approximately constant because both the wanted signal and the interference floor fade coherently. However, this argument is less valid at cell edge where the interference floor can be dominated by other-cell interference. The example link budgets presented in Table 9.5 assume that the downlink fast fade margin at the cell edge is equal to the uplink fast fade margin.

Similar to the uplink, there are three types of soft handover gain: reduction in Eb/No requirement as a result of greater diversity in the RAKE receiver; reduction of the fast fading margin as a result of not having to track all of the fast fades; and a reduction in the slow fading margin as a result of not having to track all of the slow fades. The additional diversity within the RAKE receiver is relatively significant for UE which are not capable of downlink receive diversity. The link budgets shown in Table 9.5 describe this gain as Macro Diversity Combination (MDC) gain. MDC gain is less likely to be significant for UE capable of downlink receive diversity. In this case, the downlink Eb/No requirements will be reduced as a result of increased diversity irrespective of whether or not the UE is in soft handover. The link budget entry termed soft handover gain represents the combined reduction in transmit power as a result of not having to track all of the fast and slow fading.

The combination of the soft handover and MDC gains indicates a total gain of approximately 3 dB. This figure is intended to represent an average across a range of soft handover scenarios. The gains associated with soft handover are dependent upon the relative signal strengths of the active set cells. The gains are maximised if all cells have equal strength. The gains become relatively small if the strength of one radio link is significantly greater than the strength of the other radio links. The gains tend to be a function of the relative strength of the two strongest active set cells. Other cells have relatively little impact, but effectively serve as a back-up in case propagation conditions change and one of the two strongest cells becomes weaker.

If the frequency dependence of the building penetration loss is ignored then its value should be the same as that assumed for the uplink. The indoor location probability, indoor standard deviation and slow fading margin are the same as those used for the uplink link budget. The third main result from the link budget is then the isotropic power required at the UE antenna. This power is calculated by combining the receiver sensitivity with each of the gains, losses and margins. Finally, the link budget result is defined as the difference between the transmit EIRP and the isotropic power requirement.

9.1.2 HSDPA

9.1.2.1 Uplink

The introduction of HSDPA has an impact upon both the uplink and downlink link budgets. The requirement for the UE to transmit the HS-DPCCH physical channel in addition to the DPDCH and DPCCH decreases the uplink maximum allowed path loss. Table 9.6 presents a set of uplink DPCH link budgets which include the overhead generated by the HS-DPCCH.

Both the HS-DPCCH and DPCCH represent physical channel overheads for the DPDCH. Uplink DPCH Eb/No requirements typically include the overhead generated by the DPCCH. It is also possible to include the overhead generated by the HS-DPCCH. This overhead can be calculated using the equation below.

$$\text{HS-DPCCH overhead} = 10 \times \log\left(\frac{\beta_d^2 + \beta_c^2 + \beta_c^2 \times \beta_{hs}^2}{\beta_d^2 + \beta_c^2}\right)$$

β_c and β_d represent the DPCCH and DPDCH amplitude ratios respectively. The HS-DPCCH is responsible for transferring positive acknowledgements (ACK), negative acknowledgements (NACK) and Channel Quality Indicators (CQI). Each of these can be configured with a different amplitude ratio relative to the DPCCH. These amplitude ratios are specified in 3GPP TS 25.213. The UE must be capable of transmitting ACK, NACK and CQI when located at the cell edge. This means that link

Table 9.6 Uplink DPCH link budgets when associated with an HSDPA connection

Service type	PS data	PS data	PS data
Uplink bit rate (kbps)	64	128	384
Maximum transmit power (dBm)	21.0	21.0	21.0
Terminal antenna gain (dBi)	0.0	0.0	0.0
Body loss (dB)	3.0	3.0	3.0
Transmit EIRP (dBm)	18.0	18.0	18.0
Chip rate (Mcps)	3.84	3.84	3.84
Processing gain (dB)	17.8	14.8	10.0
Required Eb/No (dB)	2.0	1.5	1.6
HS-DPCCH overhead (dB)	2.8	1.6	1.1
Target uplink load (%)	50	50	50
Rise over thermal noise (dB)	3.0	3.0	3.0
Thermal noise power (dBm)	−108.0	−108.0	−108.0
Receiver noise figure (dB)	2.0	2.0	2.0
Interference floor (dBm)	−103.0	−103.0	−103.0
Receiver sensitivity (dBm)	−116.0	−114.7	−110.3
Node B antenna gain (dBi)	18	18	18
Cable loss (dB)	0.0	0.0	0.0
Fast fading margin (dB)	1.8	1.8	1.8
Soft handover gain (dB)	2.0	2.0	2.0
Building penetration loss (dB)	12.0	12.0	12.0
Indoor location probability (%)	90	90	90
Indoor standard deviation (dB)	10	10	10
Slow fading margin (dB)	7.8	7.8	7.8
Isotropic power required (dBm)	−114.4	−113.1	−108.7
Maximum allowed path loss (dB)	132.4	131.1	126.7

Table 9.7 Example HS-DPCCH overheads

	64 kbps	128 kbps	384 kbps
β_c	11/15	7/15	7/15
β_d	15/15	15/15	15/15
β_{hs}	24/15	24/15	19/15
HS-DPCCH overhead (dB)	2.8	1.6	1.1

budgets are typically based upon the maximum of the amplitude ratios allocated to the ACK, NACK and CQI, i.e. β_{hs} represents the maximum of the three amplitude ratios. In addition, Section 6.10 explains that the HS-DPCCH amplitude ratios are usually increased when a UE is in soft handover. Link budgets assume that the UE is at cell edge so the HS-DPCCH overhead should be based upon the increased amplitude offsets. Table 9.7 presents the set of amplitude offsets used to calculate the HS-DPCCH overheads appearing within Table 9.6.

Other than the HS-DPCCH overhead, the link budgets within Table 9.6 are the same as those for an uplink DPCH without HSDPA, i.e. the same as those within Table 9.1.

9.1.2.2 Downlink

An example HSDPA link budget is presented in Table 9.8. HSDPA provides a variable bit rate which depends upon the RF channel conditions. The bit rate appearing within an HSDPA link budget should correspond to the target for the cell edge. If the network has been planned using DPCH link budgets then the maximum allowed path loss from those link budgets can be used to define the maximum allowed path loss for the HSDPA link budget, i.e. the maximum allowed path loss becomes an input rather than an output. The link budget can then be completed in reverse to identify the HSDPA bit rate which can be expected at cell edge. Alternatively, the link budget can be used to identify the HSDPA transmit power required to achieve a specific bit rate at cell edge.

The HSDPA transmit power depends upon whether a shared or dedicated RF carrier has been allocated. The transmit power available to HSDPA is greater if a dedicated RF carrier has been allocated. If a shared RF carrier has been allocated, the HSDPA transmit power should account for the power used by both the common channels and DPCH. Increasing the HSPDA transmit power improves the performance of HSDPA, but increases the downlink interference floor experienced by the other channel types.

HSDPA connections have variable Eb/No requirements because they have variable bit rates. Link level simulations can be used to generate a look-up table which associates a range of Eb/No requirements with a range of bit rates. If the link budget is completed in the forward direction then the target cell edge bit rate can be translated into a target Eb/No requirement. If the link budget is completed in the reverse direction then the Eb/No achieved at cell edge can be translated into a bit rate achieved at cell edge. The variable bit rate also means that the processing gain is variable. Nevertheless, the processing gain remains equal to the ratio between the chip rate and bit rate. The example presented in Table 9.8 is based upon an HSDPA bit rate of 384 kbps and a corresponding processing gain of $10 \times \log(3840/384) = 10$ dB.

HSDPA does not use inner loop power control so it is not necessary to include a fast fading margin. In addition, HSDPA does not use soft handover so the link budget does not include any soft handover gains. The link budget shown in Table 9.8 illustrates that the soft handover gain has been replaced by a serving cell change gain. The serving cell change gain represents a reduction in the slow fading margin as a result of the UE being able to select the best serving cell (under the control of the RNC). The serving cell change procedure can be used to switch the HSDPA connection from one active set cell to another if the current serving cell experiences a fade.

Table 9.8 HSDPA link budget

Service type	HS-PDSCH
Downlink bit rate (kbps)	384
HSDPA transmit power (dBm)	37.8
Cable loss (dB)	2.0
MHA insertion loss (dB)	0.5
Node B antenna gain (dBi)	18
Transmit EIRP (dBm)	53.3
Chip rate (Mcps)	3.84
Processing gain (dB)	10.0
Required Eb/No (dB)	2.5
Target downlink load (%)	80
Rise over thermal noise (dB)	7.0
Thermal noise power (dBm)	−108.0
Receiver noise figure (dB)	8.0
Interference floor (dBm)	−93.0
Receiver sensitivity (dBm)	−100.5
Terminal antenna gain (dBi)	0.0
Body loss (dB)	3.0
Fast fading margin (dB)	0.0
Serving cell change gain (dB)	2.0
MDC gain (dB)	0.0
Building penetration loss (dB)	12.0
Indoor location probability (%)	90
Indoor standard deviation (dB)	10
Slow fading margin (dB)	7.8
Isotropic power required (dBm)	−79.7
Max. allowed path loss (dB)	133.0

9.1.3 HSUPA

An example HSUPA link budget is presented in Table 9.9. Similar to HSDPA, the bit rate provided by an HSUPA connection depends upon the RF channel conditions. HSUPA link budgets can be completed in the forward direction to determine the maximum allowed path loss for a specific cell edge bit rate. Alternatively, link budgets can be completed in the reverse direction to identify the cell edge bit rate for a specific maximum allowed path loss. The latter tends to be used when the network has already been planned based upon a set of DPCH link budgets.

A UE with an HSUPA connection has to transmit up to five physical channels, i.e. the DPCCH, E-DPCCH, E-DPDCH, HS-DPCCH and DPDCH. The requirement for the DPDCH depends upon whether or not the SRB is transferred using the HSUPA connection (ignoring multi-service connections which could involve using the DPDCH for a CS speech connection). The Eb/No requirement within an HSUPA link budget should reflect the complete set of physical channels. HSUPA link level simulations may generate Eb/No requirements which exclude the overheads generated by one or more physical channels. The set of physical channel amplitude ratios can be used to add these overheads subsequent to the simulation process. In the case of the HS-DPCCH, the increased amplitude ratios applicable to soft handover scenarios should be used, i.e. the UE is assumed to be in soft handover when at cell edge. The example Eb/No requirement shown in Table 9.9 includes the DPCCH, E-DPCCH, E-DPDCH and HS-DPCCH. The DPDCH has been excluded based upon the assumption that it is used for the SRB which has a relatively low activity factor and has

Table 9.9 HSUPA link budget

Service type	PS data
Uplink bit rate (kbps)	128
Maximum transmit power (dBm)	21.0
Terminal antenna gain (dBi)	0.0
Body loss (dB)	3.0
Transmit EIRP (dBm)	18.0
Chip rate (Mcps)	3.84
Processing gain (dB)	14.8
Required Eb/No (dB)	1.5
Target uplink load (%)	50
Rise over thermal noise (dB)	3.0
Thermal noise power (dBm)	−108.0
Receiver noise figure (dB)	2.0
Interference floor (dBm)	−103.0
Receiver sensitivity (dBm)	−116.3
Node B antenna gain (dBi)	18
Cable loss (dB)	0.0
Fast fading margin (dB)	1.8
Soft handover gain (dB)	2.0
Building penetration loss (dB)	12.0
Indoor location probability (%)	90
Indoor standard deviation (dB)	10
Slow fading margin (dB)	7.8
Isotropic power required (dBm)	−114.7
Maximum allowed path loss (dB)	132.7

priority over the E-DPDCH. Similar to HSDPA, link level simulations can be used to generate a look-up table which associates a range of Eb/No requirements with a range of bit rates. If the link budget is completed in the forward direction then the target cell edge bit rate can be translated into a target Eb/No requirement. If the link budget is completed in the reverse direction then the Eb/No achieved at cell edge can be translated into a bit rate achieved at cell edge. The variable bit rate also means that the processing gain is variable. Nevertheless, the processing gain remains equal to the ratio between the chip rate and bit rate.

9.2 Radio Network Planning

- Radio Network Planning is used to identify the geographic locations of the Node B. It is also used to identify the associated antenna configurations in terms of antenna type, height, azimuth and tilt.
- There are two fundamental approaches to 3G radio network planning: the path loss based approach and the 3G simulation based approach.
- The path loss based approach is relatively simple and allows the radio network to be planned without modelling any subscriber traffic. This approach uses link budget results to define minimum signal strength thresholds. Results are typically presented in terms of service coverage, best server areas and C/I.
- The 3G simulation based approach is more complex and time consuming, but generates a greater quantity of information. 3G simulations are typically based upon static rather than dynamic

simulations. Results can be presented in terms of coverage, capacity, soft handover, inter-cell interference, uplink load and downlink transmit power.

- 3G simulations may be used to supplement the path loss based approach. In this case, the main planning activity is completed using path loss calculations while 3G simulations provide additional and more detailed information for specific sections of the network.

Radio Network Planning is used to identify the geographic locations of the Node B. It is also used to identify the antenna configurations in terms of antenna type, height, azimuth and tilt. The high-level interaction of radio network planning with dimensioning, site acquisition, site design and site build is illustrated in Figure 9.4.

The radio network planning process is preceded by system dimensioning. Dimensioning results are used to generate an estimate of the expected site density. This provides a guide to the number of sites which should be necessary to achieve the coverage requirements across a specific geographic area.

Site acquisition defines the actual sites which are available to radio network planners. The site acquisition team may provide radio network planners with a list of candidate sites. In this case, radio network planners are required to select the most appropriate sites from the list of candidates. Alternatively, radio network planners may identify the requirement for one or more sites at locations where there are no candidates. In this case, the site acquisition team can be requested to identify additional candidate sites. Sites represent a relatively expensive and long-term investment for an operator. Selecting sites which have a poor location or which limit good site design can lead to reduced system performance irrespective of any subsequent RF and parameter optimisation. Evaluating candidate sites should involve a site visit as well as modelling within the radio network planning tool. In practice, the interaction between site acquisition and radio network planning is likely to be a compromise between obtaining ideal site locations and the actual locations which are available.

If an operator has already deployed a 2G network then it is likely to be beneficial to re-use as many of the existing site locations as possible. Re-use of existing sites introduces the option of sharing antenna subsystems between 2G and 3G, i.e. feeder cables and antennas can be shared. In general, this requires swapping existing 2G single band antennas for new 2G/3G dual band antennas (or swapping existing 2G dual band antennas for new 2G/3G triple band antennas). The benefit of sharing the antenna subsystem is a reduced requirement for antennas and feeder cables. This can be important at sites which have limited physical space or planning restrictions. Sharing feeder cables requires a diplexor to combine the downlink signals and to separate the uplink signals. A potential drawback associated with sharing the same antenna subsystem is that it may limit the radio network planner's ability to apply independent 2G and 3G RF optimisation, e.g. if the 3G system requires mechanical

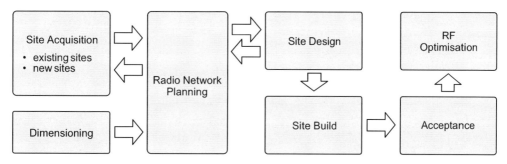

Figure 9.4 Interaction of radio network planning with site acquisition, design and build

antenna downtilt then the impact upon the 2G system must also be evaluated. However, some antennas allow independent adjustment of the 2G and 3G electrical downtilts.

Site selection should also account for site design requirements. Site design involves identifying specific locations for the Node B cabinet and each antenna. It also involves identifying the route for the feeder cable to connect the Node B cabinet to each antenna. It is necessary to account for physical space and power requirements when identifying a location for the Node B cabinet. Smaller Node B cabinets allow greater flexibility in terms of site design. The Node B location is likely to dictate the selection between an outdoor and indoor Node B cabinet. The feeder cable should be routed as directly as possible to help reduce its length and minimise feeder losses. Antenna placement represents the part of site design which has a direct impact upon radio network planning. Antennas should not be placed behind any obstacles. Roof-top site designs should ensure that antennas are either at the edge of the rooftop or sufficiently high to avoid the edge of the rooftop clipping the antenna gain pattern. In general, the height of the antenna should be increased if it is moved away from the edge of the rooftop. If antennas are mounted on the side of a building the azimuth requirements should be considered to ensure that the walls of the building do not shield the horizontal beamwidth. If the site already accommodates other antennas there is a requirement to ensure that isolation requirements are considered to avoid interference.

There are two fundamental approaches to 3G radio network planning: the path loss based approach and the 3G simulation based approach. The path loss approach is relatively simple and allows the radio network to be planned without modelling any subscriber traffic. This approach uses software tools which are relatively mature and generates results which are easy to interpret. Results are typically presented in terms of service coverage, best server areas and C/I. The 3G simulation approach is more complex and time consuming, but generates a greater quantity of information. Results are typically presented in terms of coverage, capacity, soft handover, inter-cell interference, uplink load and downlink transmit power. The path loss and 3G simulation based approaches are illustrated in Figure 9.5.

3G simulations may be used to supplement the path loss based approach. In this case, the main planning activity is completed using path loss calculations while 3G simulations are used to provide additional and more detailed information for specific sections of the network.

9.2.1 Path Loss based Approach

The path loss based approach to 3G radio network planning requires a planning tool capable of completing path loss calculations and displaying geographic areas where specific path loss thresholds

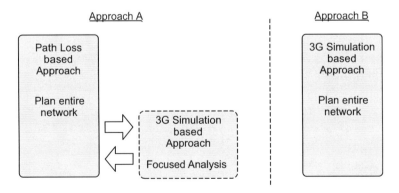

Figure 9.5   Path loss and 3G simulation approaches to radio network planning

have been exceeded. The planning tool should also be capable of displaying best server areas and ideally it should be capable of displaying downlink C/I. In each case, numerical statistics as well as graphical plots should be generated. The inputs to the planning tool for the path loss based approach are illustrated in Figure 9.6.

The 3G site data should include site locations, antenna types, antenna heights, antenna tilts, antenna azimuths, feeder types, feeder lengths, RF carrier and Node B types. The propagation model should be tuned from measurements. Accurate propagation modelling is fundamental to radio network planning. Propagation model tuning involves minimising the standard deviation of the error between the predicted and measured propagation loss while maintaining a mean error which is close to 0 dB. A different propagation model can be defined for each environment type or environment dependant correction factors can be applied to a common propagation model. Different propagation models can also be defined for different cell ranges, antenna heights and operating bands. If a propagation model has already been tuned for GSM then it is possible that the same model can be applied for 3G. GSM 1800 or 1900 MHz propagation models can be applied to 3G when operating in the 2100 MHz band. Likewise, 900 MHz GSM models can be applied to 3G when operating in the 800 or 900 MHz bands.

The digital terrain map should be adequate in terms of resolution, number of clutter categories and accuracy. The resolution should be relatively high for urban and suburban areas, but can be reduced for rural areas. It is typical to use a 20 m resolution for urban and suburban areas whereas a 50 m resolution can be used for rural areas. If the resolution is too low then the accuracy of the map and the subsequent propagation modelling becomes poor. If the resolution is too high the computer processing requirement becomes excessive and the cost of the map is more likely to be high. High resolution maps may be necessary when planning dense urban environments where cell ranges are particularly small. If microcells with below rooftop antennas are to be planned then it may be necessary to purchase a map which includes building vectors. Building vectors may have either two or three dimensions. The appropriate number of clutter categories depends upon the geographic area. It is typical to use about ten categories. Some planning tools may have a maximum number of categories which can be imported. In this case, the digital map can be processed to merge similar categories. If the number of categories is large then the propagation tuning exercise becomes more difficult. Once a digital map has been imported into the radio network planning tool, a set of checks should be completed to help validate its accuracy. Clutter types and vectors should be compared with those indicated on paper maps.

The signal strength thresholds should be based upon a link budget analysis. Different maximum allowed path loss thresholds can be generated for each service type and each environment type. In general, planning tools display contours of downlink signal strength rather than downlink path loss.

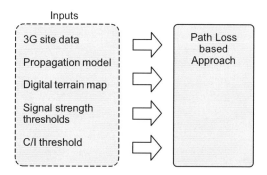

Figure 9.6   Inputs for the path loss based approach to radio network planning

This makes it necessary to translate the maximum allowed path loss thresholds into minimum downlink signal strength thresholds. A relatively arbitrary base station transmit power can be selected and the signal strength thresholds calculated by subtracting the maximum allowed path loss. It is common for the arbitrary transmit power to be chosen to equal the CPICH transmit power. This allows signal strengths calculated by the planning tool to be interpreted as CPICH RSCP. An example of the translation from a link budget maximum allowed path loss to a planning tool signal strength threshold is presented in Table 9.10.

In this example it is assumed that the maximum allowed path loss from the 3G link budget analysis is 145 dB. This figure could have originated from either an uplink or downlink link budget. In both cases, the planning tool is used to display contours of downlink signal strength, i.e. signal strength is used to quantify path loss.

The Node B antenna gain used in Table 9.10 should equal the gain assumed during the link budget analysis. It is likely that an actual radio network includes a range of different antenna types each with a different antenna gain. This does not impact the results as long as the planning tool is configured with the actual antenna types. For example, if the link budgets are based upon an antenna gain of 18 dBi whereas an actual Node B has an antenna gain of 16 dBi, the maximum allowed path loss resulting from the link budget analysis will be 2 dB more relaxed than it should be. However, applying an 18 dBi antenna gain in Table 9.10 means that the signal strength threshold is 2 dB more difficult to achieve and the two factors cancel one another. A similar argument is applicable to the UE antenna gain.

Defining an appropriate feeder loss to use in Table 9.10 is more complex. If the link budget analysis resulted in downlink limited coverage then the argument for the feeder loss is the same as the argument for the antenna gains. However, if the link budget analysis resulted in uplink limited coverage then the argument is only valid if Mast Head Amplifiers (MHA) are not used. If MHA have been used then it becomes difficult to define a single signal strength threshold which can be applied to all sites. This results from the maximum allowed path loss generated by the link budgets being independent of the feeder loss (assuming the MHA exactly compensates the feeder loss) while the signal strength calculated by the planning tool is dependant upon the feeder loss. A solution would be to assign a feeder loss value of 0 dB to all sites within the planning tool, and to the calculation within Table 9.10. However, there is often a requirement to use the planning tool as a database for site-specific feeder loss values. In this case, applying the feeder loss value assumed in the link budget (excluding the benefit of the MHA) within Table 9.10 provides an approximate signal strength threshold.

Path loss thresholds can be calculated for a range of services and bit rates, e.g. the 12.2 kbps speech service, the 64 kbps video call service and the 128 kbps packet switched data service. Thresholds can also be calculated for specific HSDPA and HSUPA bit rates. For example, HSDPA coverage contours could be plotted for 384 kbps, 500 kbps, 1 Mbps, 2.0 Mbps and 3.0 Mbps. In the case of HSDPA, the actual throughput experienced by the end-user depends upon the number of simultaneously active connections. Link budget planning thresholds illustrate the throughput which can be achieved when there is only a single active connection.

Table 9.10 Translation of maximum allowed path loss to planning tool signal strength threshold

Link budget result for maximum allowed path loss	145 dB
Downlink transmit power configured in the planning tool	33 dBm
Node B antenna gain assumed in the link budgets	18 dBi
UE antenna gain assumed in the link budgets	0 dBi
Feeder loss assumed in the link budgets	2 dB
Planning tool signal strength threshold	−96 dBm

The path loss based approach to 3G radio network planning should include an analysis of the best server areas. This helps to ensure good dominance and a relatively even distribution of network loading. Best server areas should be contiguous and should not be fragmented. Non-contiguous best server areas indicate that there is likely to be relatively poor dominance and increased levels of inter-cell interference. In general, neighbouring best server areas should be of approximately equal size. If there is a known traffic hotspot then a node B should be located as close as possible to that hotspot and the dominance area can be smaller.

A C/I analysis can also be included as part of the path loss based approach to 3G radio network planning. A C/I analysis provides an indication of cell isolation and inter-cell interference. The analysis is completed by assigning the same RF carrier to each cell and instructing the planning tool to calculate the ratio between the strongest downlink signal (interpreted as the wanted signal) and the sum of all of the other signals (interpreted as the interference). Large negative values of C/I can be interpreted as areas of poor dominance where the CPICH Ec/Io is likely to be poor. A typical threshold for the minimum allowed downlink C/I is -6 dB. This corresponds to incurring interference which is four times stronger than the desired signal, e.g. this would occur if a UE received five signals of equal strength. This type of C/I analysis typically assumes that all cells are transmitting with equal power and so are equally loaded. The analysis does not account for soft handover between cells, but this is similar to a plot of CPICH Ec/Io generated by a 3G simulation, i.e. plotting CPICH Ec/Io is equal to CPICH RSCP/RSSI, where the CPICH RSCP is measured from a single cell. A 3G simulation exercise can be completed in parallel to the first set of path loss based C/I calculations to help validate the method and to calibrate the minimum allowed C/I threshold.

9.2.2 3G Simulation based Approach

The 3G simulation based approach to radio network planning requires a 3G radio network planning tool. The majority of 3G radio network planning tools make use of Monte Carlo simulations. Monte Carlo simulations are static rather than dynamic. This means that system performance is evaluated by considering many independent instants (snap shots) in time. A dynamic simulation evaluates performance by considering a series of consecutive instants in time. In general, dynamic simulations are more time consuming than static simulations. The general principle of a static simulation is illustrated in Figure 9.7.

Each simulation snap shot is started by distributing a population of UE across the simulation area. This distribution could be based upon a uniform random distribution or a weighted random distribution. Weightings can be based upon Erlang maps generated from the traffic belonging to an existing 2G network. Alternatively, weightings could be environment type dependent so for example, urban areas could be specified to have a higher traffic density than suburban areas. Once the population of UE have been distributed, the set of uplink and downlink transmit powers are converged. Transmit power convergence requires an iterative process because the transmit power of one connection has an impact upon the transmit power of other connections. For the example of two UE, the uplink transmit power of the first UE is calculated based upon the uplink C/I requirement and the uplink interference floor. The transmit power of the second UE is then calculated based upon the uplink C/I requirement and the uplink interference floor which has now been increased by the first UE. The transmit power of the first UE then has to be recalculated because the uplink interference floor has been increased by the second UE. Repeating these calculations results in convergence of both the UE transmit powers and the uplink interference floor. A similar process can be completed in the downlink direction. The uplink and downlink transmit powers can be converged independently although there is some dependence in the case of a UE failing to maintain its connection. For example, if a UE has insufficient uplink transmit power to maintain its connection then the downlink transmit power for that UE can be cleared and made available to the remaining UE. UE which are not able to achieve their C/I requirements are categorised as being in outage. Outage may also be caused by factors such as inadequate Node B

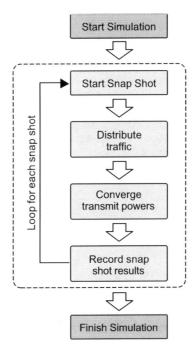

Figure 9.7   Inputs for the 3G simulation based approach to radio network planning

baseband processing resources or reaching the maximum allowed increase in uplink interference. The results are recorded at the end of a simulation snap shot and the process is repeated. This allows the simulation to generate probability distributions and to quantify the probability of certain events occurring, e.g. the probability that a UE will be able to establish a connection at a specific location. The number of snap shots necessary to generate statistically stable simulation results tends to depend upon the quantity of traffic distributed during each snap shot. Distributing relatively little traffic tends to increase the number of snap shots required.

The inputs to the planning tool for the 3G simulation based approach are illustrated in Figure 9.8. The first three inputs are similar to those required by the path loss based approach. The 3G site data may be more complex in terms of requiring greater information to describe the Node capability, e.g.

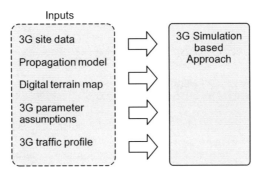

Figure 9.8   Inputs for the 3G simulation based approach to radio network planning

baseband processing capability. Propagation modelling is also more complex for a 3G simulation if slow fading is modelled. The path loss based approach makes use of a link budget threshold which includes a slow fade margin. 3G simulations typically model slow fading explicitly making it necessary to specify a standard deviation and correlation factor. The correlation factor is used to specify the coherence of the fading experienced by the signals between a UE and the set of surrounding Node B. A high correlation factor means that when one signal is experiencing a fade there is a high probability that the other signals will also be experiencing a fade. The 3G simulation tool may allow correlation factors to be configured separately for signals originating from the same Node B and signals originating from different Node B, i.e. the correlation is likely to be greater for signals originating from the same Node B because the propagation paths will be relatively similar. The slow fading correlation factor has an impact upon the soft handover gain which is maximised when soft handover radio links are uncorrelated. Uncorrelated radio links increase the probability of the receiver always having at least one signal which is not experiencing a fade.

Typical 3G parameter assumptions include maximum uplink and downlink transmit powers, common channel transmit powers, uplink and downlink noise figures, uplink and downlink Eb/No requirements, maximum allowed increase in uplink interference, orthogonality, soft handover window and soft handover gains. The soft handover gain used for 3G simulations should be interpreted differently from the soft handover gain used for link budgets. Link budgets use a soft handover gain which includes the diversity gain resulting from both fast and slow fading. 3G simulations usually model slow fading explicitly and so it is not necessary to include the slow fading diversity gain within the soft handover gain parameter. The soft handover window used for static 3G simulations tends to represent an average of the addition and deletion windows. The addition and deletion windows provide hysteresis in the live network to help avoid ping-pong in terms of cells being added and deleted from the active set. It is not possible to model this hysteresis using a static simulation tool because only snap shots in time are considered. When a simulated neighbouring cell is between the addition and deletion windows there is no historical information to indicate whether the cell is already in the active set and the signal quality is decreasing or the cell is outside the active set and the signal quality is increasing.

3G traffic profiles are specified in terms of the services used by the population of UE. The full range of services can be categorised as speech, circuit switched data and packet switched data. One or more bit rates can be associated with each of these categories. Defining accurate traffic profiles can be difficult and it is reasonable to start a simulation exercise by modelling one service at a time, e.g. generate coverage and capacity results for the 12.2 kbps speech service and then generate similar results for the 384 kbps packet switched data service. This approach provides an indication of the variance which can be expected when changing the traffic profile. Identifying the network capacity can be an iterative process, requiring the network planner to increase the quantity of traffic loading the network until the probability of blocking reaches a realistic maximum. The absolute maximum network capacity can be quantified by distributing very large quantities of traffic, but the blocking probability will become unrealistically high. The 3G traffic profile also requires the geographic distribution of the UE to be defined. This includes specifying the percentage of UE which are indoors and experience a building penetration loss.

The 3G simulation approach to radio network planning is more time consuming than the path loss based approach. Results take longer to generate and longer to interpret. Simulation time depends upon the size and resolution of the geographic area, the site density and the quantity of traffic loading the network. It is important to ensure that sufficient simulation snap shots have been completed to generate statistically stable results. 3G simulation tools may include functionality for a passive scan terminal. Passive scan terminals are used to increase the rate at which graphical geographic results are generated. At the end of each simulation snap shot a passive scan terminal is placed within each pixel across the simulated area and connection establishment is attempted without modifying the already converged uplink and downlink transmit powers. This approach allows every simulation snap shot to generate a

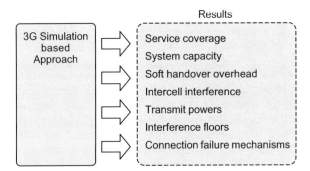

Figure 9.9   Typical results generated by a 3G simulation

result for every pixel. Otherwise, it is only possible to generate results at the pixels where UE have been distributed. The results generated by a passive scan terminal are used to update the graphical geographic results but typically are not used to update the numeric results. This means that passive scan terminals can increase the rate at which graphical results are generated, but decrease the rate at which numerical results are generated.

The main benefit of completing 3G simulations is the relatively large quantity of information which is generated. This information can help guide planning decisions as well as provide more extensive expectations of network performance. The main results generated by a 3G simulation are presented in Figure 9.9.

3G simulations can also be used for investigative studies aimed at evaluating the impact of specific network configurations. The impact of increasing sectorisation can be quantified for a range of different antenna gains and beamwidths. Some of the results can be used as an input to dimensioning exercises, e.g. the inter-cell interference ratio and soft handover overheads. Studies can also be completed to help quantify the benefit of deploying an additional RF carrier, or the impact of using mechanical rather than electrical antenna downtilts.

9.3 Scrambling Code Planning

- Scrambling code planning is usually associated with assigning a downlink primary scrambling code to each cell. It can also be associated with assigning groups of uplink scrambling codes to each RNC.
- The fundamental requirement for downlink scrambling code planning is that the isolation between cells which are assigned the same scrambling code should be sufficient to ensure that UE never simultaneously receive the same scrambling code from more than one cell.
- Scrambling code planning should be completed in combination with neighbour list planning to ensure that neighbour lists never include duplicate scrambling codes.
- Scrambling code planning can also have an impact upon the three step cell synchronisation procedure. It is possible to adopt a scrambling code planning strategy which places the emphasis upon either step 2 (identifying the scrambling code group) or step 3 (identifying the scrambling code). The impact is dependent upon UE implementation.
- Scrambling code planning should also account for future network expansion which could be the inclusion of additional Node B or increased sectorisation of existing Node B.

- If a radio network includes Node B which are configured with two or three RF carriers the same scrambling code plan can be assigned to each carrier.
- Additional rules for scrambling code planning are required at locations close to international borders where there may be another 3G operator using the same RF carrier.
- Uplink scrambling code planning involves assigning a range of uplink scrambling codes to each RNC. If these scrambling code ranges do not overlap there is no danger of neighbouring RNC assigning the same scrambling code
- Key 3GPP specifications: TS 25.213.

Scrambling code planning is usually associated with assigning a downlink primary scrambling code to each cell. It can also be associated with assigning groups of uplink scrambling codes to each RNC. The resultant uplink and downlink scrambling code plans form part of the RNC databuild.

9.3.1 Downlink

3GPP TS 25.213 specifies 512 downlink primary scrambling codes. Each primary scrambling code has 15 secondary scrambling codes. Each primary and each secondary scrambling code has a left alternative scrambling code and a right alternative scrambling code. Secondary scrambling codes may be used for beamforming whereas left and right alternative scrambling codes may be used by SF/2 compressed mode.

Each cell belonging to the radio network plan must be assigned one primary scrambling code. This defines the set of 15 secondary scrambling codes as well as the set of left and right alternative scrambling codes. The fundamental requirement for scrambling code planning is that the isolation between cells which are assigned the same scrambling code should be sufficiently great to ensure that a UE never simultaneously receives the same scrambling code from more than a single cell.

Scrambling code planning should be completed in combination with neighbour list planning to ensure that neighbour lists never include duplicate scrambling codes. This is important because RRC signalling procedures can use the scrambling code as a way to reference each neighbour. The RNC will face an ambiguity if the UE reports measurements from a neighbour which has a duplicate scrambling code within the neighbour list, i.e. the RNC will be unable to deduce from which neighbour the measurements were recorded. Figure 9.10 illustrates the scenario where multiple neighbours have been allocated the same scrambling code.

The RNC implementation may be designed to combine neighbour lists when UE are in soft handover. Neighbour list combining helps to reduce the potential for missing neighbours and so helps to improve network performance. Figure 9.11 illustrates a UE which is in soft handover with cells A and B. If neighbour list combining has been implemented the UE will be provided with a neighbour list generated from the combination of the neighbour lists belonging to cells A and B. If both cells have

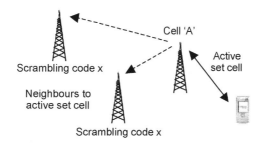

Figure 9.10   Requirement to avoid neighbour lists which include duplicate scrambling codes

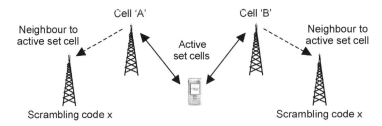

Figure 9.11   Neighbour list combining and potential duplicate scrambling codes (example 1)

neighbours with the same scrambling code, the RNC will be unable to deduce from which cell UE measurements have been recorded.

This scenario requires that when cells A and B are neighboured there should not be any duplicate scrambling codes within the neighbour lists belonging to cells A and B. A second example scenario is presented in Figure 9.12.

In this case, cell A is neighboured with cells B and C while cells B and C are not necessarily neighboured with one another. The UE could trigger an active set update which results in the active set including cells B and C. The neighbour lists belonging to cells B and C would then be combined and a duplicate scrambling code introduced. In general, neighbour list auditing should be completed after scrambling code planning to exclude the possibility of neighbour lists including duplicate scrambling codes.

Scrambling code planning can also have an impact upon the cell synchronisation procedure. This procedure is used whenever a UE needs to access a cell or measure the quality of a cell, e.g. neighbour cell measurements. 3GPP TS 25.213 specifies that the 512 primary scrambling codes are organised into 64 groups of 8. The cell synchronisation procedure is based upon this grouping and the following three steps:

1. The P-SCH is used to achieve slot synchronisation.
2. The S-SCH is used to achieve frame synchronisation and identify the primary scrambling code group.
3. The CPICH is used to identify the primary scrambling code.

The first step is relatively independent of the scrambling code plan although there is some potential to improve performance if the scrambling codes assigned to each cell of a Node B are from the same scrambling code group [51]. If all cells belong to the same scrambling code group then both the P-SCH

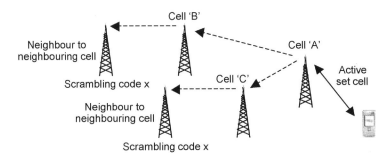

Figure 9.12   Neighbour list combining and potential duplicate scrambling codes (example 2)

and S-SCH will be identical for each cell. If all cells are configured with a Tcell of 0 chips then each cell will transmit the P-SCH and S-SCH with the same timing. This helps to improve the signal quality of the P-SCH and S-SCH in the softer handover regions and so improves the reliability of the cell synchronisation procedure. The effectiveness of this approach depends upon the UE implementation. There is a danger that after receiving a single P-SCH from a Node B the UE will assume there is only a single cell. In this case, the UE would identify only one scrambling code during step 3 instead of potentially two or more scrambling codes.

Step 2 of the synchronisation procedure involves selecting 1 scrambling code group out of 64, whereas step 3 involves selecting 1 scrambling code out of 8. Step 3 is likely to be more reliable because there are fewer alternatives. Step 3 is also likely to require greater UE processing and so have a greater impact upon UE battery life. Both of these factors are UE implementation dependent and the impact upon UE battery life needs to be kept in perspective relative to other procedures. Nevertheless, it is possible to adopt a scrambling code planning strategy which places the emphasis upon either step 2 or step 3. Placing the emphasis upon step 2 has the potential to reduce UE power consumption whereas placing the emphasis upon step 3 has the potential to improve cell synchronisation reliability. Placing the emphasis upon step 2 can be achieved by planning the scrambling codes such that each neighbour belongs to a different scrambling code group. In this case, the UE has to check for a relatively large number of different S-SCH during step 2, but once the scrambling code group has been identified the scrambling code is also known. Step 3 would serve as a check to ensure that the cell being measured is actually the cell within the neighbour list (rather than a missing neighbour belonging to the same scrambling code group). Placing the emphasis upon step 3 can be achieved by planning the scrambling codes such that neighbours tend to belong to the same scrambling code group. In this case, the UE has to check for a relatively small number of different S-SCH during step 2, but a relatively large number of scrambling codes within each group during step 3.

Figure 9.13 illustrates the concept of planning scrambling codes to minimise the number of neighbours belonging to different scrambling code groups, i.e. placing the emphasis upon step 3 of the cell synchronisation procedure. Clusters of cells are assigned scrambling codes belonging to the same group. However, it is not possible to generate a scrambling code plan in which all neighbours belong to the same scrambling code group. In practice, it is likely that neighbour lists would include cells belonging to three or four different scrambling code groups.

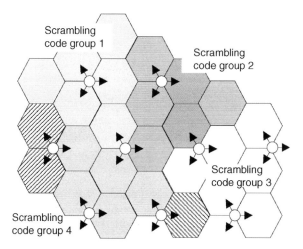

Figure 9.13   Clusters of cells which have scrambling codes belonging to the same group

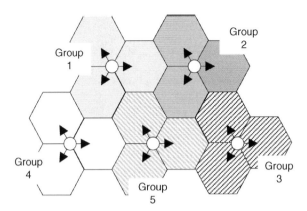

Figure 9.14 Node B with cells which have scrambling codes belonging to the same group

Figure 9.14 presents a simpler scrambling code planning strategy based upon assigning a different scrambling code group to each Node B. This approach avoids the requirement to plan clusters of cells and reduces the scrambling code planning process to the allocation of 1 out of 64 scrambling code groups, i.e. a scrambling code re-use pattern of 64. It is possible that each scrambling code group is divided into two for the purposes of scrambling code planning. This would generate 128 scrambling code subgroups and a corresponding re-use pattern of 128. Adjacent Node B could be assigned subgroups belonging to the same group to approximate clusters of cells belonging to the same code group.

The scrambling code planning strategy should also account for future network expansion. Future network expansion could mean the inclusion of additional Node B or increased sectorisation of existing Node B. Scrambling codes should be excluded from the original plan so they can be assigned when additional cells are introduced.

Additional rules for scrambling code planning are required at locations close to international borders where there may be another 3G operator using the same RF carrier. These rules are often specified by regulatory organisations. For example, in Europe the Electronic Communications Committee (ECC) within the European Conference of Postal and Telecommunications Administrations (CEPT) has specified ERC Recommendation 01-01, Border Coordination of UMTS. This recommendation prioritises the use of specific scrambling code groups on either side of an international border. It also recommends maximum allowed signal strengths for transmissions which cross international borders.

Scrambling code planning can be completed independently for different RF carriers. If a radio network includes Node B which are configured with two or three RF carriers the same scrambling code plan can be assigned to each carrier. This approach helps to reduce the quantity of work associated with scrambling code planning and reduces system complexity.

9.3.2 Uplink

Uplink scrambling codes are used by the Node B to distinguish between the radio links belonging to different UE. 3GPP TS 25.213 specifies 2^{24} long uplink scrambling codes and 2^{24} short uplink scrambling codes, i.e. there are more than 16 million of each type of scrambling code. Long scrambling codes have a length of 38 400 chips (10 ms) whereas short scrambling codes have a length of 256 chips (67 μs). Short scrambling codes are intended to be used by advanced receivers which are capable of

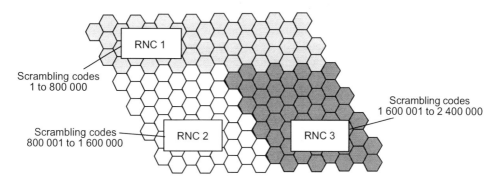

Figure 9.15   Concept of uplink scrambling code planning

interference cancellation and Multi-User Detection (MUD). Long scrambling codes represent the default for less advanced receivers. The RNC is responsible for assigning uplink scrambling codes and for ensuring that multiple UE are not assigned the same scrambling code.

An RNC does not have knowledge of the uplink scrambling codes assigned by its neighbours. If all RNC are allowed to assign any uplink scrambling code there is a danger that two UE with the same scrambling code could move into the same geographic area. Uplink scrambling code planning involves assigning a range of uplink scrambling codes to each RNC. If these scrambling code ranges do not overlap there is no danger of neighbouring RNC assigning the same scrambling code. Figure 9.15 illustrates the general concept of assigning different uplink scrambling code ranges to each RNC.

A simple approach to uplink scrambling code planning would be to equally divide the total set of uplink scrambling codes between the RNC. The large number of uplink scrambling codes means that this should be practical unless the quantity of network traffic is very high. For example, if a network includes 20 RNC then each RNC can be assigned approximately 830 000 uplink scrambling codes. If there is a risk that the set of scrambling codes assigned to an RNC becomes exhausted, the number of scrambling codes assigned to each RNC can be weighted according to the quantity of RNC traffic. Alternatively, a scrambling code re-use pattern could be introduced such that neighbouring RNC have different sets of scrambling codes, but those sets could be re-used elsewhere in the network.

9.4 Neighbour Planning

- UE rely upon neighbour lists when completing cell reselections and handovers. Maximising network performance requires high quality neighbour lists which include all necessary neighbours and exclude all unnecessary neighbours.
- Initial neighbour lists can be generated using a planning tool or they can be generated manually. Generating neighbour lists manually is practical for less dense sections of the network. Planning tools can make use of site location and antenna direction information.
- Identifying missing neighbours is often the activity which provides the greatest gains during initial RF optimisation. Neighbour lists can be optimised using either UE drive testing or network statistics.
- Intra-frequency neighbours are used for cell reselection, soft handover and intra-frequency hard handover. The RNC can instruct the UE to measure a maximum of 32 intra-frequency cells.

- It is mandatory to include the serving cell within SIB 11, i.e. it is possible to include up to 31 intra-frequency neighbours plus the serving cell. In CELL_DCH, the active set cells are included within the set of 32 intra-frequency cells.
- A UE continues to use the neighbour list information within SIB 11 after making the transition from RRC Idle mode to CELL_DCH until the RNC uses a dedicated Measurement Control message to instruct otherwise.
- Inter-frequency neighbours are used for cell reselection and inter-frequency handover. In CELL_DCH, inter-frequency neighbours are typically only signalled after triggering inter-frequency handover. A maximum of 32 inter-frequency cells can be specified.
- Uni-directional inter-frequency neighbour relations can be used to move UE from the second RF carrier to the first RF carrier for coverage reasons. Bi-directional neighbour relations can be used for load and service based handover.
- Inter-system neighbours are used for cell reselection and inter-system handover. In CELL_DCH, inter-system neighbours are typically only signalled after triggering inter-system handover. A maximum of 32 inter-system cells can be specified.
- The RNC may complete neighbour list combining when a UE is in soft handover. Neighbour list combining helps the RNC to ensure that all necessary neighbours are included within the neighbour list.
- It is preferable to have neighbour lists which are too long rather than too short. The number of neighbours is likely to depend upon the radio network plan and the associated environment. The introduction of a second RF carrier increases the total neighbour count.
- SIB 11 does not have sufficient capacity to accommodate 32 intra-frequency cells, 32 inter-frequency cells and 32 inter-system cells. The release 6 version of the 3GPP specifications introduces SIB 11bis to accommodate additional neighbours. The capacity of SIB 11 can be estimated based upon the quantity of data generated by each neighbour.
- Key 3GPP specifications: TS 25.331.

The definition of neighbouring cells represents an important part of radio network planning. UE rely upon neighbour lists when completing cell reselections and handovers. Missing neighbours can be responsible for dropped connections and UE being forced to complete cell selection rather than cell reselection. Maximising network performance requires high quality neighbour lists which include all necessary neighbours and exclude all unnecessary neighbours. The number of neighbours is likely to depend upon the radio network plan and the associated environment. Cells located in dense urban areas where the radio network is relatively dense are likely to require more neighbours than cells in rural areas where the radio network is less dense. If the radio network plan has not been optimised in terms of cell dominance the requirement for neighbours could be greater because coverage areas are more likely to be fragmented.

Initial neighbour lists can be generated using a planning tool or they can be generated manually by the radio network planner. Generating neighbour lists manually is practical for less dense sections of the radio network, e.g. motorway sections, small towns and villages. The requirement for a planning tool becomes greater for sections of the network where the site density is relatively high. Relatively simple neighbour list planning tools make use of site location and antenna direction information. These tools may propose initial neighbour lists based upon distance and direction. More complex neighbour list planning tools make use of propagation modelling and signal strength calculations. In either case it is important that neighbour lists are validated and optimised during initial RF optimisation. Identifying missing neighbours is often the activity which provides the greatest gains during initial RF optimisation. Neighbour lists can be optimised using either UE drive testing or network statistics. The former is relatively expensive as a result of requiring one or more drive test teams. It also tends to limit optimisation to vehicular routes. The benefit of this approach is that an RF scanner can be used to identify missing neighbours. It is less easy to identify neighbours which are not necessary because drive test routes are not exhaustive. Neighbour list optimisation using network counters makes use of data generated by actual end-user mobility. RNC counters can be used to identify neighbours which are not used. Missing neighbours can be identified by instructing UE to report detected set cells, i.e. cells which can be measured, but which are not included within the neighbour list.

There are three categories of neighbour lists which can be planned: intra-frequency, inter-frequency and inter-system. Intra-frequency neighbours are essential for normal operation of the 3G network. Inter-frequency neighbours become necessary if more than a single RF carrier has been deployed. Inter-system neighbours are necessary if inter-system handover has been enabled or if inter-system cell reselection is required.

9.4.1 Intra-Frequency

Intra-frequency neighbours are used for cell reselection, soft handover and intra-frequency hard handover. 3GPP TS 25.331 allows the RNC to instruct the UE to measure a maximum of 32 intra-frequency cells. In RRC Idle mode, the UE reads the set of intra-frequency neighbours from SIB 11. These neighbours can then be used for cell reselection purposes. In CELL_FACH, CELL_PCH and URA_PCH, the UE can read intra-frequency neighbour information from SIB 12. Broadcasting SIB 12 is optional and the neighbours are read from SIB 11 if SIB 12 is not broadcast. Similar to RRC Idle mode, the list of neighbours are used for cell reselection purposes. In CELL_DCH, the UE reads the set of intra-frequency neighbours from dedicated Measurement Control messages. These neighbours can then be used for soft handover and intra-frequency hard handover.

When a UE makes the transition from RRC Idle mode to CELL_DCH there is a period of time during which the UE continues to use the neighbour information read from SIB 11. The UE continues to use this information until the RNC forwards a dedicated Measurement Control to instruct otherwise. For this reason, it is possible to broadcast soft handover parameters within SIB 11, e.g. addition, drop and replace windows can be included. This allows the UE to trigger soft handover events prior to receiving the first Measurement Control message from the RNC.

It is mandatory to include the serving cell within SIB 11. This means that it is possible to include up to 31 intra-frequency neighbours plus the serving cell. The serving cell does not have to be included within SIB 12 and so there is potential to include up to 32 intra-frequency neighbours. In CELL_DCH, the active set cells are included within the 32 intra-frequency cells that the RNC is allowed to instruct the UE to measure. If the active set size is 1, the UE can be instructed to measure up to 31 intra-frequency neighbours. Likewise, if the active set size increases to 3, the UE can be instructed to measure up to 29 intra-frequency neighbours.

The RNC may complete neighbour list combining when a UE is in soft handover. This feature is not specified by 3GPP and it may not be implemented by some vendors. Neighbour list combining helps the RNC to ensure that all necessary neighbours are included within the neighbour list. The principle of neighbour list combining is that the neighbour lists belonging to each of the active set cells are combined to generate a single neighbour list. It is likely that the neighbour lists will have a relatively large number of neighbours in common. Neighbours appearing in more than a single neighbour list are typically given a high priority for inclusion within the combined list. Neighbours appearing in only one of the neighbour lists are typically given a low priority for inclusion. If the total number of intra-frequency cells exceeds 32 the neighbours with low priority are excluded.

In the case of inter-RNC soft handover, neighbour list combining requires the serving RNC to be informed of the relevant neighbour lists configured within the drift RNC. Neighbour list information can be transferred across the Iur interface using either the RNSAP Radio Link Setup Response message or the RNSAP Radio Link Addition Response message. Likewise, in an RNC anchoring scenario where all active set cells are connected to one or more drift RNC the serving RNC is informed of the relevant neighbour list information using RNSAP signalling.

It is critical to network performance that sufficient neighbours are included within the neighbour lists. Missing neighbours can cause increased levels of both uplink and downlink interference. If a UE has a missing neighbour while in CELL_DCH, the UE will be able to move relatively close to that neighbour without the RNC adding it to the active set. The downlink signal to noise ratio experienced by the UE will decrease because the power received from the missing neighbour does not include any wanted signal. The uplink interference floor belonging to the missing neighbour may increase because it is unable to power control the UE. The decrease in downlink signal to noise ratio has an impact upon the individual UE whereas the increase in uplink interference has an impact upon all UE connected to that cell. In general, the decrease in downlink signal to noise ratio dominates and the connection drops prior to being able to generate significant levels of uplink interference. This helps to avoid the scenario where a single UE can have a significant impact upon a large number of connections. However, this balance between uplink and downlink interference is scenario dependant and it is preferable to avoid the missing neighbour.

Some RNC implementations may instruct the UE to add detected set cells into the active set. Detected set cells are cells which the UE has reported, but which are outside the neighbour list. This solution can help to reduce the impact of missing neighbours, but also has an associated risk. Cells are addressed using their scrambling code when UE report CPICH Ec/Io measurements. In the case of a missing neighbour the RNC has to deduce from which cell the CPICH Ec/Io measurement has been recorded. It is likely that the scrambling code will have been re-used by the RNC at multiple cells. If the RNC selects the wrong cell then the procedure will fail. In addition, UE may be relatively slow at reporting detected set cell measurements. 3GPP TS 25.133 specifies that UE are allowed up to 30 s to identify a new detected set cell. It is preferable to avoid the missing neighbour so the RNC and UE do not have to work with detected set cell measurements.

Missing neighbours in RRC Idle mode, CELL_FACH, CELL_PCH and URA_PCH may force UE to complete a cell selection rather than a cell reselection. If inter-system neighbours have been configured UE may complete an inter-system cell reselection and temporarily move off the 3G network.

Long neighbour lists may also have an impact upon network performance although the impact is likely to be less significant and it is preferable to have neighbour lists which are too long rather than too short. If neighbour lists become too long unnecessary neighbours can be identified using network statistics, e.g. RNC counters can be used to identify which neighbour relations are not used. Long neighbour lists may reduce the rate at which a UE measures each neighbour. 3GPP TS 25.133 specifies that a UE in CELL_DCH must be capable of measuring at least eight neighbours every 200 ms (assuming the UE is not in compressed mode and the complete 200 ms is available). In general, UE should not be able to measure more than eight cells at any one location. Measuring more than eight cells indicates that it may be necessary to improve cell dominance. 3GPP TS 25.133 also specifies that a UE in CELL_DCH should be capable of identifying a new neighbouring cell within 800 ms once its CPICH Ec/Io becomes greater than -20 dB. Typical intra-frequency neighbour list lengths depend upon the site density and the level of cell dominance. It is common to have neighbour lists which include between 15 and 20 intra-frequency cells.

9.4.2 Inter-Frequency

Inter-frequency neighbours are used for cell reselection and inter-frequency handover. 3GPP TS 25.331 allows the RNC to instruct the UE to measure a maximum of 32 inter-frequency cells. In RRC Idle mode, the UE reads the set of inter-frequency neighbours from SIB 11. These neighbours can then be used for cell re-selection purposes. In CELL_FACH, CELL_PCH and URA_PCH, the UE can read inter-frequency neighbour information from SIB 12. Broadcasting SIB 12 is optional and the neighbours are read from SIB 11 if SIB 12 is not broadcast. Similar to RRC Idle mode, the list of neighbours are used for cell reselection purposes. In CELL_DCH, the UE reads the set of inter-frequency neighbours from dedicated Measurement Control messages. These neighbours can then be used for inter-frequency handover. Typically, the RNC only sends the list of inter-frequency neighbours to UE in CELL_DCH when necessary, i.e. after inter-frequency handover has been triggered. If inter-frequency handover has not been triggered then it is not usually necessary for the RNC to forward the neighbour list.

Similar to intra-frequency neighours, the RNC may complete neighbour list combining when a UE triggers inter-frequency handover while in soft handover. This helps to ensure that all necessary inter-frequency neighbours are included within the neighbour list. The principles of neighbour list combining for inter-frequency cells are the same as those for intra-frequency cells. Neighbours appearing in more than a single neighbour list are typically given a high priority for inclusion within the combined list. Neighbours appearing in only one of the neighbour lists are typically given a low priority for inclusion. If the total number of inter-frequency cells exceeds 32 the neighbours with low priority are excluded.

In the case of inter-RNC soft handover, neighbour list combining requires the serving RNC to be informed of the relevant neighbour lists configured within the drift RNC. Neighbour list information can be transferred across the Iur interface using either the RNSAP Radio Link Setup Response message or the RNSAP Radio Link Addition Response message. Likewise, in an RNC anchoring scenario where all active set cells are connected to one or more drift RNC the serving RNC is informed of the relevant neighbour list information using RNSAP signalling.

It is important that sufficient inter-frequency neighbours are included within the neighbour lists. Missing neighbours are likely to cause failed inter-frequency handover attempts and non-ideal cell reselections. Missing inter-frequency neighbours are less significant than missing intra-frequency neighbours because they do not cause the same increased levels of interference and they are only required when inter-frequency handover has been triggered. Inter-frequency neighbours can be planned using a range of different strategies. These strategies depend upon the features used to move the UE from one RF carrier to another. The simplest strategy is to limit the use of inter-frequency handover and inter-frequency cell reselection to coverage limited scenarios. In this case, inter-frequency neighbours are defined to allow the UE to move from the second RF carrier to the first RF carrier when moving out of coverage of the second RF carrier. This strategy is illustrated in the first part of Figure 9.16. It assumes that UE can be moved to the second RF carrier using directed RRC connection establishment which does not require the definition of any neighbour relations. Inter-frequency neighbour relations are uni-directional, allowing the UE to return to the first RF carrier when necessary. If the second RF carrier is introduced at individual sites or across small clusters, the levels of inter-cell interference experienced on the second RF carrier will be relatively low and the RF coverage will extend further than from the equivalent cells on the first RF carrier. This can dilute the capacity offered by the second RF carrier if it has been introduced as part of a capacity upgrade. It also means that the neighbours

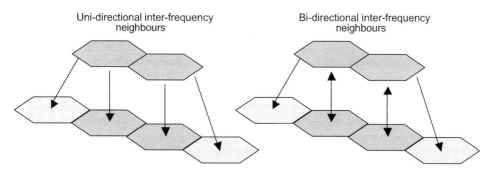

Uni-directional inter-frequency neighbours

Bi-directional inter-frequency neighbours

Figure 9.16   Example inter-frequency neighbour planning strategies

required by the second RF carrier are likely to differ from the equivalent neighbours used by the first RF carrier. Both issues can be resolved if the RNC databuild is used to reduce the extent of the coverage provided by the second RF carrier.

A more complex inter-frequency neighbour planning strategy is illustrated in the second part of Figure 9.16. Bi-directional neighbour relations have been introduced to allow the use of load-based or service-based inter-frequency handover. For example, if the load on the first RF carrier increases while the load on the second RF carrier remains relatively low, the RNC can instruct one or more UE to complete an inter-frequency handover from the first RF carrier to the second RF carrier. Service-based inter-frequency handover can be used if the operator wishes to group specific traffic types on specific RF carriers. For example, if the first RF carrier is used for speech and video call services while the second RF carrier is used for packet switched data services then the RNC can instruct inter-frequency handovers when UE establish their connections on an inappropriate RF carrier.

9.4.3 Inter-System

Inter-system neighbours are used for cell reselection and inter-system handover. 3GPP TS 25.331 allows the RNC to instruct the UE to measure a maximum of 32 inter-system cells. In RRC Idle mode, the UE reads the set of inter-system neighbours from SIB 11. These neighbours can then be used for cell reselection purposes. In CELL_FACH, CELL_PCH and URA_PCH, the UE can read inter-system neighbour information from SIB 12. Broadcasting SIB 12 is optional and the neighbours are read from SIB 11 if SIB 12 is not broadcast. Similar to RRC Idle mode, the list of neighbours are used for cell reselection purposes. In CELL_DCH, the UE reads the set of inter-system neighbours from dedicated Measurement Control messages. These neighbours can then be used for inter-system handover. Typically, the RNC only sends the list of inter-system neighbours to UE in CELL_DCH when necessary, i.e. after inter-system handover has been triggered. If inter-system handover has not been triggered then it is not usually necessary for the RNC to forward the neighbour list. The list of inter-system neighbours may be reduced during the inter-system handover procedure. GSM RSSI measurements may be recorded from a relatively large set of inter-system neighbours. The RNC may subsequently reduce the neighbour list for BSIC verification to reduce the quantity of processing required by the UE.

Similar to intra-frequency and inter-frequency neighbours, the RNC may complete neighbour list combining when a UE triggers inter-system handover while in soft handover. This helps to ensure that all necessary inter-system neighbours are included within the neighbour list. The principles of neighbour list combining for inter-system cells are the same as those for intra-frequency and inter-frequency cells. Neighbours appearing in more than a single neighbour list are typically given a high priority for inclusion within the combined list. Neighbours appearing in only one of the neighbour lists are typically given a low priority for inclusion. If the total number of inter-system cells exceeds 32 the neighbours with low priority are excluded.

In the case of inter-RNC soft handover, neighbour list combining requires the serving RNC to be informed of the relevant neighbour lists configured within the drift RNC. Neighbour list information can be transferred across the Iur interface using either the RNSAP Radio Link Setup Response message or the RNSAP Radio Link Addition Response message. Likewise, in an RNC anchoring scenario where all active set cells are connected to one or more drift RNC the serving RNC is informed of the relevant neighbour list information using RNSAP signalling.

It is important that sufficient inter-system neighbours are included within the neighbour lists. Missing neighbours are likely to cause failed inter-system handover attempts and non-ideal cell reselections. Missing inter-system neighbours are less significant than missing intra-frequency neighbours because they do not cause the same increased levels of interference and they are only required when inter-system handover has been triggered. Typical inter-system neighbour list lengths depend upon the site density. It is common to have neighbour lists which include between 15 and 20 inter-system cells.

9.4.4 Maximum Neighbour List Lengths

The preceding sections specify that a UE can be instructed to measure up to 32 intra-frequency cells, 32 inter-frequency cells and 32 inter-system cells, i.e. a total of 96 cells. In CELL_DCH, these cells are signalled to the UE using dedicated Measurement Control messages and the maximum neighbour list lengths can be used. UE are typically informed of the intra-frequency neighbours by default and conditionally informed of the inter-frequency neighbours if inter-frequency handover is triggered, or of the inter-system neighbours if inter-system handover is triggered.

In RRC Idle mode, CELL_FACH, CELL_PCH and URA_PCH, UE read the set of neighbouring cells from SIB 11 and potentially from SIB 12. The BCH transport channel is used to broadcast system information using a fixed transport block size of 246 bits and a Transmission Time Interval (TTI) of 20 ms. A single transport block is sent during each TTI so the corresponding bit rate is 12.3 kbps. The RLC and MAC layers do not add any overheads, allowing the RRC layer to use the complete set of 246 bits. The RRC layer is responsible for segmentation when an ASN.1 encoded SIB message cannot be accommodated by a single transport block. The RRC layer includes its own header information which occupies 24 bits and leaves a maximum of 222 bits for each segment of the ASN.1 encoded SIB message (the RRC header is smaller and a maximum of 226 bits are available when segmentation is not necessary and a complete SIB can be accommodated by a single transport block). The header added by the RRC layer includes the System Frame Number (SFN), the SIB type and segmentation information (when segmentation is used). 3GPP TS 25.331 specifies that a maximum of 16 segments can be used to transfer a single ASN.1 encoded SIB. This corresponds to a maximum of 3552 bits (444 bytes).

3552 bits is not sufficient to transfer information regarding 96 cells. This means that it is necessary to limit the number of neighbouring cells included within SIB 11 and SIB 12. The precise number of neighbouring cells which can be included depends upon the quantity of information associated with each neighbour, i.e. as the quantity of information increases the number of neighbours which can be included decreases. The release 6 version of the 3GPP specifications resolve this issue by introducing SIB 11bis. SIB 11bis can be used to accommodate any neighbours which have been excluded from SIB 11 as a result of its capacity limitation.

It is mandatory for SIB 11 to include information regarding the serving cell. The quantity of information signalled for the serving cell is usually less than the quantity of information signalled for non-serving cells. An example of the information signalled for the serving cell is presented in Log File 9.1. Coding this information using the ASN.1 encoding rules generates 23 bits of data. These 23 bits are illustrated in Figure 9.17.

An example of the information within SIB 11 for an intra-frequency neighbouring cell is presented in Log File 9.2. The quantity of information signalled for an intra-frequency neighbouring cell is increased by the inclusion of cell selection and reselection parameters, i.e. the maximum allowed

```
newIntraFreqCellList value 1
  intraFreqCellID: 0
  cellInfo
    modeSpecificInfo
      fdd
        primaryCPICH-Info
          primaryScramblingCode:    187
          readSFN-Indicator:        true
          tx-DiversityIndicator:    false
```

Log File 9.1 Example serving cell information within SIB 11

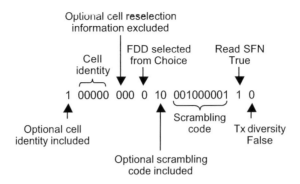

Figure 9.17   Encoding of serving cell information within SIB 11

uplink transmit power and the minimum CPICH Ec/Io and RSCP requirements. This information is not necessary for the serving cell because it is broadcast as part of SIB 3. Coding the intra-frequency neighbouring cell information using the ASN.1 coding rules generates 48 bits of data. These 48 bits are illustrated in Figure 9.18.

```
newIntraFreqCellList value 16
  intraFreqCellID: 25
  cellInfo
    modeSpecificInfo
      fdd
        primaryCPICH-Info
          primaryScramblingCode:      33
          readSFN-Indicator:          true
          tx-DiversityIndicator:      false
    cellSelectionReselectionInfo
      maxAllowedUL-TX-Power:          21
      modeSpecificInfo
        fdd
          q-QualMin :       -18
          q-RxlevMin:       -58
```

Log File 9.2   Example intra-frequency neighbouring cell within SIB 11

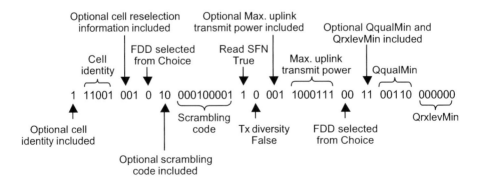

Figure 9.18   Encoding of intra-frequency neighbour cell information within SIB 11

These sections of SIB 11 illustrate that the quantity of data generated by the serving and intra-frequency neighbouring cells is defined by $23 + (n \times 48)$, where n is the number of intra-frequency neighbours.

These figures are typical, but will vary depending upon the precise set of information elements included for each intra-frequency cell. Other information elements which can be included are presented in Table 9.11.

If any of these additional parameters are included for a neighbouring cell the figure of 48 bits is increased for that neighbour, e.g. if Qoffset2 is included then the quantity of data associated with that neighbour increases to 55 bits. In the case of the reference time difference information element, the quantity of data is first increased by 2 bits to indicate which resolution is used, and then by a further 4, 8 or 10 bits, depending upon that resolution.

The examples in this chapter are based upon the Hierarchical Cell Structure (HCS) feature being disabled and the quality measure for cell selection and reselection being set to CPICH Ec/Io. The information associated with each neighbour is different when HCS is enabled or if the quality measure is set to CPICH RSCP.

An example of the information within SIB 11 for an inter-frequency neighbouring cell is presented in Log File 9.3.

Table 9.11 Additional information which can be associated with each intra-frequency cell in SIB 11

Information element	Range	Size
Cell individual offset	−20 to 20	6 bits
CPICH transmit power	−10 to 50	6 bits
Qoffset1	−50 to 50	7 bits
Qoffset2	−50 to 50	7 bits
Index for selection between reference time difference resolutions	0 to 2	2 bits
Reference time difference to cell: 2560 chips resolution	0 to 15	4 bits
256 chips resolution	0 to 150	8 bits
40 chips resolution	0 to 960	10 bits

```
newInterFreqCellList value 1
  frequencyInfo
    modeSpecificInfo
      fdd
        uarfcn-DL: 10838
  cellInfo
    modeSpecificInfo
      fdd
        primaryCPICH-Info
          primaryScramblingCode:    217
          readSFN-Indicator:        true
          tx-DiversityIndicator:    false
    cellSelectionReselectionInfo
      q-Offset2S-N:                 -50
      maxAllowedUL-TX-Power:        24
      modeSpecificInfo
        fdd
          q-QualMin:       -17
          q-RxlevMin:      -58
```

Log File 9.3   Example inter-frequency neighbouring cell within SIB 11

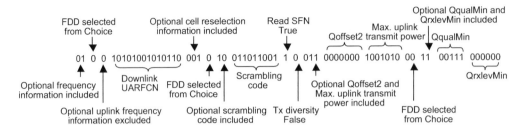

Figure 9.19   Encoding of inter-frequency neighbour cell information within SIB 11

The information describing an inter-frequency neighbouring cell is similar to that for an intra-frequency neighbouring cell. Log File 9.3 illustrates that the RF carrier is specified using the UTRA Absolute Radio Frequency Channel Number (UARFCN). It is not mandatory to include this information for every inter-frequency neighbour. If the UARFCN is excluded the UE assumes that the value associated with the previous neighbour is applicable, i.e. the UARFCN must be included for at least the first inter-frequency neighbour. This section assumes that the UARFCN is included for all inter-frequency neighbours. Log File 9.3 also illustrates that the optional Qoffset2 has been included with a large negative value. This tends to make the neighbouring cell appear more attractive for cell reselection purposes.

Coding the information presented in Log File 9.3 using the ASN.1 encoding rules generates the 67 bits of data illustrated in Figure 9.19. This example indicates that the quantity of data generated by inter-frequency neighbouring cells is defined by $(m \times 67)$, where m is the number of inter-frequency neighbours. If the optional Qoffset2 had not been included the appropriate expression would be $(m \times 60)$. In addition, if the UARFCN was only included for the first inter-frequency neighbour the expression would become $60 + ((m - 1) \times 44)$. Similar to the intra-frequency cells, other optional information elements can be included. These are presented in Table 9.12.

An example of the information within SIB 11 for an inter-system neighbouring cell is presented in Log File 9.4. This example is based upon GSM and the neighbouring cell is identified using its Network Colour Code (NCC) and Base Station Colour Code (BCC), i.e. its Base transceiver Station Identity Code (BSIC). The GSM minimum quality requirement is also specified as are the optional Qoffset1 and Qoffset2.

Table 9.12   Additional information which can be associated with each inter-frequency cell in SIB 11

Information Element	Range	Size
Inter-Frequency cell identity	0 to 31	5 bits
Uplink UARFCN	0 to 16383	14 bits
Cell individual offset	−20 to 20	6 bits
CPICH transmit power	−10 to 50	6 bits
Qoffset1	−50 to 50	7 bits
Index for selection between reference time difference resolutions	0 to 2	2 bits
Reference time difference to cell: 2560 chips resolution	0 to 15	4 bits
256 chips resolution	0 to 150	8 bits
40 chips resolution	0 to 960	10 bits

```
newInterRATCellList value 8
  technologySpecificInfo
    gsm
      cellSelectionReselectionInfo
        q-Offset1S-N:                    10
        q-Offset2S-N:                     0
        maxAllowedUL-TX-Power:           33
        modeSpecificInfo
          gsm
            q-RxlevMin:                 -53
      interRATCellIndividualOffset:  0
      bsic
        ncc: 0
        bcc: 4
      frequency-band: dcs1800BandUsed
      bcch-ARFCN: 47
```

Log File 9.4   Example inter-system neighbouring cell within SIB 11

Coding this inter-system neighbouring cell information using the ASN.1 encoding rules generates the 63 bits of data illustrated in Figure 9.20. This example indicates that the quantity of data generated by GSM inter-system neighbouring cells is defined by $(p \times 63)$, where p is the number of inter-system neighbours. If the optional Qoffset1 and Qoffset2 had been excluded the appropriate expression would be $(p \times 49)$.

In addition to individual neighbouring cell information, SIB 11 typically includes 24 bits of header information. The information within this header includes whether or not SIB 12 is broadcast, whether or not HCS is used, whether the quality measure is CPICH Ec/Io or CPICH RSCP, whether or not any inter-frequency or inter-system neighbours are included, whether or not the UE should delete any previously stored intra-frequency neighbours and the number of intra-frequency neighbours included. If any inter-frequency neighbours are included an additional 10 bits of header information is included to specify whether or not the UE should delete any previously stored inter-frequency neighbours and the number of inter-frequency neighbours included. Similarly, if any inter-system neighbours are included an additional 8 bits of header information is included to specify whether or not the UE should delete any previously stored inter-system neighbours and the number of inter-system neighbours included (the header for the inter-system neighbours is 2 bits smaller than the header for the inter-frequency neighbours because there are two less optional information elements and so two fewer flags are required).

SIB 11 can also include further information regarding the intra-frequency measurement quantity, the intra-frequency reporting quantity for the RACH, the maximum number of cells reported on the RACH

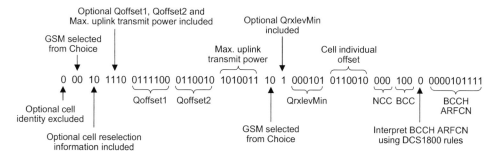

Figure 9.20   Encoding of inter-system neighbouring cell information within SIB 11

and reporting information for CELL_DCH. This additional information typically occupies a further 145 bits.

Based upon the preceding examples, the following expression can be used to calculate the size of SIB 11.

$$SIB11\_Size = Roundup\left[\frac{(24 + 23 + (n \times 48) + 145 + (m \times 67) + (p \times 63) + x + y)}{8}\right] bytes$$

where, n is the number of intra-frequency neighbours; m is the number of inter-frequency neighbours; p is the number of inter-system neighbours; $x = 10$ if there is one or more inter-frequency neighbours, otherwise $x = 0$, $y = 8$ if there is one or more inter-system neighbours, otherwise $y = 0$

This expression assumes that the UARFCN is specified individually for each inter-frequency neighbour, Qoffset2 is included for each inter-frequency neighbour and both Qoffset1 and Qoffset2 are included for each inter-system neighbour. In this case, the maximum SIB 11 size of 444 bytes is sufficient to accommodate 22 intra-frequency cells, 15 inter-frequency cells and 20 inter-system cells. This combination generates a SIB 11 size of 442 bytes.

If it is assumed that the UARFCN is only specified for the first inter-frequency neighbour and that neither Qoffset1 nor Qoffset2 is included for any inter-frequency nor inter-system neighbours the expression used to calculate the size of SIB 11 in bytes becomes:

$$SIB11\_Size = Roundup\left[\frac{(24 + 23 + (n \times 48) + 145 + 60 + ((m - 1) \times 44) + (p \times 49) + x + y)}{8}\right]$$

In this case, the maximum size of 444 bytes is sufficient to accommodate 25 intra-frequency cells, 20 inter-frequency cells and 25 inter-system cells. This combination generates a SIB 11 size of 442 bytes.

In general, these neighbour list lengths should be sufficient. If HCS is enabled then the quantity of information associated with each neighbour can increase. HCS priorities and temporary measurement offsets can be broadcast for each neighbouring cell. If the ASN.1 encoded neighbour list exceeds the capacity of SIB 11 prior to the introduction of SIB 11bis, the RNC implementation may limit the number of neighbours broadcast as part of the system information, but allow longer neighbour lists to be used within dedicated Measurement Control messages in CELL_DCH.

9.5 Antenna Subsystems

- Operators often generate a list of preferred antennas from which radio network planners are able to select. This usually provides the planners with a choice of electrical downtilts and may also provide a choice of horizontal and vertical beamwidths.
- Node B which belong to the macrocell network usually take advantage of dual-branch uplink receive diversity. This requires each cell to have two antenna elements. Cross-polar antennas accommodate two antenna elements within a single antenna housing.
- An important part of antenna selection is identifying an appropriate electrical downtilt. Downtilt provides a compromise between maximising coverage and achieving clear dominance. Electrical downtilt should be used in preference to mechanical downtilt.
- Some antennas have a fixed electrical downtilt whereas more sophisticated antennas have a variable electrical downtilt. Antennas with a variable electrical downtilt allow the optimisation engineer to adjust the delay to each of the antenna element dipoles.

- In general, macrocell deployments should be configured with high gain antennas. An increased gain helps to improve both coverage and capacity. It's necessary to balance the requirement for gain with the requirement to avoid antennas which are too directional

- A duplexor is required to allow both the transmit and receive signals to share the same antenna element. The duplexor is often included within the Node B cabinet rather than appearing as an external component.

- A Node B configured with three sectors and uplink receive diversity uses six outputs from the Node B cabinet. These six outputs are connected to six antenna elements using six parallel runs of feeder cable. Six Mast Head Amplifiers (MHA) may be included to improve the composite noise figure of the uplink receiver

- In general, feeder cable with a wide diameter has a lower loss per unit length. It also tends to be less flexible and more expensive. Flexibility can be important if the site design requires the feeder cable to follow an indirect route towards the antennas.

- Jumper cables are used to connect the RF feeder cable to the Node B cabinet and also to the MHA. Jumper cables are used for inter-connection because they are more flexible than RF feeder cables. This helps to avoid placing stress upon connectors.

- Remote RF modules avoid the requirement for long lengths of feeder cable which consequently avoids the requirement for MHA. The applicability of this design depends upon whether or not it is practical to mount the RF modules close to the antennas.

- Shared antenna subsystems require either a diplexor, triplexor or combiner to feed the separate uplink and downlink signals through a common set of RF components. The antennas, the feeder cables or both the antennas and feeder cables can be shared.

Antenna subsystems allow the downlink signals generated by the Node B cabinet to be radiated across the cell coverage area. They also allow the uplink signals generated by the population of UE to be received by the Node B cabinet. An antenna subsystem includes the antenna as well as the set of components used to connect the antenna to the Node B cabinet. These components could be relatively minimal, e.g. feeder cable and connectors, or they could be more complex, e.g. feeder cable, Mast Head Amplifier, diplexor, jumper cables and connectors. Antenna subsystems can encorporate some of the functionality usually provided by the Node B cabinet. The baseband processing and transport network interface sections of the Node B can be separated from the RF sections of the Node B. The RF sections can then be moved away to become integrated with the antenna subsystem.

9.5.1 Antenna Characteristics

Operators often generate a list of preferred antennas from which radio network planners are able to select. This usually provides the planners with a choice of electrical downtilts and may also provide a choice of horizontal and vertical beamwidths. A range of antenna types can be included to suite the requirements of different site types and different site designs, e.g. macrocell vs microcell and rooftop vs pole-mounted.

Node B which belong to the macrocell network usually take advantage of dual-branch uplink receive diversity. This requires each cell to have two antenna elements. The first can be used to transmit and receive while the second can be used to receive only. Both antenna elements transmit and receive if downlink transmit diversity is used. A duplexor is required to allow both the transmit and receive signals to share the same antenna element. A duplexor adds the signal to be transmitted in the downlink direction and filters off the signal to be received in the uplink direction. The duplexor is often included within the Node B cabinet rather than appearing as an external component. It is possible that the two antenna elements belong to two separate antennas. Two vertically polarised antennas could be used if their physical separation provides sufficient isolation. In practice, it can be

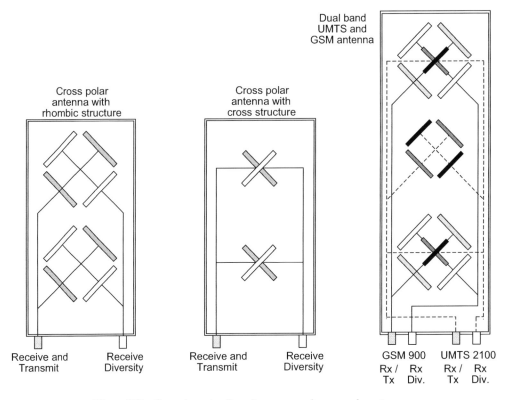

Figure 9.21   General construction of some example cross-polar antennas

difficult to create a site design which includes two antennas with sufficient isolation for every Node B sector. There may not be sufficient physical space or planning restrictions may limit the number of antennas. In addition, site owners may charge on a per antenna basis making the use of two antennas more expensive.

Cross-polar antennas accommodate two antenna elements within a single antenna housing. In this case, the two antenna elements take advantage of polarisation diversity to achieve sufficient isolation. Figure 9.21 illustrates the general construction of three example cross-polar antennas. The first two examples represent single-band antennas whereas the third example represents a dual-band antenna.

All three antenna designs are based upon elements using $\pm 45°$ dipoles. In general, it is preferable to use $\pm 45°$ rather than $0°$ and $90°$ because it allows the two antenna elements to experience similar performance. Horizontal dipoles may have weaker performance in areas where there is less scattering to change the polarisation of the signal between the UE and Node B. The two single-band designs illustrate the use of rhombic and cross structures. The rhombic structure is based upon a pair of dipoles belonging to each antenna element whereas the cross structure is based upon a single dipole per antenna element. The rhombic structure can be used to generate a narrower horizontal beamwidth than the cross structure, i.e. the rhombic and cross patterns can be used to generate horizontal beamwidths of $65°$ and $90°$ respectively. Each of the single band examples illustrate two sets of dipoles, one above the other. In practice, it is likely that the antenna design would include more than two sets of dipoles. Increasing the number of vertically separated dipoles decreases the vertical beamwidth of the antenna and increases the antenna gain. As a general rule, doubling the number of vertically separated dipoles increases the antenna gain by 3 dB and halves the vertical beamwidth. The example dual-band antenna illustrated within Figure 9.21 includes four antenna elements rather than two. There are two antenna elements for the UMTS system at 2100 MHz and two antenna elements for the GSM system at 900 MHz. The dipoles belonging to the UMTS system are smaller to reflect the smaller wavelength

An important part of antenna selection is identifying an appropriate electrical downtilt. Antenna downtilt provides a compromise between maximising coverage and achieving clear dominance. Antennas with insufficient downtilt will overshoot their intended coverage area. This is likely to result in fragmented coverage with poor dominance and increased neighbour relation requirements. In addition, the cell capacity will become diluted as a result of serving a larger geographic area and the soft handover overhead is likely to increase as a result of greater cell overlap. Antennas with too much downtilt will restrict coverage and limit the potential for soft handover. This is likely to generate either non-contiguous coverage or an increased site density. Site acceptance and subsequent RF optimisation involve drive testing to ensure that the coverage provided by an antenna reflects the expectations generated by the radio network planning tool. Electrical or mechanical tilts can be used to help optimise the coverage provided by the antenna. Electrical tilt defines the angle of the main lobe below the horizontal plane when the antenna is mounted in a vertical position. Antennas are designed to provide an electrical tilt by varying the phase of the signal reaching each of the antenna element dipoles. In practice, this corresponds to controlling the delay by using different cable lengths to each dipole. Mechanical downtilt is generated by mounting the antenna off the vertical position, i.e. angled downwards (or angled upwards if uptilt is required). Electrical downtilt should always be used in preference to mechanical downtilt. Electrical downtilt directs the complete antenna gain pattern downwards, i.e. the main lobe, side lobes and rear lobe. In contrast, mechanical downtilt directs the main lobe downwards, but directs the rear lobe upwards and does not change the angle of the side lobes.

Selecting an appropriate electrical downtilt requires the radio network planner to predict the downtilt requirement. The use of mechanical downtilt will be minimised if this can be predicted accurately, e.g. $4°$ of electrical downtilt $\pm 2°$ of mechanical downtilt. Some antennas have a fixed electrical downtilt whereas more sophisticated antennas have a variable electrical downtilt. Antennas with a variable electrical downtilt allow the optimisation engineer to adjust the delay to each of the antenna element dipoles. These antennas help to avoid the requirement for any mechanical downtilt. Site visits to adjust either electrical or mechanical downtilts can be time-consuming and expensive. Time and cost can be saved if the antenna tilts can be changed remotely. Some antenna systems support motorised components which can be controlled centrally from an operations centre.

In general, macrocell deployments should be configured with high-gain antennas. An increased gain helps to improve both coverage and capacity. However, it is necessary to balance the requirement for antenna gain with the requirement to avoid antennas which are too directional. Increasing the antenna gain is associated increasing its directivity, i.e. reducing its horizontal or vertical beamwidths. The antenna may be less able to provide uniform coverage across the cell if the beamwidths become too narrow. The horizontal beamwidth requirement tends to depend upon the sectorisation of the Node B. Six sector sites typically use a horizontal beamwidth of $33°$ whereas three sector sites may use horizontal beamwidths of either $65°$ or $90°$. The $65°$ beamwidth allows the antenna to have greater gain, but also results in greater directivity. The directivity may be less significant in urban areas where increased quantities of scattering effectively broaden the antenna gain pattern.

Microcells often use lower gain antennas with wider beamwidths. The number of vertically spaced dipoles can be reduced to generate a wider vertical beamwidth and a more compact antenna design. Indoor solutions often use low-gain compact omni-directional antennas which can be fitted to the ceiling or directional antennas which can be fitted to the walls. Repeaters use directional high-gain antennas to point back towards the donor cell. Some antennas are designed to minimise their visual impact and blend in with their local environment. Pole-mounted antennas and mast head amplifiers can be hidden within a cylinder which can be attached to the top of the pole.

9.5.2 Dedicated Subsystems

Dedicated antenna subsystems do not require the use of diplexors, triplexors nor combiners. A Node B configured with three sectors and uplink receive diversity uses six outputs from the Node B cabinet. These six outputs are connected to six antenna elements using six parallel runs of feeder cable. Six Mast Head Amplifiers (MHA) may be included to help improve the composite noise figure of the uplink receiver. An example of a dedicated antenna subsystem using MHA is illustrated in Figure 9.22.

MHA are active components which require a power supply. The feeder cables can be used to transfer a DC power supply from the Node B cabinet to the set of MHA. The power is inserted into the feeder cables using a set of bias-T units. Figure 9.22 illustrates the use of six individual MHA requiring six power supplies and six bias-T units. It is also possible to configure three dual MHA units which require only three power supplies and three bias-T units. Bias-T units are not necessary if MHA are not used. Some antennas have integrated MHA in which case the MHA does not appear as a separate unit.

In general, feeder cable with a wide diameter has a lower loss per unit length. However, feeder cable with a wide diameter tends to be less flexible and more expensive. Flexibility can be important if the

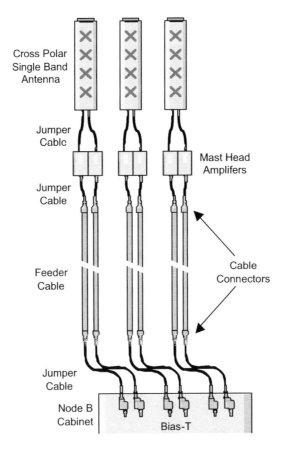

Figure 9.22   Dedicated antenna subsystem for three-sector site with receive diversity

site design requires the feeder cable to follow an indirect route towards the antennas. Typical feeder cable diameters are 1/2 inch, 7/8 inch and 1 5/8 inches. The minimum bend radius for these cables could be 8, 12 and 25 cm, respectively. Guidelines for feeder cable selection can be based upon the length of feeder cable required for a specific site design. If there is a requirement for long lengths of feeder cable there is increased justification for selecting a low-loss cable with a relatively wide diameter. Otherwise, a cable with a higher loss can be tolerated.

Jumper cables are used to connect the RF feeder cable to the Node B cabinet and also to the MHA. They are used to connect the RF feeder cable directly to the antenna if MHA are not used. Jumper cables are used for interconnection because they are more flexible than RF feeder cables. This helps to avoid placing stress upon connectors. Jumper cables typically have a bend radius of 3 cm.

The link budget analysis should encorporate the impact of all components within the antenna subsystem. Link budgets typically assume a single feeder loss figure although feeder types and feeder lengths vary from one site to another. Feeder cables with diameters of 1/2 inch, 7/8 inch and 1 5/8 inches typically have losses of 11, 6 and 4 dB per 100 m, respectively. Link budgets often assume a total cable and connector loss of 2–3 dB. This figure also includes the loss generated by the jumper cables. The connectors which join the feeder cable to the jumper cables have losses of the order of 0.1 dB. Jumper cables have losses of the order of 0.2 dB. The insertion loss of MHA must be accounted for in the downlink direction while the noise figure and gain must be accounted for in the uplink direction. The downlink insertion loss is typically 0.5 dB while the noise figure could be 2 dB and the gain between 12 and 25 dB.

Figure 9.23 illustrates an example of a dedicated antenna subsystem based upon remote RF modules. This design avoids the requirement for long lengths of feeder cable which consequently avoids the

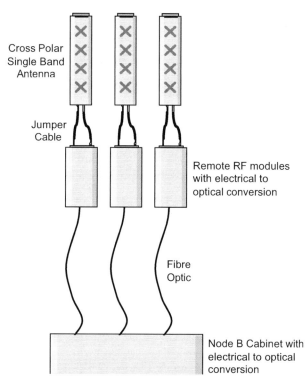

Figure 9.23 Dedicated antenna subsystem based upon remote RF modules

requirement for MHA. These factors help to reduce cost and improve performance. All RF functions can be completed within the remote unit which is connected to the Node B cabinet using lengths of fibre optic. The applicability of this design depends upon whether or not it is practical to mount the RF modules close to the antennas. This can be influenced by the size and weight of the remote RF modules as well as the type of installation, e.g. rooftop, pole-mounted or mast.

9.5.3 Shared Subsystems

Shared antenna subsystems require either a diplexor, triplexor or combiner to feed the separate uplink and downlink signals through a common set of components. Antenna subsystems can be designed to share either the antennas, the feeder cables or both the antennas and feeder cables. Figure 9.24 illustrates two shared antenna subsystems for a GSM base station and UMTS Node B. The first is based upon shared feeder cables and dedicated antennas whereas the second is based upon both shared feeder cables and a shared antenna. It is assumed that MHA are used for both the GSM and UMTS systems. Single rather than dual MHA units are shown and so there is a requirement for four power supplies at the remote end of the feeder cable. In this example, the GSM power supply is fed into the feeder cable using bias-T within the GSM base station cabinet. The UMTS system uses separate power cables which are run from the Node B cabinet to the MHA. It is possible to avoid the separate power cables if dual MHA are used and two rather than four power supplies are required. The benefit of using dedicated antennas is an increased flexibility in terms of configuring independent azimuths, heights and downtilts. Dedicated antennas may also be used for historical reasons, i.e. single-band GSM antennas have been used prior to the introduction of UMTS. Diplexors are used to combine and separate the UMTS and GSM signals at both ends of the RF feeder cable. Diplexors have insertion losses which depend upon the RF frequency. The loss for GSM900 frequencies can be 0.15 dB whereas the corresponding loss for UMTS2100 frequencies can be 0.35 dB.

Figure 9.25 illustrates a further example of a shared antenna subsystem for a GSM base station and UMTS Node B. This example does not include any MHA and the standalone diplexors at the remote end of the feeder cable are replaced with integrated diplexors within the antenna housing.

9.6 Co-siting

- Co-siting requires consideration of the isolation required to ensure that two systems do not interfere with one another. Achieving sufficient isolation should form part of the site design process.
- The most common causes of inter-system interference are spurious emissions (non-ideal transmitter filtering), receiver blocking (non-ideal receiver filtering) and intermodulation (non-linear mixing).
- A non-ideal transmitter radiates power outside its allocated channel bandwidth, whereas a non-ideal receiver does not completely reject power outside its allocated channel bandwidth. The 3GPP specifications define additional transmitter and receiver performance requirements for deployment scenarios which involve co-siting.
- An analysis of spurious emissions and receiver blocking indicates that an isolation of 45 dB would be sufficient to avoid interference between the UMTS and GSM systems.
- If the UMTS and GSM systems have separate antenna subsystems, the isolation requirement can be achieved by providing adequate physical separation between antennas.
- If the UMTS and GSM systems share the same antenna subsystem, the isolation requirement can be achieved using either a diplexor (shared feeder cables and antenna) or dual-band antenna (shared antenna only).

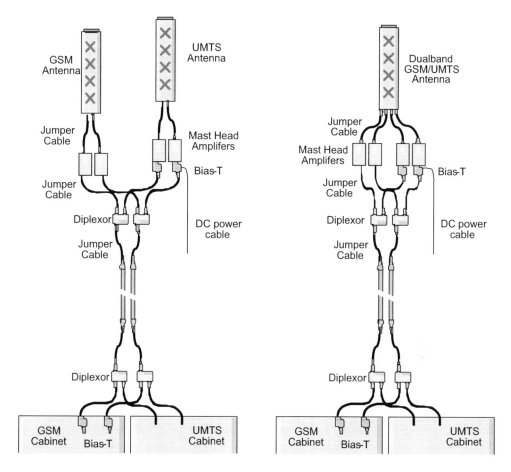

Figure 9.24 Antenna subsystems with shared RF feeder cable (single sector)

- Intermodulation products are generated by signals mixing within non-linear components, e.g. a non-linear amplifier. They can be generated at both the transmitter and receiver.
- The potential impact of intermodulation should be checked on a case-by-case basis. Intermodulation is most likely to be an issue when shared antenna sub-systems are used and there is increased potential for leakage.
- Careful frequency planning represents one way to avoid intermodulation issues although spectrum allocations may not always allow sufficient flexibility. Otherwise, increasing isolation helps to reduce the potential impact of intermodulation.
- Key 3GPP specifications: TS 05.05 and TS 25.104.

It is often necessary to select sites which already accommodate other radio equipment. That radio equipment could belong to either the same or a different operator. If the radio equipment belongs to the same operator there can be cost savings associated with co-siting. Otherwise co-siting may be necessary due to a lack of alternative sites. In either case, co-siting requires consideration of the isolation requirements between the different systems. Isolation is necessary to ensure that two systems

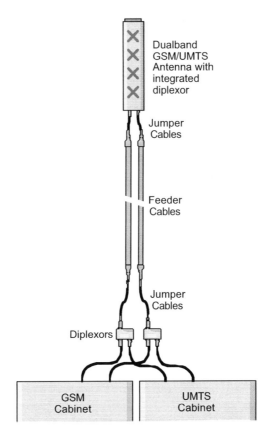

Figure 9.25   Shared antenna subsystem without MHA (single sector)

do not interfere with one another. Achieving sufficient isolation should form part of the site design process.

Isolation requirements are defined by the performance of the transmitter belonging to one system and the receiver belonging to another system. In general, the transmitter and receiver are operating in different frequency bands, e.g. a GSM base station transmitter and a UMTS Node B receiver. An example interference scenario is illustrated in Figure 9.26. This scenario is based upon a GSM 900 base station co-sited with a UMTS 2100 Node B. There is potential for interference from GSM to UMTS and from UMTS to GSM. In general, systems which have larger frequency

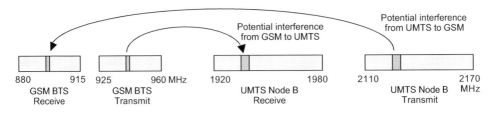

Figure 9.26   Base station to base station interference scenarios

separations require less isolation because some isolation is already provided in the frequency domain.

The most common causes of inter-system interference are spurious emissions (non-ideal transmitter filtering), receiver blocking (non-ideal receiver filtering) and intermodulation (non-linear mixing).

9.6.1 Spurious Emissions

An ideal transmitter would radiate power within the allocated channel bandwidth, but would not radiate power outside that bandwidth. An ideal transmitter would not have any isolation requirements because there would not be any potential for interference. In practice, radio systems make use of non-ideal transmitters. A non-ideal transmitter radiates power outside the allocated channel bandwidth. The concept of ideal and non-ideal transmitters is illustrated in Figure 9.27.

System specifications define the quantity of power a transmitter is allowed to radiate outside the allocated channel bandwidth. A transmit mask is often used to define the maximum allowed power at relatively small frequency offsets from the allocated channel. Spurious emission limits define the maximum allowed power at larger frequency offsets.

Figure 9.28 illustrates the concept of spurious emissions from a GSM base station causing interference to a UMTS Node B. An objective of site design is to provide an isolation between the two systems which is sufficient to ensure that the interfering power has either negligible or minimal impact.

The spurious emission limits for a GSM base station are specified within 3GPP TS 05.05. The current version of this specification includes requirements which are defined specifically for the purposes of co-siting GSM and UMTS equipment. Relatively new GSM equipment should satisfy these requirements. Older GSM equipment may have been manufactured before these additional requirements were introduced. The current version of 3GPP TS 05.05 specifies that when a GSM base station is co-sited with a UMTS Node B the GSM spurious emissions within the 1920–1980 MHz band shall not exceed the values presented in Table 9.13. This requirement is independent of the GSM operating band and is applicable to GSM 450, GSM 850, GSM 900, DCS 1800 and PCS 1900.

Figure 9.27   Concept of ideal and non-ideal transmitters

Figure 9.28   Concept of spurious emissions from a GSM BTS interfering with a UMTS Node B

Table 9.13 GSM BTS spurious emissions requirement when co-sited with a UMTS Node B

Frequency band	Maximum power	Measurement bandwidth
1920–1980 MHz	−96 dBm	100 kHz

The GSM base station spurious emissions requirement is specified using a measurement bandwidth of 100 kHz. The receiver bandwidth of a UMTS Node B is 3.84 MHz and so the interfering power becomes $-96 + 10 \times \log(3840/100) = -80$ dBm. The unloaded noise floor of a UMTS Node B is defined as $kTB \times$ Noise Figure, where k is Boltzmann's constant, T is the temperature in degrees kelvin and B is the receiver bandwidth. Assuming a noise figure of 3 dB and a temperature of 300 K generates a result of -105 dBm. If the isolation between the WCDMA Node B and GSM base station is 25 dB then the interfering power will equal the noise floor, and the noise floor will increase by 3 dB. GSM base stations are likely to transmit more than a single RF carrier and the quantity of interference accumulates. If a GSM base station is transmitting four RF carriers the spurious emissions would increase by 6 dB, i.e. $10 \times \log(4)$. The increase in interference experienced by a UMTS Node B is presented in Figure 9.29 for a range of isolations and GSM RF carriers.

An isolation of 40 dB results in a relatively small increase in interference floor. In practice GSM base stations are likely to perform better than the specifications. Nevertheless, it is common to use the figures from the specifications to represent a worse case scenario.

The GSM spurious emissions requirement presented in Table 9.13 is applicable to UMTS operating in band I, i.e. 1920–1980 MHz. Similar tables are not included for other UMTS operating bands. However, 3GPP TS 05.05 specifies an additional requirement for spurious emissions within the GSM operating bands. This is relevant because some UMTS operating bands are the same as some GSM operating bands, i.e.

- UMTS operating II ≡ PCS 1900 operating band
- UMTS operating band III ≡ DCS 1800 operating band
- UMTS operating band V ≡ GSM 850 operating band
- UMTS operating band VIII ≡ E-GSM operating band.

The spurious emissions requirement within these bands is -98 dBm when measured using a 100 kHz bandwidth, i.e. 2 dB more stringent than the co-sited requirement specified for UMTS operating band I.

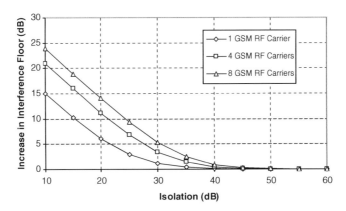

Figure 9.29 Increase in UMTS Node B noise floor generated by GSM BTS

This means that an inter-system isolation requirement of 40 dB is also applicable to these UMTS operating bands.

The spurious emission limits for a UMTS Node B are specified within 3GPP TS 25.104. Similar to the GSM specifications, there are additional requirements which are applicable to when UMTS and GSM equipment is co-sited. The spurious emissions limits for co-sited UMTS Node B are presented in Table 9.14.

A maximum power of −98 dBm measured in 100 kHz is equivalent to −95 dBm measured in the GSM receiver bandwidth of 200 kHz. If it is assumed that the GSM base station has a noise figure of 2 dB then the thermal noise floor is −119 dBm. If there is 24 dB of isolation between the UMTS Node B and the GSM base station then the noise floor of the base station would be increased by 3 dB. The UMTS Node B may be transmitting more than a single RF carrier. If the Node B is transmitting two RF carriers then its spurious emissions power will increase by 3 dB. Figure 9.30 illustrates the increase in interference floor at the GSM base station receiver for a range of isolations and either one or two UMTS RF carriers.

Similar to the previous scenario, an isolation of 40 dB results in a relatively small interference floor increase. In practice UMTS Node B are likely to perform better than the specifications. Nevertheless, it is common to use the figures from the specifications to represent a worse case scenario.

9.6.2 Receiver Blocking

An ideal receiver would accept power within the allocated channel bandwidth, but would reject power outside that bandwidth. An ideal receiver would not have any isolation requirements because there would not be any potential for interference. In practice, radio systems make use of non-ideal receivers.

Table 9.14 UMTS Node B spurious emissions requirement when co-sited with a GSM BTS

System	Frequency band (MHz)	Maximum power (dBm)	Measurement bandwidth (kHz)
GSM 900	876–915	−98	100
DCS 1800	1710–1785	−98	100
PCS 1900	1850–1910	−98	100
DCS 1800	824–849	−98	100

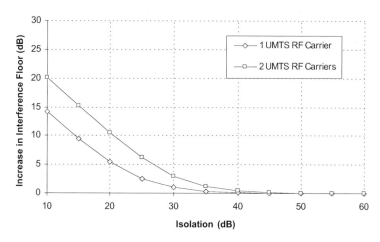

Figure 9.30 Increase in GSM BTS noise floor generated by UMTS Node B

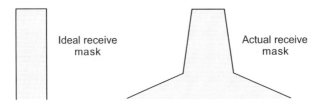

Figure 9.31   Concept of ideal and non-ideal receivers

A non-ideal receiver attenuates power outside the allocated channel bandwidth, but does not reject that power completely. The concept of ideal and non-ideal receivers is illustrated in Figure 9.31.

System specifications define the quantity of power that a receiver should be able to cope with at a specific frequency offset from the allocated channel. This receiver characteristic is known as receiver blocking performance, i.e. the ability of the receiver to block an unwanted signal outside the allocated channel bandwidth. Receiver blocking requirements are often applicable to relatively large frequency offsets from the allocated channel. The equivalent requirements for smaller frequency offsets are specified as an adjacent channel selectivity requirement.

Figure 9.32 illustrates the concept of a UMTS Node B receiver attenuating the signal power generated by a GSM base station. The UMTS Node B receiver attenuation requirement is reduced if the GSM base station signal power is relatively weak. An objective of site design is to provide an isolation between the two systems which is sufficient to ensure that the interfering signal power is at a level which has either negligible or minimal impact after being attenuated.

The receiver blocking performance requirements for a UMTS Node B are specified within 3GPP TS 25.104. There are specific requirements for when a UMTS Node B is co-sited with GSM base station. These requirements are presented in Table 9.15.

These figures indicate that a UMTS Node B should be capable of attenuating an interfering signal power of 16 dBm without having a negative impact upon the reception of a   115 dBm wanted signal. A GSM base station typically radiates a power of between 37 and 45 dBm. Assuming a GSM base station transmit power of 45 dBm means that 29 dB of isolation is required such that the GSM signal is attenuated to 16 dBm prior to reaching the WCDMA Node B. This isolation requirement may increase as a result of the GSM base station transmitting more than a single RF carrier. Nevertheless, the isolation requirement is significantly less than that associated with the spurious emissions, i.e. 29 dB compared with 40 dB.

The receiver blocking performance requirements for a GSM base station are specified within 3GPP TS 05.05. These requirements are presented in Table 9.16.

These figures indicate that a GSM 900 base station should be capable of attenuating an interfering signal power of 8 dBm without having a negative impact upon the reception of a wanted signal which is 3 dB greater than sensitivity. The same requirement exists for a GSM 850 base station except the wanted signal is only 1 dB greater than sensitivity. Assuming a UMTS Node B maximum transmit

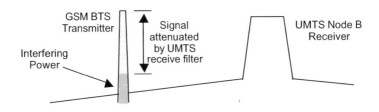

Figure 9.32   Concept of a UMTS Node B receiver attenuating the power generated by a GSM BTS

Table 9.15 WCDMA Node B receiver blocking performance requirements

System	Frequency band (MHz)	Interfering signal (dBm)	Wanted signal (dBm)	Interferer
GSM 900	921–960	16	−115	CW carrier
DCS 1800	1805–1880	16	−115	CW carrier
PCS 1900	1930–1990	16	−115	CW carrier
GSM 850	869–894	16	−115	CW carrier

Table 9.16 GSM and DCS base station receiver blocking performance requirements

System	Frequency band (MHz)	Interfering signal (dBm)	Wanted signal	Interferer
GSM 900	925–12 750	8	Sensitivity + 3 dB	CW carrier
DCS 1800	1805–12 750	0	Sensitivity + 3 dB	CW carrier
PCS 1900	1930–12 750	0	Sensitivity + 3 dB	CW carrier
GSM 850	859–12 750	8	Sensitivity + 1 dB	CW carrier

power capability of 43 dBm (20 W) results in an isolation requirement of 35 dB. This isolation ensures that the interfering UMTS signal is attenuated to 8 dBm prior to reaching the GSM base station. The isolation requirement is increased to 38 dB if a 46 dBm (40 W) UMTS transmit power capability is assumed. DCS 1800 and PCS 1900 base stations are specified to handle 0 dBm of interfering power and so require greater isolation. These base stations require 43 dB of isolation from a 43 dBm UMTS Node B and 46 dB of isolation from a 46 dBm UMTS Node B.

9.6.3 Intermodulation

Intermodulation products are generated by signals mixing within non-linear components, e.g. a non-linear amplifier. They can be generated at both the transmitter and receiver. It is possible that a signal from one base station transmitter leaks backwards into a second base station transmitter. The two downlink signals can then mix to generate a series of intermodulation products. Similarly, two base station transmit signals could leak into a base station receiver to subsequently mix and generate a series of intermodulation products. Intermodulation products can become problematic if they fall within the passband of a receiving system.

Table 9.17 presents a set of example intermodulation products generated by two DCS1800 frequencies. In this example, it is assumed that the two DCS1800 frequencies are at 1810 and 1870 MHz. The 2nd and 4th order intermodulation products are far from the UMTS operating bands. The 3rd order product $2 \times f_2 - f_1$ falls within the Node B receive band for UMTS operating band I, whereas the 3rd order product $2 \times f_1 - f_2$ falls within the Node B receive band for UMTS operating

Table 9.17 Example inter-modulation products generated by a DCS1800 base station

	Intermodulation product 1		Intermodulation product 2	
1st order	f_1	1810 MHz	f_2	1870 MHz
2nd order	$f_1 + f_2$	3680 MHz	$f_2 - f_1$	60 MHz
3rd order	$2 \times f_1 - f_2$	1750 MHz	$2 \times f_2 - f_1$	1930 MHz
4th order	$2 \times f_2 + 2 \times f_1$	7360 MHz	$2 \times f_2 - 2 \times f_1$	120 MHz
5th order	$3 \times f_1 - 2 \times f_2$	1690 MHz	$3 \times f_2 - 2 \times f_1$	1990 MHz

bands III and IV.

The signal strength of intermodulation products decreases as their order increases, i.e. 2nd order products are the strongest. In this example, the 3rd order products are potentially problematic because they are relatively strong and they fall within UMTS receive bands. In practice, the example presented within Table 9.17 may not exist because the two DCS1800 carriers have a large frequency spacing whereas regulators tend to allocate individual operators with smaller blocks of contiguous spectrum. Intermodulation products are most likely to be an issue when the isolation between the RF carriers is relatively small and there is increased potential for leakage. This could occur if the two RF carriers share the same antenna subsystem. There is less potential for interference if the two RF carriers belong to different operators using separate antenna subsystems.

In general, the potential for intermodulation problems should be checked on a case-by-case basis according to the local spectrum allocations and frequency plans. Frequency planning represents one way to avoid intermodulation issues although spectrum allocations may not always allow sufficient flexibility. The isolation between base station transmitters and receivers should be checked if the frequency plan indicates a potential for intermodulation problems.

9.6.4 Achieving Sufficient Isolation

The preceding analysis of spurious emissions and receiver blocking performance indicates that an isolation of 45 dB would be sufficient to avoid interference between the UMTS and GSM systems. Achieving this isolation requirement is dependent upon the site design. If the UMTS and GSM systems have separate antenna subsystems, the isolation can be achieved by providing adequate physical separation between the antennas. Figure 9.33 illustrates an example based upon two antennas with a horizontal separation of 1 m. The isolation achieved depends upon the horizontal beamwidth of the two antennas. An isolation of more than 40 dB can be achieved if the antennas have a separation of 1 m and a horizontal beamwidth of 90°. It may be necessary to increase the physical separation if the antennas have a wider horizontal beamwidth.

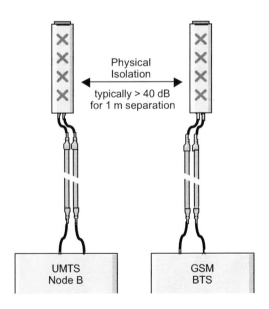

Figure 9.33   Achieving isolation between systems using the physical separation of antennas

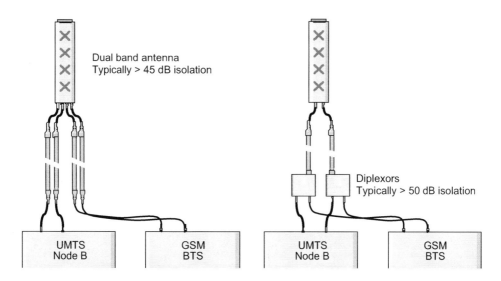

Figure 9.34   Achieving isolation between systems using the a dual-band antenna or a diplexor

Alternatively, the same isolation can be achieved by separating the two antennas vertically. A separation of 0.5 m between the top of one antenna and the bottom of the other antenna is typically sufficient to generate at least 40 dB of isolation.

There are two general scenarios if the site design allows the GSM base station and the UMTS Node B to share the same antenna subsystem. The first scenario is based upon sharing the same antenna, but using dedicated RF feeder cables. The second scenario is based upon sharing both the same antenna and the same RF feeder cables. These two scenarios are illustrated within Figure 9.34.

Sharing the same antenna while using dedicated feeder cables requires that the antenna has separate connectors for each system. Isolation is then provided by the internal antenna design. Dual-band antennas typically provide an isolation of more than 45 dB. Sharing both the same antenna and the same RF feeder cables requires the use of diplexors (also referred to as combiners). A diplexor combines the downlink signals while separating the uplink signals. Diplexors typically provide > 50 dB of isolation. The drawback associated with using a diplexor is a small insertion loss. The antenna may also have its own internal insertion loss if the GSM and UMTS signals are separated within the antenna before being connected to the antenna elements themselves. Insertion losses within the antenna are usually accounted for by specifying a slightly lower antenna gain.

9.7 Microcells

- A microcell can be categorised as a Node B with an outdoor and below rooftop antenna. The below rooftop antenna limits cell range and increases the capacity per unit area.
- A microcell can be assigned either a dedicated or shared RF carrier. Shared RF carriers require careful planning to ensure that the microcell layer has sufficient dominance. A shared RF carrier allows soft handover to be used to provide mobility between layers.
- Sectorisation of UMTS microcells is not common because it is difficult to achieve sufficient isolation between sectors. The high quantity of scattering associated with the microcell radio environment means that sectors tend to have similar coverage areas.

- The Minimum Coupling Loss (MCL) requirement defines the minimum allowed link loss between a Node B and a UE. Levels of interference can become unacceptable if the link loss becomes less than the MCL requirement.
- Modelling microcells within a radio network planning tool is more complex than modelling macrocells. The RF propagation has an increased dependence upon the neighbouring buildings and building vectors should be made available to the planning tool.
- Microcell antennas typically have a lower gain than macrocell antennas. Microcells are usually deployed with relatively short feeder lengths. Uplink Eb/No requirements increase by approximately 3 dB if uplink receive diversity is not used. Microcells are likely to have lower maximum downlink transmit powers than macrocells.
- The level of intercell interference and the quantity of soft handover between two microcells is typically less than between two macrocells. The microcell radio propagation environment is typically assumed to allow an average orthogonality factor of 0.8.
- Key 3GPP specifications: TS 25.101 and TS 25.104.

A microcell can be categorised as a Node B which has an outdoor and below rooftop antenna. Microcells are typically deployed using smaller Node B cabinets with lower transmit power capabilities. Microcells can be deployed for either coverage or capacity reasons. The below rooftop antenna placement means that coverage is relatively limited. Neighbouring buildings tend to obstruct the radio signal and limit the cell range. The limited cell range helps microcells to provide a high capacity per unit area.

9.7.1 RF Carrier Allocation

A microcell can be assigned either a dedicated or shared RF carrier. Allocating a dedicated RF carrier has the benefit of avoiding potentially significant levels of intercell interference from the macrocell layer. However, it requires the UE to complete hard handovers when moving between layers. Allocating a shared RF carrier requires careful radio network planning to ensure that the microcell layer has sufficient dominance and that levels of intercell interference from the neighbouring macrocells are not excessively high. The air-interface capacity of a microcell will be reduced if intercell interference is high. A benefit associated with using a shared RF carrier is that soft handover can be used to support mobility between layers. The longer term availability of RF carriers should be considered when defining the carrier allocation strategy. If an operator has been allocated two RF carriers then it is likely that both carriers will be allocated to the macrocell layer once the level of network traffic becomes mature. This will result in microcells having a shared RF carrier irrespective of whether or not they are allocated a dedicated carrier initially. Microcells can also be configured with two RF carriers and this is likely to represent the main capacity upgrade path.

9.7.2 Sectorisation

Sectorisation of UMTS microcells is not common because it is difficult to achieve sufficient isolation between sectors. The high quantity of scattering associated with the microcell radio environment means that sectors tend to have similar coverage areas. Directing antennas in different directions may not have a significant impact. Figure 9.35 illustrates an example of the coverage areas belonging to two sectors of a microcell. These plots were generated using a ray tracing propagation model. They illustrate that there can be significant coverage area overlap when using antennas with 120° of angular separation. This would result in high levels of intercell interference and increased soft handover overheads. It is more practical to sectorise GSM microcells because each sector can be allocated a different RF carrier. This introduces isolation in the frequency domain.

RF coverage of sector 1   RF coverage of sector 2

Figure 9.35 Overlap of the coverage areas belonging to different sectors of a microcell

9.7.3 Minimum Coupling Loss

The Minimum Coupling Loss (MCL) requirement defines the minimum allowed link loss between a Node B and a UE. Levels of interference can become unacceptable if the link loss becomes less than the MCL requirement. Large numbers of connections can be dropped if the uplink interference floor becomes unacceptable, whereas individual connections can be dropped if the downlink interference floor becomes unacceptable. The MCL requirement should be considered as part of the site design procedure.

MCL becomes significant in the uplink direction when a UE is physically close to a microcell and is transmitting at its minimum transmit power. Inner loop power control commands instruct the UE to decrease its power, but the UE is already transmitting at the lower end of its dynamic range. This results in the Node B receiving more power than necessary and generates increased levels of uplink interference. The impact of the MCL requirement depends upon the noise figure of the antenna subsystem. Assuming that microcells do not use MHA means that the uplink noise figure equals the noise figure of the Node B. A Node B noise figure of 2 dB generates an unloaded uplink interference floor of -106 dBm. The minimum UE transmit power requirement specified by 3GPP TS 25.101 is -50 dBm. In practice, UE may be able to transmit at lower powers but the figure of -50 dBm represents a worst case assumption. Figure 9.36 illustrates the increase in uplink interference floor generated by a UE transmitting at -50 dBm. The increase is plotted as a function of the MCL and is applicable to a Node B with a noise figure of 2 dB. These results indicate that an MCL requirement of at least 65 dB should be used to protect the uplink receiver from UE which are transmitting at -50 dBm.

MCL becomes significant in the downlink direction when a UE is physically close to a microcell and the microcell is transmitting at a high power. 3GPP TS 25.101 specifies that the maximum downlink

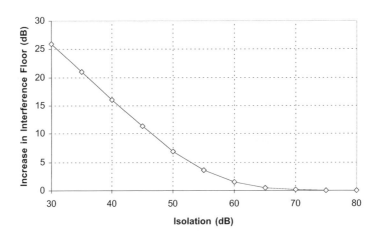

Figure 9.36   Increase in uplink interference floor generated by a UE transmitting at −50 dBm

power that a UE should be capable of receiving is −25 dBm. If a microcell Node B has a maximum downlink transmit power capability of 39 dBm then the isolation requirement should be at least 64 dB.

This analysis indicates that an MCL requirement of 65 dB should be sufficient to protect both the uplink and downlink receivers from increased levels of interference. The MCL analysis can be extended to consider UE operating on the adjacent RF carrier. These UE could belong to a different operator in which case their transmit powers cannot be controlled by the microcell. If a microcell is located towards the edge of a macrocell belonging to a different operator there could be UE transmitting at maximum power on the adjacent RF carrier. 3GPP TS 25.101 specifies an uplink Adjacent Channel Leakage Ratio (ACLR) of 33 dB for the first adjacent channel and 43 dB for the second adjacent channel. These figures indicate that a UE transmitting at 24 dBm can generate −9 dBm of interference on the first adjacent channel and −19 dBm of interference on the second adjacent channel. These interference powers remain high after 65 dB of MCL has been subtracted. A similar ACLR calculation can be completed in the downlink direction. 3GPP TS 25.104 specifies that the downlink ACLR should be at least 45 dB on the first adjacent channel and at least 50 dB on the second adjacent channel. These figures indicate that a microcell transmitting at 39 dBm can generate −6 dBm of interference on the first adjacent channel and −11 dBm of interference on the second adjacent channel. Similar to the uplink, these powers remain high after 65 dB of MCL has been subtracted. An adjacent channel analysis should also account for the Adjacent Channel Selectivity (ACS) performance of the Node B and UE receivers. 3GPP TS 25.101 specifies that the ACS performance of a UE on the first adjacent channel should be 33 dB. This means that the UE receiver should be able to attenuate adjacent channel power by at least 33 dB. If a microcell is transmitting at 39 dBm and there is 65 dB of isolation, the ACS performance of 33 dB decreases the interfering power to −59 dBm. The adjacent channel analysis presented within this paragraph is representative of a worst case scenario. In practice, there is only a small probability that this worst case scenario exists and site design requirements can be relaxed, i.e. based primarily upon the co-channel analysis rather than the adjacent channel analysis.

9.7.4 Propagation Modelling

Modelling microcells within a radio network planning tool is more complex than modelling macrocells. The RF propagation has an increased dependence upon the neighbouring buildings. Building vectors should be made available to the radio network planning tool. Building vectors can be either two or three dimensional. Two dimensional building vectors define the layout of the buildings in terms of their x–y coordinates but assume a fixed building height. Two dimensional building vectors may be sufficient for microcell modelling when all antennas are below rooftop. They may also be sufficient for areas where the building height is approximately uniform. In general there is a trade-off between the accuracy of the result, the cost of the building vectors and the associated processing time.

9.7.5 Planning Assumptions

Link budgets should be generated to identify the maximum allowed path loss associated with a microcell deployment. Microcell antennas typically have a lower gain than macrocell antennas. The reduced gain provides less directivity which allows an increase in the vertical beamwidth of the antenna gain pattern. Macrocell antennas typically have an 18 dBi gain and a 10° vertical beamwidth whereas microcell antennas typically have a 12 dBi gain and a 30° vertical beamwidth. The large quantity of scattering associated with the microcell environment means that UE receive signals originating from relatively wide sections of the antenna gain pattern. A UE could be more than 50° away from the centre of the main lobe, but still receive a signal from the main lobe if it has been reflected off a neighbouring building. This type of effect makes it difficult to predict the antenna gain experienced by a UE at a specific location and increases the requirement to use propagation modelling based upon ray tracing techniques.

It is common to assume that microcells are deployed with relatively short feeder lengths. In general, this means that the cable loss will be less than that for a macrocell unless very different types of feeder cable are used. Microcells may use feeder cable with a relatively small diameter and a correspondingly small bend radius. This allows easier installation and interconnection between the Node B and antenna. Node B cabinets which have been designed specifically for microcell deployment scenarios may include an internal antenna which does not require any feeder cable. The reduced cable loss associated with microcells means there is less potential benefit from using MHA. This removes the MHA insertion loss in the downlink direction and avoids the requirement to calculate a composite noise figure in the uplink direction.

The large quantities of scattering and multi-path associated with the microcell radio propagation environment generates relatively deep fades. Low UE speeds allow inner loop power control to track these fades. This means that the microcell link budget should include a relatively large fast fading margin. If the UE speeds were high then the inner loop power control would be unable to track the fades and the impact of the fading would become incorporated within an increased Eb/No requirement while the fast fade margin would tend to zero.

Uplink Eb/No requirements increase by approximately 3 dB if a microcell does not use uplink receive diversity. Receive diversity may be disabled to help simplify the site design and avoid the requirement to run two lengths of feeder cable between the Node B cabinet and the antenna. Increasing the uplink Eb/No requirement has an impact upon both coverage and capacity. The impact upon capacity is likely to become most significant after HSUPA has been introduced. In this case, the uplink air-interface is more likely to limit the uplink throughput performance. This provides an argument for configuring uplink receive diversity and planning microcells with a high uplink load target.

Microcells are likely to have lower maximum downlink transmit powers than macrocells. Node B cabinets designed for microcell deployments may offer lower transmit power capabilities. Otherwise, attenuators may be included within the antenna subsystem to reduce the maximum transmit power capability. High transmit powers are avoided because UE and end-users can move relatively close to microcell antennas.

If a shared RF carrier has been allocated and if a microcell is located within an area of existing macrocell coverage it is beneficial to record downlink RSSI measurements prior to completing the microcell RF planning procedure. Downlink RSSI measurements provide an indication of the interference over which the microcell must dominate. The microcell will have poor dominance and a limited coverage area if its signal strength is not sufficient relative to the downlink signal generated by the macrocell layer.

Levels of intercell interference between a microcell and macrocell sharing the same RF carrier can be relatively high. However, the level of intercell interference between two microcells is typically low. This results from the isolation provided by the neighbouring buildings. Dimensioning exercises typically assume an intercell interference ratio of 25% between a cluster of microcells compared with 65% between a cluster of macrocells. Similarly, the soft handover overhead can be assumed to reduce from 40% to 20%. This reduction results from not having softer handover (assuming a single sector microcell) and from the increased isolation between sites. The reduced delay spread associated with the microcell radio propagation environment increases the average orthogonality factor from 0.5 for a macrocell environment to 0.8.

9.8 Indoor Solutions

- It is not always practical to provide adequate indoor coverage using outdoor macrocells. In addition, some indoor locations represent traffic hot spots which require greater capacity than the macrocell layer is capable of providing.
- Assigning a shared RF carrier to an indoor solution allows soft handover between the indoor solution and neighbouring macrocells rather than inter-frequency handover. Assigning a shared carrier has the drawback of a reduced isolation.
- The indoor solution must be sufficiently dominant to overcome any interference from the neighbouring macrocells. Otherwise, UE inside the building will camp and establish connections upon the neighbouring macrocells.
- Small indoor solutions are likely to use a single sector. Sectorisation can be introduced to increase the capacity of larger indoor solutions. Indoor solutions with multiple sectors should be designed with adequate isolation between sectors.
- Passive Distributed Antenna Systems (DAS) are based upon connecting antennas to the Node B cabinet using lengths of RF feeder cable. Active DAS are based upon connecting antennas to the Node B cabinet using lengths of fibre optic or twisted pair.
- The selection between an active and passive DAS can be based upon the ability to maintain a specific downlink transmit power at each remote antenna. Alternatively, the selection can be based upon the number of antennas.
- It is not common for indoor solutions to support uplink receive diversity. The importance of uplink receive diversity is likely to increase when HSUPA is introduced.
- The minimum coupling loss requirement defines the minimum allowed link loss between a Node B and a UE. Levels of interference can become unacceptable if the link loss becomes less than the MCL requirement.
- The outdoor signal strength generated by an indoor solution should be sufficiently weak to avoid outdoor UE camping and establishing connections upon the indoor solution.
- Selecting antenna locations should account for link budget, dominance, leakage and minimum coupling loss requirements.
- Key 3GPP specifications: TS 25.101 and TS 25.104.

It is not always practical to provide adequate indoor coverage using the network of outdoor macrocells. In addition, some indoor locations represent traffic hot spots which require greater capacity than the macrocell layer is capable of providing. Indoor solutions are able to increase both the coverage and capacity of the radio access network. Example deployment scenarios include corporate customer offices, shopping centres, railway stations, airports and public event venues.

9.8.1 RF Carrier Allocation

The majority of UMTS operators have been licensed two or three RF carriers. This limits the possibility of being able to assign a dedicated RF carrier to an indoor solution in the long term. It may be possible to assign a dedicated carrier in the short term, but there will be increased pressure to assign the same carrier to the macrocell network as the quantity of traffic increases. Subsequently assigning the same carrier to the macrocell network could have an impact upon the performance of the indoor solution if its design has been based upon a dedicated carrier. Assigning a shared RF carrier allows soft handover between the macrocell layer and indoor solution. Inter-frequency hard handovers are required if a dedicated carrier is assigned. Inter-frequency handovers generally require the use of compressed mode making them less seamless than soft handovers. In addition, system design and the associated RNC databuild become more complex once inter-frequency handovers have been introduced. Assigning a shared carrier has the drawback of a reduced isolation between the indoor solution and the macrocell layer. In this case, the indoor solution design must account for the interference generated by the macrocell layer. The indoor solution must be sufficiently dominant to overcome any interference from the neighbouring macrocells. If the indoor solution is not sufficiently dominant, UE inside the building will camp and establish connections upon the macrocell layer. Ensuring indoor solution dominance may require an increased number of antennas with a corresponding increased cost.

Differences between an indoor solution link budget and a macrocell link budget can result in relatively weak soft handover connections. Indoor solutions often use a lower CPICH transmit power and are typically less sensitive in the uplink direction. The lower CPICH transmit power tends to move the soft handover area towards the indoor solution. This is beneficial from the perspective of the indoor solution sensitivity, i.e. the indoor solution requires greater uplink power. However, if the difference between the indoor solution and macrocell CPICH transmit powers is greater than the difference between the uplink sensitivities, the indoor solution will tend to instruct the UE to power down while the macrocell does not have sufficient uplink power to maintain synchronisation. This can reduce the reliability of the radio link to the macrocell.

9.8.2 Sectorisation

Small indoor solutions are likely to be deployed using a single sector. Sectorisation can be introduced to increase the capacity of larger indoor solutions. Indoor solutions with multiple sectors should be designed such that the isolation between sectors is relatively high. Intercell interference will decrease the air-interface capacity if the level of isolation is not adequate. Isolation can be achieved in multi-storey buildings by dedicating specific sectors to specific floors. Some cell overlap is required to allow time for handovers. Increasing the level of sectorisation increases the indoor solution cost. It can also influence the selection of the Node B cabinet type, i.e. some Node B cabinets may not support sectorisation.

9.8.3 Active and Passive Solutions

Passive Distributed Antenna Systems (DAS) are based upon connecting antennas to the Node B cabinet using lengths of RF feeder cable. Splitters and couplers can be used to connect more than a single antenna to the same Node B cabinet antenna connector. Splitters provide an equal distribution of downlink power whereas couplers provide an unequal distribution of power. It is usually appropriate to use a splitter when the feeder loss to each antenna is similar. Couplers can be used to maintain an equal downlink transmit power from each remote antenna when the feeder losses are unequal, e.g. a more distant antenna will have a larger feeder loss so a coupler can be used to provide the feeder connecting that antenna with an increased share of the downlink power. Splitters and couplers have losses associated with them, but indoor solution designs are usually tolerant to them, i.e. transmit powers are often higher than necessary. Additional attenuation may be required to limit the downlink transmit power radiated by an indoor solution antenna, or to achieve the minimum coupling loss requirement between the Node B cabinet and UE.

Active DAS are based upon connecting antennas to the Node B cabinet using lengths of fibre optic or twisted pair. Fibre optics and twisted pair are easier to install than RF feeder cable which can be relatively bulky. This helps to reduce the cost of installation and increase the potential for installing spare fibres or cables for future expansion of the indoor solution. Sectorisation can be introduced with relative ease if spare fibres or cables have been included as part of the initial design. In the case of using fibre optics, the downlink RF signal is converted to an optical signal at the Node B end of the DAS while the optical signal is converted back to an RF signal at the remote end of the DAS. Similarly, the uplink signal is converted to an optical signal for transmission across the DAS. The maximum downlink power which can be fed into an active DAS is often relatively low. This means that it may be necessary to install an attenuator between the Node B cabinet and active DAS.

The selection between an active and passive DAS can be based upon the ability to maintain a specific downlink transmit power at the connector of each remote antenna. An active DAS should be used if the loss associated with a passive DAS becomes so large that the target downlink transmit power at each remote antenna cannot be achieved. Alternatively, a relatively simple rule can be based upon the number of antennas belonging to the indoor solution design, e.g. if the indoor solution requires more than five antennas then its size is sufficient to justify the deployment of an active DAS.

It is not common for active or passive DAS to be designed to support uplink receive diversity. If the Node B cabinet has a receive diversity antenna connector then it is usually terminated without being connected to an antenna. This reduces the requirement for RF feeder cable in the case of a passive DAS. It reduces the requirement for fibre or cable in the case of an active DAS although that saving is less important for an active DAS. Designing an indoor solution without uplink receive diversity reduces the requirement for antennas. Indoor solution antennas have a single antenna element so uplink receive diversity requires twice as many antennas. These antennas have to be positioned with sufficient isolation to ensure that their signals are relatively uncorrelated. The importance of designing indoor solutions to include uplink receive diversity is likely to increase when HSUPA is introduced. Uplink receive diversity reduces the uplink Eb/No requirement by approximately 3 dB. This has a significant impact upon the capacity of the uplink air-interface. The throughput performance achieved by HSUPA is more likely to be limited by the air-interface if the Node B has not been configured with uplink receive diversity.

9.8.4 Minimum Coupling Loss

The Minimum Coupling Loss (MCL) requirement defines the minimum allowed link loss between a Node B and a UE. The general arguments for an indoor solution MCL are the same as those described in section 9.7.3 for a microcell. Levels of interference can become unacceptable if the link loss becomes less than the MCL requirement. An indoor solution is more likely to have a relatively small link loss between the UE and Node B cabinet because UE can position themselves relatively close to the antenna. However, the low downlink transmit power and potentially increased uplink noise figure help to reduce the MCL requirement for an indoor solution.

The uplink MCL requirement depends upon the noise figure of the Node B receiver subsystem. If the noise figure is high then the interference floor will also be high and the Node B will be less vulnerable to experiencing increases in uplink interference from UE transmitting at their minimum power. In the case of an active DAS, the uplink noise figure depends upon the number of remote units. Increasing the number of remote units increases the uplink noise figure. It is not uncommon for an active DAS to have an uplink noise figure in the order of 15 dB. This increases the usual unloaded noise

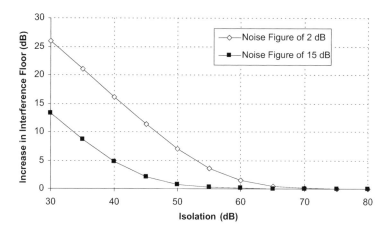

Figure 9.37   Increase in uplink interference floor generated by a UE transmitting at −50 dBm

floor from −106 dBm (assuming a noise figure of 2 dB) to −93 dBm. Figure 9.37 illustrates the increase in uplink interference floor generated by a UE transmitting at −50 dBm. The increase is plotted as a function of the MCL for Node B with noise figures of 2 and 15 dB. The noise figure of 2 dB is applicable to an indoor solution based upon a passive DAS.

These results indicate that increasing the uplink noise figure by 13 dB decreases the MCL requirement by the same amount. A passive DAS with an uplink noise figure of 2 dB has an uplink MCL requirement in the order of 65 dB whereas an active DAS with an uplink noise figure of 15 dB has an uplink MCL requirement in the order of 52 dB.

The downlink MCL requirement depends upon the transmit power of the indoor solution. A typical maximum downlink transmit power for an active DAS is 15 dBm. 3GPP TS 25.101 specifies that a UE should be capable of handling −25 dBm of downlink transmit power. This corresponds to a downlink MCL requirement of 40 dB, i.e. less than the equivalent uplink requirement. In the case of a passive DAS, the maximum downlink transmit power could be of the order of 39 dBm (depending upon the Node B transmit power capability). This corresponds to a downlink MCL requirement of 64 dB, i.e. approximately equal to the uplink requirement. The MCL requirement for a passive DAS can be at least partly achieved using the losses generated by the passive DAS feeder cables. In addition, a passive DAS may include an attenuator to limit the maximum downlink transmit power. This attenuator would also contribute towards achieving the MCL requirement.

9.8.5 Leakage Requirements

The outdoor signal strength generated by an indoor solution should be sufficiently weak to avoid the potential for outdoor UE to camp and establish connections upon the indoor solution. Allowing outdoor UE to use the indoor solution to establish connections would reduce its capacity from the perspective of the target UE population. In addition, the indoor solution could lead to increased levels of downlink intercell interference for UE connected to the macrocell network. The location of the indoor solution antennas and the maximum downlink transmit powers should account for the power received outside the target coverage area. An indoor solution design procedure may encorporate a requirement for the minimum CPICH RSCP measured from the indoor solution when outside the target coverage area. This requirement could be of the order of −110 dBm.

9.8.6 Antenna Placement

Selecting the antenna location is relatively straightforward for small indoor solutions using only a single antenna. It is common to use a small omni-directional antenna which can be mounted on the ceiling and directed down towards the target coverage area. Alternatively, a directional antenna can be mounted against one wall and directed across the target coverage area. The antenna gain of an omni-directional ceiling mounted antenna is typically 2 dBi whereas that of a directional wall mounted antenna is typically 7 dBi.

Selecting antenna locations is more complex for larger indoor solutions with multiple antennas. The results from a link budget analysis should be used to ensure that antennas provide contiguous coverage across the target coverage area. The indoor solution design procedure may take advantage of path loss measurements from test transmitters. These measurements can be used to help verify the coverage provided by the proposed antenna locations. This can be done prior to installing the active or passive DAS. Path loss measurements can be compared against link budget thresholds for each of the required service types. Antenna placement should also account for the leakage requirements and the distribution of interference from the macrocell layer. Leakage can be minimised if omni-directional antennas are positioned away from locations where the link loss to the building exterior is low, e.g. in front of windows, or if directional antennas are faced towards the building interior. Positioning antennas at locations where the signal strength from the outdoor macrocell network is relatively strong helps to ensure that the signal strength from the indoor solution remains dominant. These two requirements tend to conflict with one another, i.e. the locations where the signal strength from the macrocell network is strong tend to be the locations where the link loss to the building exterior is low. This means a balance between the two requirements must be achieved. Antenna placement should also account for the minimum coupling loss requirements. Antennas should not be positioned at locations where UE can become so close that the MCL requirements are no longer achieved.

References

[1] 3GPP TS 23.038, UMTS Alphabets and Language Specific Information.
[2] 3GPP TS 23.040, UMTS Technical Realization of Short Message Service (SMS).
[3] 3GPP TS 23.122, UMTS Non-Access-Stratum Functions related to MS in Idle Mode.
[4] 3GPP TS 23.221, UMTS Architectural Requirements.
[5] 3GPP TS 24.007, UMTS Mobile Radio Interface Signalling Layer 3, General Aspects.
[6] 3GPP TS 24.008, UMTS Mobile Radio Interface Layer 3 Specification, Core Network Protocols, Stage 3.
[7] 3GPP TS 24.011, UMTS Point-to-Point (PP) Short Message Service (SMS) support on Mobile Radio Interface.
[8] 3GPP TS 25.101, UMTS User Equipment (UE) Radio Transmission and Reception (FDD).
[9] 3GPP TS 25.104, UMTS UTRA (BS) FDD, Radio Transmission and Reception.
[10] 3GPP TS 25.133, UMTS Requirements for support of Radio Resource Management (FDD).
[11] 3GPP TS 25.211, UMTS Physical Channels and mapping of Transport Channels onto Physical Channels (FDD).
[12] 3GPP TS 25.212, UMTS Multiplexing and Channel Coding (FDD).
[13] 3GPP TS 25.213, UMTS Spreading and Modulation (FDD).
[14] 3GPP TS 25.214, UMTS Physical Layer Procedures (FDD).
[15] 3GPP TS 25.215, Physical Layer Measurements (FDD).
[16] 3GPP TS 25.301, UMTS Radio Interface Protocol Architecture.
[17] 3GPP TS 25.302, UMTS Services provided by the Physical Layer.
[18] 3GPP TS 25.304, UMTS User Equipment (UE) Procedures in Idle Mode and Procedures for Cell Reselection in Connected Mode.
[19] 3GPP TS 25.306, UMTS UE Radio Access Capabilities.
[20] 3GPP TS 25.321, UMTS Medium Access Control (MAC) Protocol Specification.
[21] 3GPP TS 25.322, UMTS Radio Link Control (RLC) Protocol Specification.
[22] 3GPP TS 25.323, UMTS Packet Data Convergence Protocol (PDCP) Specification.
[23] 3GPP TS 25.331, UMTS Radio Resource Control (RRC) Protocol Specification.
[24] 3GPP TS 25.413, UMTS UTRAN Iu Interface Radio Access Network Application Part (RANAP) Signalling.
[25] 3GPP TS 25.415, UMTS UTRAN Iu Interface User Plane Protocols.
[26] 3GPP TS 25.425, UMTS UTRAN Iur Interface User Plane Protocols for Common Transport Channel Data Streams.
[27] 3GPP TS 25.426, UMTS UTRAN Iur and Iub Interface Data Transport and Transport Signalling for DCH Data Streams.
[28] 3GPP TS 25.427, UMTS UTRAN Iur/Iub Interface User Plane Protocol for DCH Data Streams.
[29] 3GPP TS 25.430, UMTS UTRAN Iub Interface, General Aspects and Principles.
[30] 3GPP TS 25.433, UMTS UTRAN Iub Interface Node B Application Part (NBAP) Signalling.
[31] 3GPP TS 25.434, UMTS UTRAN Iub Interface Data Transport and Transport Signalling for CCH Data Streams.
[32] 3GPP TS 25.435, UMTS UTRAN Iub Interface User Plane Protocols for CCH Data Streams.
[33] 3GPP TS 25.944, UMTS Channel Coding and Multiplexing Examples.

[34] 3GPP TS 31.102, UMTS Characteristics of the USIM Application.

[35] 3GPP TS 44.018, Digital Cellular Telecommunications System (Phase 2+), Mobile Radio Interface Layer 3 Specification, Radio Resource Control Protocol.

[36] 3GPP TS 05.05, Digital Cellular Telecommunications System (Phase 2+), Radio Transmission and Reception.

[37] ITU-T Recommendation Q.2630.1, AAL type 2 Signalling Protocol - Capability Set 1.

[38] ITU-T Recommendation Q.2130, B-ISDN Signalling ATM Adaptation Layer - Service Specific Coordination Function for Support of Signalling at the User Network Interface.

[39] ITU-T Recommendation Q.2110, B-ISDN ATM Adaptation Layer - Service Specific Connection Orientated Protocol (SSCOP).

[40] ITU-T Recommendation I.363.5, B-ISDN ATM Adaptation Layer Specification: Type 5 AAL.

[41] ITU-T Recommendation I.366.1, Segmentation and Reassembly Service Specific Convergence Sublayer for the AAL type 2.

[42] ITU-T Recommendation H.223, Multiplexing Protocol for Low Bit Rate Multimedia Communication.

[43] ITU-T Recommendation H.245, Control Protocol for Multimedia Communication.

[44] ITU-T Recommendation H.324, Terminal for Low Bit Rate Multimedia Communication.

[45] AF-TM-0121.000, ATM Forum, Traffic Management Specification.

[46] AF-TM-0150.000, ATM Forum, Addendum to Traffic Management v4.1 for an Optional Minimum Desired Cell Rate Indication for UBR.

[47] IETF RFC 3985, Pseudo Wire Emulation Edge-to-Edge (PWE3) Architecture.

[48] IETF RFC 4717, Encapsulation Methods for Transport of Asynchronous Transfer Mode (ATM) over MPLS Networks.

[49] A Peak-to-Average Power Reduction Method for Third Generation CDMA Reverse Links, Kevin Laird, Nick Whinnett, Soodesh Buljore, *IEEE 49th Vehicular Technology Conference*, 1999, vol. 1, pp 551–555.

[50] Evolution of High Speed Downlink Packet Access in WCDMA by Improved ARQ Signalling, M.P.J. Baker, T.J. Moulsley, *6th IEE International Conference on 3G and Beyond*, 2005, pp 1–5.

[51] A Novel Code Planning Approach for a WCDMA Network, R.M. Joyce, T. Griparis, G.R. Conroy, B.D. Graves, I.J. Osborne, *4th IEE International Conference on 3G Mobile Communication Technologies*, 2003, pp 31–36.

Index

3GPP   10
4PAM   345, 350, 378, 385
16QAM   277, 279, 280, 328, 345, 350, 353
64QAM   277, 279, 280

Absolute grant
  scope   393, 394
  value   393, 394
Abstract Syntax Notation (ASN)   15, 27, 56
Access Class (AC)   195, 196
Access Point Name (APN)   242, 243, 483
Access Service Class (ASC)   195, 196, 198
Access Stratum (AS)   16
  release indicator   409
Acquisition Indicators (AI)   188
Activation CFN   447, 450, 468, 497, 500, 524
Active indoor solutions   594, 595
Admission control   410, 416, 441, 465, 487, 505, 509
AICH   188
  transmission timing   189, 190, 192, 195, 407
Air-interface
  capacity   36
  synchronisation   40, 61, 206, 216, 407, 423, 447, 509
ALCAP   250, 253, 255, 268
Amplitude gain factors   208, 213, 214
Antenna
  characteristics   573
  cross polar   574
  electrical downtilt   573, 575
  placement   550, 596
  dedicated subsystem   576
  shared subsystem   578
Assisted GPS   427
ATM
  AAL2   253, 257, 258, 259, 265
  AAL5   252, 253, 268

cell   259
  service categories   268
Available Bit Rate (ABR)   269, 270

BCH   158, 160, 161, 162
BCCH   43, 109, 155, 157
Beamforming   168, 226
Binding identity   415, 417, 441, 446, 467, 496
Bit Error Rate (BER)   115, 118, 129
Block Error Rate (BLER)   421, 439, 450, 467, 485, 489, 536, 538
BPSK   345, 350, 376, 378, 386
Broadcast/Multicast Control (BMC) layer   19, 20
Buffer occupancy   69, 112
Building penetration loss   541, 542

C/I requirement   538, 540, 543
Calculated Transport Format Combination (CTFC)   179, 181, 214
  for data connection   492, 493, 494
  for speech service   442, 443, 449
  for video call   466
CCCH   17, 18, 43, 77, 81, 109, 155, 157
Cell Broadcast Services (CBS)   19, 20, 33, 45
Cell Radio Network Temporary Identifier (C-RNTI)   33, 39, 110
Cell reselection   27, 32, 43, 50
Cell selection   27, 50
Cell synchronisation   172, 558, 559
Cell System Frame Number (SFN)   169, 220, 222
Cell update   42, 43, 46, 47, 49, 50, 64
CELL_DCH   33
CELL_FACH   39
CELL_PCH   44
Channel coding
  Convolutional coding   127, 129
  Turbo coding   127, 129

Radio Access Networks for UMTS   Chris Johnson
© 2008 John Wiley & Sons, Ltd

Channel estimation   36, 137, 168, 200, 205
Channel Quality Indicator (CQI)   286, 307, 309, 332, 333
Channelisation codes   8, 37, 127, 144, 146, 200, 207, 217, 219
Chase combining   282, 313, 314
Chip offset   220, 222, 223, 414, 422, 423
Ciphering   74, 77, 85, 105, 424, 427, 431, 435
Code block segmentation   129, 134, 139
Code Division Multiple Access (CDMA)   8
Code multiplexing   323, 324
Coherent detection   200, 205, 215
Common Part Convergence Sublayer (CPCS)   252, 253
Common Part Sublayer (CPS)   255, 257, 258, 259
Common transport Channel Priority Indicator (CmCH-PI)   289, 292, 293
Compressed mode   38, 44, 68, 205, 211, 216, 219, 427, 519, 521
Congestion status   401, 402
Connection establishment delay   409, 422, 428, 430, 487
Connection Frame Number (CFN)   220, 222, 223
Connection Identifier (CID)   254, 258, 259, 289, 294, 295, 416, 446, 487, 496
Connection Management (CM)   232, 238, 239
Constant Bit Rate (CBR)   269, 270
Constellation re-arrangement   327, 328
Continuous Packet Connectivity (CPC)   35, 36
Control plane protocol stack   13, 14
Controlling RNC   1
    communication context   411, 415, 424, 446
Co-siting   578
CPICH   168
    Ec/Io   170
    Ec/Io signalled value   508
    RSCP 170
CTCH   43, 45, 109, 156, 158
Cyclic Redundancy Check (CRC)   112, 116, 121, 127, 129
    indicators   115, 116, 118, 121

Data connection
    establishment   477
    NAS signalling   243, 244
    transport block processing   136, 142
    transport format combination set   165
    transport format set   161, 165
Data Description Indicator (DDI)   356, 371, 372, 400
DCH   159, 160
DCCH   18, 35, 43, 103, 110, 156, 158
Default DPCH offset   422, 423
Destination Signalling Association Identifier (DSAID)   416, 417
Digital terrain map   551

Discontinuous Reception (DRX)   27, 30, 32, 186
    cycle   45, 49, 167, 180, 182, 184, 237
Discontinuous Transmission (DTX)   25, 127
    indication bits   127, 140, 143
Doppler   138
Downlink   3, 5
Downlink transmit powers   542, 543
DPCCH   168, 204, 205, 206, 214, 215
    amplitude gain factor   148, 208, 213
    channelisation codes   207
DPCH
    downlink   214
    uplink   204
DPDCH   168, 204, 205, 206, 214, 215
    amplitude gain factor   148
    channelisation codes   207, 208
Drift RNC   2, 4
DTCH   35, 43, 53, 101, 103, 110, 156, 157, 158
Duplex spacing   5
Duplexor   573

E1   261, 262, 263
E-AGCH   348, 365, 371, 392
Eb/No   133, 147, 536, 537, 538, 539, 540
E-DCH   158, 160, 165
    data frame   400
E-DPCCH   347, 360, 376
    amplitude ratios   378
    slot format   376
E-DPDCH   344, 350, 362, 365, 367, 378
    amplitude ratios   363
    slot formats   378
E-HICH   368, 370, 387
    frame timing   389, 390
    signature sequence   387, 388
E-RGCH   348, 367, 390
    signature sequence   390, 391
E-RNTI   393, 394
E-TFC selection   346, 353, 359
Event 1a   67, 502, 505, 508, 509
Event 1b   67, 506, 511
Event 1c   67, 506, 507
Event 1d   67
Event 1e   67
Event 1f   67
Event 4a   67, 490, 491, 492
Event 6a   67, 517, 518
Event 6f   67, 510
Event 6g   67, 510

FACH   158, 159, 160
    downlink data frame   123, 124
    measurement occasions   44
    transmit power   124
    transport format set   123, 124

Fast fading margin 538, 541
F-DPCH 36, 37, 61, 214, 228
Feedback Information (FBI) 205, 206
Feeder cable 550, 577
File Transfer Protocol (FTP) 21
Filter coefficient 502, 503, 504, 517, 518
Fixed transport channel positions 226
Flexible transport channel positions 226
Flush flag 290
Forward link 3
Frame offset 220, 221, 223, 414, 422, 423, 509
Frame Protocol (FP) 14, 26, 112, 250, 257, 258, 260
   common channel control frames 126
   common channel data frames 121
   dedicated channel control frames 118
   dedicated channel data frames 113
   quality estimate 112, 115, 117
Frame Sequence Number (FSN) 289, 290, 400
Frequency Division Duplexing (FDD) 4
Frequency re-use 6
Full duplex 8

Gateway GPRS Support Node (GGSN) 242
Generalised RAKE 139
GPRS attach 30, 32, 478, 480, 481
GPRS Tunnelling Protocol (GTP) 487, 489
Gs interface 236
Guaranteed Frame Rate (GFR) 269, 270

Half duplex 7
Happy bit 347, 377, 398
HARQ 281, 313, 368
   preamble mode 336, 337
   processes 281, 315, 316, 368, 370, 394
   reordering buffer 317, 320
   round trip timing 369
   T1 timer 320
   windows 317, 319
Highest priority Logical channel Buffer Status
   (HLBS) 373
Highest priority Logical channel Identity (HLID) 373
Home PLMN 244, 245
H-RNTI 325
HSDPA 273
   Adaptive Modulation and Coding (AMC) 307
   bit rates 278
   capability 427, 428
   channelisation code occupancy 218, 329, 330
   coding rates 279, 280, 282, 283, 311, 312
   comparison with DPCH 276, 277
   flow control 301
   frame protocol 288
   in CELL_FACH 27, 35
   logical and transport channels 35
   measurement power offset 307, 309

   mobility 337
   physical channels 322
   power offsets in soft handover 341
   protocol stack 26, 277, 278
   puncturing ratios 313
   scheduler 304
   serving cell 337, 339
   slot format 329, 331, 334
   transport block 280, 282, 310, 320
   transport block sizes 310
   transport block processing 331
   transport format parameters 165, 166
   UE categories 280
HS-DPCCH 332
HS-DSCH 158, 160, 165, 274
HS-FACH 35
HS-PDSCH 329
   slot formats 331
HS-SCCH 323
HSUPA 343
   bit rates 349
   capability 409
   channelisation codes 379, 381
   comparison with DPCH and HSDPA 345
   connection uplink cell load 354
   frame protocol 399
   inner loop power control 345, 387
   logical and transport channels 35
   maximum bit rates 350, 353
   maximum transport block sizes 353
   minimum set 365, 368
   mobility 402
   physical channels 374
   protocol stack 26, 348
   reference E-TFCI 360, 363
   reference power ratios 346, 363, 367
   scheduler 395
   spreading factor 344, 350, 378
   transport block processing 383, 386
   transport block size tables 352, 353, 359, 360
   Transport Format Combination Indicator
     (E-TFCI) 347, 360, 363, 377
   transport format parameters 165, 166
   TTI 345, 351, 353, 360, 369, 378
   UE categories 351

Idle mode (RRC) 27, 29
Idle Period Downlink (IPDL) 427
IMSI attached 30, 32
Incremental redundancy 281, 282, 313, 314, 327, 370,
   377
Indoor location probability 541
Inner loop power control 72, 77, 132, 152, 153, 341,
   345, 387, 412, 415, 422, 443
Integrity protection 424, 427, 435

Interference cancellation   127, 139, 148
Interference floor   540, 544
Interference margin   345, 350, 354, 395, 397, 539
Inter-frequency
    handover   27, 36, 38, 68, 71
    reporting events   67, 68
Interleaving   127, 130, 131, 133, 134
Intermodulation   585
International Mobile Equipment Identity (IMEI)   33
International Mobile Subscriber Identity (IMSI)   33
Inter-RAT reporting events   67, 69
Inter-RNC soft handover   2
Inter-system handover   27, 29, 68
Intra-frequency reporting events   67, 68
Inverse Multiplexing for ATM (IMA)   262, 263
Iub
    architecture   260
    congestion control   401, 402
    for HSDPA   294
    overbooking   264, 271
    overheads   264
    overheads for HSDPA   296, 298
    protocol stacks   250
    transport network   249
    VCC bandwidths   263
Iu-cs protocol stack   433
Iu-ps protocol stack   486
Iur   1, 3

JT1   261, 262

Leakage requirements   595, 596
Link budget
    downlink DPCH   542
    HSDPA   546
    HSUPA   547
    uplink DPCH   535
    uplink DPCH (with HSDPA)   545
Load based handover   38
Location area   27, 30, 48, 233, 236, 237
    updates   33, 234, 235
Location Area Identity (LAI)   234, 235
Logical channels   15, 101, 155
    multiplexing   104, 106, 109

MAC-b   108
MAC-c/sh/m   101, 105
MAC-d   101, 103
    flows   287, 288, 294, 302
    for HSDPA   287
    for HSUPA   357
MAC-e
    Node B   394
    PDU   372
    re-transmissions   346, 356, 370, 377

    UE   358
MAC-es
    PDU   371
    RNC   402
    UE   358
MAC-hs   25, 106, 300
    acknowledgements   315, 317, 332, 335
    PDU   320
    re-transmissions   286, 301, 313, 317, 319
MAC layer   101
MAC Logical Channel Priority (MLP)   196, 197, 203
MAC-m   107
Mast head amplifier   540, 576
Maximum Segment Size (MSS)   21, 23
Maximum target received total wideband power   396
Maximum Transmission Unit (MTU)   23
MBMS   27, 33, 45, 55, 107, 111, 157, 167, 186
MCCH   45, 81, 111, 156, 157, 186
Messages
    Activate PDP Context Accept   490
    Activate PDP Context Request   484
    Active Set Update   224, 510, 512
    Active Set Update Complete   510
    ALCAP Establish Confirm   417
    ALCAP Establish Request   416
    Alerting   453, 459
    Attach Accept   481
    Attach Request   479
    Call Confirmed   458, 464
    Call Control   232, 239
    Call Proceeding   438, 463
    Cell Update   46, 47
    CM Service Request   432
    Compressed Mode Command   528
    Connect   453, 459
    Connect Acknowledge   454, 459
    Handover from UTRAN Command   531
    Initial Direct Transfer   17, 431
    Initial UE message   433
    Measurement Control
        BSIC verification   528
        events 1a, 1b, 1c   503
        event 1d   338, 339, 340
        events 1e, 1f   517
        events 6a, 6b   518
        GSM RSSI   527
    Measurement Report
        BSIC verification   529
        event 1a   508
        event 1b   511
        event 1f   519
        GSM RSSI   527
    Paging (RANAP)   455
    Paging Response   457
    Paging type 1   48, 456

Paging type 2   38
Physical Channel Reconfiguration   525
RAB Assignment Request
   for data   485
   for speech   440
   for video call   465
RAB Assignment Response   452, 489
Radio Bearer Reconfiguration 35, 42, 497, 499, 500
Radio Bearer Setup
   for data   488
   for speech   448, 449, 451, 452
   for video call   468
Radio Bearer Setup Complete   452
Radio Link Addition Request   513
Radio Link Reconfiguration Commit   447, 524
Radio Link Reconfiguration Prepare   442, 444, 445,
   466, 467, 493, 495
Radio Link Reconfiguration Ready   446, 524
Radio Link Restore   424
Radio Link Setup Request   411, 413
Radio Link Setup Response   415
RRC Connection Reject   410
RRC Connection Request   407
RRC Connection Setup   34, 40, 80, 418, 419, 420,
   421, 422
RRC Connection Setup Complete   425
SCCP Connection Confirm   434, 480
SCCP Connection Request   434
Service Request   17, 432, 482
Setup
   for speech   436, 457
   for video call   462, 464
Uplink Direct Transfer   437
MICH   186
MIMO   280, 281, 301
Minimum coupling loss
   indoor solutions   594
   microcells   589
Minimum Desired Cell Rate (MDCR)   270, 296
Mobility Management   16, 232, 233
Modulation   144
   constellation   148, 149, 152
Monte Carlo simulations   553
MS Classmark information   427, 432, 457
MSCH   43, 81, 111, 156, 157
MTCH   43, 45, 111, 156, 157
Multi-User Detection (MUD)   139, 148

N300   27, 64, 409
N302   42, 64
N304   64
N308   61, 64
N312   64, 423
N313   61, 64, 65
N315   61, 64, 65

NBAP   250, 251
   common   251
   dedicated   251
Neighbour planning
   inter-frequency   564
   inter-system   566
   intra-frequency   563
   maximum length   567
Network Mode of Operation (NMO)   235, 236
Network Service Access Point (NSAP)   416, 446, 447,
   482, 496
Node B   1
   communication context   415, 441, 512, 514
Node B Frame Number (BFN)   169, 170
Noise bandwidth   147
Noise figure   540, 541
Non-Access Stratum (NAS)   16, 231
Non-Scheduled Mode   373, 399
Notification Indicators   186, 187
Nrt-VBR   269, 270

Observed Time Difference of Arrival (OTDOA)
   427
OC-3   263, 264
Open loop power control   171, 197, 210, 407, 408,
   414, 422
Originating Signalling Association Identifier
   (OSAID)   417
Orthogonality   145, 146, 152, 171, 208, 308, 379
Outer loop power control   112, 116, 118, 132, 439,
   450, 467, 485
Oversampling   137

Packet Data Unit (PDU)   15
Packet Temporary Mobile Subscriber Identity
   (P-TMSI)   33, 234, 408, 454, 479
Paging   160, 180, 184, 185
   indicators   182, 183
   occasion   184, 185
   records   455
Parity bits   370, 371
Passive indoor solutions   594
PCCH   45, 77, 109, 156, 158
P-CCPCH   168, 174
PCH   159, 160, 163
   transport format set   126
PCPCH   426
PDCP 19, 20, 283, 355
PDH   261
PDP context   242, 243, 244
   activation   478, 482, 483, 489
Physical channels   15, 166
Physical layer   127
   downlink processing   139
   uplink processing   128

PICH   182
  bitmap   124, 125
PLMN Selection   29, 244
PO1   226
PO2   226
PO3   226, 412, 422, 450
Polarisation loss   540, 541
Power control   9
  preamble   211, 407, 422, 424
PRACH   168, 191
  channelisation codes   200
  message   198
  partitions   195, 198
  persistency check   198
  preamble cycle   192, 197, 198
  preamble signature   189, 193, 200
  preambles   192
  scrambling code   193, 194
  sub-channels   194, 195
Primary scrambling codes   166, 168, 173, 226
Priority queues   288, 293, 302
Processing gain   147, 536, 540
Propagation delay   71, 122, 190, 225, 227, 414
Protocol Stacks   14
Puncturing limit   209, 211, 411, 422, 442, 494
Puncturing ratio   132, 135, 140, 143

QPSK   277, 279, 280, 331, 345, 350, 378
Quality estimate selector   414
Quality of Service (QoS)   24, 439, 464, 483, 489
Quality reporting events   67
Quality target   421

RAB establishment   438, 464, 483, 487
  from CELL_DCH   40
  from CELL_FACH   41
RACH   158, 160, 162, 163
  measurements   408
  transmissions   112
  transport format combinations   204
  transport format set   122, 201, 204
Radio Access Bearer (RAB)   24
Radio Access Network (RAN)   1
Radio bearer   17, 24
Radio frame
  equalisation   130, 139
  segmentation   131, 135, 141, 143
Radio interface control plane   14
Radio interface parameter update   120
Radio link   24
Radio network control plane   250, 251
Radio network planning   548
  path loss based approach   550
  simulation based approach   553
Radio Resource Control (RRC)   27

messages   19, 56, 77, 81, 95
procedures   54
states   28
RAKE   127, 137, 138
RANAP   17
Random access procedure   153, 407
Rate matching   127, 131, 134, 140
  attributes   132, 133
Receive diversity   137, 138, 541, 573, 574, 576
Receiver blocking   583
Redundancy and constellation version   326,
    327, 331
Reference received total wideband power   396
Residual BER   439, 465, 467, 485
Restricted Digital Information (RDI)   461
Re-transmission Sequence Number (RSN)   347, 371,
    377
Reverse link   3
RF carrier   4, 6
  indoor solutions   593
  microcells   588
RFC 3095   21
Rise over thermal   539, 540
RLC   74
  acknowledged mode   81
  acknowledged mode PDU   87
  acknowledgements   92, 94
  for HSDPA   284
  for HSUPA   355
  PDU size calculations   498
  piggybacked status   88, 89
  polling   86, 88, 92, 96
  status report   83, 97
  transparent mode   75
  unacknowledged mode   77
  windows   82, 85
RNC   1
  buffer occupancy   290, 291
  communication context   411, 415, 424, 446
Robust Header Compression (ROHC)   21
Routing area   27, 32, 67
  identity   234, 235, 408, 480, 481
  update procedure   32, 33, 234, 236
RRC connection establishment   405
  cause   408
  delay   39
Rt-VBR   269, 270

S-CCPCH   176
  transport format combinations   180, 181
  transport format sets   180, 181
Scheduled grant values   362, 366, 390, 392, 394
Scheduler
  equal throughput   305, 306
  maximum throughput   305, 306

proportional fair   304, 305
round robin   304, 305, 319
Scheduling information   347, 360,
      370, 373
Scrambling   144
Scrambling code planning
   downlink   557
   uplink   560
Scrambling codes   8, 414
   alternative   217, 219, 220
   primary   168, 169, 217
   secondary   169, 217
SDH   263, 264
SDU error ratio   439, 444, 465, 485, 486
Sectorisation
   indoor solutions   593
   microcells   588
Security mode procedure   435, 436
Segmentation and Reassembly (SAR)   252, 253
Semi-static parameters   161, 165
Sensitivity   540
Served User Generated Reference (SUGR)   417, 446,
      496
Service area   433, 529
Service based handover   38
Service Data Unit (SDU)   15
Service Specific Connection Orientated Protocol
      (SSCOP)   252, 253
Service Specific Convergence Sublayer (SSCS)   252,
      257
Service Specific Coordination Function (SSCF)   252
Service Specific Segmentation and Reassembly
      (SSSAR)   257, 258
Serving cell change procedure   337, 338, 340, 403
Serving cell grant   346, 347
Serving grant value   366, 367
Serving Radio Network Subsystem (SRNS)
      relocation   3
Serving RNC   1, 3
Session management   16, 232, 239
SIB 1   235
Signalling Connection Control Part (SCCP)   434, 435,
      480
Signalling Radio Bearer (SRB)   17, 18, 19
SIR target   113, 118, 119, 132
Site acquisition   549
Site selection   550
Site Selection Diversity Transmit (SSDT)   415
Size Index Identifier (SID)   321, 495
SMS   469
   mobile originated   470
   mobile terminated   475
Soft handover   2, 15, 24, 38, 68, 71
   gain   541, 544, 555
   HSUPA   345, 403

inter-Node B   501
intra-Node B   512
Softer handover   38
SONET   263, 264
Space Time Transmit Diversity (STTD)   174, 179
Speech service
   connection establishment   97, 429
   frame protocol data frame   117
   inter-system handover   515
   Iub overheads   265, 267
   logical, transport and physical channels   24, 103
   NAS signalling   240, 241
   protocol stack   20, 23
   transport block processing   128, 134, 139, 142
   transport format combination set   164, 443
   transport format set   117, 164, 442
Spreading   144
   codes   8, 144
   gain   147
Spreading factor   167, 206, 207, 209, 215, 216
   selection   131, 132, 211
Spurious emissions   581
SRB   17, 18, 19, 412
   delay   211, 422, 423
   standalone   412, 442, 443
   transport format combination   412
   transport format combination set   412
   transport format set   413
Standardisation   10
Static simulations   553
STM-1   263, 264
Stop and Wait (SAW)   315, 368
Super Fields (SUFI)   88
Synchronisation Channels (SCH)   172
Synchronisation frame   119, 417
Systematic bits   370, 371, 383

T1   261, 262
T300   62, 64, 409
T302   42, 62, 64
T304   62, 64
T305   43, 47, 62, 64, 66
T307   62, 64, 66
T308   62, 64
T309   62
T312   62, 423
T313   55, 61, 63, 65
T314   61, 63, 65
T315   61, 63, 65
T316   63, 65, 66
T317   63, 65, 66
T318   63
T3212   235, 236, 481
T3312   235, 481
Tandem Free Operation (TFO)   77

Target Channel Type Field (TCTF) 101, 105, 109, 111
Target non-serving E-DCH to total E-DCH power
    ratio 398
T_Cell 169, 173, 224
TCP
    acknowledgements 23, 92, 93, 94
    connection establishment 491
TCP/IP
    file transfer 21, 93
    protocol stack 282
Temporary Mobile Group Identity (TMGI) 187
Temporary Mobile Subscriber Identity (TMSI) 27, 33,
        233, 234, 408
TFCS
    for data connection 494
    for speech service 442
    for video call 466
TFS
    for data connection 494
    for speech service 443
    for video call 466
Time Division Duplexing (TDD) 4
Time Division Multiple Access (TDMA) 7, 10
Time of Arrival
    Window End (toAWE) 412, 417, 495, 496
    Window Start (toAWS) 412, 417, 495, 496
Time Switched Transmit Diversity (TSTD) 174
Timing adjustment 118, 119
TNL congestion indication 118, 120
Total E-DCH Buffer Status (TEBS) 373, 374
TPC combination index 227
Traffic Handling Priority (THP) 486
Traffic volume reporting events 67, 69
Transcoder Free Operation (TrFO) 77
Transfer delay requirement 441, 465, 486, 489
Transmission Sequence Number (TSN) 317, 321, 327
Transmit Power Control (TPC) 205, 215, 224, 228
Transmit power dynamic range 10
Transport block 109, 113, 162
    BLER 421, 422, 450, 485, 486
    concatenation 129, 139
    set 113, 128, 139
Transport channels 15, 158
    multiplexing 130, 133, 135, 141
Transport Format 158, 161
    Combination (TFC) 104, 114, 127, 163
    Combination Indicator (TFCI) 140, 162, 163, 179
    Combination Set (TFCS) 163, 164
Transport Format Set (TFS) 104, 161
    for speech service 117, 442, 443
    for data connection 104, 115
    for video call 466

Transport layer address 415, 441, 446, 467, 487, 496
Transport network
    control plane 253
    user plane 257
Tunnel congestion indication control frame 398, 399,
        401

UBR 269, 270
UBR+ 270
UE
    constants 61, 64
    counters 61, 64
    measurement capability 427
    power class 210
    power headroom 347, 372, 373
    reporting events 27, 67, 68, 69
    timers 61, 62
    transmit powers 535
Unrestricted Digital Information (UDI) 461, 463, 464,
        467
Uplink 3, 5
    gating 36
    SIR target 412, 422, 439, 524
URA
    identity 51, 52
    updates 42, 52, 53, 64
URA_PCH 27, 50
U-RNTI 418, 456
User Plane 13, 19
UTRAN
    DRX Cycle length coefficient 419
    Radio Network Temporary Identity (U-RNTI) 27,
        35, 42, 46, 110
    Registration Area (URA) 50, 52
Uu 1

V300 62, 64, 409, 410
V302 62, 64
V304 62, 64
V308 62, 64
Value tag 32, 33
Video call
    connection establishment 459
Virtual Channel
    Connection (VCC) 254, 255
    Identifier (VCI) 254, 259
Virtual Path
    Connection (VPC) 254, 255, 261
    Identifier (VPI) 254, 259, 261
Voice over IP (VoIP) 20, 21, 77, 78, 228

Wideband CDMA (WCDMA) 8, 9

Printed in the USA/Agawam, MA
May 10, 2010

541189.011